T0344923

Statistics for Process Control Engineers

Statistics for Process Control Engineers

A Practical Approach

Myke King

Whitehouse Consulting, Isle of Wight, UK

This edition first published 2017

Registered Office(s)
John Wiley & Sons, Inc., 111 River Street, Hoboken, NJ 07030, USA
John Wiley & Sons Ltd, The Atrium, Southern Gate, Chichester, West Sussex, PO19 8SQ, UK

Editorial Office
The Atrium, Southern Gate, Chichester, West Sussex, PO19 8SQ, UK

For details of our global editorial offices, customer services, and more information about Wiley products visit us at www.wiley.com.

Wiley also publishes its books in a variety of electronic formats and by print-on-demand. Some content that appears in standard print versions of this book may not be available in other formats.

Library of Congress Cataloging-in-Publication Data

Names: King, Myke, 1951–
Title: Statistics for process control engineers : a practical approach / Myke King.
Description: First edition. | Hoboken, NJ : Wiley, [2017] | Includes bibliographical references and index.
Identifiers: LCCN 2017013231 (print) | LCCN 2017015732 (ebook) | ISBN 9781119383482 (pdf)
ISBN 9781119383529 (epub) | ISBN 9781119383505 (cloth)
Subjects: LCSH: Process control–Mathematical models. | Engineering–Statistical methods. | Engineering mathematics.
Classification: LCC TS156.8 (ebook) | LCC TS156.8 .K487 2018 (print) | DDC 620.001/51–dc23
LC record available at https://lccn.loc.gov/2017013231

Cover design by Wiley
Cover Images: (Background) © kagankiris/Gettyimages;
(Graph) Courtesy of Myke King

Set in 10/12pt Times by SPi Global, Pondicherry, India
Printed and bound in Singapore by Markono Print Media Pte Ltd

10 9 8 7 6 5 4 3 2 1

Contents

Preface

There are those that have a very cynical view of statistics. One only has to search the Internet to find quotations such as those from the author Mark Twain:

There are three kinds of lies: lies, damned lies, and statistics.
Facts are stubborn, but statistics are more pliable.

From the American humourist Evan Esar:

Statistics is the science of producing unreliable facts from reliable figures.

From the UK's shortest-serving prime minister George Canning:

I can prove anything by statistics except the truth.

And my personal favourite, attributed to many – all quoting different percentages!

76.3% of statistics are made up.

However, in the hands of a skilled process control engineer, statistics are an invaluable tool. Despite advanced control technology being well established in the process industry, the majority of site managers still do not fully appreciate its potential to improve process profitability. An important part of the engineer's job is to present strong evidence that such improvements are achievable or have been achieved. Perhaps one of the most insightful quotations is that from the physicist Ernest Rutherford.

If your experiment needs statistics, you ought to have done a better experiment.

Paraphrasing for the process control engineer:

If you need statistics to demonstrate that you have improved control
of the process, you ought to have installed a better control scheme.

Statistics is certainly not an exact science. Like all the mathematical techniques that are applied to process control, or indeed to any branch of engineering, they need to be used alongside good engineering judgement. The process control engineer has a responsibility to ensure that statistical methods are properly applied. Misapplied they can make a sceptical manager even more sceptical about the economic value of improved control. Properly used they can turn a sceptic into a champion. The engineer needs to be well versed in their application. This book should help ensure so.

After writing the first edition of *Process Control: A Practical Approach*, it soon became apparent that not enough attention was given to the subject. Statistics are applied extensively at every stage of a process control project from estimation of potential benefits, throughout control design and finally to performance monitoring. In the second edition this was partially addressed by the inclusion of an additional chapter. However, in writing this, it quickly became apparent that the subject is huge. In much the same way that the quantity of published process control theory far outstrips more practical texts, the same applies to the subject of statistics – but to a much greater extent. For example, the publisher of this book currently offers over 2,000 titles on the subject but fewer than a dozen covering process control. Like process control theory, most published statistical theory has little application to the process industry, but within it are hidden a few very valuable techniques.

Of course, there are already many statistical methods routinely applied by control engineers – often as part of a software product. While many use these methods quite properly, there are numerous examples where the resulting conclusion later proves to be incorrect. This typically arises because the engineer is not fully aware of the underlying (incorrect) assumptions behind the method. There are also too many occasions where the methods are grossly misapplied or where licence fees are unnecessarily incurred for software that could easily be replicated by the control engineer using a spreadsheet package.

This book therefore has two objectives. The first is to ensure that the control engineer properly understands the techniques with which he or she might already be familiar. With the rapidly widening range of statistical software products (and the enthusiastic marketing of their developers), the risk of misapplication is growing proportionately. The user will reach the wrong conclusion about, for example, the economic value of a proposed control improvement or whether it is performing well after commissioning. The second objective is to extract, from the vast array of less well-known statistical techniques, those that a control engineer should find of practical value. They offer the opportunity to greatly improve the benefits captured by improved control.

A key intent in writing this book was to avoid unnecessarily taking the reader into theoretical detail. However the reader is encouraged to brave the mathematics involved. A deeper understanding of the available techniques should at least be of interest and potentially of great value in better understanding services and products that might be offered to the control engineer. While perhaps daunting to start with, the reader will get the full value from the book by reading it from cover to cover. A first glance at some of the mathematics might appear complex. There are symbols with which the reader may not be familiar. The reader should not be discouraged. The mathematics involved should be within the capabilities of a high school student. Chapters 4 to 6 take the reader through a step-by-step approach introducing each term and explaining its use in context that should be familiar to even the least experienced engineer. Chapter 11 specifically introduces the commonly used mathematical functions and their symbology. Once the reader's initial apprehension is overcome, all are shown to be quite simple. And, in any case, almost all exist as functions in the commonly used spreadsheet software products.

It is the nature of almost any engineering subject that the real gems of useful information get buried among the background detail. Listed here are the main items worthy of special attention by the engineer because of the impact they can have on the effectiveness of control design and performance.

- Control engineers use the terms 'accuracy' and 'precision' synonymously when describing the confidence they might have in a process measurement or inferential property. As explained in Chapter 4, not understanding the difference between these terms is probably the most common cause of poorly performing quality control schemes.

- The histogram is commonly used to help visualise the variation of a process measurement. For this, both the width of the bins and the starting point for the first bin must be chosen. Although there are techniques (described in this book) that help with the initial selection, they provide only a guide. Some adjustment by trial and error is required to ensure the resulting chart shows what is required. Kernel density estimation, described in Chapter 6, is a simple-to-apply, little-known technique that removes the need for this selection. Further it generates a continuous curve rather than the discontinuous staircase shape of a histogram. This helps greatly in determining whether the data fit a particular continuous distribution.

- Control engineers typically use a few month's historical data for statistical analysis. While adequate for some applications, the size of the sample can be far too small for others. For example, control schemes are often assessed by comparing the average operation post-commissioning to that before. Small errors in each of the averages will cause much larger errors in the assessed improvement. Chapter 7 provides a methodology for assessing the accuracy of any conclusion arrived at with the chosen sample size.

- While many engineers understand the principles of significance testing, it is commonly misapplied. Chapter 8 takes the reader through the subject from first principles, describing the problems in identifying outliers and properly explaining the impact of repeatability and reproducibility of measurements.

- In assessing process behaviour it is quite common for the engineer to simply calculate, using standard formulae, the mean and standard deviation of process data. Even if the data are normally distributed, plotting the distribution of the actual data against that assumed will often reveal a poor fit. A single data point, well away from the mean, will cause the standard deviation to be substantially overestimated. Excluding such points as outliers is very subjective and risks the wrong conclusion being drawn from the analysis. Curve fitting, using all the data, produces a much more reliable estimate of mean and standard deviation. There are a range of methods of doing this, described in Chapter 9.

- Engineers tend to judge whether a distribution fits the data well by superimposing the continuous distribution on the discontinuous histogram. Such comparison can be very unreliable. Chapter 6 describes the use of quantile–quantile plots, as a much more effective alternative that is simple to apply.

- The assumption that process data follows the normal (Gaussian) distribution has become the de facto standard used in the estimation of the benefits of improved control. While valid for many datasets, there are many examples where there is a much better choice of distribution. Choosing the wrong distribution can result in the benefit estimate being easily half or double the true value. This can lead to poor decisions about the scope of an improved control project or indeed about whether it should be progressed or not. Chapter 10 demonstrates that while the underlying process data may be normally distributed, derived data may not be. For example, the variation in distillation column feed composition, as source of disturbance, might follow a normal distribution, but the effect it has on product compositions will be highly asymmetrical. Chapter 12 describes a selection of the more well-known alternative distributions. All are tested with different sets of real process data so that the engineer can see in detail how they are applied and how to select the most appropriate. A much wider range is catalogued in Part 2.

- While process control is primarily applied to continuous processes, there are many examples where statistics can be applied to assess the probability of an undesirable event. This might be important during benefit estimation, where the improvement achievable by improved control is dependent on other factors – for example, the availability of feed stock or of a key piece of process equipment. Failure to take account of such events can result in benefits being overestimated. Event analysis can also be applied to performance monitoring. For example, it is common to check an inferential property against the latest laboratory result. Estimating the probability of a detected failure being genuine helps reduce the occasions where control is unnecessarily disabled. Chapter 12 and Part 2 describe the wide range of statistical methods that are applicable to discrete events. Again, the description of each technique includes a worked example intended to both illustrate its use and inspire the engineer to identify new applications.

- One objective of control schemes is to prevent the process operating at conditions classed as extreme. Such conditions might range from a minor violation of a product specification to those causing a major hazard. Analysing their occurrence as part of the tail of distribution can be extremely unreliable. By definition the volume of data collected in this region will be only a small proportion of the data used to define the whole distribution. Chapter 13 describes techniques for accurately quantifying the probability of extreme process behaviour.

- Rarely used by control engineers, the hazard function described in Chapter 14 is simply derived from basic statistical functions. It can be beneficial in assessing the ongoing reliability of the whole control scheme or of individual items on which it depends. It can be a very effective technique to demonstrate the need to invest in process equipment, improved instrumentation, additional staff and training.

- Engineers often overlook that some process conditions have ‘memory’. It is quite reasonable to characterise the variability of a product composition by examining the distribution of a daily laboratory result. However the same methodology should not be applied to the level of liquid in a product storage tank. If the level yesterday was very low, it is extremely unlikely to be high today. The analysis of data that follow such a time series is included in Chapter 15. The technique is equally applicable to sub-second collection frequencies where it can be used to detect control problems.

- Regression analysis is primarily used by process control engineers to build inferential properties. While sensibly performed with software, there are many pitfalls that arise from not fully understanding the techniques it uses. The belief that the Pearson R coefficient is a good measure of accuracy is responsible for a very large proportion of installed inferentials, on being commissioned, causing unnoticed degradation in quality control. Chapter 16 presents the whole subject in detail, allowing the engineer to develop more effective correlations and properly assess their impact on control performance.

- Engineers, perhaps unknowingly, apply time series analysis techniques as part of model identification for MPC (multivariable predictive control). Often part of a proprietary software product, the technique is not always transparent. Chapter 17 details how such analysis is performed and suggests other applications in modelling overall process behaviour.

- Process control engineers frequently have to work with inconsistent data. An inferential property will generate a different value from that recorded by an on-stream analyser which, in turn, will be different from the laboratory result. Mass balances, required by optimisation strategies, do not close because the sum of the product flows differs from the measured feed flow. Data reconciliation is a technique, described in Chapter 18, which not only reconciles such differences but also produces an estimate that is more reliable than any of the measurements. Further, it can be extended to help specify what additional instrumentation might be installed to improve the overall accuracy.

- Much neglected, because of the perception that the mathematics are too complex to be practical, Fourier transforms are invaluable in helping detect and diagnose less obvious control problems. Chapter 19 shows that the part of the theory that is applicable to the process industry is quite simple and its application well within the capabilities of a competent engineer.

To logically sequence the material in this book was a challenge. Many statistical techniques are extensions to or special cases of others. The intent was not to refer to a technique in one chapter unless it had been covered in a previous one. This proved impossible. Many of the routes through the subject are circular. I have attempted to enter such circles at a point that requires the least previous knowledge. Cross-references to other parts of the book should help the reader navigate through a subject as required.

A similar challenge arose from the sheer quantity of published statistical distributions. Chapter 12 includes a dozen, selected on the basis that they are well known, are particularly effective or offer the opportunity to demonstrate a particular technique. The remainder are catalogued as Part 2 – offering the opportunity for the engineer to identify a distribution that may be less well known but might prove effective for a particular dataset. Several well-known distributions are relegated to Part 2 simply because they are unlikely to be applicable to process data but merit an explanation as to why not. A few distributions, which have uses only tenuously related to process control, are included because I am frequently reminded not to underestimate the ingenuity of control engineers in identifying a previously unconsidered application.

As usual, I am tremendously indebted to my clients' control engineers. Their cooperation in us together applying published statistical methods to their processes has helped hugely in proving their benefit. Much of the material contained in this book is now included in our training courses. Without the feedback from our students, putting what we cover into practice, the refinements that have improved practicability would never have been considered.

Finally, I apologise for not properly crediting everyone that might recognise, as theirs, a statistical technique reproduced in this book. While starting with the best of intentions to do so, it proved impractical. Many different statistical distributions can readily be derived from each other. It is not always entirely clear who thought of what first, and there can be dozens of papers involved. I appreciate that academics want to be able to review published work in detail. Indeed, I suspect that the pure statistician might be quite critical of the way in which much of the material is presented. It lacks much of the mathematical detail they like to see, and there are many instances where I have modified and applied their techniques in ways of which they would not approve. However this book is primarily for practitioners who are generally happy just that a method works. A simple Internet search should provide more detailed background if required.

Myke King
Isle of Wight
June 2017

About the Author

Myke King is the founder and director of Whitehouse Consulting, an independent consulting organisation specialising in process control. He has over 40 years' experience working with over 100 clients in 40 countries. As part of his consulting activities Myke has developed training courses covering all aspects of process control. To date, around 2,000 delegates have attended these courses. He also lectures at several universities. To support his consulting activities he has developed a range of software to streamline the design of controllers and to simulate their use for learning exercises. Indeed, part of this software can be downloaded from the companion web site for this book.

Myke graduated from Cambridge University in the UK with a Master's degree in chemical engineering. His course included process control taught as part of both mechanical engineering and chemical engineering. At the time he understood neither. On graduating he joined, by chance, the process control section at Exxon's refinery at Fawley in the UK. Fortunately he quickly discovered that the practical application of process control bore little resemblance to the theory he had covered at university. He later became head of the process control section and then moved to operations department as a plant manager. This was followed by a short period running the IT section.

Myke left Exxon to co-found KBC Process Automation, a subsidiary of KBC Process Technology, later becoming its managing director. The company was sold to Honeywell where it became their European centre of excellence for process control. It was at this time Myke set up Whitehouse Consulting.

Myke is a Fellow of the Institute of Chemical Engineers in the UK.

Supplementary Material

To access supplementary materials for this book please use the download links shown below.

There you will find valuable material designed to enhance your learning, including:

- Excel spreadsheets containing data used in the examples. Download directly from www.whitehouse-consulting.com/examples.xlsx
- Executable file (www.whitehouse-consulting.com/statistician.exe) to help you perform your own data analysis

For Instructor use only:

Powerpoint slides of figures can be found at

http://booksupport.wiley.com

Please enter the book title, author name or ISBN to access this material.

Part 1

The Basics

1

Introduction

Statistical methods have a very wide range of applications. They are commonplace in demographic, medical and meteorological studies, along with more recent extension into financial investments. Research into new techniques incurs little cost and, nowadays, large quantities of data are readily available. The academic world takes advantage of this and is prolific in publishing new techniques. The net result is that there are many hundreds of techniques, the vast majority of which offer negligible improvement for the process industry over those previously published. Further, the level of mathematics now involved in many methods puts them well beyond the understanding of most control engineers. This quotation from Henri Poincaré, although over 100 years old and directed at a different branch of mathematics, sums up the situation well.

> *In former times when one invented a new function it was for a practical purpose; today one invents them purposely to show up defects in the reasoning of our fathers and one will deduce from them only that.*

The reader will probably be familiar with some of the more commonly used statistical distributions – such as those described as uniform or normal (Gaussian). There are now over 250 published distributions, the majority of which are offspring of a much smaller number of parent distributions. The software industry has responded to this complexity by developing products that embed the complex theory and so remove any need for the user to understand it. For example, there are several products in which their developers pride themselves on including virtually every distribution function. While not approaching the same range of techniques, each new release of the common spreadsheet packages similarly includes additional statistical functions. While this has substantial practical value to the experienced engineer, it has the potential for an under-informed user to reach entirely wrong conclusions from analysing data.

Very few of the mathematical functions that describe published distributions are developed from a physical understanding of the mechanism that generated the data. Virtually all are empirical. Their existence is justified by the developer showing that they are better than a previously developed function at matching the true distribution of a given dataset. This is achieved by the

Statistics for Process Control Engineers: A Practical Approach, First Edition. Myke King.

inclusion of an additional fitting parameter in the function or by the addition of another non-linear term. No justification for the inclusion is given, other than it provides a more accurate fit. If applied to another dataset, there is thus no guarantee that the improvement would be replicated.

In principle there is nothing wrong with this approach. It is analogous to the control engineer developing an inferential property by regressing previously collected process data. Doing so requires the engineer to exercise judgement in ensuring the resulting inferential calculation makes engineering sense. He also has to balance potential improvements to its accuracy against the risk that the additional complexity reduces its robustness or creates difficult process dynamics. Much the same judgemental approach must be used when selecting and fitting a distribution function.

2

Application to Process Control

Perhaps more than any other engineering discipline, process control engineers make extensive use of statistical methods. Embedded in proprietary control design and monitoring software, the engineer may not even be aware of them. The purpose of this chapter is to draw attention to the relevance of statistics throughout all stages of implementation of improved controls – from estimation of the economic benefits, throughout the design phase, ongoing performance monitoring and fault diagnosis. Those that have read the author's first book *Process Control: A Practical Approach* will be aware of the detail behind all the examples and so most of this has been omitted here.

2.1 Benefit Estimation

The assumption that the variability or standard deviation (σ) is halved by the implementation of improved regulatory control has become a de facto standard in the process industry. It has no theoretical background; indeed it would be difficult to develop a value theoretically that is any more credible. It is accepted because post-improvement audits generally confirm that it has been achieved. But results can be misleading because the methodology is being applied, as we will show, without a full appreciation of the underlying statistics.

There are a variety of ways in which the benefit of reducing the standard deviation is commonly assessed. The *Same Percentage Rule*[1,2] is based on the principle that if a certain percentage of results already violate a specification, then after improving the regulatory control, it is acceptable that the percentage violation is the same. Halving the standard deviation permits the average giveaway to be halved.

$$\Delta\bar{x} = 0.5\left(x_{\text{target}} - \bar{x}\right) \tag{2.1}$$

This principle is illustrated in Figure 2.1. Using the example of diesel quality data that we will cover in Chapter 3, shown in Table A1.3, we can calculate the mean as 356.7°C and the standard deviation as 8.4°C. The black curve shows the assumed distribution. It shows that the probability of the product being on-grade, with a 95% distillation point less than 360°C, is 0.65.

Statistics for Process Control Engineers: A Practical Approach, First Edition. Myke King.

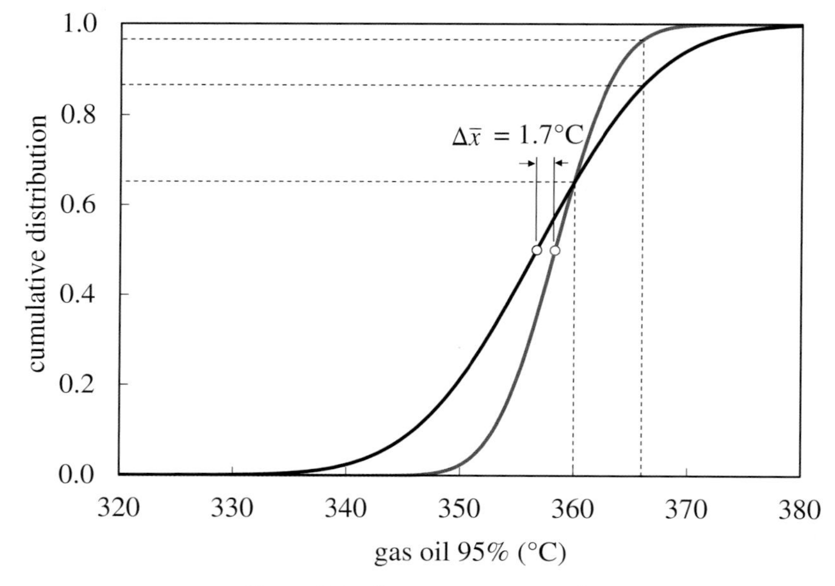

Figure 2.1 Same percentage rule

In other words, we expect 35% of the results to be off-grade. Halving the standard deviation, as shown by the coloured curve, would allow us to increase the mean while not affecting the probability of an off-grade result. From Equation (2.1), improved control would allow us to more closely approach the specification by 1.7°C.

We will show later that it is not sufficient to calculate mean and standard deviation from the data. Figure 2.2 again plots the assumed distribution but also shows, as points, the distribution

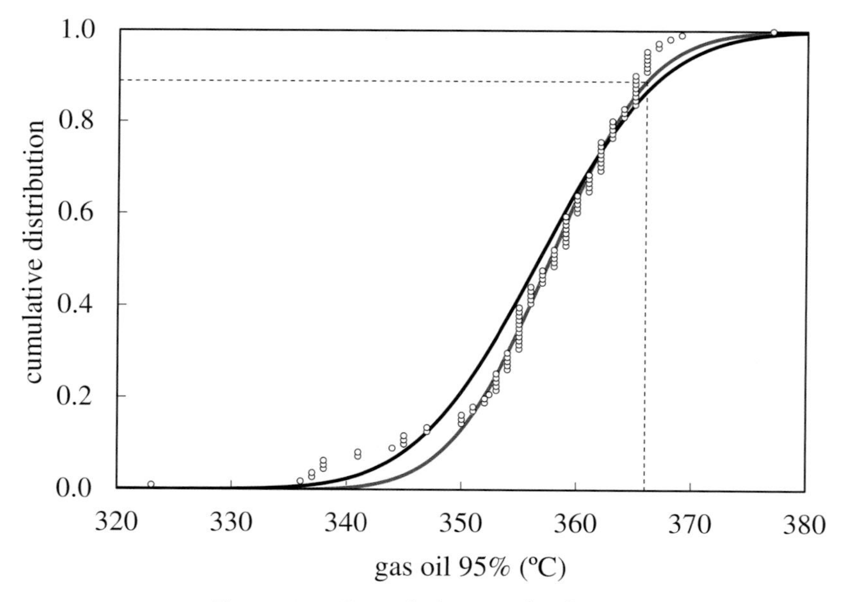

Figure 2.2 Properly fitting a distribution

of the actual data. The coloured curve is the result of properly fitting a normal distribution to these points, using the method we will cover in Chapter 9. This estimates the mean as 357.7°C and the standard deviation as 6.9°C. From Equation (2.1), the potential improvement is now 1.2°C. At around 30% less than the previous result, this represents a substantial reduction in the benefit achievable.

A second potential benefit of improved control is a reduction in the number of occasions the gasoil must be reprocessed because the 95% distillation point has exceeded 366°C. As Figure 2.1 shows, the fitted distribution would suggest that the probability of being within this limit is 0.888. This would suggest that, out of the 111 results, we would then expect the number of reprocessing events to be 12. In fact there were only five. It is clear from Figure 2.2 that the assumed distribution does not match the actual distribution well – particularly for the more extreme results. The problem lies now with the choice of distribution. From the large number of alternative distributions it is likely that a better one could be chosen. Or, even better, we might adopt a discrete distribution suited to estimation of the probability of events. We could also apply an extreme value analytical technique. Both these methods we cover in Chapter 13.

2.2 Inferential Properties

A substantial part of a control engineer's role is the development and maintenance of inferential property calculations. Despite the technology being well established, not properly assessing their performance is the single biggest cause of benefits not being fully captured. Indeed, there are many examples where process profitability would be improved by their removal.

Most inferentials are developed through regression of previously collected process data. Doing so employs a wide range of statistical techniques. Regression analysis helps the engineer identify the most accurate calculation but not necessarily the most practical. The engineer has to apply other techniques to assess the trade-off between complexity and accuracy.

While there are 'first-principle' inferentials, developed without applying statistical methods, once commissioned both types need to be monitored to ensure the accuracy is maintained. If an accuracy problem arises, then the engineer has to be able to assess whether it can be safely ignored as a transient problem, whether it needs a routine update to its bias term or whether a complete redesign is necessary. While there is no replacement for relying on the judgement of a skilled engineer, statistics play a major role in supporting this decision.

2.3 Controller Performance Monitoring

Perhaps the most recent developments in the process control industry are process control performance monitoring applications. Vendors of MPC packages have long offered these as part of a suite of software that supports their controllers. But more recently the focus has been on monitoring basic PID control, where the intent is to diagnose problems with the controller itself or its associated instrumentation. These products employ a wide range of statistical methods to generate a wide range of performance parameters, many of which are perhaps not fully understood by the engineer.

2.4 Event Analysis

Event analysis is perhaps one of the larger opportunities yet to be fully exploited by process control engineers. For example, they will routinely monitor the performance of advanced control – usually reporting a simple service factor. Usually this is the time that the controller is in service expressed as a fraction of the time that it should have been in service. While valuable as reporting tool, it has limitations in terms of helping improve service factor. An advanced control being taken out of service is an example of an event. Understanding the frequency of such events, particularly if linked to cause, can help greatly in reducing their frequency.

Control engineers often have to respond to instrument failures. In the event of one, a control scheme may have to be temporarily disabled or, in more extended cases, be modified so that it can operate in some reduced capacity until the fault is rectified. Analysis of the frequency of such events, and the time taken to resolve them, can help justify a permanent solution to a recurring problem or help direct management to resolve a more general issue.

Inferential properties are generally monitored against an independent measurement, such as that from an on-stream analyser or the laboratory. Some discrepancy is inevitable and so the engineer will have previously identified how large it must be to prompt corrective action. Violating this limit is an event. Understanding the statistics of such events can help considerably in deciding whether the fault is real or the result of some circumstance that needs no attention.

On most sites, at least part of the process requires some form of sequential, rather than continuous, operation. In an oil refinery, products such as gasoline and diesel are batch blended using components produced by upstream continuous processes. In the polymers industry plants run continuously but switch between grades. Downstream processing, such as extrusion, has to be scheduled around extruder availability, customer demand and product inventory. Other industries, such as pharmaceuticals, are almost exclusively batch processes. While most advanced control techniques are not applicable to batch processes, there is often the opportunity to improve profitability by improved scheduling. Understanding the statistical behaviour of events such as equipment availability, feedstock availability and even the weather can be crucial in optimising the schedule.

Many control engineers become involved in alarm studies, often following the guidelines[3] published by Engineering Equipment and Materials Users' Association. These recommend the following upper limits per operator console:

- No more than 10 *standing alarms*, i.e. alarms which have been acknowledged
- No more than 10 *background alarms* per hour, i.e. alarms for information purposes that may not require urgent attention
- No more than 10 alarms in the first 10 minutes after a major process problem develops

There are also alarm management systems available that can be particularly useful in identifying repeating *nuisance* and *long-standing* alarms. What is less common is examination of the probability of a number of alarms occurring. For example, if all major process problemshave previously met the criterion of not more than 10 alarms, but then one causes 11, should this prompt a review? If not, how many alarms would be required to initiate one?

2.5 Time Series Analysis

Often overlooked by control engineers, feed and product storage limitations can have a significant impact on the benefits captured by improved control. Capacity utilisation is often the major source of benefits. However, if there are periods when there is insufficient feed in storage or insufficient capacity to store products, these benefits would not be captured. Indeed, it may be preferable to operate the process at a lower steady feed rate rather than have the advanced control continuously adjust it. There is little point in maximising feed rate today if there will be a feed shortage tomorrow.

Modelling the behaviour of storage systems requires a different approach to modelling process behaviour. If the level in a storage tank was high yesterday, it is very unlikely to be low today. Such levels are *autoregressive*, i.e. the current level (L_n) is a function of previous levels.

$$L_n = a_0 + a_1 L_{n-1} + a_2 L_{n-2} + \ldots \tag{2.2}$$

The level is following a *time series*. It is not sufficient to quantify the variation in level in terms of its mean and standard deviation. We need also to take account of the sequence of levels.

Time series analysis is also applicable to the process unit. Key to effective control of any process is understanding the process dynamics. *Model identification* determines the correlation between the current process value (PV_n), previous process values (PV_{n-1}, etc.) and previous values of the manipulated variable (MV) delayed by the process deadtime (θ). If ts is the data collection interval, the *autoregressive with exogenous input* (*ARX*) model for a single MV has the form

$$PV_n = a_0 + a_1 PV_{n-1} + a_2 PV_{n-2} + \ldots + b_1 MV_{n-\theta/ts} + b_2 MV_{n-1-\theta/ts} \cdots \tag{2.3}$$

For a *first order* process, this model will include only one or two historical values. Simple formulae can then be applied to convert the derived coefficients to the more traditional parametric model based on process gain, deadtime and lag. These values would commonly be used to develop tuning for basic PID controllers and for advanced regulatory control (ARC) techniques. Higher order models can be developed by increasing the number of historical values and these models form the basis of some proprietary MPC packages. Other types of MPC use the time series model directly.

There is a wide range of proprietary model identification software products. Control engineers apply them without perhaps fully understanding how they work. Primarily they use *regression analysis* but several other statistical techniques are required. For example, increasing the number of historical values will always result in a model that is mathematically more accurate. Doing so, however, will increasingly model the noise in the measurements and reduce the robustness of the model. The packages include statistical techniques that select the optimum model length. We also need to assess the reliability of the model. For example, if the process disturbances are small compared to measurement noise or if the process is highly nonlinear, there may be little confidence that the identified model is reliable. Again the package will include some statistical technique to warn the user of this. Similarly statistical methods might also be used to remove any suspect data before model identification begins.

3

Process Examples

Real process data has been used throughout to demonstrate how the techniques documented can be applied (or not). This chapter simply describes the data and how it might be used. Where practical, data are included as tables in Appendix 1 so that the reader can reproduce the calculations performed. All of the larger datasets are available for download.

The author's experience has been gained primarily in the oil, gas and petrochemical industries; therefore much of the data used come from these. The reader, if from another industry, should not be put off by this. The processes involved are relatively simple and are explained here. Nor should the choice of data create the impression that the statistical techniques covered are specific to these industries. They are not; the reader should have no problem applying them to any set of process measurements.

3.1 Debutaniser

The debutaniser column separates C_{4-} material from naphtha, sending it to the de-ethaniser. Data collected comprises 5,000 hourly measurements of reflux (R) and distillate (D) flows. Of interest is, if basic process measurements follow a particular distribution, what distribution would a derived measurement follow? In Chapter 10 the flows are used to derive the reflux ratio (R/D) to demonstrate how the ratio of two measurements might be distributed.

3.2 De-ethaniser

The overhead product is a gas and is fed to the site's fuel gas system, along with many other sources. Disturbances to the producers cause changes in fuel gas composition – particularly affecting its molecular weight and heating value. We cover this later in this chapter.

The bottoms product is mixed LPG (propane plus butane) and it routed to the splitter. The C_2 content of finished propane is determined by the operation of the de-ethaniser. We cover, later in this chapter, the impact this has on propane cargoes.

Statistics for Process Control Engineers: A Practical Approach, First Edition. Myke King.

3.3 LPG Splitter

The LPG splitter produces sales grade propane and butane as the overheads and bottoms products respectively. Like the debutaniser, data collected includes 5,000 hourly measurements of reflux and distillate flows. These values are used, along with those from the debutaniser, to explore the distribution of the derived reflux ratio.

The reflux flow is normally manipulated by the composition control strategy. There are columns where it would be manipulated by the reflux drum level controller. In either case the reflux will be changed in response to almost every disturbance to the column. Of concern on this column are those occasions where reflux exceeds certain flow rates. Above 65 m^3/hr the column can flood. A flow above 70 m^3/hr can cause a pump alarm. Above 85 m^3/hr, a trip shuts down the process.

Figure 3.1 shows the variation in reflux over 5,000 hours. Figure 3.2 shows the distribution of reflux flows. The shaded area gives the probability that the reflux will exceed 65 m^3/hr. We will show, in Chapter 5, how this value is quantified for the normal distribution and, in subsequent chapters, how to apply different distributions.

Alternatively, a high reflux can be classed as an event. Figure 3.1 shows 393 occasions when the flow exceeded 65 m^3/hr and 129 when it exceeded 70 m^3/hr. The distribution can then be based on the number of events that occur in a defined time. Figure 3.3 shows the distribution of the number of events that occur per day. For example, it shows that the observed probability of the reflux not exceeding 70 m^3/hr in a 24 hour period (i.e. 0 events per day) is around 0.56. Similarly the most likely number of violations, of the 65 m^3/hr limit, is two per day, occurring on approximately 29% of the days. We will use this behaviour, in Chapter 12 and Part 2, to show how many of the discrete distributions can be applied. Another approach is to analyse the variation of time between high reflux events. Figure 3.4 shows the observed distribution of the interval between exceeding 65 m^3/hr. For example, most likely is an interval of one hour – occurring

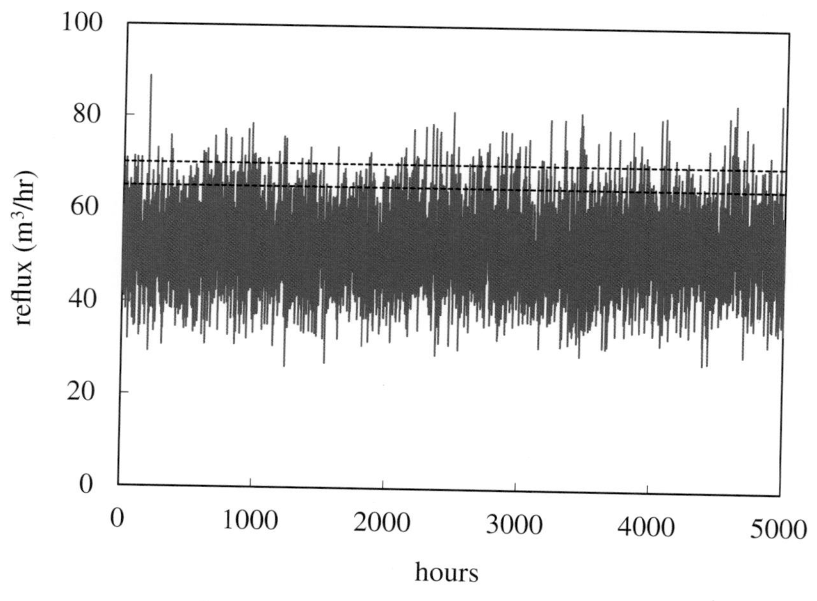

Figure 3.1 Variation in LPG splitter reflux

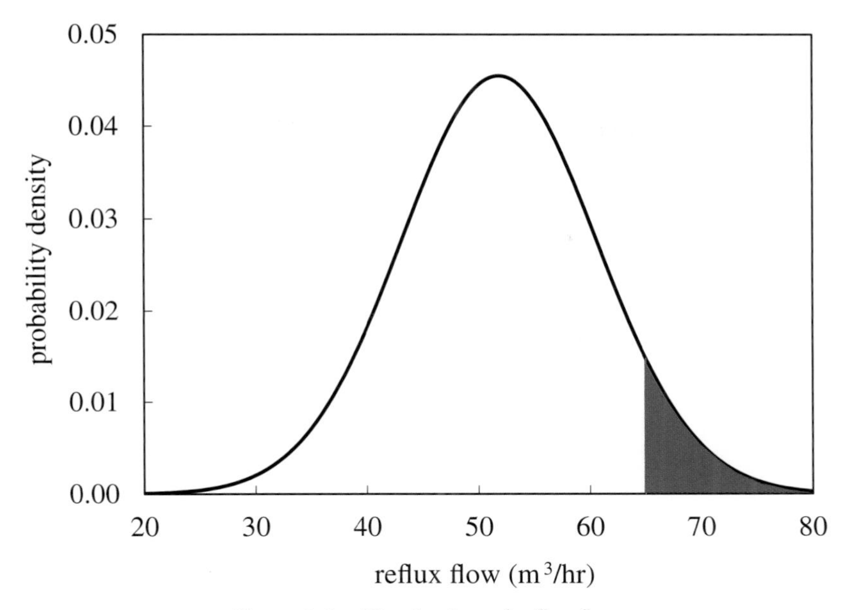

Figure 3.2 *Distribution of reflux flow*

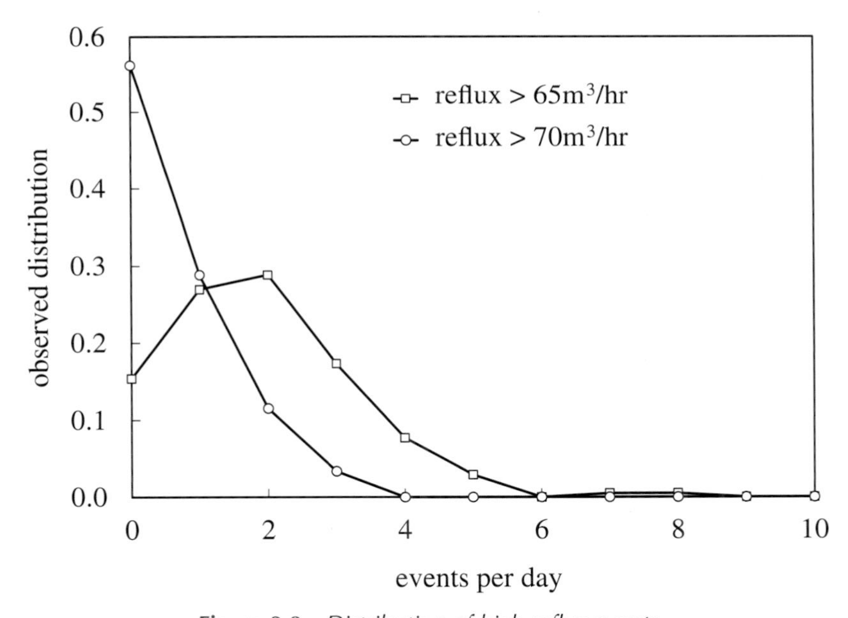

Figure 3.3 *Distribution of high reflux events*

on about 9% of the occasions. In this form, continuous distributions can then be applied to the data.

Table A1.1 shows the C_4 content of propane, not the finished product but sampled from the rundown to storage. It includes one year of daily laboratory results, also shown in Figure 3.5. Of interest is the potential improvement to composition control that will increase the C_4 content

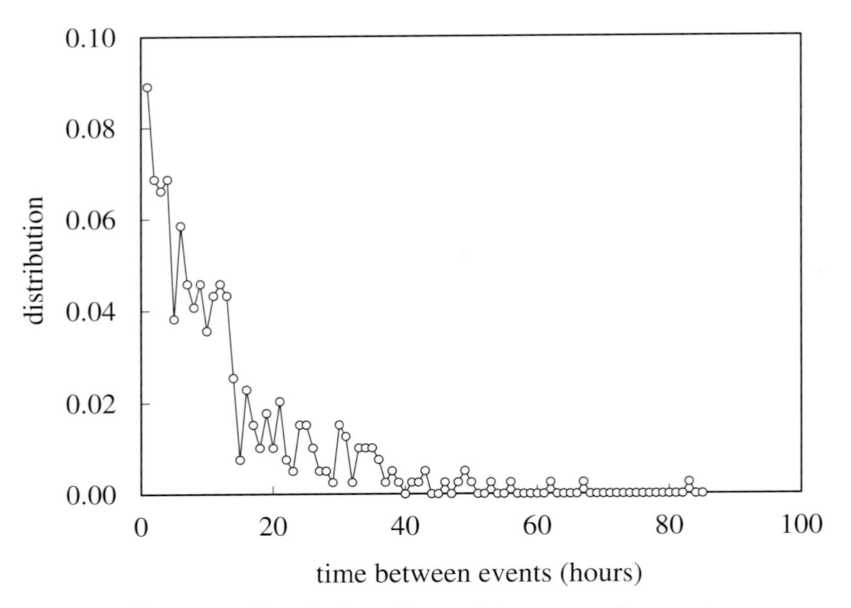

Figure 3.4 *Distribution of intervals between reflux events*

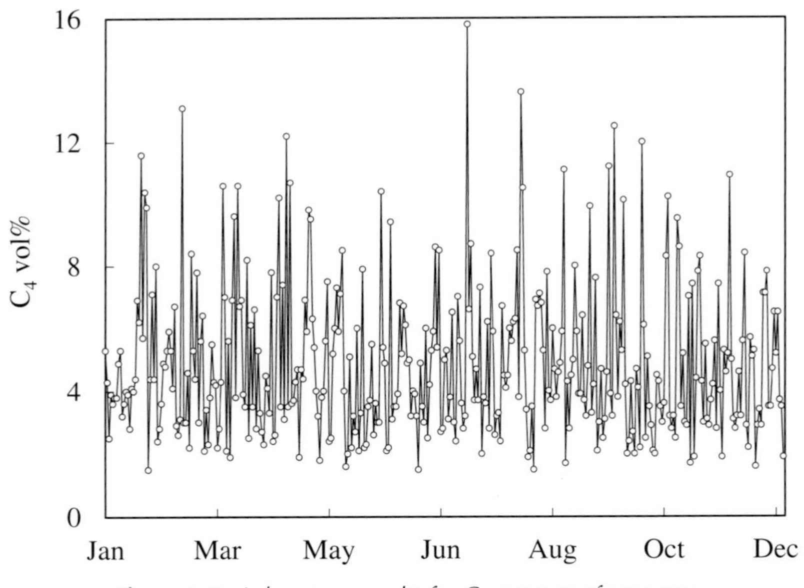

Figure 3.5 *Laboratory results for C_4 content of propane*

closer to the specification of 10 vol%. To determine this we need an accurate measure of the current average content and its variation. The key issue is choosing the best distribution. As Figure 3.6 shows, the best fit normal distribution does not match well the highly skewed data. Indeed, it shows about a 4% probability that the C_4 content is negative. We will clearly need to select a better form of distribution from the many available.

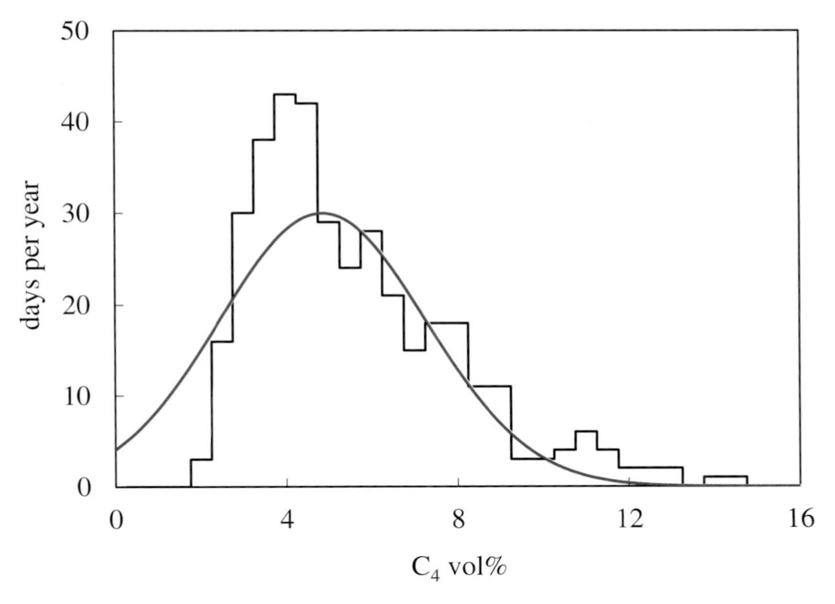

Figure 3.6 Skewed distribution of C_4 results

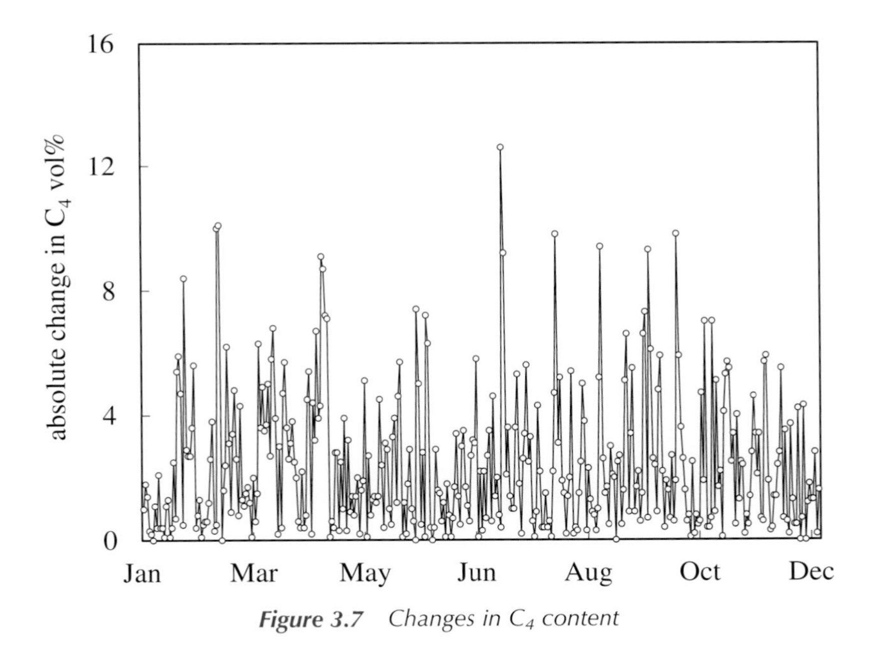

Figure 3.7 Changes in C_4 content

Also of concern are the occasional very large changes to the C_4 content, as shown by Figure 3.7, since these can cause the product, already in the rundown sphere, to be put off-grade. We will show how some distributions can be used to assess the impact that improved control might have on the frequency of such disturbances.

There is an analyser and an inferential property installed on the bottoms product measuring the C_3 content of butane. Figure 3.8 shows data collected every 30 minutes over 24 days, i.e.

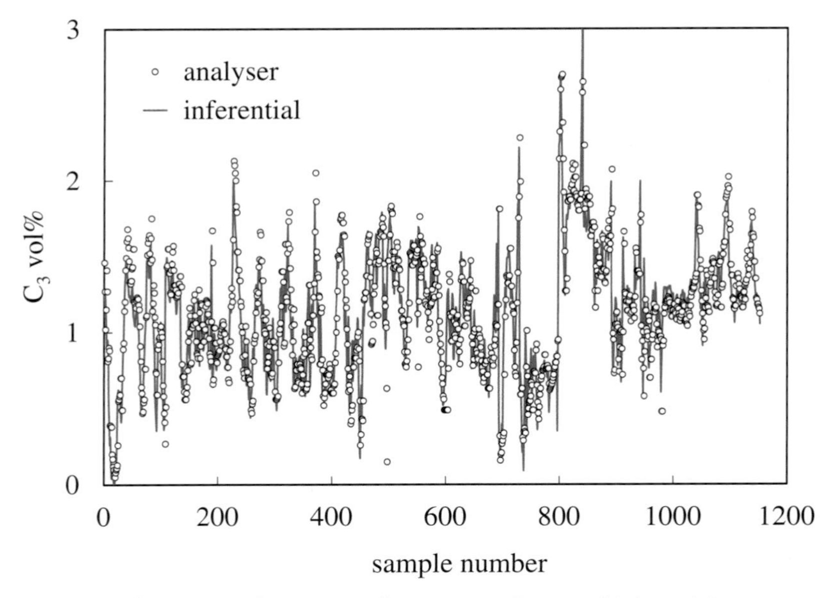

Figure 3.8 Comparison between analyser and inferential

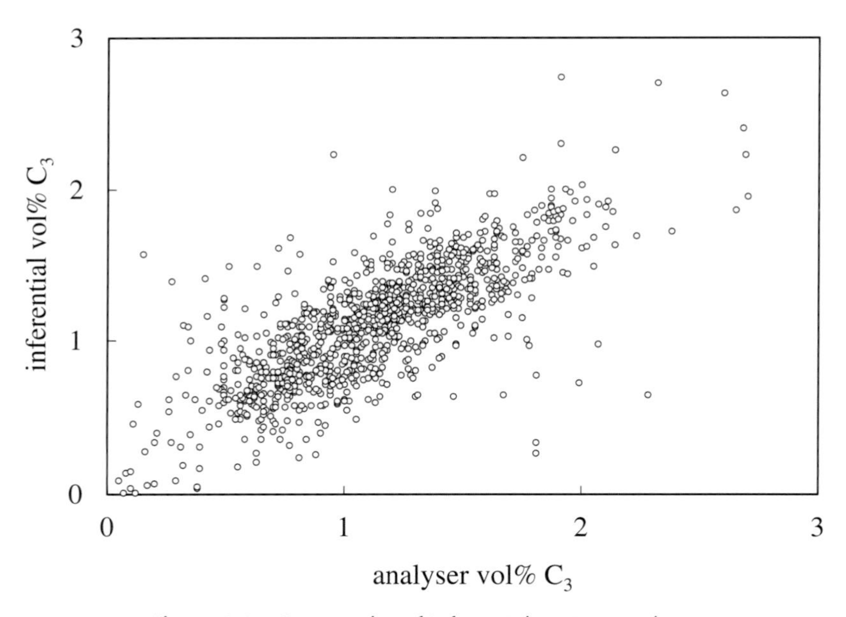

Figure 3.9 Scatter plot of inferential against analyser

1,152 samples. Such line plots are deceptive in that they present the inferential as more accurate than it truly is. Figure 3.9 plots the same data as a scatter diagram showing that, for example, if the analyser is recording 1 vol%, the inferential can be in error by ±0.5 vol%. Further, there is tendency to assume that such errors follow the normal distribution. Figure 3.10 shows the best fit normal distribution. The actual frequency of small errors is around double that suggested by

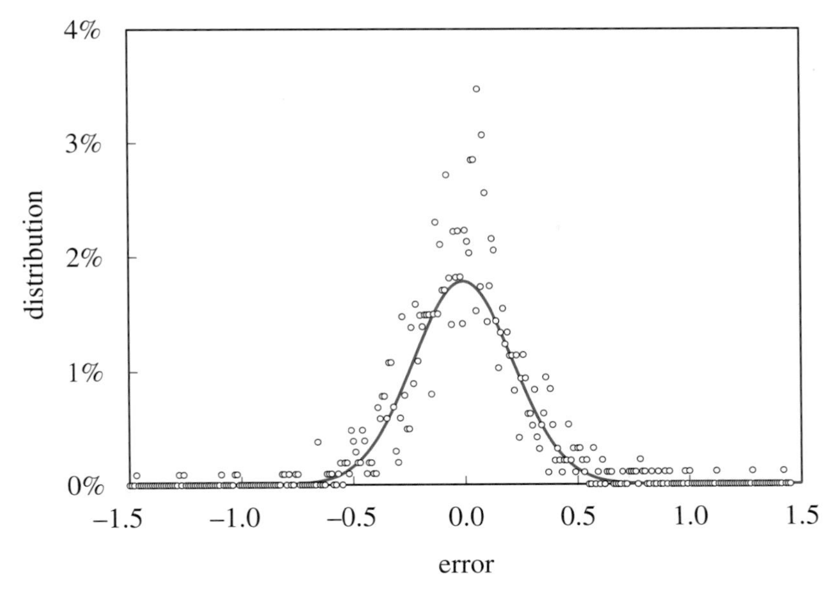

Figure 3.10 *Actual distribution very different from normal*

the normal distribution. We will look at how other distributions are better suited to analysing this problem.

3.4 Propane Cargoes

Propane from the LPG splitter is routed to one of two spheres. Once a sphere is full, production is switched to the other sphere. Cargoes are shipped from the filled sphere once the laboratory has completed a full analysis for the certificate of quality. The maximum C_2 content permitted by the propane product specification is 5 vol%. Table A1.2 includes the results, taken from the certificates of quality, for 100 cargoes. While primarily used to illustrate, in Chapter 6, methods of presenting data, it is also used to draw attention to the difference between analysing finished product data as opposed to data collected from the product leaving the process.

3.5 Diesel Quality

A property commonly measured for oil products is the *distillation point*. Although its precise definition is more complex, in principle it is the temperature at which a certain percentage of the product evaporates. Gasoil, for example, is a component used in producing diesel and might have a specification that 95% must evaporate at a temperature below 360°C. There is an economic incentive to get as close as possible to the specification.

Table A1.3 shows the results of 111 daily laboratory samples taken from the gasoil rundown. The product is routed to a storage tank which is well-mixed before being used in a blend. A certain amount of off-grade production is permissible provided it is balanced by product in giveaway, so that the filled tank is within specification. Indeed, as Figure 3.11 shows, 40 of the results violate the specification. But the simple average, represented by the coloured

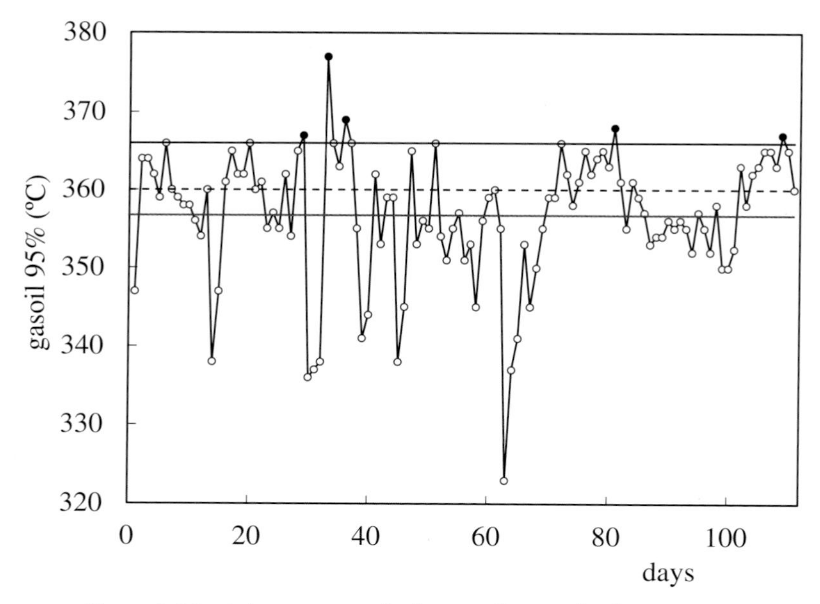

Figure 3.11 *Laboratory results for gasoil 95% distillation point*

line, shows giveaway of 3.3°C. Improving composition control would reduce the variation and so allow closer approach to the specification and an increase in product yield. We will use these data to show how to more properly estimate the average operation and its variation.

Of course there are industries where no amount of off-grade production is permitted. Most notable are the paper and metals industries where the off-grade material cannot be blended with that in giveaway. Our example has a similar situation. At a distillation point above 366°C undesirable components can be present in the product that cannot be sufficiently diluted by mixing in the tank. Any material this far off-grade must be reprocessed. The figures highlighted in Table A1.3 and Figure 3.11 show the five occasions when this occurred. Improved control would reduce the number of occasions and so reduce reprocessing costs. We will use these data to explore the use of discrete distributions that might be used to determine the savings.

3.6 Fuel Gas Heating Value

In common with many sites all fuel gas from producers, such as the de-ethaniser, is routed to a header from which all consumers take their supply. A disturbance on any producer causes a change in NHV (net heating value) that then upsets a large number of fired heaters and boilers. Some consumers of fuel gas are also producers; a disturbance to the gas supplied to these units propagates through to further disturb the header NHV.

The data, included as Table A1.4, comprises laboratory results collected daily for a period of six months. It was collected to identify the best approach to handling variations to reduce the process disturbances. There are several solutions. One might be to install densitometers, use these to infer NHV and effectively convert the flow controllers on each consumer to duty controllers. Another might be to switch from conventional orifice plate type flowmeters to coriolis types on the principle that heating value measured on a weight basis varies much less than that

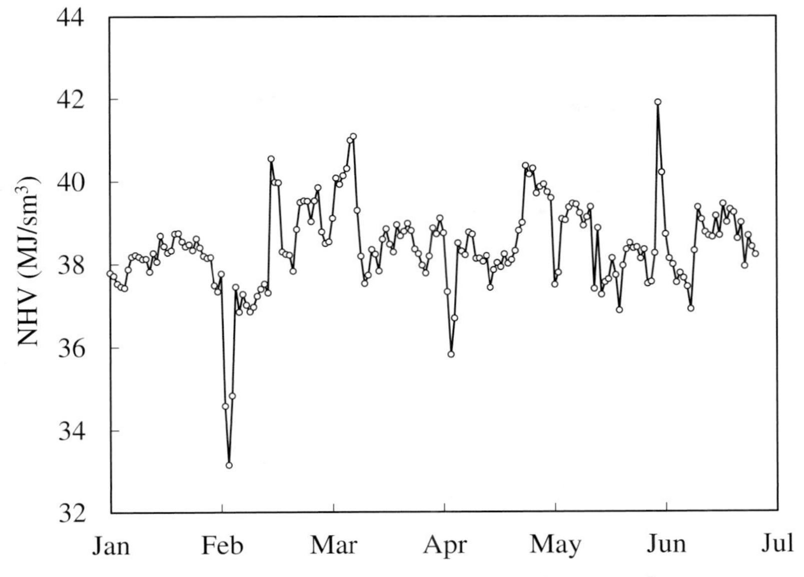

Figure 3.12 Variation in site's fuel gas heating value

measured on a volume basis. Understanding the variation in NHV permits each solution to be assessed in terms of the achievable reduction in disturbances.

Figure 3.12 shows the variation of the NHV of fuel gas routed to a fired heater. The disturbances (x) are determined from Table A1.4 as

$$x_n = \frac{NHV_n - NHV_{n-1}}{NHV_{n-1}} \times 100 \qquad n > 1 \tag{3.1}$$

These disturbances are plotted as Figure 3.13. Figure 3.14 shows that the best fit normal distribution is unsuitable since it has tails much fatter than the distribution of the data. The data will therefore be used to help assess the suitability of alternative distributions, some of which do not accommodate negative values. For this type of data, where it would be reasonable to assume that positive and negative disturbances are equally likely, one approach is to fit to the absolute values of the disturbances.

Table A1.5 comprises analyses of 39 of the samples of fuel gas showing the breakdown by component. These we will use to illustrate how a multivariate distribution might be applied.

3.7 Stock Level

While the control engineer may feel that there is little application of process control to product storage, understanding its behaviour can be crucial to properly estimating benefits of improved control and in assessing what changes might be necessary as a result of a control improvement. For example, the final product from many sites is the result of blending components. Properly controlling such blending can substantially improve profitability but, in estimating the benefits of a potential improvement we need to assess the availability not only of the blend components but also available capacity for finished product. There are many projects that have been justified

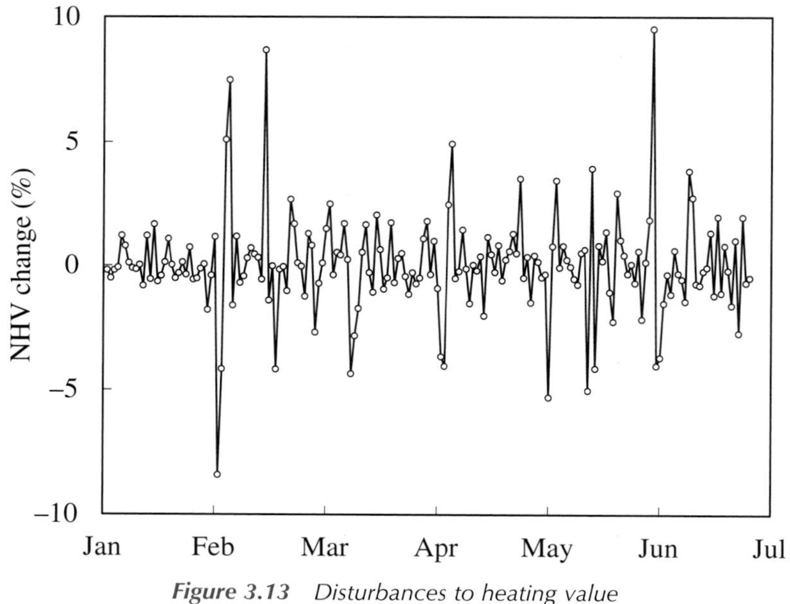

Figure 3.13 *Disturbances to heating value*

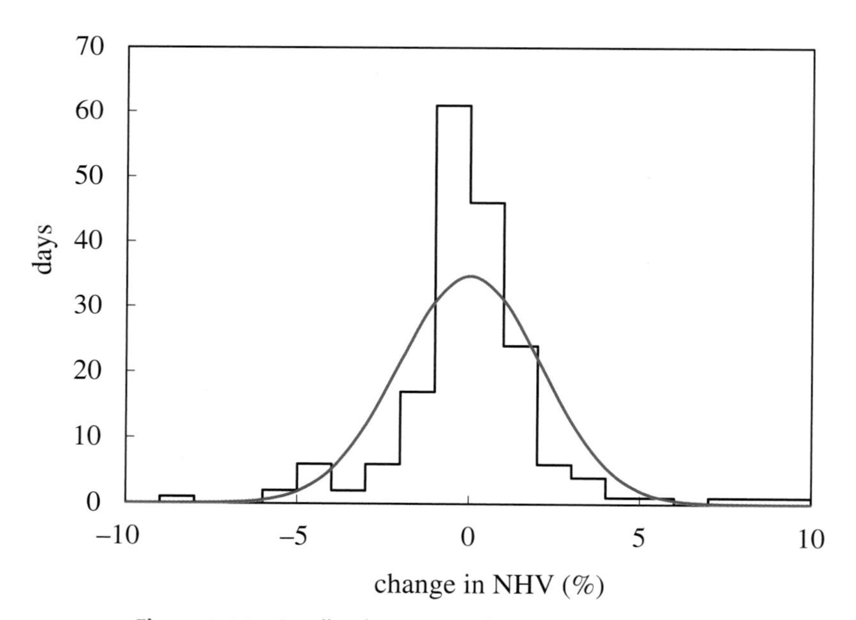

Figure 3.14 *Small tails compared to normal distribution*

on increasing production rates only to find that this cannot be fully accommodated by the storage facilities.

The data, included in Table A1.6, is the daily stock level of a key component collected over a seven month period or 220 days. The variation is also shown in Figure 3.15. Figure 3.16 shows

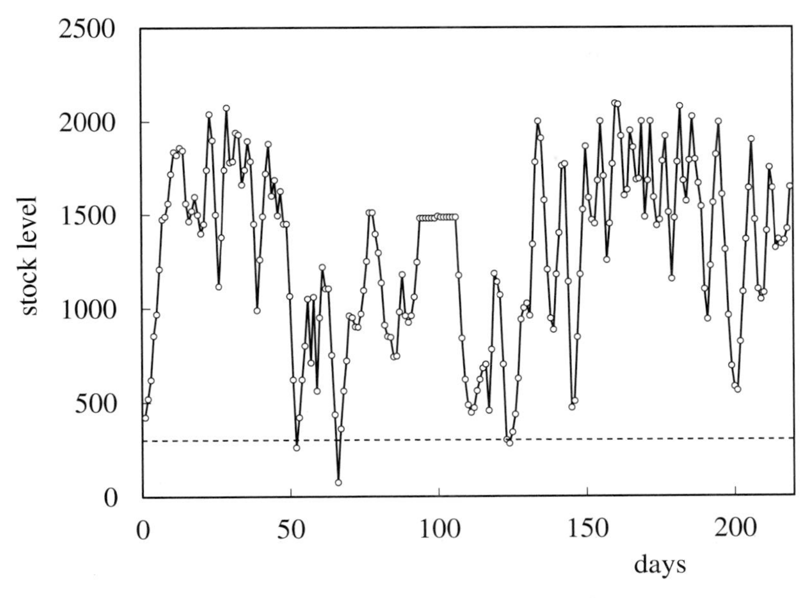

Figure 3.15 *Variation in stock level*

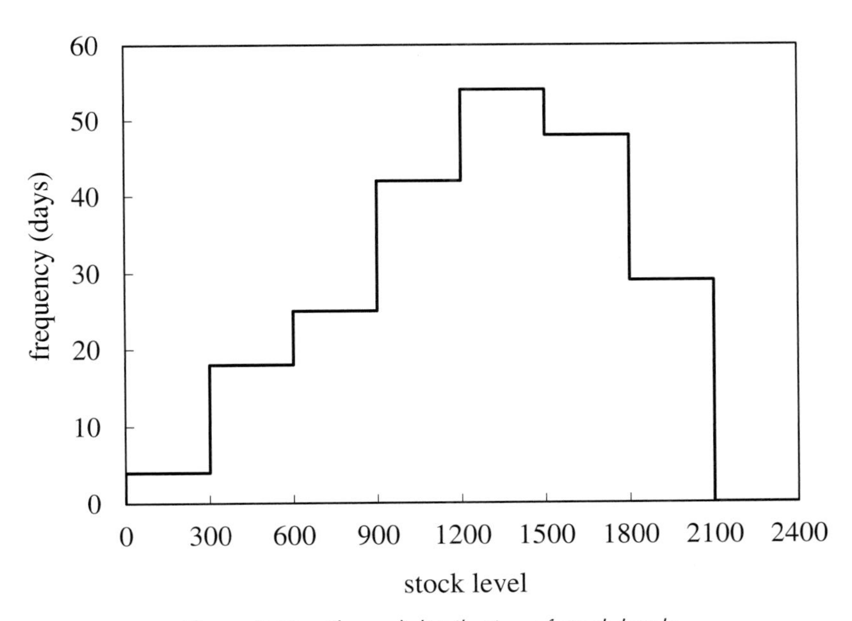

Figure 3.16 *Skewed distribution of stock levels*

the non-symmetrical distribution of the data. The main concern is those occasions when the inventory drops to 300, below which a blend cannot be started. There are three occasions, highlighted as bold in Table A1.6, when this occurred. The data will be used to demonstrate how discrete distributions might estimate the probability of such an occurrence in the future. We will also apply a time series technique to predict the future variation in level.

3.8 Batch Blending

Batch blending is common to many industries, including quite large continuous processes such as oil refineries. In this particular example, a batch is produced by blending components and then taking a laboratory sample of the product. The key specification is 100; if the laboratory result is less than this, a corrective trim blend is added to the product and it is then retested. Each blend takes a day, including the laboratory analysis of its quality. This is repeated as necessary until the product is on-grade. 100 m^3 is then routed to a downstream process and then to sales. The material remaining in the blend vessel forms part of the next batch.

There is large incentive to improve the control and so reduce the number of trim blends. This would permit an increase in production and reduction in storage requirements. Table A1.7 and Figure 3.17 show the intermediate and finished laboratory results for the 78 blends resulting in 44 finished batches. This example is used to explore the use of both continuous and discrete distributions in assessing the improvement that might arise from improved control. We will also show how to use some distributions to explore what change in storage facilities might be required if production is increased.

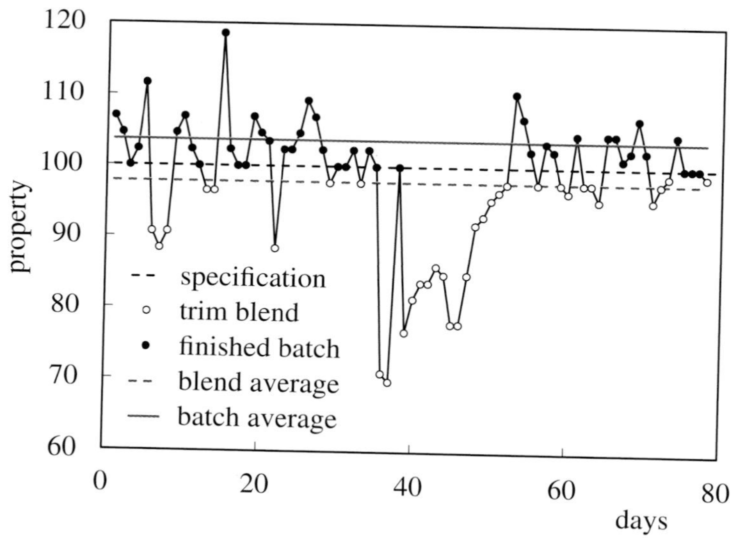

Figure 3.17 Variation in blend property

4

Characteristics of Data

4.1 Data Types

Data fall into one of three types:

Dichotomous data can have one of only two values. In the process industry examples might include pass/fail status of a check against a product specification. Similarly it might be the pass/fail status from validating a measurement from an on-stream analyser or inferential. Such data can be averaged by assigning a numerical value. For example, we might assign 1 if a MPC application is in use and 0 if not. Averaging the data would then give the service factor of the application.

Nominal data have two or more categories but cannot be ranked sensibly. For example, the oil industry produces multiple grades of many products. Specifications can vary by application, by season and by latitude. For example, a data point might have the value 'summer grade European regular gasoline'. Only limited mathematical manipulation is possible with such data. For example, it would be possible to determine what percentage of cargoes fell into this category.

Cardinal data have two or more categories that can be ranked. Most process data fall into this category. Within this type, data can be *continuous* or *discrete*. Process measurements might generally be considered as continuous measurements. Strictly a DCS can only generate discrete values although, for statistical purposes, the resolution is usually such that they can be treated as continuous. Laboratory results are usually discrete values. This arises because testing standards, for example, those published by the ASTM, specify the resolution to which they should be reported. This is based on the accuracy of the test. For example, the flash point of products like jet fuel is quoted to the nearest 0.5°C, whereas the cloud point of diesel is quoted to the nearest 3°C. Another common example of a discrete variable is the number of events that occur in a fixed time. In principle, the reciprocal of this, which is the elapsed time between events, is a continuous variable. However, real-time databases historise data at a fixed interval, for example one minute, and so even time can then be treated as a discrete variable.

Statistics for Process Control Engineers: A Practical Approach, First Edition. Myke King.

4.2 Memory

Data can arise from a process that has *memory*. This occurs when the current measurement is in any way dependent on the preceding measurement. For example, we might wish to assess the probability of violating the maximum permitted inventory in a storage tank. It is extremely unlikely that the level in the tank will be very low today if it was very high yesterday. Today's inventory will be largely influenced by what it was yesterday.

The same might apply to assessing the likelihood of equipment failure. As equipment ages it might become more prone to failure. The time to failure is no longer an independent variable; it becomes shorter as the length of memory increases.

Process events, such as alarms, can also show memory. The condition that activates the alarm might also trigger several others. Analysis of alarm frequency would show that, rather than follow a random pattern, they tend to occur in batches. In other words the likelihood of an alarm increases if one has just occurred.

Most process data do not have memory. For example, product composition can change quite rapidly on most processes. While the composition now will be closely related to what it was a few seconds ago, it will show little correlation with what it was an hour ago. If composition data were collected as daily laboratory measurements, the process will almost certainly appear *memoryless* or *forgetful*. However measuring a property that changes slowly over time, such as catalyst activity, will show memory.

4.3 Use of Historical Data

We have seen that the process control engineer requires historical process data for a wide range of applications. Primarily these fall into two categories – assessing current performance and predicting future performance.

There are three basic methods of using the data for prediction:

- The simplest is to assume that future operation will be identical to the past. Process data are used directly. For example, in studying how a new control scheme might react to changes in feed composition, it is assumed that it will have to deal with exactly the same pattern of changes as it did in the past.

- The second method is to analyse historical data to identify parameters that accurately describe the distribution of the data as a *probability density function* (*PDF*) or *cumulative density function* (*CDF*). This distribution is then used in assessing future process behaviour. This is perhaps the most common method and a large part of this book presents these density functions in detail.

- The third approach is *Monte Carlo simulation*. This uses the derived distribution to generate a very large quantity of data that have the same statistics as the historical data. The synthesised data is then used to study the likely behaviour of a process in the future. For example, it is commonly used in simulating imports to and exports from product storage to determine what storage capacity is required. Provided the simulated imports and exports have the same statistical distribution as the real situation then the *law of large numbers* tells us the average of the results obtained from a large enough number of trials should be close to the real result.

Key to the success of the latter two methods is accurately defining the shape of the distribution of historical data.

4.4 Central Value

A dataset requires two key parameters to characterise its properties. Firstly, data generally show *central tendency* in that they are clustered around some central value. Secondly, as we shall see in the next section, a parameter is needed to describe how the data is dispersed around the central value. The most commonly used measure of the central value is the *mean* – more colloquially called the *average*. There are many versions of the mean. There are also several alternative measures of the central value. Here we define those commonly defined and identify which might be of value to the control engineer.

The *arithmetic* mean of the population (μ) of a set of N values of x is defined as

$$\mu = \frac{\sum_{i=1}^{N} x_i}{N} \tag{4.1}$$

We will generally work with samples of the population. A sample containing n values will have the mean

$$\bar{x} = \frac{\sum_{i=1}^{n} x_i}{n} \tag{4.2}$$

For example, as part of benefits study, we might examine the average giveaway against the maximum amount of C_4 permitted in propane product. If propane attracts a higher price than butane, we would want to maximise the C_4 content. We would normally take a large number of results to calculate the mean but, as an illustration, let us assume we have only three results of 3.9, 4.7 and 4.2. From Equation (4.2) we calculate the mean as 4.27. If the maximum permitted content is 5, the average giveaway is 0.73. Knowing the annual production of propane, we could use this to determine how much additional C_4 could be sold at the propane price rather than at the lower butane price.

However we should more properly use the *weighted mean*. Imagine that the three results were collected for three cargoes, respectively sized 75, 25 and 50 km^3. If w is the *weighting factor* (in this case the cargo size) then the mean butane content is

$$\bar{x} = \frac{\sum_{i=1}^{n} w_i x_i}{\sum_{i=1}^{n} w_i} \tag{4.3}$$

The true mean C_4 content is therefore 4.13 and the giveaway 0.87. Calculating how much more C_4 could be included in propane would, in this example, give a result some 19% higher than that based on the simple mean. Equation (4.3) is effectively the total C_4 contained in all the cargoes divided by the total propane shipped. The additional C_4 that could have been included in propane is therefore 1.3 km^3 – given by

$$\frac{5\sum_{i=1}^{n} w_i - \sum_{i=1}^{n} w_i x_i}{100} \tag{4.4}$$

We might to keep track of the mean C_4 content over an extending period. Imagine that the three results are the first in a calendar year and we want to update the mean as additional cargoes are shipped to generate a year-to-date (YTD) giveaway analysis. We could of course recalculate the mean from all the available results. Alternatively we can simply update the previously determined mean. In the case of the simple mean, the calculation would be

$$\bar{x}_{n+1} = \frac{n\bar{x}_n + x_{n+1}}{n+1} = \bar{x}_n + \frac{x_{n+1} - \bar{x}_n}{n+1} \tag{4.5}$$

For example, the fourth cargo of the year contains 3.7 vol% C_4. The year-to-date mean then becomes

$$\bar{x}_{n+1} = 4.27 + \frac{3.7 - 4.27}{3+1} = 4.13 \tag{4.6}$$

Note that this is different from a *rolling average*, in which the oldest result is removed when a new one is added. If is the number of values in the average (m) is 3, then

$$\bar{x}_{n+1} = \bar{x}_n + \frac{x_{n+1} - x_{n-m+1}}{m} = 4.27 + \frac{3.7 - 3.9}{3} = 4.20 \tag{4.7}$$

While not normally used as part of improved control studies, the rolling average can be applied as a means of filtering out some of the random behaviour of the process and measurement. Indeed, in some countries, finished product specifications will permit wider variation in the property of a single cargo, provided the rolling average is within the true specification.

To update a weighted average

$$\bar{x}_{n+1} = \frac{\bar{x}_n \sum_{i=1}^{n} w_i + w_{n+1} x_{n+1}}{\sum_{i=1}^{n} w_i + w_{n+1}} = \bar{x}_n + \frac{w_{n+1}(x_{n+1} - \bar{x}_n)}{\sum_{i=1}^{n} w_i + w_{n+1}} \tag{4.8}$$

So, if the fourth cargo was 60 km^3

$$\bar{x}_{n+1} = 4.13 + \frac{60(3.7 - 4.13)}{150 + 60} = 4.01 \tag{4.9}$$

Table 4.1 shows the result of applying the calculations described as additional cargoes are produced.

In addition to the arithmetic mean there is the *harmonic mean*, defined as

$$\bar{x}_h = \frac{n}{\sum_{i=1}^{n} \frac{1}{x_i}} \tag{4.10}$$

The *weighted harmonic mean* is

$$\bar{x}_h = \frac{\sum_{i=1}^{n} w_i}{\sum_{i=1}^{n} \frac{w_i}{x_i}} \tag{4.11}$$

Using heavy fuel oil as an example, its maximum permitted density is 0.991. Giveaway is undesirable because density is reduced by adding diluent, such as gasoil, that would otherwise be sold at a price higher than heavy fuel oil. Consider three cargoes sized 80, 120 and 100 kt with densities of 0.9480, 0.9880 and 0.9740. The weighted average, if calculated from Equation (4.3), would be 0.9727. However, density does not blend linearly on a weight basis.

Table 4.1 Averaging C_4 content of propane cargoes

C_4 vol%	YTD mean	rolling average $n = 3$	cargo km^3	YTD weighted mean
3.9	3.90		75	
4.7	4.30		25	
4.2	4.27	4.27	50	4.13
3.7	4.13	4.20	60	4.01
4.7	4.24	4.20	60	4.16
4.4	4.27	4.27	80	4.22
4.7	4.33	4.60	75	4.30
4.0	4.29	4.37	25	4.29
4.8	4.34	4.50	70	4.35
4.2	4.33	4.33	25	4.35
4.1	4.31	4.37	80	4.32
4.0	4.28	4.10	65	4.29
4.0	4.26	4.03	25	4.28
4.4	4.27	4.13	75	4.29
4.8	4.31	4.40	80	4.34
3.9	4.28	4.37	60	4.31
4.3	4.28	4.33	25	4.31
4.0	4.27	4.07	80	4.28
4.8	4.29	4.37	60	4.31
4.1	4.29	4.30	75	4.30

To properly calculate the mean we should first convert each of the cargoes to km^3. Volume is mass (w) divided by density (x), so Equation (4.11) effectively divides the total mass of the cargoes by their total volume and gives the mean density of 0.9724. While an error in the fourth decimal place may seem negligible, it may be significant when compared to the potential improvement. For example, if improved control increased the mean density to 0.9750, the increase would be about 13% higher than that indicated by Equation (4.3) and so too would be the economic benefit.

Table 4.2 shows how the harmonic mean changes as additional cargoes are produced.

There is also the *geometric mean* derived by multiplying the n values together and taking the nth root of the result, i.e.

$$\bar{x}_g = \sqrt[n]{\prod_{i=1}^{n} x_i} \tag{4.12}$$

The geometric mean can be useful in determining the mean of values that have very different ranges. For example, in addition to density, heavy fuel oil is subject to a maximum viscosity specification of 380 cSt and a maximum sulphur content of 3.5 wt%. If only one diluent is available then it must be added until all three specifications are met. The most limiting specification will not necessarily be the same for every cargo since the properties of the base fuel oil and the diluent will vary. To assess giveaway we would have to divide cargoes into three groups, i.e. those limited on density, those limited on viscosity and those limited on sulphur. In principle we can avoid this by calculating, for each cargo, the geometric mean of the three properties. If all three specifications were exactly met, the geometric mean of the properties would be 10.96. If any property is in giveaway, for example by being 10% off target, then the geometric mean would be reduced by 3.2% – no matter which property it is. However, this should be considered

Table 4.2 Averaging SG of heavy fuel oil cargoes

SG	cargo kt	YTD harmonic mean
0.9480	80	0.9480
0.9880	120	0.9716
0.9740	100	0.9724
0.9830	120	0.9754
0.9510	100	0.9706
0.9520	80	0.9681
0.9830	80	0.9698
0.9560	100	0.9680
0.9640	80	0.9677
0.9560	120	0.9662
0.9840	100	0.9678
0.9700	100	0.9680
0.9600	80	0.9675
0.9770	100	0.9682
0.9550	80	0.9675
0.9700	120	0.9676
0.9730	80	0.9679
0.9730	80	0.9681
0.9770	100	0.9686
0.9840	100	0.9694
0.9830	80	0.9699
0.9580	120	0.9693

as only an indicative measure of the potential to reduce giveaway. The amount of diluent required to reduce density by 10% is not the same as that required to reduce viscosity by 10%. A more precise approach would be to calculate, for each cargo, exactly how much less diluent could have been used without violating any of the three specifications.

Laws of heat transfer, vaporisation and chemical reaction all include a logarithmic function. Taking logarithms of Equation (4.12)

$$\log(\bar{x}_g) = \frac{\sum_{i=1}^{n} \log(x_i)}{n} \tag{4.13}$$

Rather than taking the logarithm of each measurement before calculating the mean, we could instead calculate the geometric mean.

Not to be confused with this definition of the geometric mean, there is also the *logarithmic mean*. It is limited to determining the mean of two positive values. If these are x_1 and x_2 then it is defined as

$$\bar{x}_l = \frac{x_1 + x_2}{\ln(x_1) - \ln(x_2)} = \frac{x_1 + x_2}{\ln\left(\dfrac{x_1}{x_2}\right)} \tag{4.14}$$

It has limited application but is most notably used in calculating the *log mean temperature difference* (*LMTD*) used in heat exchanger design. While Equation (4.14) uses the natural logarithm, the logarithm to any base (e.g. 10) may be used.

Table 4.3 *Analyses of propane cargoes*

vol% C_2	vol% C_3	vol% C_4
1.3	97.8	0.9
3.2	95.2	1.6
1.1	96.2	2.7
2.9	96.0	1.1
0.8	95.8	3.4
0.5	95.1	4.4
1.5	95.2	3.3
0.3	97.7	2.0
2.4	95.4	2.2
2.5	96.9	0.6

The definition of mean can be extended to multidimensional space. Table 4.3 shows the composition of 10 cargoes of propane. Propane must be at least 95 vol% pure and so the total of the C_2 content and the C_4 content must not exceed 5 vol%. Both these components have a lower value than propane and so their content should be maximised. To quantify what improvement is possible we can determine the *centroid*. This is the point at which the *residual sum of the squares* (*RSS*) of the distances from it to the data points is a minimum. In our example we have three variables: vol% C_2 (x_1), vol% C_3 (x_2) and vol% C_4 (x_3). We therefore adjust the coordinates a, b and c to minimise the function

$$RSS = \sum_{i=1}^{n}(x_{1i}-a)^2 + \sum_{i=1}^{n}(x_{2i}-b)^2 + \sum_{i=1}^{n}(x_{3i}-c)^2 \tag{4.15}$$

To identify the minimum we set the partial derivatives of this function to zero. For example, partially differentiating with respect to x_1 gives

$$\frac{\partial RSS}{\partial x_1} = 2\sum_{i=1}^{n}(x_{1i}-a) = 0 \tag{4.16}$$

$$\therefore \sum_{i1}^{n} x_{1i} - na = 0 \quad \text{and so} \quad a = \frac{\sum_{i=1}^{n} x_{1i}}{n} = \bar{x}_1 \tag{4.17}$$

Therefore *RSS* will be at a minimum when the coordinates of the centroid are the arithmetic means of x_1, x_2 and x_3, i.e. (1.65, 96.13, 2.22). The centroid is therefore mathematically no different from calculating the means separately. Its main advantage, for the two-dimensional case, is the way it presents the data. This is shown by Figure 4.1, which plots two of the three dimensions. The coloured points are the compositions of the cargoes; the white point is the centroid. It shows that giveaway could be eliminated by increasing the C_2 content to 2.96 or increasing the C_4 content to 3.35. Of course any combination of increases in the two components is possible, provided they sum to 1.13. Indeed one of the components could be increased beyond this value, provided the other is reduced. The strategy adopted would depend on the relative values of C_2 and C_4 when routed to their next most valuable alternative disposition.

In the same way we can add weighting factors to the arithmetic mean, we can do so to the calculation of the centroid. Indeed its alternative name, *centre of mass*, comes from using it to calculate the position of the centre of gravity of distributed weights.

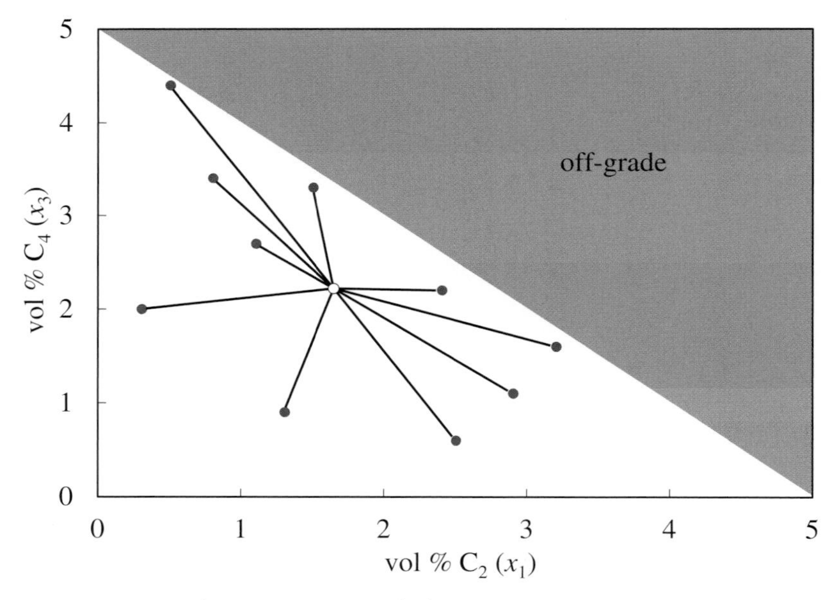

Figure 4.1 *Centroid of propane composition*

While using the mean makes good engineering sense in assessing opportunities for process improvements, it can give a distorted measure of performance. The inclusion of a single measurement that is very different from the others can, particularly if the sample size is small, significantly affect the estimate of the mean. An example might be a measurement collected under very unusual conditions, such as a process upset, or simply an error. We will present later how such *outliers* might be excluded. Doing so results in a *trimmed* (or *truncated*) mean. For example, we could simply exclude the lowest and highest values. So, after ranking the values of x from the lowest (x_1) to the highest (x_n), the trimmed arithmetic mean becomes

$$\bar{x}_t = \frac{\sum_{i=2}^{n-1} x_i}{n-2} \tag{4.18}$$

The *Winsorised mean* is based on a similar approach but the most outlying values are replaced with the adjacent less outlying values. For example, if we Winsorise the two most outlying values, the mean would be calculated as

$$\bar{x}_w = \frac{x_2 + x_{n-1} + \sum_{i=2}^{n-1} x_i}{n} \tag{4.19}$$

If n is 10, Equation (4.19) describes the *10% Winsorised mean*; we have removed 10% of the lower and upper outliers. If the sample size were increased to 20 then, to maintain the same level of Winsorisation, we would replace x_1 and x_2 with x_3, and x_{19} and x_{20} with x_{18}.

We will show later that exclusion of outliers carries the risk that an important aspect of process behaviour will be overlooked. An alternative approach is to define the centre of our sample using the *median*. In order to determine the median we again rank the dataset. If there is an odd number of measurements in the set then the median is the middle ranked value. If there is an even number, it is the average of the two middle values. The addition of an outlier (no matter what its value) to the dataset will therefore have very little effect on the median – shifting it half the distance between two adjacent values ranked in the middle. The median can also be described as the 50 percentile, i.e. 50% of the values lie below (or above) the median.

We can also define *quartiles* – 25% of the values lie below the *first quartile* (Q_1) and 75% lie below the *third quartile* (Q_3). While the principle of determining the quartiles is clear, there are two different ways they can be calculated. These are known as the *inclusive* and *exclusive* methods. The inclusive method is based on the intervals between the ranked data. If there are n samples in the dataset, then there are $n - 1$ intervals. The first quartile is then the value $(n - 1)/4$ intervals from the start, i.e. the value ranked $(n - 1)/4 + 1$. The third quartile is the value ranked $3(n - 1)/4 + 1$. For example, in the dataset containing the values 10, 20, 30, 40 and 50, the first quartile is 20 and the third is 40. It is quite likely, however, that the calculated rank will not be an integer. If so, then we obtain the quartile by interpolating between the adjacent values. For example, if we were to add the value 60 to the dataset, the ranking for the first quartile becomes 2.25. Since this is nearer 2 than 3, we interpolate as $(0.75 \times 20) + (0.25 \times 30)$ or 22.5. The third quartile is 47.5.

The exclusive approach effectively adds a value ranked 0 to the data. The first quartile is then the value ranked $(n + 1)/4$. For example, if we were to increase n to 7 by adding 70 to the dataset, the first quartile would be the value ranked at 2, i.e. 20. The third quartile would be the value ranked $3(n + 1)/4$, i.e. 60. Non-integer results would similarly be dealt with by interpolation. For example, if we revert to the dataset without the 70, the first quartile is given by $(0.25 \times 10) + (0.75 \times 20)$, or 17.5. The third quartile is 52.5.

While each method gives very different results, this is primarily because the dataset is very small. The difference would be negligible if the dataset is large and well dispersed.

The median can also be described as the *second quartile* (Q_2). The average of the first and third quartile is known at the *midhinge* – another parameter representing the central value. There is also the *trimean*, defined as $(Q_1 + 2Q_2 + Q_3)/4$. Finally there is the *midrange*, which is simply the average of the smallest and largest values in the dataset.

There is a multidimensional equivalent to the median, the *geometric median*. It is similar to the centroid but is positioned to minimise the sum of the distances, not the sum of their squares. For the three-dimensional case the penalty function (F) is described by

$$F = \sum_{i=1}^{n} \sqrt{(x_{1i} - a)^2 + (x_{2i} - b)^2 + (x_{3i} - c)^2} \tag{4.20}$$

Unlike the centroid, the coordinates of the geometric median cannot be calculated simply. An iterative approach is necessary that adjusts a, b and c to minimise F. Applying this to the data in Table 4.3 gives the coordinates as (1.62, 96.02, 2.36). This is very close to the centroid determined from Equation (4.17). However, like its one-dimensional equivalent, the geometric median is less sensitive to outliers. For example, if we increase the value of the last C_2 content in Table 4.3 from 2.5 to 12.5, the mean increases from 1.65 to 2.65. The geometric median moves far less to (1.74, 95.77, 2.49).

Unlike the median, the geometric median will not be one of the data points. The data point nearest to the geometric mean is known as the *medoid*.

In process engineering, the mean value of a parameter has a true engineering meaning. For example, daily production averaged over a year can be converted to an annual production (simply by multiplying by 365) and any cost saving expressed per unit production can readily be converted to an annual saving. The same is not true of the median. While of some qualitative value in presenting data, it (and its related parameters) cannot be used in any meaningful calculation.

4.5 Dispersion

Once we have defined the central value, we need some measure of the dispersion of the data around it – often described as *variability* or *spread*. There are several simple parameters that we might consider. They include the *range*, which is simply the difference between the highest and lowest values in the dataset. It is sensitive to outliers, but we can deal with this in much the same way as we did in determining the mean. For example, we can use the *trimmed range* or *truncated range* where some criterion is applied to remove outliers. We can similarly *Winsorise* the range (by replacing outliers with adjacent values nearer the central value), although here this will give the same result as truncation. However these measures of dispersion, while offering a qualitative view, cannot readily be used in engineering calculations, for example, to assess potential improvements.

We can base measures of dispersion on the distance from the median. The most common of these is the *interquartile range* – the difference between the third and first quartile. It is equivalent to the 25% Winsorised range. There is also the *quartile deviation*, which is half of the interquartile range. There are measures based on *deciles*. For example, 10% of the values lie below the first decile (D_1) and 10% lie above the ninth decile (D_9). The difference between the two is another measure of dispersion. However, as described in the previous section, all such parameters are largely qualitative measures.

A better approach is to use the mean as the central value. We then calculate the deviation from the mean of each value in the dataset. In principle we could then sum these deviations as a measure of dispersion. However, from the definition of the arithmetic mean given by Equation (4.2), they will always sum to zero.

$$\sum_{i=1}^{n}(x_i-\bar{x}) = \sum_{i=1}^{n}x_i - n\bar{x} = \sum_{i=1}^{n}x_i - \sum_{i=1}^{n}x_i = 0 \tag{4.21}$$

A possible solution is to use the absolute value (D) of the deviation from the mean. This suffers the problem that increasing the number of data points in the set will increase the sum of the absolute deviations, even if there is no increase in dispersion. To overcome this we divide by n to give the *mean absolute deviation*

$$\bar{D} = \frac{\sum_{i=1}^{n}|x_i-\bar{x}|}{n} \tag{4.22}$$

However, this does lend itself to use in further mathematical analysis. For example, if we know only the mean absolute deviation for each of two datasets, we cannot derive it for the combined set.

A more useful way of removing the sign of the deviation is to square it. The average sum of the squares of the deviations is the *variance*. Its positive square root is the *standard deviation* (σ).

$$\sigma^2 = \frac{\sum_{i=1}^{n}(x_i-\bar{x})^2}{n} \tag{4.23}$$

Expanding gives

$$\sigma^2 = \frac{\sum_{i=1}^{n}x_i^2 - 2\bar{x}\sum_{i=1}^{n}x_i + n\bar{x}^2}{n} \tag{4.24}$$

Combining with Equation (4.2) gives an alternative way of calculating the variance.

$$\sigma^2 = \frac{1}{n}\sum_{i=1}^{n} x_i^2 - \bar{x}^2 \tag{4.25}$$

The main advantage of variances is that they are additive. If, in addition to the series of values for x, we have one for values of y then from Equation (4.25)

$$\sigma_{x+y}^2 = \frac{\sum_{i=1}^{n} (x_i + y_i)^2}{n} - (\bar{x} + \bar{y})^2 \tag{4.26}$$

$$= \frac{\sum_{i=1}^{n} x_i^2 + 2\sum_{i=1}^{n} x_i y_i + \sum_{i=1}^{n} y_i^2}{n} - (\bar{x}^2 + 2\bar{x}\bar{y} + \bar{y}^2) \tag{4.27}$$

$$= \left(\frac{1}{n}\sum_{i=1}^{n} x_i^2 - \bar{x}^2\right) + \left(\frac{1}{n}\sum_{i=1}^{n} y_i^2 - \bar{y}^2\right) + 2\left(\frac{1}{n}\sum_{i=1}^{n} x_i y_i - \bar{x}\bar{y}\right) \tag{4.28}$$

$$= \sigma_x^2 + \sigma_y^2 + 2\sigma_{xy} \tag{4.29}$$

Following the same method we can determine the variance of the difference between two variables.

$$\sigma_{x-y}^2 = \sigma_x^2 + \sigma_y^2 - 2\sigma_{xy} \tag{4.30}$$

The term σ_{xy} is the *covariance*. We cover this in more detail in Section 4.9 but, if x and y are independent variables, their covariance will be zero. The variance of the sum of (or difference between) two variables will then be the sum of their variances.

Variance, because of the squaring of the deviation from the mean, is very sensitive to outliers. For example, the variance of C_2 vol% in Table 4.3 is 1.05. Increasing just the last value from 2.5 to 5.0 increases the variance to 2.11.

We may wish to compare dispersions that have different ranges or engineering units. Some texts suggest we modify the measure of dispersion to make it dimensionless. For example, dividing standard deviation by the mean gives the *coefficient of variation*. Other commonly documented dimensionless measures include the *coefficient of range* (the range divided by the sum of the lowest and highest values) and the *quartile coefficient* (the interquartile range divided by the sum of the first and third quartiles). In practice these values can be misleading. Consider a product that comprises mainly a component A and an impurity B. Assume the percentage of B in the product has a mean of 5 vol% and a standard deviation of 1%. The coefficient of variation is therefore 0.20. It follows therefore that the mean percentage of component A is 95 vol%. Its standard deviation will also be 1 vol% – giving a coefficient of variation of 0.01. This would appear to suggest that control of product purity is 20 times better than control of impurity where, in a two-component mixture, they must be the same.

4.6 Mode

Another measure commonly quoted is the *mode*. In principle it is the value that occurs most frequently in the dataset. However, if the values are continuous, it is quite possible that no two are the same. Instead we have to partition the data into ranges (known as *bins*); the mode is then the centre of the most populated range. While in most cases the mode will be towards the centre of the dataset, there is no guarantee that it will be. It has little application in the statistics associated with process control. However it is important that we work with distributions that are *unimodal*, i.e. they have a single mode. A distribution would be *bimodal*, for example, if a second grade of propane was produced (say, with a minimum purity target of 90%) and the analyses included in the same dataset as the higher purity grade. A distribution can also be *multimodal* – having two or more modes.

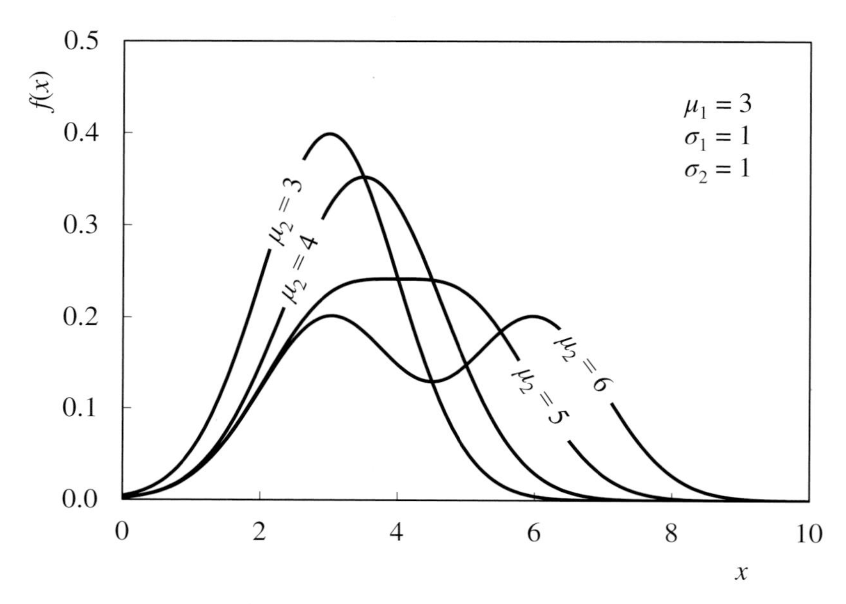

Figure 4.2 *Multimodal distributions*

A multimodal distribution can be easily mistaken for a unimodal one. Figure 4.2 illustrates this. Two distributions, with different means, have been combined. Both have a standard deviation of 1. In general, if the difference between the means is greater than double the standard deviation, the distribution will be clearly bimodal. In our example, where μ_2–μ_1 is 2, a second mode is not visible. Calculating the standard deviation without taking account of the modality would substantially overestimate the dispersion of the data. In this example it would be around 40% higher than the true value.

If a product is produced to different specifications, then the distribution will certainly be multimodal, even if not visibly so. A number of the distributions described in this book can be bimodal. However, their use is better avoided by segregating the data so that the standard deviation is determined separately for each grade of product. Alternatively, instead of obtaining the statistics for the propane purity, we obtain them for the deviation from target. Indeed this is exactly how we might assess the performance of a PID controller. We determine the standard deviation of the error, not the measurement. We need, however, to be more careful in assessing the performance of an inferential property. We might do so by monitoring the standard deviation of the difference between it and the laboratory. However, a change in operating mode might cause a bias error in the inferential. A bias error only contributes to standard deviation when it changes. Frequent changes in mode will therefore increase the standard deviation, suggesting that the inferential is performing poorly. To properly check its performance we should segregate the data for each operating mode.

Although not generally encountered in the process industry, a distribution can pass through a minimum, rather than a maximum. The minimum is known as the *anti-mode* – the value that occurs least often in the dataset. The distribution would then be described as *anti-modal.*

4.7 Standard Deviation

Standard deviation can be considered a measure of *precision*. Indeed, some texts define precision as the reciprocal of the variance (often using the term τ). Others define it as the reciprocal of the standard deviation (using the term τ'). Control engineers, of course, use τ to represent process lag. To avoid confusion, we avoid using precision as a statistical parameter. It does however have meaning. For example, the standard deviation of a variable that we wish to control is a measure of how precisely that variable is controlled. Indeed, reducing the standard deviation is often the basis of benefit calculations for process control improvements. If control were perfect the standard deviation would be zero. We similarly assess the performance of an inferential property from the standard deviation of the prediction error. However precision is not the same as *accuracy*. For example, if an inferential property consistently misestimates the property by the same amount, the standard deviation would be zero but the inferential would still be inaccurate. We have to distinguish between bias error and random error. Standard deviation is a measure only of random error.

We need to distinguish between *population* and *sample*. The population includes every value. For example, in the process industry, the population of daily production rates includes every measured rate since the process was commissioned until it is decommissioned. We clearly have no values for production rates between now and decommissioning. Records may not be available as far back as commissioning and, even if they are, the volume of data may be too large to retrieve practically. In practice, we normally work with a subset of the population, i.e. a sample. We need, of course, to ensure that the sample is representative by ensuring it includes values that are typical and that the sample is sufficiently large.

The standard deviation (σ_p) of the whole population of N values, if the *population mean* is μ, is given by

$$\sigma_p^2 = \frac{\sum_{i=1}^{N} (x_i - \mu)^2}{N} \tag{4.31}$$

When executing process control benefits studies we select a sample – a period for analysis comprising n data points. Those performing such analysis will likely have noticed that, to account for this, μ in Equation (4.31) is replaced by the *sample mean* and the denominator is replaced by $n - 1$. The following explains why.

From the data points collected in the period we estimate the sample mean.

$$\bar{x} = \frac{\sum_{i=1}^{n} x_i}{n} \tag{4.32}$$

Applying Equation (4.31) to a sample of the population will underestimate the true standard deviation. This is because the sum of the squared deviations of a set of values from their sample mean ($\bar{x}$) will always be less than the sum of the squared deviations from a different value, such as the population mean (μ). To understand this consider the trivial example where we have a sample of two data points with values 1 and 5. The sample mean is 3 and the sum of the deviations from the mean is 8 ($2^2 + 2^2$). If, instead of using the sample mean, we choose to use a higher value of 4, the sum of the deviations will then be 10 ($3^2 + 1^2$). We would get the same result if we had chosen a lower value of 2 for the mean. The nonlinearity, caused by squaring, results in the increase in squared deviation in one direction being greater than the decrease in the other.

We do not know the mean of the whole population (μ). Applying Equation (4.32) to different samples selected from the population will give a number of possible estimates of the true mean. These estimates will have a mean of μ. Similarly we do not know the standard deviation of the whole population. Imagine the sample mean being determined by, before summing all the data points, dividing each by n. The standard deviation of the resulting values will therefore be n times smaller, i.e. σ_p/n, giving a variance of $(\sigma_p/n)^2$. We have seen that variances are additive. The sum of the n values, which will now be the sample mean, will therefore have a variance n times larger, i.e. $\sigma_p{}^2/n$. The square root of this value is sometimes described as the *standard error*.

The variance of the sample (σ^2) is

$$\sigma^2 = \frac{\sum_{i=1}^{n}(x_i - \bar{x})^2}{n} \tag{4.33}$$

The variance of the population will be the variance of the sample plus the variance of the sample mean.

$$\sigma_p^2 = \sigma^2 + \frac{\sigma_p^2}{n} \quad \text{or} \quad \sigma_p^2 = \frac{n}{n-1}\sigma^2 \tag{4.34}$$

Substituting for σ^2 from Equation (4.33) gives

$$\sigma_p^2 = \frac{\sum_{i=1}^{n}(x_i - \bar{x})^2}{n-1} \tag{4.35}$$

We use Equation (4.35) to generate a *sample-adjusted* or *unbiased* variance. This technique is known as *Bessel's correction*. The denominator is often described as the number of *degrees of freedom*. We will see it is used in number of distributions. By definition it is the number of values in the dataset that can be freely adjusted while retaining the value of any quantified statistical parameters. In this case we have quantified only one parameter – the mean. We can freely adjust $n - 1$ of the values, provided we adjust the nth value to keep the total, and hence the mean, constant.

In practice, if the number of data points is sufficiently large, the error introduced is small. For example, with a value of 50 for n, the effect on σ^2 will be to change it by about 2%, with a change in σ of less than 1%.

To remove the need to first calculate the sample mean, Equation (4.35) can be rewritten as

$$\sigma_p^2 = \frac{\sum_{i=1}^{n} x_i{}^2 - 2\bar{x}\sum_{i=1}^{n} x_i + n\bar{x}^2}{n-1} = \frac{n\sum_{i=1}^{n} x_i{}^2 - \left(\sum_{i=1}^{n} x_i\right)^2}{n(n-1)} \tag{4.36}$$

4.8 Skewness and Kurtosis

Like variance, *skewness* (γ) and *kurtosis* (κ) are used to describe the shape of the distribution. To mathematically represent the distribution of the data we first have to choose the form of the distribution. That chosen is known as the *prior distribution*. It will contain parameters (such as mean and variance) that are then adjusted to fit the real distribution as close as possible. The main use of skewness and kurtosis is to assess whether the actual distribution of the data is then accurately represented. They are examples of *moments*. The kth *raw* moment (m) is defined as

$$m_k = \frac{\sum_{i=1}^{N} x_i^k}{N} \tag{4.37}$$

Although of little use, the *zeroth raw moment* ($k = 0$) has a value of 1. The *first raw moment* ($k = 1$) is the population mean (μ). *Central moments* (m') are calculated about the population mean

$$m'_k = \frac{\sum_{i=1}^{N} (x_i - \mu)^k}{N} \tag{4.38}$$

The *first central moment* will evaluate to zero.

$$m'_1 = \frac{\sum_{i=1}^{N} (x_i - \mu)}{N} = \frac{\sum_{i=1}^{N} x_i - N\mu}{N} = \mu - \mu = 0 \tag{4.39}$$

The second central moment is the population variance; replacing k in Equation (4.38) with 2 gives Equation (4.31). Higher moments are generally *normalised* or *standardised*, by dividing by the appropriate power of standard deviation of the population, so that the result is dimensionless.

$$m_k = \frac{\sum_{i=1}^{N} (x_i - \mu)^k}{N\, \sigma_p^k} \quad k > 2 \tag{4.40}$$

Skewness (γ) is the third central moment. If the number of data points in the sample is large then we can calculate it from Equation (4.40). Strictly, if calculated from a sample of the population, the formula becomes

$$\gamma = \frac{n}{(n-1)(n-2)} \frac{\sum_{i=1}^{n} (x_i - \bar{x})^3}{\sigma_p^3} \quad n > 2 \tag{4.41}$$

If the skewness is greater than zero, it might indicate there are more values higher than the mean than there are below it. Or it might indicate that the values higher than the mean are further from it than the values below it. The value of skewness does not indicate the cause. It does not tell us whether the mean is less than or greater than the median.

As a simple example, consider a dataset containing the values 98, 99, 100, 101 and 107. The majority of the values are less than the mean of 101, indicating that the skew might be negative.

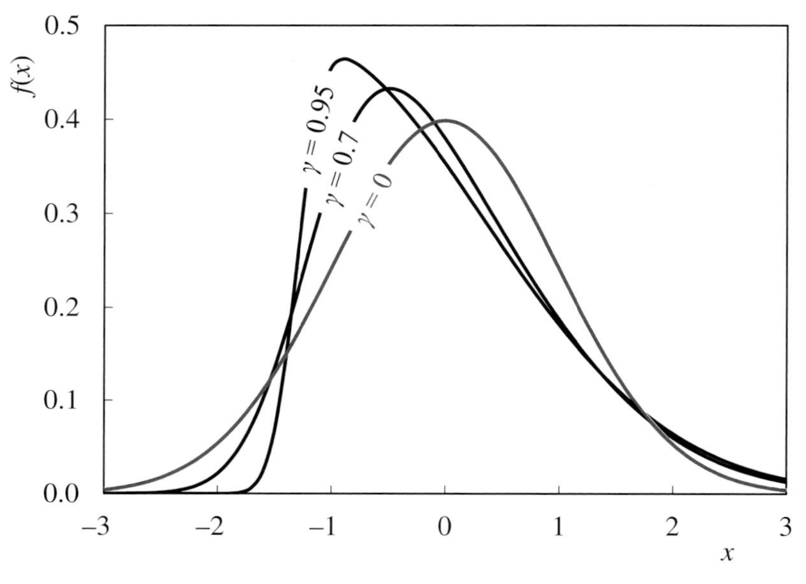

Figure 4.3 *Skewness*

However, the one value greater than the mean is far more distant from it than the others – possibly indicating a positive skew. This is confirmed by Equation (4.41), which gives the skewness as 1.7.

Figure 4.3 shows (as the coloured line) the normal distribution with a mean of 0 and standard deviation of 1. A source of confusion is that a positive skewness indicates a skew to the right. The black lines show increasing skewness while keeping the mean and standard deviation constant. The mode has moved to the left but, as skewness increases, values below the mean have approached the mean while some of values above the mean now form a more extended tail.

A normal distribution is symmetrical about the mean. Some texts therefore suggest that skewness lying between –0.5 and +0.5 is one of the indications that we can treat the distribution as normal. However, while a symmetrical distribution has a skewness of zero, the converse is not necessarily true. For example, a large number of values a little less than the mean might be balanced by a small number much higher than the mean. Skewness will be zero, but the distribution is clearly not symmetrical.

In any symmetrical distribution, the mean, median and mode will all have the same value.

Kurtosis (κ) is the fourth central moment given, for a sample of the population, by

$$\kappa = \frac{n(n+1)}{(n-1)(n-2)(n-3)} \frac{\sum_{i=1}^{n}(x_i - \bar{x})^4}{\sigma_p^4} - \frac{3(3n-5)}{(n-2)(n-3)} \qquad n > 3 \tag{4.42}$$

The kurtosis of a normal distribution is 3; for this reason many texts use the parameter γ_1 as skewness and γ_2 as *excess kurtosis*.

$$\gamma_1 = \gamma \qquad \gamma_2 = \kappa - 3 \tag{4.43}$$

Kurtosis is a measure of how flat or peaked is the distribution. It is a measure of how much of the variance is due to infrequent extreme deviations from the mean, rather than more frequent small deviations. This is apparent from examining Equation (4.42). If the deviation from the

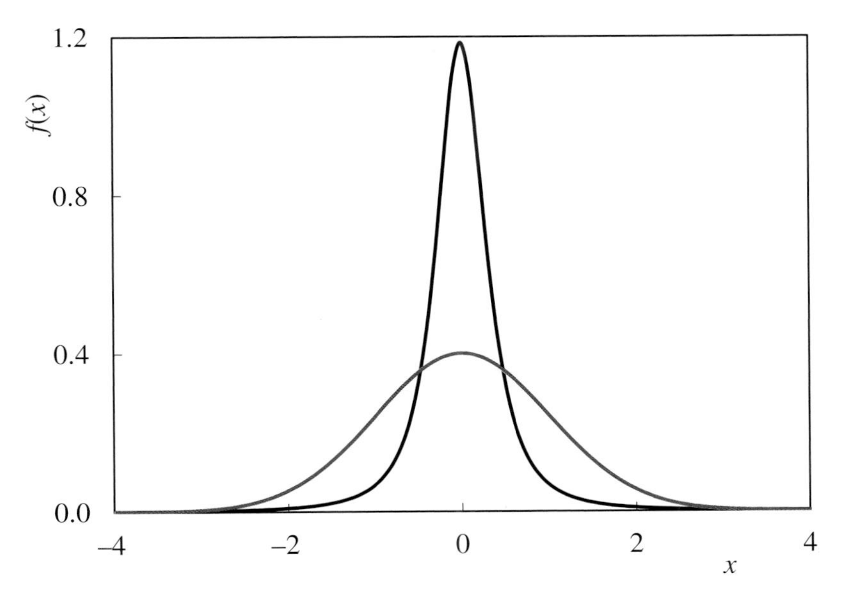

Figure 4.4 *Increasing kurtosis*

mean is less than the standard deviation then $(x_i - \bar{x})/\sigma_P$ will be less than 1. Raising this to the fourth power will make it considerably smaller and so it will contribute little to the summation. If the deviations are predominantly larger than the standard deviation (i.e. the distribution has long tails) then kurtosis will be large. Specifically, if excess kurtosis is positive ($\kappa > 3$) then the distribution is *leptokurtic*, i.e. it has a higher peak with long tails. If negative ($\kappa < 3$) then it is *platykurtic*, i.e. more flat with short tails. If excess kurtosis is zero ($\kappa = 3$) the distribution is described as *mesokurtic*. Indeed, if excess kurtosis is outside the range –0.5 to +0.5, we should not treat the distribution as normal. Many commonly used distributions, as we will see later, are leptokurtic and can be described as *super-Gaussian*. Platykurtic distributions can be described as *sub-Gaussian*.

Most spreadsheet packages and much statistical analysis software use excess kurtosis. To avoid confusion, and to keep the formulae simpler, this book uses kurtosis (κ) throughout.

Kurtosis is quite difficult to detect simply by looking at the distribution curve. Figure 4.4 shows (as the coloured line) the normal distribution – with a mean of 0 and a variance of 1. The black line has the same mean and variance but with kurtosis increased (to around 20). Figure 4.5 also shows the same normal distribution but this time the kurtosis is kept at zero and the variance reduced (to around 0.115). The dashed line looks almost identical to the solid line – although close inspection of the tails of the distribution shows the difference. Figure 4.6 shows the same distributions plotted on a cumulative basis and shows much the same difficulty. If, instead of plotting the function, the distribution were plotted from the data, it is even less likely that kurtosis could be seen. To detect it reliably (and to quantify it) kurtosis should at least be calculated as above, but preferably estimated from curve fitting.

Higher order moments can be defined but their interpretation is difficult and are rarely used. The fifth moment is *hyperskewness* and the sixth *hyperflatness*.

In addition to calculating the skewness and kurtosis from the data, we also need to determine them for the chosen distribution function. As we will see later, many distributions are documented with simple formulae but for some the calculations are extremely complex. Under these

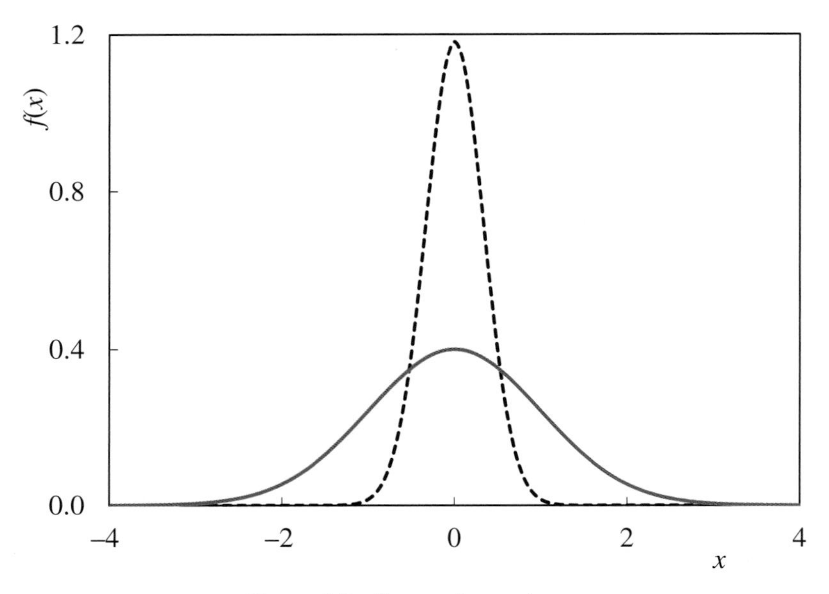

Figure 4.5 Decreasing variance

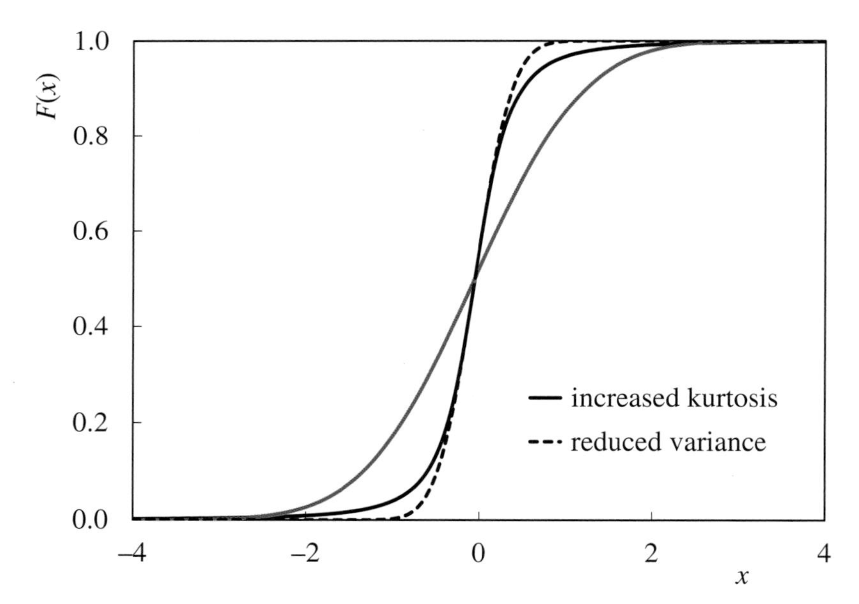

Figure 4.6 Comparison between kurtosis and variance

circumstances it is common to instead give a formula for the raw moments. We will see in the next chapter that a distribution can usually be described by a *probability density function* (*PDF*), $f(x)$. Central moments are then defined as

$$m'_n = \int_{-\infty}^{+\infty} (x-\mu)^n f(x).dx \tag{4.44}$$

Raw moments defined as

$$m_n = \int_{-\infty}^{+\infty} x^n f(x).dx \tag{4.45}$$

The term 'moment' is used in other areas of engineering. In general it is a physical quantity multiplied by distance. For example, in mechanical engineering, where it is also known as *torque*, it is force multiplied by the perpendicular distance to where the force is applied. Equation (4.44) explains why the term is applicable to statistics. If n is 1, it is the probability density multiplied by the distance from the mean. The mechanical engineer might use the first moment to determine the centre of gravity; the statistician uses it to determine the mean.

Provided Equation (4.45) can be integrated then we can develop formulae for the raw moments (m) and use these to derive the required parameters. The mean (μ) is

$$\mu = m_1 \tag{4.46}$$

Variance, by definition, is

$$\sigma_P^2 = \frac{\sum_{i=1}^{N}(x_i - \mu)^2}{N} \tag{4.47}$$

Expanding gives

$$\sigma_P^2 = \frac{\sum_{i=1}^{N}(x_i)^2}{N} - \frac{2\mu\sum_{i=1}^{N}x_i}{N} + \mu^2 = m_2 - \mu^2 \tag{4.48}$$

So, given the formulae for m_1 and m_2, we can calculate the standard deviation

$$\sigma_P^2 = m_2 - m_1^2 \tag{4.49}$$

A similar approach can be developed for skewness; by definition it is

$$\gamma = \frac{\sum_{i=1}^{N}(x_i - \mu)^3}{N\sigma_P^3} \tag{4.50}$$

Expanding gives

$$\gamma = \frac{1}{N\sigma_P^3}\left[\sum_{i=1}^{N}(x_i)^3 - 3\mu\sum_{i=1}^{N}(x_i)^2 + 3\mu^2\sum_{i=1}^{N}x_i - \mu^3\right] \tag{4.51}$$

$$\therefore \ \gamma = \frac{m_3 - 3m_1m_2 + 2\,m_1^3}{\left(m_2 - m_1^2\right)^{3/2}} \tag{4.52}$$

The formulae for some distributions is documented in terms of γ^2, suggesting that γ might be positive or negative. While a negative skewness is quite possible, it does not arise from choosing the negative square root. Mathematically Equation (4.47) might suggest that σ_p can be negative, but we always take the positive root of variance to give standard deviation. The denominator of Equation (4.50) is therefore positive. Skewness will only be negative if the numerator is negative.

For kurtosis

$$\kappa = \frac{\sum_{i=1}^{N} (x_i - \mu)^4}{N\sigma_P^4} \tag{4.53}$$

Because its definition includes only even powers, kurtosis cannot be less than zero. Excess kurtosis however, if kurtosis is less than 3, will be negative.

Expanding Equation (4.53) gives

$$\kappa = \frac{1}{N\sigma_P^4}\left[\sum_{i=1}^{N}(x_i)^4 - 4\mu\sum_{i=1}^{N}(x_i)^3 + 6\mu^2\sum_{i=1}^{N}(x_i)^2 - 4\mu^3\sum_{i=1}^{N}x_i + \mu^4\right] \tag{4.54}$$

$$\therefore \quad \kappa = \frac{m_4 - 4m_1m_3 + 6\,m_1^2m_2 - 3\,m_1^4}{\left(m_2 - m_1^2\right)^2} \tag{4.55}$$

Instead of a formula for the moments, the *moment generating function* may be documented. This is normally written as $M(t)$. Moments are determined by successively differentiating $M(t)$ and setting t to zero.

We will see later that the formulae for skewness and kurtosis for the normal distribution are trivial; skewness is 0 and kurtosis is 3. However, as an example, we will derive them from the published moment generating function that, for the normal distribution, is

$$M(t) = \exp\left(\mu t + \frac{\sigma^2}{2}t^2\right) \tag{4.56}$$

Differentiating

$$\frac{dM(t)}{dt} = \left(\mu t + \sigma^2 t\right)\exp\left(\mu t + \frac{\sigma^2}{2}t^2\right) \tag{4.57}$$

Setting t to zero gives

$$m_1 = \mu \tag{4.58}$$

Differentiating again

$$\frac{d^2M(t)}{dt^2} = \left[\left(\mu t + \sigma^2 t\right)^2 + \sigma^2\right]\exp\left(\mu t + \frac{\sigma^2}{2}t^2\right) \tag{4.59}$$

Setting t to zero gives

$$m_2 = \mu^2 + \sigma^2 \tag{4.60}$$

From Equations (4.49) and (4.58)

$$\sigma_p^2 = \sigma^2 \tag{4.61}$$

Differentiating again

$$\frac{d^3M(t)}{dt^3} = \left[\left(\mu t + \sigma^2 t\right)^3 + 3\sigma^2\left(\mu t + \sigma^2 t\right)\right]\exp\left(\mu t + \frac{\sigma^2}{2}t^2\right) \tag{4.62}$$

Setting t to zero gives

$$m_3 = \mu^3 + 3\mu\sigma^2 \tag{4.63}$$

From Equation (4.51)

$$\gamma = \frac{\mu^3 + 3\mu\sigma^2 - 3\mu(\mu^2 + \sigma^2) + 2\mu^3}{\sigma^3} = 0 \tag{4.64}$$

Differentiating again

$$\frac{d^4 M(t)}{dt^4} = \left[\left(\mu t + \sigma^2 t\right)^4 + 6\sigma^2\left(\mu t + \sigma^2 t\right)^2 + 3\sigma^4\right]\exp\left(\mu t + \frac{\sigma^2}{2}t^2\right) \tag{4.65}$$

Setting t to zero gives

$$m_4 = \mu^4 + 6\mu^2\sigma^2 + 3\sigma^4 \tag{4.66}$$

From Equation (4.54)

$$\kappa = \frac{\mu^4 + 6\mu^2\sigma^2 + 3\sigma^4 - 4\mu(\mu^3 + 3\mu\sigma^2) + 6\mu^2(\mu^2 + \sigma^2) - 3\mu^4}{\sigma^4} = 3 \tag{4.67}$$

If Equation (4.45) cannot readily be integrated, it is possible to calculate the raw moments using the trapezium rule.

$$m_n = \frac{1}{2}\sum_{i=-\infty}^{\infty}\left[x_i^n f(x_i) + x_{i-1}^n f(x_{i-1})\right](x_i - x_{i-1}) \tag{4.68}$$

We can again use the normal distribution to demonstrate this method. Figure 4.7 plots $f(x)$ against x for the normal distribution that has a mean of 0 and a variance of 1. We will show in the next chapter how $f(x)$ is determined.

Table 4.4 shows how Equation (4.68) is applied. The range of i must be large enough so that $f(x)$ is very close to zero at the extremes. Because of the multiplying effect of the x^n term, this is particularly important for the higher moments. In this example, we achieve this by ranging x

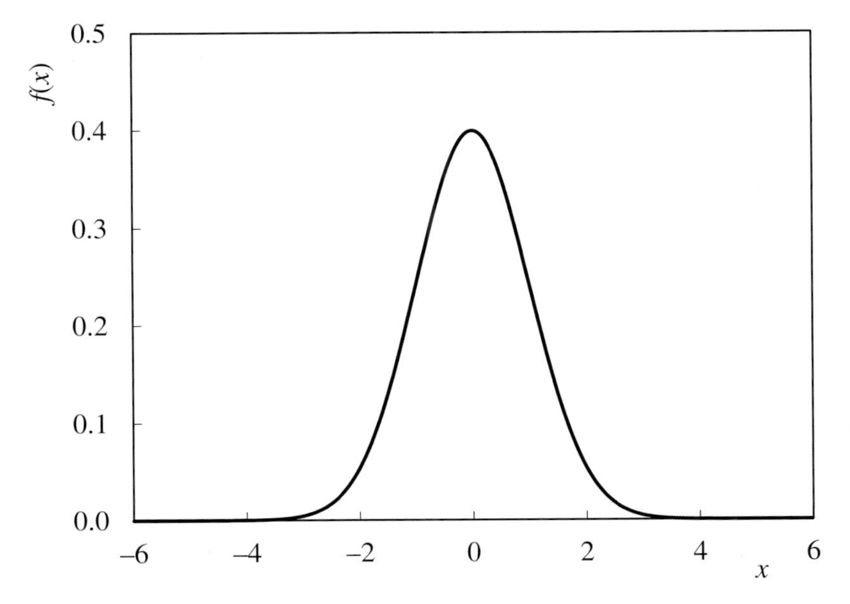

Figure 4.7 Normal distribution

Table 4.4 *Applying trapezium rule to determine moments*

x	$f(x)$	$xf(x)$	m_1	$x^2f(x)$	m_2	$x^3f(x)$	m_3	$x^4f(x)$	m_4	μ	σ	γ	κ
−6.0	0.000	0.000	0.000	0.000	0.000	0.000	0.000	0.000	0.000	0.000	0.000		
−5.9	0.000	0.000	0.000	0.000	0.000	0.000	0.000	0.000	0.000	0.000	0.000	−34211	1170468119
−5.8	0.000	0.000	0.000	0.000	0.000	0.000	0.000	0.000	0.000	0.000	0.000	−20442	417928670
−5.7	0.000	0.000	0.000	0.000	0.000	0.000	0.000	0.000	0.000	0.000	0.000	−13954	194749909
−5.6	0.000	0.000	0.000	0.000	0.000	0.000	0.000	0.000	0.000	0.000	0.001	−10011	100267085
−5.5	0.000	0.000	0.000	0.000	0.000	0.000	0.000	0.000	0.000	0.000	0.001	−7365	54268904
.	.	.	.	.	.	.	.	.	.	.	.	.	.
−0.5	0.352	−0.176	−0.352	0.088	0.484	−0.044	−0.792	0.022	1.498	−0.352	0.600	−1.699	5.361
−0.4	0.368	−0.147	−0.368	0.059	0.492	−0.024	−0.795	0.009	1.499	−0.368	0.597	−1.657	5.303
−0.3	0.381	−0.114	−0.381	0.034	0.496	−0.010	−0.797	0.003	1.500	−0.381	0.593	−1.636	5.306
−0.2	0.391	−0.078	−0.391	0.016	0.499	−0.003	−0.798	0.001	1.500	−0.391	0.588	−1.632	5.343
−0.1	0.397	−0.040	−0.397	0.004	0.500	0.000	−0.798	0.000	1.500	−0.397	0.585	−1.636	5.385
0.0	0.399	0.000	−0.399	0.000	0.500	0.000	−0.798	0.000	1.500	−0.399	0.584	−1.640	5.404
0.1	0.397	0.040	−0.397	0.004	0.500	0.000	−0.798	0.000	1.500	−0.397	0.586	−1.631	5.376
0.2	0.391	0.078	−0.391	0.016	0.501	0.003	−0.798	0.001	1.500	−0.391	0.590	−1.602	5.290
0.3	0.381	0.114	−0.381	0.034	0.504	0.010	−0.797	0.003	1.500	−0.381	0.599	−1.546	5.144
0.4	0.368	0.147	−0.368	0.059	0.508	0.024	−0.795	0.009	1.501	−0.368	0.611	−1.466	4.949
0.5	0.352	0.176	−0.352	0.088	0.516	0.044	−0.792	0.022	1.502	−0.352	0.626	−1.364	4.720
.	.	.	.	.	.	.	.	.	.	.	.	.	.
5.5	0.000	0.000	0.000	0.000	1.000	0.000	0.000	0.000	3.000	0.000	1.000	0.000	3.000
5.6	0.000	0.000	0.000	0.000	1.000	0.000	0.000	0.000	3.000	0.000	1.000	0.000	3.000
5.7	0.000	0.000	0.000	0.000	1.000	0.000	0.000	0.000	3.000	0.000	1.000	0.000	3.000
5.8	0.000	0.000	0.000	0.000	1.000	0.000	0.000	0.000	3.000	0.000	1.000	0.000	3.000
5.9	0.000	0.000	0.000	0.000	1.000	0.000	0.000	0.000	3.000	0.000	1.000	0.000	3.000
6.0	0.000	0.000	0.000	0.000	1.000	0.000	0.000	0.000	3.000	0.000	1.000	0.000	3.000

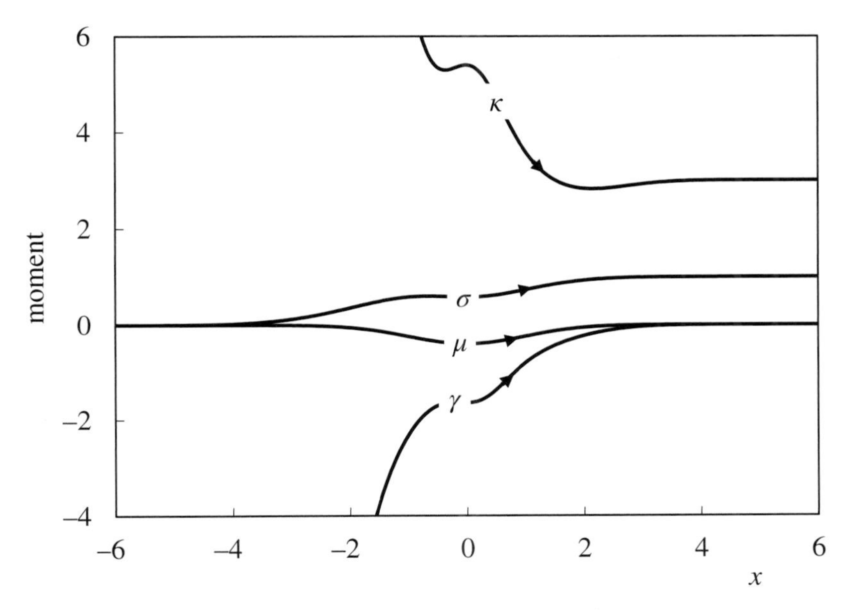

Figure 4.8 Convergence of calculations of moments

from −6 to +6 (i.e. 6 standard deviations) where the value of $f(x)$ is less than 10^{-8}. As usual, with the trapezium rule, accuracy is improved by making $(x_i - x_{i-1})$ as small as practical. In our example, an interval of 0.1 is used. Due to space restrictions, only selected rows are included in the table. Each of the raw moments (m_1, m_2, m_3 and m_4) is calculated from Equation (4.68) as cumulative sums. Equations (4.46), (4.49), (4.52) and (4.55) are then used to calculate μ, σ, γ and κ. Figure 4.8 plots these values against x. It shows that the graphs for all four parameters converge quickly to the values we expect.

Not all distributions have moments that converge; the areas of trapezia can increase as i increments, so giving an infinite value for the total area. In other words, the result of Equation (4.45) can be infinite.

There are other measures of skewness. One is the *Pearson first skewness coefficient* which is based on the mode (x_{mode}) and is defined as

$$\gamma_P = \frac{\mu - x_{\text{mode}}}{\sigma} \tag{4.69}$$

The *Pearson second skewness coefficient* is defined as three times the first coefficient. There is the *non-parametric skew* that replaces the mode in Equation (4.69) with the median. Others include *Bowley's measure of skewness*, also known as *Galton's measure of skewness*, which uses the first, second and third quartiles.

$$\gamma_B = \frac{(Q_3 - Q_2) - (Q_2 - Q_1)}{Q_3 - Q_1} \tag{4.70}$$

There is also *Kelly's measure of skewness* that uses the first, fifth and ninth deciles.

$$\gamma_K = \frac{(D_9 - D_5) - (D_5 - D_1)}{D_9 - D_1} \tag{4.71}$$

While such measures are indicative, there is no guarantee that they relate to the skewness, as defined by Equations (4.41) and (4.52). For example, it is possible for them to have a different sign.

There are similar alternative measures of kurtosis – again which have limited use.

4.9 Correlation

The calculation of *covariance* (σ_{xy}) is a useful first step in determining whether there is any correlation between two variables (x and y). It is an example of a *mixed moment*. For a sample, it is defined as

$$\sigma_{xy} = \frac{\sum_{i=1}^{n} (x_i - \bar{x})(y_i - \bar{y})}{n-1} \tag{4.72}$$

If there is no correlation between x and y then the covariance will be close to zero. However if y tends to be above its mean when x is above its, then it will have a positive value. Similarly if one variable tends to be below its mean when the other is above, then it will have a negative value. This can be a useful step in dynamic model identification. Determining the covariance between PV and MV will tell us whether the process gain is positive or negative.

The variance of a value derived from two or more measurements, for example the rate of change of inventory determined by subtracting the flow out from the flow in to a process, is normally determined by summing the variances of the two measurements. This is correct if the error in one measurement is not influenced by the other, i.e. they are truly independent variables. Measurement errors might be correlated if, for example, they have a common cause. For example, a change in fluid properties might affect both flow measurements similarly. If they are correlated then calculation of the combined variance must take account of the covariance(s).

$$\sigma_{x-y}^2 = \left(\sigma_x^2 - \sigma_{xy}\right) + \left(\sigma_y^2 - \sigma_{yx}\right) = \sigma_x^2 + \sigma_y^2 - 2\sigma_{xy} \tag{4.73}$$

We can see from Equation (4.72) that σ_{yx} is the same as σ_{xy}. By convention we write it in this way because we subtract, from the variance of y, its covariance with respect to x. Note that, depending on the value of σ_{xy}, the variance of the difference in flows can be smaller than the variance of either flow. If x and y are perfectly correlated, for example governed by $y = x + c$, then the variance of the flow difference will be zero.

The variance of the sum of correlated measurements is determined by modifying Equation (4.73) so that the covariances are added rather than subtracted. For example, estimating the feed to a process by adding correlated measurements of its three product flows would have a variance calculated from

$$\begin{aligned} \sigma_{x+y+z}^2 &= \left(\sigma_x^2 + \sigma_{xy} + \sigma_{xz}\right) + \left(\sigma_y^2 + \sigma_{yx} + \sigma_{yz}\right) + \left(\sigma_z^2 + \sigma_{zx} + \sigma_{zy}\right) \\ &= \sigma_x^2 + \sigma_y^2 + \sigma_z^2 + 2\left(\sigma_{xy} + \sigma_{xz} + \sigma_{yz}\right) \end{aligned} \tag{4.74}$$

Note that, while it is true that independent variables have a covariance of zero, the converse is not true. Covariance measures only linear correlations. For example, if $y = \sin(x)$, then x and y are clearly correlated but, provided x varies over a range greater than 2π, the covariance would be close to zero.

A limitation of covariance is that it is difficult to determine the significance of its value since it depends on the magnitude of the variables. When x and y are identical the covariance will reach its maximum possible value. This will be the variance of x (which is also the product of the standard deviations of x and y). Dividing the covariance by the standard deviations of x and y gives the dimensionless *Pearson coefficient* (R).

$$R = \frac{\sigma_{xy}}{\sigma_x \sigma_y} = \frac{\sum_{i=1}^{n} (x_i - \bar{x})(y_i - \bar{y})}{\sqrt{\sum_{i=1}^{n} (x_i - \bar{x})^2} \sqrt{\sum_{i=1}^{n} (y_i - \bar{y})^2}} \tag{4.75}$$

To avoid having to first calculate $\bar{x}$ and $\bar{y}$, this can be rewritten as

$$R = \frac{n\sum_{i=1}^{n} x_i y_i - \sum_{i=1}^{n} x_i \sum_{i=1}^{n} y_i}{\sqrt{\left(n\sum_{i=1}^{n} x_i^2 - \left(\sum_{i=1}^{n} x_i\right)^2\right)\left(n\sum_{i=1}^{n} y_i^2 - \left(\sum_{i=1}^{n} y_i\right)^2\right)}} \tag{4.76}$$

R will be in the range −1 to +1. If there is an exact correlation (such that $y_i = mx_i + c$) then R will be +1 if m is positive and −1 if m is negative. Often R^2 is used to remove the sign. A value for R of zero means that there is no relationship between x and y. It is important to appreciate that a nonzero value of R does not indicate that the values of x are necessarily close to the corresponding values of y. It only tells us they are correlated. It is therefore of limited value, for example, in assessing whether an inferential property closely matches the corresponding laboratory result. It is a measure of precision, not accuracy.

4.10 Data Conditioning

Data conditioning is a mathematical method that may be required to modify the data in some way to make it more suitable for use. Process control engineers will be familiar with techniques, such as linearisation and filtering, to condition a process measurement to improve control. Here we use techniques to enable a distribution to be well fitted.

In selecting which distribution to fit it is important to consider the range of the data. Process data is generally *bounded*. If subject to either a minimum or maximum limit, but not both, it is *single-bounded*. If subject to both it is described as *double-bounded*. The same terms apply to distributions. Engineers most commonly choose to fit the normal distribution. This is *unbounded*. It assumes that there is a finite (albeit small) probability that the process measurement can have any value.

Some process measurements are very clearly bounded. For example, the level in a product storage tank must lie between zero and its maximum capacity. Further it is reasonable to expect the level to be anywhere in this range. Some double-bounded distributions assume that the variable x is ranged from 0 to 1. We would therefore scale the tank level (L) accordingly.

$$x_i = \frac{L_i - L_{min}}{L_{max} - L_{min}} \tag{4.77}$$

If we wish to fit a double-bounded distribution to such data, the choice of L_{min} and L_{max} would be clear. However, as another example, the C_4 content of propane must lie between 0 and 100 vol% but in practice we would never see laboratory results anywhere near 100%.

The same would apply, more so, to concentrations measured in ppm (parts per million). We would not have chosen such units if the concentration could approach 10^6 ppm. Clearly the range must be chosen to cover all possible results but making it too large can reduce the accuracy of the fitted distribution.

Conditioning the data in this way will change the value of some of the location and dispersion parameters determined by curve-fitting. The mean will now be a fraction of the range and must be converted back to engineering units.

$$\bar{L} = (L_{max} - L_{min})\bar{x} + L_{min} \tag{4.78}$$

Standard deviation and variance will be similarly affected and need conversion.

$$\sigma_L = (L_{max} - L_{min})\sigma_x \tag{4.79}$$

$$\sigma_L^2 = (L_{max} - L_{min})^2 \sigma_x^2 \tag{4.80}$$

Parameters such as skewness and kurtosis are dimensionless and so not affected by scaling.

Since an unbounded distribution covers an infinite range of possible values, a single-bounded distribution can also be described as *semi-infinite*. They are usually *lower-bounded* in that they assume a minimum value for x, usually zero, but no upper limit. They will, however, usually include a location parameter to move the lower bound. Its value can be either chosen by the engineer or adjusted as part of curve-fitting. Such a distribution might be better suited to our example of the C_4 content of propane. We are sure that a result cannot be below zero but we are not entirely sure of the upper limit.

There are, of course, many examples where we would prefer to use an *upper-bounded* distribution. If, instead of C_4 content, the laboratory reported purity (i.e. the C_3 content), then the maximum is clearly 100% and we might be unsure about the minimum. However there are far fewer upper-bounded distribution functions and so, to widen the choice, we might convert the purity (P) results to impurities (x).

$$x_i = 100 - P_i \tag{4.81}$$

This form of conditioning only affects the resulting mean; it does not affect the standard deviation.

It is common to use basic controller outputs in constraint controllers as an indication that a hydraulic constraint has been reached. Usually the requirement is to run at the maximum output. In assessing how well this is being achieved would therefore require an upper-bounded distribution. To avoid this we can convert the output (OP) to a deviation from target (SP), remembering that we must then choose a distribution that can include negative measurements.

$$x_i = SP - OP_i \tag{4.82}$$

Such an approach is also helpful if the target changes. For example, it is common for some operators to be more comfortable than others in running the process closer to its limits, and so increase SP. In principle we could partition the data by SP and fit a distribution for each operator. By converting each result to a deviation from SP we can fit a single distribution to all the results.

Data conditioning might also be considered if the variation in the dataset is small compared to the mean. This can reduce the accuracy of the fitted distribution. As a rule of thumb, the coefficient of variation (as described in Section 4.5) should be at least 0.02. If this is not the case then the data should be conditioned by subtracting a fixed bias from every value. The bias can be any sensible value, close to that defined by Equation (4.83).

$$bias > \bar{x} - 50\sigma \tag{4.83}$$

Such a problem can arise when applying Equation (4.81). While P may show sufficient variation with respect to its mean, x may not. Applying the bias effectively subtracts P from a value less than 100.

Once the distribution has been fitted, the bias must be added back to the mean, quartiles and to any bounds. Parameters derived from moments, such as standard deviation, require no correction.

5

Probability Density Function

The probability density function (PDF) is a mathematical function that represents the distribution of a dataset. For example, if we were to throw a pair of unbiased six-sided dice 36 times, we would expect (on average) the distribution of the total score to be that shown by Figure 5.1. This shows the *frequency distribution*. To convert it to a *probability distribution* we divide each frequency by the total number of throws. For example, a total score of 5 would be expected to occur four times in 36 throws and so has a probability of 4/36 (about 0.111 or 11.1%). Figure 5.2 shows the resulting probability distribution.

Throwing dice generates a *discrete* distribution; in this case the result is restricted to integer values. Probability should not be plotted as continuous line. The probability of a non-integer result is zero. But we can develop an equation for the line. In this case, if x is the total scored, the probability of scoring x is

$$p(x) = \frac{6-|x-7|}{36} \qquad 1 \leq x \leq 13 \tag{5.1}$$

Because x is discrete this function is known as the *probability mass function (PMF)*. If the distribution were continuous we can convert the probability distribution to a *probability density function (PDF)* by dividing the probability by the range over which it applies.

A condition of both functions is that the area they contain must be unity (or 100%) – in other words we are certain to be inside the area. So, in general

$$\int_{x=-\infty}^{\infty} p(x)dx = 1 \tag{5.2}$$

Provided this condition is met then any function can be described as a PDF (or PMF). We will show later that there are many functions that have a practical application. Unfortunately there are a larger number that appear to have been invented as a mathematical exercise and are yet to be shown that they describe any real probability behaviour.

Statistics for Process Control Engineers: A Practical Approach, First Edition. Myke King.

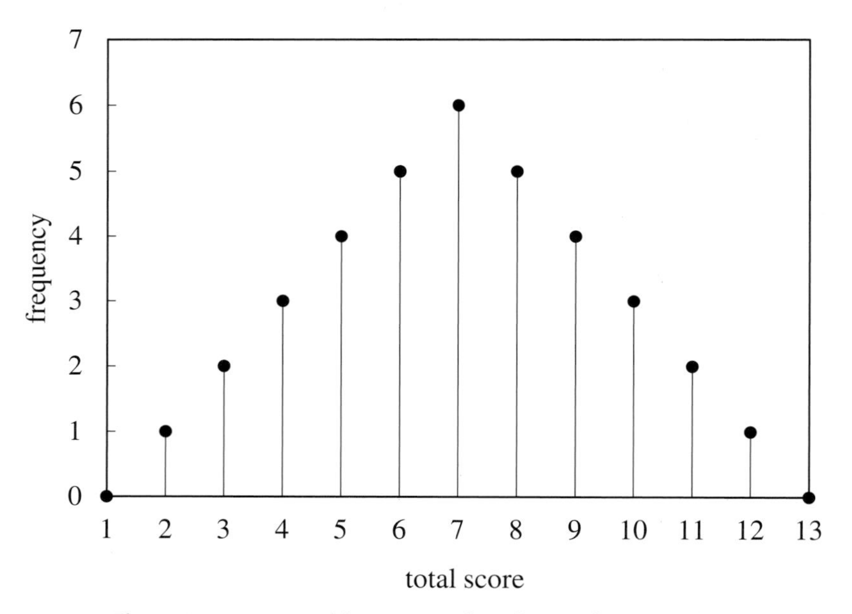

Figure 5.1 Expected frequency of total score from two dice

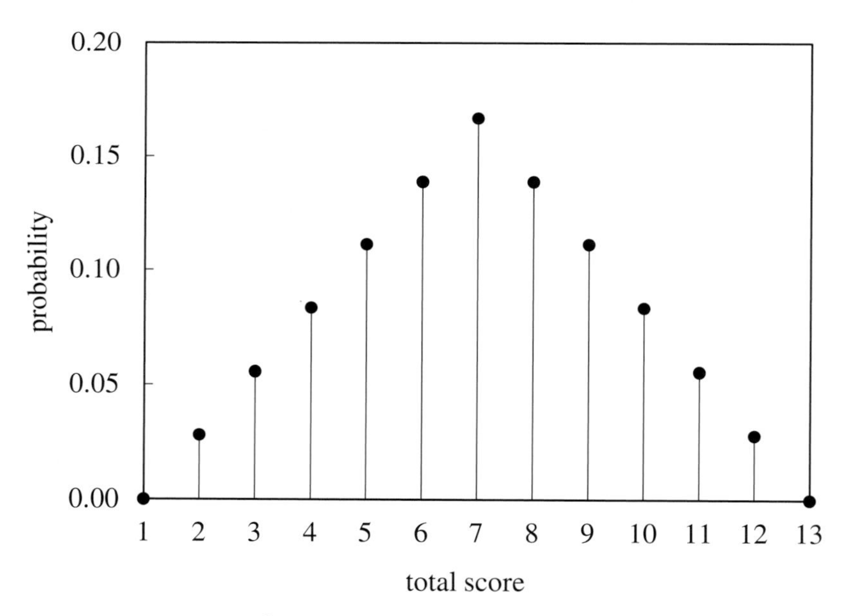

Figure 5.2 Expected distribution

While the PMF allows us to estimate the probability of x having a certain value, the PDF does not. It only tells us the probability of x falling within a specified range. The probability of x being between a and b is

$$P(a \leq x \leq b) = \int_a^b p(x)dx \tag{5.3}$$

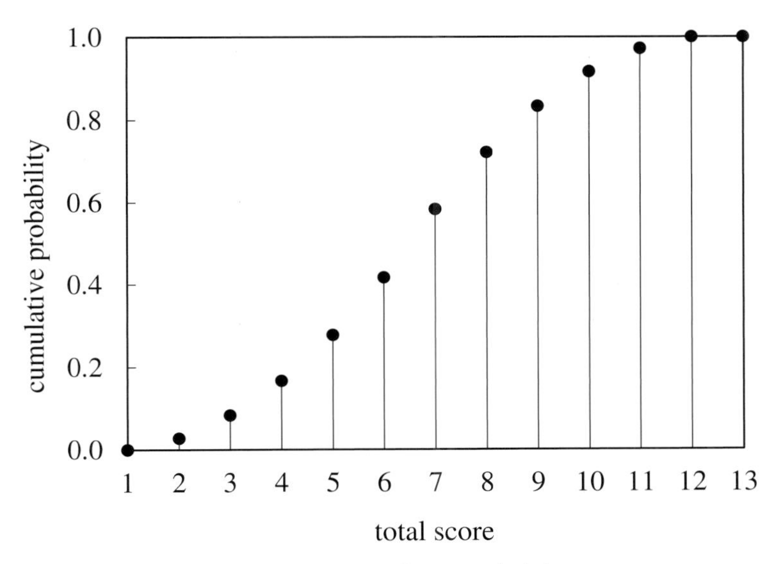

Figure 5.3 *Cumulative probability*

In fact, if x is a continuous parameter, the probability of it having a particular value is zero. To understand this, let us assume we have calculated the probability of it lying, for example, between 99 and 100 as 0.1. The probability of it lying between 99.9 and 100 will be approximately 0.01; between 99.99 and 100 it will be about 0.001, etc. To approach 100 the interval $(b - a)$ must be zero, as will the probability. Indeed, setting a equal to b in Equation (5.3), would give the same result.

The integral of the PDF (or PMF) is known as the *cumulative distribution function (CDF)*. It is obtained, as the name suggests, by cumulatively summing the distribution. For example, Figure 5.3 is derived by summing the probabilities in Figure 5.2.

Figure 5.4 plots the PDF of each of two continuous distributions that we will cover later. The black curve is that of a uniform distribution and the coloured curve (often described at the *bell curve*) that of a normal distribution. The area under both is unity.

So far, we have used $p(x)$ as the notation for the PMF and $P(x)$ for the discrete CDF. It is common for $f(x)$ to be used for the PDF and $F(x)$ for the continuous CDF. In Figure 5.4 the normal distribution, $f(x)$, exceeds 1 for a range of values for x. This can be confusing; it does not mean that the probability at this point is greater than 1. To illustrate this, a point on the curve is highlighted at $x = 2.1$. Figure 5.5 shows the corresponding CDF for each distribution. It shows that the probability of x being less than or equal to 2.1 (i.e. x being between $-\infty$ and 2.1) is about 0.63.

$F(x)$ must range from 0 to 1 as x varies from its minimum possible value (x_{min}) to its maximum (x_{max}). If it did not, then the underlying $f(x)$ could not be described as a PDF. Provided $F(x)$ passes through the points (x_{min},0) and (x_{max},1) and is *monotonic* (i.e. increases over the range without ever decreasing), then the CDF can be virtually any shape. That for the normal distribution is described as an *ogive*.

If the CDF is the integral of the PDF, it follows that the PDF can be derived by differentiating the CDF. Similarly the PDF can be derived as the slope of the CDF at the value x.

$$f(x) = \frac{F(x_i) - F(x_{i-1})}{x_i - x_{i-1}} \tag{5.4}$$

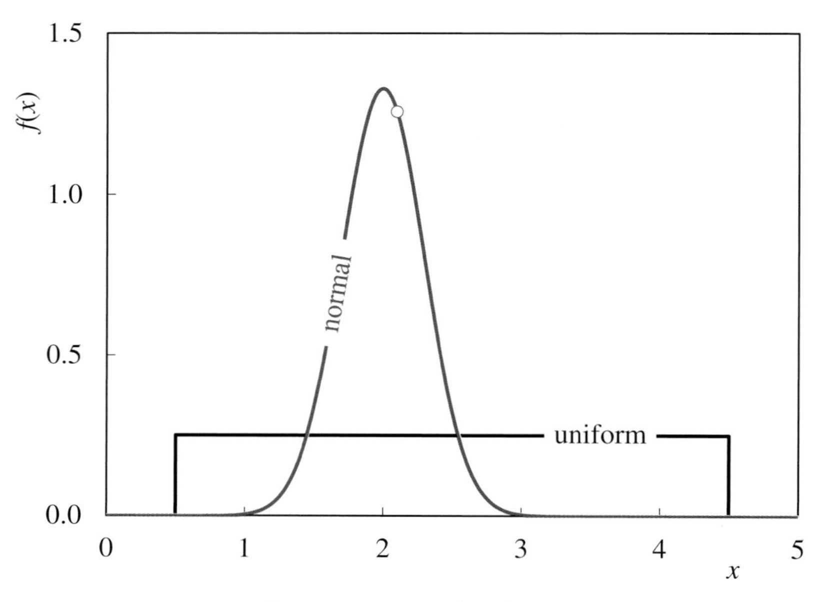

Figure 5.4 Examples of PDF

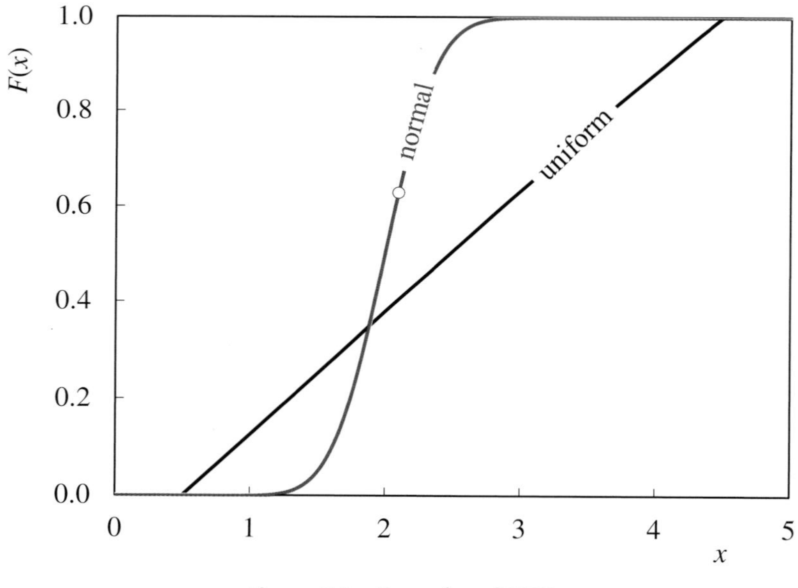

Figure 5.5 Examples of CDF

The PDF is therefore measured as distribution per unit of x – explaining why it is described as a 'density' function.

The CDF permits us to determine the probability of being within part of the feasible range. While we may be able to integrate a PDF, integrating a PMF will give the wrong result. To illustrate this, we can integrate Equation (5.1) over the range 11 to 12, expecting the result to be 3/36.

$$\int_{11}^{12} \frac{6-|x-7|}{36} dx = \int_{11}^{12} \frac{13-x}{36} dx = \frac{1}{36}\left[13x - \frac{x^2}{2}\right]_{11}^{12} = \frac{1.5}{36} \tag{5.5}$$

Because Equation (5.1) is a PMF it must be integrated as a summation.

$$P(a \leq x \leq b) = \frac{1}{36}\sum_{x=a}^{b} 6-|x-7| \tag{5.6}$$

Figure 5.3 plots this function; as required it varies from 0 (at $x = 1$) to 1 (at $x = 12$). While essential to convert a PMF to a CDF, similar numerical methods are also applied to convert any PDF that is mathematically too complex to be integrated.

Functions that do not integrate to unity over the feasible range require a *standardisation* (or *normalisation*) constant to force them to do so. Indeed, in this example, the factor 1/36 in Equation (5.6) could be thought of as this constant.

Another term occasionally used is the *survival function*, usually written as $S(x)$. The name comes from the application of statistics to demographic studies; $S(x)$ would be the probability of an individual surviving beyond a certain age. It can also be described as the *reliability function*, the *reverse distribution*, or the *complementary CDF*. It is used, for example, in situations where we may wish to estimate the probability that equipment will continue to operate beyond a specified time. $F(x)$ will tell us the probability of survival being less than the specified time so it follows that

$$S(x) = 1 - F(x) \tag{5.7}$$

5.1 Uniform Distribution

The *uniform distribution* is also known as the *rectangular distribution.* While process data is unlikely to be uniformly distributed, it has an important role in generating statistical data for studies – most commonly in Monte Carlo simulation.

An example of a physical system that shows a uniform distribution is the throwing of an unbiased six-sided dice. Each of the six possible outcomes is equally likely. If we throw the dice a sufficient number of times then the distribution of outcomes will be uniform. This is described by the discrete form of the uniform distribution. Including the minimum integer value (x_{min}) and the maximum (x_{max}), there are n possible values of x, where

$$n = x_{max} - x_{min} + 1 \tag{5.8}$$

The PMF is the probability of any of these values occurring and, since the distribution is uniform, is

$$p(x) = \frac{1}{n} \tag{5.9}$$

The CDF is the probability of being less than or equal to x. The number of values included is $x - x_{min} + 1$. As usual, the CDF is obtained by summing $p(x)$ from x_{min} to x.

$$P(x) = \sum_{i=1}^{x-x_{min}+1} \frac{1}{n} \tag{5.10}$$

This can be written in the form

$$P(x) = \frac{\lfloor x \rfloor - x_{min} + 1}{n} \tag{5.11}$$

The reader may be unfamiliar with the open-topped brackets $\lfloor x \rfloor$. They represent the *floor of x*, i.e. the largest integer that is less than or equal to x. Their purpose is to allow the distribution to be plotted as a continuous staircase-shaped line, rather than as a series of discrete points. In effect, x is treated as a continuous variable. For simplicity we will omit the brackets in later examples of discrete distributions. The reader may be interested to know that there is a converse symbol $\lceil x \rceil$ which is the *ceiling of x* or the smallest integer that is greater than or equal to k. Further, placing x between a floor bracket and a ceiling bracket means that x should be *rounded* to the nearest integer.

Key parameters of the discrete uniform distribution are

$$\mu = \frac{x_{min} + x_{max}}{2} \tag{5.12}$$

$$\sigma^2 = \frac{n^2 - 1}{12} \tag{5.13}$$

$$\gamma = 0 \tag{5.14}$$

$$\kappa = \frac{3(7n^2 - 3)}{5(n^2 - 1)} \tag{5.15}$$

Most spreadsheet packages include a random number function. Typically it would generate a number in the range 0 to 1. This distribution is described as U(0,1). In the absence of this function, pseudo-random numbers can be generated using a modified multiplicative congruential method. One example is

$$x_n = \{[\alpha(\beta x_{n-1} + n)] \bmod \beta + x_{n-2}\} \bmod 1 \tag{5.16}$$

For readers unfamiliar with *modulus*, 'a mod b' is the remainder when a is divided by b. The special case of 'a mod 1' is also known as the *mantissa* of a. The generator is *seeded* with values (between 0 and 1) for x_{n-2} and x_{n-1} and started by setting n to 1. It will then generate the next value in the range 0 to 1. Theoretically at least β values can be generated before any cycling occurs. The terms α and β are prime numbers (e.g. 30341 and 628051), large enough so that the cycle length is sufficiently large. Choosing different seed values produces different sequences of random numbers. If this is required, the seeds can be derived as some function of the system clock.

The result is described by the continuous form of the distribution. The plot of the PDF for U (0,1) is a rectangle of unity width. Since the area it contains must be unity, the PDF must be

$$f(x) = 1 \tag{5.17}$$

More generally, if x is ranged from x_{min} to x_{max}, the PDF of U(x_{min}, x_{max}) is

$$f(x) = \frac{1}{x_{max} - x_{min}} \tag{5.18}$$

To obtain the CDF we integrate the PDF between x_{min} and x

$$F(x) = \int_{x_{min}}^{x} \frac{dx}{x_{max} - x_{min}} = \left[\frac{x}{x_{max} - x_{min}}\right]_{x_{min}}^{x} = \frac{x - x_{min}}{x_{max} - x_{min}} \tag{5.19}$$

This is plotted as the black line in Figure 5.5.

We describe, later in this chapter, the role of the *quantile function (QF)*. Generally, it is obtained by inverting the CDF. From it, given the probability (F) that a measurement lies between $-\infty$ and x, we can determine x. In this case it would be

$$x(F) = x_{min} + F(x_{max} - x_{min}) \quad 0 \le F \le 1 \tag{5.20}$$

Key parameters of the continuous uniform distribution are

$$\mu = \frac{x_{min} + x_{max}}{2} \tag{5.21}$$

$$\sigma^2 = \frac{(x_{max} - x_{min})^2}{12} \tag{5.22}$$

The distribution is symmetrical and so

$$\gamma = 0 \tag{5.23}$$

Not surprisingly, with no peak or long tails in the distribution, it is platykurtic ($\kappa < 3$).

$$\kappa = 1.8 \tag{5.24}$$

5.2 Triangular Distribution

While one might think that the triangular distribution is too simplistic to be applicable to process data, it can often fit extremely well. Further, the simplicity of the mathematics involved helps with the general understanding of the PDF, the CDF and the relationship between them.

Figure 5.6 shows a continuous triangular distribution. To be a PDF the area it encloses must be unity.

$$\frac{(x_{max} - x_{min})h}{2} = 1 \quad \therefore h = \frac{2}{(x_{max} - x_{min})} \tag{5.25}$$

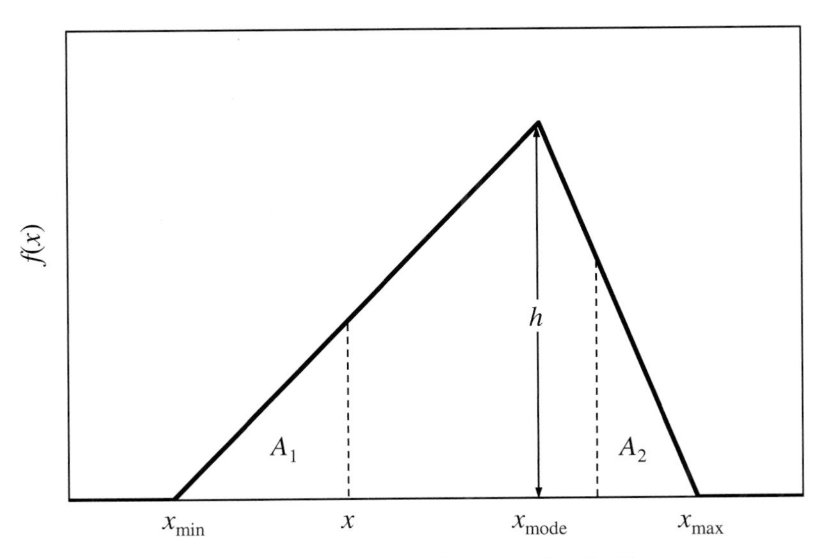

Figure 5.6 *Development of the triangular distribution*

The PDF can therefore be developed. Firstly

$$f(x)=0 \quad \text{for } x<x_{min} \text{ and } x \geq x_{max} \tag{5.26}$$

Secondly, from the similarity of triangles

$$f(x)=\frac{x-x_{min}}{x_{mode}-x_{min}} \times \frac{2}{x_{max}-x_{min}} \quad \text{for } x_{min} \leq x<x_{mode} \tag{5.27}$$

$$f(x)=\frac{x_{max}-x}{x_{max}-x_{mode}} \times \frac{2}{x_{max}-x_{min}} \quad \text{for } x_{mode} \leq x<x_{max} \tag{5.28}$$

The CDF is, by definition, the area under the PDF between $-\infty$ and x.

$$F(x)=0 \quad \text{for } -\infty<x \leq x_{min} \tag{5.29}$$

$$F(x)=1 \quad \text{for } x_{max} \leq x<\infty \tag{5.30}$$

For values of x greater than the minimum and less than the mode, $F(x)$ is the area labelled A_1, i.e.

$$F(x)=\frac{(x-x_{min})f(x)}{2}=\frac{(x-x_{min})^2}{(x_{mode}-x_{min})(x_{max}-x_{min})} \quad \text{for } x_{min} \leq x<x_{mode} \tag{5.31}$$

For values of x greater than the mode and less than the maximum $F(x)$ is $(1-A_2)$, i.e.

$$F(x)=1-\frac{(x_{max}-x)^2}{(x_{max}-x_{mode})(x_{max}-x_{min})} \quad \text{for } x_{mode} \leq x<x_{max} \tag{5.32}$$

Note that differentiating the equations for $F(x)$ gives the corresponding equations for $f(x)$.

As an example of a triangular distribution we can plot $f(x)$ and $F(x)$ for the total score from throwing two unbiased six-sided dice. The minimum score (x_{min}) is 2, the maximum (x_{max}) is 12 and the most frequent (x_{mode}) is 7. Figure 5.7 plots Equations (5.26) to (5.28). Figure 5.8

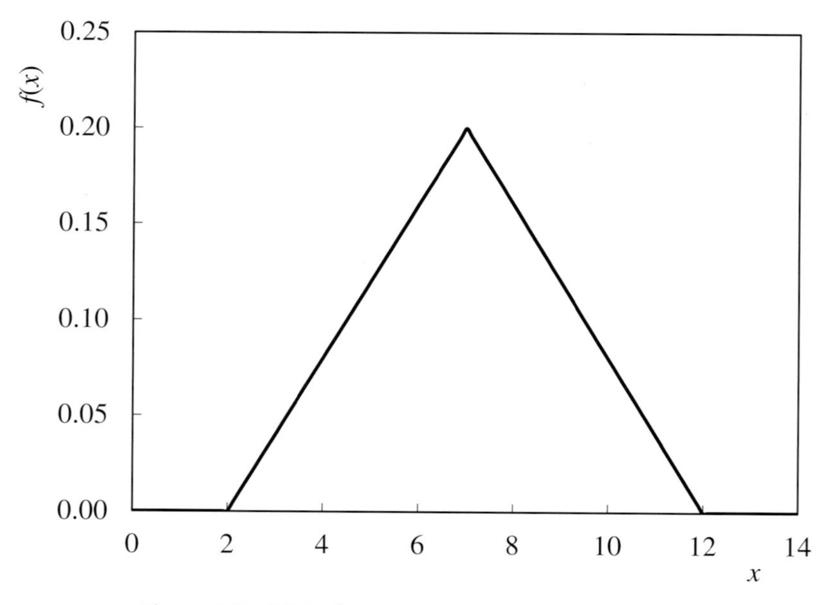

Figure 5.7 PDF of continuous triangular distribution

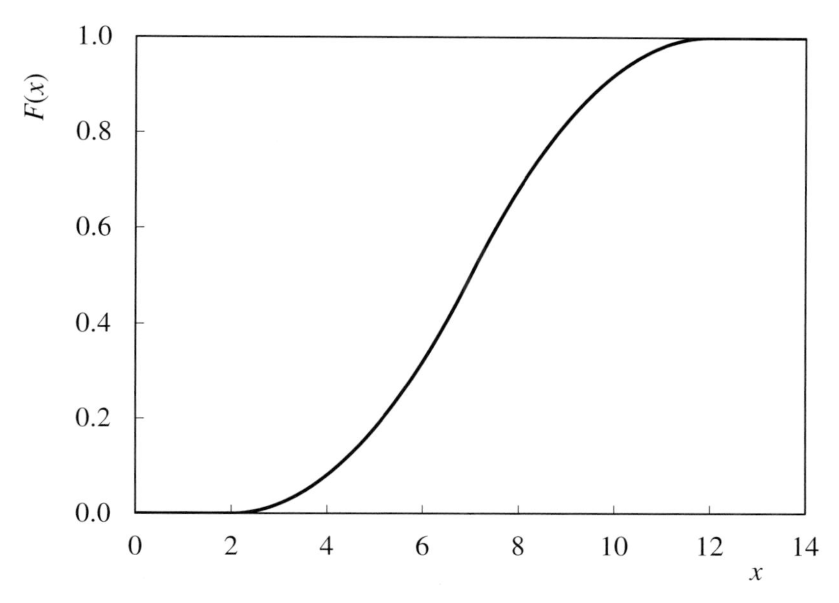

Figure 5.8 *CDF of continuous triangular distribution*

plots Equations (5.29) to (5.32). Note that the plots assume that x is a continuous variable that, in the case of dice, it is clearly not. We cannot use Figure 5.8 to determine the probability of the total score being within a specified range.

Historically the triangular distribution has been used where very little data are available. Indeed, it has been called the *lack of knowledge distribution* – requiring only a minimum, a maximum and a 'best guess' mode. We will cover later its use (now discredited) in the *PERT distribution*.

The mean of the triangular distribution is

$$\mu = \frac{x_{min} + x_{mode} + x_{max}}{3} \tag{5.33}$$

The variance is

$$\sigma^2 = \frac{x_{min}^2 + x_{mode}^2 + x_{max}^2 - x_{min}x_{mode} - x_{min}x_{max} - x_{mode}x_{max}}{18} \tag{5.34}$$

Skewness is restricted to the range −0.566 (when $x_{mode} = x_{min}$) to 0.566 (when $x_{mode} = x_{max}$). Its formula is too large to present here. In the absence of long tails, the distribution is platykurtic. Kurtosis is fixed at 2.4.

5.3 Normal Distribution

The *normal distribution* is by far the most commonly used. Also known as the *Gaussian distribution*, it has a mean of μ and a standard deviation of σ. The PDF is

$$f(x) = \frac{1}{\sigma\sqrt{2\pi}} \exp\left[\frac{-(x-\mu)^2}{2\sigma^2}\right] \tag{5.35}$$

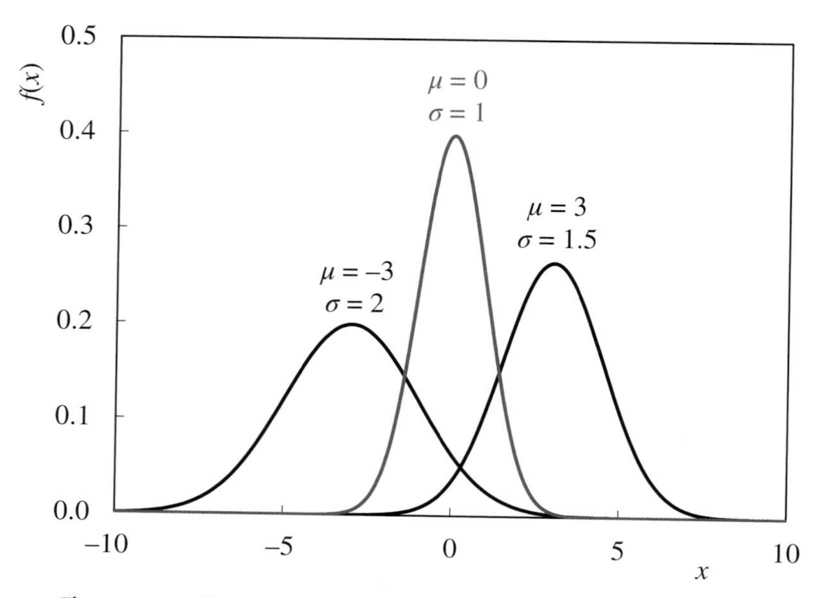

Figure 5.9 *Effect of varying μ and σ on PDF of normal distribution*

This is described (as it will be in this book) as N(μ,σ^2). Confusingly, N(μ,σ) will appear in other texts. Figure 5.9 shows the effect of changing μ and σ.

This equation is converted to the *standard normal function* by setting the mean to 0 and the variance to 1. Described as N(0,1), it will have the formula

$$\phi(z) = \frac{1}{\sqrt{2\pi}} \exp\left(\frac{-z^2}{2}\right) \tag{5.36}$$

Figure 5.9 shows this as the coloured curve. The term z is known as the *z score*, defined by

$$z = \frac{x-\mu}{\sigma} \tag{5.37}$$

Although rarely used in this book, the term $\phi[f(z)]$ is used commonly in other texts as an abbreviation, i.e.

$$\phi[f(z)] = \frac{1}{\sqrt{2\pi}} \exp\left[-\frac{[f(z)]^2}{2}\right] \tag{5.38}$$

To obtain the CDF we would normally integrate Equation (5.36)

$$F(z) = \frac{1}{\sqrt{2\pi}} \int_{-\infty}^{x} \exp\left(\frac{-z^2}{2}\right) dz \tag{5.39}$$

However the function does not integrate simply. Instead we define the *error function (erf)*. This has nothing to do with errors. It is the probability that a random variable, chosen from N(0,0.5), falls between $-x$ and x.

$$\text{erf}(z) = \frac{1}{\sqrt{\pi}} \int_{-x}^{x} \exp(-z^2).dz = \frac{2}{\sqrt{\pi}} \int_{0}^{x} \exp(-z^2).dz \tag{5.40}$$

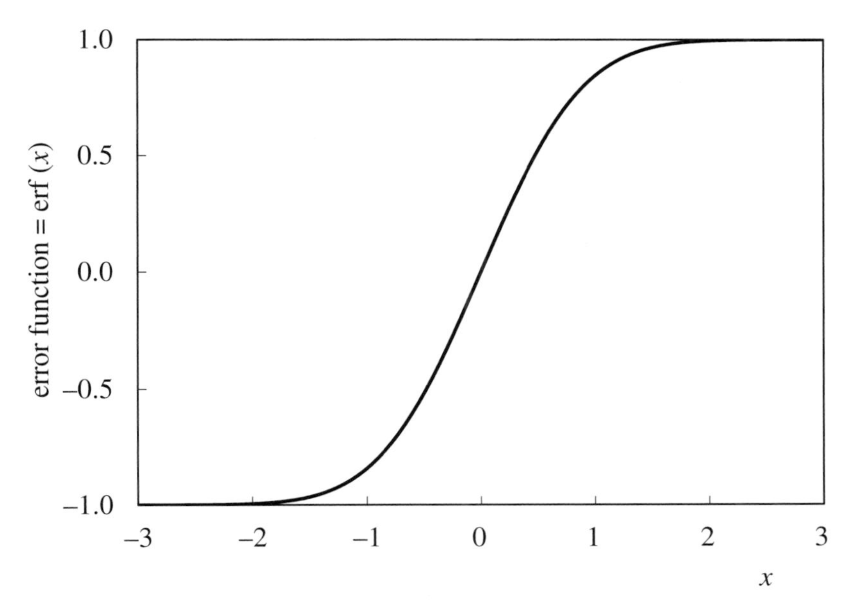

Figure 5.10 *Error function*

A useful property of the error function is

$$\operatorname{erf}(-x) = -\operatorname{erf}(x) \tag{5.41}$$

Using the trapezium rule, Figure 5.10 plots Equation (5.40), applying Equation (5.41) for negative values of x. Using a numerical approximation, the erf function is provided in most spreadsheet packages.

The CDF of the standard normal distribution is then

$$\Phi(z) = F(z) = \frac{1}{2}\left[1 + \operatorname{erf}\left(\frac{z}{\sqrt{2}}\right)\right] \tag{5.42}$$

Like the term for the PDF, $\phi[f(z)]$, $\Phi[f(z)]$ is commonly used as an abbreviation for the CDF.

Including μ and σ gives the CDF for the normal distribution

$$F(x) = \Phi\left(\frac{x-\mu}{\sigma}\right) = \frac{1}{2}\left[1 + \operatorname{erf}\left(\frac{x-\mu}{\sigma\sqrt{2}}\right)\right] \tag{5.43}$$

Figure 5.11 shows how this distribution is affected by the values of μ and σ.

The *complementary error function (erfc)* is defined as

$$\operatorname{erfc}(x) = 1 - \operatorname{erf}(x) \quad \text{or} \quad 1 + \operatorname{erf}(-x) \tag{5.44}$$

Not to be confused with the function, there is the *erf distribution*, described by

$$f(x) = \frac{\alpha}{\sqrt{\pi}}\exp\left(-\alpha^2 x^2\right) \tag{5.45}$$

Comparison with Equation (5.35) shows that it is a form of the normal distribution with

$$\mu = 0 \quad \text{and} \quad \sigma = \frac{1}{\alpha\sqrt{2}} \tag{5.46}$$

Using it offers no advantage over the normal distribution.

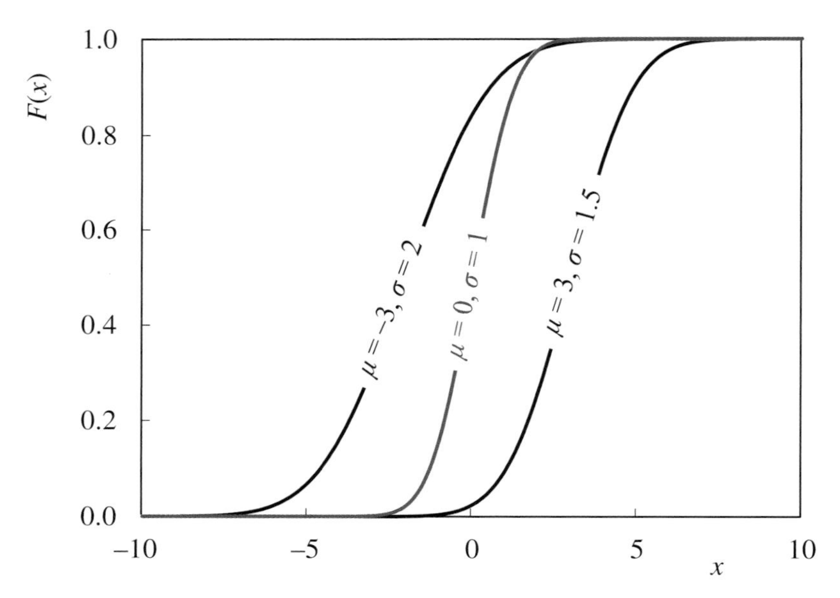

Figure 5.11 Effect of varying μ and σ on CDF of normal distribution

5.4 Bivariate Normal Distribution

All the distributions included in this book are described in their *univariate* form, i.e. they describe the distribution of a single variable. The mathematics of *multivariate* distributions are quiet complex and better handled with a software product. However, we can use a *bivariate* example to demonstrate the key features. In the same way that there are multiple forms of distribution for a single variable, there are a large number for multiple variables. As an example of a *joint distribution* we will use the *bivariate normal distribution*.

The PDF for two variables, x and y, is

$$f(x,y) = \frac{1}{2\pi\sigma_x\sigma_y\sqrt{1-R^2}}\exp\left[-\frac{z}{2(1-R^2)}\right] \quad \sigma_x, \sigma_y > 0 \tag{5.47}$$

where

$$z = \frac{(x-\bar{x})^2}{\sigma_x^2} + \frac{(y-\bar{y})^2}{\sigma_y^2} - \frac{2R(x-\bar{x})(y-\bar{y})}{\sigma_x\sigma_y} \tag{5.48}$$

R is the Pearson coefficient, as defined by Equation (4.75).

Using the analyses of a site's fuel gas, shown Table A1.5, we could develop a full multivariate distribution for all 11 components. However, as a bivariate example, we will choose x as mol% hydrogen (H_2) and y as mol% methane (CH_4).

By calculation from the data

$$\bar{x} = 57.36 \qquad \bar{y} = 12.69 \tag{5.49}$$

$$\sigma_x = 6.63 \qquad \sigma_y = 1.87 \qquad \sigma_{xy} = -10.35 \tag{5.50}$$

$$R = -0.835 \tag{5.51}$$

Remembering that R is ranged from −1 to 1, the result shows that there is strong reverse correlation between x and y. Not surprisingly, if the concentration of the major component

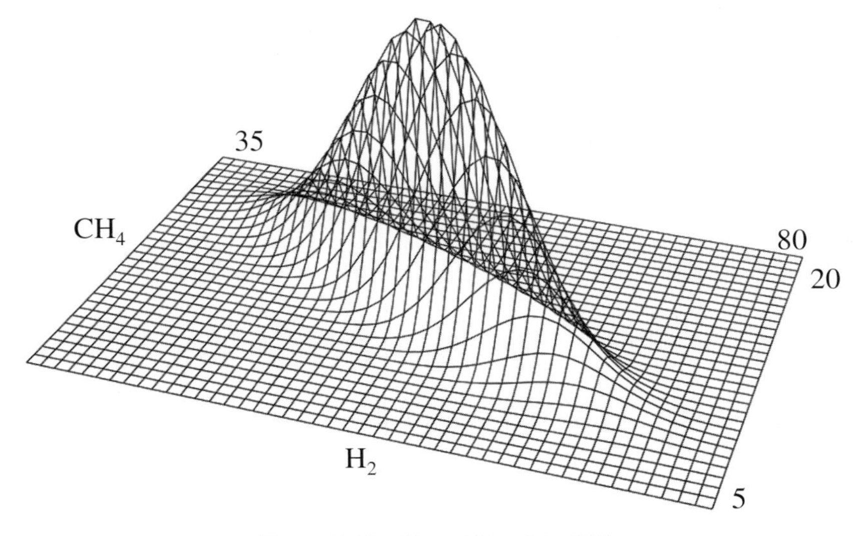

Figure 5.12 Plot of bivariate PDF

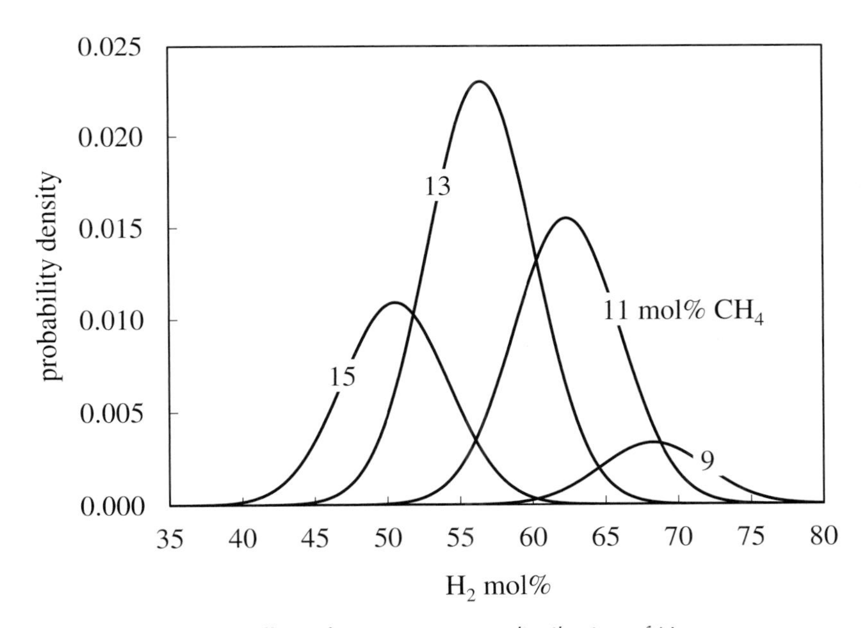

Figure 5.13 Effect of CH_4 content on distribution of H_2 content

increases, the concentration of the next largest component will likely decrease. In effect, the sum of the two components is approximately constant.

Figure 5.12 plots Equation (5.47) against x and y. The diagonal orientation confirms the correlation – high H_2 content corresponds to low CH_4 content and vice versa. Figure 5.13 is a two-dimensional representation of the same distribution. It comprises four vertical slices along the CH_4 axis, showing how the distribution of H_2 content varies as the CH_4 content varies. Figure 5.14 comprises five slices along the H_2 axis showing the distribution of CH_4 content as H_2 content varies. Figure 5.15 plots the cumulative distribution. In the absence of a CDF this is derived by applying the trapezium rule twice – for both x and y.

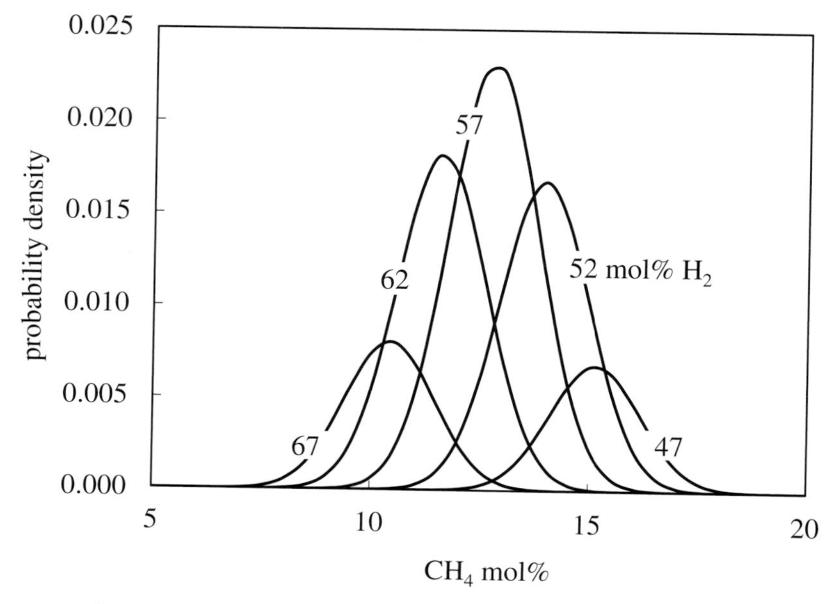

Figure 5.14 Effect of H_2 content on distribution of CH_4 content

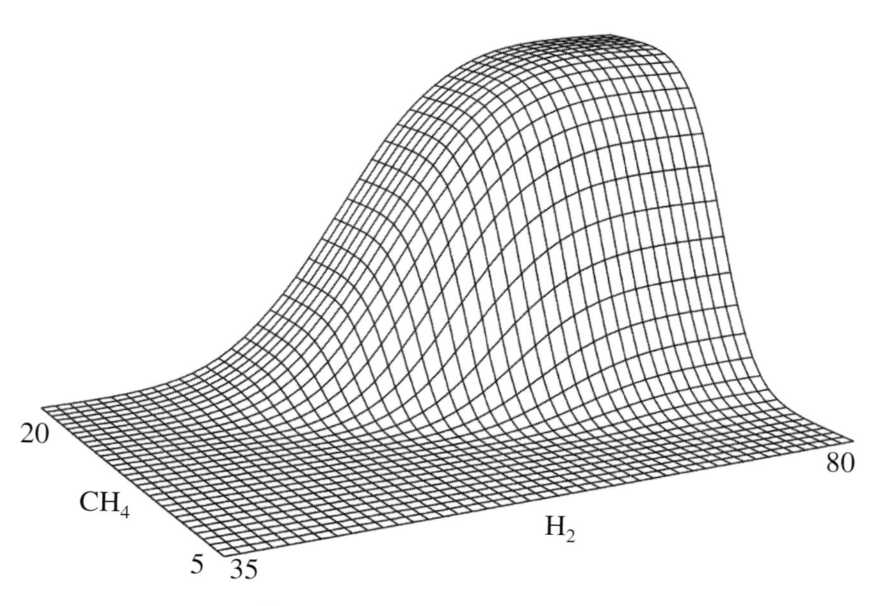

Figure 5.15 Plot of bivariate CDF

While useful for display purposes, bivariate distributions probably have little application to the role of process control engineers. Further, our example violates a key aspect of this book. We should fit the distribution's shape parameters to the data not, as we have done here, calculate them from the data. While it would be possible to fit the five parameters, listed in Equations (5.49) and (5.50), it would likely offer little advantage over the use of a univariate distribution.

5.5 Central Limit Theorem

Proof of the Central Limit Theorem involves mathematics beyond the scope of this book. Nevertheless its conclusion is of great importance – particularly in supporting the assumption that the variation in key process performance parameters follows a normal distribution. Fundamentally the theorem states that summing a large number of independent values, selected randomly, will result in a total that is normally distributed – no matter what the shape of the distribution of the values selected. In effect a dependent process parameter, such as weekly butane production, is determined by the sum of a large number of independent flow measurements recorded every minute. Each of these independent measurements does not have to be normally distributed but the dependent variable will be.

As a simple example consider the results of throwing a single unbiased six-sided dice. There is an equal probability (of 1/6) of throwing any number from 1 to 6. Throwing the dice enough times would result in a uniform distribution of the score (x) with a mean (μ) and variance (σ^2).

$$\mu = \frac{\sum_{i=1}^{6} x_i}{6} = 3.5 \tag{5.52}$$

$$\sigma^2 = \frac{\sum_{i=1}^{6} (x_i - \mu)^2}{6} = \frac{35}{12} = 2.9 \tag{5.53}$$

Figure 5.16 shows the normal distribution, with the same mean and variance, superimposed on the true distribution. Clearly the distributions are very different. However if we sum two values each selected from the uniform distribution, the equivalent of throwing two dice 36 times, we expect the distribution shown in Figure 5.17. The mean and variance of this distribution are obtained by summing the means and variances of the source distributions, giving

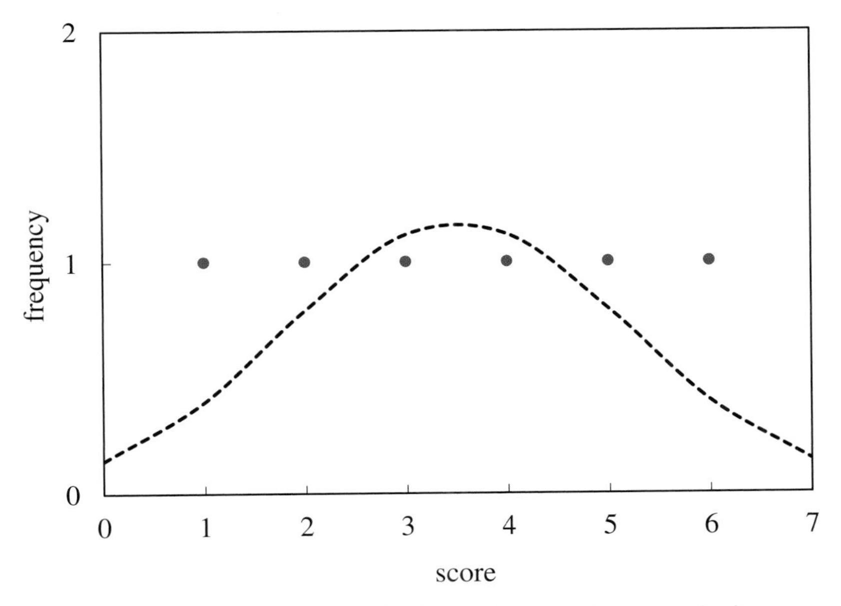

Figure 5.16 *Frequency distribution of the score from a single dice*

values of 7 and 5.8 respectively. Figure 5.17 shows that the now triangular distribution is closer to the normal distribution. Closer still is the distribution for throwing five dice 7,776 times, as shown in Figure 5.18. It will have a mean of 17.5 and a variance of 14.6.

We have demonstrated how summing values selected randomly from a uniform distribution will generate a normal distribution, but the Central Limit Theorem goes further than this.

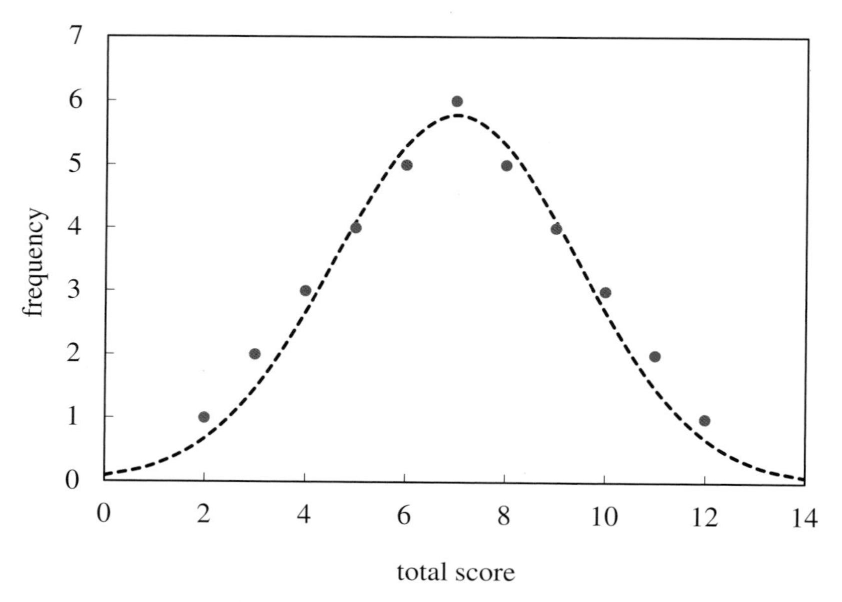

Figure 5.17 *Frequency distribution of the total score from two dice*

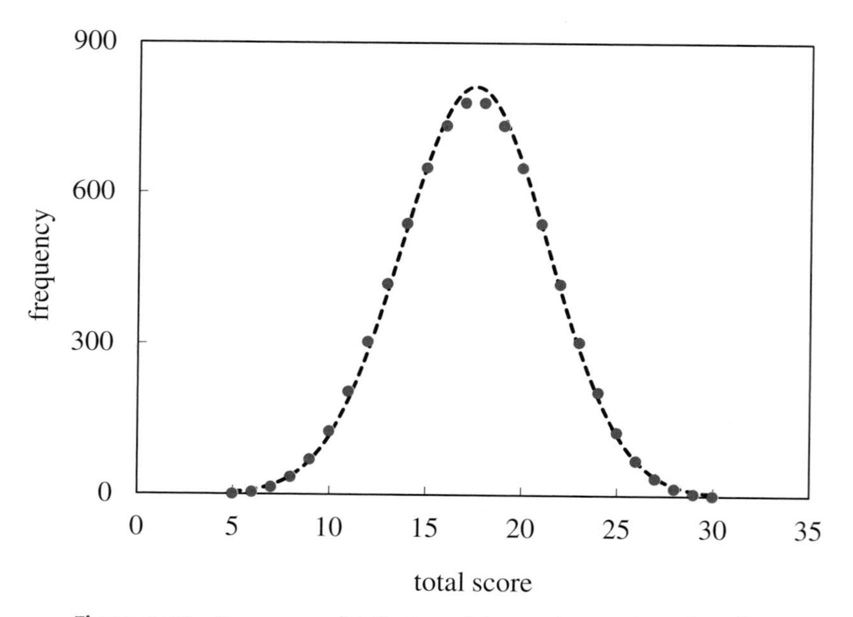

Figure 5.18 *Frequency distribution of the total score from five dice*

Consider a six-sided dice with the sides having values of 1, 1, 2, 5, 6 and 6. Throwing such a dice will generate the distribution, shown in Figure 5.19, which is far from uniform. Throwing a large enough number of such dice will generate a normal distribution. For example, as shown in Figure 5.20, throwing seven such dice 279,936 times approaches it.

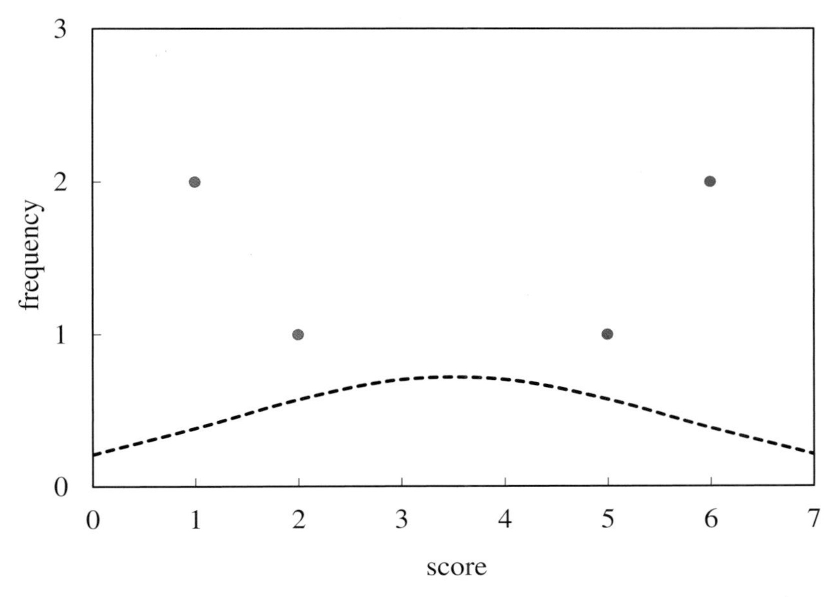

Figure 5.19 Frequency distribution of the score from a single modified dice

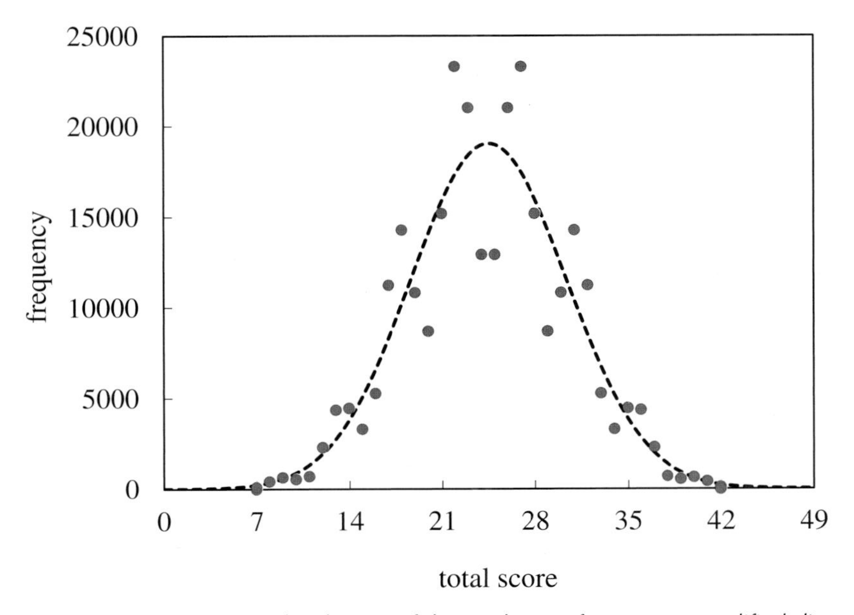

Figure 5.20 Frequency distribution of the total score from seven modified dice

The Central Limit Theorem applies only to the addition (or subtraction) of random variables. For example, instead of summing the scores from three six-sided dice, we multiply them. The expected distribution of the result is shown in Figure 5.21. Highly skewed, it is far from a normal distribution. However, multiplication is equivalent to summing logarithms. The Central Limit Theorem would therefore suggest that the logarithm of the combined score would be normally distributed. Indeed, as Figure 5.22 shows, this would appear to be the case. Rather

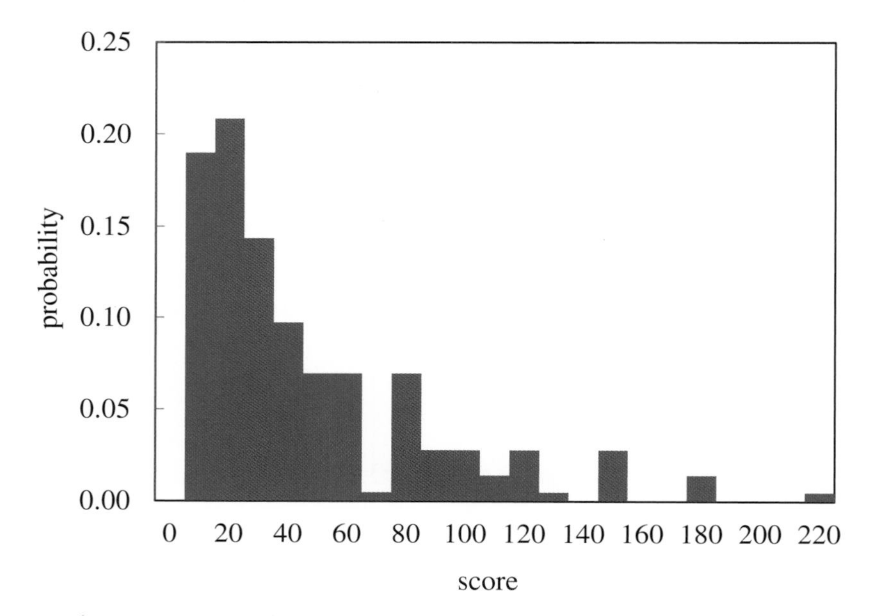

Figure 5.21 Probability distribution of the product of three dice throws

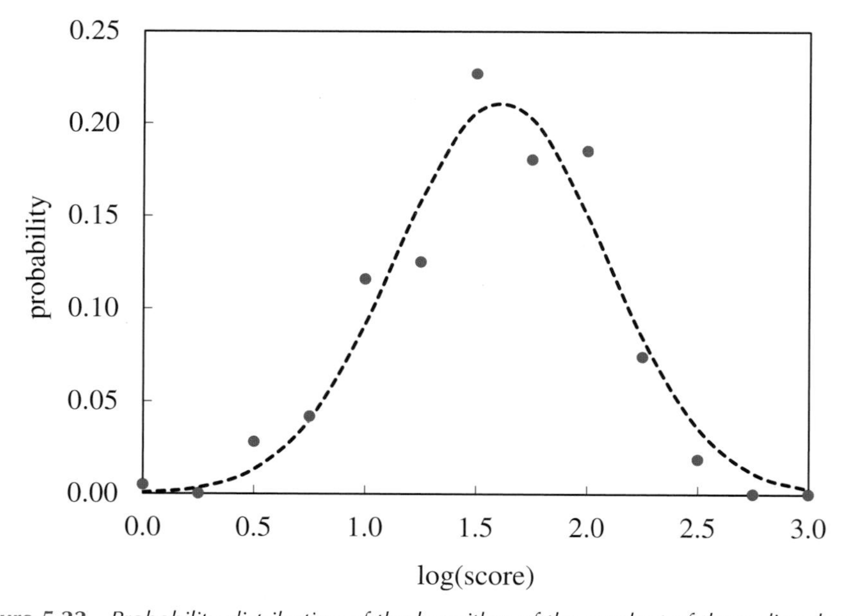

Figure 5.22 Probability distribution of the logarithm of the product of three dice throws

than fit the normal distribution to the logarithm of a measurement, we will see later that the lognormal distribution fits to the measurements directly.

Process measurements are the result of far more complex mathematical functions that describe process behaviour. These include those describing reaction kinetics, heat and mass balancing, heat transfer, vaporisation, equipment behaviour, etc. It is not practical to design a distribution function based on an understanding of the mathematics involved. Instead we take the far more pragmatic approach of selecting a distribution largely on the basis that it fits the data well.

5.6 Generating a Normal Distribution

With dice, since only integer scores are possible, the mean and standard deviation must be calculated from discrete values. For continuous functions a slightly different approach is required. Consider a number randomly generated from a uniform distribution in the range 0 to 1. Frequency, on the vertical axis, is replaced with probability. Since we are certain to generate a value in this range, the area under the distribution curve must be unity – as shown in Figure 5.23.

The mean of the distribution is 0.5. Since $f(x) = 1$, from Equations (4.45) and (4.48), the variance is

$$\sigma^2 = \int_0^1 x^2.dx - \mu^2 = \left[\frac{x^3}{3}\right]_0^1 - \left(\frac{1}{2}\right)^2 = \frac{1}{12} \tag{5.54}$$

As we saw from the Central Limit Theorem, if we add together 12 values (x_1 to x_{12}) randomly selected from the same uniform distribution, we would obtain a value chosen from a distribution very close to normal. The total would have a mean of 6, a variance of 1 and thus a standard

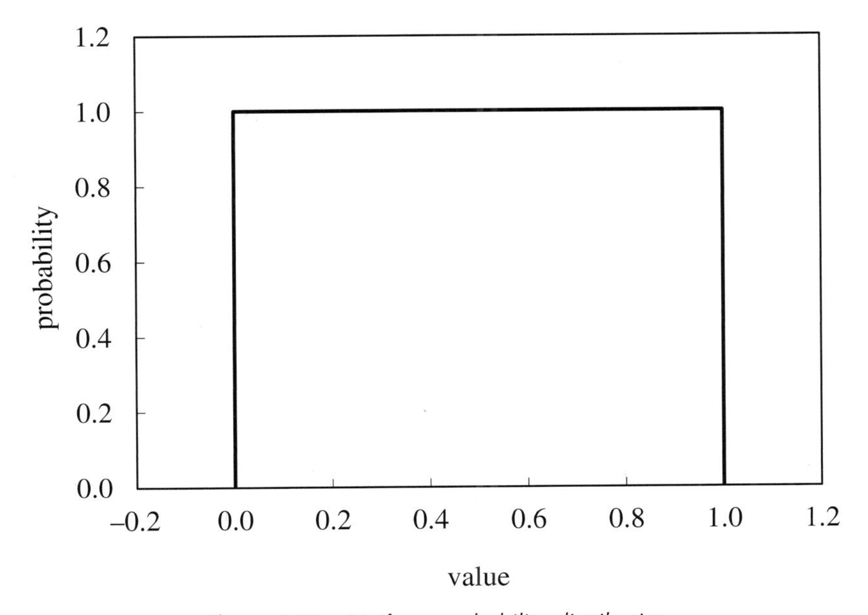

Figure 5.23 *Uniform probability distribution*

deviation of 1. This gives us a method of generating a value (x) from a normal distribution required to have a mean of μ and a standard deviation of σ.

$$x=\mu+\sigma\left[\sum_{i=1}^{12}x_i-6\right] \tag{5.55}$$

5.7 Quantile Function

We previously defined quartiles – the three points that divide the ranked data into four equal groups. Similarly deciles divide the data into 10 groups; *centiles* divide the data into 100 groups. *Quantiles* divide the data into n equal groups, where n is the number of data points and so each group will contain one data point. We will make use of these later in *quantile plots*. The *quantile function (QF)* is related to these. Also known as the *per cent point function*, it is the inverse of the CDF. For example, the CDF of the standard normal distribution is given by Equation (5.43). Inverting it gives

$$x(F)=\mu+\sigma\sqrt{2}\mathrm{erf}^{-1}(2F-1) \qquad 0\le F\le 1 \tag{5.56}$$

This allows us, given the probability (F), to determine x. While the erf^{-1} may not be included as a function in a spreadsheet package, the inverse of the normal distribution is likely to be. Should it not be, then the CDF of the normal distribution can be closely approximated by

$$F(x)\approx\frac{1}{2}\left\{1+\frac{|x-\mu|}{x-\mu}\left[1-\exp\left(-\frac{2(x-\mu)^2}{\pi\sigma^2}\right)\right]^{\frac{1}{2}}\right\} \tag{5.57}$$

The advantage of this approximation is that it can be inverted to give a QF that is a close approximation to Equation (5.56).

$$x(F)\approx\mu+\sigma\sqrt{-\frac{\pi}{2}\ln\left[1-(2F-1)^2\right]} \qquad 0\le F\le 1 \tag{5.58}$$

Figure 5.24 shows a plot of Equation (5.56) for the case where μ is 0 and σ is 1. It is the inverse of the standard normal function developed from rearranging Equation (5.42), i.e.

$$z(F)=\sqrt{2}\mathrm{erf}^{-1}(2F-1) \qquad 0\le F\le 1 \tag{5.59}$$

As Figure 5.24 shows, by selecting 0.25 as the value for F, we obtain the first quartile as –0.674. We could have obtained the median ($F=0.5$) and the third quartile ($F=0.75$). The QF can also be used to obtain the confidence interval that we will cover this in more detail in Section 8.2. As shown the single tailed 95% confidence value is 1.64.

Another important use of the QF is the generation of data that follow the specified PDF, e.g. for use in Monte Carlo simulation. We first determine F by sampling from the uniform distribution U(0,1) and use the QF to generate the corresponding x.

Some CDF are readily invertible. For others the function may exist in a spreadsheet package. Failing this, then an iterative approach can be used – adjusting x in the CDF until the required value for F is reached. Alternatively F can be plotted against x. The ease with which x can be obtained from F might be a criterion in selecting the best distribution.

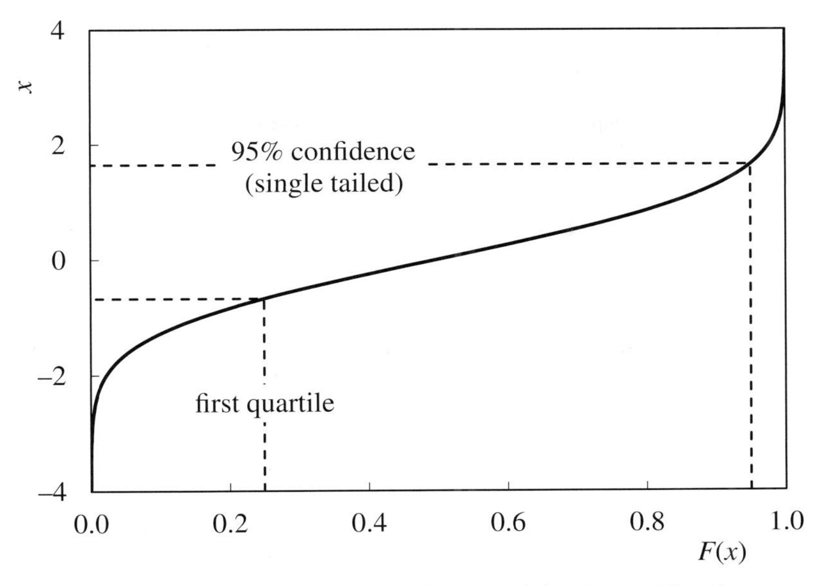

Figure 5.24 *Quantile function for normal distribution N(0,1)*

Although relatively unusual, there are distributions that can only be expressed as a QF. As we will see later, this does not prevent them being used; we simply have to take a slightly different approach to fitting them to data.

5.8 Location and Scale

We have seen, for the normal distribution, that μ defines the central value (or *location*) of the data and σ the dispersion (or *scale*). Most other distributions require these parameters but, for simplicity, they are often documented without them. The distribution is then in *standard form*. Including the location parameter (e.g. μ) in any standard distribution changes it to a *shifted* distribution. Including the scaling parameter (e.g. σ) changes it to a *scaled* distribution. There are other less used definitions of a standard distribution. For example, Gauss defined his as having a variance of 0.5. The word 'standard' is also used by authors to describe a distribution that is written as the developer first specified it. Perhaps worse is the use of 'normal' for the same purpose. To avoid ambiguity, in this book we instead use the word 'classic'.

Only in a few distributions are the location and scale parameters the same as the mean and the standard deviation. Further there is no standardisation on what symbols are used. Throughout this book the terms μ and σ have been reserved specifically for mean and standard deviation. Wherever possible we have used α as a more general measure of location and β as a more general measure of scale. Where available, formulae are given so that μ and σ can be calculated from α and β. If there is no simple formula for σ, in most distributions it is proportional to β. We can exploit this, for example, when determining the improvement achievable by improved control. We typically base this on halving the standard deviation. We can instead halve β. Provided we can then determine what impact this can have on the average operation, we do not need to quantify σ. We will illustrate this with examples later.

In some distributions, where it makes better sense to do so, β is replaced with $1/\lambda$. For example, β might represent the mean time between events occurring. Its reciprocal (λ) would then be the frequency of events and might be described as the *rate parameter*. We have used δ when an additional shape parameter is included. If used to change skewness it might be described as the *asymmetry parameter*. If to change kurtosis it might be the *tail index* or *tail heaviness parameter*. We have used δ_1, δ_2, etc. if there is more than one. Symbols, such as θ and τ, which have other meanings specific to process control, have been replaced.

Some distributions include in their names words, such as 'beta', 'gamma', 'kappa' and 'lambda', indicating the use of these characters in the originally published PDF. Standardising results in these symbols being largely replaced. However, as will become clear, naming of distributions is already very ambiguous and not a reliable way of distinguishing one from another. While we have used the names under which distributions are published, the reader need not be concerned too much about their origin – using instead the PDF or CDF to identify a distribution unambiguously.

Care must be taken if a scale parameter is added to a PDF expressed in standard form. In standard form it might be expressed as $f(z)$ and we wish to convert it to $f(x)$, where

$$z = \frac{x - \alpha}{\beta} \tag{5.60}$$

Firstly, we have to compensate the normalisation constant to ensure the PDF still integrates to 1 over the range $-\infty$ to $-\infty$. Remembering that the PDF is the derivative of the CDF, so

$$f(x) = \frac{dF(x)}{dx} \tag{5.61}$$

$$f(z) = \frac{dF(z)}{dz} \tag{5.62}$$

$$\frac{dF(x)}{dx} = \frac{dF(z)}{dz}\frac{dz}{dx} \tag{5.63}$$

Differentiating Equation (5.60)

$$\frac{dz}{dx} = \frac{1}{\beta} \tag{5.64}$$

So

$$f(z) = \frac{1}{\beta} f(x) \tag{5.65}$$

For example, comparing Equations (5.35) and (5.43) for the normal distribution shows the inclusion of $1/\sigma$ outside of the exponential term.

Equation (5.65) explains why β is described as the scale parameter in some distributions. Changing it changes the vertical scale of the plot of the PDF.

Secondly, from rearranging Equation (5.60)

$$x = \alpha + \beta z \tag{5.66}$$

Therefore

$$\mu_x = \alpha + \beta \mu_z \tag{5.67}$$

Thus any calculation of the mean must be modified to include α and β. Similarly any calculation of the standard deviation must be multiplied by β or, more commonly, the calculation of variance must be multiplied by β^2. Skewness and kurtosis, since they are dimensionless, are unaffected by the inclusion of α and β.

Many distributions require z to be positive. For example, they might include functions such as z^k, where k is not an integer. Similarly, some distributions include $\ln(z)$ or $1/z$ and so z cannot be 0. These are one reason why a distribution may be lower-bounded at zero. Adding the location parameter moves the lower bound to α. If fitting a distribution that includes values of x that are less than α then, for these values, $f(x)$ and $F(x)$ are set to zero. Indeed, in some distributions, α is replaced with x_{min} – the minimum permitted value of x.

5.9 Mixture Distribution

There are several ways in which distributions can be combined. A distribution can even be combined with another version of itself. But we need to be careful with terminology. There are several words that are synonymous with 'mixture' but describe quite different distributions. For example, the *compound distribution* we cover later is quite different from the *mixture distribution* we describe here. Similarly, *joint distribution* is another name for a *multivariate distribution*. As covered in Section 5.4, it describes the distribution of more than one variable. A *combined distribution* is again something different, as explained in the next section.

A mixture distribution is the weighted sum of several distributions. The weights are known as *mixture weights* and the distributions as *mixture components*. As a relatively simple example we might produce multiple grades of the same product. For example, many polymers are manufactured to meet a specification on *MFI (melt flow index)* but the target value for the index varies by product grade. Most plants will produce several grades, not simultaneously but by switching from one to another as product demand requires. If we were to attempt to fit a distribution to *MFI*, we need to separate the variation caused by poor control from that caused by changing target. One way of doing this is to determine the distribution for each grade and then 'mix' the distributions to produce a PDF known as the *mixture density*.

For example, two grades of polymer might have mean *MFI* of 55 and 65. We might find the standard deviation of the first grade is 2 and that of the second is 3. Production might be split 30/70 per cent. If each is normally distributed then, from Equation (5.35), the PDF is therefore

$$f(MFI) = \frac{0.3}{2\sqrt{2\pi}}\exp\left[\frac{-(MFI-55)^2}{2\times 2^2}\right] + \frac{0.7}{3\sqrt{2\pi}}\exp\left[\frac{-(MFI-65)^2}{2\times 3^2}\right] \tag{5.68}$$

Figure 5.25 plots this function. We could further refine it by choosing different distributions for each grade. For example, the data for one might be highly skewed and therefore cannot be accurately represented by the normal distribution.

While a mixture distribution might be an effective way of presenting control performance, it is inadvisable to combine the data in this way for the purpose of quantifying control improvement. Ideally this should be calculated by grade and the weighting factors applied to sum the benefits.

5.10 Combined Distribution

While a mixture distribution is a *combined distribution*, this term is more usually reserved for one that has shape parameters derived from the parameters of two or more distributions of the

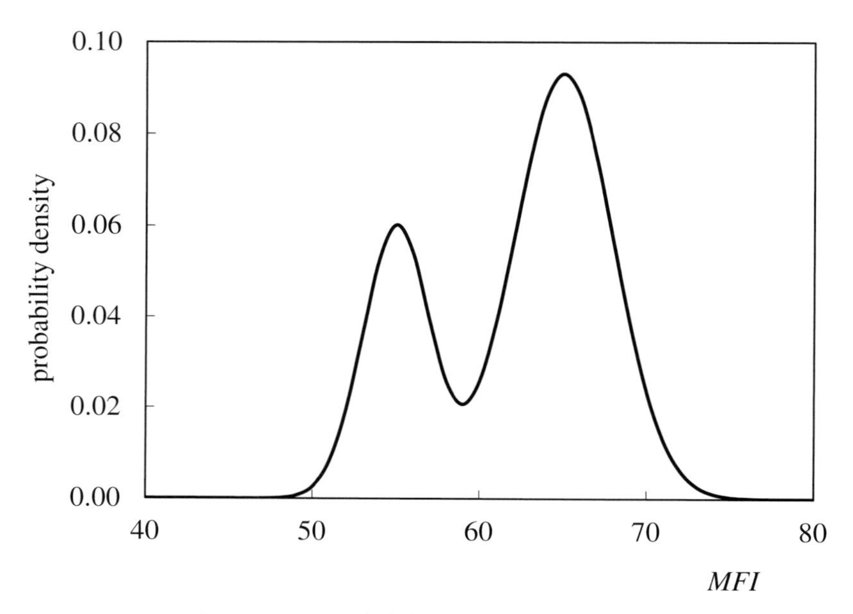

Figure 5.25 Probability density of MFI results

same type. For example, we can two combine normal distributions by taking a weighted average of their means and variances to produce another normal distribution.

We might calculate product loss by subtracting what has been delivered from what has been produced. The result will have a distribution that reflects the confidence we have in the measure of loss. Its mean is the difference between the two measurements. We can assign variances to both measurements that reflect their accuracy. The variance of the loss is then the sum of the two variances. We have developed a combined normal distribution.

Such an approach is used in *data reconciliation*. This is covered in detail in Chapter 18. Most process measurements have multiple measurements of the same property, none of which are completely reliable. For example, a composition might be measured by an on-stream analyser and by the laboratory. Or the flow of feed to a process might be measured and also derived by totalling the measured product flows.

To take the simplest example of two measurements of the same property, we might assume the measurements (μ_1 and μ_2) are means of two normal distributions. From instrument specifications we can estimate the reliability of each measurement and convert these to standard deviations (σ_1 and σ_2). We show in Chapter 18 that the best estimate (μ) is the weighted average of the two 'opinions', where the weights are

$$w_1 = \frac{\sigma_1^2}{\sigma_1^2 + \sigma_2^2} \qquad w_2 = \frac{\sigma_2^2}{\sigma_1^2 + \sigma_2^2} \tag{5.69}$$

$$\mu = w_1\mu_1 + w_2\mu_2 \tag{5.70}$$

The resulting variance will be the weighted sum of the variances.

$$\sigma^2 = w_1\sigma_1^2 + w_2\sigma_2^2 \qquad \therefore \sigma = \frac{\sigma_1\sigma_2}{\sqrt{\sigma_1^2 + \sigma_2^2}} \tag{5.71}$$

Other combined distributions arise from combining variables randomly selected from other distributions. For example, if we select x_1 from N(0,1) and x_2 from U(0,1), the ratio x_1/x_2 will follow the slash distribution. Included in this book are many distributions derived in this way. While they probably originated as mathematical curiosities and bear no relation to the mechanism by which process conditions arise, this does not mean they should be discounted as impractical. If the end result proves consistent with process behaviour, they should at least be considered as a choice for prior distribution. Indeed, we will show that the slash distribution will often represent well the distribution of process disturbances.

5.11 Compound Distribution

We will cover, with worked examples in Part 2, a large number of *compound distributions*. These are developed by assuming that a shape parameter in one distribution is not a constant, but varies according to some other distribution. For example, if we assume the mean of the normal distribution itself follows the *exponential distribution*, we obtain the *exponentially modified Gaussian (EMG) distribution*. If we assume the standard deviation follows the *inverse gamma distribution*, we obtain the *Student t distribution*.

The mathematics involved in developing compound distribution can be very complex. It is not important we know exactly how any distribution is derived. Indeed, we do not need to know even what 'parent' distributions are used. As with any distribution, we test its worth by exploring how well it fits the data.

5.12 Generalised Distribution

It is common practice to *generalise* a distribution by including additional shape parameters. The CDF of a distribution, $F(x)$, by definition ranges from 0 to 1 as x ranges from $-\infty$ to $+\infty$. It will remain a CDF if it raised to any power (δ). Doing so produces a *generalised distribution*. By setting δ to 1, it still includes the original distribution but adjusting it means it also includes many other distributions.

However, this is not the only method by which a distribution can be generalised. Many distributions shift and scale the variable x to produce the variable z which is ranged 0 to 1. Raising z to the power δ does not affect its range and therefore offers another way a distribution might be generalised.

$$z^\delta = \left(\frac{x-\alpha}{\beta}\right)^\delta \tag{5.72}$$

Other approaches include making a shape parameter a variable, dependent on x. For example, the scale parameter (β) might be replaced with a simple function

$$z = \frac{x-\alpha}{\beta-\delta(x-\alpha)} \tag{5.73}$$

Since this changes the range of z, some compensating change will also be made to the normalisation constant so that the CDF still ranges from 0 to 1. Using the method described in Section 5.8, the CDF must be multiplied by A, where

$$A = \frac{dz}{dx} = \frac{\beta}{[\beta-\delta(x-\alpha)]^2} \tag{5.74}$$

The problem is that many of the common distributions have had several of these methods applied. For example, there are at least three distributions sharing the name *generalised Gaussian distribution* but that are entirely different from each other.

5.13 Inverse Distribution

There is considerable confusion about the use of the term 'inverse'. As we saw in Section 5.7, the inverse of a CDF is known as the QF. It is simply the CDF rearranged to make the variable (x) the subject of the equation. But there are also distributions that can be described as 'inverse'. One common definition of an *inverse distribution* is the distribution of the reciprocals of the values in the dataset.

As an example we can use the exponential distribution that we cover later in Section 12.7. In its simplest form, its CDF is

$$F(x) = 1 - e^{-z} \quad \text{where} \quad z = \frac{x-\alpha}{\beta} \tag{5.75}$$

$$\therefore F(x) = 1 - \exp\left(-\frac{x-\alpha}{\beta}\right) \tag{5.76}$$

As usual, we obtain its PDF by differentiating its CDF.

$$f(x) = \frac{1}{\beta}\exp\left(-\frac{x-\alpha}{\beta}\right) \tag{5.77}$$

The CDF of the reciprocal of z is therefore

$$F\left(\frac{1}{z}\right) = 1 - \exp\left(-\frac{\beta}{x-\alpha}\right) \tag{5.78}$$

Differentiating

$$f(x) = \frac{\beta}{(x-\alpha)^2}\exp\left(-\frac{\beta}{x-\alpha}\right) \tag{5.79}$$

This is quite logically named the *inverse exponential distribution* and we will cover it in detail in Part 2. However it might well have been named the 'reciprocal exponential distribution'. Indeed there are distributions that have adopted this convention. There are also distributions described as 'inverse' that appear to have no connection with any reciprocal. There are even 'reciprocal inverse' distributions.

5.14 Transformed Distribution

In addition to z^δ and $1/z$, many other transformations can be applied. Most commonly, these include $\ln(z)$ and $\exp(z)$. These can be useful if the data contains very large or very small values. Other transformations include $|z|$ which, like z^2, can be useful if we have no need to distinguish between positive and negative values. While distributions are often named to reflect the transformation, there is no agreed convention. Differentiating a CDF using the transformed z can produce a very complex PDF; often its origin may not therefore be immediately obvious.

5.15 Truncated Distribution

Truncation is a method of converting an unbounded distribution into one that is bounded. When fitting it to data, values less than x_{min} or greater than x_{max} are effectively removed. This will reduce the area under the PDF curve. To keep it at unity, we adjust the normalisation of $f(x)$ to produce $f_t(x)$ – the PDF for the truncated distribution.

$$f_t(x) = \frac{f(x)}{F(x_{max}) - F(x_{min})} \tag{5.80}$$

We have to do the same to the CDF.

$$F_t(x) = \frac{F(x)}{F(x_{max}) - F(x_{min})} \tag{5.81}$$

Figures 5.26 and 5.27 show an example of the normal distribution N(5,1) being truncated so that it is bounded between 3 and 6.

To determine the mean and standard deviation, we first convert x to standard form (z).

$$z_{min} = \frac{x_{min} - \mu}{\sigma} \tag{5.82}$$

$$z_{max} = \frac{x_{max} - \mu}{\sigma} \tag{5.83}$$

We then use the distribution in its standard form. For the normal distribution this is described by Equations (5.36) and (5.42). The mean and variance are then

$$\mu_t = \mu + \frac{\phi(z_{min}) - \phi(z_{max})}{\Phi(z_{max}) - \Phi(z_{min})}\sigma \tag{5.84}$$

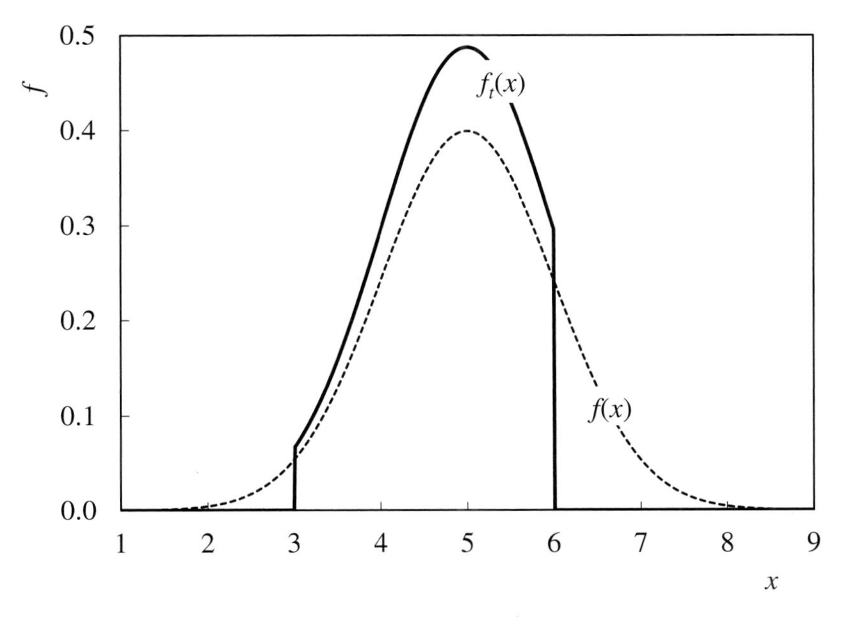

Figure 5.26 Truncated PDF

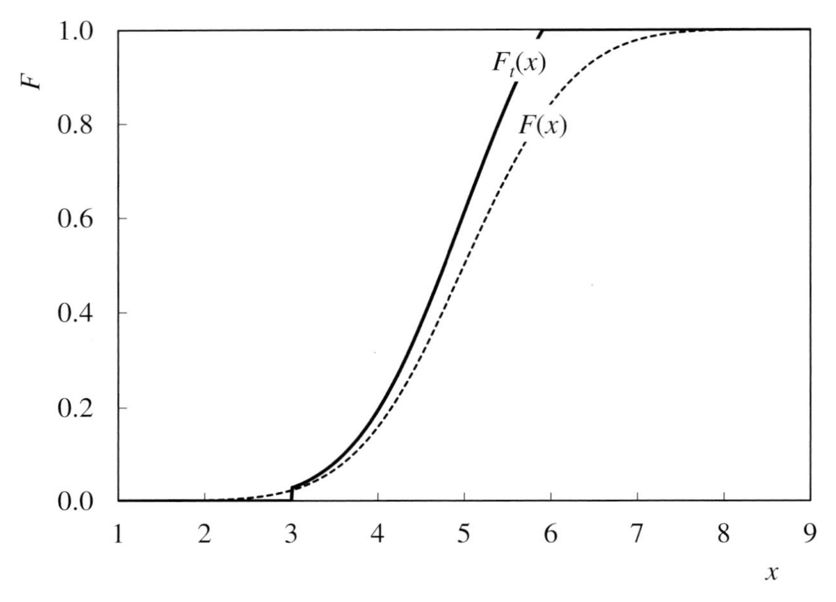

Figure 5.27 *Truncated CDF*

$$\sigma_t^2 = \left[1 + \frac{z_{\min}\phi(z_{\min}) - z_{\max}\phi(z_{\max})}{\Phi(z_{\max}) - \Phi(z_{\min})} - \left(\frac{\phi(z_{\min}) - \phi(z_{\max})}{\Phi(z_{\max}) - \Phi(z_{\min})}\right)^2\right]\sigma^2 \tag{5.85}$$

For the example above, the mean (μ_t) of the truncated distribution is therefore 4.77 and the standard deviation (σ_t) 0.72.

5.16 Rectified Distribution

Rectification sets, to zero, all negative values in a dataset. Figure 5.28 shows the result of doing so to N(1,1). The spike at $x = 0$ includes all the values set to zero. Theoretically it has width of zero so, to contribute to the area under the PDF, it must be infinitely tall. This is known as the *Dirac delta function* or *δ function*. It is defined as

$$\frac{1}{\sigma\sqrt{\pi}}\exp\left(\frac{x^2}{\sigma^2}\right) \to \delta_\sigma(x) \quad \text{as} \quad \sigma \to 0 \tag{5.86}$$

Perhaps fortunately, rectification of process data is unlikely ever to be required.

5.17 Noncentral Distribution

Although this may not be immediately obvious from the form of the PDF, many of the distributions covered in this book are transformations of the normal distribution. These include the beta, chi, F and Student *t* distributions. We omit the complex mathematics of such transformations on the basis that omission does not inhibit successful application. The resulting PDFs are presented in their *central* form. This simply means that they were derived from the standard normal distribution N(0,1). If developed from the more general normal

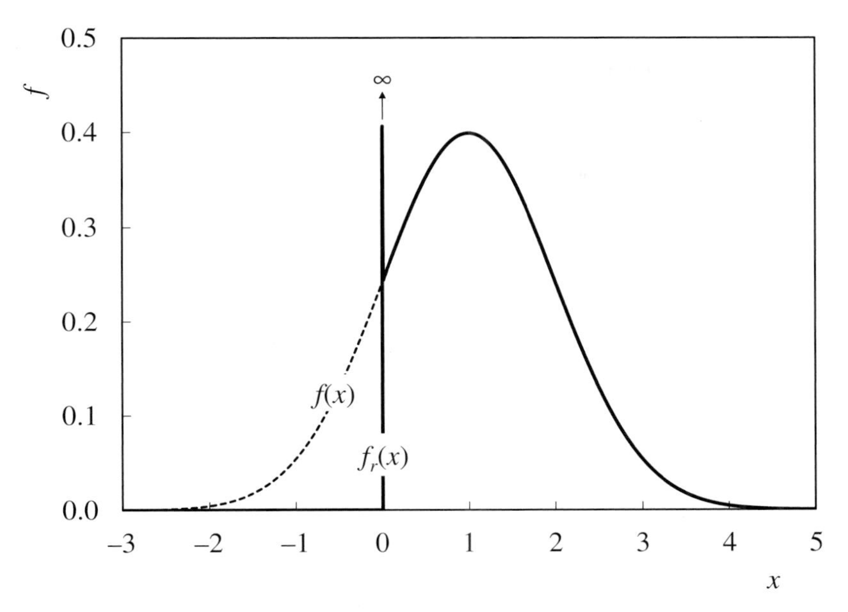

Figure 5.28 Rectified PDF

distribution, with an arbitrary mean and variance, they become *noncentral* versions. While this makes them a little more versatile, the level of mathematics involved in their application is complex and often only feasible with custom software. We will, in Part 2, apply some of them. The remainder are included only in sufficient detail to demonstrate their impracticality.

5.18 Odds

Those familiar with gambling will be aware of probability being expressed as *odds*. For example, if the probability of a horse winning a race is 0.25, out of four races we would expect it to lose three and win one. In the UK and Ireland, odds would then be quoted as '3-to-1 against' (a win). These are known as *fractional odds*. In general it is the probability of failure expressed as a ratio to the probability of success. If p is the probability and the odds against are p_a, conversion is performed by

$$p_a = \frac{1-p}{p} \quad \therefore p = \frac{1}{p_a + 1} \quad p < 0.5 \tag{5.87}$$

Should the probability of success be greater than that of failure, then the ratio is quoted as 'odds on' (p_o) rather than 'odds against'. For example, if the probability was 0.8, then the odds would be quoted '4-to-1 on', not '1-to-4 against'.

$$p_o = \frac{p}{1-p} \quad \therefore p = \frac{p_o}{p_o + 1} \quad p > 0.5 \tag{5.88}$$

The ratio would always be expressed using integers and not necessarily is the simplest form. For example, by tradition, 100-to-8 is preferred to 25-to-2. Both convert to a probability close to 0.074. If p is 0.5, then the odds are quoted as 'evens'.

In most other parts of the world *decimal odds* are used. These are simply the reciprocal of the probability. So '3-to-1 against' would translate to 4; '4-to-1 on' would be 1.25; '100-to-8 against' would be 13.5 and 'evens' would be 2.

In practice, odds quoted by bookmakers do not equate to real probabilities. They are *shortened*, i.e. the probability increased, so that the bookmaker makes a profit. So, if the odds of all the horses in a race were converted to probabilities and summed, the total would be greater than 1. For example, in a race of three equally matched horses, we would expect each horse to have odds of 2-to-1. The bookmaker might offer odds of 3-to-2. If £100 was then bet on each horse, those who bet on the winner would collect £250 and the bookmaker has made a profit of £50.

While odds are not commonly used in statistics, there are a small number of distributions that use Equation (5.87) as a transformation to convert a double-bounded distribution ($0 < p < 1$) to a lower-bounded one ($0 < p_a < \infty$). For example, we will cover later the minimax distribution. With this conversion it becomes the *minimax odds distribution*.

Another commonly used transformation removes both bounds ($-\infty < p' < \infty$).

$$p' = \frac{2p-1}{2p(1-p)} \qquad 0<p<1 \tag{5.89}$$

5.19 Entropy

Entropy is a term with which the control engineer may already be familiar, usually from its use in thermodynamics – where it is loosely defined as the 'randomness' of a process. As an example, consider a heater burning a hydrocarbon fuel. If we were interested in a particular hydrogen atom we can be sure that, before combustion, it is somewhere within the fuel system. After combustion, now part of water molecule, it could literally be anywhere on the planet. There has been a massive increase in the unpredictability of its location.

Entropy has a very similar meaning in statistics, where it is also a measure of unpredictability. However, it is primarily a theoretical concept with no immediately obvious application to the process industry. It is included here primarily for those readers who have come across it elsewhere and are curious about its meaning.

As an example, consider a single toss of an unbiased coin. Because it is equally likely to land heads or tails, there is no predictability. Consider now a double-headed coin. The outcome of the toss is now completely predictable.

In general, for a discrete variable, entropy (H) is defined as

$$\mathrm{H} = \sum_{i=1}^{n} p_i \log_2\left(\frac{1}{p_i}\right) \tag{5.90}$$

The number of possible outcomes (n) in tossing an unbiased coin is 2. One outcome is landing heads with a probability (p_1) of 0.5. The other is landing tails with a probability (p_2) of 0.5. Entropy is therefore

$$\mathrm{H} = 0.5\log_2\left(\frac{1}{0.5}\right) + 0.5\log_2\left(\frac{1}{0.5}\right) = 1 \tag{5.91}$$

By convention, logarithm to the base 2 is used, in which case the units of entropy will be *bits* or sometimes *Shannon*. But any base can be used; base 3 gives the result in *trits*, while base 10 gives *Hartley*. Using the natural logarithm gives *nats*.

For the double-headed coin, entropy is

$$H = \log_2(1) = 0 \tag{5.92}$$

Figure 5.29 shows the relationship between entropy and the probability of the coin landing heads. As expected, there is zero unpredictability if p is 0 or 1 and maximum unpredictability if p is 0.5.

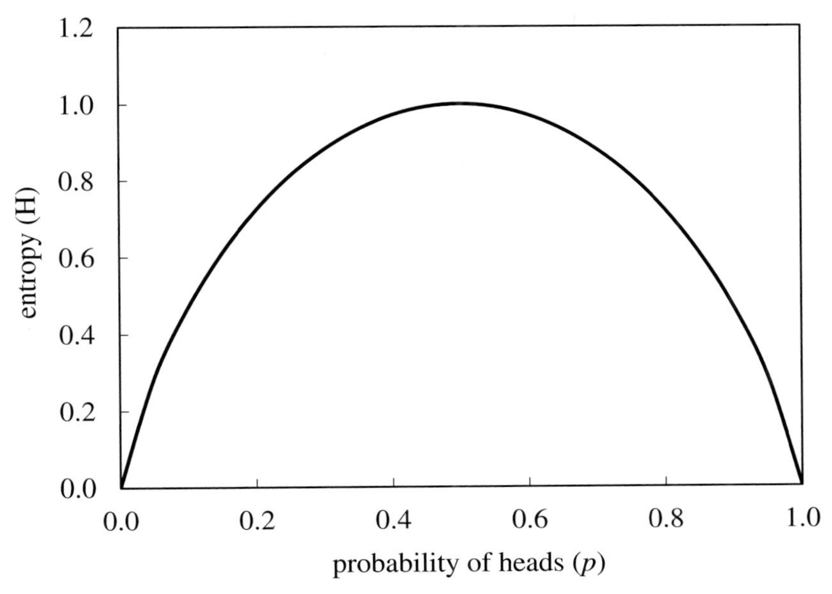

Figure 5.29 Variation of predictability of coin toss

The entropy for a continuous variable is defined as

$$H = \int_{-\infty}^{\infty} f(x)\log_2\left(\frac{1}{f(x)}\right)dx \tag{5.93}$$

For example, from the PDF of the uniform distribution, described by Equation (5.18)

$$\begin{aligned} H &= \int_{x_{min}}^{x_{max}} \frac{1}{x_{max}-x_{min}} \log_2(x_{max}-x_{min})dx \\ &= \frac{\log_2(x_{max}-x_{min})}{x_{max}-x_{min}} \int_{x_{min}}^{x_{max}} dx \\ &= \log_2(x_{max}-x_{min}) \end{aligned} \tag{5.94}$$

Since x is equally likely to be anywhere in its range, Equation (5.94) gives the maximum uncertainty for any double-bounded variable.

We can apply the same methodology to the normal distribution, giving

$$H = \frac{1+\log_2(2\pi\sigma^2)}{2} \tag{5.95}$$

This can be shown to be the maximum uncertainty for an unbounded variable of known variance (σ^2).

From the exponential distribution we get

$$H = 1 + \log_2(\mu) \tag{5.96}$$

This can be shown to be the maximum uncertainty for a lower-bounded variable of known mean (μ).

Theoretically, when fitting a distribution to data, we should choose the one that has the maximum entropy. This minimises the amount of prior information built into the distribution. This is more a philosophical argument; in practice we simply choose the distribution that fits the best.

6

Presenting the Data

Many of the ways that data are analysed for presentation do nothing to help the engineer design and support control schemes. A pie chart, for example, cannot readily be used to estimate the benefits captured by improved control or to assess the accuracy of an inferential property. But one of the key responsibilities of a process control engineer is to present data to others that have little understanding of the technology. This might be to senior management in order to obtain approval of major expenditure on an advanced control project. It might be to the maintenance department to highlight how poor performing instrumentation is restricting the benefits being captured by improved control. It might be to the laboratory drawing attention to unexplained differences between their results and an inferential property or an on-stream analyser. With a little imagination, even a simple pie chart can be used to great effect to illustrate a key issue to others.

This chapter covers techniques that might be used for such presentation, along with those that can also be used for engineering.

6.1 Box and Whisker Diagram

Figure 6.1 is *box and whisker diagram* drawn for the 100 results for C_2 in propane cargoes, shown as Table A1.2. The width of the box is the interquartile range – in this case 0.925. The division is the median – in this case 3.5. In this form of the diagram, the ends of whiskers represent the range of data – 1.9 to 5.0. There are other forms in which the whiskers are used to represent different aspects of the distribution. One is to set their length so that points outside their range become outliers; it is common then to also show the outliers as points.

The diagram can also of course be drawn vertically. It is also possible to have multiple diagrams drawn on the same axis to permit comparison between datasets. There are various formulae for determining the width of the *notch* cut into each box at the median. If a notch in one diagram overlaps that in another, some conclusion might be drawn concerning similarity of the distributions.

While perhaps a useful visual aid, the box and whisker diagram offers little scope for further analysis of the data.

Statistics for Process Control Engineers: A Practical Approach, First Edition. Myke King.

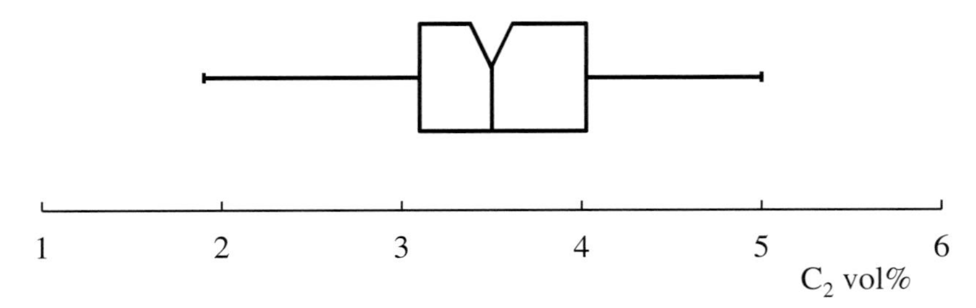

Figure 6.1 Box and whisker diagram

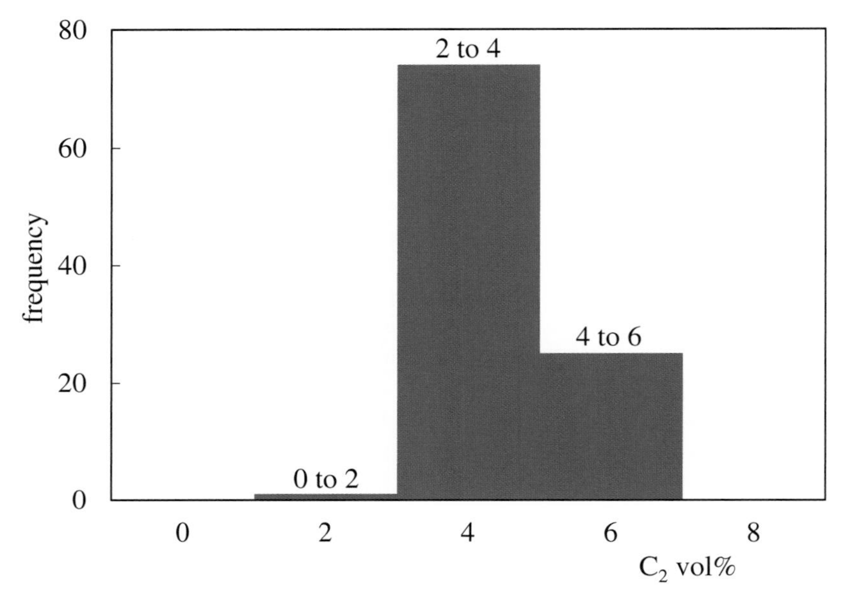

Figure 6.2 Bins too wide

6.2 Histogram

Of the many ways of presenting data in chart form, the most commonly used by control engineers is likely to be the *histogram*. Strictly it shows the probability density at which values occur in the dataset. But it can also be drawn to show observed probability or, in its simplest form, frequency.

The histogram is a special form of a *bar graph*. A bar graph can be used to display any type of data. If used to display cardinal data, it becomes a histogram. Further the bars in a bar graph are usually spaced, while those in a histogram touch. Bar graphs can be drawn vertically or horizontally, whereas histograms are generally vertical.

To plot the results in Table A1.2 as a histogram, we first split the range of the data into *bins*. Figure 6.2 shows the result of choosing a bin width of 2. It shows one result is in the interval 0 to 2, 74 between 2 and 4 and the remaining 25 are between 4 and 6. Note that the horizontal axis labels show the upper limit of each bin. The chosen bin is clearly too large to help us understand the distribution. Figure 6.3 shows the result of choosing too small a bin, in this case 0.1.

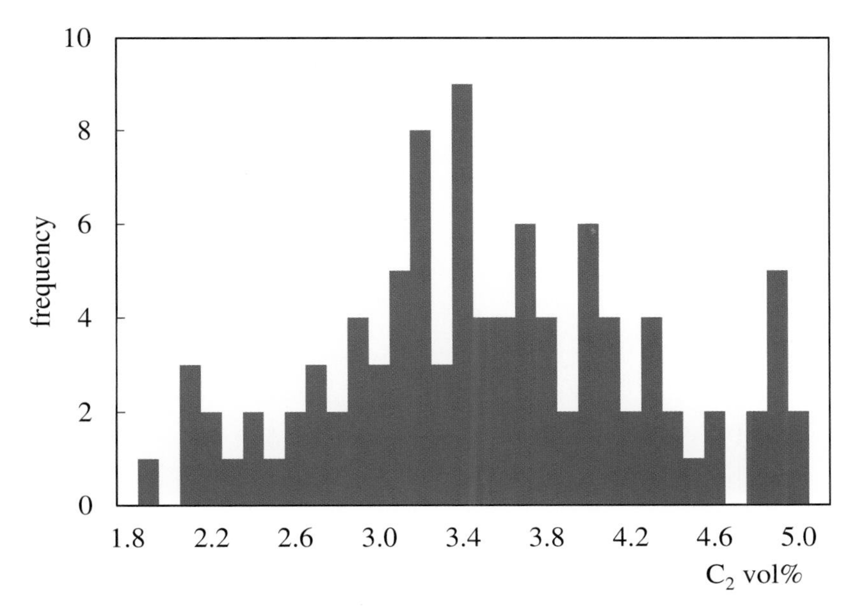

Figure 6.3 *Bins too narrow*

There are several published methods for choosing the optimum number of bins (n_{bins}) based on the number of data points (n). From this we can determine the bin width (w_{bin}) based on the range of the data, although it is likely that we would need to round w_{bin} to some sensible value.

$$w_{\text{bin}} = \frac{x_{\max} - x_{\min}}{n_{\text{bins}}} \tag{6.1}$$

The simplest technique for estimating n_{bins} is

$$n_{\text{bins}} = \sqrt{n} \tag{6.2}$$

In our example this would suggest 10 bins. We clearly have to cover the full range of the data, in this case 1.9 to 5.0; this would give an interval of 0.31. Rounding this off to a more sensible 0.3 gives Figure 6.4.

Other techniques for choosing the number of bins include *Sturge's formula*.

$$n_{\text{bins}} = \log_2(n) + 1 \approx 3.32\log_{10}(n) + 1 \tag{6.3}$$

This is applicable if the distribution is close to normal and n is at least 30. The rule states that the result of applying Equation (6.3) should always be rounded up to the nearest integer. For our example it would suggest 8 bins with a (rounded) width of 0.4. This is shown in Figure 6.5, which is considerably better at showing the distribution than Figure 6.4. However, in addition to choosing the width of the bin, we also have to choose where the first one starts. In Figure 6.5 the first bin is ranged from 1.4 to 1.8. In Figure 6.6 it has been moved to be ranged from 1.5 to 1.9. The distribution is now significantly less clear.

To be reliable any method needs to take account of the dispersion of the data. Those that suggest the number of bands do so indirectly since the width of the band is calculated from the range of the data using Equation (6.1). There are methods that suggest the width of the bin directly. *Scott's normal reference rule* uses the standard deviation (σ) of the data.

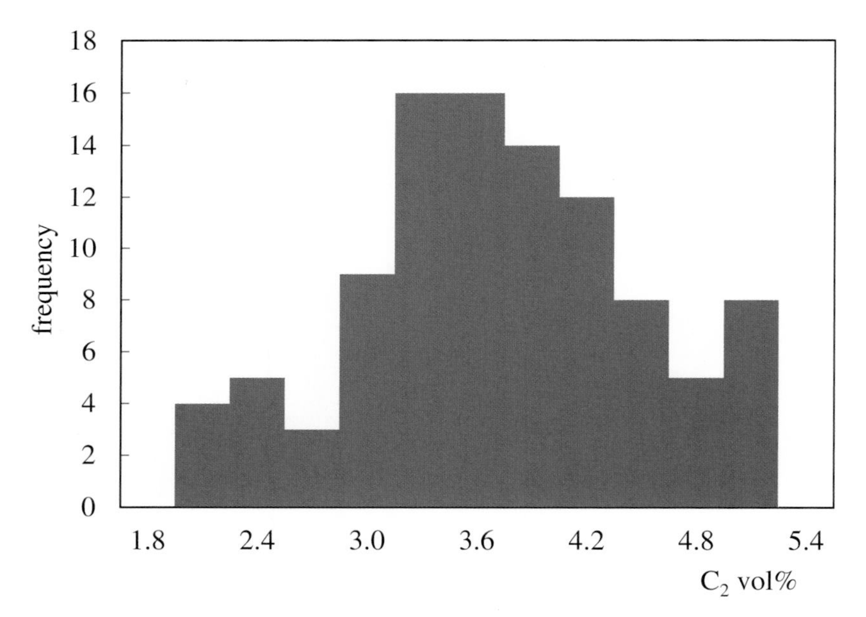

Figure 6.4 *Using $\sqrt{n}$ bins*

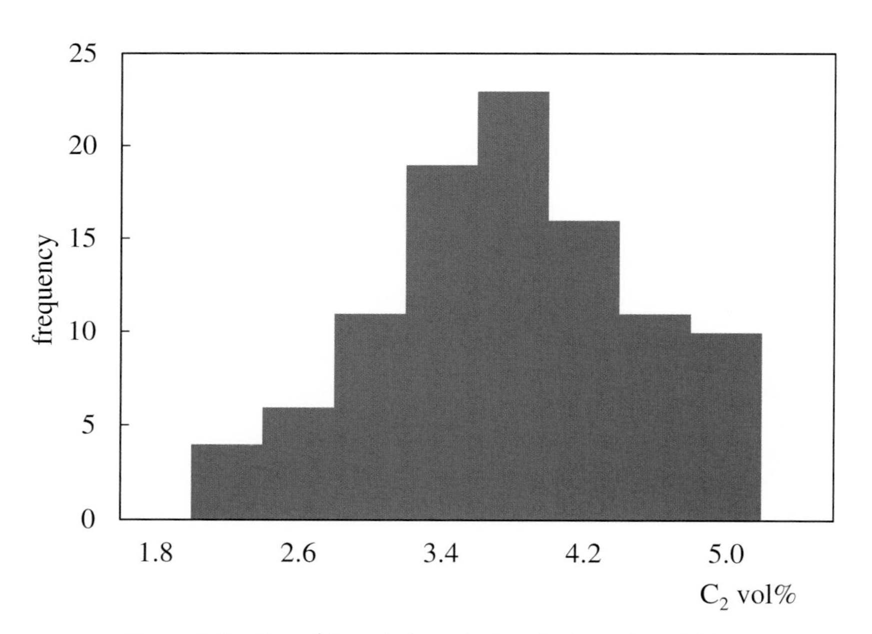

Figure 6.5 *Use of Surge's formula to select number of bins*

$$w_{\text{bin}} = \frac{3.49\sigma}{\sqrt[3]{n}} \tag{6.4}$$

In our example σ is 0.74, suggesting that the width of the bins should be 0.56. Rounding to 0.6 gives Figure 6.7, which is also probably an acceptable view of the distribution.

Similar to Scott's technique is the *Freedman–Diaconis rule*, which relies on the interquartile range (*IQR*), rather than the standard deviation, in order to reduce the sensitivity to outliers.

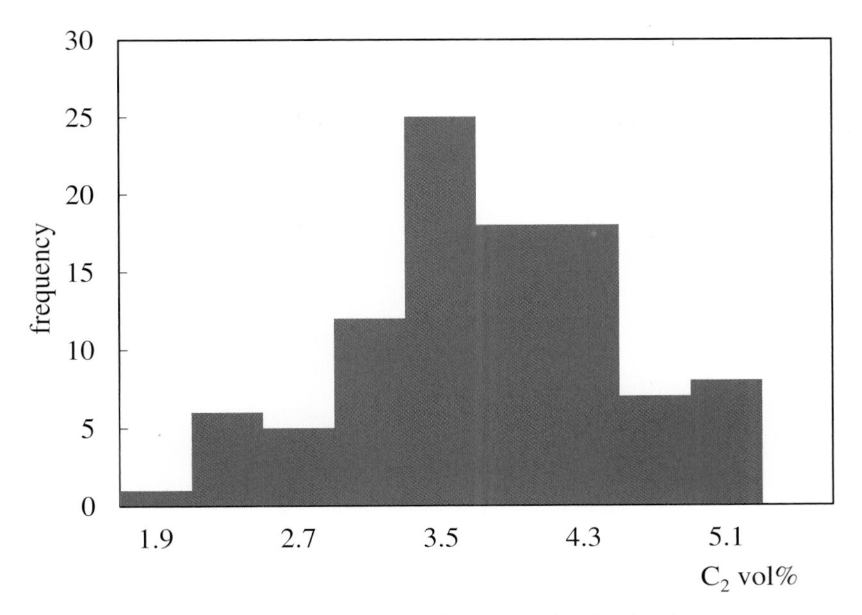

Figure 6.6 Poor choice of starting value for first bin

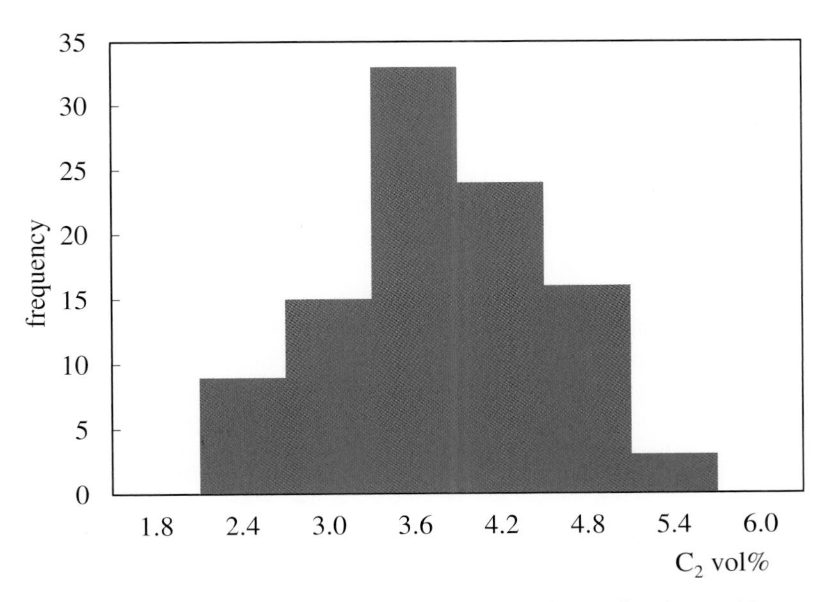

Figure 6.7 Use of Scott's normal reference rule to select bin width

$$w_{\text{bin}} = \frac{2 \times IQR}{\sqrt[3]{n}} \tag{6.5}$$

In our example IQR is 2.9, suggesting that w_{bin} should be 1.25. In this case, this is well above any value we might choose.

The calculated bin width may be smaller than the smallest difference between ranked values in the dataset. This is undesirable because some bins will be then empty and the distribution poorly represented by the histogram.

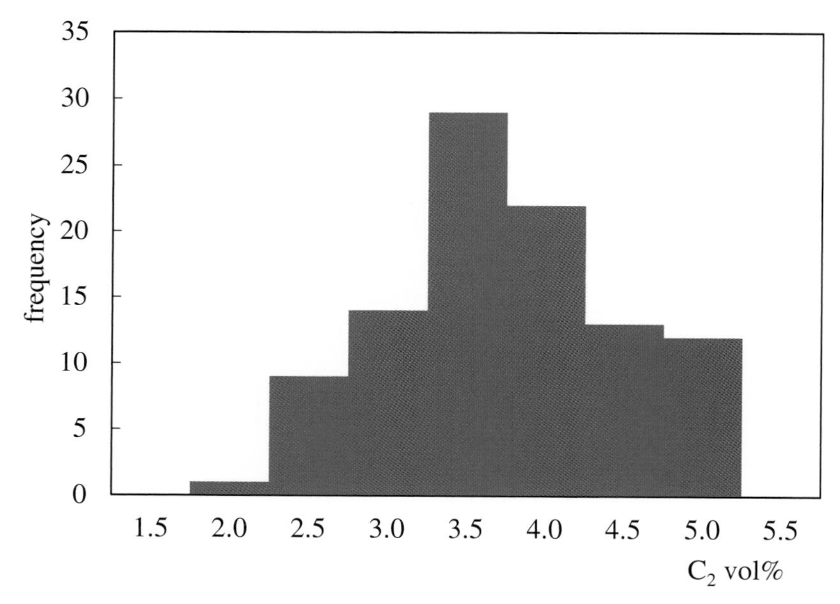

Figure 6.8 Optimised bin width

What is clear from the techniques described above is that they should only be considered as a starting point for adjustment by trial and error to determine both bin width and the start point. For example, Figure 6.8 provides probably the best view. It is based on a bin of 0.5 – not suggested by any of the methods described above.

It is common to use the same bin width for the whole histogram – in which case the frequency is represented by the height of the bar. But probability is proportional to the area of the bar. This becomes important if we vary the bin width. This is useful if the distribution has a long tail. The few data points in the tail can be scattered intermittently and the variation in bar heights will disguise the true distribution. Making the bins wider in this region helps resolve this.

Histograms can be plotted based on frequency, as we have done so far, or based on a dimensionless distribution. We do this by dividing the number of data points in each bin by the total number of points. The height of each bar then represents the observed probability and the sum of all the bar heights will then be unity. However, this does mean the area of the histogram is unity. To achieve this we have to divide the height of each bar by its width to give the probability density. This technique is known as *normalisation* and the result is shown as Figure 6.9. This, and the mid-points of each bin, are necessary if we want to fit a PDF and then superimpose the result onto the histogram. Alternatively, we can convert the probability density $f(x)$, calculated from the PDF, to probability – by multiplying by the bin width. We could then multiply this by the number of results in the sample to give the expected frequency.

We can convert the data used to plot a histogram into a *cumulative distribution* chart. The solid black line in Figure 6.10 shows the actual probability plot based on individual results rather than intervals. It happens, because the laboratory reports to the nearest 0.1 vol%, that most of the intervals are 0.1. Figure 6.10 is effectively the cumulative version of Figure 6.3. If the results were from an on-stream analyser reporting more decimal places, then intervals would vary. The coloured points and the line joining them are determined from the

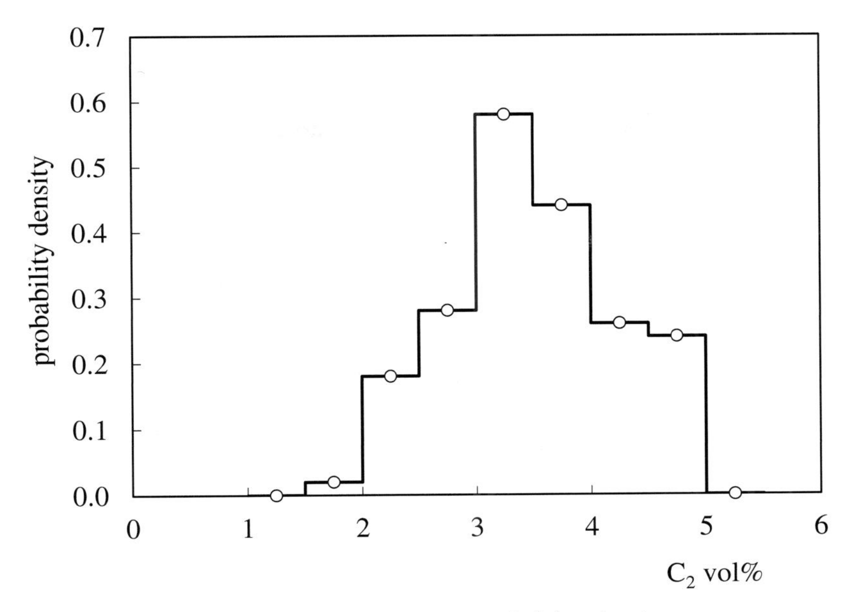

Figure 6.9 Conversion to probability density

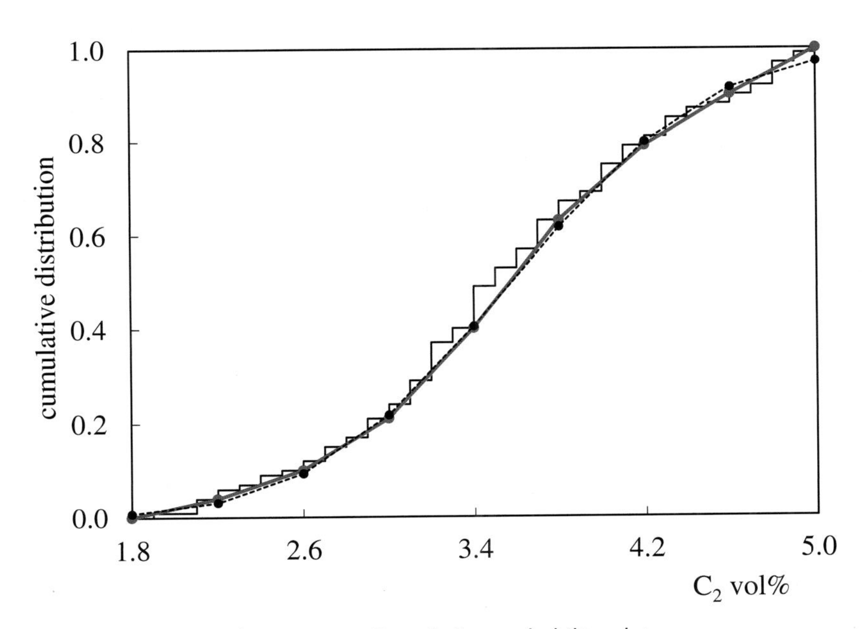

Figure 6.10 Cumulative probability plot

data used to prepare Figure 6.5. Using the method that we will cover in Chapter 9, the dashed line is the best fit normal distribution with a mean of 3.58 and a standard deviation of 0.74. The close match between this and the coloured line confirms that the chosen interval of 0.4 works well.

6.3 Kernel Density Estimation

Kernel density estimation is a technique for building a *non-parametric* PDF. For example, the PDF of the normal distribution given by Equation (5.35), is *parametric* in that it includes the location and shape parameters μ and σ. To fit a parametric distribution we first have to choose the prior distribution and then adjust the parameters to give the best fit. The advantage of a non-parametric PDF is that it makes no assumption about the form of the distribution.

A histogram is a form of non-parametric PDF. Kernel density estimation is effectively a histogram in which the right-angled corners have been smoothed out. Its advantage is that there is no need to identify, by trial and error, the width of the bins or their starting point.

The *kernel* of a parametric PDF is only those parts that are a function of the variable (x). It omits therefore any normalisation constant, even if it includes any of the distribution shape parameters. For example, the kernel of the normal distribution is derived from Equation (5.35) by omitting any part of $f(x)$ which is not a function of x. This gives

$$K(x) = \exp\left[\frac{-(x-\mu)^2}{2\sigma^2}\right] \tag{6.6}$$

In general the kernel, $K(x)$, of a non-parametric PDF is any mathematical function that satisfies two conditions. The first is that integrates to unity over the range to which is applies. If its range is unrestricted then

$$\int_{-\infty}^{\infty} K(x)dx = 1 \tag{6.7}$$

The second condition is that it must be symmetrical about zero, i.e.

$$K(x) = K(-x) \tag{6.8}$$

A convenient choice for $K(x)$ is the PDF of the normal distribution, known as the *Gaussian kernel*, derived from Equation (5.35).

$$K(x) = \frac{1}{h\sqrt{2\pi}}\exp\left[-\left(\frac{x-x_i}{h}\right)^2\right] \tag{6.9}$$

The inclusion of h, which is known as either the *smoothing factor* or the *bandwidth*, makes $K(x)$ a *scaled kernel*. In this example h is set at 0.5. Equation (6.9) is applied to every data point (x_i) in the sample. For example, from Table A1.2, x_1 is 3.6 and so the kernel is plotted with x_i set to 3.6. This is shown as the solid black line in Figure 6.11. We include kernels for all x_i; for example, the next three are at 3.4, 4.8 and 2.7. Their kernels are plotted as dashed lines. The coloured line is the kernel density estimate calculated as the average of the four kernels. When completed for all 100 results, it will approximate to the overall PDF of the data.

The two conditions placed on $K(x)$ ensure that firstly, since the area under each kernel is unity, the area under the overall PDF will be unity – as required of any PDF. Secondly the symmetry of each kernel about zero ensures that the mean of the overall PDF will be the same as the mean of the data.

Figure 6.12 illustrates the effect of the smoothing factor (h). Too small and the resulting distribution will appear noisy – more closely resembling a histogram. Too large a choice will flatten the distribution. There are several techniques published for choosing the optimum value. For example, the *normal distribution approximation* (sometimes known as *Silverman's rule of thumb*) suggests

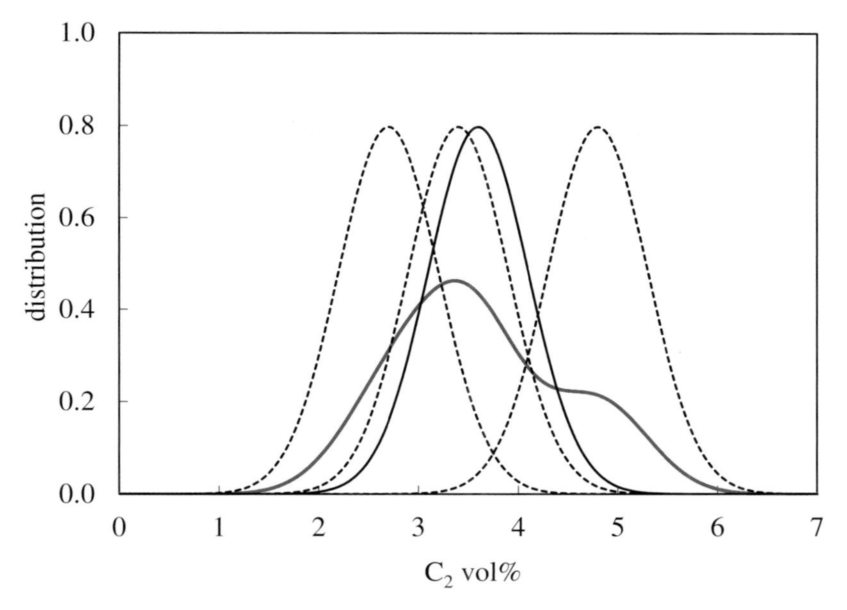

Figure 6.11 *Construction of kernel density plot*

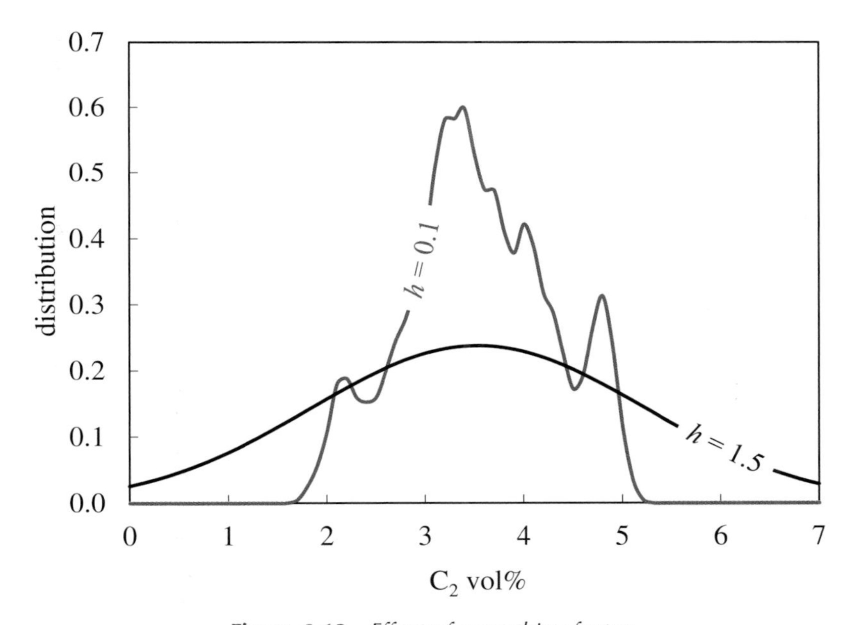

Figure 6.12 *Effect of smoothing factor*

$$h = \left(\frac{4}{3n}\right)^{\frac{1}{5}} \sigma \tag{6.10}$$

The standard deviation (σ) of the C_2 results is 0.74, leading to a value for h of 0.31. Figure 6.13 plots, as the coloured line, the resulting distribution. The dashed line shows the classic normal distribution.

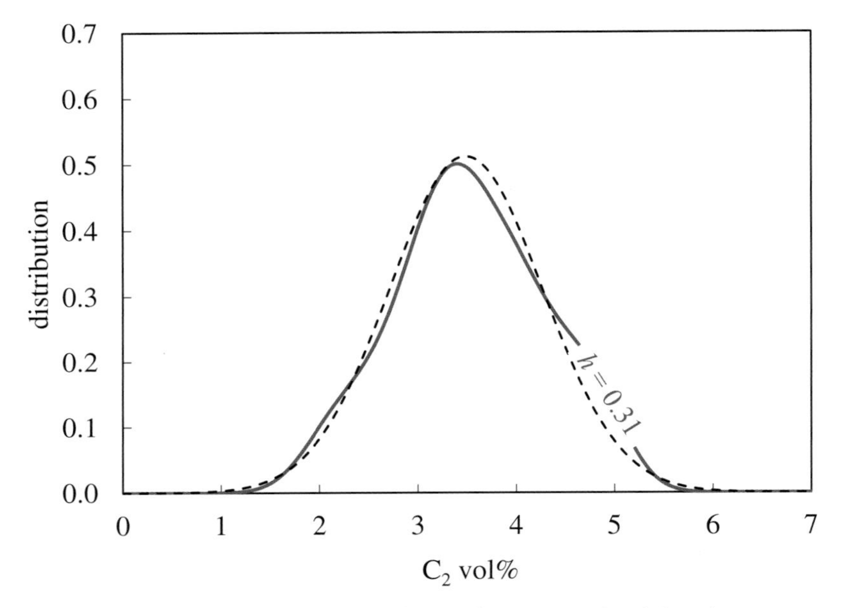

Figure 6.13 Result of applying Silverman's rule of thumb

In addition to the Gaussian kernel, there are many others that satisfy the two conditions. Theoretically the most accurate is the *Epanechnikov kernel*.

$$K(x) = \frac{3}{4h}\left[1 - \left(\frac{x - x_i}{h}\right)^2\right] \quad |x - x_i| < h \tag{6.11}$$

$$K(x) = 0 \quad |x - x_i| \geq h \tag{6.12}$$

However, the choice of kernel is very much secondary to the choice of smoothing factor. As the coloured line in Figure 6.14 shows, adopting the Epanechnikov kernel with h chosen as 0.64 results in a distribution almost identical to that developed from the Gaussian kernel (shown as the dashed line).

While a non-parametric distribution is useful pictorially, the absence of simple parameters makes impractical any mathematical analysis. Its main application in the field of process control is to assess whether the distribution of the data closely matches the prior distribution. For example, Figure 6.13 would indicate that the data closely follows the normal distribution. Figure 6.15 is a *P–P plot*. The black line plots the probability density of the normal distribution against that of the Epanechnikov kernel distribution – as they increase from zero to their peak at 0.5 and then back to zero. Closely following the coloured line shows that that the two distributions are virtually identical. In other words, our choice of prior distribution is excellent.

Table A1.1 shows a year of daily laboratory results for C_4 content of the same propane product – except that these are from the product rundown leaving the LPG splitter, not the cargoes. Figure 6.16 shows the distribution. The results for the propane leaving the site as cargoes are likely to be normally distributed, no matter what the distribution of the rundown results, as suggested by the Central Limit Theorem. However the distribution of rundown results is likely to be non-symmetrical. Process disturbances cannot reduce the C_4 content below zero but the occasional large disturbance can take it well over the average content. In our example the mean is around 4.5 but the largest value is around 16.

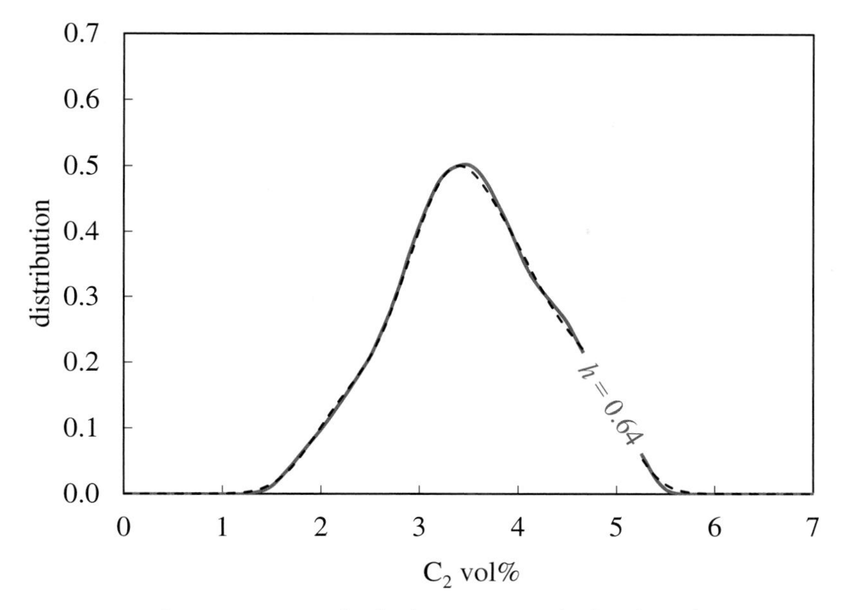

Figure 6.14 *Result of selecting Epanechnikov kernel*

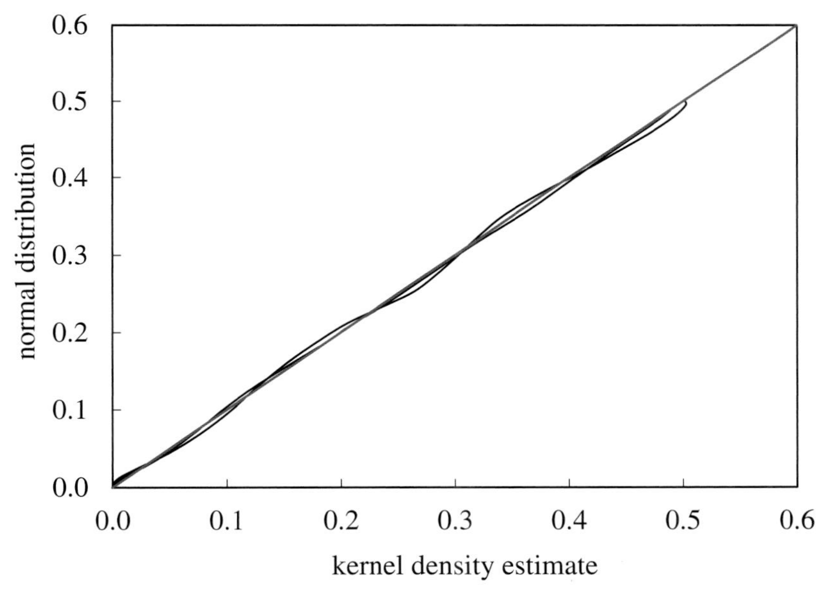

Figure 6.15 *P–P plot demonstrating accuracy of fit*

Plotting the mid-range of the bins in Figure 6.16, Figure 6.17 shows the comparison between the kernel density estimate (coloured line) and a fitted normal distribution (black line). It shows clearly that the real distribution is highly skewed. The *frequency–frequency plot* in Figure 6.18 is effectively a P–P plot but based on days/year rather than probability density. It confirms that the distribution is far from normal.

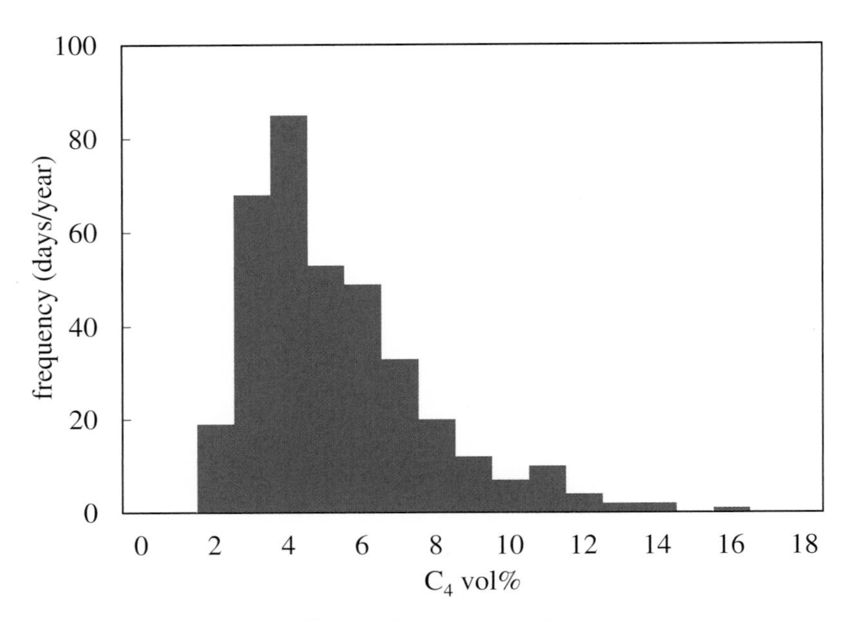

Figure 6.16 Distribution of C4 content of propane rundown

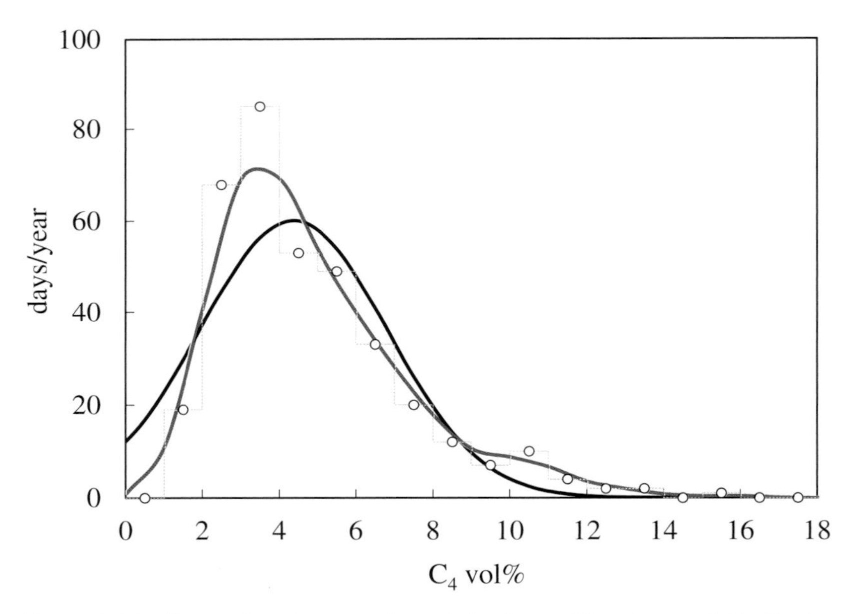

Figure 6.17 Comparison between kernel density and fitted normal distribution

From Equation (4.41) the skewness of the C_4 content is calculated as 1.27; this is significantly higher than the limit of 0.5 for treating the distribution as normal. Its standard deviation is 2.42. If treated as normal, from Equation (5.43) we would estimate that there is a probability of 0.022 that the C_4 content would be negative. We clearly do not expect eight such results per year.

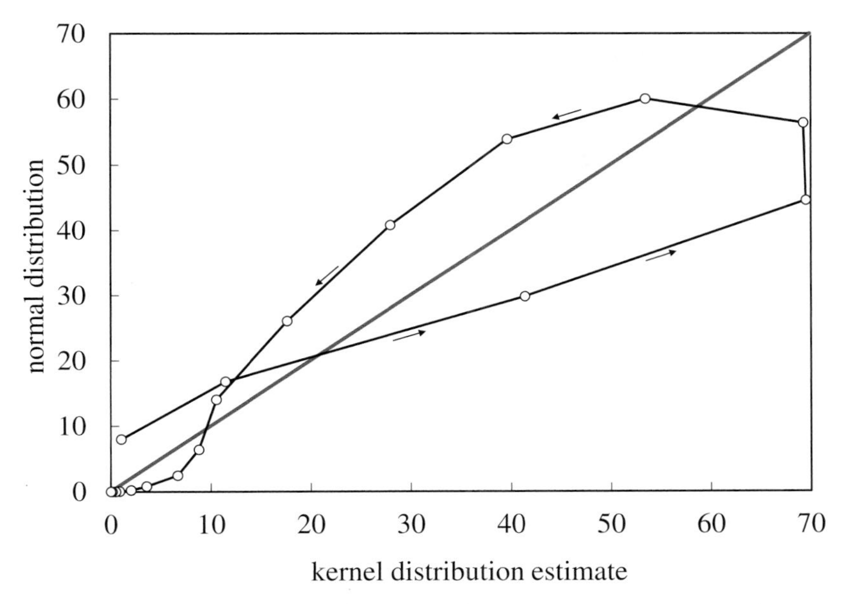

Figure 6.18 *Frequency–frequency plot demonstrating very poor fit*

6.4 Circular Plots

There are a variety of ways of devising circular plots. In a histogram the columns are arranged vertically. They can also be arranged radially – effectively like the spokes of a wheel. This approach might be used if there is no reason for columns to be ranked. For example, while the months of the year follow a sequence, they are not ranked. There may be no reason for starting with any particular month. If we wanted to display the distribution of days by month we would arrange 12 spokes, at 30° intervals, with their lengths proportional to the number of days in the month. It would make no difference if we wanted to show a calendar year, from January to December, or the UK tax year, from April to March. Joining the ends of the spokes with straight lines results in a *radar plot*. For example, taking the data from Table A1.1, Figure 6.19 shows the number of days, by month, that the C_4 content of propane was off-grade. Because of its resemblance to a spider's web, the radar plot is occasionally described as a *spider chart*. However this term is more commonly used for a method of organising ideas that results in a chart that resembles a spider's legs. An extended form of this chart is used for *mind mapping*.

A radar plot can also be used to present the behaviour of a multivariable controller, as illustrated by Figure 6.20. Each spoke would be a selected MV or CV. On each spoke, three points would be located corresponding to the LO limit, the HI limit and the current value. Joining the LO points and joining the HI points would display the area in which the process must operate. Joining the current values will generate an irregular polygon. Process operators quickly learn to recognise changes in the shape of the polygon and can use this to explore why the controller has made unexpected moves.

We will see later that there are distribution functions that are specifically designed for circular plots. They produce something similar to a radar plot, except that the points are joined by a continuous function, rather than by straight lines. It can be thought of a plot based on polar,

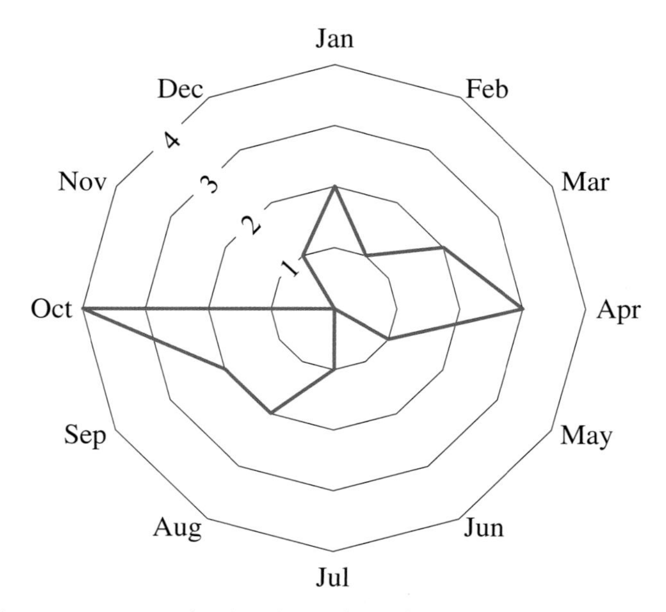

Figure 6.19 Use of radar plot to show days of off-grade production

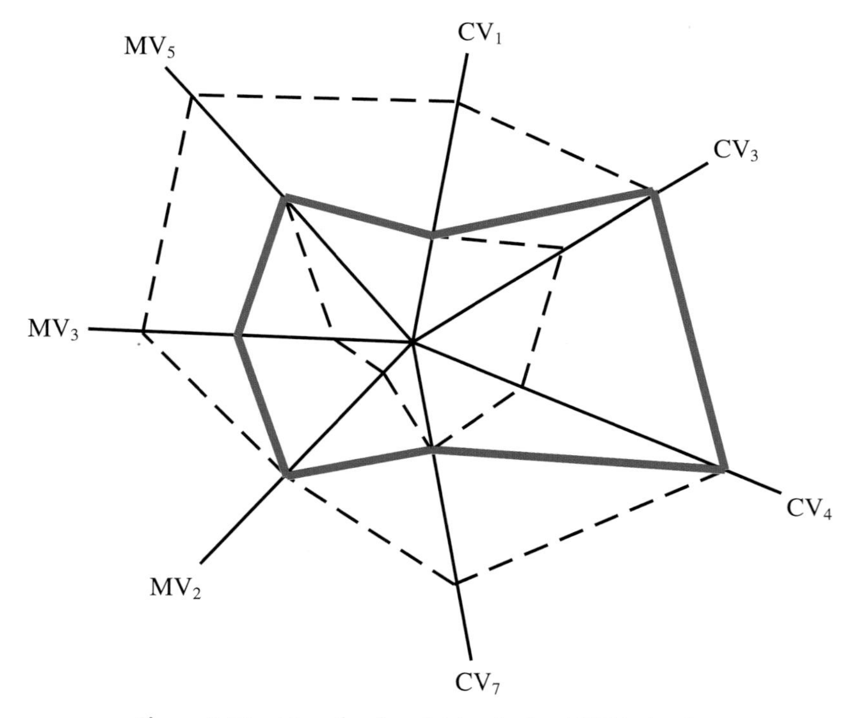

Figure 6.20 Use of radar plot to display MPC behaviour

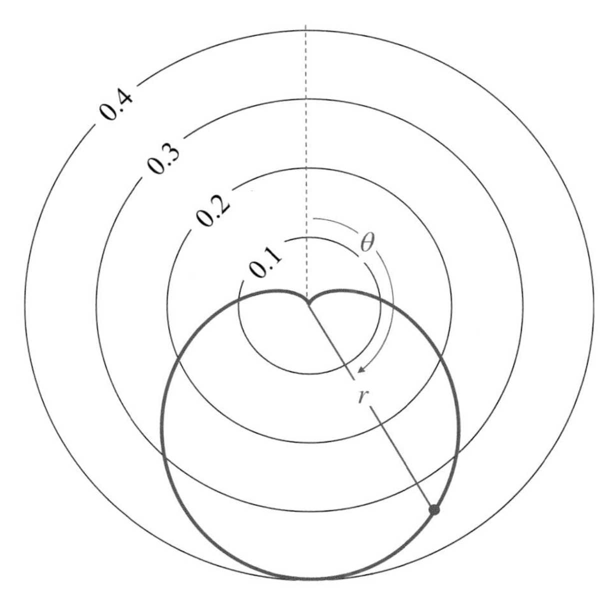

Figure 6.21 Circular plot based on polar coordinates

rather than Cartesian, coordinates except that the angle is measured in a clockwise direction. For example, Figure 6.21 shows the plot of the standard normal distribution.

$$r = \frac{1}{\sqrt{2\pi}} \exp\left(\frac{-\theta^2}{2}\right) \quad -\pi \leq \theta \leq \pi \tag{6.13}$$

The point marked illustrates how the curve is produced. Such an approach might be applicable if we wished to show, for example, the distribution of wind speeds by direction.

6.5 Parallel Coordinates

Taking a radar plot and returning the spokes to align vertically results in a *parallel coordinates* plot. These too can be very useful in assessing the behaviour of MPC or, indeed, any aspect of process operation. The plot is a two-dimensional graphical method for representing multidimensional space. In the example shown in Figure 6.22, a point in seven-dimensional space is represented by the coordinates (x_1, x_2, x_3, x_4, x_5, x_6, x_7). Since we cannot visualise space of more than three dimensions, the value of each coordinate is plotted on vertical parallel axes. The points are then joined by straight lines.

The technique is well suited to predicting the behaviour of a multivariable controller, even before step-testing has been started. Plant history databases comprise a number of instrument tag names with measurements collected at regular intervals. If we imagine the data arranged in a matrix so that each column corresponds to either a MV or a CV in the proposed controller and each row is a time-stamped snapshot of the value of each parameter. To this we add a column in which we place the value of the proposed MPC objective function (C) derived from the values in the same row (where P are the objective coefficients for the m CVs and Q the objective coefficients for the n MVs), i.e.

$$C = \sum_{i=1}^{m} P_i CV_i + \sum_{j=1}^{n} Q_j MV_j \tag{6.14}$$

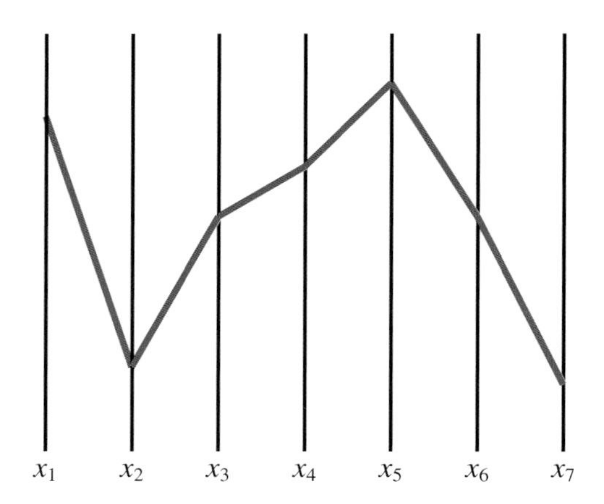

Figure 6.22 Parallel coordinates

Each row in the database is then plotted as a line on the parallel coordinates chart. The result will initially appear very confused with a large number of lines superimposed. The next step is to add the HI/LO constraints on each vertical axis. If a line violates any constraint on any axis then the whole line is deleted. The lines remaining will each represent an occasion in the past when all the process conditions satisfied all the constraints. The final step is to choose the line for which the value on the cost axis is the lowest. Since this axis is the MPC cost function, the line with the lowest value will represent the operation that MPC would select. Provided that the process has at some stage operated close to the optimum (as defined by MPC) then the chosen dataset will give some idea of the operating strategy that MPC will implement. If different from the established operating strategy, this approach gives an early opportunity to explore why. Any difference should be seen as an opportunity to adopt a more profitable way of operating the process rather than an error that should be corrected by adjusting the individual objective coefficients.

6.6 Pie Chart

Instead of making the length of the spokes in a radar plot proportional to the values, we make the angles between the spokes proportional to them. This results in a *pie chart*, usually applied if data is sorted into non-sequential categories and we want to display the proportion of the data that falls into each category. Figure 6.23 illustrates how the C_4 in propane data might be displayed. The product is off-grade if the C_4 content is greater than 10%. It is considered 'on spec' if it is between 8 and 10%. Results in giveaway are split into two groups, depending on the severity. While not lending itself to any statistical analysis, it does make clear just how poorly the composition control is performing.

6.7 Quantile Plot

A *quantile plot* (or *Q–Q plot*) is similar to a P–P plot and also used to compare the distributions of two sets of values. It requires each set of values to be ranked in increasing order. While changing the sequence of the values has no effect on mean or standard deviation, there is no

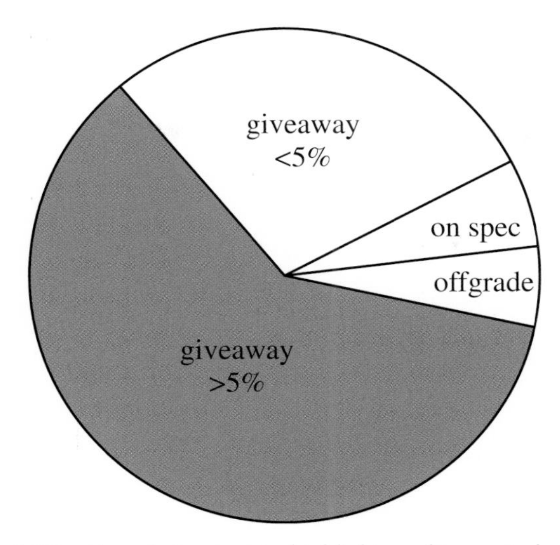

Figure 6.23 *Use of pie chart to highlight quality control problem*

guarantee that values in one set are sorted into the same time order as the other. For this reason the technique cannot be used to determine whether the two variables are correlated. For example, it cannot be used to validate an inferential property against its corresponding laboratory result.

The use that the control engineer might make of it is to confirm that data collected for statistical analysis are distributed according to an assumed distribution. Commonly this would be applied to data assumed to be normally distributed, but any assumed distribution can be tested. The QF is used to determine the values expected, if the data follow the assumed distribution. These are then plotted against the corresponding observed values.

The data points are first ranked, where k is the ranking of each point. The kth smallest value is known as the kth *order statistic*. The quantile (q) for each point is then most simply determined from

$$q = \frac{k}{n} \tag{6.15}$$

We will use this definition throughout this book when fitting distributions to data. It will appear when we cover, in Chapter 9, fitting a CDF to data and in many of the worked examples following this. There are however many other formulae, mainly of the form

$$q = \frac{k-a}{n+1-2a} \tag{6.16}$$

The term a is chosen (usually in the range 0.0 to 0.5) although, if n is large, the choice has negligible impact on the conclusion. Depending on its choice, each quantile is therefore calculated from

$$a = 0.0 \quad q = \frac{k}{n+1} \tag{6.17}$$

$$a = 0.5 \quad q = \frac{k-0.5}{n} \tag{6.18}$$

The expected value (x) corresponding to each quantile is calculated using the mean (μ) and standard deviation (σ) of the observed data. If we are checking whether the distribution is

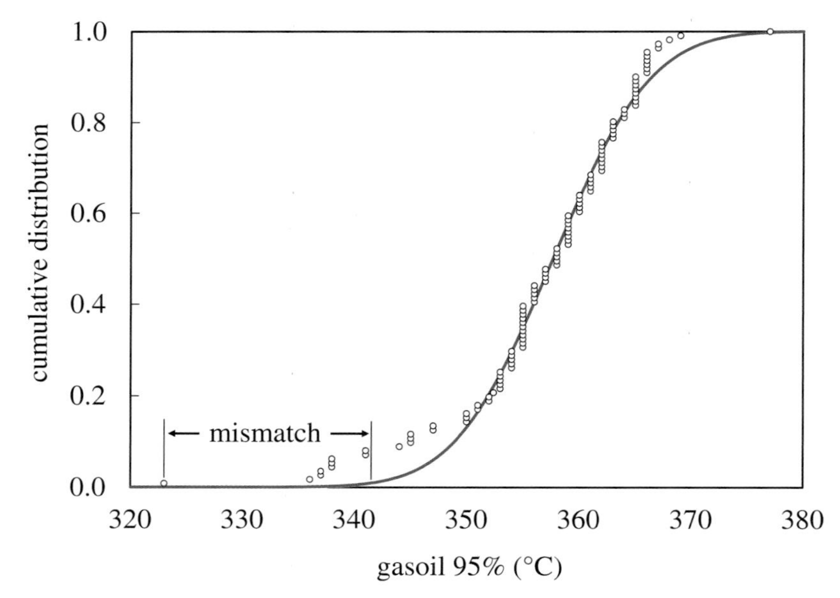

Figure 6.24 Cumulative distribution of gasoil 95% with normal distribution fitted

normal, we use the CDF derived by integrating the PDF as described by Equation (5.35) and solve for x.

$$\frac{1}{\sigma\sqrt{2\pi}}\int_{-\infty}^{x} e^{\frac{-(x-\mu)^2}{2\sigma^2}}dx = q \tag{6.19}$$

This is the QF and, in this case, would have to be solved by iteration. Fortunately most spreadsheet packages include the inverse normal function. Equation (6.19) is the same as (5.56).

In fact, we can choose any value for the mean and standard deviation. For example, we might choose μ as 0 and σ as 1 – effectively applying Equation (5.59). Under these circumstances the expected values can be described as *rankits*. If the data are normally distributed then the plot of actual values against those expected will be a straight line. For the line to pass through the origin and have a slope of 1, we must use the mean and standard deviation derived from the data. To highlight any deviation from normal distribution we can add a reference line – commonly drawn through the first and third quartiles.

Figure 6.24 shows the normal distribution fitted to the gasoil 95% data from Table A1.3. If we consider the value at 323°C, it would appear to lie very close to the fitted distribution. However, it is the horizontal separation that is important. There is a mismatch of about 19°C between the true value and what the distribution curve predicts

Figure 6.25 is the P–P plot but, unlike previous examples, based on cumulative probability rather than probability density. While it may appear that the fit is reasonable, it is a plot of the vertical separation between actual and predicted values shown in Figure 6.24.

The Q–Q plot, shown as Figure 6.26, plots the horizontal separation. As can be seen from the mismatch at 323°C, it now emphasises the poor fit to the tails. Since the intent of improved control would be to reduce these tails, it is important that they be properly represented by the chosen distribution. Examination of skewness, calculated from Equation (4.41) as

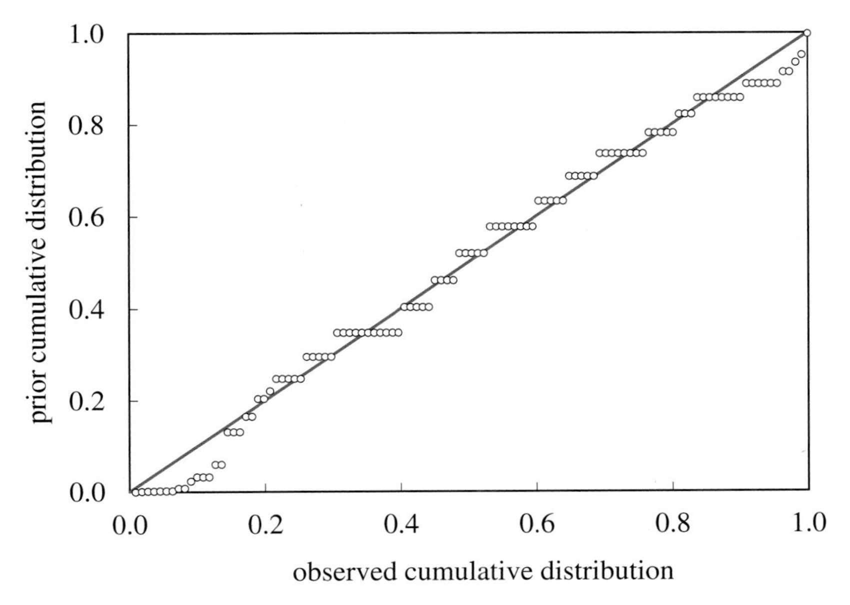

Figure 6.25 P–P plot for gasoil 95%

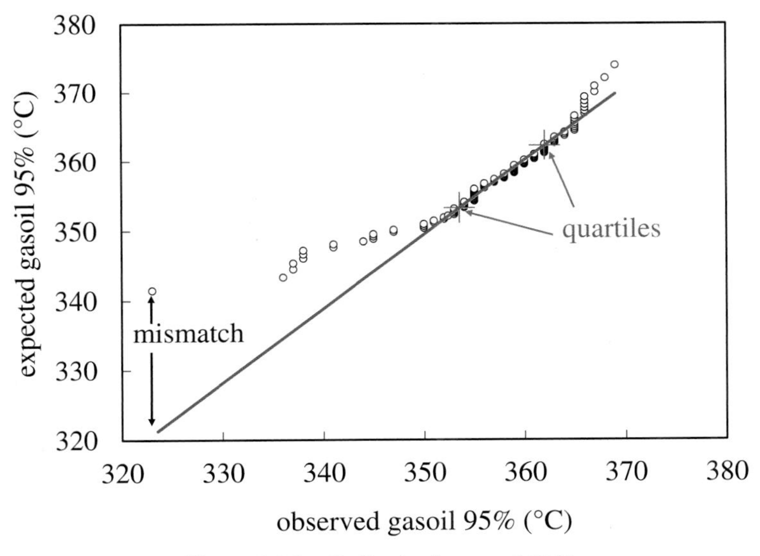

Figure 6.26 Q–Q plot for gasoil 95%

−1.13, and kurtosis, calculated from Equation (4.42) as 5.09, confirms that these are well outside the criteria for the distribution to be normal.

Rather than comparing a set of values to an expected distribution, we may wish to compare two sets of values to determine whether they have the same distribution. For example, we might want to compare two contending inferentials. The datasets are the prediction errors for each. Both datasets need to contain the same number of values. The C_3 in butane inferential has 1,152

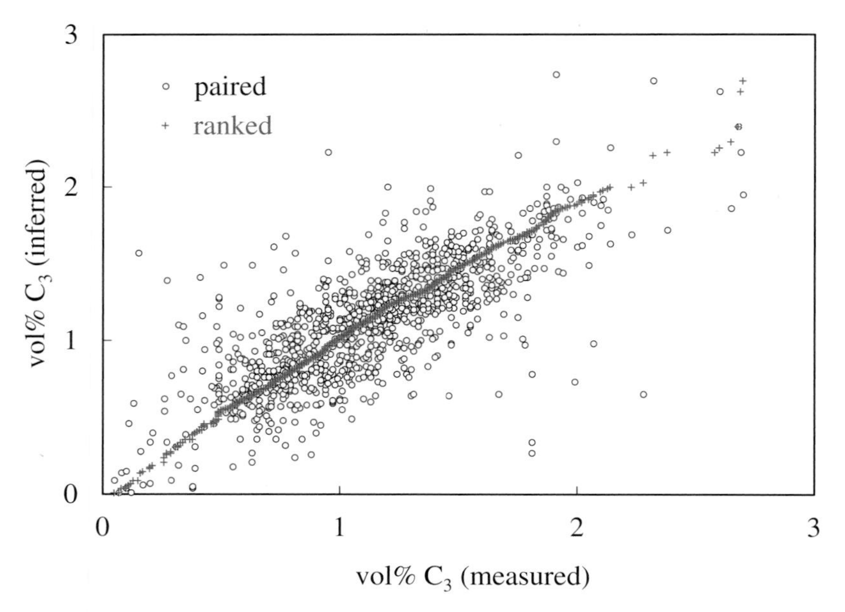

Figure 6.27 *Ranked versus paired points comparing inferential to on-stream analyser*

inferred results with the same number of corresponding analyser measurements. To compare these we would conventionally simply plot each inferred result against its corresponding analyser measurement. These are shown as the 'paired' points in Figure 6.27. For a Q–Q plot we rank each dataset and then plot the lowest value in one set against the lowest in the other, the second lowest against the second, etc. The result of doing so is shown as the 'ranked' points. Perhaps surprisingly the correlation appears to be better. However, we have lost the timestamp correspondence. The resulting, almost straight, line tells us only that the two datasets have very similar distributions. It tells us nothing about the accuracy of the inferential. However, if the resulting line were more curved, we might conclude that there was a problem with the inferential. But we cannot quantify the problem in terms of, for example, prediction error.

If the two datasets do not contain the same number of values then the quantiles in one set, which are missing from the other, need to be added by interpolation – and vice versa. To do so, we first identify the ranking (k and $k+1$) of the two values to be used for interpolation. If we have used Equation (6.17) to determine q, then

$$k = \text{int}\left(q_{missing}(n+1)\right) \tag{6.20}$$

We then determine the weighting factor (w) that should be applied.

$$w = q_{missing}(n+1) - k \tag{6.21}$$

The interpolated value (x) is then

$$x = (1-w)x_k + wx_{k+1} \tag{6.22}$$

To illustrate simply how this approach works we can take two very small datasets that have been ranked. The first (series 1) comprises the four values 2, 4, 7 and 8; the second (series 2) comprises the three values 3, 7 and 9. If we apply Equation (6.17), from the first set we have

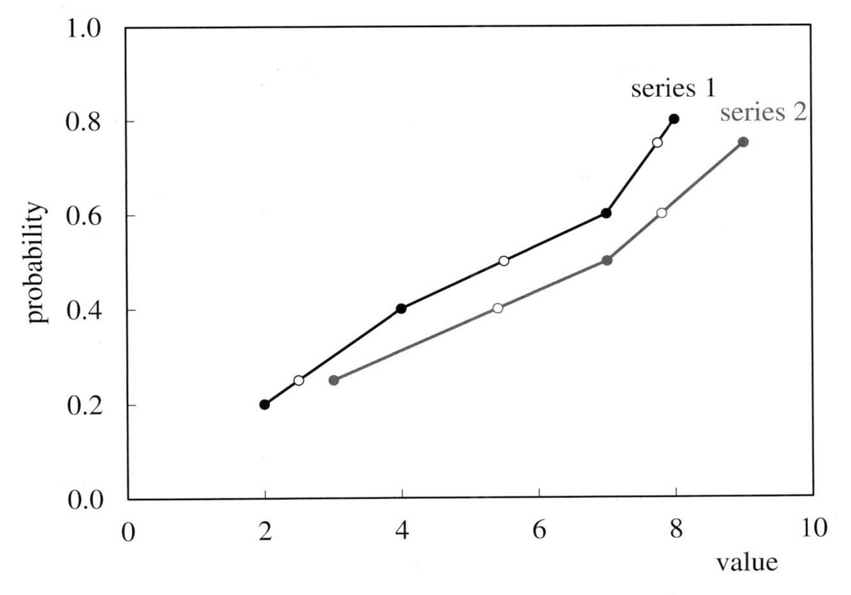

Figure 6.28 Procedure for adding missing quantiles

the quantiles 0.2, 0.4, 0.6 and 0.8. From the second we have 0.25, 0.5 and 0.75. We follow the procedure:

1. We need to add, to the first set, the missing quantile 0.25. Applying Equation (6.20), with n as 4, gives k as 1. Equation (6.21) then gives w as 0.25. From Equation (6.22), x is therefore 2.5 $(0.75 \times 2 + 0.25 \times 4)$.
2. We similarly need to add the quantile 0.5. From Equation (6.20), k is now 2. From Equation (6.19), w is 0.5 and so, from Equation (6.22), x is 5.5 $(0.5 \times 4 + 0.5 \times 7)$.
3. Following the same procedure to add the quantile 0.75 gives x as 7.75. The first dataset now comprises the values 2, 2.5, 4, 5.5, 7, 7.75 and 8.
4. We also have to add the missing quantiles to the second dataset. The first of these is 0.2. From Equation (6.21), with n now 3, we find that k is 0. There is no x_0 and so we have to ignore the 0.2 quantile.
5. We can add the 0.4 quantile. From Equation (6.20) we obtain k as 1. From Equation (6.21) we find that w is 0.6 and, from Equation (6.22), x is 5.4 $(0.4 \times 3 + 0.6 \times 7)$.
6. Adding the 0.6 quantile gives x as 7.8.
7. We cannot add the 0.8 quantile because there is no x_4. We therefore ignore this. The first set therefore now includes 2.5, 4, 5.5, 7 and 7.75; the second is 3, 5.4, 7, 7.8 and 9.

Figure 6.28 shows, perhaps more clearly, what the procedure above has achieved. The original sets are shown as the solid points. The missing quantiles are added by linearly interpolating between the two adjacent points. Plotting gives the Q–Q plot shown as Figure 6.29. Although there are far too few points to reach a firm conclusion, the distance of the points from the straight line would indicate that the two datasets have dissimilar distributions.

As we saw with the gasoil 95% example, the Q–Q plot was best at drawing attention to the differences between two distributions. If the number of values is not the same in each dataset,

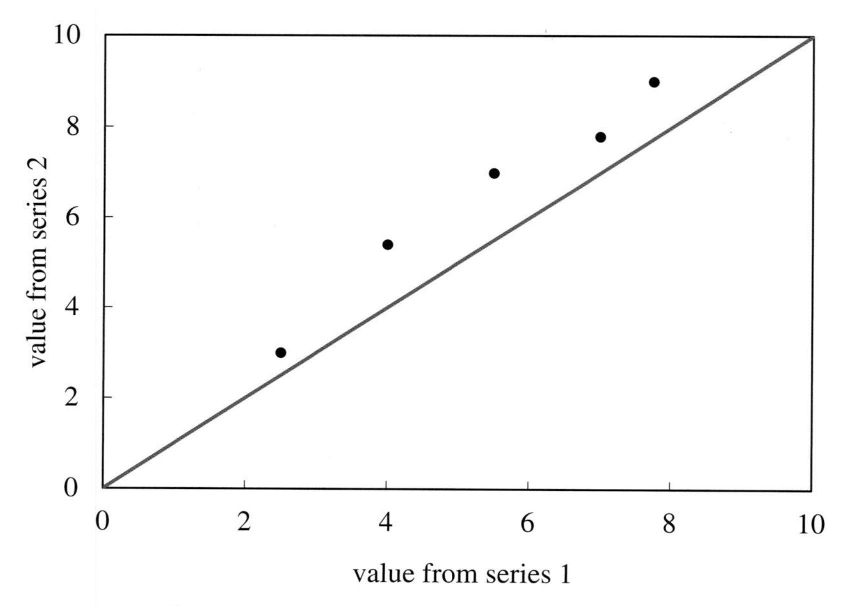

Figure 6.29 Q–Q plot including missing quantiles

the Q–Q plot is more complex to prepare than the P–P plot, and perhaps not as intuitively understood. Nevertheless, the effort is likely to be justified – particularly so if the calculation above is computerised.

7

Sample Size

7.1 Mean

To have confidence in the conclusion of any statistical analysis, we must ensure that the number of samples (n) is sufficiently large. As an example, consider the batch blending quality data in Table A1.7. By calculation, the 78 results have a mean of 97.9 and a standard deviation of 9.1. To have confidence in our estimate of benefits from improved control, we need the confidence interval of the estimate of the mean. We showed in Section 4.7 that estimates of the sample mean, calculated using different selection of n results from the population, will have a variance of σ_p^2/n. Choosing a confidence interval from Table A1.8, for example 95%, gives a value of 1.96 for z. The margin of error (ε) is therefore

$$\varepsilon = \frac{z\sigma_p}{\sqrt{n}} = \frac{1.96 \times 9.1}{\sqrt{78}} = 2.0 \tag{7.1}$$

In other words we are 95% confident that the population mean lies between 95.9 and 99.9. If we wanted to improve on this we can determine the required sample size by rearranging Equation (7.1).

$$n = \left(\frac{z\sigma_p}{\varepsilon}\right)^2 \tag{7.2}$$

So, for example, if we wanted to improve the accuracy of the estimate of mean to ±1.0, we would need 318 results.

It becomes more complex if we want to check the confidence that the mean has changed, for example, as a result of implementing control improvements. The confidence we have in the estimate of mean before the improvement is

$$\mu_1 = \bar{x}_1 \pm \frac{z(\sigma_p)_1}{\sqrt{n}} \tag{7.3}$$

Statistics for Process Control Engineers: A Practical Approach, First Edition. Myke King.

Assuming we use the same number of results in the calculation, after implementation, it is

$$\mu_2 = \bar{x}_2 \pm \frac{z(\sigma_p)_2}{\sqrt{n}} \tag{7.4}$$

We showed in Section 4.6 that the variance of the difference between two values is the sum of the variances. Assuming the intent of the control improvement is to increase the mean, then the improvement is

$$\mu_2 - \mu_1 = \bar{x}_2 - \bar{x}_1 \pm z\left(\frac{(\sigma_p)_1 + (\sigma_p)_2}{\sqrt{n}}\right) \tag{7.5}$$

So, if we want to be 95% sure that the mean has increased, then

$$\bar{x}_2 - \bar{x}_1 > 1.96\left(\frac{(\sigma_p)_1 + (\sigma_p)_2}{\sqrt{n}}\right) \tag{7.6}$$

The weakness of this method of course is that it assumes we know the standard deviations of the population – both before and after the improvement. In practice these will also be estimated from the same n values. This reduces the confidence we have in σ_p and so widens the confidence interval of the increase in mean. In reality the control engineer will use judgement in collecting sufficient results to represent typical before and after operations. Any doubt that this may not have been achieved can be alleviated by repeating the exercise with a different set of results and checking the estimate of improvement changes little. This is a simple example of *bootstrapping* that we cover later in this section.

7.2 Standard Deviation

To determine, for example, the 95% confidence interval we first calculate p and $1 - p$.

$$p = \frac{1}{2}\left(1 - \frac{95}{100}\right) = 0.025 \quad \therefore 1 - p = 0.975 \tag{7.7}$$

The confidence interval for the estimate of the standard deviation of the population is then

$$\frac{n-1}{\chi^2_{p,n-1}} \le \frac{\sigma_p^2}{\sigma^2} \le \frac{n-1}{\chi^2_{1-p,n-1}} \tag{7.8}$$

We will cover the chi-squared (χ^2) distribution in Section 12.9. It has no simple CDF and hence no QF, but this is available in most spreadsheet packages. It gives the value corresponding to a chosen probability (e.g. p) for a defined number of degrees of freedom. In our batch blending example this is 77 ($n - 1$). So

$$\sqrt{\frac{77}{103.2}} \le \frac{\sigma_p}{\sigma} \le \sqrt{\frac{77}{54.6}} \tag{7.9}$$

The standard deviation of the sample (σ) is 9.1 and so we can be 95% certain that the standard deviation of the population (σ_p) lies between 7.9 and 10.8.

If we perform the same calculation after the implementation of the control improvement, we would have expected the standard deviation to halve to 4.6. Assuming we use another 78 results then Equation (7.9) tells us the 95% confidence interval on this estimate is from 3.9 to 5.4.

Using the highest and lowest likely values of standard deviations then, from Equation (7.6), the 95% confidence interval on the measured increase is given by

$$1.96\left(\frac{10.8+5.4}{\sqrt{78}}\right)-1.96\left(\frac{7.9+3.9}{\sqrt{78}}\right)=1.0 \tag{7.10}$$

According to Equation (2.1), if the specification is 100, the mean should have increased from 97.9 to 99.0, i.e. an increase of 1.1. With the number of samples taken, we can be 95% confident that the increase is somewhere between 0.6 and 1.6.

7.3 Skewness and Kurtosis

Sample size also affects the confidence we have in the estimates of other parameters. For example, if the data follow the normal distribution, the variance of the estimates of skewness (γ) and kurtosis (κ) are

$$\sigma_\gamma^2=\frac{6n(n-1)}{(n-2)(n+1)(n+3)} \qquad n>2 \tag{7.11}$$

$$\sigma_\kappa^2=\frac{24n(n-1)^2}{(n-3)(n-2)(n+3)(n+5)} \qquad n>3 \tag{7.12}$$

From Table A1.8 the 95% confidence interval (ε) is 1.96σ. This is plotted against n in Figure 7.1. Skewness and kurtosis are key to deciding whether a chosen distribution fits well to the data.

For example, if n is 100, we can be 95% sure that the estimate of skewness is accurate within ±0.47. We typically reject the proposition that a distribution is normal if skewness is outside the range −0.5 to +0.5. So, with 100 results, the skewness must be in the range −0.03 to +0.03 for us

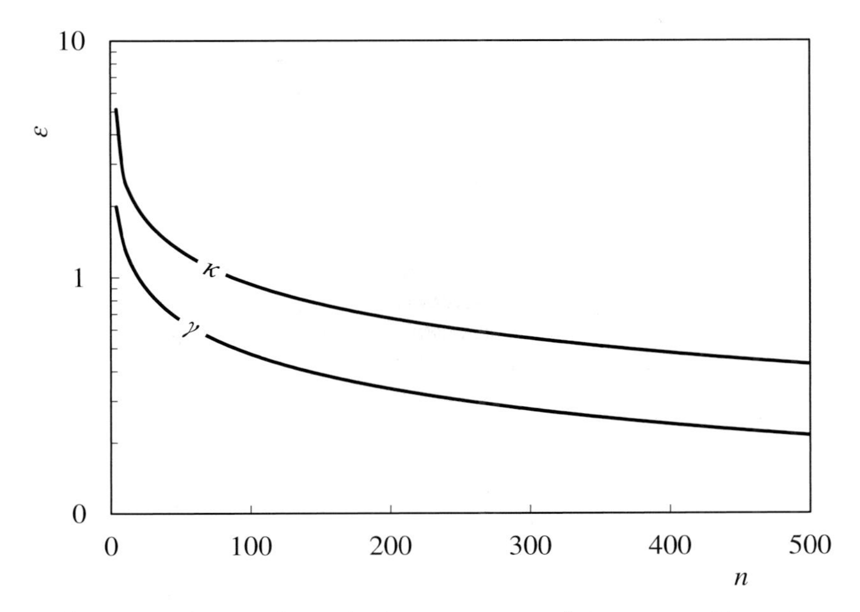

Figure 7.1 Impact of sample size on accuracy of skewness and kurtosis

not to reject the proposition. By increasing the number of results to 500, ε falls to 0.21 and so the distribution would not be rejected as normal provided skewness is between −0.29 and +0.29.

Figure 7.1 shows that the estimate of kurtosis is far less reliable. We reject a distribution as normal if kurtosis is outside the range 2.5 to 3.5. Even if calculated exactly at 3, we therefore need ε to be less than 0.5. This requires n to be at least 365.

7.4 Dichotomous Data

A different approach to selecting sample size is required if the data is dichotomous. For example, we might wish to assess the probability that a blend is produced on-grade. Of the 78 results 44 are on-grade. We could therefore estimate the probability (p) as 0.564 or 56.4%. But, again, we need to know how much confidence we have that this represents the likelihood of success over a much longer period.

To determine this we again choose a margin of error, here expressed as fraction. For example, if we wish to determine the probability of an on-grade blend to within an accuracy of ±5% then ε would be 0.05. If the total number of blends over a long period is N then the minimum sample size is

$$n = \frac{z^2 p(1-p)}{\varepsilon^2 + \left(\frac{z^2 p(1-p)}{N}\right)} \tag{7.13}$$

Since we do not yet have a value for p we take the worst case. The value of $p(1-p)$ is largest when p is 0.5. If, for example, we want to check how much confidence we have that the estimate of p would be applicable to a year's operation then N is 365. Figure 7.2 plots Equation (7.13) for

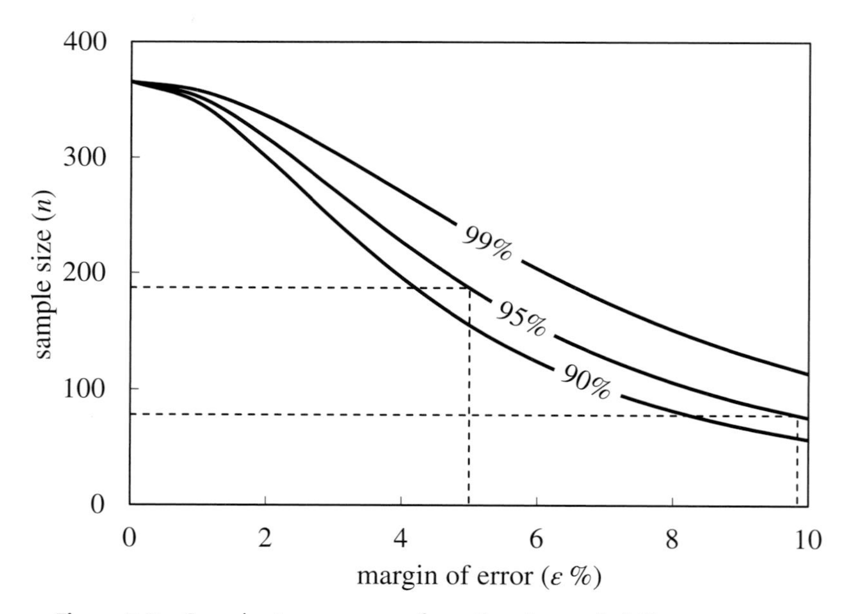

Figure 7.2 Sample size necessary for estimating probability over a year

a range of confidence intervals. In our example it shows we would need to calculate p based on at least 187 results.

Rearranging Equation (7.13) gives

$$\varepsilon = z\sqrt{p(1-p)\left(\frac{1}{n}-\frac{1}{N}\right)} \tag{7.14}$$

Restricting the sample size to 78 means that we are 95% confident that our estimate of the probability of success over a year is correct within ± 9.8%.

Examination of Equation (7.13) shows that as $N \rightarrow \infty$

$$n \rightarrow \frac{z^2 p(1-p)}{\varepsilon^2} \tag{7.15}$$

Figure 7.3 plots Equation (7.15); if we wanted to be 95% confident that our probability is accurate to ±5% for all blends then we would have to calculate p based on at least 384 results. Rearranging Equation (7.15) with p set to 0.5 gives

$$\varepsilon = z\sqrt{\frac{0.25}{n}} \tag{7.16}$$

The probability derived from 78 results means that we are 95% confident that our estimate of the probability of success for all blends is 56.4 ± 11.1%.

Equation (7.16) is known as the *Wald method*. It is one of several published. For example, the *Agresti–Coull method* includes the term n_1 – the number of successful outcomes in the sample of size n.

$$\varepsilon = z\sqrt{\frac{\tilde{p}(1-\tilde{p})}{n+z^2}} \quad \text{where} \quad \tilde{p} = \frac{n_1 + 0.5z^2}{n+z^2} \tag{7.17}$$

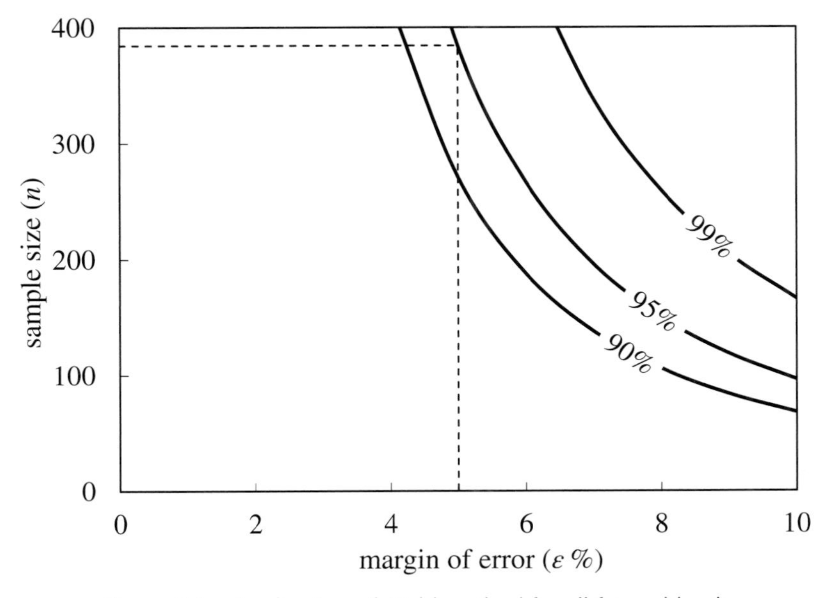

Figure 7.3 Application of Wald method for all future blends

Applying Equation (7.17) to our example of 44 successes in 78 blends indicates that we can be 95% sure that the probability of success for all blends is 56.4 ± 10.8%. This is very close to the accuracy derived using the Wald method. Rearranging Equation (7.17) gives

$$n = z^2\left[\frac{\tilde{p}(1-\tilde{p})}{\varepsilon^2} - 1\right] \tag{7.18}$$

If we were to assume the worst case value for $\tilde{p}$ (of 0.5) we need 380 results to be 95% sure that the accuracy of the estimate of p is within ±5%. This conclusion is also very close to that derived using the Wald method. Further, as Figure 7.4 shows, plotting Equation (7.18) gives results very similar to Figure 7.3. However, because it takes no account of the size of the population (N), the Agresti–Coull method cannot be applied to small populations. Equation (7.18) tells us that n must be at least 380 – even if the population is less than this number.

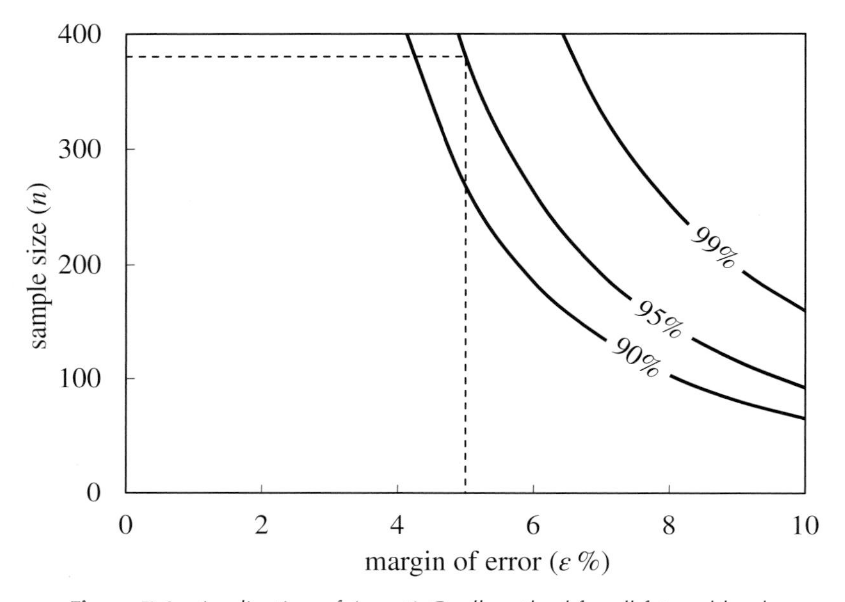

Figure 7.4 Application of Agresti–Coull method for all future blends

7.5 Bootstrapping

Bootstrapping is another technique for determining the confidence we have in estimating, from a sample, the statistical parameters of the population. The principle is to first make a random selection of measurements from the process data. The number of randomly selected measurements is the same as the sample size but they are selected *with replacement*. The selected measurement remains in the dataset and can be selected again. Indeed, the same measurement may appear many times within any sample. We then use the selected measurements to determine the parameter we require. We then repeat this many times to generate multiple estimates of the parameter. Fitting a distribution to the estimates then permits us to determine the best estimate of the parameter and its confidence interval.

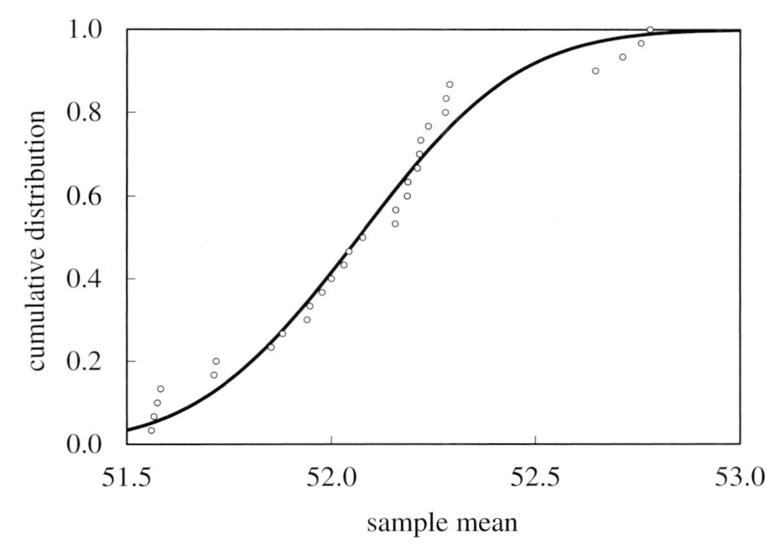

Figure 7.5 *Use of bootstrapping to determine confidence in estimating* μ

As an example, instead of using all 5,000 measurements of the LPG splitter reflux flow, we assume we have only the first 1,000. From these, we select 1,000 measurements randomly, remembering that many are likely to be selected more than once. We then determine the mean of this sample. Repeating this 30 times gives 30 estimates of the population mean.

Fitting a normal distribution to the 30 estimates of the population mean, as shown by Figure 7.5, gives a mean of 52.1 and standard deviation of 0.31. We can therefore be 95% confident that the population mean is within ±0.6 (1.96 × 0.31) of the estimate. If we consider the original 5,000 measurements as the population, the population mean is actually 52.1, well within the confidence interval.

The technique is not restricted to estimating the mean. Using the same data, the standard deviation is estimated as 8.83 ± 0.43 versus 8.73 for the population. Skewness is 0.17 ± 0.32 versus 0.23 for the population. Kurtosis is 3.34 ± 1.45 versus 2.98 for the population. Remembering we would reject the distribution as normal if skewness is outside 0 ± 0.5 or kurtosis outside 3 ± 0.5, skewness is just within the limits but kurtosis is not. This may be because the 1,000 recorded measurements do not comprise a sufficiently large sample. Indeed, repeating the exercise with all 5,000 values estimates kurtosis as 3.10 ± 0.36 – just meeting the criterion to not reject the hypothesis that the data are normally distributed.

Due to the randomisation of the data selection, bootstrapping can give different results, simply by repeating the same test. For this reason, the reader will not be able to exactly reproduce this result. But, more importantly, we should validate any conclusions by increasing the number of trials. Indeed, some judgement is required in interpreting whether the results justify collection of additional data.

Bootstrapping can be extended to any form of curve fitting. For example, it is common when developing an inferential property to split the data into two sets – one for developing the inferential and the other for testing. This is a trivial form of the technique. We could instead develop the inferential from randomly selected data points randomly and repeat this process many times. We would obtain multiple estimates for each coefficient in the inferential and could use these to determine how much confidence we have in their values. From these, we can then determine how much confidence we have in the resulting inferential.

Assume the inferential takes the form

$$y = a_0 + a_1 x_1 + a_2 x_2 \tag{7.19}$$

If σ_0 is the standard deviation of the estimate for a_0, σ_1 is that for a_1, etc., then the standard deviation for the estimate of y is given by

$$\sigma_y = \sqrt{\sigma_0^2 + x_1 \sigma_1^2 + x_2 \sigma_2^2} \tag{7.20}$$

As usual, if we required the 95% confidence interval, this would be $1.96\sigma_y$.

8

Significance Testing

One of the key reasons we might want to fit a statistical distribution to process data is to explore whether an observed result is expected from normal process variations or whether it is exceptional. *Significance testing* enables us to establish whether our expectation of process variations is correct. Related to this, the *confidence interval* gives the range over which we might expect a process condition to vary.

8.1 Null Hypothesis

The probability (p) of an unbiased coin landing heads when tossed is 0.5. If we toss it 100 times we would expect it to land heads 50 times. We will show later how the binomial distribution is used to calculate the probability of this outcome but we should be quite surprised if it occurred since, as Figure 8.1 shows, the probability is only 8%. If it landed heads 49 times we would accept that this is just chance and would not doubt that the coin is unbiased. Clearly, if it landed heads 99 times, we would. The question is how many times must it land heads before we conclude that the coin is biased? For example, the probability of 60 heads is small at 0.011. Is this sufficiently low for us to conclude that p is actually closer to 0.6?

The approach is to first make the *null hypothesis*, sometimes given the symbol H_0. In this example this might be that the coin is unbiased. Less commonly, we make the *alternative hypothesis (H_a)*. In this example it would be that the coin is biased. Choosing the null hypothesis, we calculate the *P-value* – the probability that the observed, or more extreme, result occurs if the null hypothesis is true. It can also be described as the probability of making a *Type I error*, i.e. wrongly rejecting a true hypothesis – a *false positive*. A *Type II error* (a *false negative*) occurs if we were to accept that a hypothesis is correct, when it is not.

If the P-value is below a chosen *significance level* we reject the null hypothesis and conclude that the coin is biased. The key is choosing the significance level. Commonly 0.05 (5%) is used. Figure 8.2 shows the cumulative probability; from this we can calculate that the probability of there being 58 or fewer heads is 0.956. The probability of 59 or more heads is therefore below 0.05 at 0.044; so if there were 60 heads, we would reject the null hypothesis.

Statistics for Process Control Engineers: A Practical Approach, First Edition. Myke King.

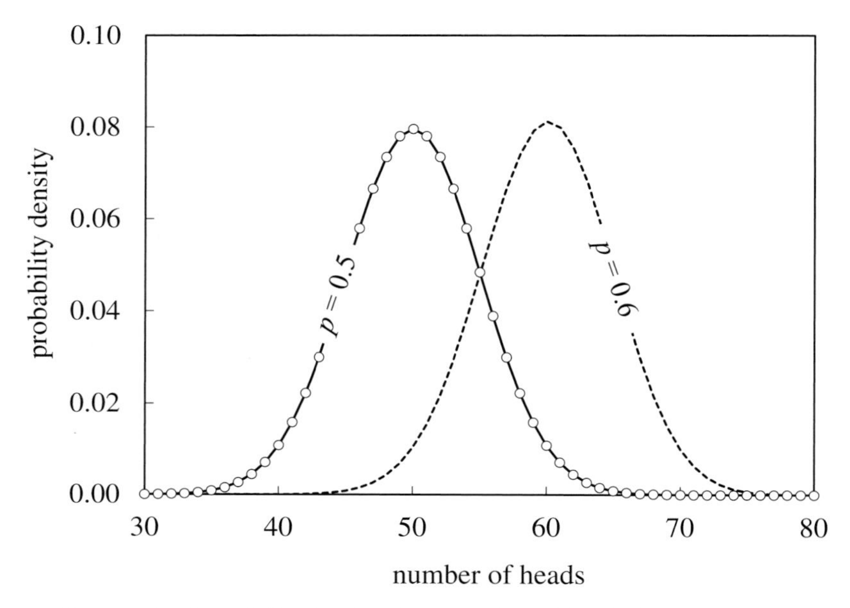

Figure 8.1 Probability density of the number of heads from 100 tosses

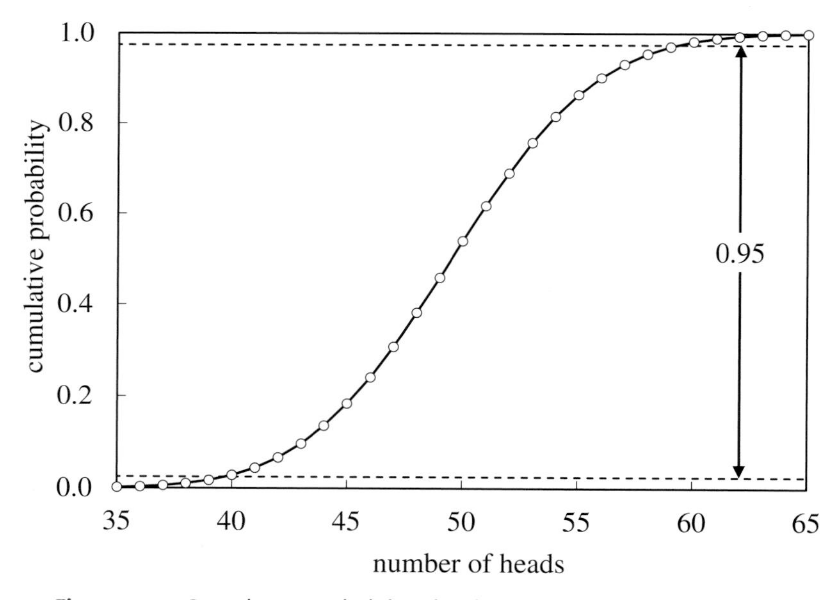

Figure 8.2 Cumulative probability distribution of the number of heads

It is important to remember that this approach only allows us to reject the null hypothesis. It would be wrong to accept that the null hypothesis is true if the P-value is greater than the significance level. In this example, if the number of heads was less than 60, this does not prove that the coin is unbiased. For example, if we wanted to be more certain that our decision to reject the hypothesis is correct, we would choose a lower significance level of, say, 0.01. Again from

Figure 8.2, the probability of there being 61 or fewer heads is 0.990. If there were 60 heads we would not reject the null hypothesis. This does not mean we are sure the null hypothesis is correct; it merely means we have found no evidence to reject it.

The test above is *one-tailed* in that we have only considered the case when the coin lands as heads more times than we might expect. We would of course be equally suspicious about a coin that landed excessively as tails or, in other words, if it landed too few times as heads. The distribution is symmetrical; so if we were concerned that there are 60 heads, we should be equally concerned if there were 40. This would then entail applying the *two-tailed* test. If we wanted to be 95% certain that a coin is unbiased then we need to determine what outcomes lie between the cumulative probabilities of 0.025 and 0.975. These limits are shown in Figure 8.2; to not reject the null hypothesis, the number of heads must be between 40 and 59.

Although the tossing of a coin follows the binomial distribution, the curves in Figure 8.1 are very close to the normal distribution with a mean and standard deviation of

$$\mu = np \qquad \sigma = \sqrt{\frac{p(1-p)}{n}} \tag{8.1}$$

We will show later that this is because p is close to 0.5 and n is large. It means that we can use the normal distribution to plot the curves shown in Figure 8.3. From Equation (5.43), for the one-tailed case, the probability of throwing 60 or more heads is

$$P = 1 - \frac{1}{2}\left[1 + \operatorname{erf}\left(\frac{60-\mu}{\sigma\sqrt{2}}\right)\right] \tag{8.2}$$

For the two-tailed case, the probability of throwing fewer than 40 or more than 60 heads is

$$P = 1 - \left\{\frac{1}{2}\left[1 + \operatorname{erf}\left(\frac{60-\mu}{\sigma\sqrt{2}}\right)\right] - \frac{1}{2}\left[1 + \operatorname{erf}\left(\frac{40-\mu}{\sigma\sqrt{2}}\right)\right]\right\} \tag{8.3}$$

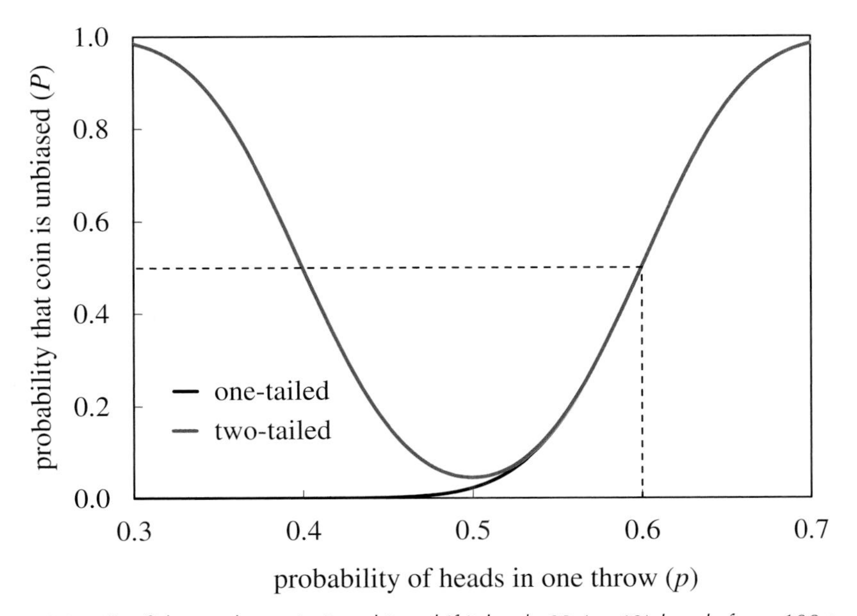

Figure 8.3 *Confidence that coin is unbiased if it lands 60 (or 40) heads from 100 tosses*

In effect we have replaced the P-value with the *z-test*. We will describe later the use of the *t-test*, which arises from using the Student t distribution. This is more applicable when n is small.

The black curve in Figure 8.3 is the one-tailed case. It plots the probability (P), if the coin lands heads for 60 or times out of 100 tosses, that it is unbiased ($p = 0.5$). We can see that, if p is 0.6, the probability that the coin is unbiased is 0.5. In other words, it is equally probable that it is biased. We would have no reason to change our opinion. However, if we hypothesise that p is 0.5, the probability that the coin is unbiased drops to 0.018. This is well below the 5% significance level at which we would accept that 60 heads could be due to chance and so we would reject the hypothesis.

The coloured curve is the two-tailed case. It shows the probability (P) that the coin is unbiased if it lands heads fewer than 40 times or more than 60 times. The probability that p is 0.5 is now 0.046 – only just below the 5% significance level.

Whether we apply the one-tailed or two-tailed test depends on the application. For example, if we were checking the prediction error of an inferential property, it is equally likely that it overestimates or underestimates the true value and we would want to detect either problem. We would therefore apply the two-tailed test. However, if we were analysing the impact that a control scheme had on reducing the number of plant trips caused by exceeding, for example, a high pressure limit, we would be unlikely to be concerned about low pressure and so would apply the one-tailed test.

8.2 Confidence Interval

The confidence interval is used to indicate the reliability of an estimate. It is an example of a two-tailed test. For example, a 95% confidence interval means that there is a probability of 0.95 that the true value is in the quoted range. If we assume that the estimate is normally distributed then we can derive the interval by integrating Equation (5.35) between the limits of the range.

We often define the confidence interval in terms of a multiple (n) of the standard deviation. Figure 8.4 illustrates the case when n is 1. The probability of lying within one standard deviation of the mean is the shaded area under the probability density function. Figure 8.5 is the plot of cumulative probability, given by Equation (5.43). The shaded area is thus determined as 0.683 and confidence interval in this example is therefore 68.3%. In general the probability (P) of the value (x) being between $\mu - n\sigma$ and $\mu + n\sigma$ is

$$P = \int_{\mu - n\sigma}^{\mu + n\sigma} y.dx \tag{8.4}$$

To perform the integration we first make, into Equation (5.35), the substitution

$$x = \mu + \sigma\sqrt{2}z \tag{8.5}$$

Therefore

$$P = \frac{1}{\sqrt{\pi}} \int_{-n/\sqrt{2}}^{n/\sqrt{2}} \exp\left(-z^2\right).dz \tag{8.6}$$

$$\therefore \quad P = \text{erf}\left(\frac{n}{\sqrt{2}}\right) \tag{8.7}$$

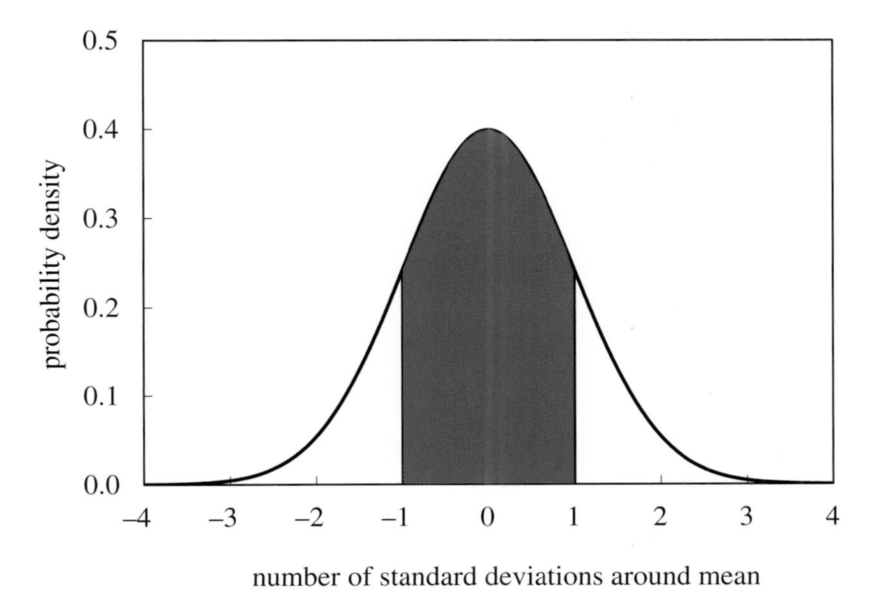

Figure 8.4 *PDF showing probability of being within one standard deviation of the mean*

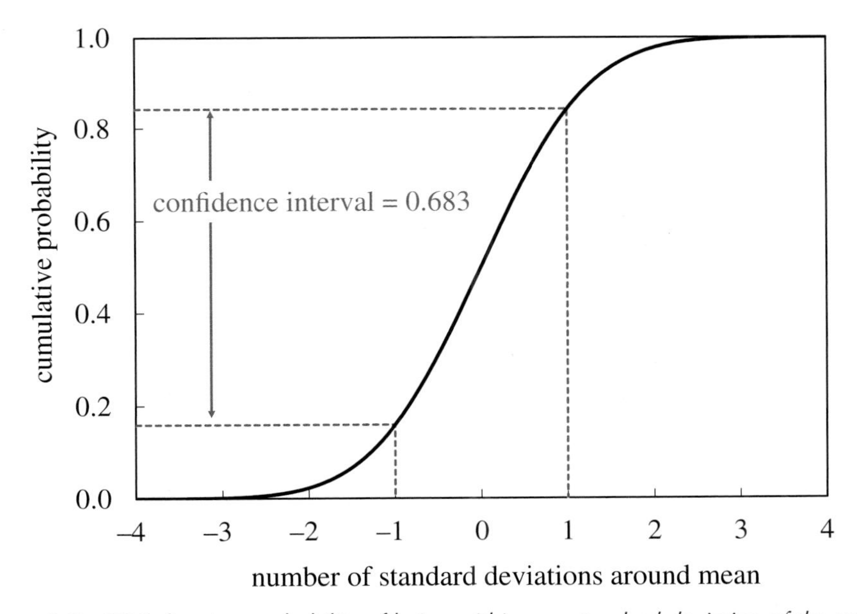

Figure 8.5 *CDF showing probability of being within one standard deviation of the mean*

This function is plotted as curve A in Figure 8.6. So, for example, the 95% confidence interval is often quoted as 2σ (strictly 1.96σ). Table A1.8 shows some commonly used values. It should be remembered that these have been derived from the function for the normal distribution. There are other methods of determining confidence interval in situations where the distribution is not normal.

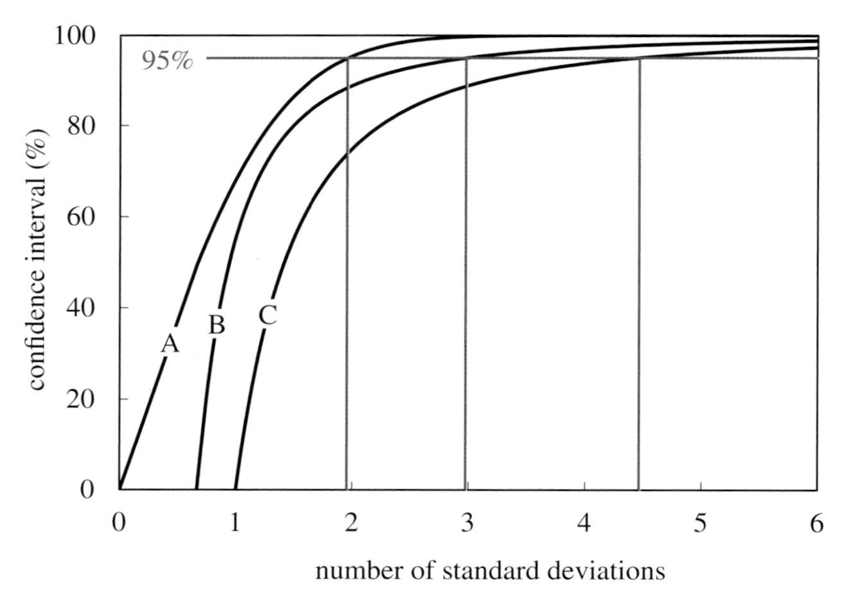

Figure 8.6 Confidence interval as a function of the number of standard deviations

- For any unimodal distribution the *Vysochanskij–Petunin inequality* states that the probability of a result lying more than $n\sigma$ from the mean will be less than $4(3n)^{-2}$. This is plotted as curve B in Figure 8.6 – showing the 95% confidence interval increases to 3σ, whereas 2σ now corresponds to a confidence interval of 89%.
- If not unimodal then *Chebyshev's inequality*, which applies to any distribution, states that the probability will be less than n^{-2}. This is plotted as curve C in Figure 8.6 – showing a further increase to 4.5σ, whereas 2σ now corresponds to a confidence interval of 75%.

As we shall, there are other distributions that will generate different results. In certain circumstances these may be more applicable.

For the one-tailed significance level, n is determined from

$$P = 1 - \frac{1}{2}\left[1 - \text{erf}\left(\frac{n}{\sqrt{2}}\right)\right] \tag{8.8}$$

$$n = \sqrt{2}\,\text{erf}^{-1}[1 - 2(1 - P)] \tag{8.9}$$

Figure 8.7 plots both the one- and two-tailed curves. For example, as described above, the 95% significance level for the two-tailed case is 1.96σ. For the one-tailed case it is 1.64σ. Other commonly used values are included in Tables A1.8 and A1.9.

8.3 Six-Sigma

Six-sigma is a much publicised method used in quality control. However rather than, as it name suggests, it being based on six standard deviations, it is actually based on 4.5. A 1.5σ *shift* is included to reflect long-term changes thought to occur in the variation. The choice of this value

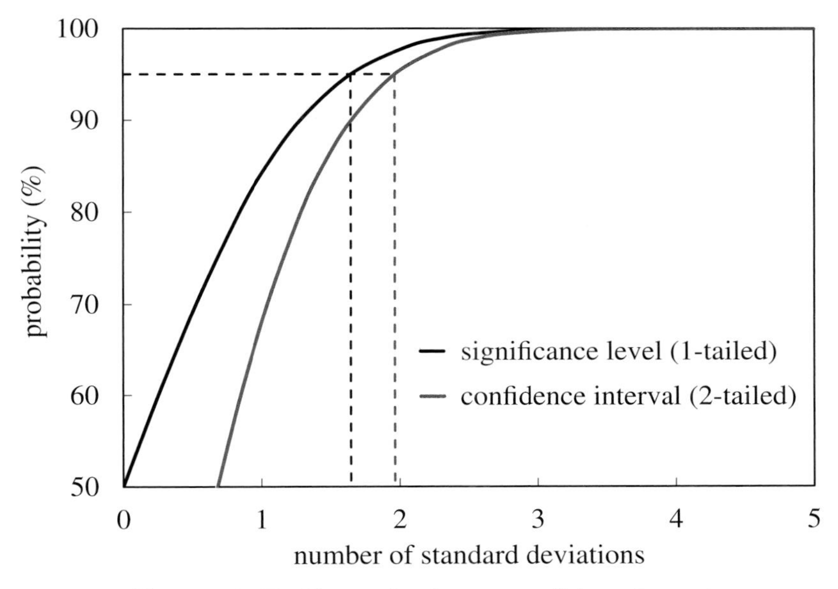

Figure 8.7 Significance level versus confidence interval

is entirely empirical and often criticised by statisticians. However it now forms part of the definition of the procedure.

Six-sigma only considers off-grade results and so uses the probability of values exceeding $\mu + 4.5\sigma$ or falling below $\mu - 4.5\sigma$, but not both. We therefore use the single-tailed confidence interval. From Table A1.9 we can see that this is 0.9999966. There is a probability 0.0000034 that the result is off-grade, i.e. we expect fewer than 3.4 defects per million.

8.4 Outliers

Deciding which data points to exclude from statistical analysis is very subjective. If the scatter is poor then the position of a single outlier, for example on a XY plot, will have a significant impact on the slope of the line of best fit. Basing any conclusion on the effect of a single point might be unwise and the point better treated as an outlier. However, if there is evidence that the point is representative, then excluding it would also result in a suspect conclusion.

The most frequently published method of defining outliers is to treat as *mild* any value that falls outside 1.5 times the interquartile range and as *extreme* any value outside 3 times the range. By definition there is a 50% probability of a value lying in the interquartile range. From Table A1.8 we can see that this corresponds to a range of $\pm 0.6745\sigma$. Multiplying this by 1.5 gives a value close to σ and by 3 a value close to 2σ. We could therefore adopt a policy of excluding all values more than 2σ from the mean and reviewing those that lie between σ and 2σ away from the mean. We are effectively choosing to exclude any value that we are 95% certain is an outlier.

Excluding outliers will change the mean and reduce the standard deviation of the values retained. Thus data points that were mild outliers may now fail the 2σ criterion. Excluding these, and any others that arise from the next recalculation of μ and σ, could severely undermine the confidence we have in any regression analysis.

A similar approach is based on directly on the quartiles. If the first quartile is Q_1 and third Q_3, we exclude any value that is less than $2.5Q_1–1.5Q_3$ or greater than $2.5Q_3–1.5Q_1$. Again, if the quartiles are recalculated following removal of the outliers, it is likely that additional data points would become outliers.

An approach, similar to that used to identify outliers, can be taken in identifying suspect results from an inferential once in operation. The development of the inferential will have quantified the expected σ_{error}. If, for example, the error between the inferential and the laboratory result exceeds three times this value then, from Table A1.8, we can determine that the probability that the inferential is incorrect is 99.73%. The probability that it is correct is therefore very low at 0.27%. This would prompt an investigation of the reliability of the inferential and a possible update. This is the principle behind the *Shewhart statistical process control (SPC) chart*. This includes a centre line, drawn at the mean (where the mean would be zero if the chart is being used to assess the error in an inferential) and the lower and upper *control limits*, usually drawn at $\pm 3\sigma_{error}$. The inferential error is then plotted against sample number.

This simple approach is unlikely however to detect small shifts in the accuracy. A more sophisticated approach is based on the *Westinghouse Electric Company (WECO) rules* that were originally developed to improve the reliability of Shewhart charts. If any of these four rules are violated then the inferential would be considered suspect.

1. The error exceeds the $3\sigma_{error}$ limit – as described above.
2. Two out of three consecutive errors exceed the $2\sigma_{error}$ limit but are within the $3\sigma_{error}$ limit. The errors exceeding the limit must do so on the same side of the mean. When the inferential was developed the mean error would have been zero and so we need either consecutive positive errors or consecutive negative errors. From Table A1.8, the probability of being within the lower limit is 95.4%. One of the three errors must be in this region. The probability for being between the limits on the same side of the mean is (99.73–95.4)/2 or 2.17%. Two of the errors must be in this region. There are three combinations of results that meet the criteria and each can occur on either side of the mean. The probability of violating this rule is therefore $0.954 \times 0.0217^2 \times 3 \times 2$ or 0.27%.
3. Four out of five consecutive errors (with same sign) exceed the σ_{error} limit but are within the $2\sigma_{error}$ limit. Taking the same approach, the probability of violating this rule is $0.683 \times 0.1355^4 \times 5 \times 2$ or 0.23%.
4. Eight consecutive points fall on the same side of the mean. The probability of violation of this rule is 0.5^8 or 0.39%.

Applying only the first rule would result in a false indication of a problem once in every 371 instances. While increasing the reliability of detecting a problem, applying all four rules will increase the false indication rate to once in every 92 instances.

These rules all assume that the prediction error is normally distributed; so it would be wise to confirm this before applying them. For example, it is possible to define asymmetric rules for skewed distributions.

8.5 Repeatability

Repeatability is a measure of *precision*. It has little to do with *accuracy*. A measurement can be inaccurate and precise. For example, a measurement that is consistently wrong by the same amount is precise. The error is caused by a fixed bias. Repeatability is a measure of random error. The smaller the random error, the better the repeatability.

Repeatability of a laboratory result, as defined by the ASTM, is determined from test results that are obtained by applying the same test method in the same laboratory by the same operator with the same equipment in the shortest practicable period of time using test specimens taken at random from a single quantity of homogeneous material. It can be defined as the maximum difference between two measurements that can be expected 95% of the time.

Imagine that the laboratory tests the same material twice, under these conditions, and obtains results x_1 and x_2. Both these results are taken from a normal distribution of mean μ and standard deviation σ. The average of the two results is

$$\bar{x} = \frac{x_1}{2} + \frac{x_2}{2} \tag{8.10}$$

Since we halve x_1 and x_2, we also halve their standard deviation. The variance of the mean will be the sum of variances.

$$\sigma_{mean}^2 = \left(\frac{\sigma}{2}\right)^2 + \left(\frac{\sigma}{2}\right)^2 \quad \text{or} \quad \sigma_{mean} = \frac{\sigma}{\sqrt{2}} \tag{8.11}$$

The 95% confidence interval is $1.96\sigma_{mean}$. So, if the repeatability (r) is given, the standard deviation of the test result can be derived. The approximation is that made by the ASTM.

$$\sigma = \frac{r}{1.96\sqrt{2}} \approx \frac{r}{2.8} \tag{8.12}$$

8.6 Reproducibility

Reproducibility is another measure of precision. Like repeatability, it says nothing about accuracy. The ASTM defines reproducibility as the expected degree of agreement between results obtained by two different laboratories employing the same procedures but different equipment when analysing splits of the same gross sample. It is determined from an *interlaboratory study* involving between 8 and 12 laboratories. It will always be greater than repeatability since it includes the variation within individual laboratories as well as the variation between them. It will typically be double the repeatability and, as with repeatability, it is determined from 2.8σ.

Normally the control engineer is more concerned with repeatability. However reproducibility becomes important, for example, if a study involves laboratory results provided by a customer testing product that is also routinely tested by the manufacturer.

For tests that measure concentration, Horwitz[4,5] identified that the accuracy of a laboratory result was related to the concentration being measured. In particular, accuracy becomes very poor at very low concentrations. He studied the results of over 10,000 interlaboratory studies and developed *Horwitz's curve* as shown as Figure 8.8. The *relative standard deviation (RSD)* is defined as percentage of the measurement (C), where C is the concentration expressed as a weight fraction.

$$RSD = 2^{[1-0.5\log(C)]} \tag{8.13}$$

From this equation

$$\log(RSD) = \log(2)(1-0.5\log(C)) \quad \text{or} \quad \log\left(\frac{RSD}{2}\right) = \log\left(C^{-0.5\log(2)}\right) \tag{8.14}$$

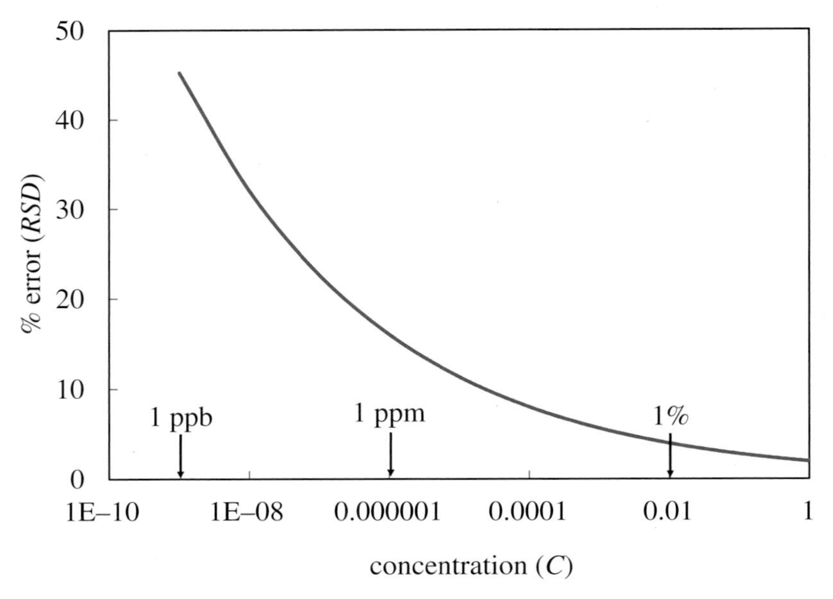

Figure 8.8 *Horwitz's curve*

This gives rise to the alternative definition

$$RSD = 2C^{-0.15} \tag{8.15}$$

If we want the standard deviation (σ_H) expressed in the same units as C then the equation is further modified.

$$RSD = 100\frac{\sigma_H}{C} \quad \text{and so} \quad \sigma_H = \frac{RSD.C}{100} = 0.02C^{0.85} \tag{8.16}$$

Laboratory testing methods are generally accepted if their reproducibility is less than half σ_H. Although there are those that will accept a method if it performs better than $0.67\sigma_H$ and there are others who believe that the Horwitz method is unreliable.

8.7 Accuracy

A measurement can be accurate without being precise. If the average of multiple estimates of the same property is equal to the true measurement, then the estimates are deemed accurate. If the variation between the estimates is large, then they would be considered imprecise. From a statistical point of view, high accuracy requires the mean of the estimates to be close to the true value. High precision requires that the standard deviation of the error is small. Accuracy requires a small bias error, while precision requires a small random error. Another term, *trueness*, defined in ISO 5725 is synonymous with our use of the word 'accuracy'.

For field instrumentation, precision is usually quoted as $\pm\varepsilon$, where ε is the maximum error. In some cases, the overestimate ($\varepsilon_{\text{high}}$) may not be the same as the underestimate (ε_{low}). The standard deviation of the measurement error can be determined from

$$\sigma = \sqrt{\frac{\varepsilon_{\text{high}}^2 - \varepsilon_{\text{high}}\varepsilon_{\text{low}} + \varepsilon_{\text{low}}^2}{3}} \tag{8.17}$$

8.8 Instrumentation Error

Control engineers are generally more concerned that process instrumentation gives a precise rather than an accurate measurement. For example, the SP of a reactor temperature controller might be set by the operator to meet the required product quality. Any inaccuracy in the temperature measurement would result in the quality being different from what might be expected but this would be dealt by making a once-off change in the temperature SP. However, a lack of precision will cause the controller to take unnecessary corrective action, responding to the random variation in the temperature measurement, and hence disturb quality. If the operator were to respond to this variation, the SP changes would cause further disturbances.

Precision is also of importance when estimating the potential benefit of improving the control of product quality. This would commonly involve assessing the standard deviation of the deviation of the quality from its target. It is generally assumed that improved control will halve the standard deviation. However, if the variation is due entirely to the lack of precision of the temperature measurement and the resulting operator action, then improved control would have no impact. Instead, what is required is an upgrade of the instrumentation to something more precise.

Where accuracy becomes important is in the comparison of two (or more) measurements of the same property. This occurs, for example, when the engineer wants to confirm that an on-stream analyser agrees with the laboratory result. Knowing the precision of both measurements we can decide whether the difference between the two is within what might be expected or not.

The repeatability of an on-stream analyser is determined by keeping the composition constant and observing how much the measurement varies. Reproducibility is determined by changing the composition and is a measure of the ability of the instrument to reproduce the measurement when a predefined set of conditions is recreated. If we were selecting an analyser for installation we would be interested in both. When checking its measurement against the laboratory we should use repeatability.

For example, the ASTM quoted repeatability for the laboratory measurement of the flash point of kerosene is 4.1°C. From Equation (8.12), this is equivalent to the measurement error having a standard deviation of 1.5°C. Analyser manufacturers usually quote a repeatability that is better than the laboratory test method. Let us assume that the standard deviation for the analyser measurement error is 1°C.

We saw in Section 4.5 that the variance of the difference between the laboratory and the analyser is the sum of the variances of the two measurements.

$$\sigma_{error} = \sqrt{1.5^2 + 1^2} = 1.8°\text{C} \tag{8.18}$$

From Table A1.8, we can be 95% sure that there is no problem with either measurement, provided that the difference between them is less than 1.96σ, or 3.5°C. With the flash point of kerosene typically around 40°C, such a discrepancy is substantial. The product could be seriously off-grade or there could a substantial loss of product yield. In practice, it would not be ignored. It suggests that the repeatabilities are much better than quoted.

Another common example is validating an inferential property against the laboratory measurement. In Chapter 16 we will, using regression analysis, develop an inferential for the C_4 content of propane.

$$\%C_4 = -20.54 - 0.2092T_1 + 0.6544T_2 \tag{8.19}$$

T_1 and T_2 are tray temperatures. Typically measured with K-type thermocouples, the quoted accuracy is ±2.2°C or 0.75% of range, whichever is the greater. The temperatures are around

60°C and so would typically have a range of 0 to 100°C. Applying Equation (8.17) gives a standard deviation of the measurement error as 1.3°C. The standard deviation of the error this causes in the inferential is therefore

$$\sigma_{infer} = \sqrt{0.2092 \times 1.3^2 + 0.6544 \times 1.3^2} = 1.2 \text{ vol\%} \tag{8.20}$$

We also require the standard deviation of the laboratory error. The C_4 content is typically 5 vol%. This is close to a weight fraction of 0.05. From Equation (8.16) we obtain σ_H as 0.0016 – close to 0.16 vol%. This has negligible impact on the standard deviation of the difference between the laboratory and the inferential.

$$\sigma_{error} = \sqrt{1.2^2 + 0.16^2} = 1.2 \text{ vol\%} \tag{8.21}$$

We can be 95% certain that the inferential agrees with the laboratory, provided that the difference between them is less than 1.96σ, or 2.4 vol%. In other words, if the inferential is controlling the composition at 5, we should not be concerned with any discrepancy provided the laboratory result is in the range 2.6 to 7.4. As with the kerosene example, this range covers a significant loss of product revenue through to the product being seriously off-grade. Such imprecision in an inferential (or analyser) would not be tolerated.

We will study the behaviour of a similar inferential later in Section 16.1. For this, σ_{error} is far less than predicted at 0.28. This number was obtained, during the development of the inferential, by calculating the standard deviation of the difference between it and the laboratory. The overestimation is a result of the temperature indicators being far more reliable than quoted. This probably arises from the manufacturer including both random and bias error in the quoted accuracy. When developing the inferential, regression takes account of any bias error (whether it is in the temperature measurements or in the laboratory result). To give a consistent result, in this case, we would have to assume a thermocouple precision of about ±0.3°C.

Of course, measurements errors can be worse than expected. They arise not only from the instrument but also from the way it is installed and the operating conditions. For example, contact resistance on a poorly connected signal cable can severely undermine the accuracy of a resistance type temperature device (RTD). Not compensating for changes in pressure and density can generate far greater errors in the gas flow reported by an orifice type flowmeter. Poor design of sampling conditioning for an on-stream analyser can cause large measurement errors.

Poor estimates of precision can arise when assessing the economic incentive to improve control. For example, the 'true' variation in product quality should theoretically be derived from the variation reported by the laboratory and, from Equation (8.12), the repeatability (r) of the test method.

$$\sigma_{true} = \sqrt{\sigma_{reported}^2 - \left(\frac{r}{2.8}\right)^2} \tag{8.22}$$

As with field instrumentation, the laboratory repeatability is often better than that quoted. Indeed, applying Equation (8.22) can show that there is no variation in the 'true' result and so remove the incentive to improve control. But, as with field instrumentation, the error in the laboratory result can also be much worse than expected. This can arise from poor time-stamping, poor sampling procedures or simple transcription mistakes.

While control engineers must be aware of the sources of error, for the reasons outlined above, it would be unusual for them to use vendor-provided accuracy and precision information. For field instrumentation this is far more in the realm of the instrument engineer, firstly during the selection of new instruments and secondly in resolving any apparent measurement problem.

While many confuse the roles of the control engineer and the instrument engineer, this is one of the skills that distinguishes one from the other.

Often, instrument selection is a compromise between instrument range, accuracy, maintainability and cost. To complete this properly, considerable skill is required in understanding the performance claimed by the instrument vendors. Accuracy can be quoted on an absolute basis, as a percentage of range (*span*) or as a percentage of full scale. In addition to the performance measures covered so far, there are several others. *Drift*, as the name suggests, is a measure of how the measurement of a fixed property might change over time. *Resolution* is the smallest detectable change in property. Some instrumentation is routinely calibrated. For example, many on-stream analysers are routinely checked with a known sample and custody transfer flow meters checked with provers. The quoted accuracy can be meaningless since it will also depend on the calibration method.

9

Fitting a Distribution

The process control engineer is likely to be familiar with the uniform and normal distributions, and perhaps one or two others, but these form only part of a bewildering array of over 250 published. Very few documented PDF have a solid engineering basis. Primarily they were developed empirically and justified on the basis that they fit published datasets better than their predecessors. Indeed the approach we take in this book is to select the most appropriate distribution for each dataset – irrespective of the purpose for which the distribution might originally have been developed. Indeed, the control engineer has little choice in adopting this approach. Development of distributions has been largely focussed on pharmaceutical, meteorological and demographic studies. More recently, new techniques have been developed for analysis of financial investments. But there are none that are specifically designed for the process industry.

We will see that each PDF (or PMF), CDF and QF includes at least one parameter that affects its shape. For example, we have seen that the shape of a normal distribution is defined by its mean (μ) and standard deviation (σ). Presented with a dataset we choose a CDF (or PDF or QF) that should describe its distribution and then estimate the shape parameters based on the observed distribution. This is known as the *Bayesian* approach and, as defined earlier, the chosen function is known as the prior distribution.

The approach an engineer takes to statistical analysis needs to be somewhat different to that taken by a theoretician. As the reader progresses through this book it will become apparent that there will almost always be a more accurate way of mathematically representing process behaviour. However the skill of the engineer is in deciding whether the fitted distribution is 'good enough'. The warning, usually associated with financial investments, that 'past performance is no guarantee of future performance' applies to all statistics. Simply because a process has followed some statistical behaviour in the past does not imply it will, if no changes are made, follow the same behaviour in the future. Control engineers seeking management approval to invest in control improvements are concerned with future behaviour. Their argument is generally that, without improvement, the average operation will be the same and that, with improved control, it will be better. However, how much better is a matter of judgement. The outcome is unlikely to exactly match the prediction. The error in predicting the improvement could be considerably larger than that caused by not choosing the best fitting distribution. There is little

Statistics for Process Control Engineers: A Practical Approach, First Edition. Myke King.

point in choosing a very precise distribution, mathematically very complex to apply, if its precision is undermined by other assumptions.

So what should the engineer look for is selecting the prior distribution?

- Data can be discrete or continuous. We need therefore to first choose a distribution suitable for the type of data. This decision is not always quite as simple as it might first appear. Time, for example, would normally be considered a continuous variable but when data is historised it becomes discontinuous – particularly if the data collection interval is large. We might similarly consider that product quality is a continuous measurement but there are many laboratory tests that give a simple pass/fail result or report the result as an integer.
- We also need to decide whether the process has memory or not and choose an appropriate distribution. This too is not always straightforward. For example, a product composition measured now will be highly correlated to the value recorded one second ago but probably has no relation to one recorded on the previous day. Would the process exhibit memory if the interval was one minute? What if it was one hour?
- There are distributions designed for specific situations. For example, a distribution might have been developed to deal with very small datasets or maybe to compare datasets. Clearly, if one of these situations arises, the appropriate distribution should be used.
- We need to accurately represent current process behaviour. We need to know what this is on average, by determining the mean (μ). And we need to understand the variability of the data by determining the standard deviation (σ) or an equivalent parameter. While this may seem obvious, the reader will discover that there are many distributions that fit the data extremely well but that do not permit these parameters to be calculated easily. They can, of course, be calculated from the data but we will show later that this is often unreliable.
- We should also look at whether the distribution has simple calculations for skewness (γ) and kurtosis (κ). While we are not normally interested in their actual values, we can calculate what combinations of the two can be represented. Indeed, many of the distributions described in this book have charts showing what is feasible. By first calculating γ and κ from the data we can shortlist those distributions that are capable of matching these parameters.
- Care needs to be taken if the distribution is bounded. It is often the case that fitting such a distribution results in the more extreme data points being excluded. Or it might include all the current data but exclude the possibility of more extreme, but quite feasible, values occurring in the future.
- Related to the previous point, we must be careful about what statistical behaviour we want to represent. For example, we might want to estimate the probability of an unusual and highly undesirable event. We should then be primarily concerned that our distribution properly represents extreme behaviour. Fitting even an unbounded distribution can give little priority to accurately representing operating regions where very little data exist.
- Another consideration is the ability to invert the CDF to a give the QF. Firstly this requires that the CDF exists; not all distributions have one that is easily defined. Secondly it has to be in a form that can be inverted; often this is not the case. A QF, or something equivalent, is essential if the engineer plans on using Monte Carlo simulation, as described in Section 5.7. It is also useful should the engineer wish to explore the value of x corresponding to a probability. Indeed, it can be informative to check whether the 1st and 99th centiles of the fitted distribution match the data – particularly so if we are primarily interested in modelling extreme behaviour. We have also seen that a Q–Q plot is by far the most sensitive in highlighting any mismatch. While quantiles can be determined from a plot of the CDF or by

iteration, it is far simpler to use the QF. The converse can also apply. Albeit rare, there are distributions that can only be expressed as a QF. The absence of a CDF and a PDF can be at least an inconvenience.

When selecting the prior distribution, many can be ruled out as inapplicable but there will remain many that are. Selection is then based on which fits the data best, tempered by ease of application and the robustness of the end result. One of the purposes of this book is to document candidate distributions, identify under what circumstances each might be considered and present worked examples of their application. Distributions that have no obvious application to the process industry have largely been excluded. However, some are included simply on the basis that they are well known and therefore require at least an explanation as to why they should not be used. Those that include complex mathematics and offer no advantage over simpler versions have also largely been omitted – although some of those better known are used to demonstrate the trade-off between accuracy and complexity. The main criterion for inclusion is ease of use within a spreadsheet environment.

There are a wide range of commercially available software products, some of which include virtually all published distributions. While these can cover most of what is covered in this book, they do so at a (sometimes substantial) cost. Often targeted at other applications, none are specifically designed to support process control design and maintenance. Further, they do not excuse the engineer from properly understanding the techniques they employ. Indeed the wise engineer will first evaluate the techniques in the spreadsheet environment and use this experience to decide whether the cost of the product is justified and, if so, which would be the best choice.

While every attempt has been made to properly title each distribution, it is almost impossible to be unambiguous. Many distributions have multiple names, perhaps because they were published by developers unaware that they already existed. We will see later that there are several examples of the same distribution being known under five different names. One distribution can be a special case of another and so also share its name. Many are strongly linked and can be derived from each other. Many are numbered (e.g. as Type I, Type II) but the numbering may have been changed by the developer as the range expanded or misused by others not adhering to the original publication. Many are documented by authors that have not checked the original source, assuming that the source they are using has accurately reproduced the distribution. Further that source may have done the same. It would be very time-consuming to prove, but it is highly likely that many distributions have passed along a long chain of authors – each making a minor modification or introducing an error. The end result can bear little resemblance to the original.

As we saw in Chapter 5, there is no consistency in including in a distribution's title, words such as 'generalised' and 'standard'. Similarly there is little consistency in the symbols used by developers. In this book we have standardised on μ for the mean, σ for the standard deviation and γ for skewness. We use κ for kurtosis, largely avoiding the use of excess kurtosis for which we use γ_2. We have generally used α as the location parameter, β as the scale parameter and δ as an additional shape parameter that typically affects skewness and/or kurtosis. Complete standardisation is impractical and so symbols are largely defined as they are used and can have different meanings in different formulae. Those documenting the same distribution can make other changes. For example, β will appear in one version, replaced by the reciprocal $1/\lambda$ in another. And there are many changes that have a less trivial impact on the look of a PDF or CDF. For example, replacing β with $\ln(\lambda)$ will change $e^{\beta x}$ to λ^x and generate what might first appear to be a very different distribution. Parameters can also be omitted completely, perhaps to simplify the look of the equations. The developer may assume that the reader is an experienced statistician who knows that, for example, variable z should more fully be written as $(x-\mu)/\sigma$. Further, as we will see in Chapter 11, some of the mathematical functions used in documenting distributions can be written in several forms.

In general, the more parameters that can be adjusted in the distribution, the more closely it will fit the data. However this has to be balanced against the disadvantages this brings. Firstly, it can make fitting the distribution difficult. Some of the parameters may influence the shape of the distribution in very similar ways. As a result the engineer may find that multiple sets of parameters give an almost identical fit. The concern would then be about the robustness of the result. Would we be confident that it truly represents the behaviour of the process – particularly if it moves into the region where there are few historically collected data points? Secondly, calculation of key parameters, such as mean and standard deviation, can be extremely complex if there are more parameters – often to the point where they can only be performed by computer code. Finally, process data follows an underlying distribution but with random 'noise' superimposed. Such noise can arise from errors in measurement and from poor timestamping. The more parameters in a distribution, the more likely it is to model the noise in addition to the true distribution. A distribution that fits recently collected data may not therefore reliably predict future behaviour.

Distributions are often developed to handle published datasets. The developer will usually specify ranges for any parameter used in the PDF. While generally these limits should be adhered there are examples where fitting them to an unrelated dataset violates these limits. While the engineer should be cautious if this occurs, in itself it is not a reason for enforcing the limits or choosing another distribution. However, in some cases, the calculation of μ, σ, γ and κ is possible only over a range of values for a shape parameter. A fitted parameter falling outside this range might be a reason for choosing an alternative distribution.

Finally, publications contain a significant number of errors. When typing a complex equation, using an equation editor, it is very easy to omit or mistype a term. From personal experience the author of this book can confirm that proofreading, even by someone not involved in writing the text, can easily miss such errors. The result is that a PDF or CDF might appear to be a version of the distribution not previously published. A simple check is to plot the PDF (or PMF) and check that the area under the curve is unity and is always so as shape parameters are changed. Often visually approximating the area as a triangle and calculating its area is enough to highlight any error. A CDF can be checked by ensuring it generates zero when the variable is at its minimum and unity when it is at its maximum.

9.1 Accuracy of Mean and Standard Deviation

A common approach in the process industry is to use Equations (4.2) and (4.35) to calculate the mean and standard deviation, assume that data are normally distributed and then analyse the results – effectively using Equations (5.35) and (5.43). The calculation of the mean can have a true engineering meaning. For example, the (weighted) mean of the C_4 content of cargoes of propane truly represents the quantity of C_4 sold at propane prices. But averaging the laboratory results of the same product as it leaves the LPG splitter, even if weighted for variation in flow, gives only an approximation of the C_4 sold as propane. We have effectively assumed that the composition (and flow) remained unchanged between laboratory samples. This is clearly not the case and the less frequently we sample the stream, the less confidence can we have that mean composition truly represents what was sold.

Accurate estimation of mean is important from two aspects. Firstly we need to know reliably what is being achieved by the process currently so that we can accurately quantify any improvement. The process control engineer, however, is primarily interested in the variation and relies heavily on standard deviation as a measure of this. As Equation (4.35) shows, to calculate standard deviation, we first need the sample mean. Any error in estimating the mean will always cause the standard deviation to be overestimated. To understand this, consider the trivial example where we have two measurements of 90 and 110. The mean is clearly 100 and so the

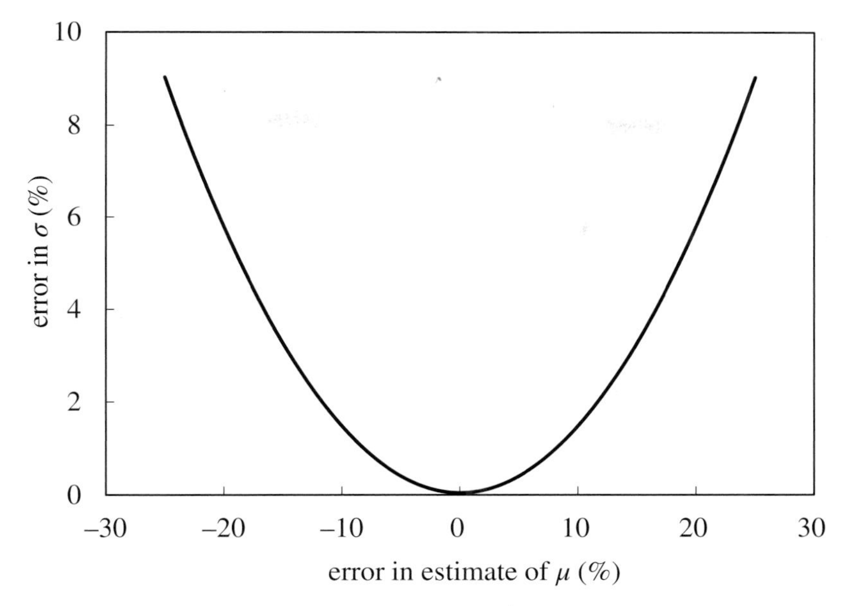

Figure 9.1 *Error in estimating σ arising from error in estimating μ*

variance is $(10^2 + 10^2)/2$, or 100, giving a standard deviation of 10. If the mean is overestimated as 105 the variance will be calculated as $(15^2 + 5^2)/2$, or 125, giving a standard deviation of 11.2. Exactly the same would result from underestimating the mean at 95. A 5% error in the mean has caused a 12% error in the standard deviation. If being used as part of calculating the benefit of improved control, the benefit would then be overstated. Fortunately, for large samples, the error introduced is less, as illustrated by Figure 9.1. Drawn for data following a uniform distribution, it shows what error is introduced by misestimating the mean. For data that are approximately normally distributed, an error in the mean has much less of an effect. However, it is always a wise precaution to ensure the mean is estimated as accurately as practically possible.

The standard deviation forms the basis of many other techniques used by the control engineer. Identifying whether correlations exist, for example in the development of inferential properties, relies on the estimate of standard deviation of the prediction error. Something similar is required for monitoring the performance of MPC and inferentials. There are two concerns with simply applying Equation (4.35). Firstly, as we showed in Section 4.5, standard deviation is very sensitive to outliers. A single point can give an entirely incorrect conclusion about the degree of dispersion. We showed in Section 8.4 that excluding outliers is not an exact science. Instead we here describe the better approach of fitting the chosen distribution to all the data. Secondly, although it is likely that the distribution of the data will be close to normal, this is not guaranteed. Closer examination of process data can reveal that this assumption can cause large errors in the resulting conclusions.

9.2 Fitting a CDF

Column 1 of Table 9.1 shows 30 process measurements (x). We will later describe methods for checking whether a normal distribution can be used and, if not, what should replace it. But, for this example, we will assume that the data are normally distributed. As the first step in the curve fitting procedure the data has been ranked – smallest to largest. Column 2 is the

Table 9.1 *Fitting the CDF and quantile function of a normal distribution*

			calculated		fitted CDF		fitted quantile function	
x	ranking (k)	observed EDF	$F(x)$ $\mu = 47.2$ $\sigma = 8.46$	RSS	$F(x)$ $\mu = 46.0$ $\sigma = 8.63$	RSS	x $\mu = 46.5$ $\sigma = 8.72$	RSS
34.4	1	0.0333	0.0645	0.0010	0.0895	0.0032	30.5	15.48
36.9	2	0.0667	0.1108	0.0019	0.1459	0.0063	33.4	12.47
37.1	3	0.1000	0.1153	0.0002	0.1512	0.0026	35.3	3.30
37.3	4	0.1333	0.1200	0.0002	0.1567	0.0005	36.8	0.28
38.1	5	0.1667	0.1399	0.0007	0.1800	0.0002	38.0	0.01
38.4	6	0.2000	0.1480	0.0027	0.1893	0.0001	39.1	0.52
40.2	7	0.2333	0.2026	0.0009	0.2508	0.0003	40.1	0.01
40.5	8	0.2667	0.2127	0.0029	0.2620	0.0000	41.0	0.28
40.7	9	0.3000	0.2197	0.0065	0.2696	0.0009	41.9	1.42
42.2	10	0.3333	0.2756	0.0033	0.3298	0.0000	42.7	0.26
43.1	11	0.3667	0.3122	0.0030	0.3684	0.0000	43.5	0.15
43.5	12	0.4000	0.3291	0.0050	0.3860	0.0002	44.3	0.57
44.5	13	0.4333	0.3729	0.0037	0.4309	0.0000	45.0	0.25
44.6	14	0.4667	0.3774	0.0080	0.4355	0.0010	45.7	1.29
44.9	15	0.5000	0.3909	0.0119	0.4492	0.0026	46.5	2.45
47.0	16	0.5333	0.4885	0.0020	0.5460	0.0002	47.2	0.04
47.7	17	0.5667	0.5215	0.0020	0.5780	0.0001	47.9	0.05
48.4	18	0.6000	0.5544	0.0021	0.6094	0.0001	48.7	0.08
48.5	19	0.6333	0.5590	0.0055	0.6138	0.0004	49.4	0.88
48.5	20	0.6667	0.5590	0.0116	0.6138	0.0028	50.2	2.96
50.6	21	0.7000	0.6542	0.0021	0.7028	0.0000	51.0	0.19
51.0	22	0.7333	0.6715	0.0038	0.7187	0.0002	51.9	0.81
52.4	23	0.7667	0.7289	0.0014	0.7707	0.0000	52.8	0.17
53.6	24	0.8000	0.7738	0.0007	0.8106	0.0001	53.8	0.04
57.5	25	0.8333	0.8873	0.0029	0.9085	0.0057	54.9	6.74
57.7	26	0.8667	0.8917	0.0006	0.9123	0.0021	56.2	2.39
57.7	27	0.9000	0.8917	0.0001	0.9123	0.0002	57.6	0.00
60.0	28	0.9333	0.9342	0.0000	0.9475	0.0002	59.6	0.19
64.4	29	0.9667	0.9787	0.0001	0.9835	0.0003	62.5	3.75
65.9	30	1.0000	0.9863	0.0002	0.9894	0.0001		
total				0.0871		0.0271		41.53

ranking itself (k). Column 3 is the ranking divided by the number of measurements (n). This now represents the cumulative distribution we wish to represent. The result can be represented by what is sometimes described as the *ogive distribution*. If each process measurement is x_k, with k ranging from 1 to n, then the *observed* distribution is

$$f_n(x) = \frac{1}{n(x_k - x_{k-1})} \qquad k > 1 \tag{9.1}$$

$$F_n(x_k) = \frac{k}{n} \tag{9.2}$$

Figure 9.2 shows the result of applying Equation (9.2) to the data collected. This is known as the *empirical distribution function (EDF)*. It is a discontinuous function. It is the cumulative version of a histogram that has a bin for each value that occurs in the dataset. Conventionally we add the suffix n to make clear that f_n and F_n are discontinuous.

As starting values for the fit we calculate the mean (μ), using Equation (4.2), as 47.2 and the standard deviation (σ), using Equation (4.35), as 8.46. These are known as the *maximum likelihood estimates*. We will see later, if fitting another form of distribution, calculating these can more be complex. Indeed for many distributions it can be much easier to make an educated guess.

There are two basic approaches to fitting the prior distribution. It is not always possible to define a continuous equation, $F(x)$, for the CDF. Our first method assumes that it is. Column 4 of Table 9.1 is the result of applying Equation (5.43). If our assumption that the distribution is normal is correct, as is choice of μ and σ, then columns 3 and 4 would be identical. Most spreadsheet packages include a *solver* that can be set up to obtain the best fit, adjusting μ and σ to minimise the *residual sum of the squares (RSS)* of the prediction error. This is calculated in each row of column 5 as $(\text{predicted} - \text{actual})^2$ and summed, in this example, to 0.0871. By adjusting μ to 46.0 and σ to 8.63, *RSS* is reduced to the minimum of 0.0271. Figure 9.3 plots both the initial estimate (as the dashed line) and the best fit (as the solid line).

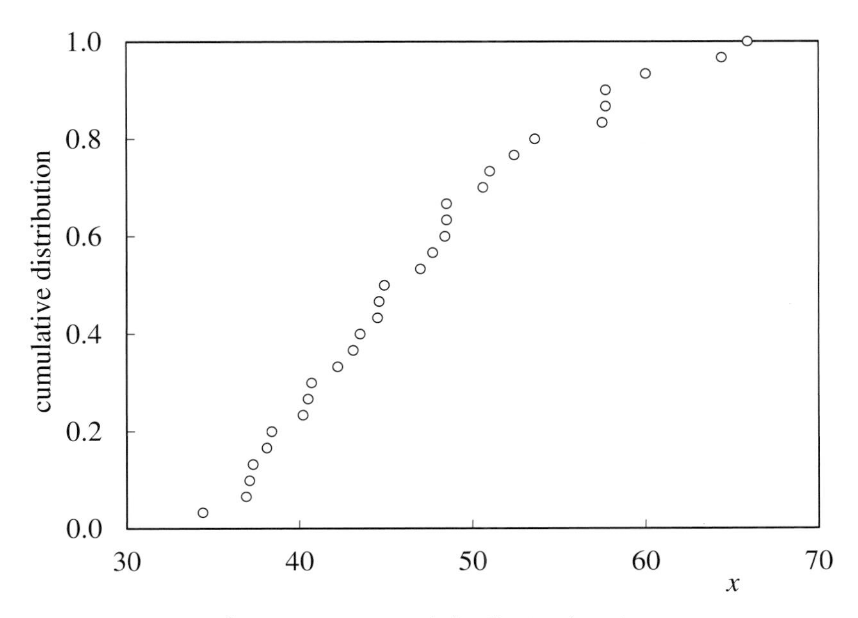

Figure 9.2 *Empirical distribution function*

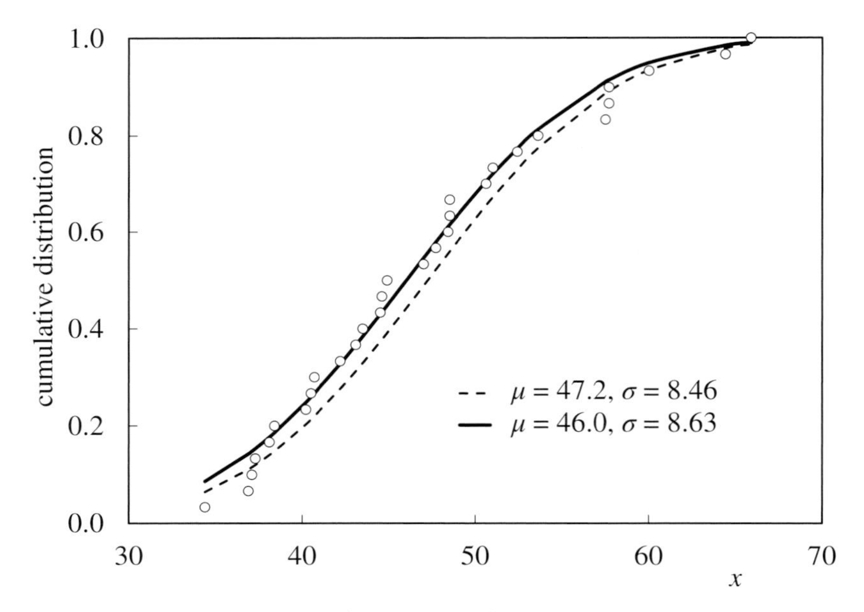

Figure 9.3 *Comparison between fitted and calculated distributions*

Those trying to reproduce this example as a learning exercise may find they obtain slightly different results. This can be caused by the precision of the software being used. Small round-off errors can accumulate to produce a value for *RSS* noticeably different and so affect the fitted values. The higher the moment, the more noticeable the change. Standard deviation will show it more than the mean. We will see later that skewness shows it more than standard deviation and kurtosis more than skewness.

While the difference between the calculated and fitted values of μ and σ might appear, in this case, to be relatively small, we can demonstrate that the fitted approach gives a much more robust result. If, for example, the first process measurement is reduced by 10 to 24.4 and the last increased by 10 to 75.9, the mean is unchanged but the calculated standard deviation increases to 9.84. The fitted value however remains virtually unchanged at 8.65. There is now no need to decide whether the first and last measurements are outliers.

Figure 9.4 is a P–P plot of the calculated and fitted distributions (columns 4 and 6 of Table 9.1) versus the observed distribution (column 3). It provides visual demonstration that the fitted distribution is noticeably more accurate than that using the calculated values of μ and σ.

9.3 Fitting a QF

Fitting a QF requires a slightly modified approach. Since we cannot determine F from x, to fit it to data, we use the function to predict x from each of the values of F in the EDF. In our example, the QF for the normal distribution cannot be expressed as a simple equation but most spreadsheet packages include the function. Column 8 in Table 9.1 is the result of using this.

We then adjust the coefficients to minimise the sum of the squares of the differences between the predicted x and the actual x (column 9).

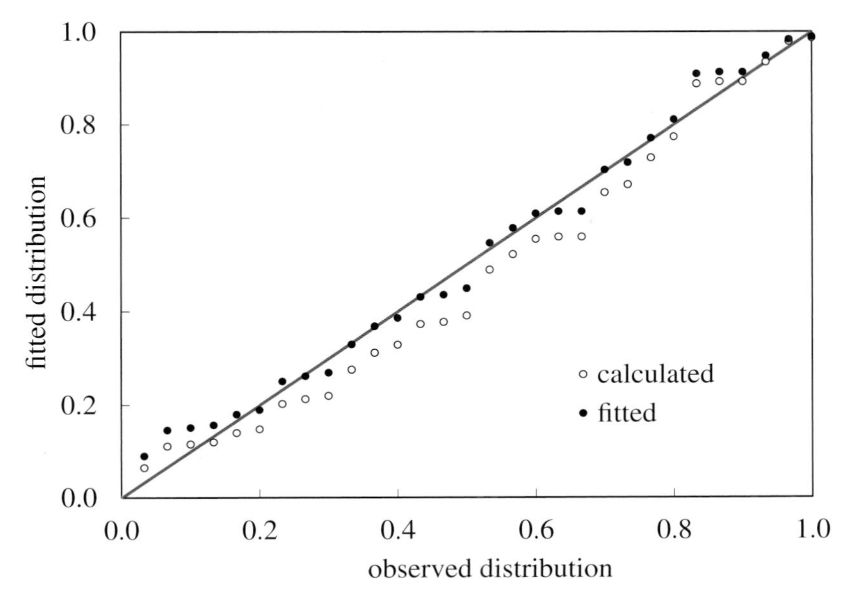

Figure 9.4 *P–P plots of fitted and calculated distributions*

Many distributions can be expressed as both a CDF and a QF, offering the engineer the choice of which to use. It is important to recognise that fitting them will not give the same result. For example, as Table 9.1 shows, both the mean and standard deviation from fitting the QF are slightly different from those obtained by fitting the CDF. In this example, the difference is small. In other, less robust functions, the effect can be much larger. Fitting a QF will generally result in a better fit to the tails of a distribution, sacrificing a little accuracy elsewhere. This would be noticeable in a Q-Q plot and would be the preferred approach if we are primarily interested in the more extreme process behaviour.

When comparing different distributions, the value of *RSS* is used to help select the one that most accurately represents the data. It is not possible to do this if one distribution has been fitted as a CDF and the other as a QF. *RSS* will have the units of probability in the first case and the units of the data in the second. Numerically, even if the two fits are identical, their *RSS* will be very different. Scaling the data so that it has the range 0 to 1 will bring the values closer but not entirely solve the problem. This issue is common to regression analysis and is addressed in detail in Chapter 16.

When comparing the accuracy of fits of contending CDF, the units of *RSS* will be the same – even if x has been scaled to fit within the bounds of the distribution. *RSS* will be in the units of $F(x)$, not x. However, when comparing the fit of different QFs, *RSS* will be in the units of x. If x has been scaled in one of the distributions, but not the other, it is important to transform the predicted values of x back to their original units before calculating *RSS*.

9.4 Fitting a PDF

Let us imagine that, for the normal distribution, $F(x)$ does not exist as a continuous function. Instead we have to use the equation of the PDF, $f(x)$, and apply the trapezium rule

$$F(x_i) = F(x_{i-1}) + \frac{f(x_i) + f(x_{i-1})}{2}(x_i - x_{i-1}) \tag{9.3}$$

Table 9.2 contains the same data as Table 9.1 but column 4 is now the result of applying Equation (5.35). Equation (9.3) is known as the *ascending cumulative distribution*. To apply it we have to choose a value for $F(x_1)$. If we are sure that future measurements will not include a value less than x_1 then we can assume $F(x_1)$ is 0. Alternatively, if we are sure that future measurements will not include a value greater than x_n then we can assume $F(x_n)$ is 1 and rearrange Equation (9.3) to give the *descending cumulative distribution*.

$$F(x_{i-1}) = F(x_i) - \frac{f(x_i) + f(x_{i-1})}{2}(x_i - x_{i-1}) \tag{9.4}$$

But we can choose any point at which to start. For example, we can assume that the median will not be changed by the addition of future measurements and assume that $F(x_{n/2})$ is 0.5. We would then apply Equation (9.3) to obtain values of i above $n/2$ and Equation (9.4) when below $n/2$.

Figure 9.5 shows the result of curve fitting from making each of these three assumptions. In each case the curve passes through the chosen point shown as shaded. The resulting estimates of μ and σ vary significantly. The problem, in this case, is that we want to fit an unbounded distribution, i.e. x can be anywhere between $-\infty$ and ∞. While the probability of additional measurements falling well away from the median is very small, the possibility does exist. A better approach is to make the chosen $F(x_1)$ one of the parameters that is adjusted to achieve the best fit to the data. Whichever approach is adopted, the limits below should be included so that fitting does not take $F(x)$ outside its feasible range.

$$F(x_{\min}) \geq 0 \quad \text{and} \quad F(x_{\max}) \leq 1 \tag{9.5}$$

$$\text{i.e.} \quad F(x_1) \geq 0 \quad \text{and} \quad F(x_n) \leq 1 \tag{9.6}$$

Column 5 has been calculated assuming that $F(x_1)$ is 0. Column 6 is the calculation of *RSS*. Fitting gives μ as 43.7 and σ as 7.27, with *RSS* as 0.0415. Columns 7 to 9 show the impact of minimising *RSS* to 0.0201 by adjusting $F(x_1)$ to 0.0562, μ to 44.7 and σ to 8.27. The resulting curve is shown in Figure 9.6. The P–P plot is shown in Figure 9.7.

In this example, fitting the PDF gives a result marginally better than fitting the CDF. When fitting the CDF we were able to adjust two parameters – the mean, which varies the horizontal position of the curve, and the standard deviation, which varies the shape. When fitting the PDF we can additionally adjust $F(x_1)$, which varies the vertical position of the curve. This extra degree of freedom permits a closer fit.

For example, fitting the CDF of the normal distribution to the C_4 in propane data gives μ as 4.47 and σ as 2.09. $F(x_1)$ is fixed by the CDF at 0.0776. The resulting *RSS* is 0.6138. Alternatively, fitting the PDF gives μ as 3.94, σ as 1.96 and $F(x_1)$ as 0.0221. The resulting *RSS* is reduced substantially to 0.2303.

However, inaccuracy can arise because the area under the PDF curve does not comprise exactly trapezia. This can be illustrated by fitting the CDF and PDF of the normal distribution to the NHV disturbance data. The CDF gives μ as –0.0206, σ as 1.19 and *RSS* as 0.1437. Fitting the PDF gives μ as –0.0107, σ as 1.19 and $F(x_1)$ at its minimum of 0. The resulting *RSS* is slightly higher at 0.1479.

If the process data is continuous then the accuracy of fitting the PDF would be improved by increasing the number of values analysed. This reduces the width of each trapezium. The number of straight-line segments that are used to approximate the distribution curve are increased in number and therefore shortened in length – more closely matching the curve.

			RSS minimised with $F(x_1) = 0$			*RSS* minimised with $F(x_1)$ fitted		
x	ranking	observed (EDF)	$f(x)$ $\mu = 43.7$ $\sigma = 7.27$	$F(x)$ by trapezium rule	*RSS*	$f(x)$ $\mu = 44.7$ $\sigma = 8.27$	$F(x)$ by trapezium rule	*RSS*
34.4	1	0.0333	0.0243	0.0000	0.0011	0.0222	0.0562	0.0005
36.9	2	0.0667	0.0356	0.0749	0.0001	0.0309	0.1227	0.0031
37.1	3	0.1000	0.0365	0.0821	0.0003	0.0316	0.1290	0.0008
37.3	4	0.1333	0.0374	0.0895	0.0019	0.0323	0.1354	0.0000
38.1	5	0.1667	0.0409	0.1208	0.0021	0.0351	0.1624	0.0000
38.4	6	0.2000	0.0422	0.1332	0.0045	0.0361	0.1730	0.0007
40.2	7	0.2333	0.0490	0.2153	0.0003	0.0416	0.2430	0.0001
40.5	8	0.2667	0.0499	0.2301	0.0013	0.0424	0.2556	0.0001
40.7	9	0.3000	0.0505	0.2401	0.0036	0.0429	0.2641	0.0013
42.2	10	0.3333	0.0538	0.3183	0.0002	0.0461	0.3309	0.0000
43.1	11	0.3667	0.0547	0.3671	0.0000	0.0474	0.3730	0.0000
43.5	12	0.4000	0.0549	0.3890	0.0001	0.0477	0.3920	0.0001
44.5	13	0.4333	0.0545	0.4437	0.0001	0.0482	0.4400	0.0000
44.6	14	0.4667	0.0544	0.4491	0.0003	0.0482	0.4448	0.0005
44.9	15	0.5000	0.0541	0.4654	0.0012	0.0482	0.4593	0.0017
47.0	16	0.5333	0.0494	0.5741	0.0017	0.0464	0.5586	0.0006
47.7	17	0.5667	0.0471	0.6079	0.0017	0.0452	0.5906	0.0006
48.4	18	0.6000	0.0444	0.6399	0.0016	0.0436	0.6217	0.0005
48.5	19	0.6333	0.0440	0.6443	0.0001	0.0434	0.6261	0.0001
48.5	20	0.6667	0.0440	0.6443	0.0005	0.0434	0.6261	0.0016
50.6	21	0.7000	0.0348	0.7271	0.0007	0.0374	0.7109	0.0001
51.0	22	0.7333	0.0330	0.7407	0.0001	0.0361	0.7256	0.0001
52.4	23	0.7667	0.0267	0.7825	0.0002	0.0312	0.7727	0.0000
53.6	24	0.8000	0.0216	0.8114	0.0001	0.0270	0.8076	0.0001
57.5	25	0.8333	0.0090	0.8711	0.0014	0.0145	0.8887	0.0031
57.7	26	0.8667	0.0085	0.8728	0.0000	0.0140	0.8915	0.0006
57.7	27	0.9000	0.0085	0.8728	0.0007	0.0140	0.8915	0.0001
60.0	28	0.9333	0.0044	0.8877	0.0021	0.0087	0.9176	0.0002
64.4	29	0.9667	0.0009	0.8995	0.0045	0.0028	0.9430	0.0006
65.9	30	1.0000	0.0005	0.9006	0.0099	0.0018	0.9464	0.0029
total					0.0415			0.0201

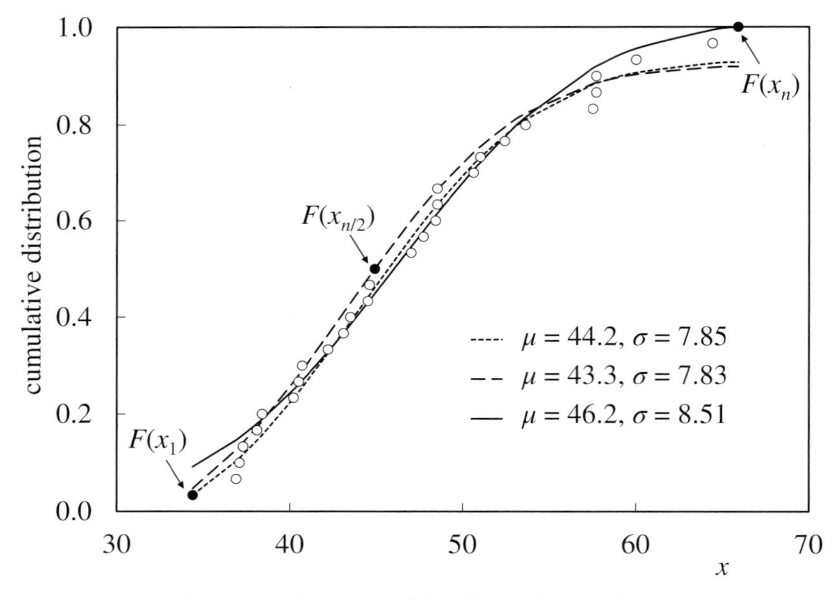

Figure 9.5 *Impact of fixing* $F(x_1)$, $F(x_n)$ *or* $F(x_{n/2})$

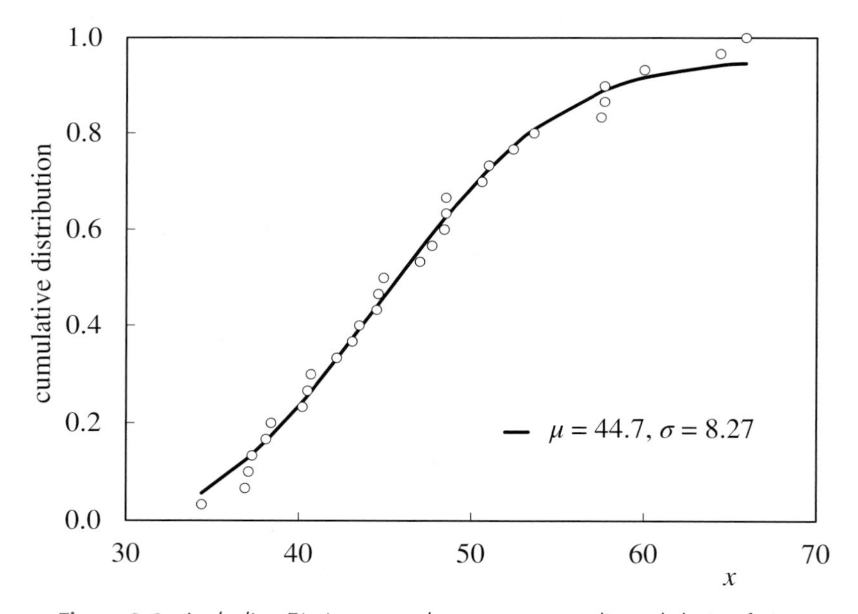

Figure 9.6 *Including* $F(x_1)$ *among the parameters adjusted during fitting*

9.5 Fitting to a Histogram

Figure 9.8 shows the same data presented as histogram. It is possible to fit a distribution from the information presented in the drawing. Table 9.3 demonstrates this. Column 2 is the mid-point of the ranges plotted in the histogram. Column 3 is the frequency. We can calculate the maximum likelihood estimate of the mean by applying Equation (4.3). The weighting (w) is the

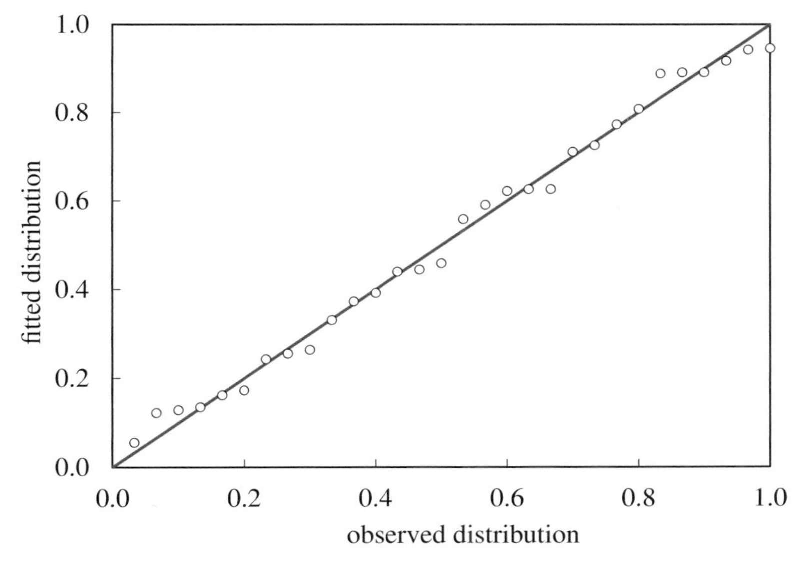

Figure 9.7 *P–P plot from fitting the PDF*

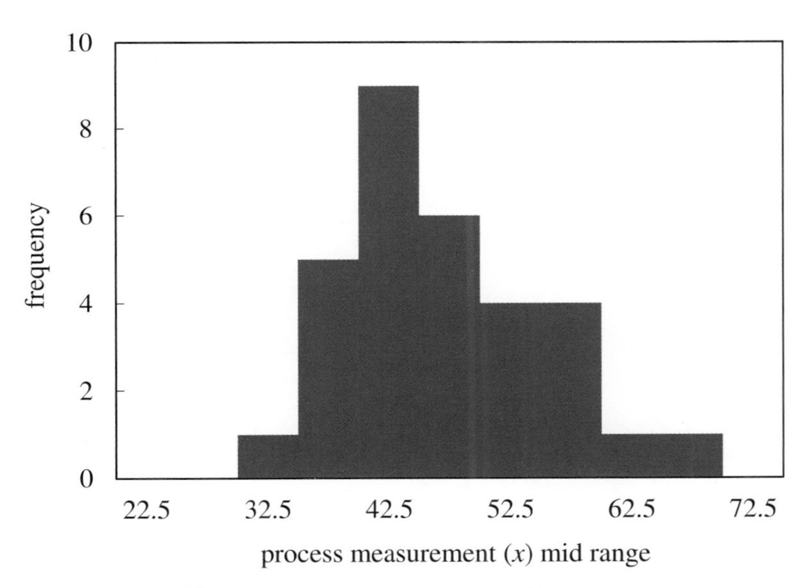

Figure 9.8 *Data presented as a histogram*

number of measurements falling within the range. Multiplying column 2 by column 3, dividing by the total number of measurements and summing gives the weighted mean of 47.0. We can similarly calculate the maximum likelihood estimate for the variance using the equation

$$\sigma^2 = \frac{\sum_{i=1}^{n} w_i (x_i - \bar{x})^2}{\sum_{i=1}^{n} w_i} \tag{9.7}$$

Table 9.3 *Fitting a normal distribution to a histogram*

ranges	mid-range	frequency	mean	variance	probability	EDF	CDF $\mu = 46.1$ $\sigma = 8.60$	*RSS*	PDF $\mu = 44.1$ $\sigma = 7.69$	*RSS*	CDF derived from PDF
25	22.5	0			0.0000	0.0000	0.0071	0.0000			0.0000
30	27.5	0			0.0000	0.0000	0.0306	0.0000	0.0256	0.0000	0.0128
35	32.5	1	1.05	6.80	0.0323	0.0323	0.0984	0.0044	0.0839	0.0027	0.0675
40	37.5	5	6.05	14.61	0.1613	0.1935	0.2391	0.0104	0.1804	0.0018	0.1997
45	42.5	9	12.34	5.92	0.2903	0.4839	0.4493	0.0108	0.2541	0.0118	0.4169
50	47.5	6	9.19	0.05	0.1935	0.6774	0.6751	0.0000	0.2346	0.0101	0.6613
55	52.5	4	6.77	3.88	0.1290	0.8065	0.8498	0.0075	0.1419	0.0007	0.8495
60	57.5	4	7.42	14.18	0.1290	0.9355	0.9471	0.0005	0.0563	0.0212	0.9486
65	62.5	1	2.02	7.73	0.0323	0.9677	0.9861	0.0003	0.0146	0.0003	0.9841
70	67.5	1	2.18	13.54	0.0323	1.0000	0.9973	0.0000	0.0025	0.0009	0.9926
75	72.5	0			0.0000	1.0000	0.9996	0.0000	0.0003	0.0000	0.9940
total		31	47.0	66.70	1.0000			0.0340		0.0494	

Column 5 gives a value for the standard deviation of 8.17. Column 3 is converted to the probability distribution, in column 6, by dividing by the total number of measurements. Column 7 shows the cumulative distribution derived from summing the values in column 6. Column 8 is determined using Equation (5.43). The values in column 9 are now the square of the prediction error, between columns 7 and 8, multiplied by the number of measurements (column 3) falling in the band. Without this weighting a very large number of measurements in one band would only influence the curve fitting by the same amount as a single measurement in another. This would make the accuracy of fitting very vulnerable to outliers.

Although the starting value of *RSS* is not included in the table, by adjusting μ to 46.1 and σ to 8.60, it has been minimised from 0.0724 to 0.0340. This result is similar to that obtained previously, when fitting the CDF to the EDF. It is however an approximation since it is based on the assumption that measurements within each band are uniformly distributed. We know that this is not the case. Reducing the size of the bands will give a poorly presented histogram but will reduce the impact of this assumption. There is no requirement that the bands be of equal size. The most accurate fit is achieved by reducing the size of each band (theoretically to zero) so that each contains only elements of the dataset that have exactly the same value. This is exactly the procedure described by Table 9.1.

If a CDF cannot be fitted directly then, as in the previous section, we can fit a PDF. The result of doing so is also presented in Table 9.3. Column 10 is derived by applying Equation (5.35), remembering that this gives probability density. To convert to probability we must multiply by the width of the interval. In this example all the intervals are 5. Column 11 is the weighted square of the difference between column 6 and 10. Although the starting value of *RSS* is not included in the table, by adjusting μ to 47.0 and σ to 8.17, it has been minimised from 0.1009 to 0.0494. Figure 9.9 shows how well the derived distribution matches the histogram. By its nature, comparing the curve of a continuous distribution to a histogram will appear not to show a good match – even if only the midpoints of each column are considered. However, we can derive the CDF in column 12 from the cumulative sum of the fitted PDF in column 10. Figure 9.10 plots this against the observed EDF, showing that the fit is good.

While minimising *RSS* tells us that we have the best fit that the method can deliver, we should be cautious using it to compare methods. As we might expect, fitting to a histogram results in a higher *RSS* than fitting to every point. However *RSS* depends not only on the deviation between fitted and observed distributions but also on the number of points used in its calculation. We might therefore consider normalising *RSS* by dividing by n. This leads us on to the next topic.

9.6 Choice of Penalty Function

So far, in identifying the best fit (of a CDF), we have chosen to minimise *RSS* where

$$RSS = \sum_{i=1}^{n} [F(x_i) - F_n(x_i)]^2 \tag{9.8}$$

This *least squares* approach is common to all forms of curve fitting. It is used, for example, in developing an inferential property from historical data using regression analysis to identify the coefficients in the chosen function. We are not normally concerned with actual value of *RSS*;

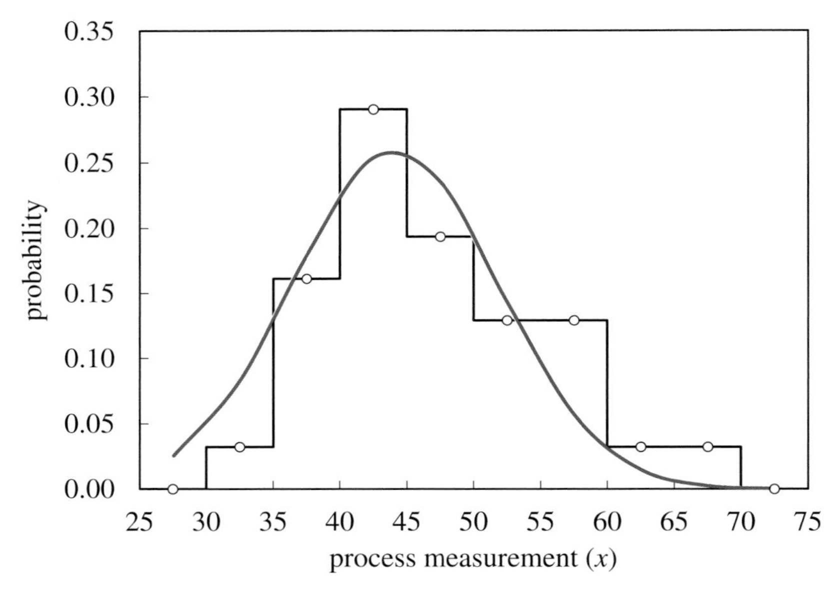

Figure 9.9 Comparing derived probability distribution to histogram

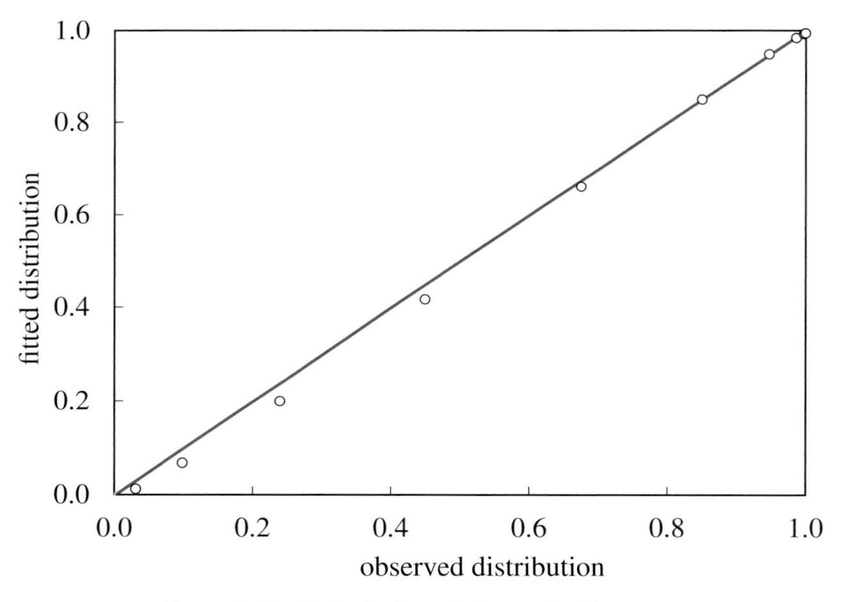

Figure 9.10 P–P plot from fitting to the histogram

we only require that it has been minimised. The value can however be used more systematically when fitting a statistical distribution function. It can help us decide whether the fit is good enough. Specifically the *Cramér–von Mises test* involves the use of published tables to determine whether *RSS* is below the value we would expect at a chosen confidence level and the number (n) of data points used. If *RSS* is greater than this value then we would reject the hypothesis that the data follow the prior distribution.

There are many other similar tests. For example, the *Anderson–Darling test* is based on a weighted *RSS*.

$$RSS_w = \sum_{i=1}^{n} w[F(x_i) - F_n(x_i)]^2 \tag{9.9}$$

The weight (w) is defined as

$$w = \frac{1}{F(x)[1-F(x)]} \tag{9.10}$$

Figure 9.11 shows how w varies. Its principle is to place a higher weighting on values in the tails of the distribution to compensate for the fact that there are far fewer of them. There are tables published to cover the most commonly considered distributions that give the maximum expected value for RSS_w.

Another commonly used test is that described as *Kolmogorov–Smirnov*. This was originally designed to enable two continuous functions to be compared to determine whether any difference might be accounted for by chance. Quite simply it defines the parameter D_n as the largest mismatch between the two curves. A modification to the test was later developed to permit comparison between a continuous function and a discontinuous one.

As an example, Table 9.4 shows the method involved. The first column contains the data, ranked in increasing value. The second is simply a counter. Column 3 is calculated according to

$$F_n(x_{k-1}) = \frac{k-1}{n} \tag{9.11}$$

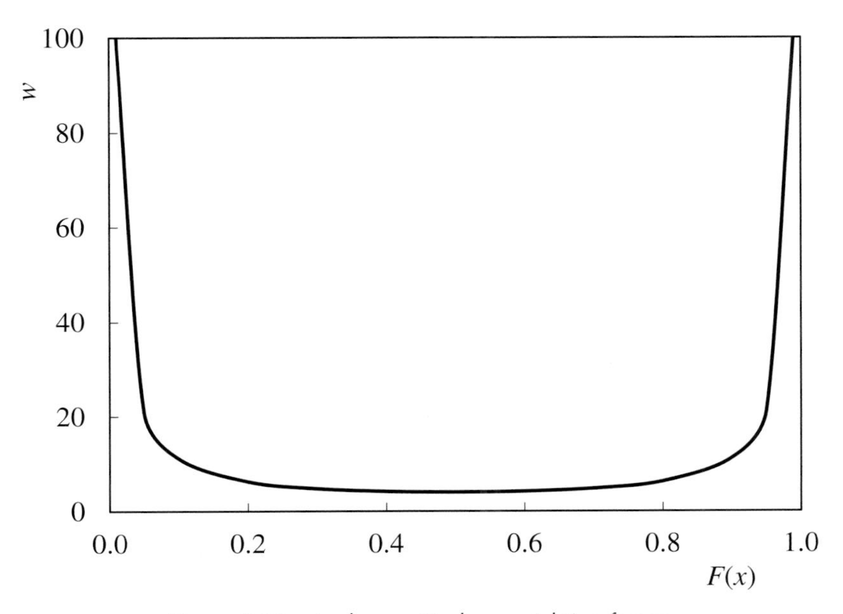

Figure 9.11 *Anderson–Darling weighting factor*

Table 9.4 Kolmogorov–Smirnov test

x_k	k	$F_n(x_{k-1})$	$F(x)$	$F_n(x_k)$	$\lvert F_n(x_{k-1}) - F(x)\rvert$	$\lvert F(x) - F_n(x_k)\rvert$
34.4	1	0.0	0.0869	0.1	0.0869	0.0131
37.3	2	0.1	0.1423	0.2	0.0423	0.0577
40.2	3	0.2	0.2177	0.3	0.0177	0.0823
42.2	4	0.3	0.2810	0.4	0.0190	0.1190
44.5	5	0.4	0.3632	0.5	0.0368	0.1368
47.0	6	0.5	0.4602	0.6	0.0398	0.1398
48.5	7	0.6	0.5199	0.7	0.0801	**0.1801**
57.7	8	0.7	0.8340	0.8	0.1340	0.0340
57.7	9	0.8	0.8340	0.9	0.0340	0.0660
65.9	10	0.9	0.9633	1.0	0.0633	0.0367

Column 4 is calculated from the prior distribution. In this example the distribution is assumed to be normal, using Equation (5.43) with a mean (μ) of 48 and a standard deviation (σ) of 10. Column 5 is the EDF calculated from Equation (9.2). Column 6 is the absolute value of the difference between columns 3 and 4. Column 7 is the absolute value of the difference between columns 4 and 5. We choose the largest value that appears in columns 6 and 7, in this case 0.1801. This is the maximum distance between the prior distribution (coloured line) and the EDF (black line) shown by Figure 9.12. Again there is a published table that is used to determine whether D_n is below the maximum expected.

Strictly the Kolmogorov–Smirnov test should not be used to identify whether a distribution that has been fitted to data has been selected correctly. The assumption behind the test is that the distributions being compared are independent. More properly it should be used, for example, if the engineer wishes to check that data collected more recently fits a distribution that was

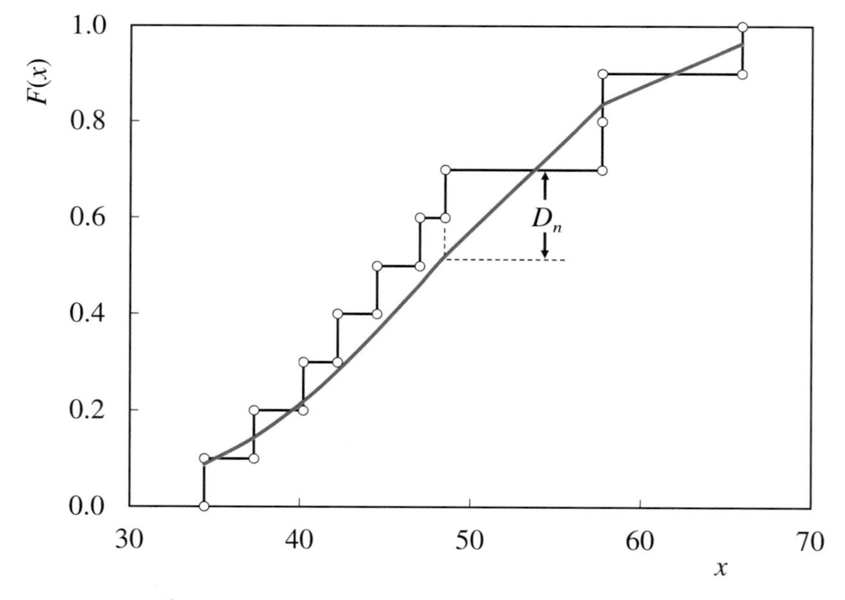

Figure 9.12 Kolmogorov–Smirnov penalty function

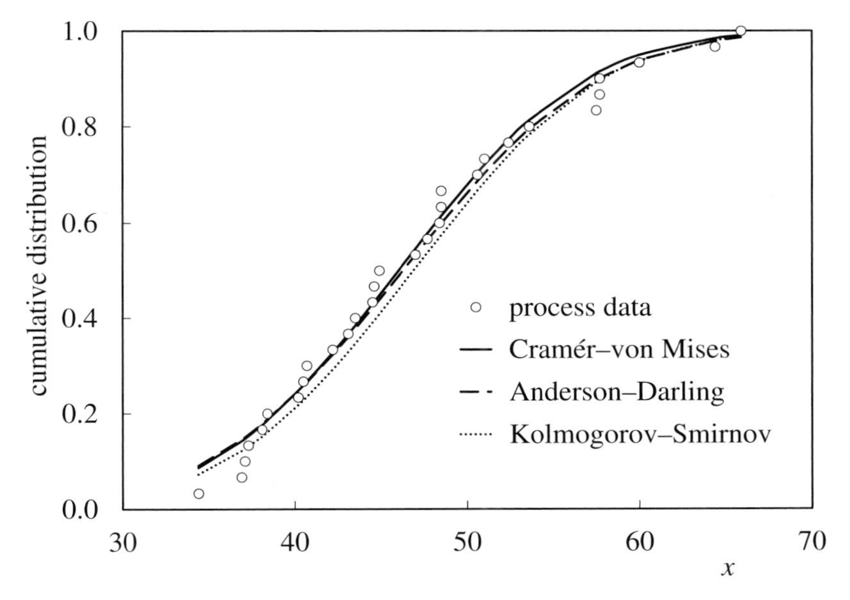

Figure 9.13 *Impact of choice of penalty function*

developed from older data. Similarly it would permit comparison between two distributions developed from different datasets to determine if they are significantly different.

While these techniques might be applicable to other sciences, in practice control engineers typically rely on more visual techniques, such as Q–Q plots, to identify whether a distribution has been well chosen. However there is another use for the penalty functions like RSS_w and D_n. We are free to use these parameters, instead of RSS, as part of curve fitting. Using the same data as in the previous section, Figure 9.13 shows how the fitted normal distribution changes depending on the choice of penalty function. For example, the estimate of the mean is 46.0 when RSS is used. Using RSS_w gives 46.3 and using D_n gives 46.9. Similarly the estimates of standard deviation are 8.53, 8.88 and 8.57 respectively. Again there is no absolute measure of which is the best choice; as ever selection is based on engineering judgement. For example, if extreme values are not well modelled by the chosen distribution, the use of RSS_w might help improve the fit. Similarly, if the largest mismatch is in the region of the distribution being used for further study, the use of D_n may resolve this.

10

Distribution of Dependent Variables

Many distributions result from manipulating data that follow another distribution. For example, if we collect data from a normal distribution, square each value and sum together a defined number of the squares, the results will follow a chi-squared distribution. The control engineer might rightly be unconcerned by such a result. It is unlikely that process data would be manipulated in such a way. However, it highlights an important principle. While the raw data collected from a process might be normally distributed, any property derived from the data is likely not to be so.

For example, an inferential property is derived from independent variables. The form of calculation can influence the form of distribution. But most of the process variables we analyse are derived from others. A product composition, for example, depends on a number of independently set process parameters such as flows, temperatures and pressures. A process simulation would contain a complex calculation that derives composition from basic measurements. We can think of the process as this calculation.

In fact, by definition, a variable we wish to control must be dependent on others. So, while the independent variables might reasonably be expected to be normally distributed, the nonlinearity inherent to the process can significantly alter the form of distribution of the dependent variable.

10.1 Addition and Subtraction

We have seen that the Central Limit Theorem tells us that deriving a value by summing two others will result in a distribution closer to normal, no matter what the distribution of the base variables. For example, total daily production might be determined by adding a large number of flow measurements recorded during the day. It would be reasonable to assume that variation in production would follow a normal distribution.

The same would apply to values derived by subtraction. For example, we might calculate losses by subtracting total product from total feed. Again we would reasonably expect the variation in losses to follow a normal distribution.

Statistics for Process Control Engineers: A Practical Approach, First Edition. Myke King.

10.2 Division and Multiplication

The product of two variables can be determined by first summing their logarithms. The Central Limit Theorem would therefore tell us that the logarithms of the products will more closely follow the normal distribution. For example, we might derive the duty of a heat exchanger by multiplying the flow through it by the change in temperature. The same will apply to division, since this can be achieved by subtracting logarithms. We might determine the reflux ratio on a distillation column by dividing reflux flow by the flow of distillate. Such derived measurements are commonly analysed to support benefit estimation or design of inferential properties. In order to fit a normal distribution to the results, it is likely that the fit would be improved by taking the logarithm of each result. However, as we will see later, there is a modified form of the normal distribution, known as the lognormal distribution, which effectively does the same.

Figure 10.1 shows the distribution of a variable derived by dividing values (x_1 and x_2) both selected from a uniform distribution between 0 and 1. The distribution is clearly far from normal. However, real process data are unlikely to be uniformly distributed. For example, Figure 10.2 shows the distribution of reflux and distillate flows on the LPG splitter that operates with a reflux ratio close to 1. Figure 10.3 shows a trend over time of the calculated reflux ratio. It displays some of the behaviour expected in that the distribution is clearly not symmetrical. Positive deviations from the mean are significantly greater than negative deviations. Indeed, it shows a skewness of 0.67 – greater than the limit of 0.5 normally applied in assuming a distribution is normal. Its kurtosis of 3.95 is similarly outside the limit of 3.5. Figure 10.4 shows the distribution of the reflux ratio. The solid line is the result of fitting a normal distribution giving a mean of 1.09 and standard deviation of 0.21. The dashed line is the result of fitting a lognormal distribution giving a mean of 1.10 and standard deviation of 0.22. While visually the lognormal is a much better fit, its choice in this case has relatively little impact on the statistics.

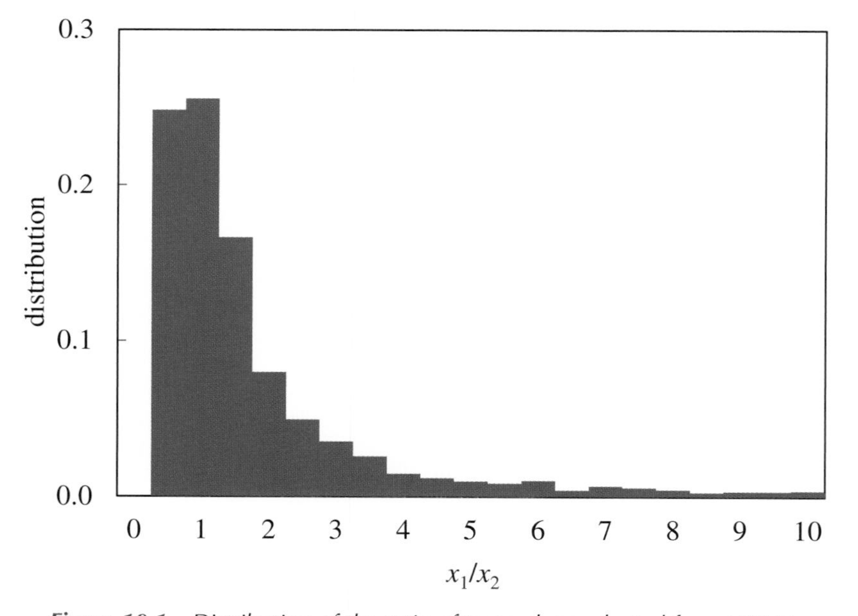

Figure 10.1 Distribution of the ratio of two values selected from U(0,1)

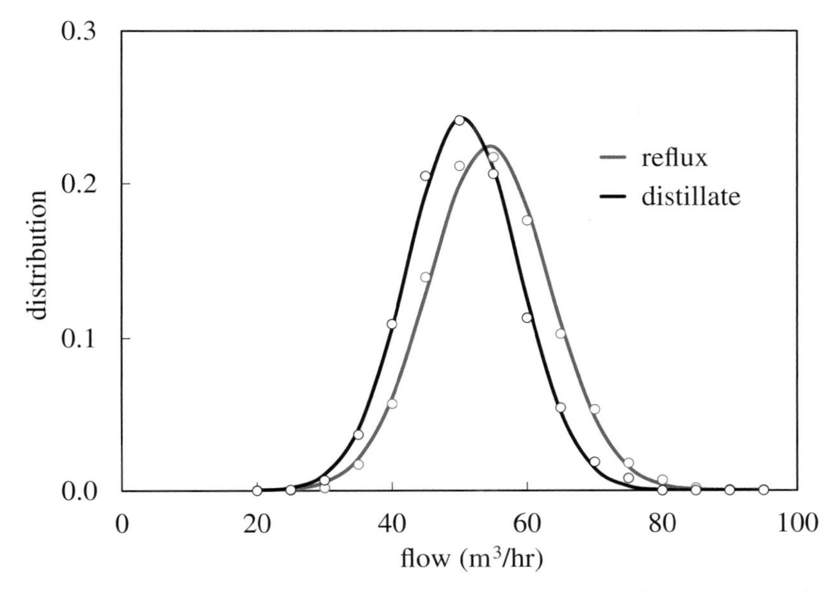

Figure 10.2 Normal distribution fitted to reflux and distillate flows on LPG splitter

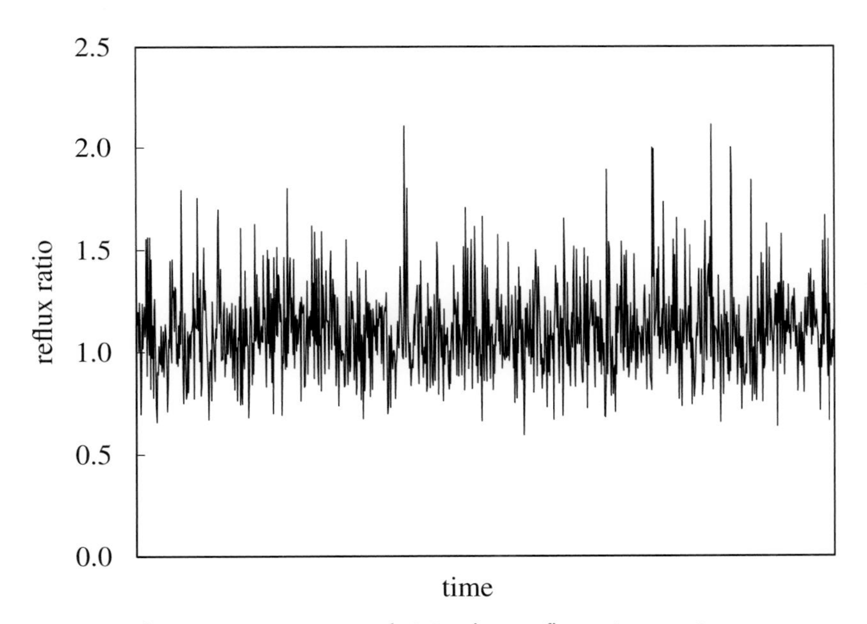

Figure 10.3 Variation of LPG splitter reflux ratio over time

Figure 10.5 shows the distribution of reflux and distillate flows on the debutaniser – this time operating with a reflux ratio of around 0.3. Figure 10.6 shows the distribution of the reflux ratio. Its skewness is 0.20 and kurtosis 3.02 – both well within the limits of the assumption that the distribution is normal. Visually the fit is almost perfect. Although the lognormal distribution

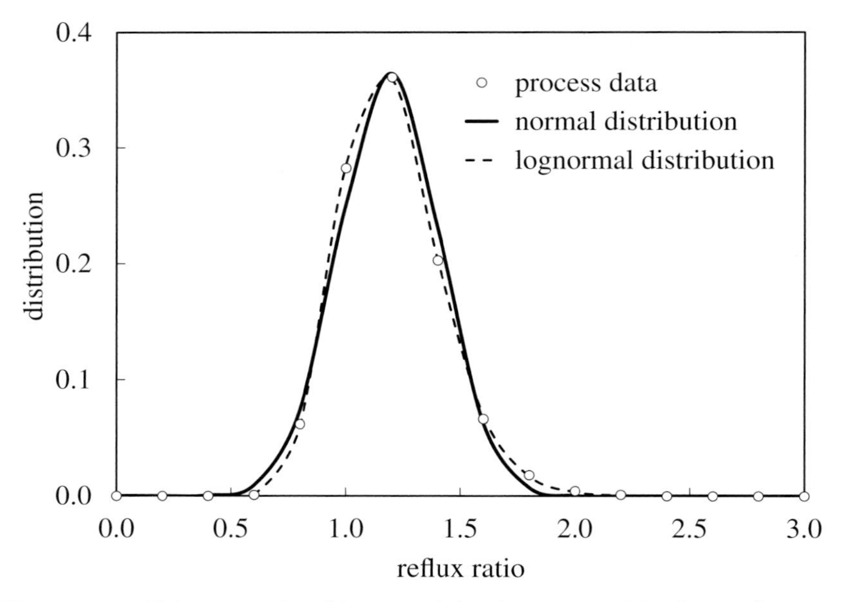

Figure 10.4 *Fitting normal and lognormal distributions to LPG splitter reflux ratio*

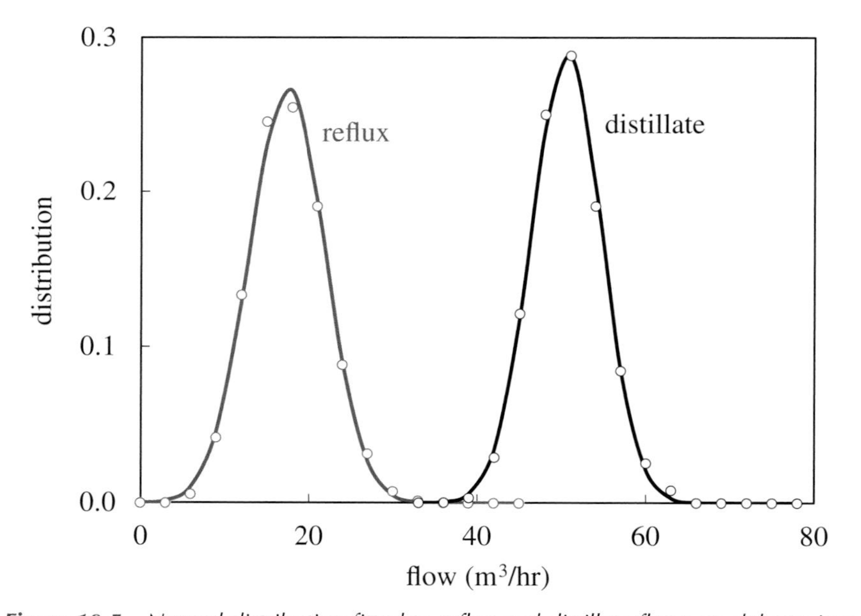

Figure 10.5 *Normal distribution fitted to reflux and distillate flows on debutaniser*

would be a marginally better choice, there would be no reason to choose another type of distribution. In general, as we have seen here, the worst case is when the two measurements used to calculate a ratio are numerically virtually equal. But, even under these circumstances, the assumption that the distribution of the ratio is normal would introduce relatively little error in the statistics.

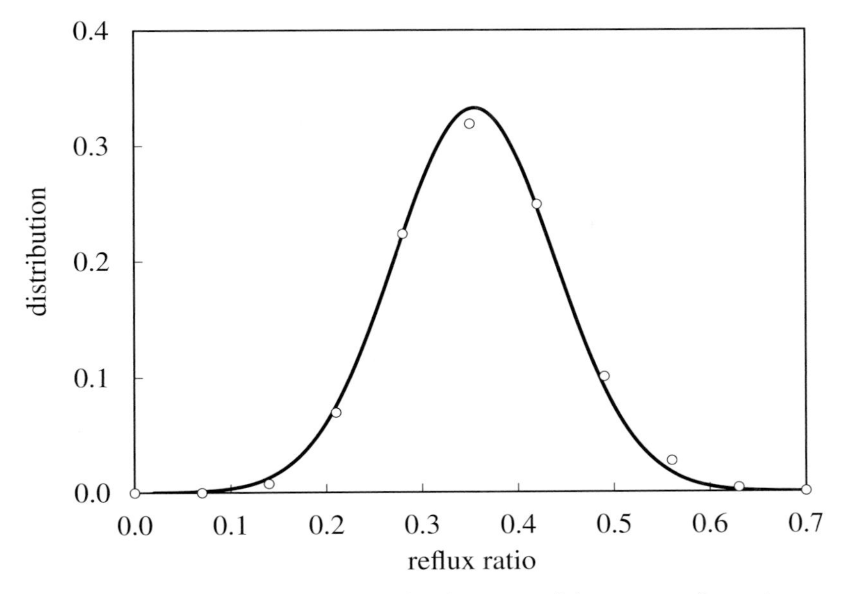

Figure 10.6 Fitting the normal distribution to debutaniser reflux ratio

Figure 10.7 shows the distribution resulting from multiplying x_1 by x_2 – again both selected from a uniform distribution between 0 and 1. Figures 10.8 and 10.9 show the distribution of the two process measurements used to calculate heat exchanger duty – the flow and the temperature difference between inlet and outlet. Figure 10.10 shows the distribution of the product of these two measurements. It has a skewness of 0.42 and a kurtosis of 3.13. It should therefore be

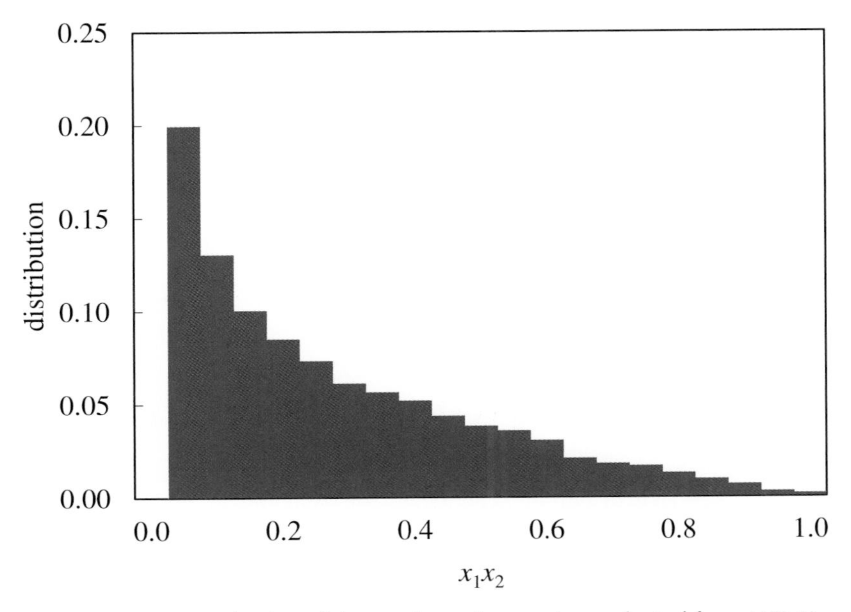

Figure 10.7 Distribution of the product of two values selected from U(0,1)

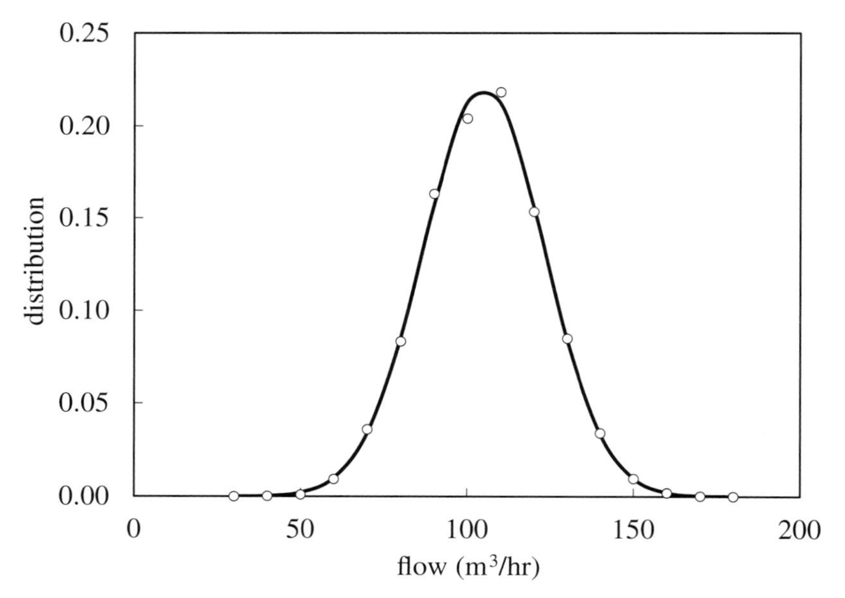

Figure 10.8 Normal distribution fitted to heat exchanger flow

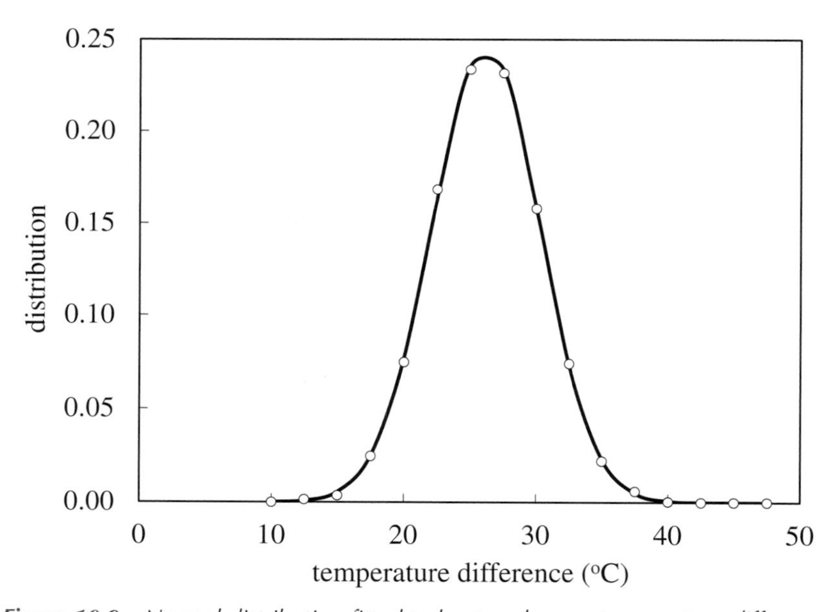

Figure 10.9 Normal distribution fitted to heat exchanger temperature difference

reasonable to fit a normal distribution, as shown by the solid line, giving a mean of 2476 and a standard deviation of 608. Fitting a lognormal distribution (the dashed line) gives the mean as 2513 and the standard deviation as 627. As with division, in this case, multiplication of normally distributed variables gives a result that closely follows a normal distribution but which is noticeably better represented by a lognormal distribution.

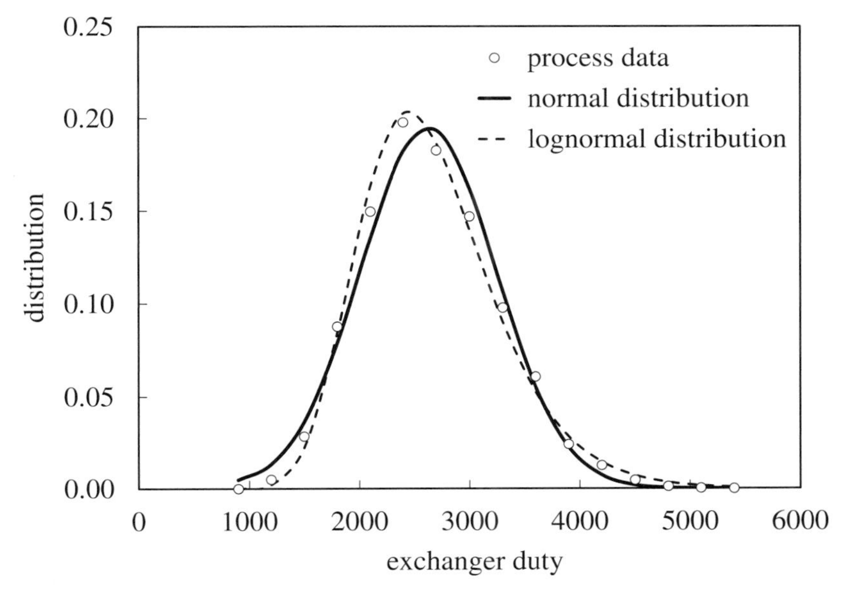

Figure 10.10 *Fitting normal and lognormal distributions to exchanger duty*

10.3 Reciprocal

While it may not be immediately obvious, control engineers will often fit a distribution to the reciprocal of a measurement. For example, the time taken to produce a batch of gasoline thorough a recipe blender is inversely proportional to the total flow rate of the components. If we were to study the statistics of batch duration, we are effectively fitting a distribution to the reciprocal of the flow. If, for example, records show that the flow is normally distributed then the batch duration will not be. Chemical engineers are concerned with residence time in a reactor, since this can affect conversion. Residence time is the reciprocal of the reactor feed rate, multiplied by reactor volume. The same applies to events. Events can be quantified either as the number that occur in a defined time or as the time between events. One is the reciprocal of the other.

The propane from the LPG splitter is routed to a sphere. The time taken to fill the sphere will be inversely proportional to the product flow rate. Figure 10.11 shows the distribution of the reciprocal of this flow. We know that the flow is normally distributed, but its reciprocal is clearly not. Skewness for the distillate flow is 0.22; for its reciprocal it is 0.89. Similarly kurtosis changes from 3.0 to 4.7. A reciprocal can be determined by subtracting the logarithm of the measurement from zero (the logarithm of 1). So, again, the lognormal distribution would provide a considerably improved fit to the data.

10.4 Logarithmic and Exponential Functions

More complex calculations may or may not significantly distort the distribution away from normal. Figure 10.12 shows the distribution of each of the temperatures around an exchanger that uses hot oil to heat up a product. Key to exchanger design and monitoring is the *log mean temperature difference*.

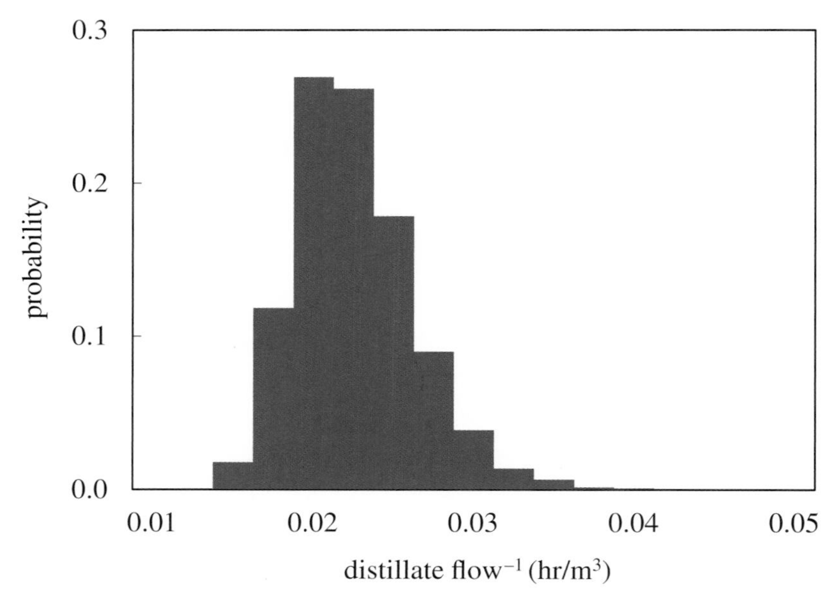

Figure 10.11 *Distribution of reciprocal of propane flow from LPG splitter*

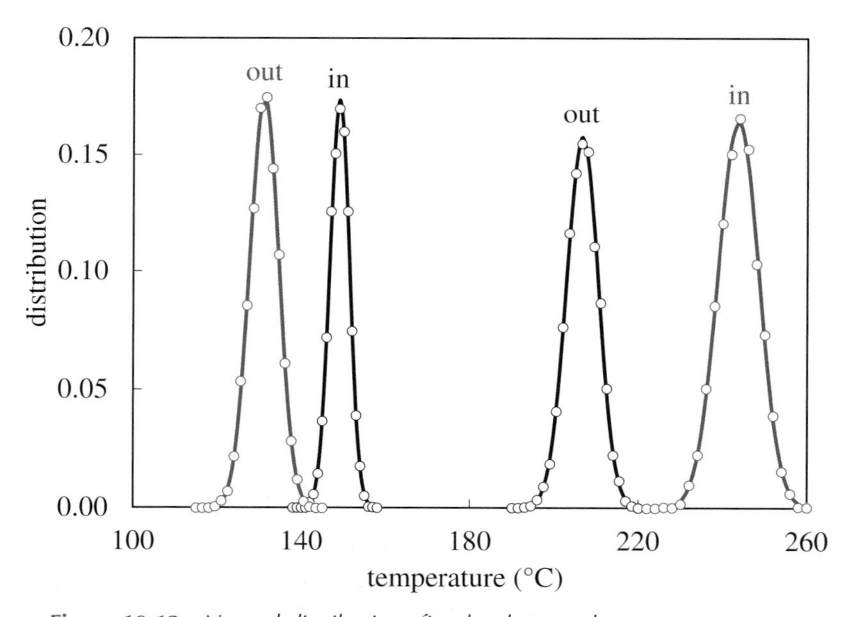

Figure 10.12 *Normal distributions fitted to heat exchanger temperatures*

$$LMTD = \frac{\Delta T_{in} - \Delta T_{out}}{\ln\left(\Delta T_{in} / \Delta T_{out}\right)} \tag{10.1}$$

Figure 10.13 shows the distribution of *LMTD*. In this case assuming the distribution is normal (solid line) gives a mean of 26.29 and a standard deviation of 3.58. The lognormal (dashed line), with a mean of 26.41 and a standard deviation of 3.30, is a slightly poorer fit. This is because the slight negative skewness of the LMTD results cannot be represented by the lognormal distribution. It is almost certain that a better choice of distribution could be found.

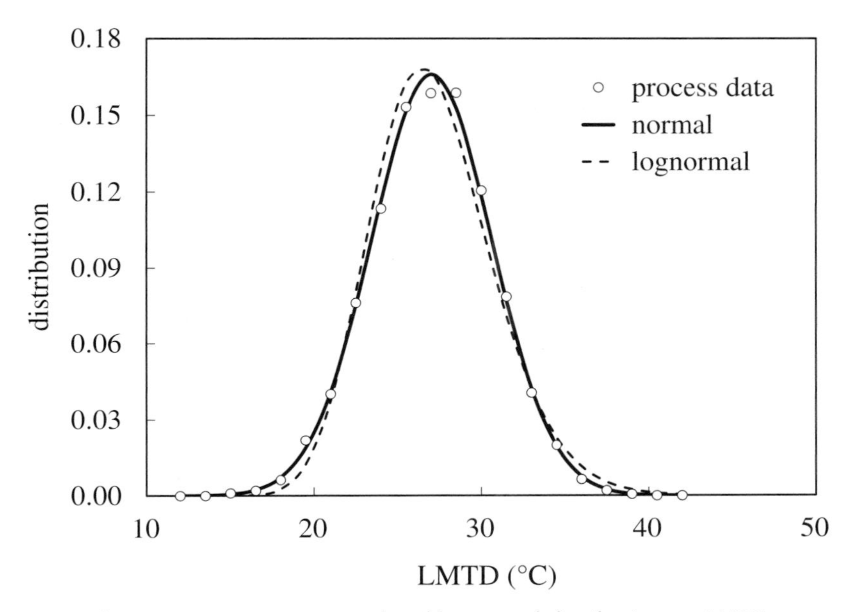

Figure 10.13 *Fitting normal and lognormal distributions to LMTD*

As another example, Figure 10.14 shows the variation in pressure in the LPG splitter. Distillation is governed by the *Antoine Equation* which relates absolute vapour pressure (P bara) to temperature (T°C).

$$\ln(P) = A - \frac{B}{T + C} \tag{10.2}$$

Among other applications it can be used to determine the boiling point of a liquid at a given pressure (P barg) and hence the tray temperature (T°C) required to meet the product purity target. Rearranging Equation (10.2) and using the coefficients for butane.

$$T = \frac{2154.897}{9.05800 - \ln(P + 1.01325)} - 238.730 \tag{10.3}$$

Figure 10.15 shows the distribution of the calculated temperature. The solid line is the fitted normal distribution – showing an almost perfect match. This is explained by Figure 10.16 which shows that, despite the reciprocal logarithmic function in Equation (10.3), the relationship between temperature and pressure is close to linear over the operating range. In fact this is the key to the linear PCT (pressure-compensated temperature) often used in composition control schemes.

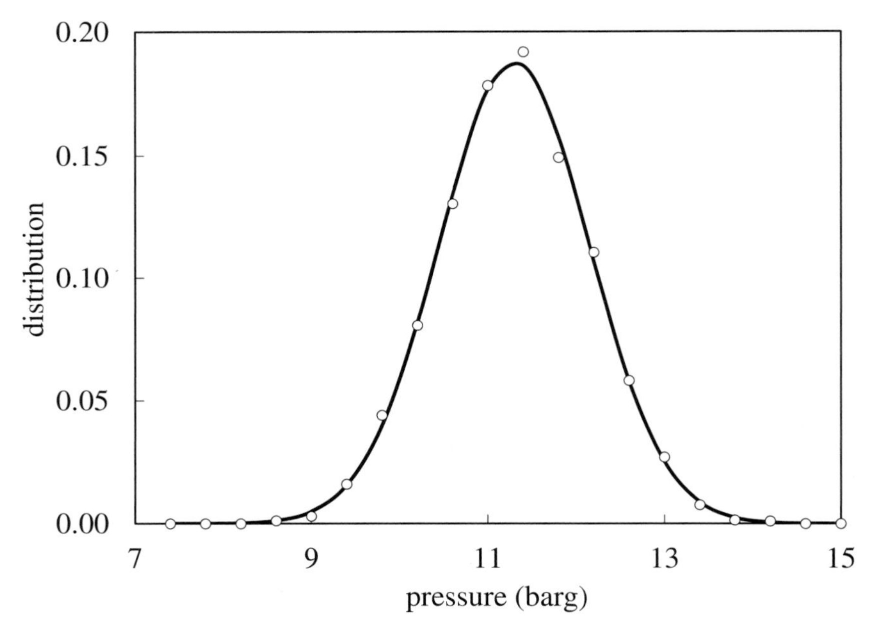

Figure 10.14 *Normal distribution fitted to LPG splitter pressure*

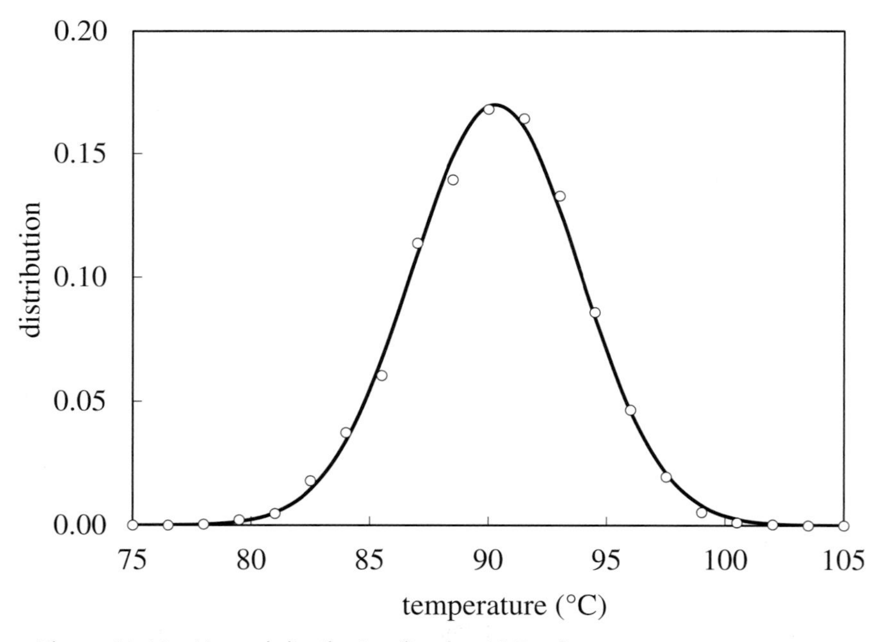

Figure 10.15 *Normal distribution fitted to LPG splitter tray temperature target*

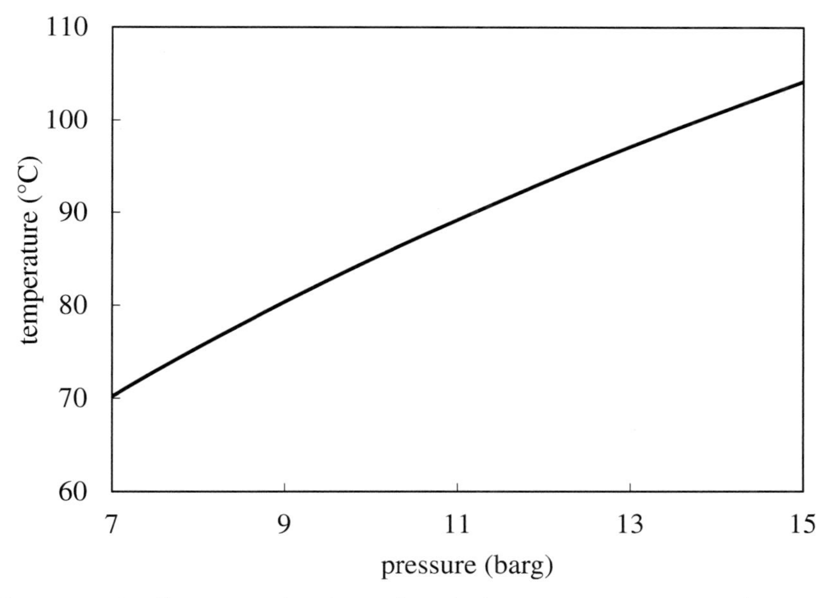

Figure 10.16 Close approximation to linearity between temperature and pressure

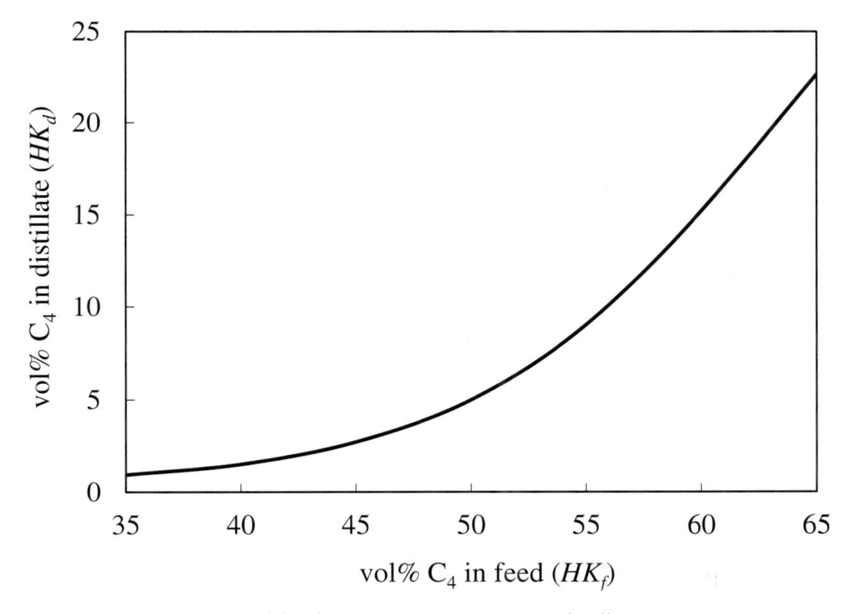

Figure 10.17 Impact of feed composition (HK_f) on distillate composition (HK_d)

However, distillation is a highly non-linear process and can distort the distribution of disturbances. Figure 10.17 shows the relationship, developed from a simulation of the LPG splitter, between the C_4 content of the overhead propane product (HK_d) and that of the feed (HK_f). Figure 10.18 shows that this relationship is close to logarithmic. The dashed line is the plot of

$$\log(HK_d) = 0.0475HK_f + 1.69 \tag{10.4}$$

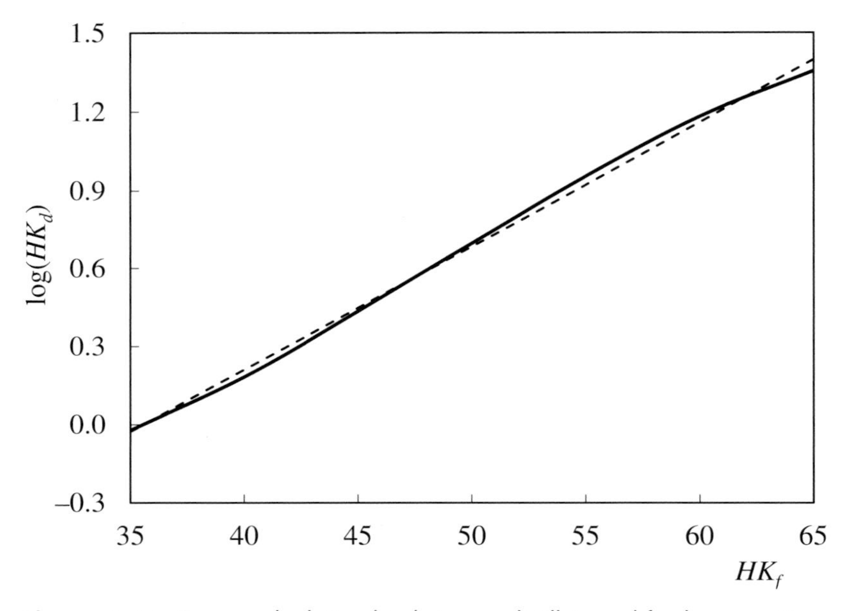

Figure 10.18 Regressed relationship between distillate and feed compositions

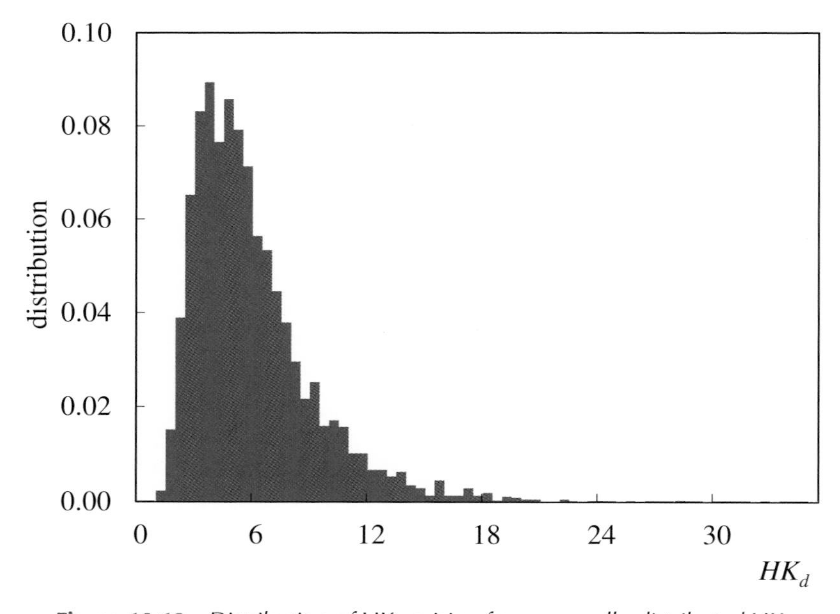

Figure 10.19 Distribution of HK_d arising from normally distributed HK_f

If the feed is routed directly from an upstream process then we might find that HK_f follows a normal distribution. Figure 10.19 shows the distribution of HK_d resulting from a Monte Carlo simulation of a large number of measurements of HK_f that have a mean of 50 vol% and standard deviation of 5 vol%. The result is far from normal with a skewness of 1.9 and a kurtosis of 9.9. In fact, by definition, the results will follow a lognormal distribution. If this product were routed directly to another column, for example a propene/propane splitter, then the distribution of the finished propane product composition will be even further skewed, possibly to the point where a lognormal distribution can no longer describe the behaviour.

Blending components to meet a viscosity target is a highly nonlinear process. The *Refutas Equation* is used to convert viscosity in cSt (ν) to a linear blending number (*VBN*).

$$VBN = 14.534\ln[\ln(\nu + 0.8)] + 10.975 \tag{10.5}$$

Its inverse is

$$\nu = \exp\left[\exp\left(\frac{VBN - 10.975}{14.534}\right)\right] - 0.8 \tag{10.6}$$

The result is then used determine the relative amounts of each component required to meet the target viscosity. Figure 10.20 shows the variation in *VBN* for a fuel oil blend. From applying Equation (10.6), Figure 10.21 shows the resulting variation in viscosity.

Again, despite the apparently highly non-linear function, there appears only a little advantage in selecting the lognormal distribution. The data has a skewness of 0.2 and a kurtosis of 2.9, both well within the limits. The normal distribution estimates the mean as 369.0 cSt and the standard deviation as 27.0. The lognormal gives values within 1% of these, at 369.4 and 27.3. Figure 10.22 illustrates that, over the operating range, the non-linearity is considerably less than that might be anticipated from the double logarithm in the Refutas Equation.

Reactor conversion is governed by the residence time (t) and the rate of reaction (k) which varies with absolute temperature (T) according to the *Arrhenius Equation*.

$$\text{conversion}\,(\%) = 100\left(1 - e^{-kt}\right) \quad \text{where} \quad k = A\exp\left(\frac{-E_a}{RT}\right) \tag{10.7}$$

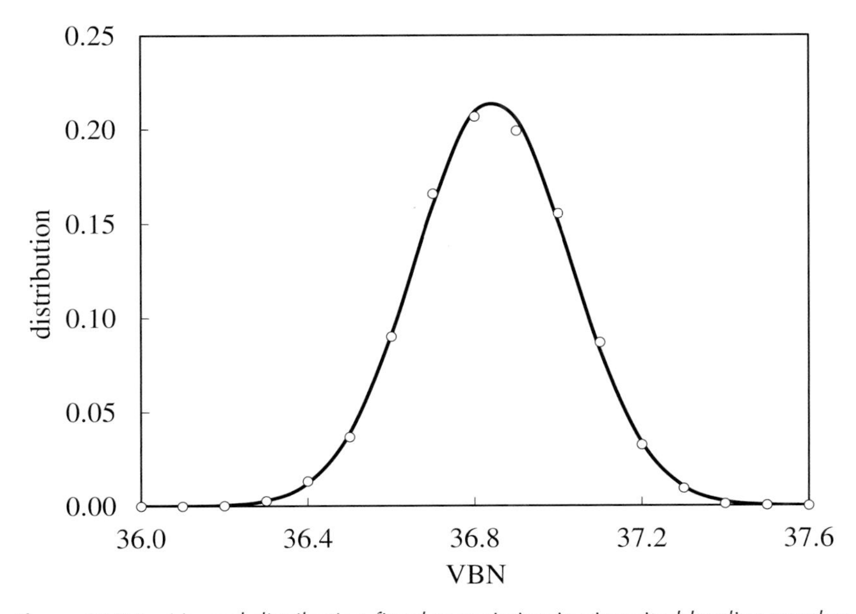

Figure 10.20 *Normal distribution fitted to variation in viscosity blending number*

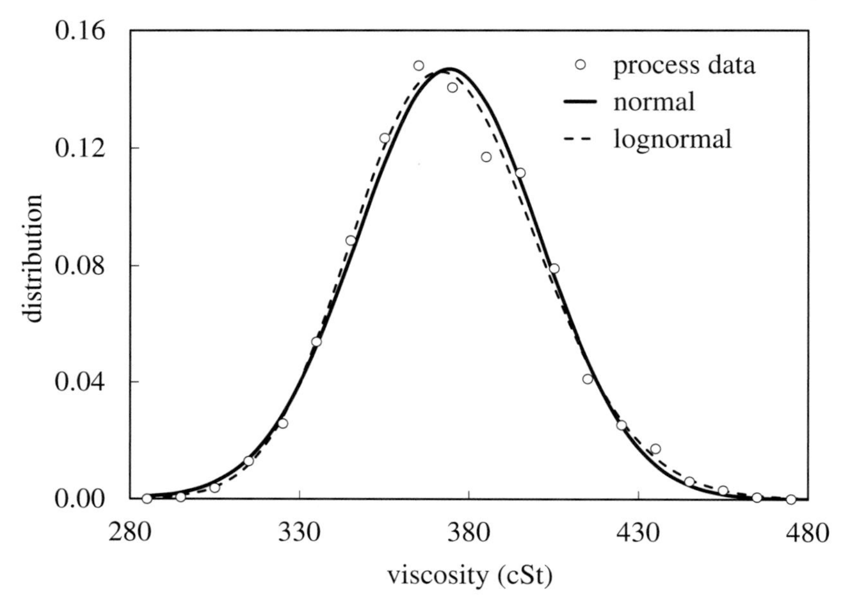

Figure 10.21 Fitting normal and lognormal distributions to viscosities

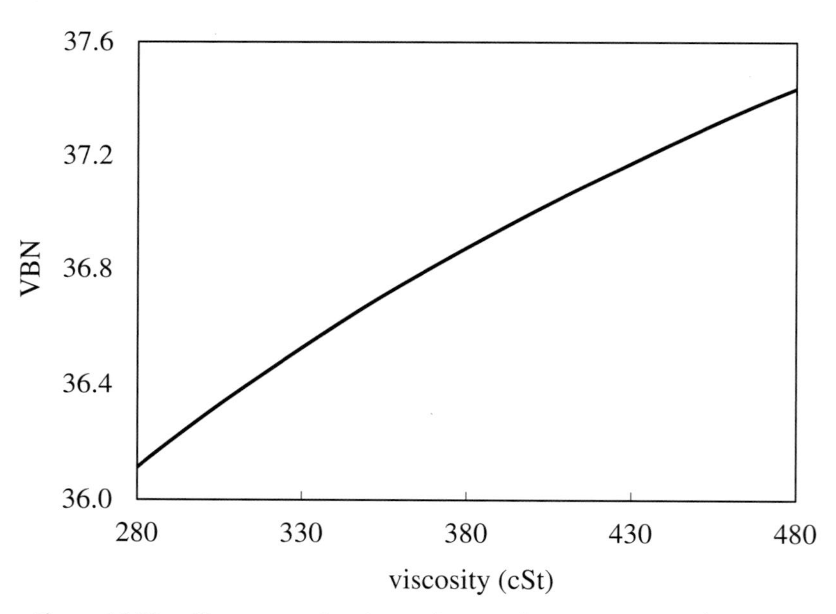

Figure 10.22 Close approximation to linearity between VBN and viscosity

In this example the *frequency factor* (A) has a value of 45,000 hr^{-1} and the *activation energy* (E_a) is 49.5 kJ/mol. The *Universal Gas Constant* (R) is 8.314 J/mol/K. Figure 10.23 shows the variation in reactor temperature closely follows the normal distribution. Applying Equation (10.7), assuming a residence time of 20 minutes, gives the variation in conversion shown in Figure 10.24.

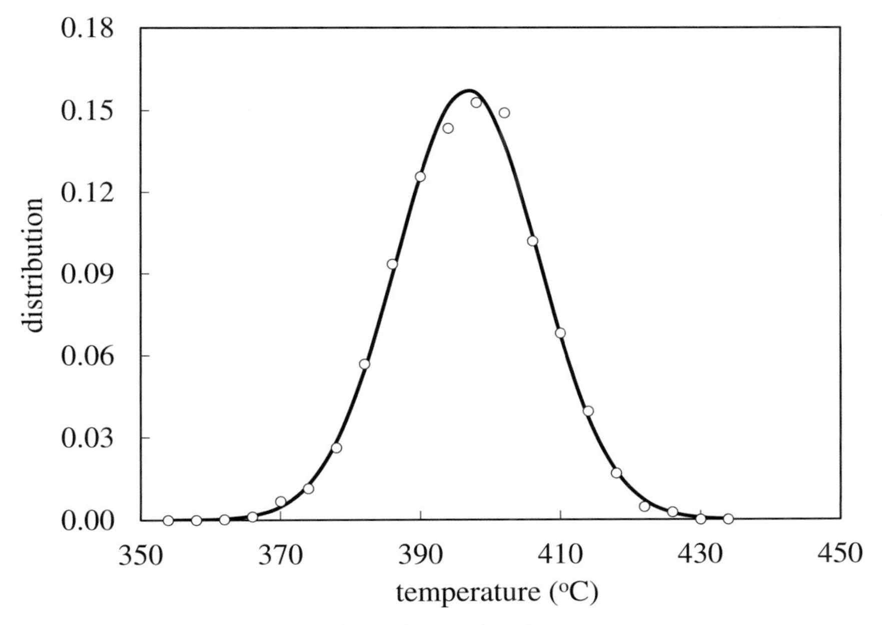

Figure 10.23 Normal distribution fitted to reactor temperature

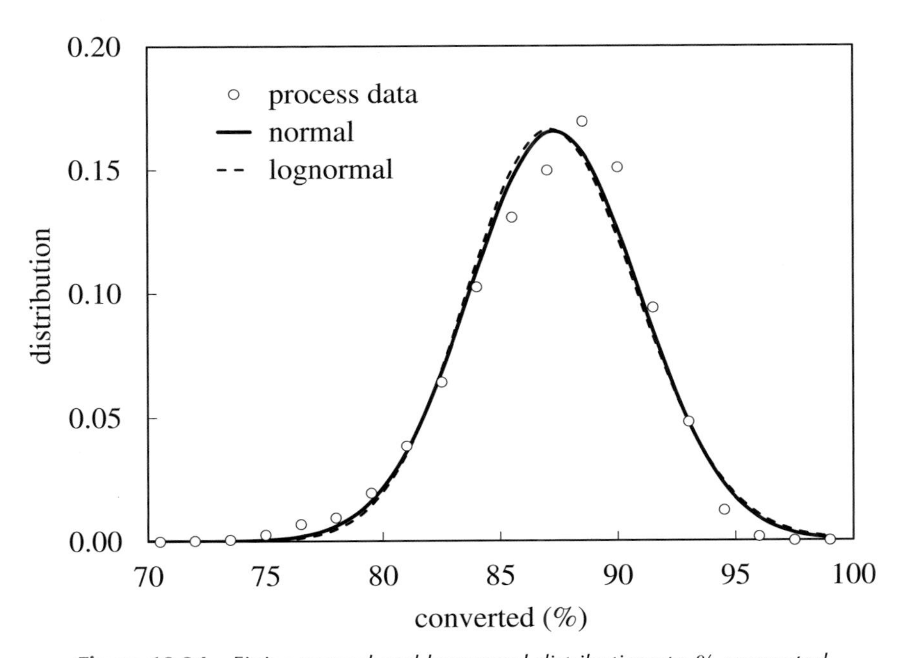

Figure 10.24 Fitting normal and lognormal distributions to % converted

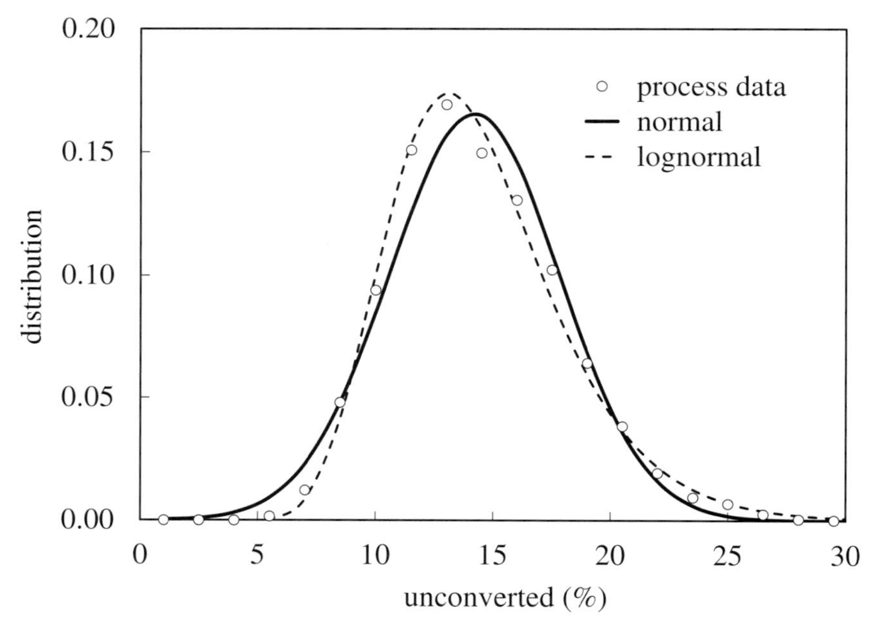

Figure 10.25 Fitting normal and lognormal distributions to % unconverted

Fitting the normal and lognormal distributions give very similar results – although both appear not to closely match the actual distribution. The reason for this is that the skewness is negative (at −0.5). Under these circumstances the lognormal will be no better a fit than the normal distribution. Indeed, as we will see later, many of the non-symmetrical distributions cannot represent data that is negatively skewed.

Rather than attempt to identify a distribution that is skewed to the left, a better solution may be to apply some transformation to the data. In this case we can change from % converted to % unconverted – simply by subtracting each measurement from 100%. Figure 10.25 shows the result of fitting the normal and lognormal distributions to the transformed data. Being symmetrical, the normal distribution is no different. Its mean has changed simply from 86.56% converted to 13.44% unconverted. The standard deviation of 3.59% is also unchanged.

The lognormal distribution, however, now fits the data very well – with a mean of 13.68% and a standard deviation of 3.74%. Figure 10.26 shows that the nonlinearity is significant – in this case justifying the use of a non-symmetrical distribution.

10.5 Root Mean Square

Figure 10.27 shows the distribution of a variable (x) derived as the root mean square of two variables (x_1 and x_2) both selected from a normal distribution with a mean of 0 and a variance of 1.

$$x_1 = \mathrm{N}(0,1) \quad x_2 = \mathrm{N}(0,1) \quad x = \sqrt{x_1^2 + x_2^2} \tag{10.8}$$

We will describe later the Rayleigh distribution. Theoretically it is followed by the root mean square of two variables with zero mean and the same standard deviation. While in the process

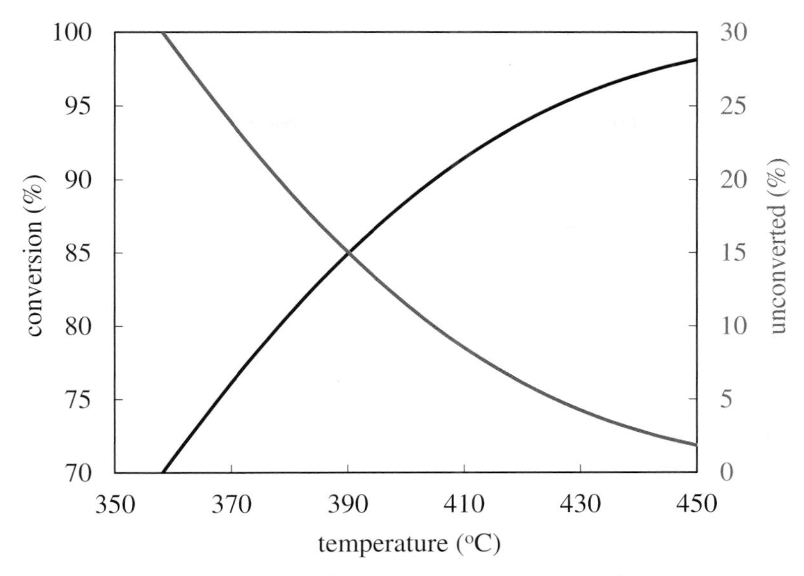

Figure 10.26 Nonlinear relationship between conversion and reactor temperature

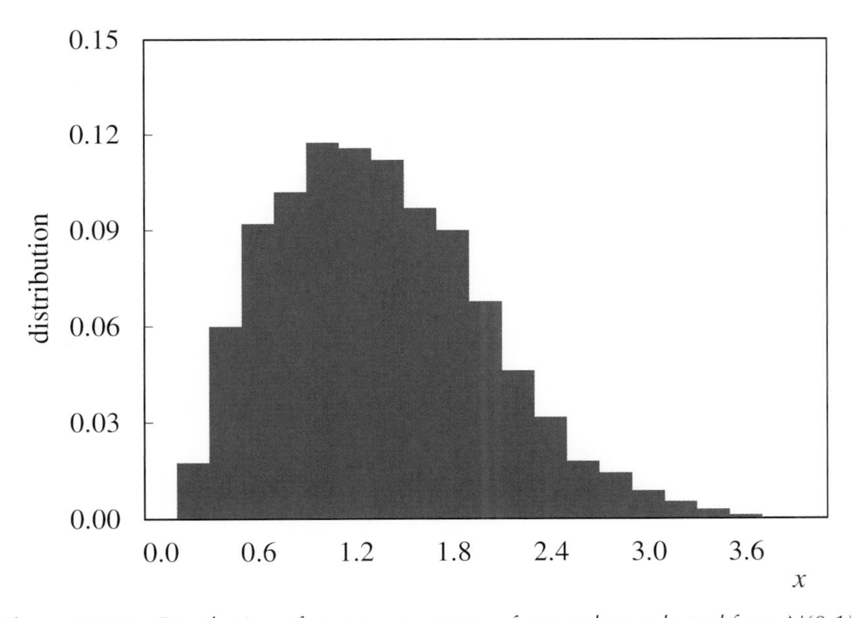

Figure 10.27 Distribution of root mean square of two values selected from N(0,1)

industry there are situations where the root mean square of two variables might be used, it is highly unlikely that the mean of both variables will be zero. Figure 10.28 shows the distribution of the root mean square of the two flows shown in Figure 10.2. Despite the flows having almost the same mean and standard deviation, the root mean square closely follows the normal

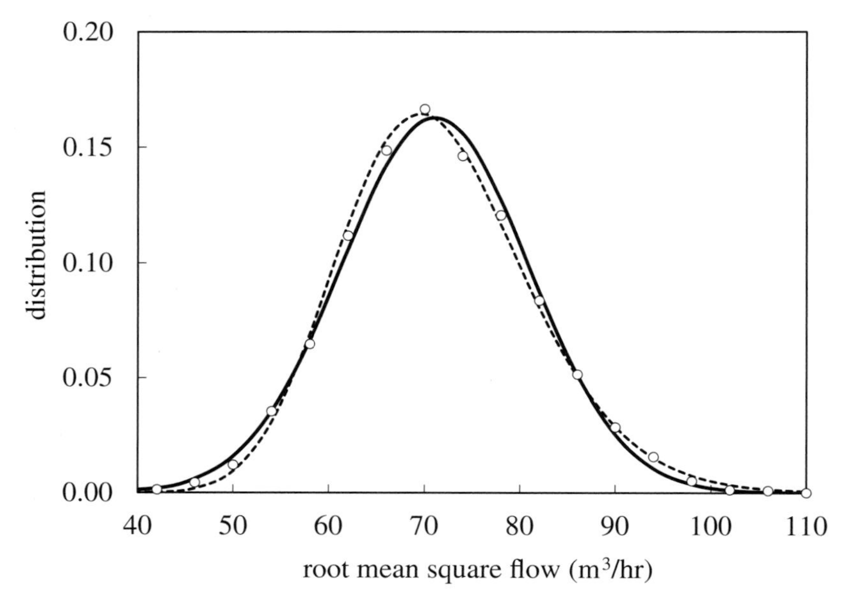

Figure 10.28 Fitting normal and lognormal distributions to root mean square flow

distribution (solid line). But, because the data are slightly skewed to the right ($\gamma = 0.21$), the lognormal distribution (dashed line) fits a little better.

10.6 Trigonometric Functions

Data derived by applying, for example, trigonometric functions is highly unlikely to fit either the normal or lognormal distributions. Such situations are rare in the process industry but, if required, there are distributions that can be adapted to fit such data.

11

Commonly Used Functions

There is a wide range of mathematical functions used in statistics. The reader may not be familiar with many of them. The symbols used and the language involved can be daunting. Their value is that they allow us to simply document a complex combination of basic mathematical operations. They are analogous to using $n!$ to replace $1 \times 2 \times 3 \times \ldots \times n$. This chapter includes those most commonly used. With a little practice, the reader will quickly become comfortable using them. Indeed, they are becoming commonplace in spreadsheet packages.

11.1 Euler's Number

The reader is likely to be familiar with a number of mathematical constants used frequently in statistical formulae. For example, we know π is the ratio of the circumference of a circle to its diameter. Another is *Euler's number (e)* but less well-known is its definition, which is

$$\left(1+\frac{1}{n}\right)^n \rightarrow e \quad \text{as} \quad n \rightarrow \infty \tag{11.1}$$

It is important because similar functions can appear in the derivation of a PDF or CDF and so lead to the inclusion of an exponential function. For example,

$$\text{as } n \rightarrow \infty \quad \left(1+\frac{\lambda}{n}\right)^n \rightarrow e^{\lambda} \quad \text{and} \quad \left(1-\frac{\lambda}{n}\right)^n \rightarrow e^{-\lambda} \tag{11.2}$$

This convergence is illustrated in Figure 11.1.

Equation (11.2) is often the basis for the limit of other functions. For example, if we replace λ with x and n with $1/\delta$, we get

$$(1+\delta x)^{1/\delta} \rightarrow e^x \quad \text{and} \quad (1-\delta x)^{1/\delta} \rightarrow e^{-x} \quad \text{as} \quad \delta \rightarrow 0 \tag{11.3}$$

Statistics for Process Control Engineers: A Practical Approach, First Edition. Myke King.

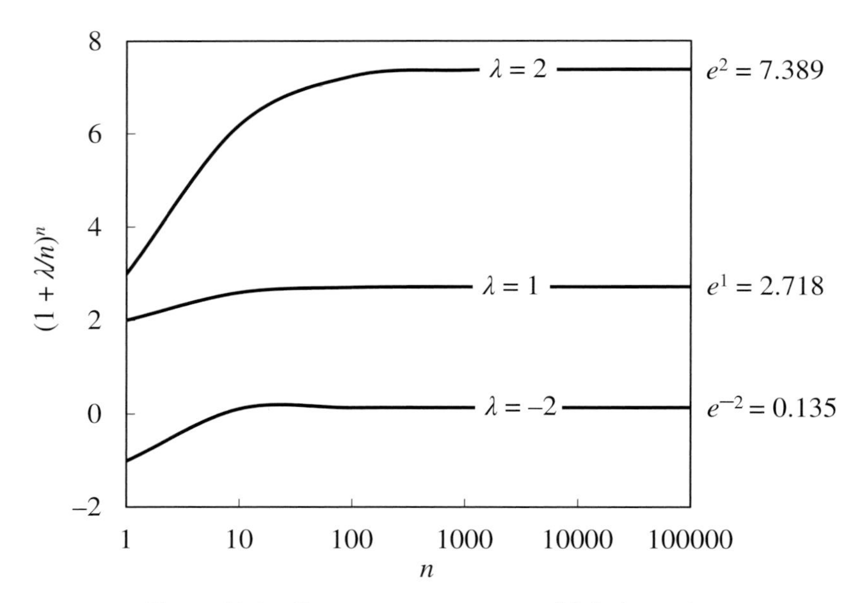

Figure 11.1 *Convergence to powers of Euler's number*

And so, for example, we will see terms in distributions of the form

$$\left[1-\delta\left(\frac{x-\alpha}{\beta}\right)\right]^{1/\delta} \rightarrow \exp\left(-\frac{x-\alpha}{\beta}\right) \quad \text{as} \quad \delta \rightarrow 0 \tag{11.4}$$

11.2 Euler–Mascheroni Constant

Sometimes confusingly described as *Euler's constant*, it is not the same as Euler's number. By convention it is given the symbol γ, which has nothing in common with its use to represent skewness. It is defined by

$$\sum_{i=1}^{n}\left(\frac{1}{i}\right) - \ln(n) \rightarrow \gamma \quad \text{as} \quad n \rightarrow \infty \tag{11.5}$$

Figure 11.2 plots Equation (11.5) showing that γ has a value of about 0.5772. The constant appears in a number of situations. Those of interest to the control engineer include its use in the gamma function described later in this chapter. Its only other use in this book, we shall see later, is in determining the mean of the Gumbel distributions.

11.3 Logit Function

The *logit function* gives its name to the logit-normal distribution. It is defined by

$$\text{logit}(x) = \ln\left(\frac{x}{1-x}\right) = \ln(x) - \ln(1-x) \quad \text{for} \quad 0 < x < 1 \tag{11.6}$$

Figure 11.3 plots the function.

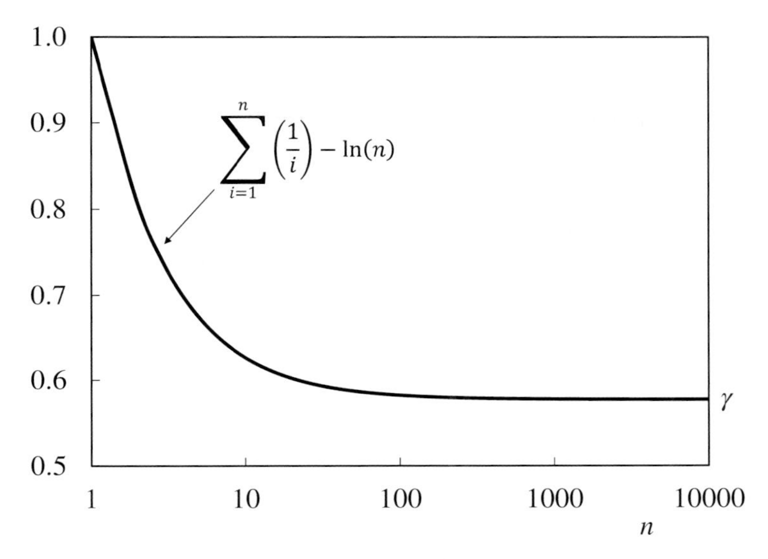

Figure 11.2 Euler–Mascheroni constant

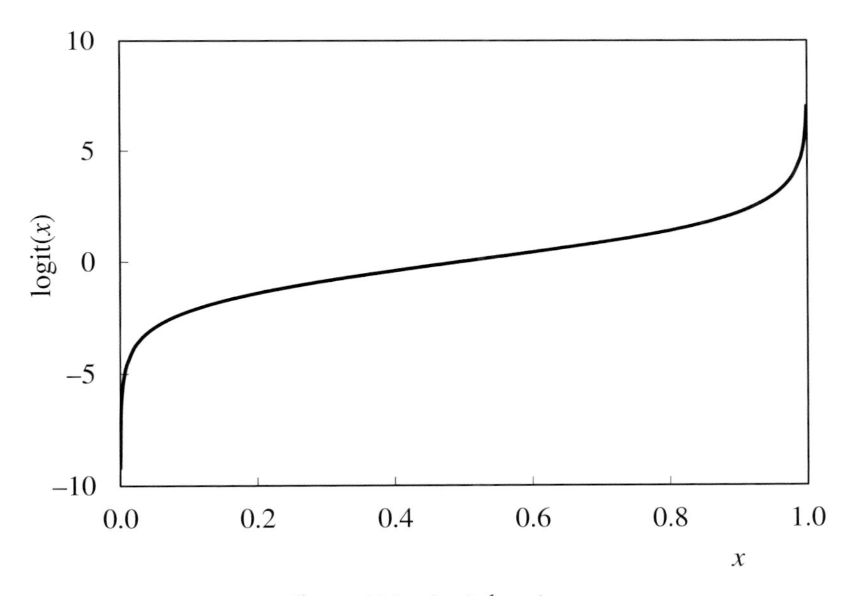

Figure 11.3 Logit function

11.4 Logistic Function

The *logistic function* gives its name to the logistic distribution. It is the inverse of the logit function.

$$\text{logit}^{-1}(x) = \text{logistic}(x) = \frac{1}{1+e^{-x}} = \frac{e^{x}}{1+e^{x}} \tag{11.7}$$

Unlike the logit function, there is no restriction on the value of x.

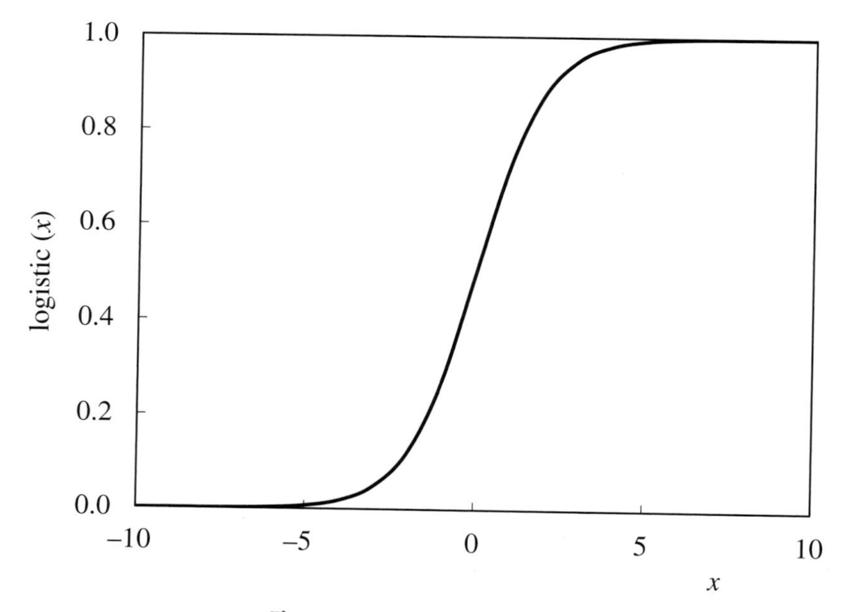

Figure 11.4 Logistic function

$$x \to -\infty \qquad \text{logistic}(x) \to 0 \tag{11.8}$$

$$x \to \infty \qquad \text{logistic}(x) \to 1 \tag{11.9}$$

Figure 11.4 plots the function. We will come across it again in Section 16.4; when used in an artificial neural network, it is more commonly known as the *sigmoid function*.

11.5 Gamma Function

The *gamma function (Γ)* was introduced by Euler as a way of more simply presenting lengthy multiplication of a series of values that arise in defining probability distributions. It is defined as

$$\Gamma(x) = \int_0^\infty t^{x-1} e^{-t} dt \tag{11.10}$$

So

$$\Gamma(1) = \int_0^\infty e^{-t} dt = [-e^{-t}]_0^\infty = 1 \tag{11.11}$$

Remembering the general formula for integration by parts,

$$\int u.dv = u.v - \int v.du \tag{11.12}$$

Let

$$u = t^{x-1} \qquad \text{and} \qquad dv = e^{-t} dt \tag{11.13}$$

Therefore

$$du = (x-1)t^{x-2} \quad \text{and} \quad v = -e^{-t} \tag{11.14}$$

Substituting into Equation (11.12)

$$\Gamma(x) = \left[-t^{x-1}e^{-t}\right]_0^{\infty} + \int_0^{\infty} (x-1)t^{x-2}e^{-t}dt = (x-1)\int_0^{\infty} t^{x-2}e^{-t}dt \tag{11.15}$$

which gives the *reduction formula*

$$\Gamma(x) = (x-1)\Gamma(x-1) \tag{11.16}$$

Therefore

$$\Gamma(x) = (x-1)\Gamma(x-1) = (x-1)(x-2)\Gamma(x-2) = \ldots \tag{11.17}$$

The series terminates when the last term $(x-r)$ is less than or equal to 1. So, if x is an integer, the last term in the multiplication will be $\Gamma(1)$ – which we have shown is 1. So, for example,

$$\Gamma(4) = (4-1)(3-1)(2-1)\Gamma(1) = 6 \tag{11.18}$$

Extending Equation (11.16) gives a formula used frequently to simplify PDF and CDF.

$$\Gamma(x+1) = x\Gamma(x) \tag{11.19}$$

For integers, $\Gamma(x)$ is simply $(x-1)!$ – shown as the round points on the curve in Figure 11.5. So, for example, Equation (11.19) can be written as

$$x! = x(x-1)! = x(x-1)(x-2)! = \ldots \tag{11.20}$$

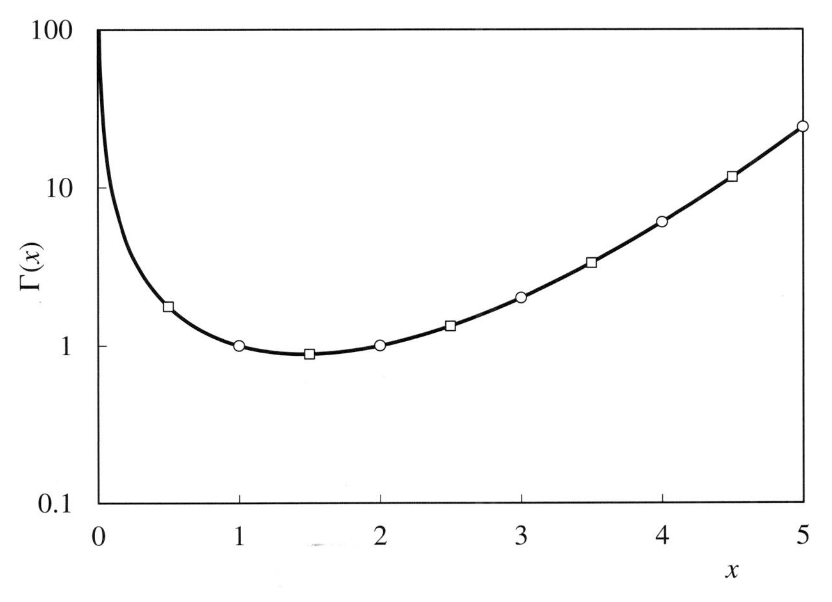

Figure 11.5 Gamma function

But x need not be an integer; for example if x is an odd multiple of 0.5, then the last term in Equation (11.17) will be Γ(0.5). We can quantify this by using *Euler's reflection formula*

$$\Gamma(x)\Gamma(1-x)=\frac{\pi}{\sin(\pi x)} \tag{11.21}$$

By setting x to 0.5 we obtain

$$\Gamma(0.5)=\sqrt{\pi} \tag{11.22}$$

For example,

$$\Gamma(3.5)=(3.5-1)(2.5-1)(1.5-1)\Gamma(0.5)=1.875\sqrt{\pi} \tag{11.23}$$

In general

$$\Gamma(x+0.5)=\frac{(2x-1)!!}{2^x}\sqrt{\pi} \tag{11.24}$$

The !! symbol is the *double factorial*, defined as

$$n!!=n(n-2)(n-4)(n-6)\ldots(n-r) \tag{11.25}$$

The last term in Equation (11.25) is either 1 or 2 – depending on whether n is odd or even. The square points in Figure 11.5 are plotted from Equation (11.24).

Most spreadsheet packages include the gamma function and so can provide a result for any value of x. Since the function increases rapidly with x, it is common to plot its logarithm. Figure 11.6 does this – taking advantage of this to extend the x axis well beyond that in Figure 11.5. It is used in distributions, where x can be large, to avoid computational overflow problems.

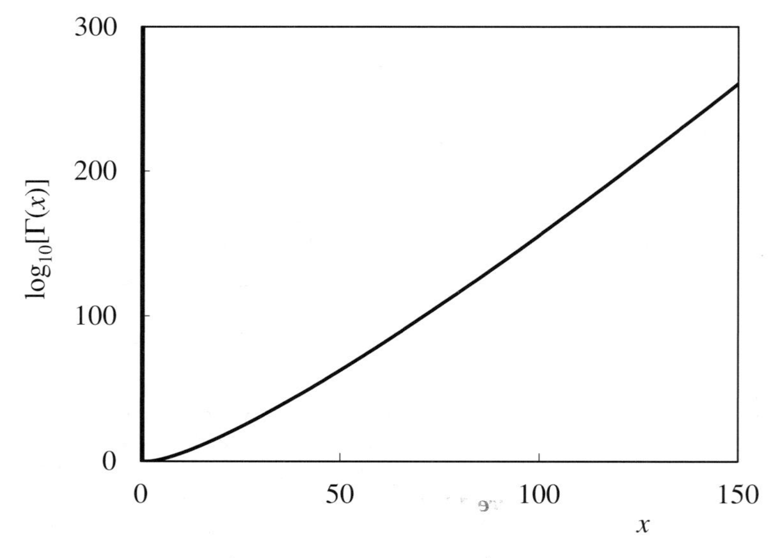

Figure 11.6 Log-gamma function

A common appearance of the gamma function is in distributions that involve calculating the number of combinations possible when r items are selected from a total of n. This is known as the *binomial index* and is commonly denoted by $^{n}C_{r}$ where

$$^{n}C_{r}=\binom{n}{r}=\frac{n!}{r!(n-r)!}=\frac{\Gamma(n+1)}{\Gamma(r+1)\Gamma(n-r+1)}=\frac{n}{r(n-r)}\frac{\Gamma(n)}{\Gamma(r)\Gamma(n-r)} \tag{11.26}$$

Many distributions that include the gamma function can be written in several different ways that, at first glance, make them appear different. We have seen how Equations (11.16) and (11.19) might be used. Other alternatives are not so immediately obvious. For example, Equation (11.21) might be used to replace a gamma function with a trigonometric one.

The gamma function of a negative number requires a different approach to its calculation. Rearranging Equation (11.19) gives

$$\Gamma(x)=\frac{\Gamma(x+1)}{x} \tag{11.27}$$

So

$$\Gamma(-1)=\frac{\Gamma(0)}{-1}=\frac{\infty}{-1}=-\infty \tag{11.28}$$

$$\Gamma(-2)=\frac{\Gamma(-1)}{-2}=\frac{-\infty}{-2}=\infty \tag{11.29}$$

We can see that the gamma function for odd negative integers is $-\infty$, while that for even negative integers is ∞. An example of a non-integer value for x is

$$\Gamma(-0.5)=\frac{\Gamma(0.5)}{-0.5}=-2\sqrt{\pi} \tag{11.30}$$

And so

$$\Gamma(-1.5)=\frac{\Gamma(-0.5)}{-1.5}=\frac{4}{3}\sqrt{\pi} \tag{11.31}$$

In general

$$\Gamma(0.5-x)=\frac{(-2)^{x}}{(2x-1)!!}\sqrt{\pi} \tag{11.32}$$

Figure 11.7 plots the gamma function for negative numbers. The square points are those calculated from Equation (11.32).

There are methods of obtaining the gamma function for other non-integer values but there is no analytical method that can use Equation (11.10) to determine it for any value. Instead a numerical approximation is used, the most common of which is the *Lanczos approximation*. Too complex to merit inclusion here, it is almost certainly the method used by spreadsheet packages and other proprietary software.

There are occasions when we need the derivative of the gamma function, $\Gamma'(x)$. Differentiating Equation (11.19)

$$\Gamma'(x+1)=x\Gamma'(x)+\Gamma(x) \tag{11.33}$$

Dividing by $x\Gamma(x)$

$$\frac{\Gamma'(x+1)}{x\Gamma(x)}=\frac{\Gamma'(x)}{\Gamma(x)}+\frac{1}{x} \tag{11.34}$$

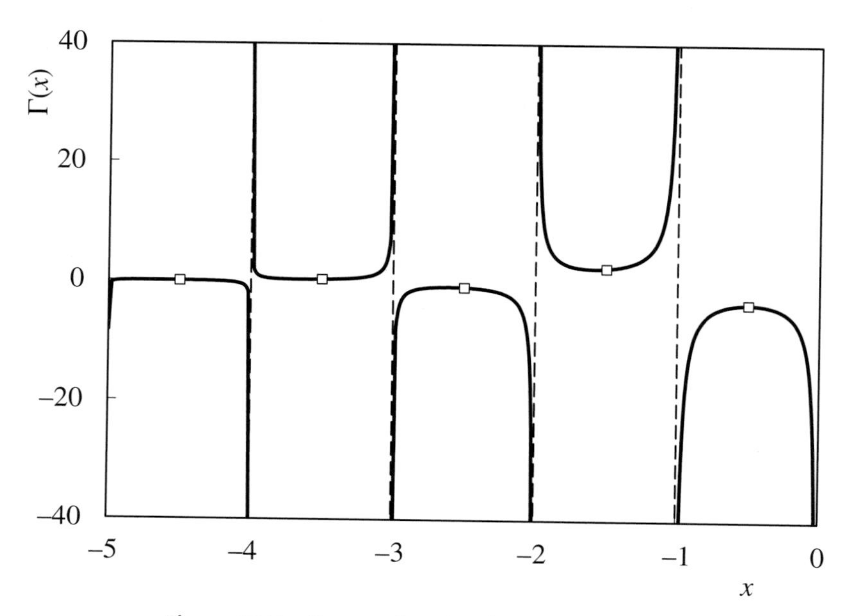

Figure 11.7 Gamma function for negative values

Replacing $x\Gamma(x)$ from Equation (11.19)

$$\frac{\Gamma'(x+1)}{\Gamma(x+1)} = \frac{\Gamma'(x)}{\Gamma(x)} + \frac{1}{x} \tag{11.35}$$

Thus

$$\frac{\Gamma'(x+2)}{\Gamma(x+2)} = \frac{\Gamma'(x+1)}{\Gamma(x+1)} + \frac{1}{x+1} = \frac{\Gamma'(x)}{\Gamma(x)} + \frac{1}{x} + \frac{1}{x+1} \tag{11.36}$$

In general

$$\frac{\Gamma'(x+n-1)}{\Gamma(x+n-1)} = \frac{\Gamma'(x)}{\Gamma(x)} + \frac{1}{x} + \frac{1}{x+1} + \ldots + \frac{1}{x+n-2} \tag{11.37}$$

So, if $x = 1$

$$\frac{\Gamma'(n)}{\Gamma(n)} = \frac{\Gamma'(1)}{\Gamma(1)} + \frac{1}{1} + \frac{1}{2} + \ldots + \frac{1}{n-1} \tag{11.38}$$

An alternative definition of the Euler–Mascheroni constant (γ), defined in Section 11.2, is

$$\gamma = -\frac{\Gamma'(1)}{\Gamma(1)} \tag{11.39}$$

So (replacing n with x), generally

$$\Gamma'(x) = \Gamma(x)\left[\sum_{i=1}^{x-1}\frac{1}{i} - \gamma\right] \tag{11.40}$$

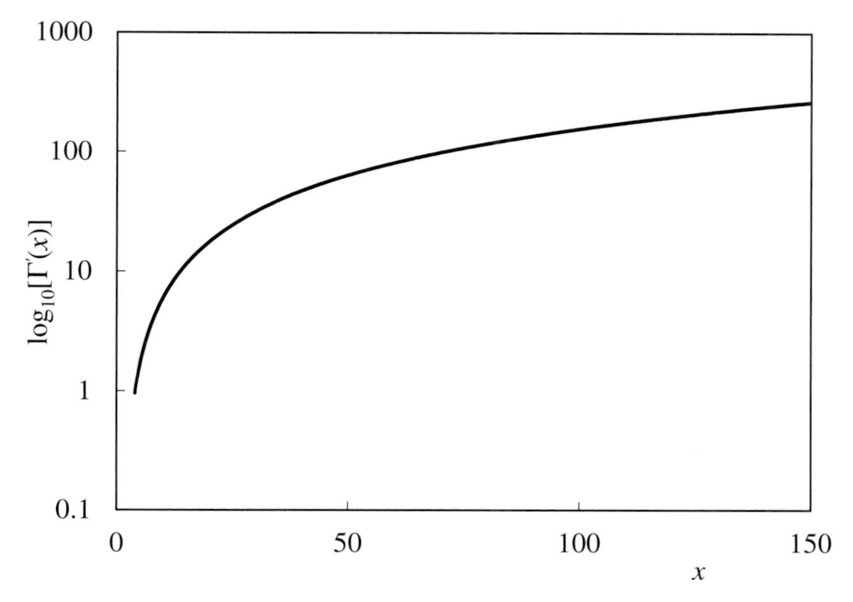

Figure 11.8 Derivative of gamma function

Like the gamma function itself, we can readily calculate its derivative if x is an integer. There are software-based techniques, beyond the scope of this book, for which this restriction does not apply. Figure 11.8 plots the result.

Some CDF involve the use of the *incomplete gamma function*. This is the gamma function described by Equation (11.10), but integrated over only part of the range. The *lower* incomplete gamma function is defined as

$$\gamma(\delta,x) = \int_0^x t^{\delta-1} e^{-t} dt \tag{11.41}$$

The *upper* incomplete gamma function is defined as

$$\Gamma(\delta,x) = \int_x^\infty t^{\delta-1} e^{-t} dt \tag{11.42}$$

Adding Equations (11.41) and (11.42) and then comparing with Equation (11.10) gives

$$\gamma(\delta,x) + \Gamma(\delta,x) = \Gamma(\delta) \tag{11.43}$$

The *regularised* lower and upper incomplete gamma functions are obtained by dividing by the gamma function. These sum to unity.

$$\frac{\gamma(\delta,x)}{\Gamma(\delta)} + \frac{\Gamma(\delta,x)}{\Gamma(\delta)} = 1 \tag{11.44}$$

The CDF of the gamma and chi-squared distributions involve these functions.

Not included in this book, because of their complexity, are any of the published numerical techniques that give very close approximation to the incomplete functions. But they are used in spreadsheet packages and other proprietary software that support these distributions.

Similarly excluded are distributions that include the gamma function of a complex number – the result of which may be real, but still not offer a worthwhile improvement over distributions that are easier to apply.

11.6 Beta Function

The definition of the *beta function* is

$$B(\alpha,\beta)=\int_0^1 t^{\alpha-1}(1-t)^{\beta-1}dt \tag{11.45}$$

In practice it is calculated from the gamma function.

$$B(\alpha,\beta)=\frac{\Gamma(\alpha)\Gamma(\beta)}{\Gamma(\alpha+\beta)} \tag{11.46}$$

It is often used to provide the normalisation constant in a PDF, as we will see later in the distribution that takes its name. As with the gamma function, distributions using the beta function can be documented in different ways. For example,

$$B(\alpha+1,\beta)=\frac{\alpha}{\alpha+\beta}B(\alpha,\beta) \tag{11.47}$$

$$B(\alpha-1,\beta)=\frac{\alpha+\beta-1}{\alpha-1}B(\alpha,\beta) \tag{11.48}$$

$$B(1,\beta)=\frac{\Gamma(1)\Gamma(\beta)}{\Gamma(\beta+1)}=\frac{(\beta-1)!}{\beta!}=\frac{1}{\beta} \tag{11.49}$$

In the same way that there is an incomplete gamma function, there is an *incomplete beta function*. It is used, for example, in the CDF of the beta-I and Fisher distributions. There is no analytical method of directly integrating Equation (11.45). Numerical methods include those based on a converging infinite series known as a *continued fraction*. Almost certainly, this method is used in spreadsheet functions and other statistical software.

11.7 Pochhammer Symbol

There are actually two *Pochhammer symbols*. They are each abbreviations of another form of factorial. The first can be described as a *rising* or *ascending factorial*. Strictly, because it does not start at 1, it is not a factorial. In fact, it is the ratio of two factorials. More properly it is a *sequential product*. It is defined as

$$x^{(n)}=x(x+1)(x+2)\ldots(x+n-1)=\frac{\Gamma(x+n)}{\Gamma(x)}=\frac{(x+n-1)!}{(x-1)!} \tag{11.50}$$

Figure 11.9 shows the effect of changing n.

The *falling* or *descending* version is defined as

$$(x)_n=x(x-1)(x-2)\ldots(x-n+1)=\frac{\Gamma(x+1)}{\Gamma(x-n+1)}=\frac{x!}{(x-n)!} \tag{11.51}$$

It is shown in Figure 11.10.

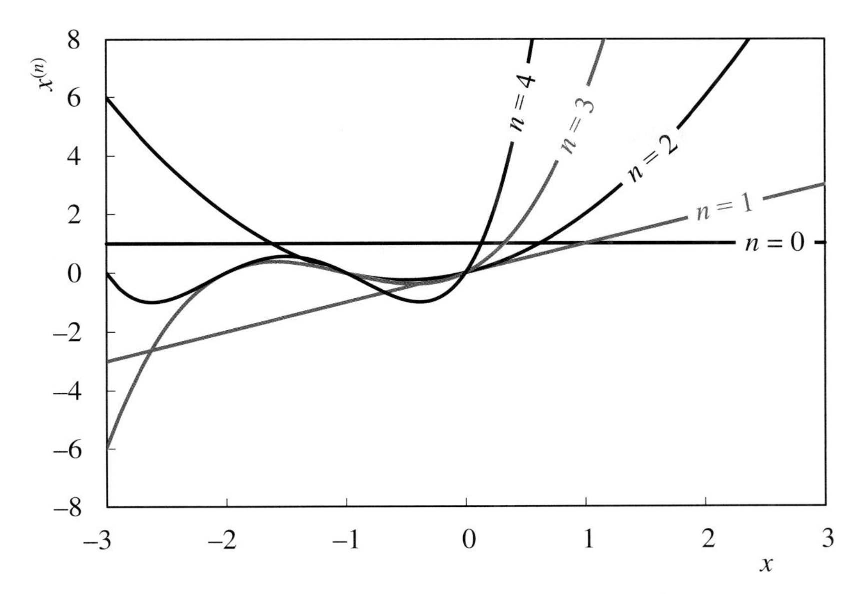

Figure 11.9 *Pochhammer ascending factorial*

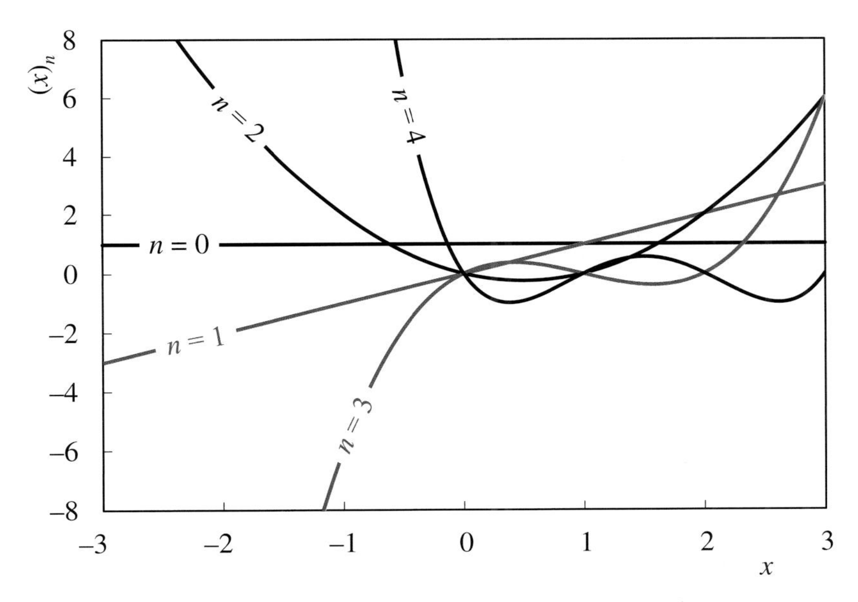

Figure 11.10 *Pochhammer descending factorial*

Pochhammer originally used the symbol for the binomial index, as described by Equation (11.26). There are also other forms of the symbols. Care must therefore be taken by checking the definition used in each case. For this reason, in this book, the notation is avoided by instead including the full calculation in any function that includes Pochhammer.

11.8 Bessel Function

The *Bessel function* (J), of *order* (*k*), is defined as

$$J_k(x) = \sum_{i=0}^{\infty} \frac{(-1)^i}{i!\Gamma(k+i+1)} \left(\frac{x}{2}\right)^{2i+k} \tag{11.52}$$

Of more interest is the *modified Bessel function* (I), which is used by the von Mises, Rician and Skellam distributions. It can be derived from J_k but this involves the use of $\sqrt{-1}$.

$$I_k(x) = \left(\sqrt{-1}\right)^{-k} J_k\left(x\sqrt{-1}\right) \tag{11.53}$$

It can, however, also be determined directly from

$$I_k(x) = \sum_{i=0}^{\infty} \frac{1}{i!\Gamma(k+i+1)} \left(\frac{x}{2}\right)^{2i+k} \tag{11.54}$$

Figure 11.11 shows this function plotted for a number of different values of *k*.

A summation of a series to infinity may not appear a practical function but in fact it converges quite quickly, especially for small values of *x*. Figure 11.12 shows how the function converges as the number of terms in the series is increased. It is drawn with *k* set to 0. This is the worst case; higher values of *k* require fewer terms to achieve the same accuracy. Figure 11.13 shows how many terms are necessary for the error to fall below 0.01%. More conveniently, the function is included in some spreadsheet packages. However, many spreadsheet packages permit only integer values of *k*. Of those that appear to accept non-integers, some in fact round off to the nearest integer. Figure 11.14 plots the same values as Figure 11.11. The points correspond to integer values of *k* but, as shown, the function can be treated as continuous.

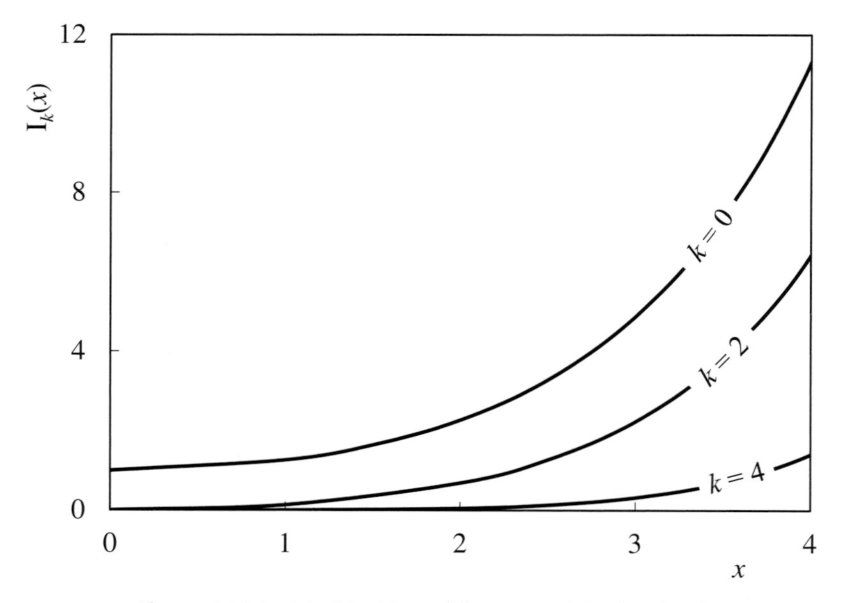

Figure 11.11 Modified Bessel function of the first kind

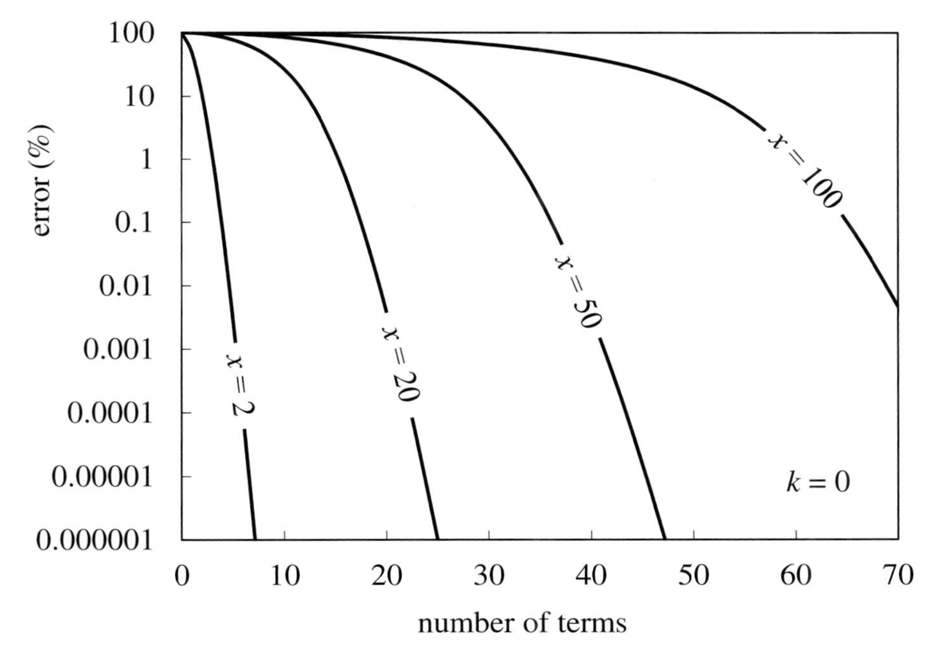

Figure 11.12 *Convergence of Bessel function*

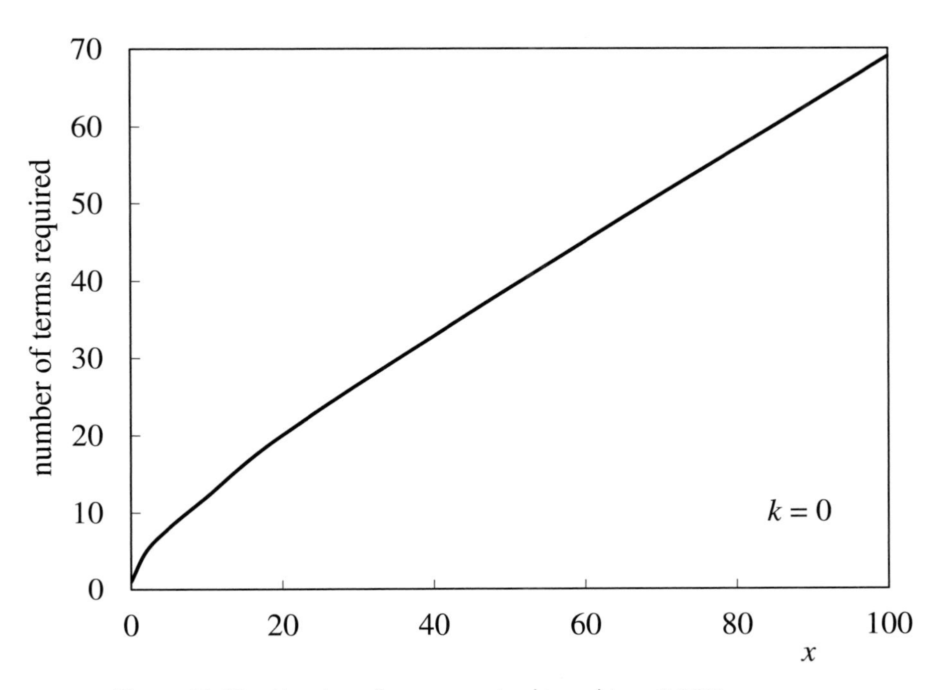

Figure 11.13 *Number of terms required to achieve 0.01% accuracy*

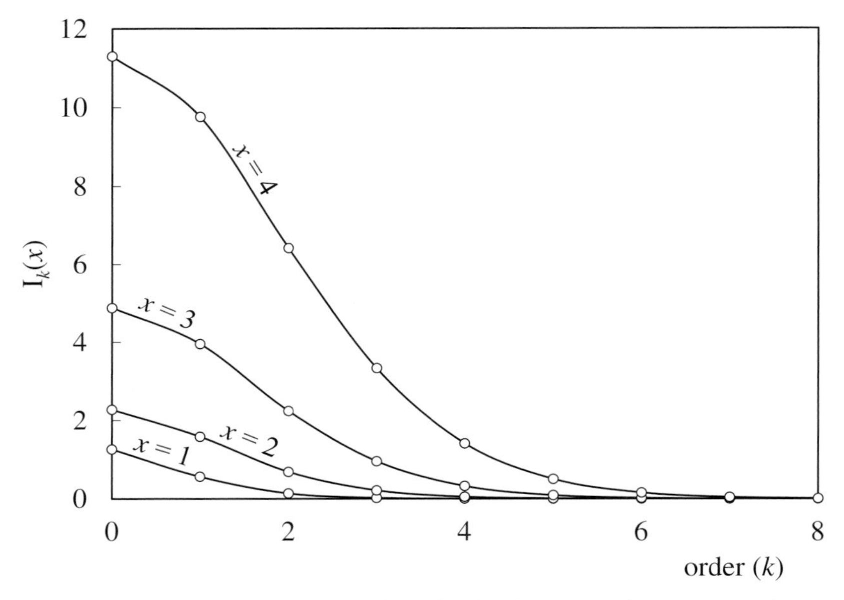

Figure 11.14 *Bessel function (of the first kind) restricted to integer orders*

More fully, Equations (11.52) and (11.54) describe Bessel functions of the *first kind*. The *Bessel function of the second kind* is

$$\mathrm{Y}_k(x) = \frac{\mathrm{J}_k \cos(kx) - \mathrm{J}_{-k}(x)}{\sin(k\pi)} \tag{11.55}$$

The *modified Bessel function of the second kind* is used by the generalised inverse Gaussian distribution. Its definition is

$$\mathrm{K}_k(x) = \frac{\pi}{2} \frac{\mathrm{I}_{-k}(x) - \mathrm{I}_k(x)}{\sin(k\pi)} \tag{11.56}$$

The formulae involve Bessel functions with a negative order, which is usually not supported by spreadsheet packages. However, by using different mathematics, they do support the functions of the second kind. Figure 11.15 shows the function plotted against x. Again, most spreadsheet packages treat the function as discrete although, as Figure 11.16 shows, it can be considered continuous.

11.9 Marcum Q-Function

The *Marcum Q-function* is used in the CDF of the noncentral chi-squared and Rician distributions. It is defined as

$$\mathrm{Q}_k(a,b) = \exp\left(-\frac{a^2 + b^2}{2}\right) \sum_{i=1-k}^{\infty} \left(\frac{a}{b}\right)^i \mathrm{I}_i(ab) \tag{11.57}$$

It is not commonly supported by spreadsheet packages. While the calculation could be built, it relies on the modified Bessel function of the first kind (I_i) which is only supported for integer values of i. We will therefore not rely on it; instead we apply the trapezium rule to the PDF.

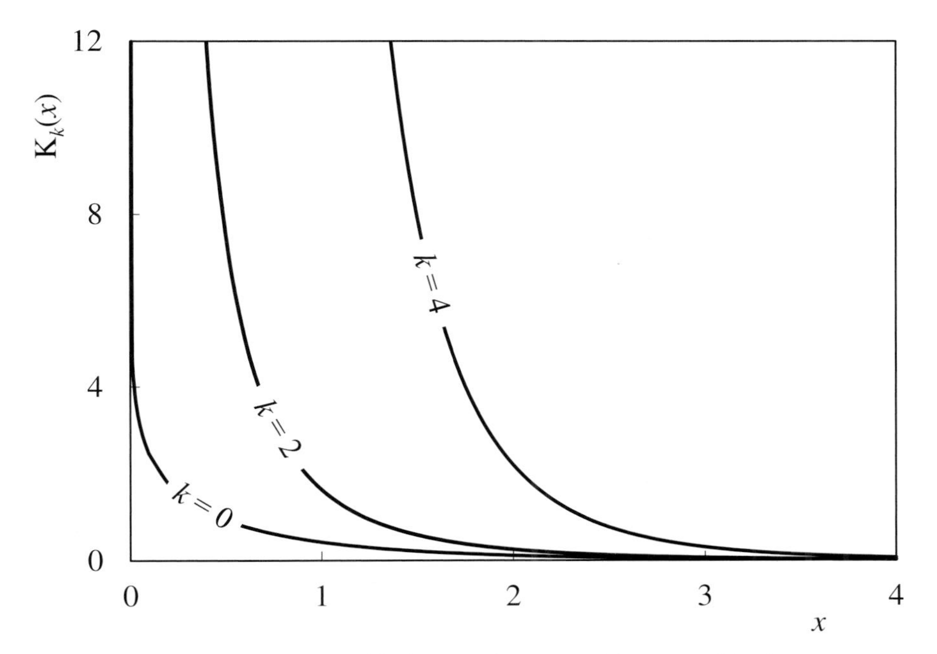

Figure 11.15 *Modified Bessel function of the second kind*

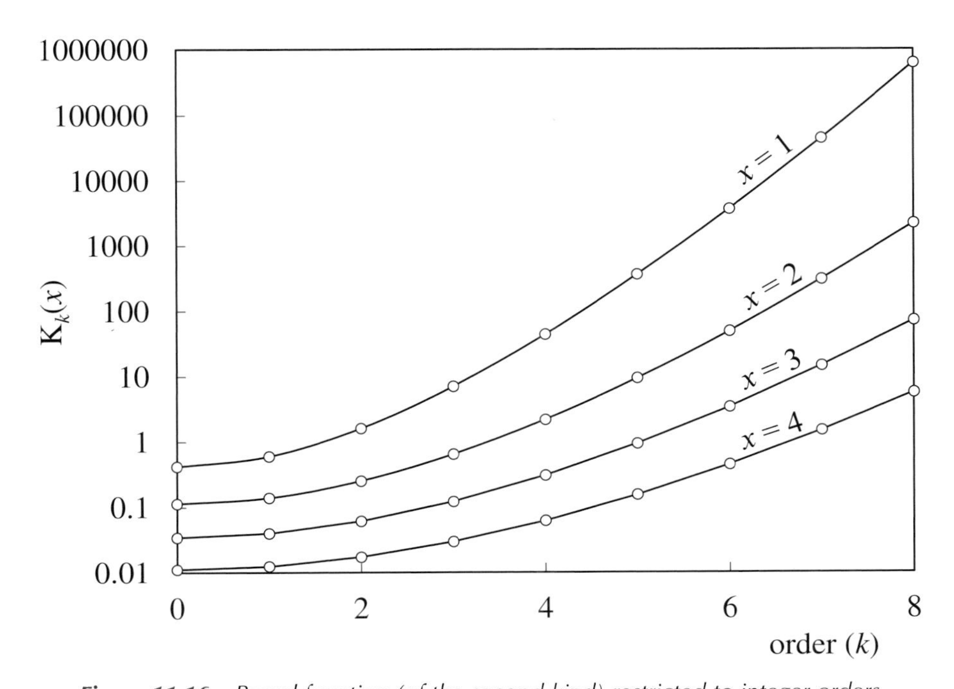

Figure 11.16 *Bessel function (of the second kind) restricted to integer orders*

11.10 Riemann Zeta Function

Named after the Greek letter zeta (ζ), not to be confused with the damping factor used in control theory, the *Riemann zeta function* is defined as

$$\zeta(x) = \sum_{i=1}^{\infty} i^{-x} \tag{11.58}$$

It gives its name to the zeta distribution. It converges only if x is greater than 1. $\zeta(1)$ is infinite. For values of x only slightly greater than 1 it converges very slowly. Fortunately, for higher values, convergence is much quicker. Figure 11.17 illustrates this.

In the Gumbel distribution $\zeta(3)$ is required to calculate the skewness. This has a value of about 1.2025, which is accurate to within 0.01% once i exceeds 64. As Figures 11.17 and 11.18 show, for large values of x, $\zeta(x)$ is approximately 1.

The *Hurwitz zeta function* includes an additional term (α).

$$\zeta(x,\alpha) = \sum_{i=1}^{\infty} i^{-(x+\alpha)} \tag{11.59}$$

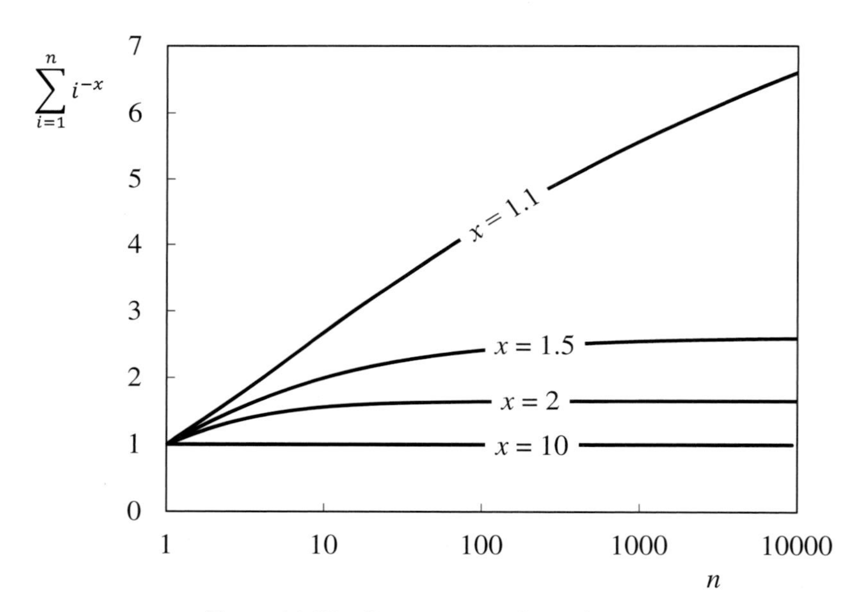

Figure 11.17 Convergence of zeta function

11.11 Harmonic Number

Related to the Riemann function is the *harmonic number* (H_n)

$$H_n = 1 + \frac{1}{2} + \frac{1}{3} \ldots + \frac{1}{n} = \sum_{i=1}^{n} \frac{1}{i} \tag{11.60}$$

A more general harmonic number appears in the Zipf distribution.

$$H_{n,\lambda} = 1 + \frac{1}{2^{\lambda}} + \frac{1}{3^{\lambda}} \ldots + \frac{1}{n^{\lambda}} = \sum_{i=1}^{n} \frac{1}{i^{\lambda}} \tag{11.61}$$

Figure 11.19 shows that, for high values of λ, harmonic numbers tend towards a constant value. Indeed, as $n \to \infty$, the harmonic number approaches the zeta function. Neither function is commonly available in spreadsheet packages.

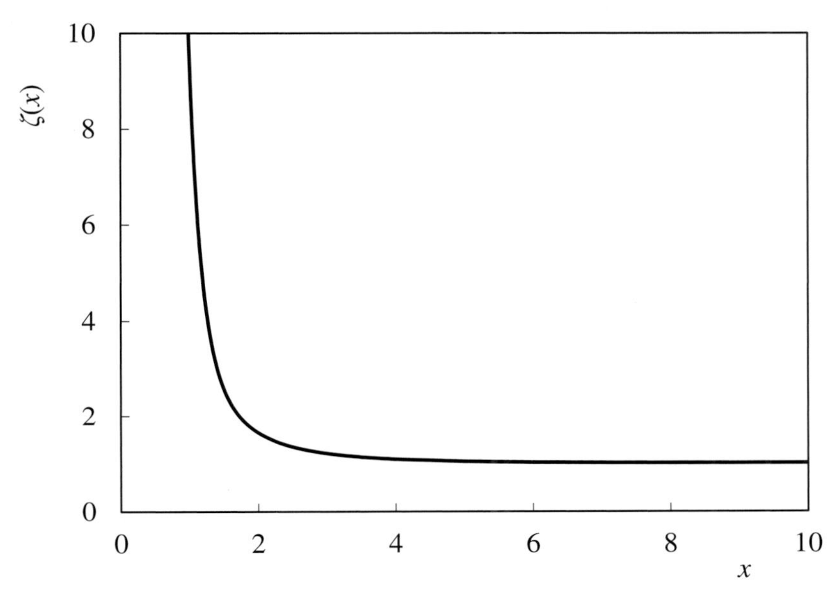

Figure 11.18 Zeta function

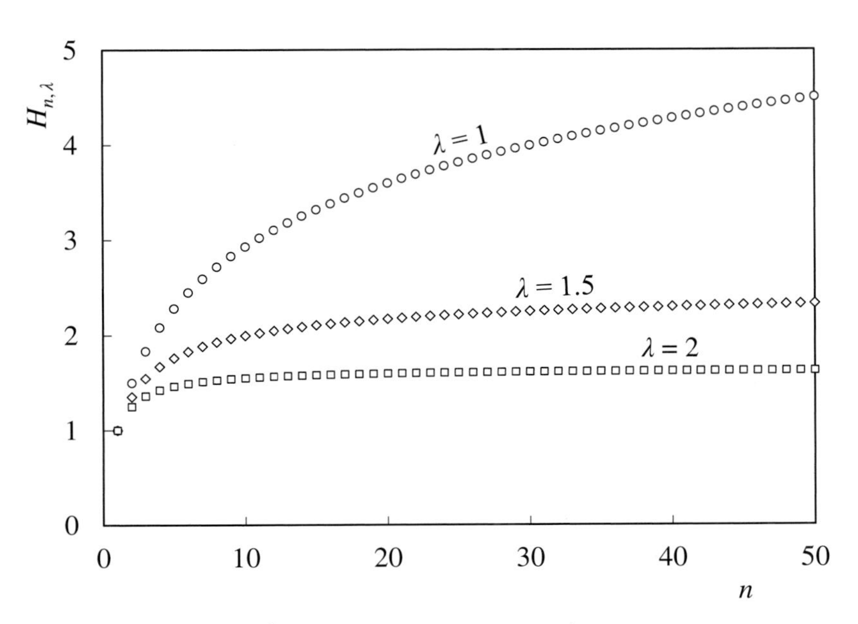

Figure 11.19 Harmonic number

Harmonic numbers are used in the Zipf–Mandelbrot distribution as $H_{n,\alpha,\lambda}$, where

$$H_{n,\alpha,\lambda} = \frac{1}{(1+\alpha)^{\lambda}} + \frac{1}{(2+\alpha)^{\lambda}} + \frac{1}{(3+\alpha)^{\lambda}} \ldots + \frac{1}{(n+\alpha)^{\lambda}} = \sum_{i=1}^{n} (i+\alpha)^{-\lambda} \tag{11.62}$$

11.12 Stirling Approximation

The *Stirling approximation* provides a reasonably accurate estimate of the factorial of a number (n).

$$n! \approx \sqrt{2\pi n}\left(\frac{n}{e}\right)^{n} \tag{11.63}$$

As Figure 11.20 shows, as n exceeds 83, the error in the approximation falls to less than 0.1%. Also shown are the less well-known, higher order, versions of the approximation in which the error is many orders of magnitude smaller.

$$n! \approx \sqrt{2\pi n}\left(1 + \frac{1}{12n}\right)\left(\frac{n}{e}\right)^{n} \tag{11.64}$$

$$n! \approx \sqrt{2\pi n}\left(1 + \frac{1}{12n} + \frac{1}{288n^{2}}\right)\left(\frac{n}{e}\right)^{n} \tag{11.65}$$

While once a convenient method, particularly for non-integer values of n, most spreadsheet packages now include the factorial function for integers and, for non-integers, the gamma function $\Gamma(n+1)$.

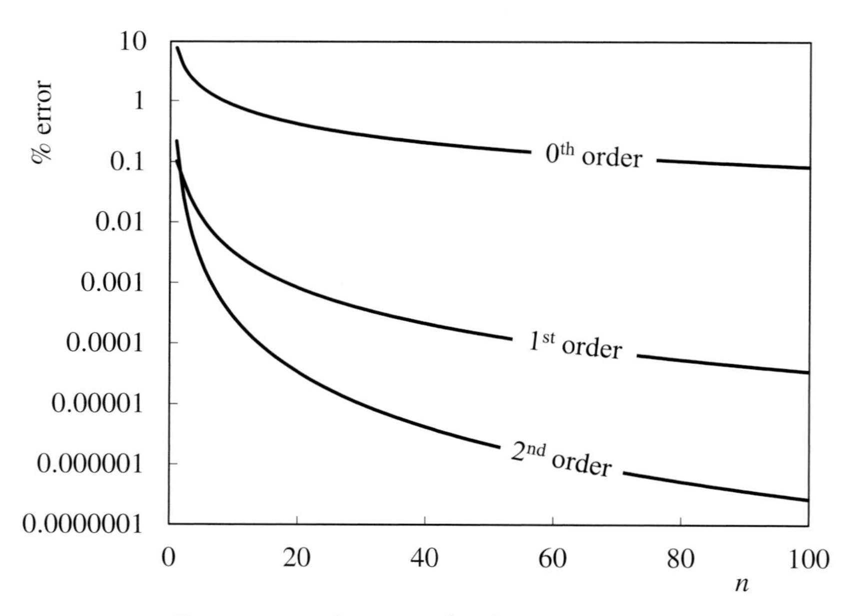

Figure 11.20 Accuracy of Stirling approximation

11.13 Derivatives

There are a wide range of mathematical functions used in CDF that need to be differentiated to generate the corresponding PDF. Many of the derivatives will be familiar to the reader. For those that may not be able to immediately recall them, they are summarised in Table 11.1.

Table 11.1 *Common derivatives*

$F(x)$	$\frac{dF(x)}{dx}$	$F(x)$	$\frac{dF(x)}{dx}$
x^n	nx^{n-1}	a^x	$\ln(a)a^x$
e^x	e^x	$\ln(x)$	$\frac{1}{x}$
$\sin(x)$	$\cos(x)$	$\sin^{-1}\left(\frac{x}{a}\right)$	$\frac{1}{\sqrt{a^2-x^2}}$
$\cos(x)$	$-\sin(x)$	$\cos^{-1}\left(\frac{x}{a}\right)$	$\frac{-1}{\sqrt{a^2-x^2}}$
$\tan(x)$	$\sec^2(x)$	$\tan^{-1}\left(\frac{x}{a}\right)$	$\frac{a}{x^2+a^2}$
$\mathrm{cosec}(x)$	$-\mathrm{cosec}(x)\cot(x)$	$\mathrm{cosec}^{-1}\left(\frac{x}{a}\right)$	$\frac{-a}{x\sqrt{x^2-a^2}}$
$\sec(x)$	$\sec(x)\tan(x)$	$\sec^{-1}\left(\frac{x}{a}\right)$	$\frac{a}{x\sqrt{x^2-a^2}}$
$\cot(x)$	$-\mathrm{cosec}^2(x)$	$\cot^{-1}\left(\frac{x}{a}\right)$	$\frac{-a}{x^2+a^2}$
$\sinh(x)$	$\cosh(x)$	$\sinh^{-1}\left(\frac{x}{a}\right)$	$\frac{1}{\sqrt{x^2+a^2}}$
$\cosh(x)$	$\sinh(x)$	$\cosh^{-1}\left(\frac{x}{a}\right)$	$\frac{1}{\sqrt{x^2-a^2}}$
$\tanh(x)$	$1-\tanh^2(x)$	$\tanh^{1}\left(\frac{x}{a}\right)$	$\frac{a}{a^2-x^2}$
$\mathrm{cosech}(x)$	$-\mathrm{cosech}(x)\coth(x)$	$\mathrm{cosech}^{-1}\left(\frac{x}{a}\right)$	$\frac{-a}{x\sqrt{x^2+a^2}}$
$\mathrm{sech}(x)$	$-\mathrm{sech}(x)\tanh(x)$	$\mathrm{sech}^{-1}\left(\frac{x}{a}\right)$	$\frac{-a}{x\sqrt{a^2-x^2}}$
$\coth(x)$	$1-\coth^2(x)$	$\coth^{-1}\left(\frac{x}{a}\right)$	$\frac{a}{x^2-a^2}$

12

Selected Distributions

We have seen that it is a common error to assume that data are normally distributed. This can even be done unknowingly. The control engineer may not appreciate that a statistical formula being applied was derived from the PDF or CDF of the normal distribution. This can result in seriously erroneous conclusions, of which there are many examples. They include proceeding with an improved control project that appears to be economically attractive when, in fact, it is not. Another is deducing an inferential property is reliable when, in fact, including it in a control scheme actually worsens control of the property.

There are many reasons why process data may not be normally distributed. One of the aims of this book is to highlight this issue and present a wide range of alternative choices for the prior distribution. This chapter focusses on those that are more commonly presented in textbooks. They represent a small fraction of many hundred published. In some cases their popularity is justified; they work well for the purpose for which they were developed. However, popularity is rarely a reliable measure of effectiveness. Take, as an example, the Ziegler–Nichols controller tuning method. It is almost universally included in university courses on process control, despite it now being virtually useless as a tuning method. Of the dozens of other published methods, perhaps one or two are effective and these are among the least well known. The same is true of statistical distributions. Indeed, the best choice for any dataset may not be among those included in this chapter. For this reason, a summary of the many alternatives is included as Appendix 2.

To provide a benchmark against which to assess alternatives we can fit a normal distribution to some of our example datasets:

- We showed in Chapter 9 that Equation (5.35) can be fitted to the C_4 in propane data more accurately than Equation (5.43). The better choice gives a mean (μ) of 3.94, a standard deviation (σ) of 1.96 and *RSS* as 0.2303. Skewness, calculated from Equation (4.41), is 1.27 and kurtosis, from Equation (4.42), is 4.79 – both well outside the acceptable range.
- The NHV disturbance data has relatively little skewness ($\gamma = 0.61$) but a very high kurtosis ($\kappa = 8.87$). In Chapter 9 we showed that Equation (5.43) gives the better fit, with μ as −0.0206, σ as 1.19 and *RSS* as 0.1437.

Statistics for Process Control Engineers: A Practical Approach, First Edition. Myke King.

At the end of this chapter it will become apparent which of the distributions described best fit these two datasets. This, of course, does not imply that they will always be the best choice. With experience, the engineer will be able to reject some distributions as being unsuitable but will, for every dataset, need to test each of the remaining contenders. Indeed, as will become apparent in Part 2, there might be several dozen contenders.

12.1 Lognormal

The *lognormal distribution* is one of many derived from the normal distribution; others are included in Part 2. It is the probability distribution of a variable whose logarithm is normally distributed. Its advantage is its ability to represent the distribution of positively skewed data.

We saw in Section 10.2 how the Central Limit Theorem shows that values, derived by multiplying or dividing data, tend to follow a lognormal distribution. Many of the parameters that affect process behaviour are the result of such calculations. It is not therefore surprising that the lognormal distribution often fits process data well.

Equation (5.35) describes the normal distribution; the PDF of the lognormal distribution is derived from it.

$$f(x)=\frac{1}{\beta x\sqrt{2\pi}}\exp\left[\frac{-(\ln(x)-\alpha)^2}{2\beta^2}\right] \qquad x>0;\ \beta>0 \tag{12.1}$$

The mean (α) and standard deviation (β) are those for $\ln(x)$. Raw moments for x are given by

$$m_n=\exp\left(n\alpha+\frac{n^2\beta^2}{2}\right) \tag{12.2}$$

Hence

$$\mu=\exp\left(\alpha+\frac{\beta^2}{2}\right) \tag{12.3}$$

$$\sigma^2=\left(\exp(\beta^2)-1\right)\exp(2\alpha+\beta^2) \tag{12.4}$$

Rearranging

$$\alpha=\ln(\mu)-\ln\left(\sqrt{1+\frac{\sigma^2}{\mu^2}}\right) \tag{12.5}$$

$$\beta=\sqrt{\ln\left(1+\frac{\sigma^2}{\mu^2}\right)} \tag{12.6}$$

Figure 12.1 shows the effect of changing σ, with μ fixed at 5. As σ gets smaller the distribution becomes increasingly symmetrical.

The CDF cannot be obtained by integrating Equation (12.1). So, like the normal distribution, it uses the error function.

$$F(x)=\frac{1}{2}+\frac{1}{2}\operatorname{erf}\left[\frac{\ln(x)-\alpha}{\beta\sqrt{2}}\right] \tag{12.7}$$

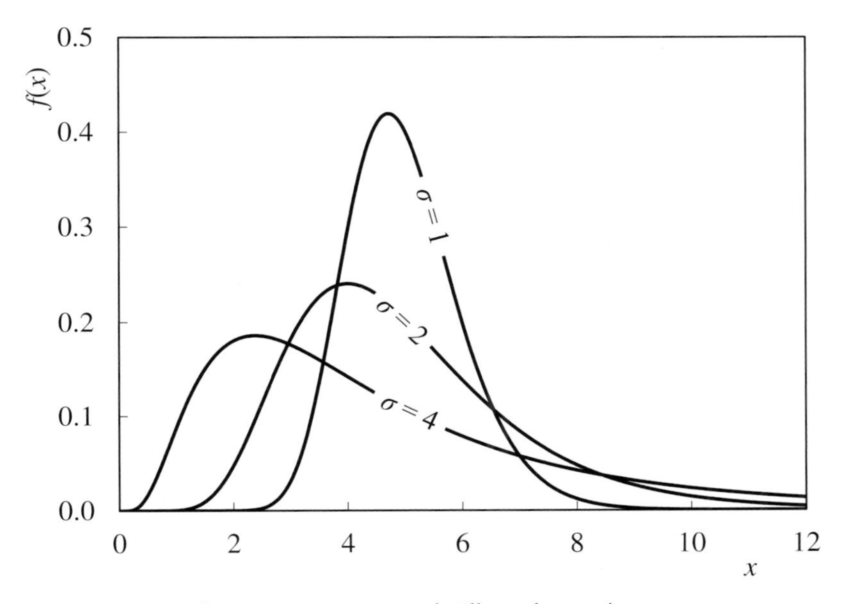

Figure 12.1 *Lognormal: Effect of σ on shape*

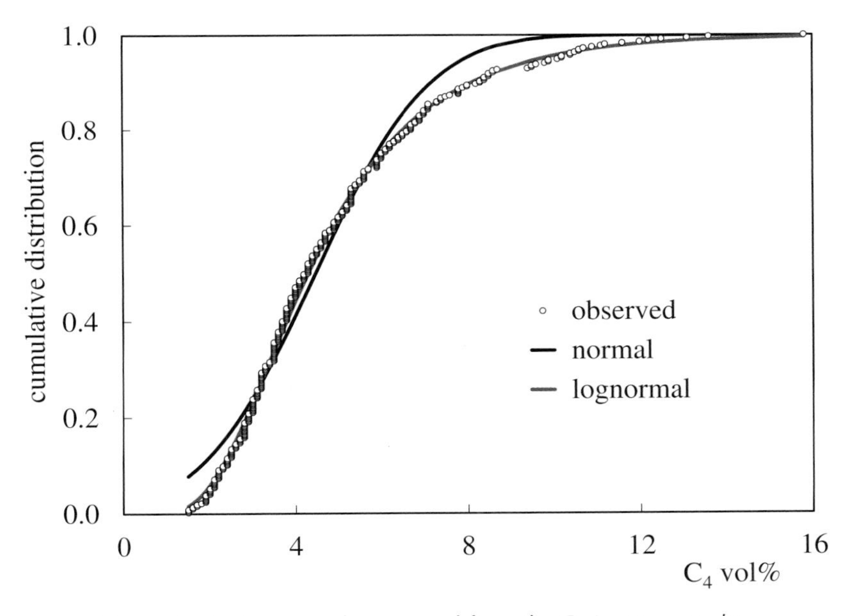

Figure 12.2 *Lognormal: Improved fit to the C_4 in propane data*

From the C_4 in propane data we can calculate μ as 4.87 and σ as 2.43. We can, from Equation (12.5), calculate the maximum likelihood estimate for α as 1.47 and, from Equation (12.6), that for β as 0.47. Adjusting these values to 1.46 and 0.49 respectively minimises *RSS* to 0.0477. The fit is considerably better than that achieved by the normal distribution, as illustrated by Figures 12.2 and 12.3. Figure 12.4 further shows a close fit of the PDF to the observed distribution plotted as a histogram.

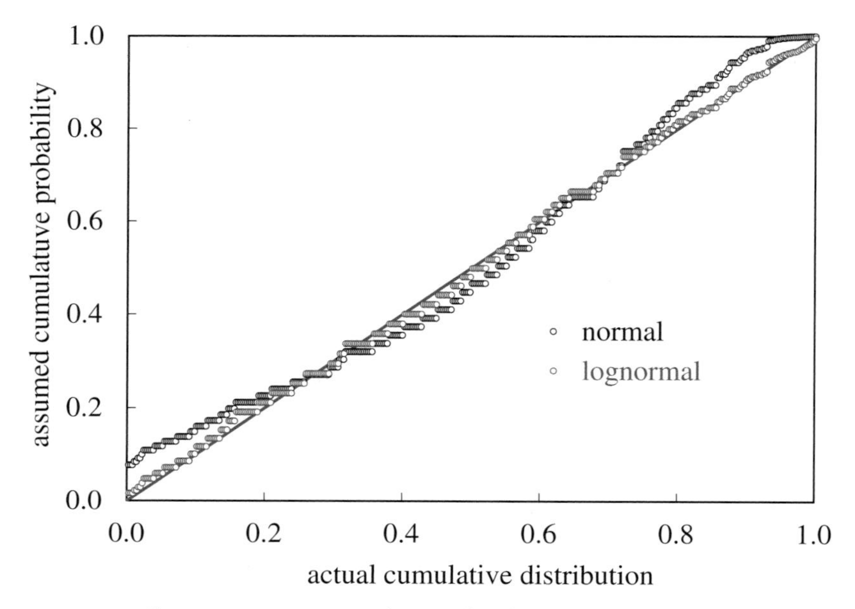

Figure 12.3 Lognormal: P–P plot showing improved fit

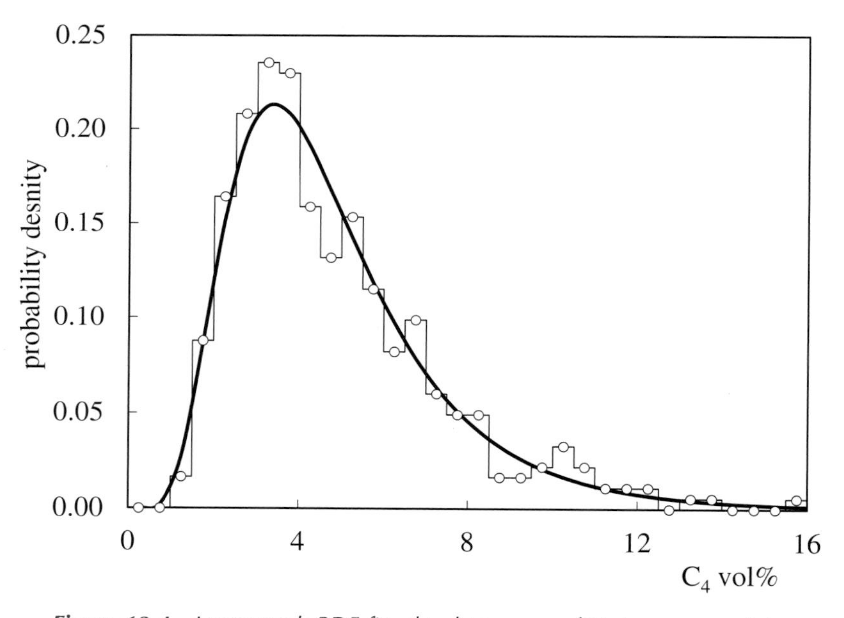

Figure 12.4 Lognormal: PDF fitted to histogram of C_4 in propane data

From Equation (12.3), we obtain the mean as 4.85 and, from Equation (12.4), the standard deviation as 2.53.

Skewness and kurtosis are

$$\gamma = \left[\exp\left(\beta^2\right) + 2\right]\sqrt{\exp\left(\beta^2\right) - 1} \quad (12.8a)$$

$$\kappa = \exp(4\beta^2) + 2\exp(3\beta^2) + 3\exp(2\beta^2) - 3 \tag{12.8b}$$

Figure 12.5 shows the feasible combinations of skewness and kurtosis. The values calculated from the data are close to this line – further indicating a good fit. But, as we saw in some of the examples in Chapter 10, the lognormal distribution cannot represent negative skewness. For example, the distribution of the C_3 content of propane (approximating to 100 – C_4) would not be well represented.

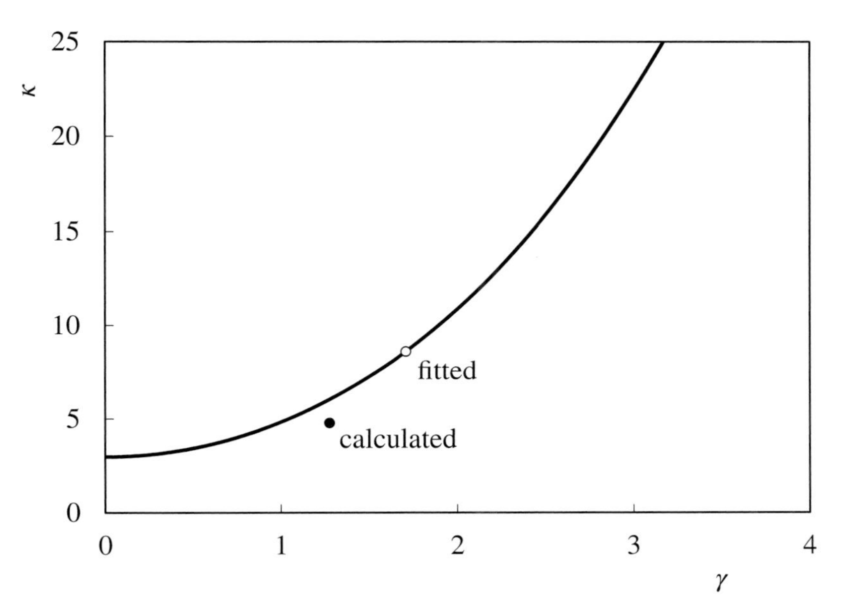

Figure 12.5 *Lognormal: Feasible combinations of γ and κ*

12.2 Burr

More fully entitled the *Burr-XII distribution* is also known as the *Pareto-IV distribution* or the *Singh–Maddala distribution*. Its PDF is

$$f(x) = \frac{\frac{\delta_1 \delta_2}{\beta}\left(\frac{x-\alpha}{\beta}\right)^{\delta_2 - 1}}{\left(1 + \left(\frac{x-\alpha}{\beta}\right)^{\delta_2}\right)^{\delta_1 + 1}} \qquad x \geq \alpha;\ \beta, \delta_1, \delta_2 > 0 \tag{12.9}$$

Its CDF is

$$F(x) = 1 - \left[1 + \left(\frac{x-\alpha}{\beta}\right)^{\delta_2}\right]^{-\delta_1} \tag{12.10}$$

We will see later, if δ_2 is set to 1, the distribution becomes the *Pareto-II distribution*.

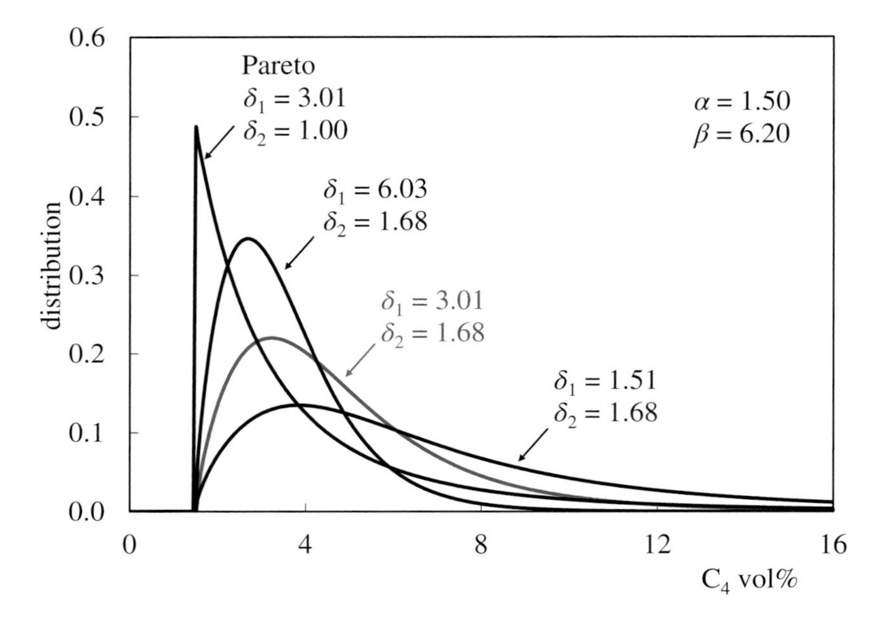

Figure 12.6 *Burr-XII: Fitted to the C_4 in propane data*

Fitting Equation (12.10) to the C_4 in propane data gives α as 1.50, β as 6.20, δ_1 as 3.01 and δ_2 as 1.68. Figure 12.6 shows, as the coloured line, this distribution and also the impact of changing these values. They minimise *RSS* to 0.0350. We shall see that this is one of the best fits among those distributions that permit simple calculation of mean and standard deviation.

The QF can be obtained by inverting the CDF.

$$x(F) = \alpha + \beta \left\{ [1-F]^{-1/\delta_1} - 1 \right\}^{1/\delta_2} \quad 0 \leq F \leq 1 \tag{12.11}$$

Raw moments are given by

$$m_n = \delta_1 \mathrm{B}\left(\frac{\delta_1\delta_2 - n}{\delta_2}, \frac{\delta_2 + n}{\delta_2} \right) \beta^n = \frac{n}{\delta_2 \Gamma(\delta_1)} \Gamma\left(\frac{n}{\delta_2}\right) \Gamma\left(\delta_1 - \frac{n}{\delta_2}\right) \beta^n \quad \delta_1\delta_2 > n \tag{12.12}$$

Setting n to 1 in Equation (12.12) gives the mean (μ) as 4.94. Setting n to 2 and putting the results into Equation (4.48) gives the standard deviation (σ) as 2.77. The third moment and Equation (4.52) give the skewness (γ) as 2.68. The fourth moment and Equation (4.55) give the kurtosis (κ) as an unrealistic 25.7. Figure 12.7 shows the feasible combinations of skewness and kurtosis derived from varying δ_1 and δ_2. Also shown are the fitted and calculated values. While *RSS* might indicate a good fit, the large discrepancy in kurtosis places some doubt on the applicability of this distribution.

Fitting Equation (12.10) to the NHV disturbance data gives −8.42 for α, 2.20 for β, 0.837 for δ_1 and 12.7 for δ_2 – resulting in a value of 0.0942 for *RSS*. Figure 12.8 compares the CDF against the EDF. We will see later that this is substantially bettered by other distributions. This means, unfortunately, that the Burr-XII distribution does not outperform others as a general-purpose choice. In fact, for many datasets, fitting will result in extremely large values of δ_1 and δ_2. Indeed, if no restriction is placed on the number of iterations, the values can approach infinity and cause overflow problems. These may occur during fitting or, afterwards, when calculating the moments.

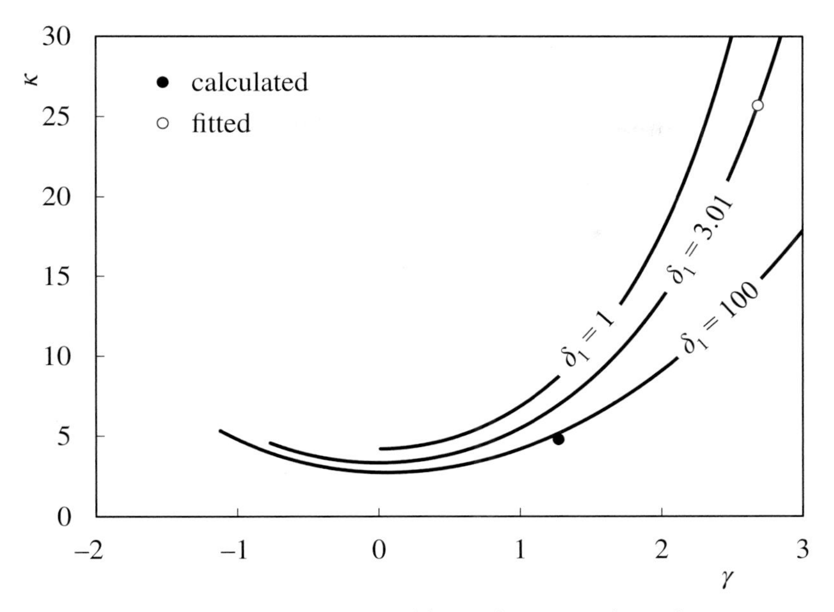

Figure 12.7 Burr-XII: Feasible combinations of γ and κ

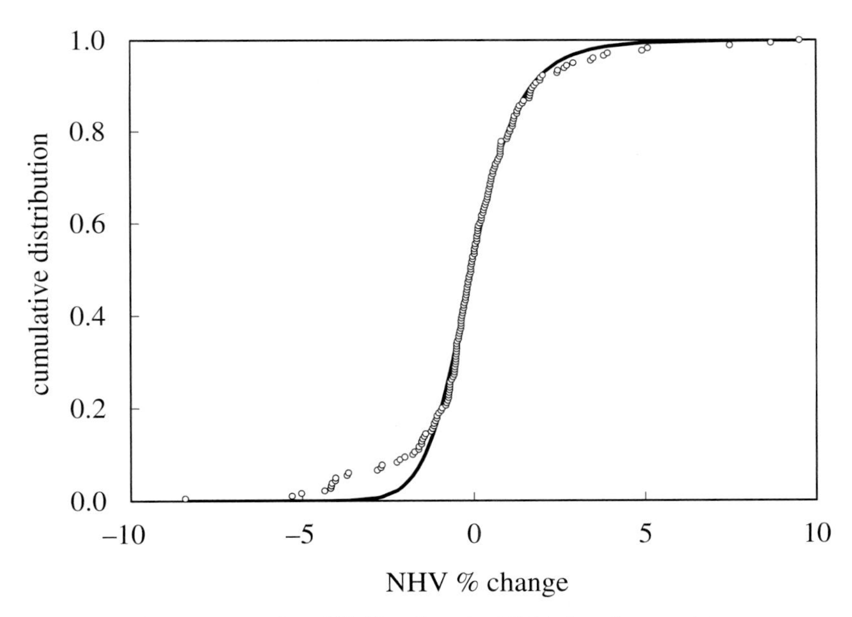

Figure 12.8 Burr-XII: Fitted to the NHV disturbance data

12.3 Beta

There are several versions of the beta distribution. The *beta-I distribution* is also known as the *Feller–Pareto distribution*. It is non-symmetrical, taking its usual name from the use of the beta function.

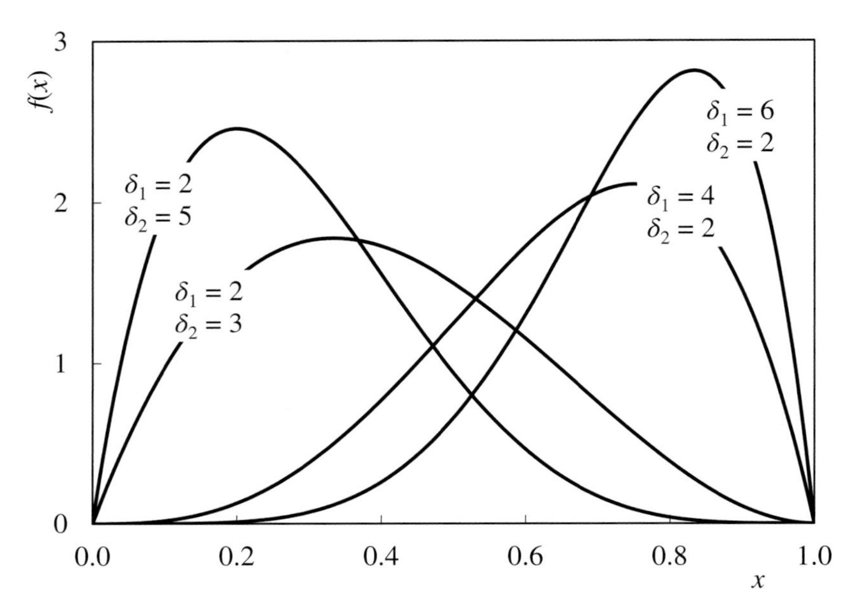

Figure 12.9 *Beta-I: Effect of* δ_1 *and* δ_2 *on shape*

Its PDF is

$$f(x) = \frac{x^{\delta_1 - 1}(1-x)^{\delta_2 - 1}}{\mathrm{B}(\delta_1, \delta_2)} \qquad 0 \leq x \leq 1;\ \delta_1, \delta_2 > 0 \tag{12.13}$$

The reciprocal of the beta function, $\mathrm{B}(\delta_1,\delta_2)$, is the normalisation constant, chosen so that the area under the probability distribution curve is 1.

The coefficients δ_1 and δ_2 are shape parameters. Figure 12.9 shows the effect of varying these. If set equal to each other, we obtain the *symmetric beta distribution* – as shown by Figure 12.10. If δ_1 and δ_2 are both 1, we obtain the uniform distribution U(0,1). While mathematically δ_1 and δ_2 can both approach zero, if both are less than 1, the resulting anti-modal distribution has little application to process data.

The CDF is the regularised incomplete beta function, as described in Section 11.6, and is supported by most spreadsheet packages. Alternatively, as we saw in Chapter 9, the trapezium rule can be applied to the PDF.

The key statistical parameters are

$$\mu = \frac{\delta_1}{\delta_1 + \delta_2} \tag{12.14}$$

$$\sigma^2 = \frac{\delta_1 \delta_2}{(\delta_1 + \delta_2)^2 (\delta_1 + \delta_2 + 1)} \tag{12.15}$$

$$\gamma = \frac{2(\delta_2 - \delta_1)}{\delta_1 + \delta_2 + 2} \sqrt{\frac{\delta_1 + \delta_2 + 1}{\delta_1 \delta_2}} \tag{12.16}$$

$$\kappa = \frac{\left(6(\delta_1 - \delta_2)^2 + 3\delta_1\delta_2(\delta_1 + \delta_2 + 2)\right)(\delta_1 + \delta_2 + 1)}{\delta_1 \delta_2 (\delta_1 + \delta_2 + 2)(\delta_1 + \delta_2 + 3)} \tag{12.17}$$

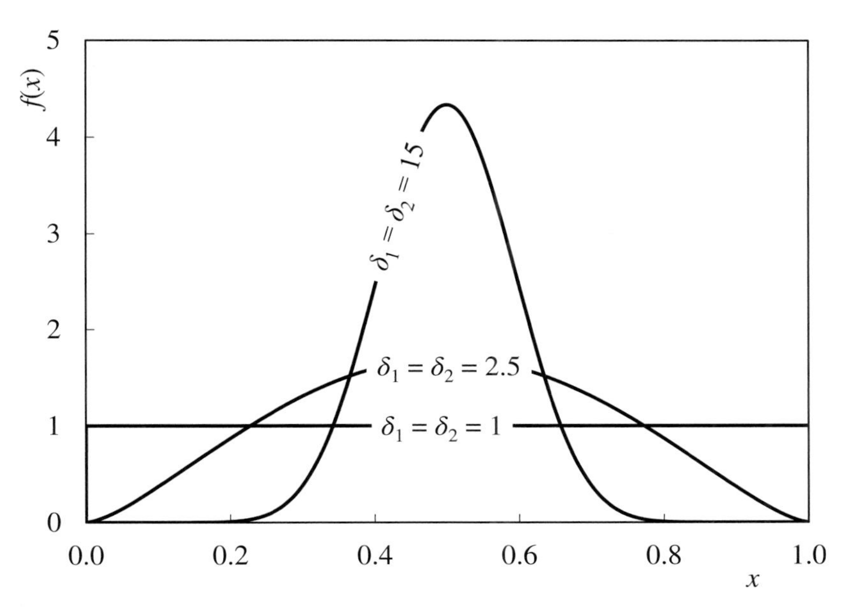

Figure 12.10 *Beta-I: Symmetric-beta ($\delta_1 = \delta_2$)*

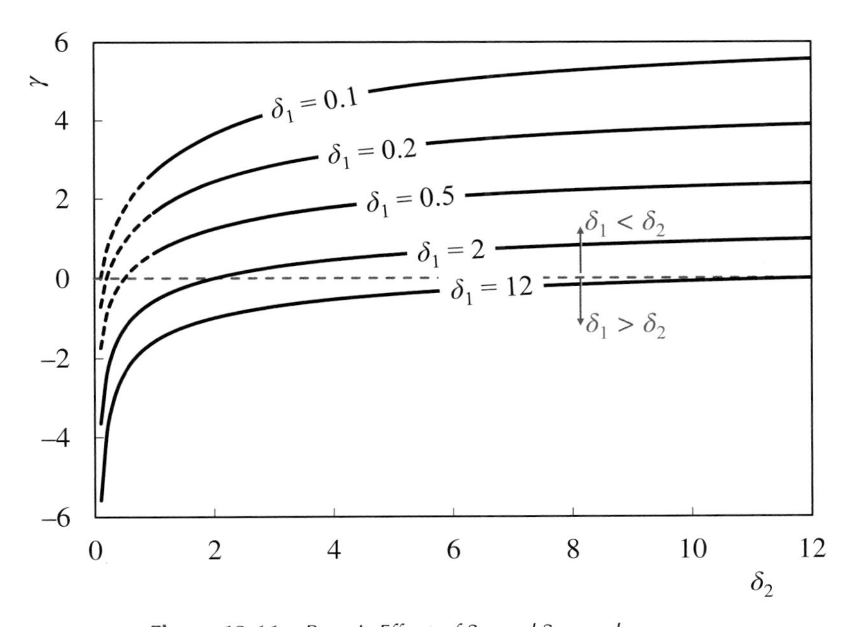

Figure 12.11 *Beta-I: Effect of δ_1 and δ_2 on skewness*

Figures 12.11 and 12.12 show the effect of δ_1 and δ_2 on skewness and kurtosis, demonstrating the wide range of shapes that can be represented by the distribution. One of the advantages of the distribution is that skewness can be introduced without affecting kurtosis and vice versa. As Figure 12.11 shows, if δ_1 is set less than δ_2, the distribution is skewed to the right (γ is positive). If δ_1 is set greater than δ_2 then γ is negative. Examination of Equation (12.16) shows that if

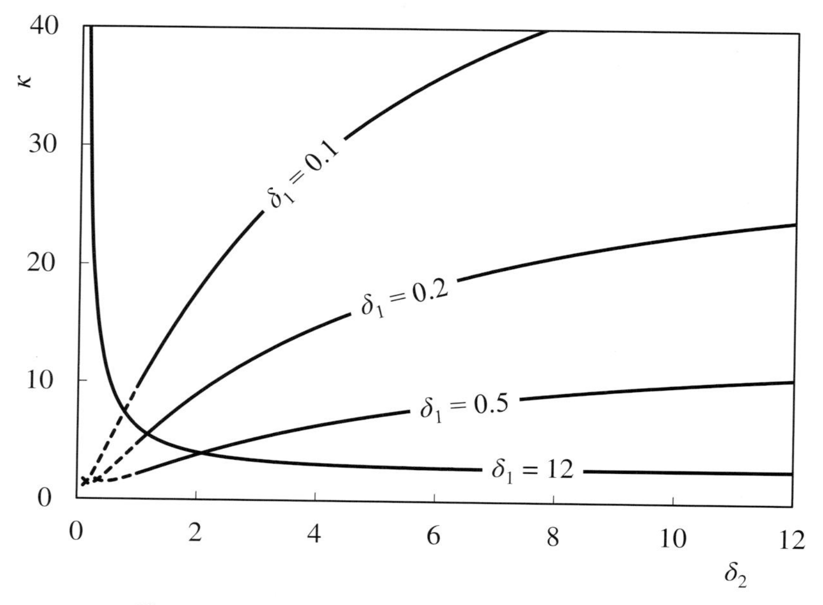

Figure 12.12 *Beta-I: Effect of δ1 and δ2 on kurtosis*

δ_1 and δ_2 are equal then skewness will be zero – as already shown by Figure 12.10. As Figure 12.12 shows, there are multiple combinations of δ_1 and δ_2 that give the same kurtosis.

From Equation (12.17), the condition for no kurtosis ($\kappa = 3$) is

$$\delta_1^3 - (2\delta_2 - 1)\delta_1^2 - 2\delta_2(\delta_2 + 2)\delta_1 + \delta_2^2(\delta_2 + 1) = 0 \tag{12.18}$$

As a cubic this equation has three solutions. One is trivial in that δ_1 and δ_2 are both zero. The other two are given by Figure 12.13. If the distribution is skewed to the right then δ_1 can be derived from δ_2 (or vice versa) using the upper line. If the distribution is skewed to the left, the relationship is given by the lower line. While the lines appear straight, they are not exactly so. The equations given are approximations valid only over the range plotted.

These relationships can of course be used to fit a distribution to data that, by calculation, has either zero skewness or zero kurtosis but this is not the intent. It merely demonstrates the versatility of the beta-I distribution. The approach, as usual, should be to adjust δ_1 and δ_2 to give the best fit.

Rearranging Equations (12.14) and (12.15) gives

$$\delta_1 = \frac{\mu(\mu - \mu^2 - \sigma^2)}{\sigma^2} \tag{12.19}$$

$$\delta_2 = \frac{(1-\mu)(\mu - \mu^2 - \sigma^2)}{\sigma^2} \tag{12.20}$$

Using the C_4 in propane data as an example, because x must be between 0 and 1, we first choose a range for the data – in this case from 1 to 16 vol%. Each result is then converted to a fraction of this range. A preliminary estimate of the mean, calculated from Equation (4.2), is 0.258. From Equation (4.35) we obtain an estimate of 0.162 for the standard deviation. Applying Equations (12.19) and (12.20) gives preliminary estimates for δ_1 and δ_2 of 1.63 and 4.69 respectively. Curve fitting gives better values as 2.01 and 6.53 respectively.

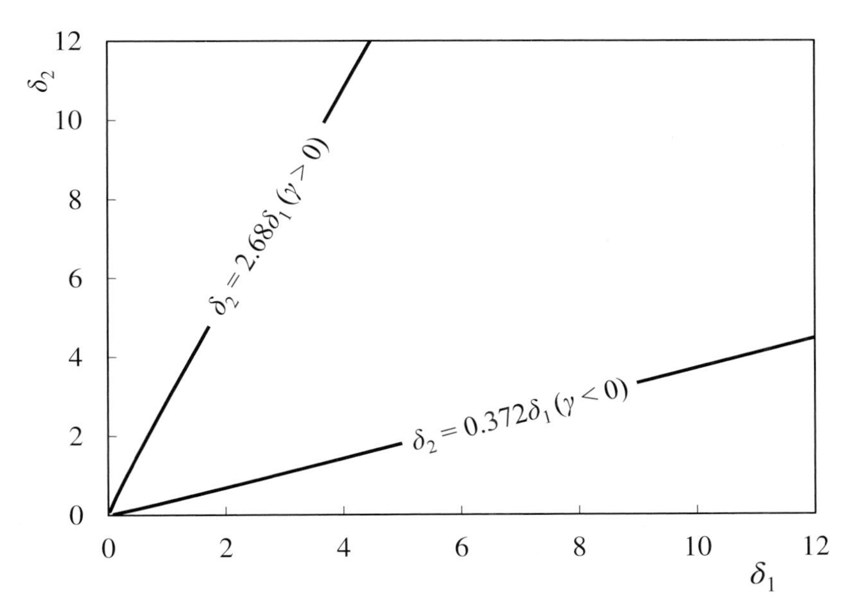

Figure 12.13 *Beta-I: Conditions for zero kurtosis*

Equations (12.14) and (12.15) give the mean and standard deviation as 0.235 and 0.137. Remembering that these are for x, the scaling must be used to convert them to measures of product composition.

$$\mu = 0.235(16 - 1) + 1 = 4.52 \tag{12.21}$$

$$\sigma = 0.137(16 - 1) = 2.06 \tag{12.22}$$

Although we fit the distribution to scaled variables, *RSS* can still be used to compare the accuracy of fit to an alternative distribution. *RSS* is based on the accuracy with which we predict $f(x)$ – not x. In this example it is 0.0841. While the beta-I distribution fits the data less well than many of distributions we cover, it has the advantage that it can model skewness in either direction. For example, if the composition data was presented as purity (i.e. vol% C_3) rather than as impurity, then the skew would have been reversed. As we have seen, we could have handled this by obtaining the distribution of (100 – C_3) vol% but we would then have to assume that the content of other components, such as C_2, does not vary.

The accuracy of fit is also influenced by the choice of range. Figure 12.14 shows the distribution developed above, using the range of 1 to 16%, compared to that based on a range determined by fitting. Of note is the large change in δ_2 and the significant reduction in *RSS*. The mean, standard deviation, skewness and kurtosis match more closely the values calculated from the data. The slight concern is that the range excludes the three occasions where the C_4 content fell to 1.5 vol% – effectively saying that this is impossible.

12.4 Hosking

More fully entitled the *Hosking four-parameter kappa distribution*,[6] it perhaps will not be as well known to the reader as others in this chapter. It is included because it will fit a broad range of datasets. It also includes, as special cases, many other distributions that therefore need not

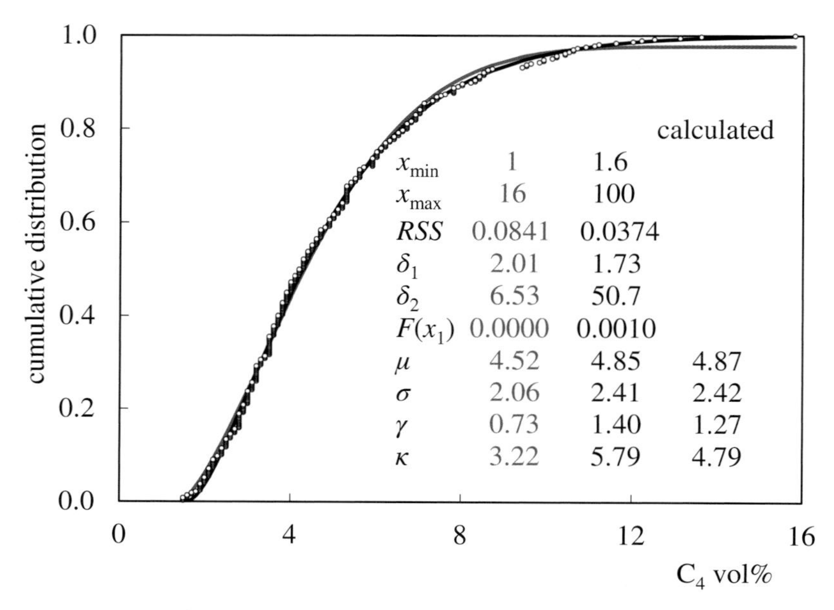

Figure 12.14 *Beta-I: Impact of choice of range*

be considered separately. In particular some are used in extreme value analysis that we will cover in Chapter 13. In the absence of any formulae for moments, it also gives an opportunity to demonstrate how they can be calculated using the trapezium rule.

The distribution is described by the PDF

$$f(x)=\frac{1}{\beta}\left[1-\frac{\delta_1(x-\alpha)}{\beta}\right]^{\frac{1}{\delta_1}-1}\left\{1-\delta_2\left[1-\frac{\delta_1(x-\alpha)}{\beta}\right]^{\frac{1}{\delta_1}}\right\}^{\frac{1}{\delta_2}-1} \qquad \beta>0 \tag{12.23}$$

To ensure a negative number is not raised to a non-integer power, the bounds described by Table 12.1 are placed on x; outside of these $f(x)$ is zero.

Table 12.1 *Bounds for the Hosking distribution*

δ_1	δ_2	minimum	maximum
>0	>0	$\alpha+\frac{\beta(1-\delta_2^{-\delta_1})}{\delta_1}$	$\alpha+\frac{\beta}{\delta_1}$
0	>0	$\alpha+\beta\ln(\delta_2)$	∞
<0	>0	$\alpha+\frac{\beta(1-\delta_2^{-\delta_1})}{\delta_1}$	∞
>0	≤0	$-\infty$	$\alpha+\frac{\beta}{\delta_1}$
0	≤0	$-\infty$	∞
<0	≤0	$\alpha+\frac{\beta}{\delta_1}$	∞

The CDF is

$$F(x)=\left\{1-\delta_2\left[1-\frac{\delta_1(x-\alpha)}{\beta}\right]^{\frac{1}{\delta_1}}\right\}^{\frac{1}{\delta_2}} \tag{12.24}$$

It is possible to invert the CDF to give

$$x(F)=\alpha+\frac{\beta}{\delta_1}\left[1-\left(\frac{1-F^{\delta_2}}{\delta_2}\right)^{\delta_1}\right] \qquad 0\leq F\leq 1 \tag{12.25}$$

Fitting the distribution to the C_4 in propane data gives 3.26 for α, 2.04 for β, −0.0604 for δ_1 and 0.365 for δ_2. At 0.0319 *RSS* is one of the lowest of the distributions considered.

There are no published formulae for mean, standard deviation, skewness or kurtosis. Instead we can apply the trapezium rule, described by Equation (4.68), to determine the moments. From Table 12.1, the distribution is lower-bounded at 1.26. So, although i is ranged from zero, only values of x greater than this are actually used. Choosing 200 as the maximum value of x ensures that $f(x)$ is, not surprisingly, very close to zero. This value is around 80 standard deviations from the mean. Choosing the interval as 0.1 gives the upper limit of i as 2000. Table 12.2 includes selected rows of the calculations. For the first and last groups of six, $f(x)$ is effectively zero. As accumulating sums, the raw moments increase by row until they reach their asymptotic values of

$$m_1=4.91 \qquad m_2=31.0 \qquad m_3=254 \qquad m_4=2707 \tag{12.26}$$

From Equations (4.46), (4.49), (4.52) and (4.55) the properties of the distribution are therefore asymptotic to

$$\mu=4.91 \qquad \sigma=2.62 \qquad \gamma=1.90 \qquad \kappa=9.65 \tag{12.27}$$

Calculated from the data, the corresponding properties are

$$\mu=4.87 \qquad \sigma=2.43 \qquad \gamma=1.27 \qquad \kappa=4.79 \tag{12.28}$$

This illustrates how higher moments are more prone to error. For example, the calculation of kurtosis involves the fourth raw moment and hence shows the largest discrepancy. In contrast, the mean is the first raw moment and is closest to the estimate. This result is typical of even a well-fitting distribution.

If we were using this distribution to assess the improvement arising from improved control we might, as described in Chapter 2, apply the same percentage rule. If, before implementation, it is acceptable to violate the specification 5% of the time then we should accept the same afterwards. From Equation (12.25) we currently expect the C_4 content, for 95% of the time, to be 9.95 or less. If improved control halves the standard deviation, this is equivalent to halving β to 1.02. Rearranging Equation (12.25)

$$\alpha=x-\frac{\beta}{\delta_1}\left[1-\left(\frac{1-F^{\delta_2}}{\delta_2}\right)^{\delta_1}\right]=9.95-\frac{1.02}{-0.0604}\left[1-\left(\frac{1-0.95^{0.365}}{0.365}\right)^{-0.0604}\right]=6.59 \tag{12.29}$$

Improved control would therefore enable the average C_4 content to be increased by 3.33 (6.59 − 3.26) as shown by Figure 12.15.

Fitting the distribution to the NHV disturbance data gives −0.0143 for α, 0.689 for β, −0.0906 for δ_1 and 1.10 for δ_2. *RSS* is 0.0959 – showing one of the poorest fits. As Figure 12.16 shows the fit is poor in the region of downward disturbances in the range 1 to 5.

Table 12.2 *Calculation of μ, σ, γ and κ for Hosking distribution*

x	$f(x)$	$xf(x)$	m_1	$x^2f(x)$	m_2	$x^3f(x)$	m_3	$x^4f(x)$	m_4	μ	σ	γ	κ
0.0	0.00	0.00	0.00	0.00	0.00	0.00	0.00	0.00	0.00	0.00	0.00	∞	∞
0.1	0.00	0.00	0.00	0.00	0.00	0.00	0.00	0.00	0.00	0.00	0.00		
0.2	0.00	0.00	0.00	0.00	0.00	0.00	0.00	0.00	0.00	0.00	0.00		
0.3	0.00	0.00	0.00	0.00	0.00	0.00	0.00	0.00	0.00	0.00	0.00		
0.4	0.00	0.00	0.00	0.00	0.00	0.00	0.00	0.00	0.00	0.00	0.00		
0.5	0.00	0.00	0.00	0.00	0.00	0.00	0.00	0.00	0.00	0.00	0.00		
.	.	.	.	.	.	.	.	.	.	.	.		
.	.	.	.	.	.	.	.	.	.	.	.		
2.0	0.13	0.26	0.08	0.52	0.14	1.03	0.25	2.06	0.45	0.08	0.36	4.57	22.03
2.1	0.15	0.31	0.10	0.64	0.20	1.35	0.37	2.83	0.70	0.10	0.43	3.90	16.36
2.2	0.16	0.35	0.14	0.78	0.27	1.71	0.52	3.76	1.03	0.14	0.50	3.38	12.58
2.3	0.17	0.40	0.18	0.92	0.35	2.12	0.71	4.87	1.46	0.18	0.57	2.97	9.95
2.4	0.19	0.44	0.22	1.07	0.45	2.56	0.95	6.15	2.01	0.22	0.64	2.63	8.04
2.5	0.20	0.49	0.26	1.22	0.57	3.05	1.23	7.62	2.70	0.26	0.70	2.34	6.62
.	.	.	.	.	.	.	.	.	.	.	.	.	.
.	.	.	.	.	.	.	.	.	.	.	.	.	.
199.5	0.00	0.00	4.91	0.00	31.0	0.00	254	0.00	2707	4.91	2.62	1.90	9.65
199.6	0.00	0.00	4.91	0.00	31.0	0.00	254	0.00	2707	4.91	2.62	1.90	9.65
199.7	0.00	0.00	4.91	0.00	31.0	0.00	254	0.00	2707	4.91	2.62	1.90	9.65
199.8	0.00	0.00	4.91	0.00	31.0	0.00	254	0.00	2707	4.91	2.62	1.90	9.65
199.9	0.00	0.00	4.91	0.00	31.0	0.00	254	0.00	2707	4.91	2.62	1.90	9.65
200.0	0.00	0.00	4.91	0.00	31.0	0.00	254	0.00	2707	4.91	2.62	1.90	9.65

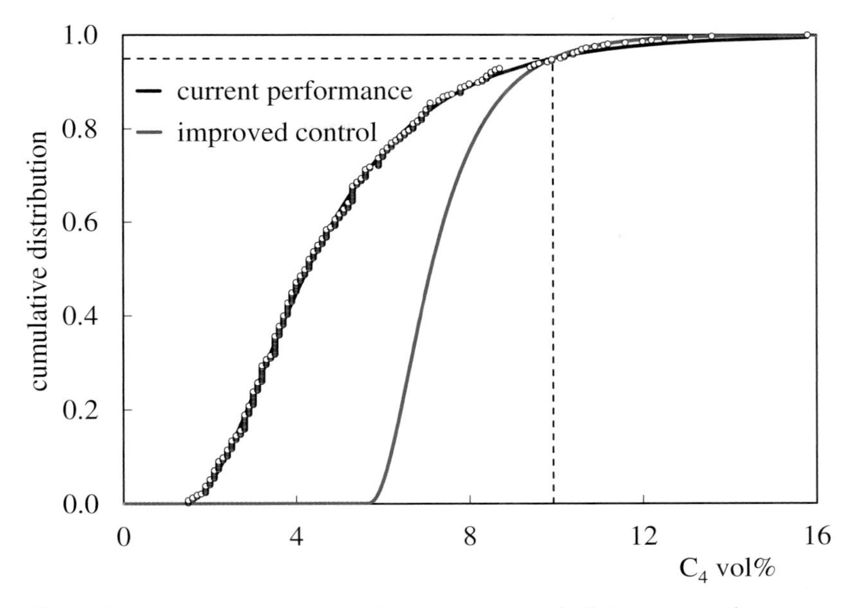

Figure 12.15 Hosking: Impact of improved control of C_4 content of propane

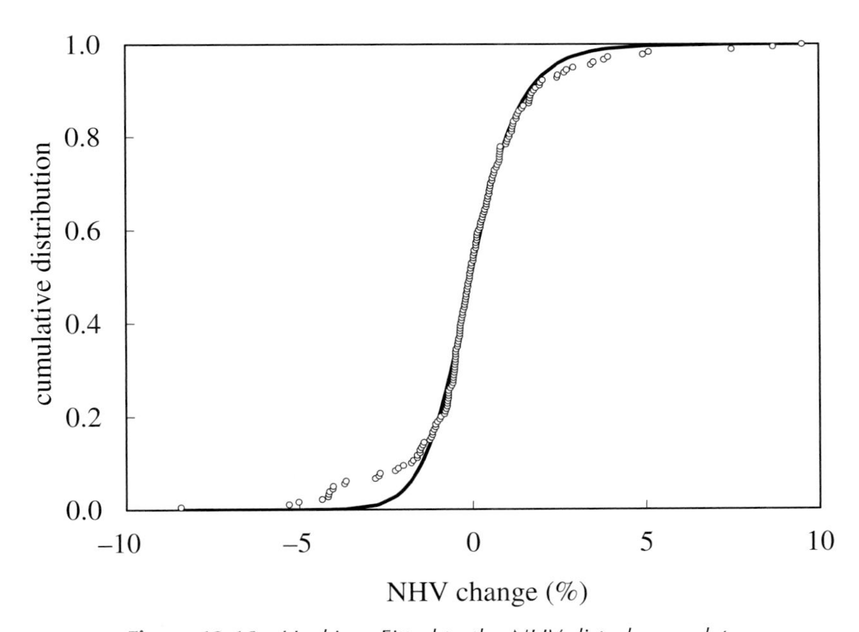

Figure 12.16 Hosking: Fitted to the NHV disturbance data

Figure 12.17 presents the same data, but based on absolute changes in NHV. The fit gives 0.416 for α, 0.591 for β, −0.558 for δ_1 and 0.265 for δ_2. With RSS at 0.0272, it is the best of the distributions considered but, more importantly, demonstrates the flexibility of Equation (12.24) in representing very different distributions.

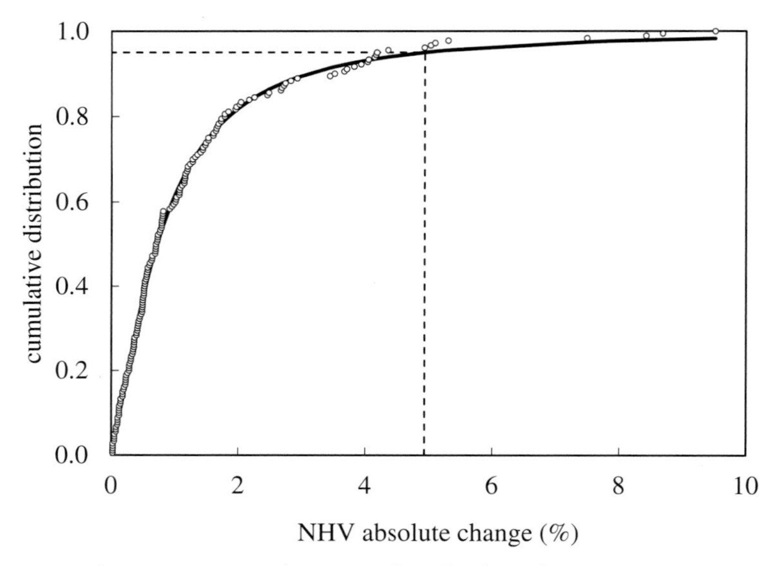

Figure 12.17 *Hosking: Fitted to absolute changes in NHV*

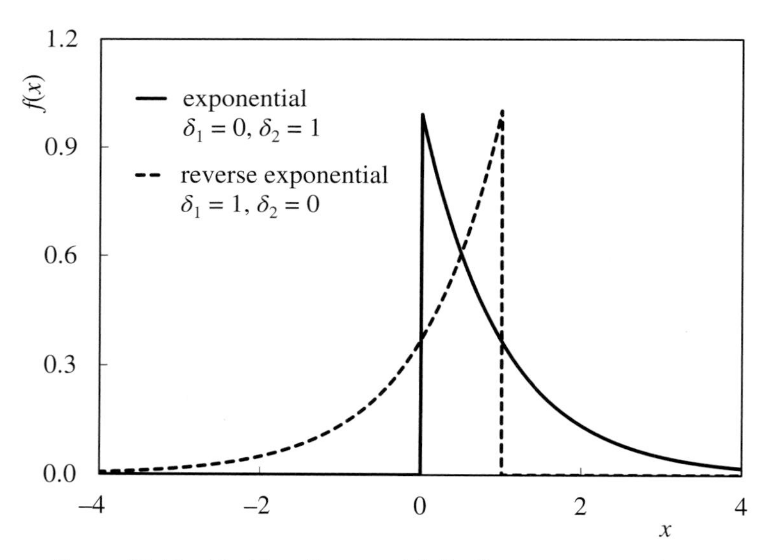

Figure 12.18 *Hosking: Exponential distributions as special case*

Several of the distributions we cover later are special cases of the Hosking distribution. These are illustrated by Figures 12.18 to 12.21. In all cases α is fixed at 0 and β at 1.

In the case of the *Gumbel distribution*, because both δ_1 and δ_2 are set to 0, we apply Equation (11.4) to generate the PDF

$$f(x) = \frac{1}{\beta}\exp\left[-\frac{x-\alpha}{\beta}\right]\exp\left\{-\exp\left[-\frac{x-\alpha}{\beta}\right]\right\} = \frac{1}{\beta}\exp\left[-\frac{x-\alpha}{\beta}-\exp\left(-\frac{x-\alpha}{\beta}\right)\right] \quad \beta>0 \tag{12.30}$$

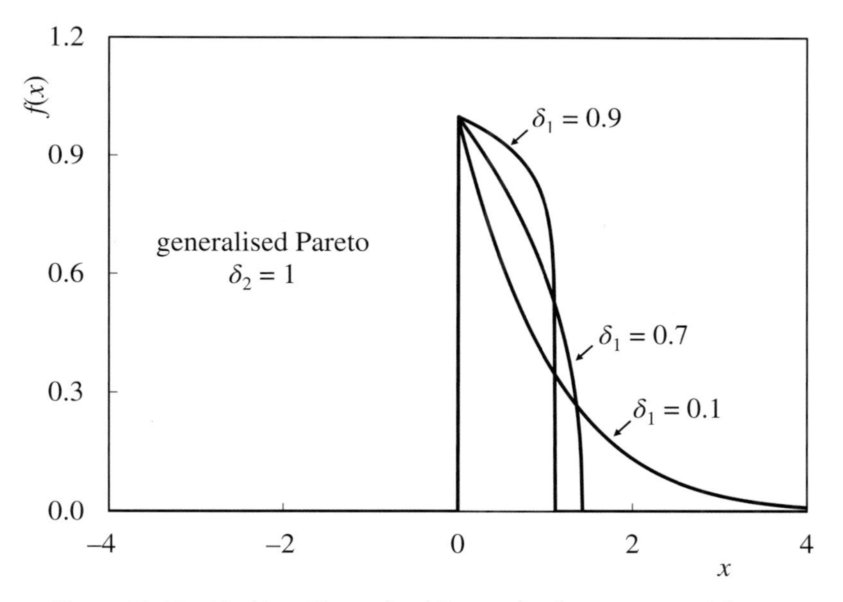

Figure 12.19 *Hosking: Generalised Pareto distribution as special case*

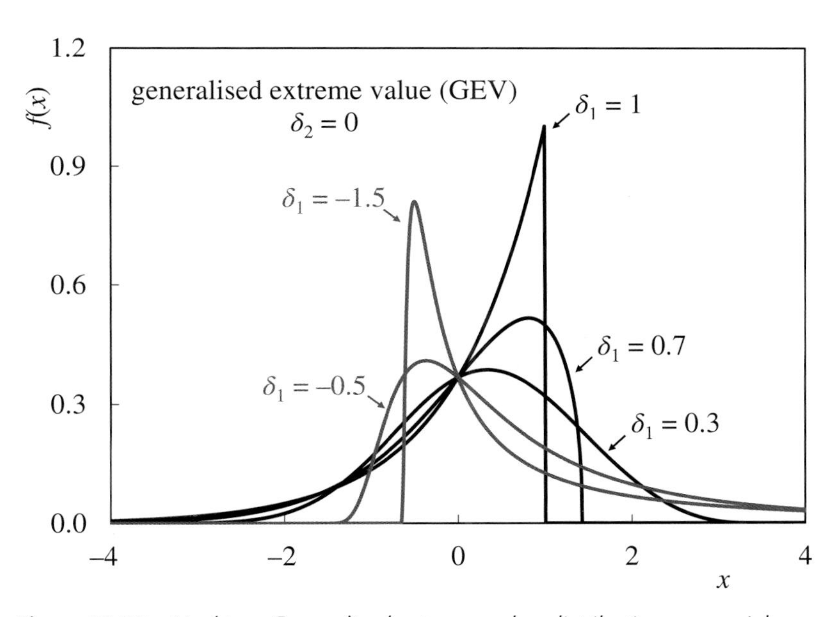

Figure 12.20 *Hosking: Generalised extreme value distribution as special case*

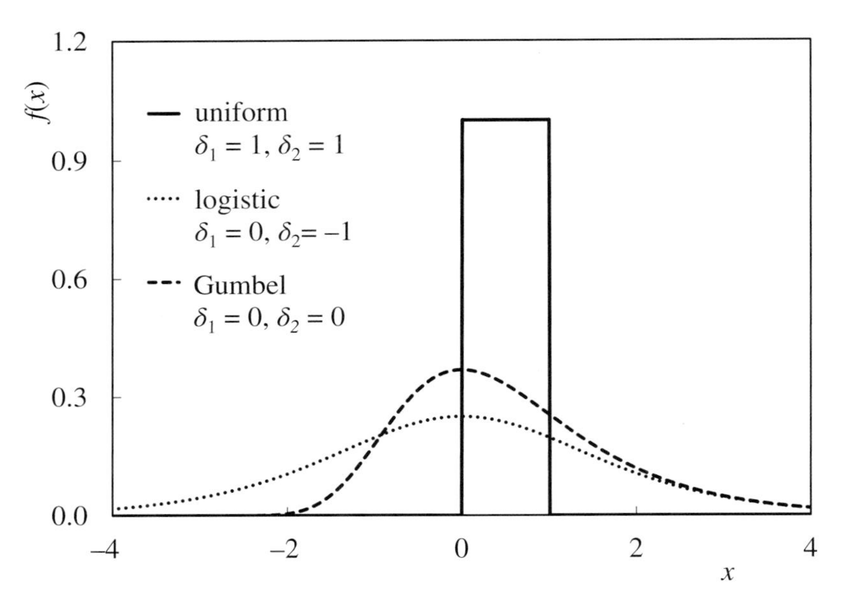

Figure 12.21 *Hosking: Other special cases*

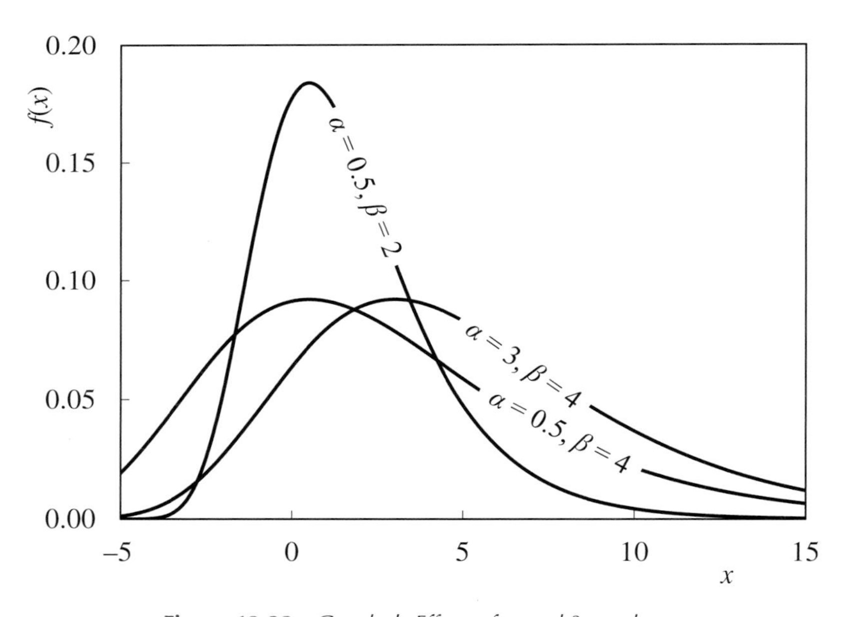

Figure 12.22 *Gumbel: Effect of α and β on shape*

It can also be described as the *extreme value-I distribution*, reflecting its use in extreme value analysis that we will cover in Chapter 13.

Figure 12.22 shows the effect of varying α and β.

Its CDF and QF are

$$F(x) = \exp\left[-\exp\left(-\frac{x-\alpha}{\beta}\right)\right] \tag{12.31}$$

$$x(F) = \alpha - \beta \ln(-\ln F) \quad 0 \le F \le 1 \tag{12.32}$$

The calculation of the mean uses the Euler–Mascheroni constant as defined in Section 11.2.

$$\mu \approx \alpha + 0.5772\beta \tag{12.33}$$

$$\sigma^2 = \frac{\pi^2}{6}\beta^2 \approx 1.64\beta^2 \tag{12.34}$$

The calculation of skewness uses the Riemann zeta function described in Section 11.7, where $\zeta(3)$ is approximately 1.2025.

$$\gamma = \frac{12\sqrt{6}\zeta(3)}{\pi^3} \approx 1.14 \tag{12.35}$$

$$\kappa = 5.4 \tag{12.36}$$

Noting the convention to change the sign of δ, the PDF of the *generalised extreme value (GEV) distribution* is

$$f(x) = \frac{1}{\beta}\left[1 + \frac{\delta(x-\alpha)}{\beta}\right]^{-\frac{1}{\delta}-1} \exp\left\{-\left[1 + \frac{\delta(x-\alpha)}{\beta}\right]^{-\frac{1}{\delta}}\right\} \quad \delta x \ge \delta\alpha - \beta;\ \beta > 0;\ \delta \ne 0 \tag{12.37}$$

Note that, because δ can be negative, the lower bound on x cannot generally be expressed as $\alpha - \beta/\delta$. Indeed, if δ is negative, this is the upper bound.

This distribution also has a specific application to extreme value analysis. Its CDF and QF are

$$F(x) = \exp\left\{-\left[1 + \frac{\delta(x-\alpha)}{\beta}\right]^{-\frac{1}{\delta}}\right\} \tag{12.38}$$

$$x(F) = \alpha + \frac{\beta}{\delta}\left\{[-\ln(F)]^{-\delta} - 1\right\} \tag{12.39}$$

Mean and variance are

$$\mu = \alpha + \frac{\Gamma(1-\delta) - 1}{\delta}\beta \quad \delta < 1 \tag{12.40}$$

$$\sigma^2 = \frac{\Gamma(1-2\delta) - \Gamma^2(1-\delta)}{\delta^2}\beta^2 \quad \delta < \frac{1}{2} \tag{12.41}$$

Skewness and kurtosis are

$$\gamma = \frac{\Gamma(1-3\delta) - 3\Gamma(1-\delta)\Gamma(1-2\delta) + 2\Gamma^2(1-\delta)}{\left[\Gamma(1-2\delta) - \Gamma^2(1-\delta)\right]^{\frac{3}{2}}}\frac{\delta}{|\delta|} \quad \delta < \frac{1}{3} \tag{12.42}$$

$$\kappa = \frac{\Gamma(1-4\delta) - 4\Gamma(1-\delta)\Gamma(1-3\delta) + 6\Gamma(1-2\delta)\Gamma^2(1-\delta) - 3\Gamma^4(1-\delta)}{\left[\Gamma(1-2\delta) - \Gamma^2(1-\delta)\right]^2} \quad \delta < \frac{1}{4} \tag{12.43}$$

While most of the special cases are named, no name has been given to the distribution if δ_1 is set to 0 and δ_2 to δ. The PDF becomes

$$f(x) = \frac{1}{\beta}\exp\left[-\frac{(x-\alpha)}{\beta}\right]\left\{1 - \delta\exp\left[-\frac{(x-\alpha)}{\beta}\right]\right\}^{\frac{1}{\delta}-1} \tag{12.44}$$

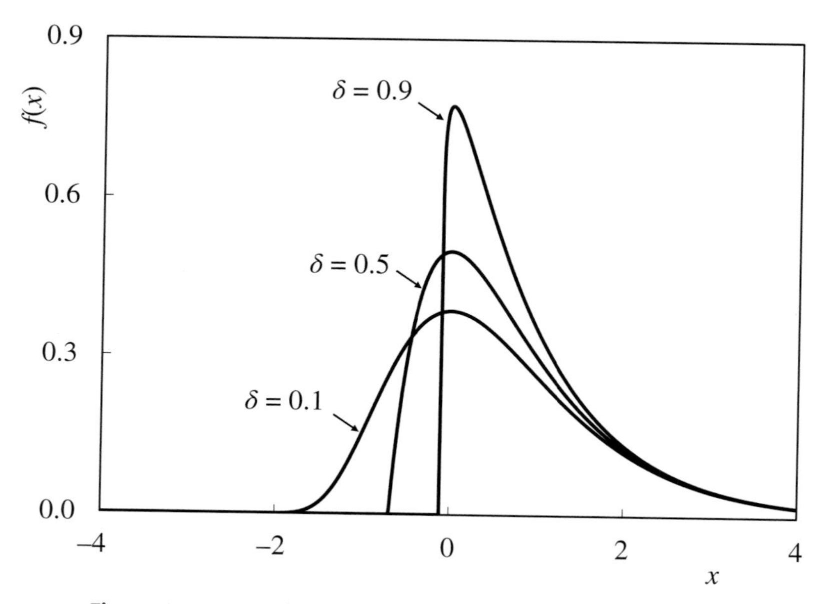

Figure 12.23 *Hosking: Unnamed distribution as special case*

Figure 12.23 shows the effect of varying δ. The CDF and its inverse are

$$F(x) = \left[1-\delta\left(1-\frac{x-\alpha}{\beta}\right)^{\frac{1}{\delta}}\right] \tag{12.45}$$

$$x(F) = \alpha - \beta\ln\left(\frac{1-F^{\delta}}{\delta}\right) \tag{12.46}$$

Similarly no name has been given to the distribution if δ_1 is set to 1 and δ_2 to δ. However, other than the special cases already covered ($\delta_2 = 0$ and $\delta_2 = 1$), it is unlikely that the distribution would have any application in the process industry.

Figure 12.24 maps the range of values for δ_1 and δ_2, summarising the range of distributions that the Hosking distribution can represent. Also shown are the values fitted to the C_4 in propane and NHV disturbance data.

12.5 Student *t*

The *standard Student t distribution* was actually proposed by William Sealy Gosset using the pseudonym Student. It is a special case of the Pearson-VII distribution described later in Section 27.7. It was developed specifically for use when the mean of the population is estimated from a very small number of results. The method gives the reliability of the estimate. This is useful, for example, in determining whether a few unusual results are representative of the expected behaviour of the process or whether they indicate a problem.

As we showed previously in defining the calculation of standard deviation (Section 4.7), if σ_p is the standard deviation of the population containing a sample of n values then estimate of

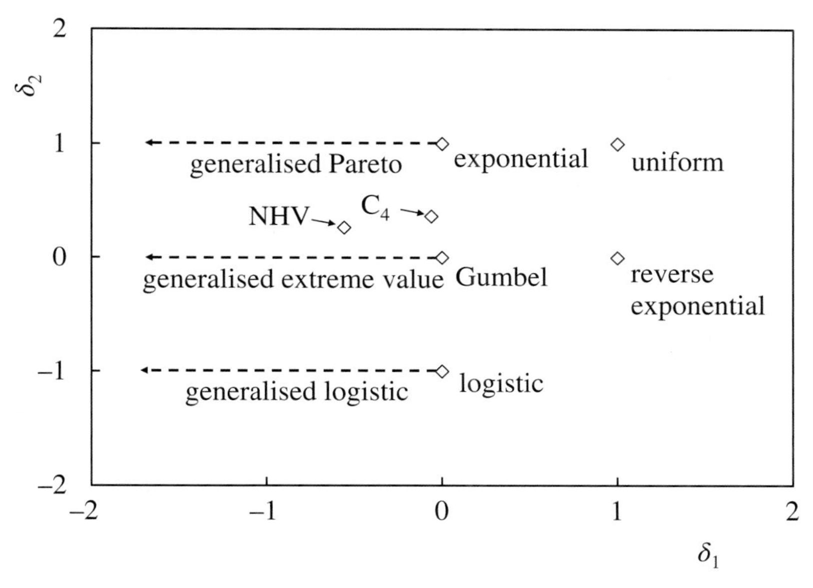

Figure 12.24 *Hosking: Summary of distributions represented*

the mean ($\bar{x}$) of the sample will have a standard deviation of $\sigma_p/\sqrt{n}$. If μ is the mean of the population then Student's t is defined by

$$t = \frac{\bar{x} - \mu}{\sigma_p/\sqrt{n}} \tag{12.47}$$

The PDF is

$$f(t) = \frac{\Gamma\left(\frac{f+1}{2}\right)}{\Gamma\left(\frac{f}{2}\right)\sqrt{\pi f}}\left(1 + \frac{t^2}{f}\right)^{-\frac{f+1}{2}} \quad f > 0 \tag{12.48}$$

To apply this function we first need to determine the number of degrees of freedom (f). We defined this in Section 4.7. As usual, since we have used n values to calculate the mean, the number degrees of freedom is $n - 1$. The shape of the distribution depends on f and hence on the sample size.

Figure 12.25 shows how the distribution for $f = 1$ compares to the normal distribution with a variance of 1. This is also the standard Cauchy distribution, described in Section 27.4 as a special case of the Pearson-IV distribution. As f increases, the t distribution becomes closer to the normal distribution. The two distributions are identical for an infinite number of degrees of freedom. For values of f greater than 15, the t distribution will be close enough to the normal distribution to make this approximation.

For example, imagine we are making a product that has an inferential property that, over the last 24 hours, shows that the property has been well controlled at 100 and so μ is 100. In the same period three 8-hourly laboratory tests gave the results 98, 100 and 96. The mean of these

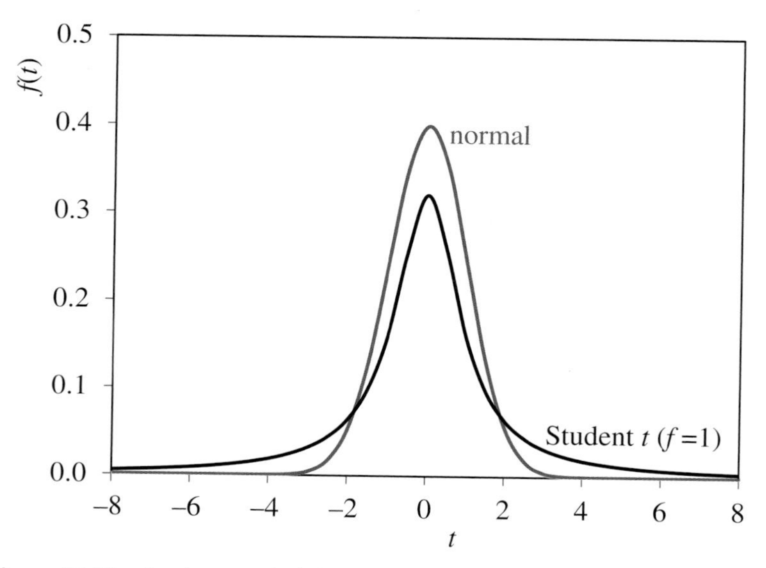

Figure 12.25 Student: Probability density compared to normal distribution

results ($\bar{x}$) is 98 and the standard deviation (σ), from Equation (4.35), is 2. From Equation (12.47), t is

$$t = \frac{98 - 100}{2/\sqrt{3}} = -1.732 \tag{12.49}$$

We have calculated one mean and so two degrees of freedom remain. Figure 12.26 shows the cumulative t distribution, drawn by applying the trapezium rule to Equation (12.48), for different values of f. For comparison the normal distribution is included as the coloured line. Figure 12.27 zooms in at the lower end of the plot range. For $f = 2$ (3 samples) it shows that, if the inferential is accurate, there is a probability of 0.113 that the true mean could be less than 98. One could thus conclude that this is within the 95% confidence that the inferential is correct. Figure 12.27 shows that, had there been only two samples ($f = 1$) with the same mean and standard deviation then, again using Equation (12.48), the probability that the mean could be less than 98 rises to 0.196. Similarly had there been five samples ($f = 4$), using Equation (12.48) shows it would fall to less than 0.05 – the level of probability that would typically prompt a redesign of the inferential. The t distribution therefore reflects our natural inclination to place more trust in a conclusion that is based on more results.

Another application of the t distribution is comparing the means of two small datasets to determine whether they are significantly different. One set has n_1 measurements, a mean of $\bar{x}_1$ and a standard deviation of σ_1. The other has n_2 measurements, a mean of $\bar{x}_2$ and a standard deviation of σ_2. The variance of the difference between the two means (σ_d) is

$$\sigma_d^2 = \sigma_{\bar{x}_1}^2 + \sigma_{\bar{x}_2}^2 = \frac{\sigma_1^2}{n_1} + \frac{\sigma_2^2}{n_2} \tag{12.50}$$

And so t can be derived using Equation (12.47).

$$t = \left| \frac{\bar{x}_1 - \bar{x}_2}{\sigma_d} \right| \tag{12.51}$$

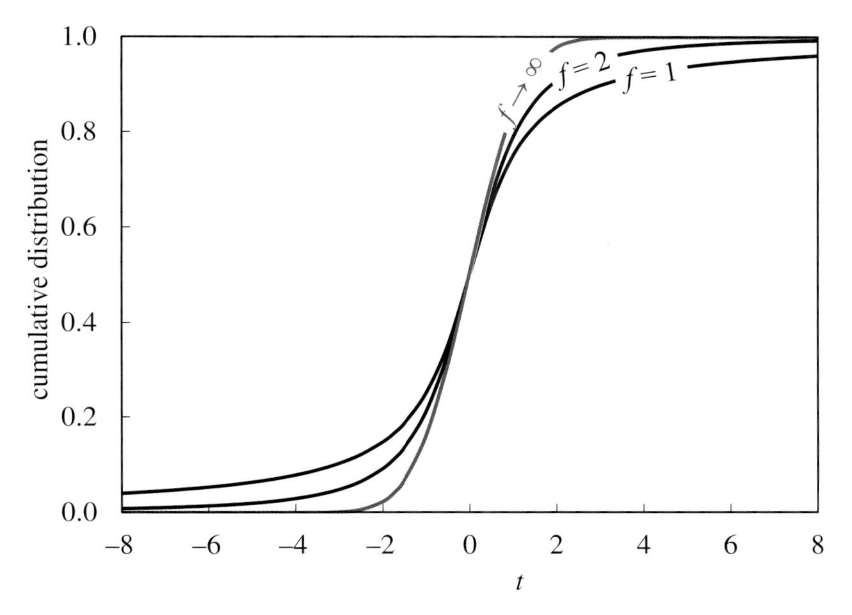

Figure 12.26 *Student: Effect of* f *on cumulative distribution*

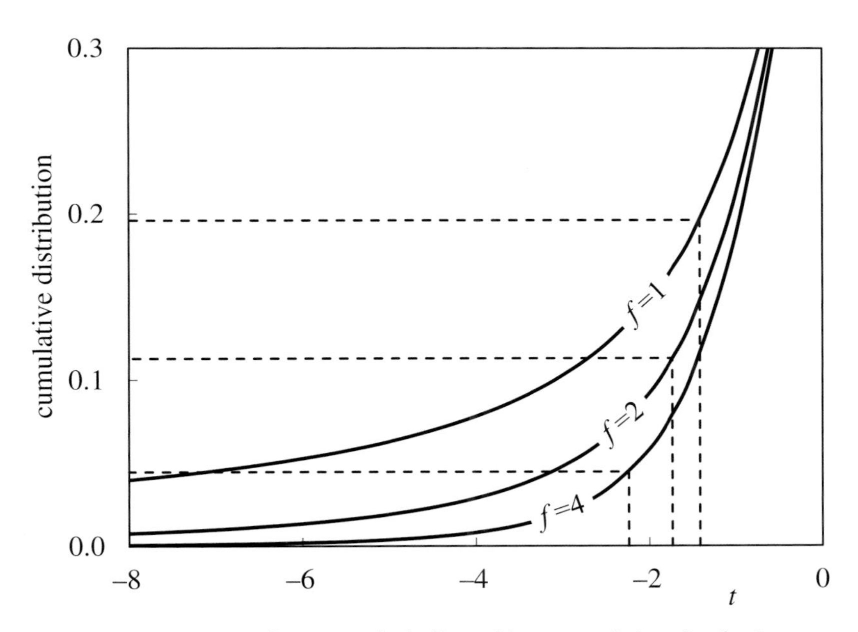

Figure 12.27 *Student: Detail of effect of* f *on cumulative distribution*

On the basis that we have derived two values (the two means) the number of degrees of freedom (f) is

$$f = n_1 + n_2 - 2 \tag{12.52}$$

This might be used to determine, from a small number of samples, whether an inferential has developed a significant bias error. One dataset will be laboratory results and the other the corresponding estimates made by the inferential.

The mean (μ) and the skewness (γ) of the distribution are both zero. Other key parameters, although unlikely ever to be required, can be determined from

$$\sigma^2 = \frac{f}{f-2} \quad \text{for } f > 2 \tag{12.53}$$

$$\kappa = \frac{3(f-2)}{f-4} \quad \text{for } f > 4 \tag{12.54}$$

12.6 Fisher

Named in honour of Ronald Fisher, this *F distribution* is also known as the *Fisher–Snedecor distribution* or the *Snedecor F distribution*. It is a special case of the Pearson-VI distribution that we will cover in Section 27.6. It is the distribution of the ratio of the estimates of the variance of a normal distribution. While at first this might seem a purely theoretical function it has, as we shall see later, a range of applications in the *analysis of variance (ANOVA)* and forms the basis of the *F test*. In particular we can use it to determine whether a measured change in variance is significant.

The mathematical function that describes its shape is quite complex, involving two values for the number of degrees of freedom – f_1 and f_2. Fortunately, it has been converted by others into tables and spreadsheet functions.

$$f(F) = \frac{\left(\frac{f_1}{f_2}\right)^{\frac{f_1}{2}} F^{\frac{f_1}{2}-1}}{\mathrm{B}\left(\frac{f_1}{2}, \frac{f_2}{2}\right)\left(1 + \frac{f_1}{f_2}F\right)^{\frac{f_1+f_2}{2}}} \quad f_1, f_2 > 0 \tag{12.55}$$

To show the function graphically is also more complex since it depends on the values of two parameters (f_1 and f_2). Figures 12.28 to 12.30 give some indication of how the shape of the curve varies. The distribution's main application is in the area of regression analysis, as we will demonstrate in Chapter 16.

The CDF is a regularised incomplete beta function, as described in Section 11.6, and is commonly available as a spreadsheet function and in proprietary software. Alternatively the trapezium rule can be applied to the PDF.

If the distribution were considered for fitting to process data, F in Equation (12.55) would be replaced with $(x - \alpha)/\beta$. Remembering, when doing so, as explained by Equations (5.60) to (5.65), the resulting PDF must then have β included in the denominator. The parameters f_1 and f_2 are assumed, by most proprietary software, to be integers. This makes fitting more complex. It is also common that the best fit is achieved by impractically large values for f_2. Nevertheless, there are occasions where it provides an accuracy of fit better than many of the contending distributions.

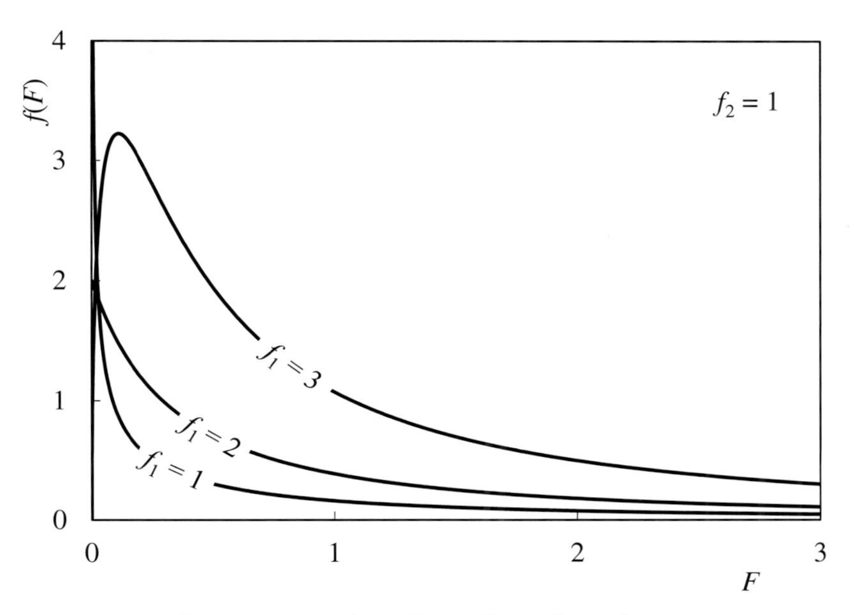

Figure 12.28 *Fisher: Effect of* f_1 *on shape* $(f_2 = 1)$

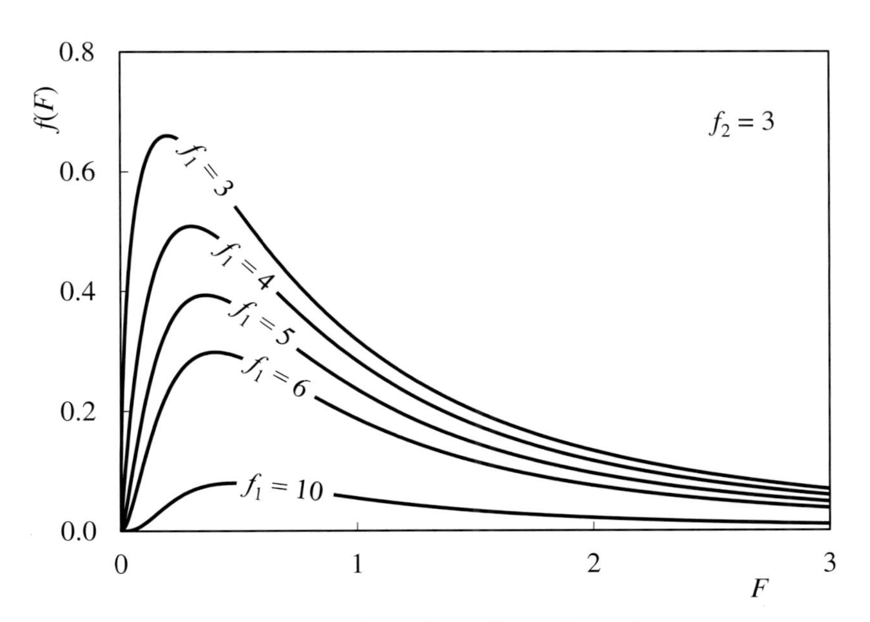

Figure 12.29 *Fisher: Effect of* f_1 *on shape* $(f_2 = 3)$

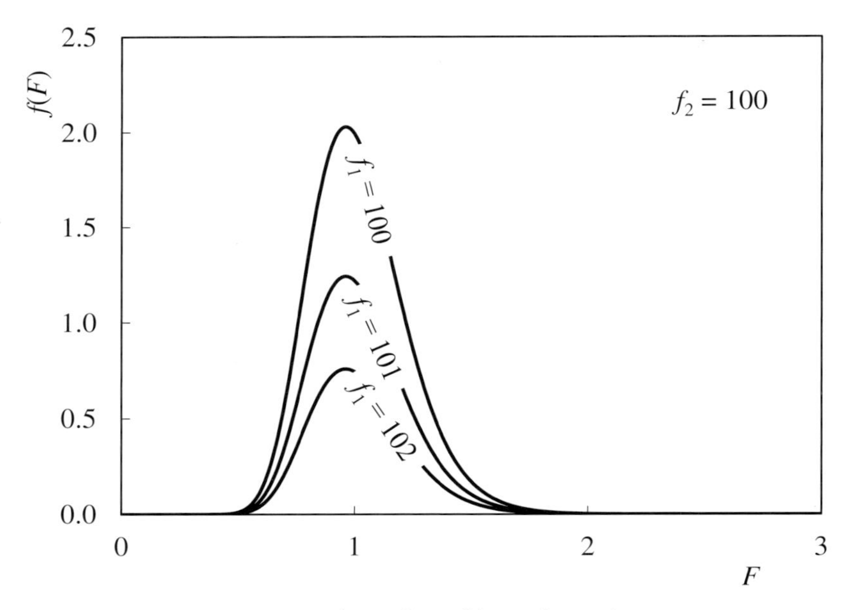

Figure 12.30 *Fisher: Effect of f_1 on shape (f_2 = 100)*

Moments are given by

$$m_n = \left(\frac{f_2}{f_1}\right)^n \frac{\Gamma\left(\frac{f_1}{2}+n\right)}{\Gamma\left(\frac{f_1}{2}\right)} \frac{\Gamma\left(\frac{f_2}{2}-n\right)}{\Gamma\left(\frac{f_2}{2}\right)} \beta^n \quad f_2 > 2n \tag{12.56}$$

Hence

$$\mu = \alpha + \frac{f_2}{f_2-2}\beta \quad f_2 > 2 \tag{12.57}$$

$$\sigma^2 = \frac{2f_2^2(f_1+f_2-2)}{f_1(f_2-2)^2(f_2-4)}\beta^2 \quad f_2 > 4 \tag{12.58}$$

$$\gamma = \frac{2(2f_1+f_2-2)}{f_2-6}\sqrt{\frac{2(f_2-4)}{f_1(f_1+f_2-2)}} \quad f_2 > 6 \tag{12.59}$$

$$\kappa = \frac{6(f_2-4)}{(f_2-6)(f_2-8)}\left[\frac{2(f_2-2)^2}{f_1(f_1+f_2-2)} + \frac{f_2}{2} + 5\right] \quad f_2 > 8 \tag{12.60}$$

Figure 12.31 shows feasible combinations of skewness and kurtosis.

12.7 Exponential

The *exponential distribution* is a special case of several other distributions. For example, if in the Hosking distribution, described by Equation (12.23), we set δ_1 to 0 and δ_2 to 1 we obtain

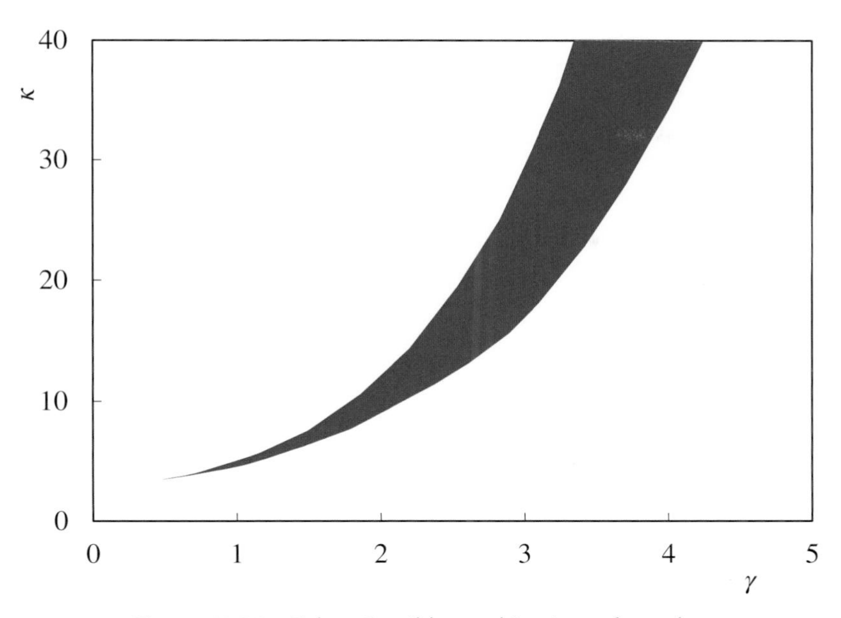

Figure 12.31 *Fisher: Feasible combinations of γ and κ*

$$f(x)=\frac{1}{\beta}\exp\left[-\frac{x-\alpha}{\beta}\right] \qquad x\geq\alpha;\ \beta>0 \tag{12.61}$$

$$F(x)=1-\exp\left[-\frac{x-\alpha}{\beta}\right] \tag{12.62}$$

This is sometimes described as the *shifted exponential distribution*, because the x axis is shifted by α. Setting α to 0 and replacing β with $1/\lambda$ gives what is normally accepted as the *standard exponential distribution*, although 'standard' would usually suggest that β is 1.

$$f(x)=\lambda e^{-\lambda x} \qquad x\geq 0;\ \lambda>0 \tag{12.63}$$

$$F(x)=1-e^{-\lambda x} \tag{12.64}$$

It is used to estimate the probability of the interval between events (x) based on the expected rate of events (λ). It is applicable when events occur independently at a constant average rate. The probability of an event must not be affected by previous occurrences. The distribution is therefore one of the few that are memoryless. It should not therefore be applied to processes that have memory.

Inverting the CDF gives the QF

$$x(F)=\alpha-\beta\ln(1-F)=-\frac{\ln(1-F)}{\lambda} \qquad 0\leq F\leq 1 \tag{12.65}$$

Key parameters are

$$\mu=\alpha+\beta=\frac{1}{\lambda} \tag{12.66}$$

$$\sigma^2=\beta^2=\frac{1}{\lambda^2} \tag{12.67}$$

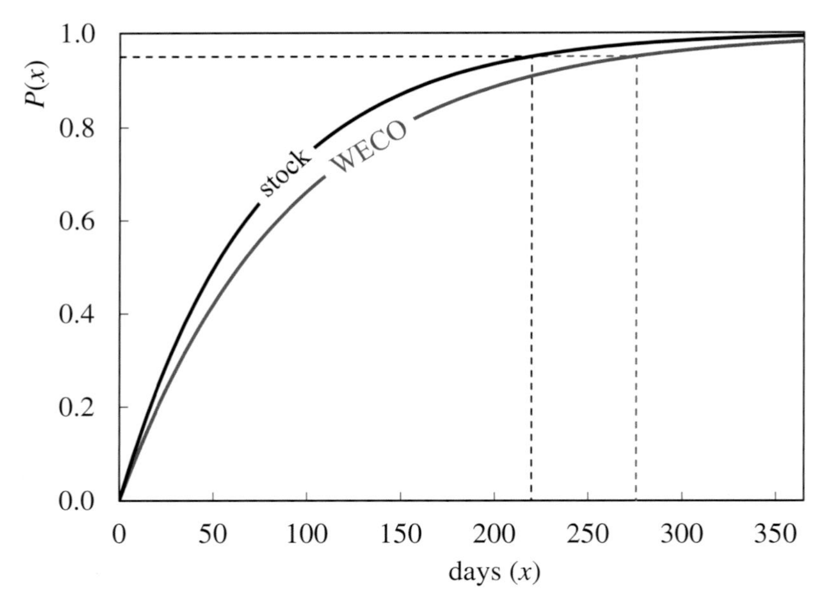

Figure 12.32 *Exponential: Application to WECO rules and stock level example*

$$\gamma = 2 \tag{12.68}$$

$$\kappa = 9 \tag{12.69}$$

Taking the WECO rules example from Section 8.4, the expected number of false indications (λ) is 1/92 per day. Equation (12.64) is plotted using this value as Figure 12.32. If F is 0.95, then, from Equation (12.65), x is very close to $3/\lambda$. In other words we are 95% sure that the interval between false indications will not exceed $3/\lambda$ or 276 days. Should it do so then we should conclude that the value of σ_{error} (used in assessing the inferential) has changed.

Similarly, for the stock level example, the expected number of low stock events (λ) is 3/220 per day. As Figure 12.32 shows, we can therefore be 95% certain that there will be problem with component availability within 220 days of the previous occasion. However, some caution should be used in interpreting this result. Stock level is a process with memory. We will cover later more reliable methods of assessing the likelihood of such events.

Other applications include assessing process availability based on the MTBF (mean time between failures) of critical process equipment. For example, the feed pump may be known to have a recurring problem with a MTBF of 100 days ($\lambda = 0.01$ failures per day). We could instigate a preventative maintenance program under which the pump was serviced at the time when the chance of failure exceeded, for example, 50%. From Equation (12.65) such work would need to take place every 69 days.

We described, in Section 3.3, the distribution of times between events of the LPG splitter reflux exceeding 65 m^3/hr. Over the 5,000 hours of data, there were 393 such events. The first occurred after 10 hours, and the last after 4,968 hours. By calculation the mean time between events is therefore 12.6 hours. Rather than calculate λ from the data we can fit the exponential distribution to the event distribution. Fitting Equation (12.62) gives α as 0.429 and β as 12.1. Figure 12.33 shows the fit; *RSS* is 0.1190. From Equation (12.66), the mean time between events is 12.5 hours. Applying Equation (12.65), with F chosen as 0.01, we can be 99% confident that a high reflux event will occur within 56 hours.

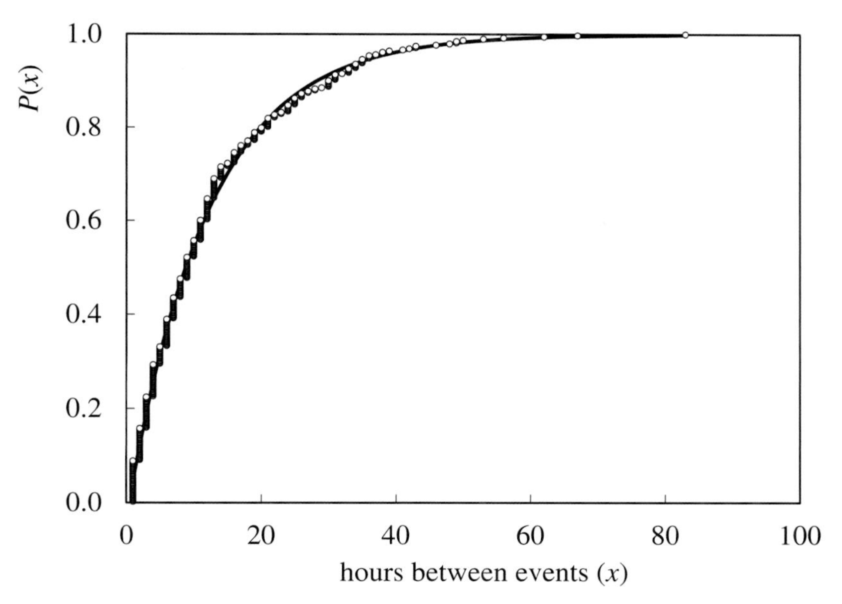

Figure 12.33 Exponential: Fitted to interval between high reflux flow events

12.8 Weibull

The *Weibull distribution* is used, like the exponential distribution, to estimate the probability of the time between events, but for processes that have memory.

It is described by the PDF

$$f(x) = \delta x^{\delta-1} \exp\left(-x^{\delta}\right) \quad x > 0;\ \delta > 0 \tag{12.70}$$

This is plotted for a range of values of the shape parameter (δ) as Figure 12.34. It is strictly the *Weibull-I distribution*. More commonly a scale parameter (β) is included to produce the *Weibull-II distribution* – also known as the *Rosin–Rammler distribution*. Its PDF is

$$f(x) = \frac{\delta x^{\delta-1}}{\beta^{\delta}} \exp\left[-\left(\frac{x}{\beta}\right)^{\delta}\right] \quad x > 0;\ \delta, \beta > 0 \tag{12.71}$$

This is plotted, for different values of δ and β, as Figure 12.35. Its CDF is

$$F(x) = 1 - \exp\left[-\left(\frac{x}{\beta}\right)^{\delta}\right] \tag{12.72}$$

The CDF can be inverted to give the QF.

$$x(F) = \beta\left\{\ln\left[\frac{1}{1-F}\right]\right\}^{\frac{1}{\delta}} \quad 0 \leq F \leq 1 \tag{12.73}$$

The *Weibull-III distribution* also includes a location parameter (α).

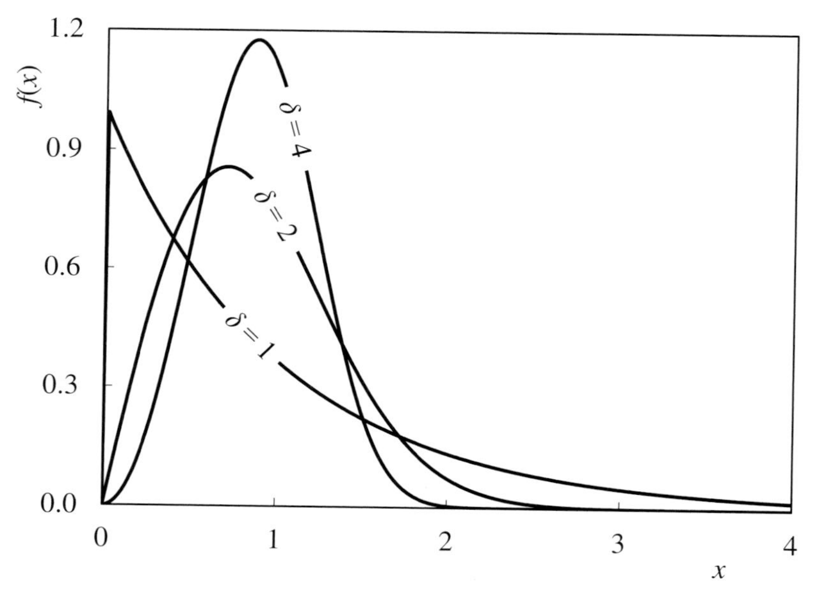

Figure 12.34 *Weibull-I: Effect of δ on shape*

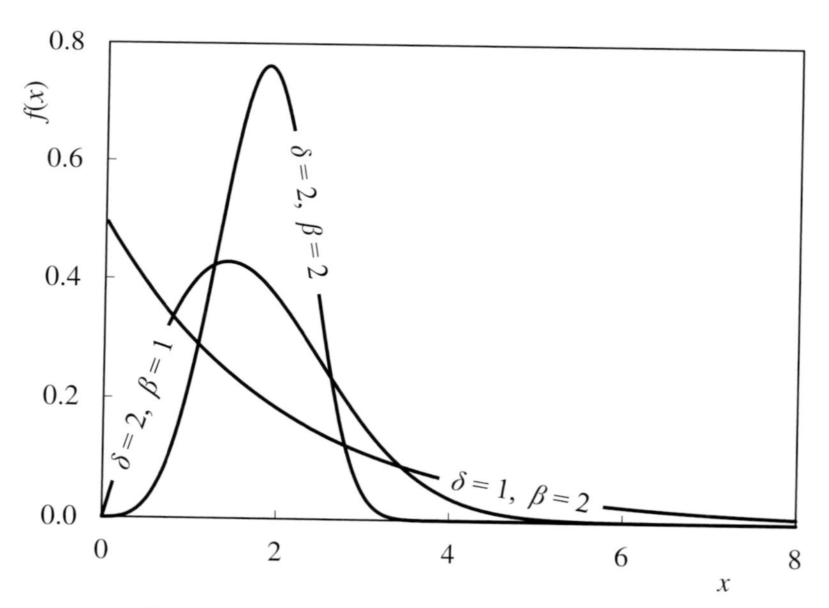

Figure 12.35 *Weibull-II: Effect of δ and β on shape*

$$f(x) = \frac{\delta(x-\alpha)^{\delta-1}}{\beta^{\delta}} \exp\left[-\left(\frac{x-\alpha}{\beta}\right)^{\delta}\right] \qquad x>\alpha;\ \delta,\beta>0 \tag{12.74}$$

$$F(x) = 1-\exp\left[-\left(\frac{x-\alpha}{\beta}\right)^{\delta}\right] \tag{12.75}$$

$$x(F) = \alpha+\beta\left\{\ln\left[\frac{1}{1-F}\right]\right\}^{\frac{1}{\delta}} \qquad 0 \le F \le 1 \tag{12.76}$$

Fitting Equation (12.72), to the absolute value of the changes in the C_4 content of the propane rundown, gives values for δ and β of 0.974 and 2.51 respectively. But the Weibull-II distribution lends itself to another approach. Rearranging Equation (12.73) gives

$$\ln\{-\ln[1-F(x)]\} = \delta\ln(x) - \delta\ln(\beta) \tag{12.77}$$

$F(x)$ is now the actual distribution, determined as if we were to fit a prior distribution. We then plot $\ln\{-\ln[1-F(x)]\}$ against $\ln(x)$. This is only possible if all values of x are greater than zero. Since we have defined changes in C_4 content as absolute values, this is the case for most of the results. Changes that are zero are included in the total for the purpose of calculating the actual $F(x)$ but are excluded from the plot.

If the distribution follows Weibull then we would expect the plot to be a straight line of slope δ and intercept $-\delta\ln(\beta)$. Figure 12.36 shows the result. The slope (δ) is 0.999 – compared to 0.974 derived by curve fitting. When δ is 1 the Weibull-II distribution becomes the exponential distribution. In this example the process has been shown to be memoryless; the timing of a disturbance is not influenced by the timing of the last. We would therefore expect the behaviour to follow the exponential distribution. The intercept is –0.898, which is $-0.999\ln(\beta)$, so β is 2.46 – a value close to that derived from curve fitting.

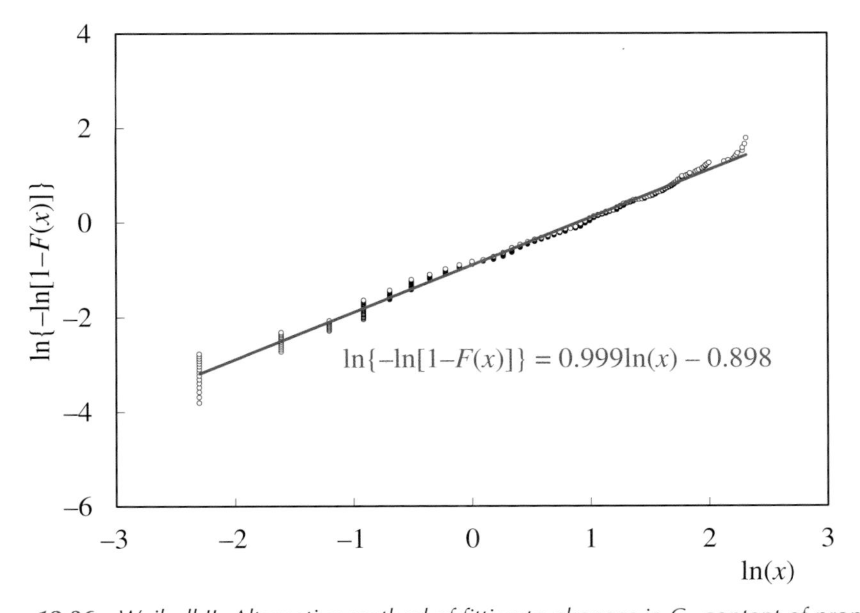

Figure 12.36 Weibull-II: Alternative method of fitting to changes in C_4 content of propane

Using Equation (12.73), with the values derived for δ and β, we can be 95% certain that x will not exceed 7.4. Alternatively we might expect disturbances to be larger than this on 18 days in a year. This might be enough of a concern to justify some improvement to the control scheme.

The raw moments are given by

$$m_n = \Gamma\left(1+\frac{n}{\delta}\right)\beta^n = \frac{n}{\delta}\Gamma\left(\frac{n}{\delta}\right)\beta^n \tag{12.78}$$

Hence the mean and variance are

$$\mu = \alpha + \beta\Gamma\left(1+\frac{1}{\delta}\right) = \alpha + \frac{\beta}{\delta}\Gamma\left(\frac{1}{\delta}\right) \tag{12.79}$$

$$\sigma^2 = \left[\Gamma\left(1+\frac{2}{\delta}\right) - \Gamma^2\left(1+\frac{1}{\delta}\right)\right]\beta^2 = \left[\frac{2}{\delta}\Gamma\left(\frac{2}{\delta}\right) - \frac{1}{\delta^2}\Gamma^2\left(\frac{1}{\delta}\right)\right]\beta^2 \tag{12.80}$$

Because δ is 1, the mean and standard deviation have the same value (β) of 2.46. Simply calculating these parameters from the data gives μ as 2.43 and σ as 2.25. The true variation in the size of disturbances is thus about 9% larger than simple calculation suggests.

There are a variety of ways in which the formulae for skewness and kurtosis can be presented; deriving them from the raw moments gives

$$\gamma = \frac{3\delta^2\Gamma\left(\frac{3}{\delta}\right) - 6\delta\Gamma\left(\frac{2}{\delta}\right)\Gamma\left(\frac{1}{\delta}\right) + 2\Gamma^3\left(\frac{1}{\delta}\right)}{\left[2\delta\Gamma\left(\frac{2}{\delta}\right) - \Gamma^2\left(\frac{1}{\delta}\right)\right]^{\frac{3}{2}}} \tag{12.81}$$

$$\kappa = \frac{4\delta^3\Gamma\left(\frac{4}{\delta}\right) - 12\delta^2\Gamma\left(\frac{3}{\delta}\right)\Gamma\left(\frac{1}{\delta}\right) + 12\delta\Gamma^2\left(\frac{1}{\delta}\right)\Gamma\left(\frac{2}{\delta}\right) - 3\Gamma^4\left(\frac{1}{\delta}\right)}{\left[2\delta\Gamma\left(\frac{2}{\delta}\right) - \Gamma^2\left(\frac{1}{\delta}\right)\right]^2} \tag{12.82}$$

Feasible combinations of skewness and kurtosis are shown as Figure 12.37.

Figure 12.38 is the result of applying the same technique, as described by Equation (12.77), to the stock level example. While the result is not exactly a straight line, it gives an estimate for δ of 2.51 – much higher than the value of 1 which would result from a memoryless process. This behaviour might have been expected, since today's inventory will be determined substantially by the quantity the tank contained yesterday.

Curve fitting, as shown in Figure 12.39, gives a higher value of 3.00 for δ and 1489 for β. While perhaps not a good fit it underlines the principle of fitting a distribution that could, for example, be used to estimate the probability of the inventory falling below the minimum required for downstream processing or of exceeding the maximum storage capacity.

12.9 Chi-Squared

The *chi-squared (χ^2) distribution* is defined by the PDF

$$f(x) = \frac{\left(\frac{x-\alpha}{\beta}\right)^{\frac{f}{2}-1}\exp\left(-\frac{x-\alpha}{2\beta}\right)}{2^{\frac{f}{2}}\beta\Gamma\left(\frac{f}{2}\right)} \qquad \beta, f > 0 \tag{12.83}$$

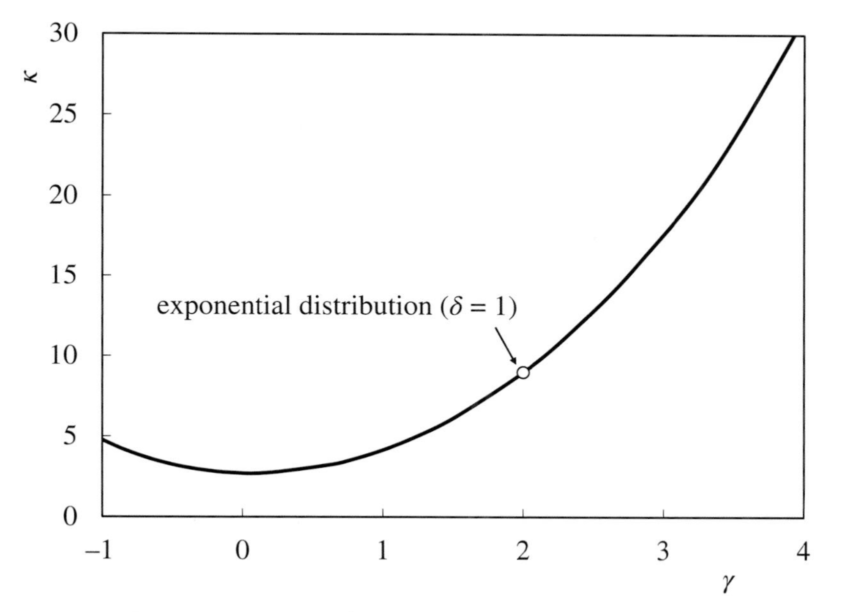

Figure 12.37 Weibull: Feasible combinations of γ and κ

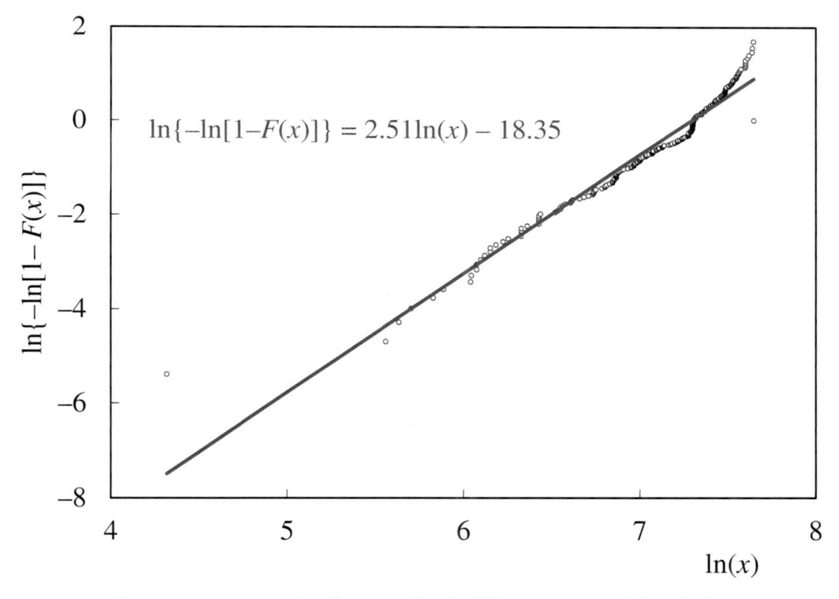

Figure 12.38 Weibull-II: Method of fitting to stock level data

It is useful when we wish to compare a set of observed results with those predicted and to determine the probability that our method of predicting the results is correct. Under these circumstances α is set to 0 and β to 1 to give the *standard chi-squared distribution*. The PDF then becomes

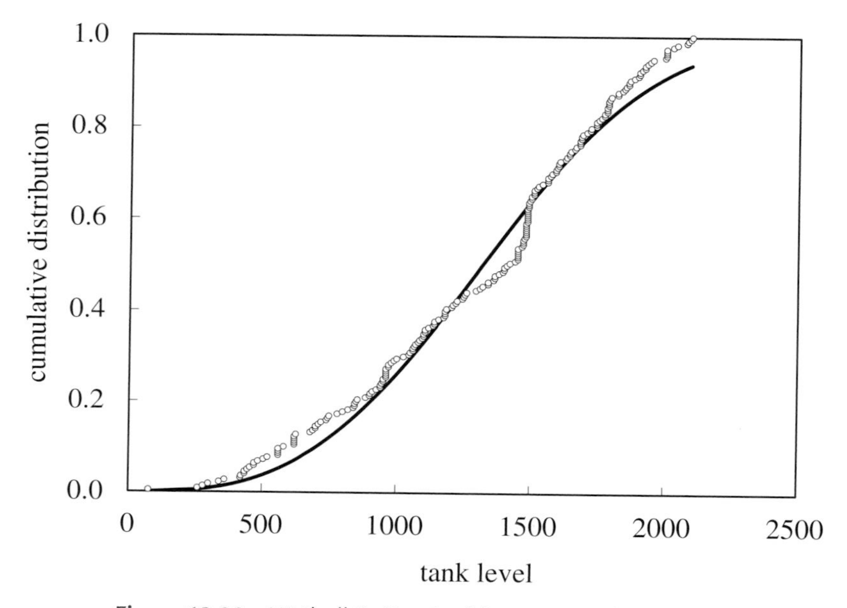

Figure 12.39 Weibull-II: Result of fitting to stock level data

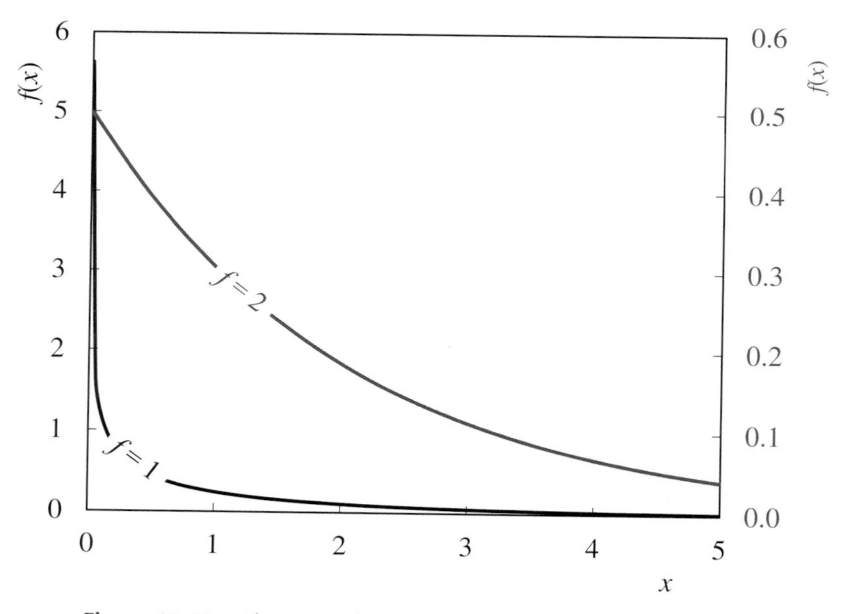

Figure 12.40 Chi-squared: Probability density for f = 1 and f = 2

$$f(x) = \frac{x^{\frac{f}{2}-1} \exp\left(-\frac{x}{2}\right)}{2^{\frac{f}{2}} \Gamma\left(\frac{f}{2}\right)} \tag{12.84}$$

Figures 12.40 and 12.41 show this function for a range of values of f. The CDF uses the regularised incomplete gamma distribution, described in Section 11.5, and is supported by most spreadsheet packages.

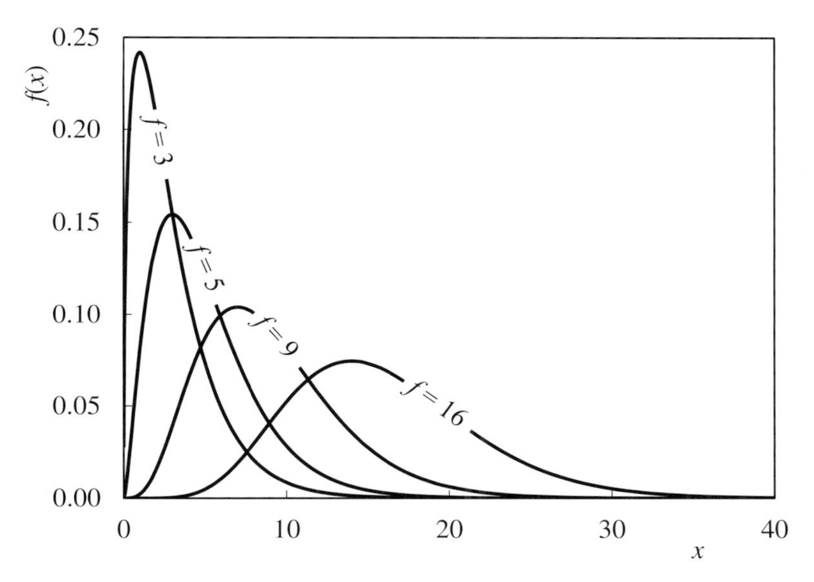

Figure 12.41 Chi-squared: Probability density for f ≥ 3

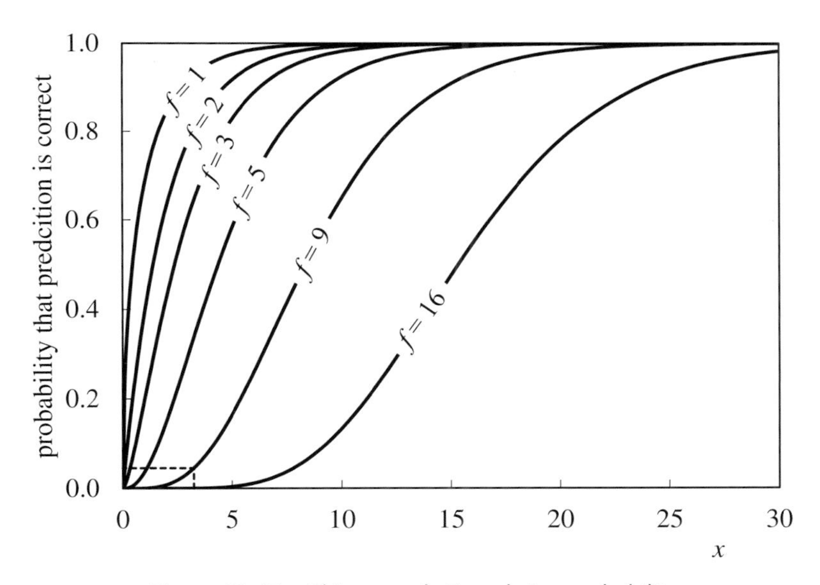

Figure 12.42 Chi-squared: Cumulative probability

To use the distribution we first calculate χ^2 from the predicted values ($\hat{y}$) and the observed values (y).

$$\chi^2 = \sum_{i=1}^{n} \frac{(y_i - \hat{y}_i)^2}{\hat{y}_i} \tag{12.85}$$

Figure 12.42 shows the cumulative probability curves, plotted by applying the trapezium rule to Equation (12.84), for a range of values of f. From the appropriate curve we identify the probability corresponding to the value when x is χ^2.

As an example we will use a correlation developed using regression analysis later in Chapter 16. The control engineer might think of this as an inferential property calculation.

$$\hat{y} = 25.76 + 1.468x_1 \tag{12.86}$$

Table 12.3 shows daily measurements of the true property (y) versus those predicted by the inferential. The same information is portrayed in Figure 12.43. We can see that every measurement is greater than that predicted and so we wish to determine whether the inferential calculation should be updated. If we consider all 10 data points then f is 9 and χ^2 is 3.24. We can see from Figure 12.42 that the probability that our inferential property calculation is correct is 0.046. This means we are 95.4% sure that the inferential has been incorrect and so we would work to resolve the problem. Had we performed the analysis a day sooner, when

Table 12.3 Measured and predicted values for y

x_1	y	$\hat{y}$	$\frac{(y-\hat{y})^2}{\hat{y}}$	$\sum\frac{(y-\hat{y})^2}{\hat{y}}$	f	P
11	45	42	0.23	0.23	0	
13	49	45	0.39	0.61	1	0.566
17	56	51	0.55	1.16	2	0.441
27	71	65	0.48	1.64	3	0.351
35	80	77	0.11	1.75	4	0.218
49	105	98	0.55	2.30	5	0.193
19	56	54	0.10	2.40	6	0.120
52	109	102	0.47	2.87	7	0.103
65	127	121	0.28	3.15	8	0.075
66	126	123	0.09	3.24	9	0.046

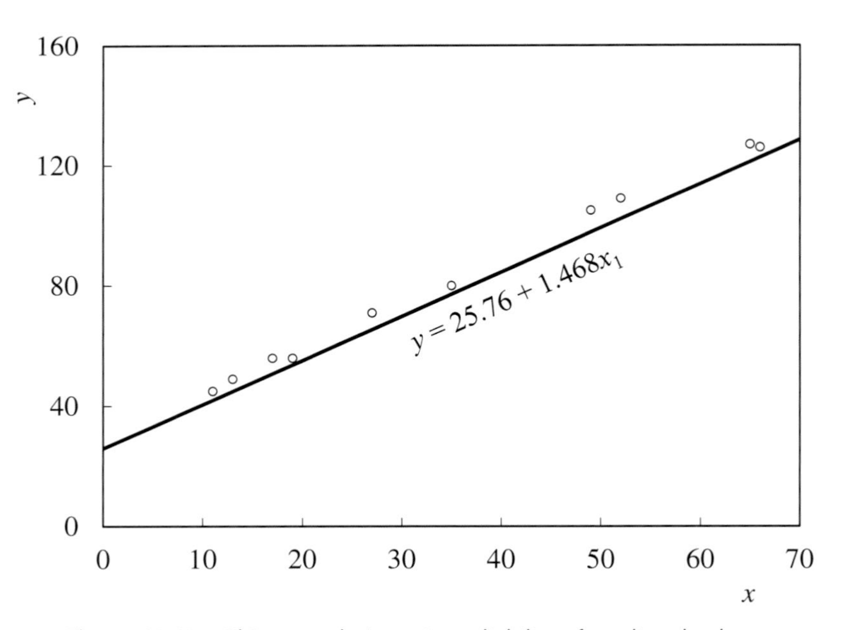

Figure 12.43 Chi-squared: Assessing reliability of predicted values

f was 8 and χ^2 was 3.15, the probability that the inferential was incorrect was 92.5% – just below the point where we would investigate.

Following the fourth Westinghouse rule, described in Section 8.4, we would have taken action two days sooner – once 8 consecutive errors had the same sign. But this criterion takes no account of the size of the error and so is perhaps too crude.

Although not generally used for this purpose, Equation (12.83) can be fitted to process data. While it will fit some datasets well, for others f must be restricted to avoid overflow errors.

The moment generating function is

$$M(t) = (1-2t)^{-\frac{f}{2}} \tag{12.87}$$

leading to

$$\mu = \alpha + f\beta \tag{12.88}$$

$$\sigma^2 = 2f\beta^2 \tag{12.89}$$

$$\gamma^2 = \frac{8}{f} \tag{12.90}$$

$$\kappa = \frac{12}{f} + 3 \tag{12.91}$$

12.10 Gamma

The *gamma distribution* is a continuous distribution but is strongly connected to the discrete Poisson distribution that we cover later. Indeed, we will show later that the PDF can be derived from that of the discrete Poisson distribution. It is also a special case of the Pearson-III distribution that we will cover in Section 27.3.

The PDF is

$$f(x) = \frac{\lambda^k x^{k-1} e^{-\lambda x}}{(k-1)!} = \frac{\lambda^k x^{k-1} e^{-\lambda x}}{\Gamma(k)} \qquad x \geq 0;\ k, \lambda > 0 \tag{12.92}$$

Replacing k with $f/2$ and x with $x/2$ shows that the chi-squared distribution, described by Equation (12.84), is a special case of the gamma distribution.

It is common for texts to replace λ (the expected frequency of events or *rate parameter*) with β (the mean time between events $1/\lambda$); hence

$$f(x) = \frac{x^{k-1}}{\beta^k \Gamma(k)} \exp\left(-\frac{x}{\beta}\right) \qquad x \geq 0;\ k, \beta > 0 \tag{12.93}$$

If k is restricted to integers then the distribution can also be called the *Erlang distribution* and can be used to estimate the probability of the waiting time (x) until a specified number (k) events occur.

Figure 12.44 shows the effect of k and β. If k is 1 (coloured curve), we obtain the exponential distribution. As k is increased the distribution becomes increasingly symmetrical and approaches the normal distribution.

The CDF uses the regularised incomplete gamma function, described in Section 11.5. It is commonly available as a spreadsheet function.

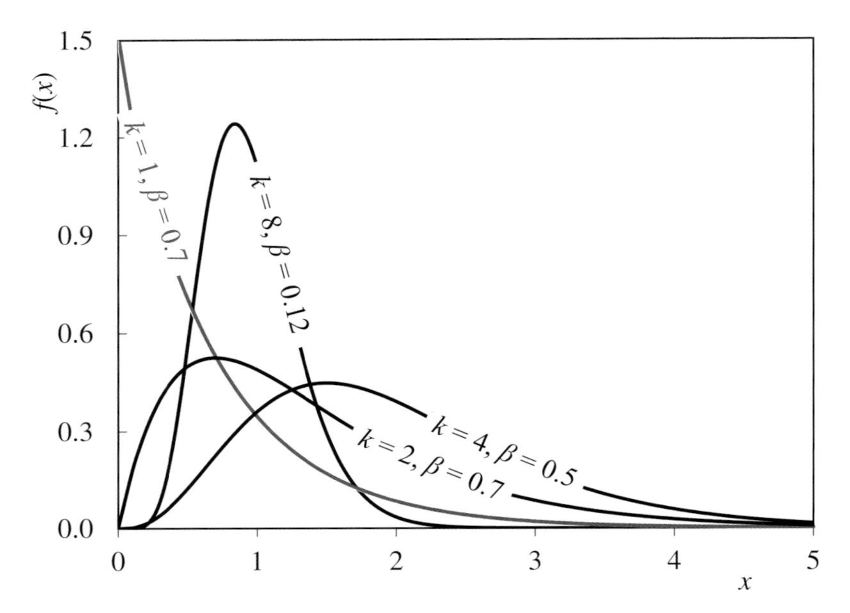

Figure 12.44 *Gamma: Effect of* k *and* β *on shape*

The raw moments are given by

$$m_n = \frac{\Gamma(k+n)}{\Gamma(k)\lambda^n} = \frac{\Gamma(k+n)}{\Gamma(k)}\beta^n \tag{12.94}$$

leading to

$$\mu = \frac{k}{\lambda} = k\beta \tag{12.95}$$

$$\sigma^2 = \frac{k}{\lambda^2} = k\beta^2 \tag{12.96}$$

$$\gamma = \frac{2}{\sqrt{k}} \tag{12.97}$$

$$\kappa = \frac{6}{k} + 3 \tag{12.98}$$

Figure 12.45 shows the feasible values for γ and κ.

Using the batch blending example, we apply Equation (12.92) in which k will be 1 and λ will be 44/78 or 0.564. The probability of x being greater than a day is thus

$$P(x>1) = 1 - e^{-0.564}\frac{0.564^0}{0!} = 0.431 \tag{12.99}$$

The probability of completing a batch within a day is therefore 0.569 (1 – 0.431). We can now explore the impact of making blends of half the size. Assuming this then permits two blends per day and that the probability of being on-grade remains at 0.564, then k will be 2

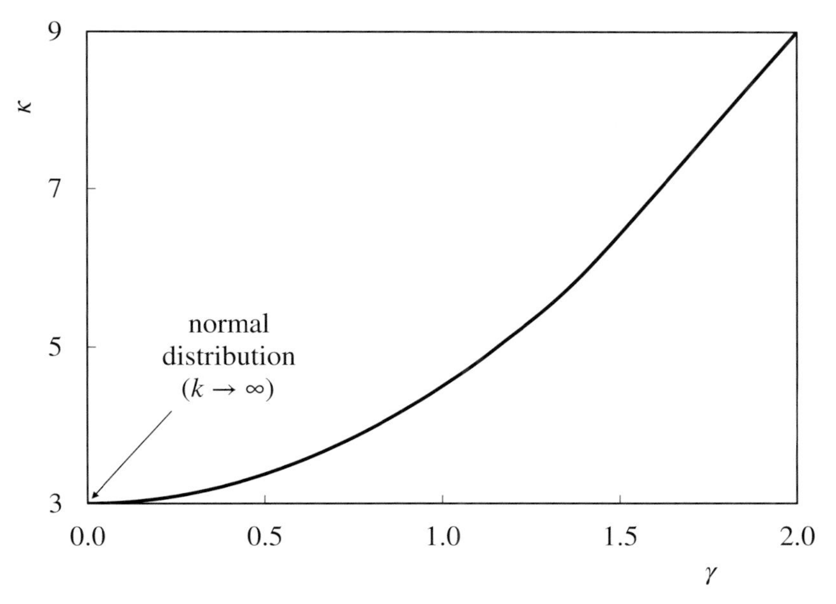

Figure 12.45 *Gamma: Feasible combinations of γ and κ*

and the expected frequency of success (λ) will be doubled to 1.128. The probability of x now being more than a day is

$$P(x>1)=1-e^{-1.128}\left(\frac{1.128^0}{0!}+\frac{1.128^1}{1!}\right)=0.311 \tag{12.100}$$

The probability of completing a batch within a day has increased to 0.689. This is equivalent to increasing the number of batches in a year from 208 to 252 – effectively increasing capacity by 21%. The cumulative distribution curves are included as Figure 12.46 and illustrate how further reductions in blend size will reduce the probability of the batch taking longer than a day to complete. There will of course be a practical limit on the number of blends that can be completed in a day.

The gamma distribution can also be used to study whether a site has sufficient product storage capacity. Each completed batch of product is 100 m^3 so, for 208 batches per year, production is an average 57 m^3/day. Assuming this is withdrawn as a continuous flow (say, to another process), if a batch is completed within a day, we have to store the surplus 43 m^3. If a batch takes four days then we need to have previously stored 128 m^3 (57 × 4 – 100) to keep the downstream unit running at 57 m^3/day.

From Equation (12.92) we can determine the probability of the number of days required for a batch. These are plotted (in black) in Figure 12.47. We might choose to size intermediate storage on the basis that it is sufficient 99% of the time. We therefore need to allow for a batch taking seven days to complete. This would require 299 m^3 of previously stored product. Figure 12.48 plots probability against storage requirements. If we now consider the proposal of performing two blends per day, to achieve the 99% target, we now need to allow for a batch taking six days. We can exploit this in one of two ways. We could maintain the same downstream processing rate and reduce the inventory required by 57 to 242 m^3. Alternatively, since we can produce 252 batches per year, we can operate at a higher rate of 69 m^3/day. This case is

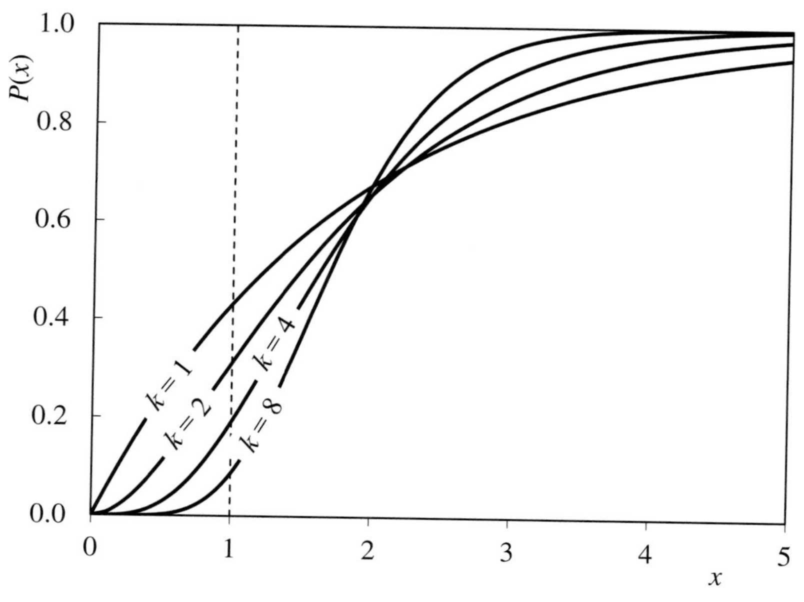

Figure 12.46 Gamma: Cumulative distribution

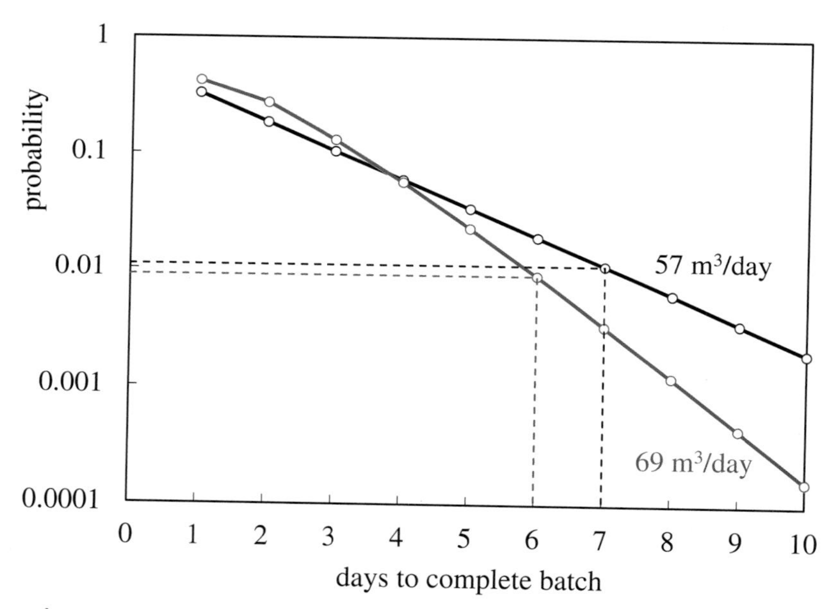

Figure 12.47 Gamma: Probability distribution of time to produce a batch

shown as the coloured points in Figures 12.47 and 12.48. The required inventory would increase to 314 m^3 – although keeping it at 299 m^3 would reduce the availability very slightly below 99%. In other words, there would probably be no need to invest in additional storage to handle the increased production.

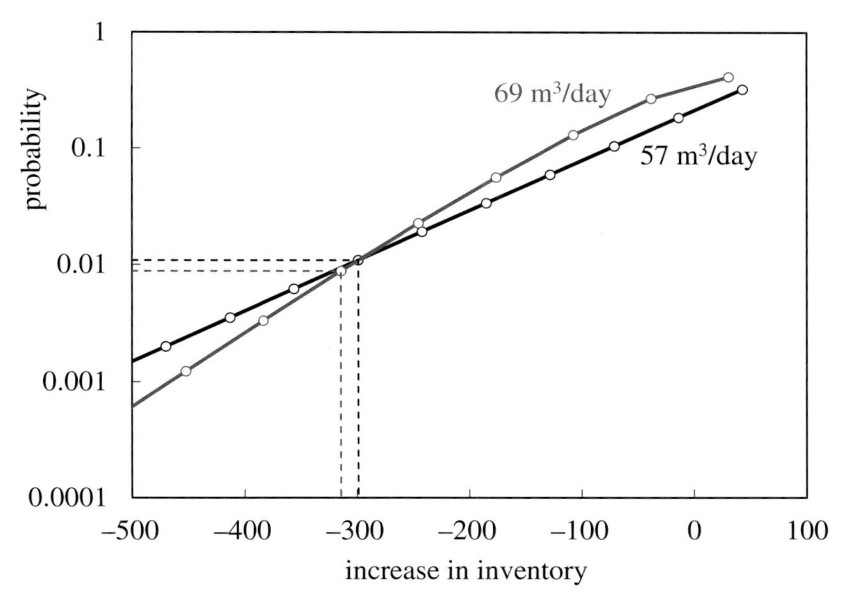

Figure 12.48 *Gamma: Probability distribution of required inventory*

12.11 Binomial

The *binomial distribution* is perhaps the most well known of the discrete distributions. It can be used to estimate the probability, $p(x)$, of a given number of successes (x) occurring in a number (n) of success/failure trials. This is based on the probability (p) of success in a single trial. The PMF is

$$p(x) = \frac{n!}{x!(n-x)!} p^x (1-p)^{n-x} \quad 0 \le x \le n;\ 0 \le p \le 1 \tag{12.101}$$

For large values of n, a problem can arise from calculating n! For example, the maximum number supported by a pocket calculator might be 10^{100} and so cause a problem if n is greater than 69. Similarly a spreadsheet package, with a limit of 2^{1024}, restricts n to 170. Spreadsheets, which include the binomial function, avoid this by first computing each fraction before multiplying.

$$\frac{n!}{x!(n-x)!} = \frac{n}{x} \times \frac{n-1}{x-1} \ldots \times \frac{n-x+1}{1} = \prod_{i=0}^{x-1} \frac{n-i}{x-i} \tag{12.102}$$

This problem will then only arise if n is much larger than x.

The binomial distribution is a special case of the *multinomial distribution* in which n possible outcomes are possible. If the probability of outcome 1 is p_1, the probability of outcome 2 is p_2 and so on, then

$$\sum_{i=1}^{n} p_i = 1 \tag{12.103}$$

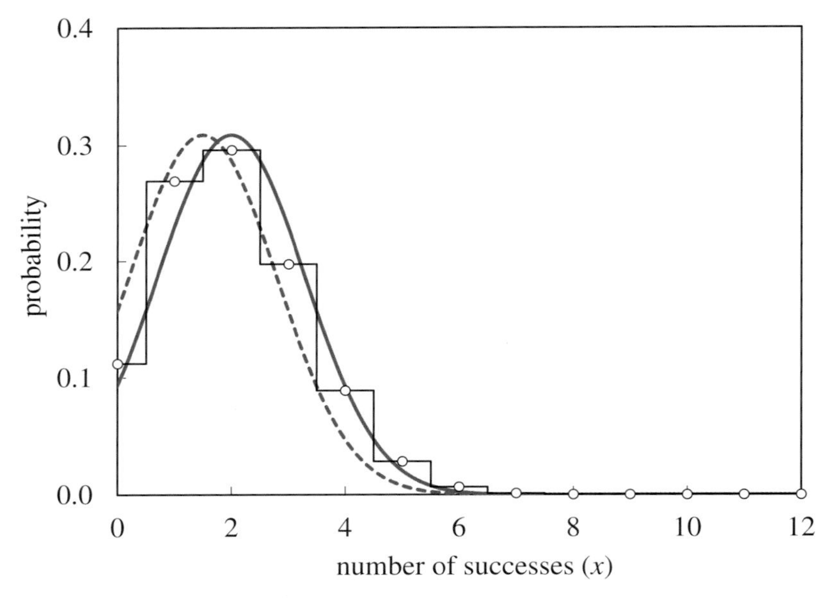

Figure 12.49 *Binomial: Fitting a normal distribution to the number of sixes from 12 throws*

The probability that there will be x_1 of outcome 1, x_2 of outcome 2 and so on is given by the PMF

$$P_{k_1 k_2 \ldots k_n} = \frac{n!}{x_1! x_2! \ldots x_n!} p_1^{x_1} p_2^{x_2} \ldots p_n^{x_n} \quad \text{where} \quad \sum_{i=1}^{n} x_i = n \tag{12.104}$$

If the probability of all the events is the same ($p_1 = p_2 = \ldots = p_n$), we obtain the *Poisson binomial distribution*.

The *Bernoulli distribution* is a special form of the binomial distribution in which n is 1 and so x can be either 0 or 1. Its PMF is

$$p(x) = p^x (1-p)^{1-x} \tag{12.105}$$

If x is 0 then $p(0)$ will be $1-p$. If x is 1, $p(1)$ will be p. As an example consider throwing once ($n = 1$) an unbiased six-sided dice. Success is defined as throwing a six ($x = 1$), for which p is 1/6. The probability of not throwing a six ($x = 0$) is 5/6. Figure 12.49 shows the probable outcome if the dice is thrown 12 times ($n = 12$). The black points are plotted from Equation (12.101) and show, as one might expect, that the most likely outcome is 2 successes – but with a probability of only 0.296.

The moment generating function of the binomial distribution is

$$M(t) = (1 - p + pe^t)^n \tag{12.106}$$

This leads to

$$\mu = np \tag{12.107}$$

$$\sigma^2 = np(1-p) \tag{12.108}$$

$$\gamma^2 = \frac{n^2(1-2p)^2}{p(1-p)} \tag{12.109}$$

$$\kappa = \frac{n[1-3p(1-p)]}{p(1-p)} \tag{12.110}$$

In order to apply many of the techniques covered, we need to be able to assume that the distribution of possible outcomes is normal. The solid coloured curve in Figure 12.49 shows a normal distribution, plotted from Equation (5.35) using the parameters derived from Equations (12.107) and (12.108). Part of the difference arises because x is an integer. The binomial distribution is a histogram with bins defined as $x \pm 0.5$. The closest continuous function would then pass through x at the top of each column. It might therefore be considered more accurate to base the normal distribution not on x but on $x + 0.5$. This is known as the *Yates adjustment* and the result is included as the dashed coloured curve. However, as it has in this case, the adjustment is known to overcompensate.

How close the distribution is to normal depends on n and p. Indeed we will show it is virtually identical when p is 0.5. As p moves away from this value, the approximation requires larger values of n. Provided that n is greater than 5,[7] then the minimum value of n can be determined from

$$n > 11\left[\sqrt{\frac{1-p}{p}} - \sqrt{\frac{p}{1-p}}\right]^2 \tag{12.111}$$

This function is plotted as Figure 12.50. In our example, with $p = 1/6$ (shown by the dashed line), the distribution can be assumed to be normal provided n is greater than 36. Figure 12.51 illustrates this for n set at 36.

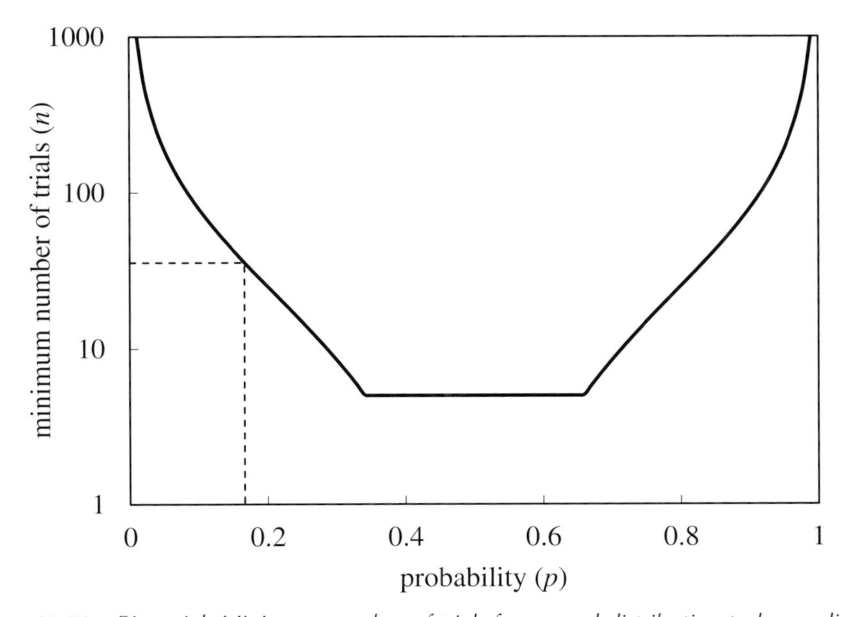

Figure 12.50 *Binomial: Minimum number of trials for normal distribution to be applicable*

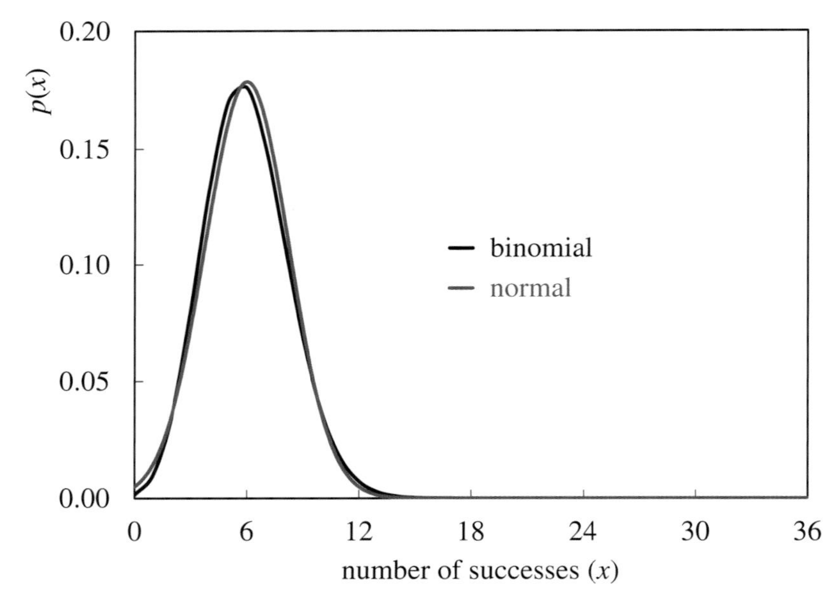

Figure 12.51 *Binomial: Demonstration that normal distribution can be used for* n = 36

Because x is discrete we cannot integrate Equation (12.101) to obtain the CDF. Instead it must be determined using the discrete function

$$P(x) = \sum_{i=0}^{x} \frac{n!}{i!(n-i)!} p^i (1-p)^{n-i} \tag{12.112}$$

We might make a daily check on an inferential to ensure its deviation from the laboratory is acceptable. If, when the inferential was first developed, the standard deviation of the prediction error was σ_{error}, then, from Table A1.8, we are 95% sure that the inferential is reliable if the error is less than $1.96\sigma_{error}$. Figure 12.52 is a plot of Equation (12.112) with $n = 365$ and $p = 0.95$. The dashed lines show that we are 95% sure that the inferential will be deemed accurate on no less than 338 days per year and on no more than 354. If the actual number falls outside this range then we are 95% sure that σ_{error} has changed. So, if it fails on more than 27 days, the inferential performance has degraded sufficiently to prompt further investigation. If it fails on fewer than 11 days then the check on accuracy should be made more demanding by reducing the value used for σ_{error}.

A very similar result would be obtained by assuming the distribution is normal. By considering a year of daily results, the number of trials is greater than the minimum of 188 specified by Equation (12.111) when p is 0.95. Equation (12.107) gives μ as 346.8 and Equation (12.108) gives σ as 4.2. The 95% confidence interval ($\mu \pm 1.96\sigma$) would therefore be from 338 to 355.

If we were to validate the performance monitoring over a much smaller period, say 30 days, then such an approximation would be unwise. Use of Equation (12.112) would give a 95% confidence interval of 26 to 29 days while assuming a normal distribution would give 26 to 31 days. Clearly we cannot expect 31 successes out of 30 trials. The problem arises because the binomial distribution, unlike the normal distribution, is only symmetrical about the mean if p is 0.5. In general, use of the normal distribution will suggest the inferential is more reliable than it is.

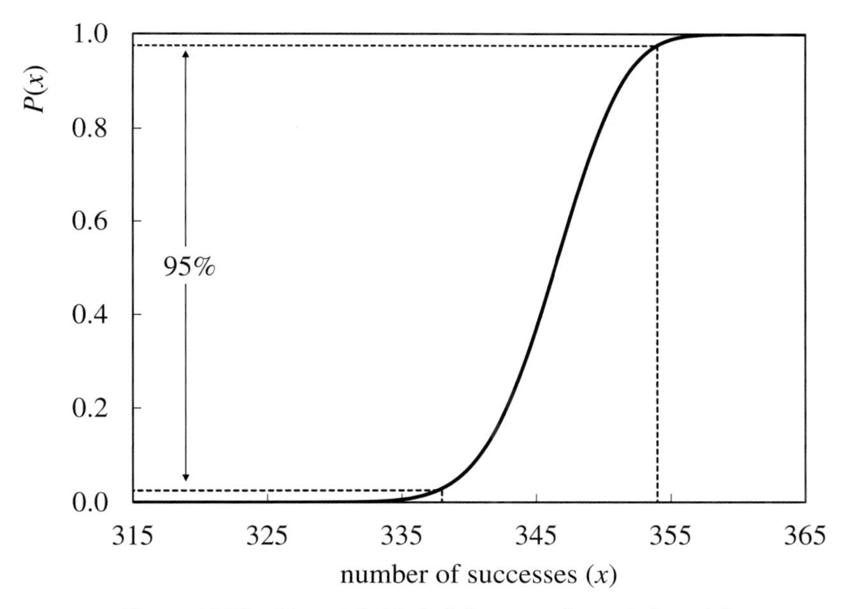

Figure 12.52 *Binomial: Likely failure rate for an inferential*

Process control benefit calculations are typically based on halving the standard deviation of key variables, such as finished product properties. If p is the current probability that a batch of product will be off-grade and p_{new} the probability after implementation of the control improvement then, from Equation (12.108)

$$\sqrt{np_{new}(1-p_{new})}=0.5\sqrt{np(1-p)} \quad \text{or} \quad p_{new}=\frac{1-\sqrt{1-p(1-p)}}{2} \tag{12.113}$$

$$\mu_{new}=\frac{1-\sqrt{1-p(1-p)}}{2p}\mu \tag{12.114}$$

In improved control studies, it is common to exploit the reduction in standard deviation by operating, on average, closer to the product specification – keeping the number of violations of the specification the same. This approach is acceptable if the off-grade product is blended with some that is in giveaway so that what is supplied to the customer is within specification. This is usually the situation in the oil and bulk petrochemical industries. In others however, no amount of off-grade product is permitted. For example, in the paper and metal industries, the problem cannot be resolved by blending. Off-grade material has to be downgraded or reprocessed. Under these circumstances we would capture benefits by reducing the number of occasions when this is necessary.

Using the diesel quality example, of the 111 daily laboratory results there are five occasions when the product is deemed off-grade. From this we estimate p as 0.0450. Choosing n as 365 then, using Equation (12.101), we can plot the black curve in Figure 12.53. Equation (12.107) tells us that, on average, the current operation results in 16 off-grade results per year. From Equation (12.114) we determine that p_{new} will be 0.0108. This is used to plot the coloured curve in Figure 12.53 and shows that improved control will reduce the number of off-grade batches to an average of four per year. Knowing the cost of dealing with an off-grade batch would allow us to estimate the benefit of producing 12 fewer per year. Further, if correcting an off-grade batch

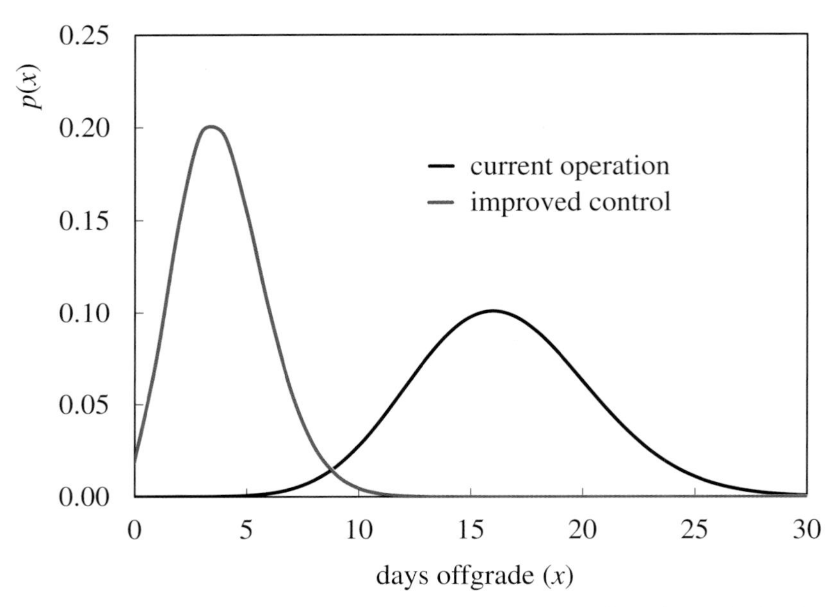

Figure 12.53 *Binomial: Reducing the average number of off-grade results for diesel*

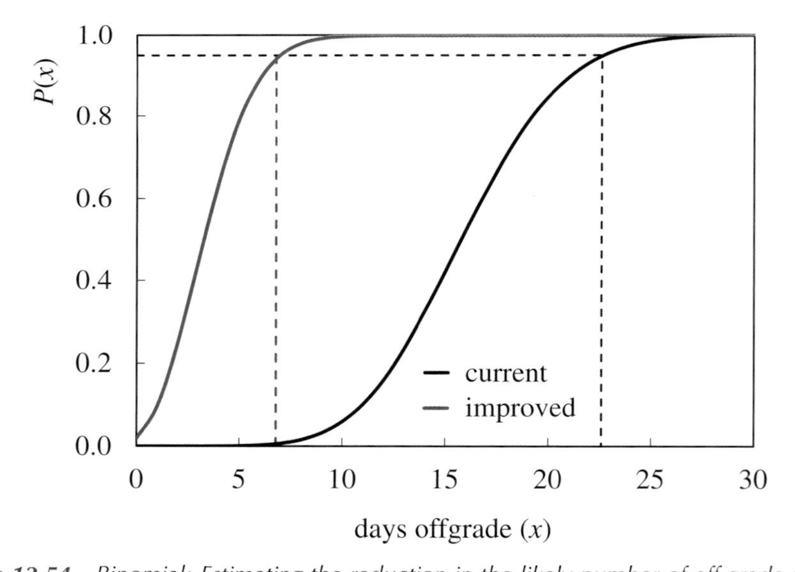

Figure 12.54 *Binomial: Estimating the reduction in the likely number of off-grade results*

uses the same process equipment as that used to produce the batch in the first place, the economic benefit can be far greater. Reprocessing might be using valuable capacity that would be better used to increase total production by about 3%.

The reduction in standard deviation also gives us greater confidence in the operation. Figure 12.54 shows the corresponding cumulative probabilities. Currently we can be 95% sure that there will be no more than 23 off-grade batches per year. With improved control this number reduces to six.

12.12 Poisson

The *Poisson distribution* is used to estimate the probability, $p(x)$, of a number of events (x) occurring in a fixed interval of time. It can be derived from several distributions; here we derive it from the binomial distribution.

The expected number of successes (λ) in a given interval is given by the number of trials (n) multiplied by the probability of success of a single trial (p), i.e. np. Replacing p, in Equation (12.101), with λ/n gives

$$p(x) = \frac{n!}{x!(n-x)!}\left(\frac{\lambda}{n}\right)^x\left(1-\frac{\lambda}{n}\right)^{n-x} \tag{12.115}$$

Cancelling by $(n-x)!$

$$p(x) = \frac{n(n-1)(n-2)\ldots(n-x+1)}{x!}\left(\frac{\lambda}{n}\right)^x\left(1-\frac{\lambda}{n}\right)^{n-x} \tag{12.116}$$

Rearranging

$$p(x) = \frac{n(n-1)(n-2)\ldots(n-x+1)}{n^x}\frac{\lambda^x}{x!}\left(1-\frac{\lambda}{n}\right)^{-x}\left(1-\frac{\lambda}{n}\right)^{n} \tag{12.117}$$

The Poisson distribution is given by $n \to \infty$, so

$$\frac{n(n-1)(n-2)\ldots(n-x)}{n^k} \to 1 \tag{12.118}$$

$$\left(1-\frac{\lambda}{n}\right)^{-x} \to 1 \tag{12.119}$$

Remembering, from Section 11.1, the definition of Euler's number

$$\left(1-\frac{\lambda}{n}\right)^{n} \to e^{-\lambda} \tag{12.120}$$

Equation (12.117) therefore becomes the PMF of the Poisson distribution.

$$p(x) = \frac{\lambda^x}{x!}e^{-\lambda} \qquad x \geq 0;\ \lambda \geq 0 \tag{12.121}$$

The effect of λ is shown in Figure 12.55. Mean and variance are

$$\mu = \lambda \tag{12.122}$$

$$\sigma = \sqrt{\lambda} \tag{12.123}$$

The CDF is

$$P(x) = e^{-\lambda}\sum_{i=0}^{x}\frac{\lambda^i}{i!} \tag{12.124}$$

The distribution is only applicable if the events are independently timed; one occurrence should not influence the timing of another, i.e. the process is memoryless. For example, as part of process operator manning level study, it might be used to estimate the probability of a large number of process alarms occurring in a short time. However, since the circumstances causing one alarm are likely to cause others, one might argue that the alarms will not be independently

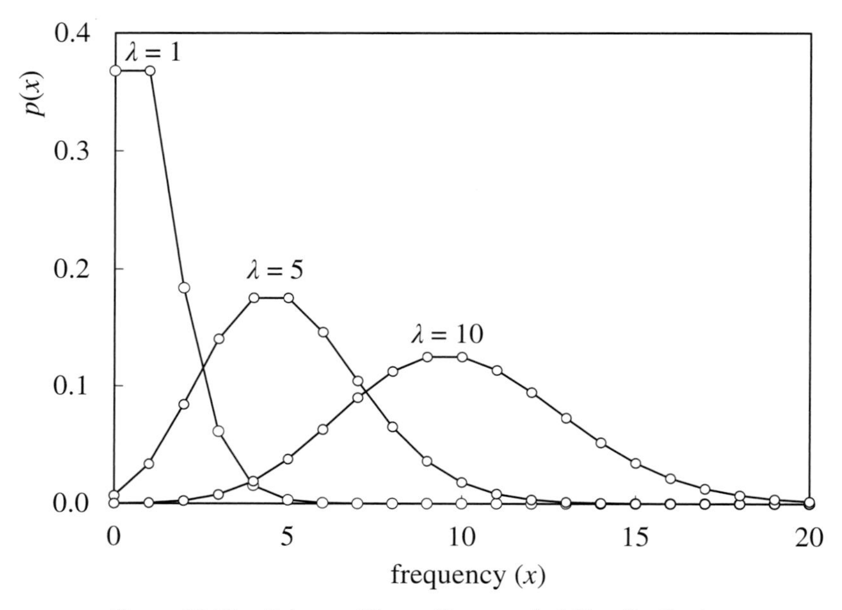

Figure 12.55 *Poisson: Effect of λ on probability distribution*

timed. While one alarm may not directly cause another, the analysis might be better performed by including only *first-up* alarms.

We showed, in Section 8.4, that using the WECO rules to detect a suspect inferential measurement would be expected to result in a false indication once in every 92 instances. If the inferential is checked against a daily laboratory result, we would therefore expect four false indications per year. Equation (12.124) is plotted, for a range of values of λ, as Figure 12.56. The coloured lines show that, for $\lambda = 4$, we might expect the number of false indications in a year to be between one and seven. So, if there were zero or more than eight occurrences, we would be 95% sure that the value of σ_{error} (used in assessing the inferential) has changed.

Using the stock level example, on three of the 220 days, the inventory was too small to permit a blend to be started. Based on this past performance, we might therefore expect this problem five times per year. From Figure 12.56, for $\lambda = 5$, we can therefore be 95% sure that the number of future occasions will be no more than nine per year.

The Poisson distribution leads directly to the gamma distribution, covered in Section 12.10, which is used to estimate the probability of the waiting time (x) until a specified number (k) events occur. The number of events expected in the time period x is λx. The frequency (λ) used in Equation (12.121) can be defined for any period and so

$$f(x) = \frac{(\lambda x)^k}{k!} e^{-\lambda x} \tag{12.125}$$

The waiting time will be greater than x if there are fewer than k events in the interval.

$$F(x) = e^{-\lambda x} \sum_{i=0}^{k-1} \frac{(\lambda x)^i}{i!} \tag{12.126}$$

Hence we can determine the probability of there being more than k events.

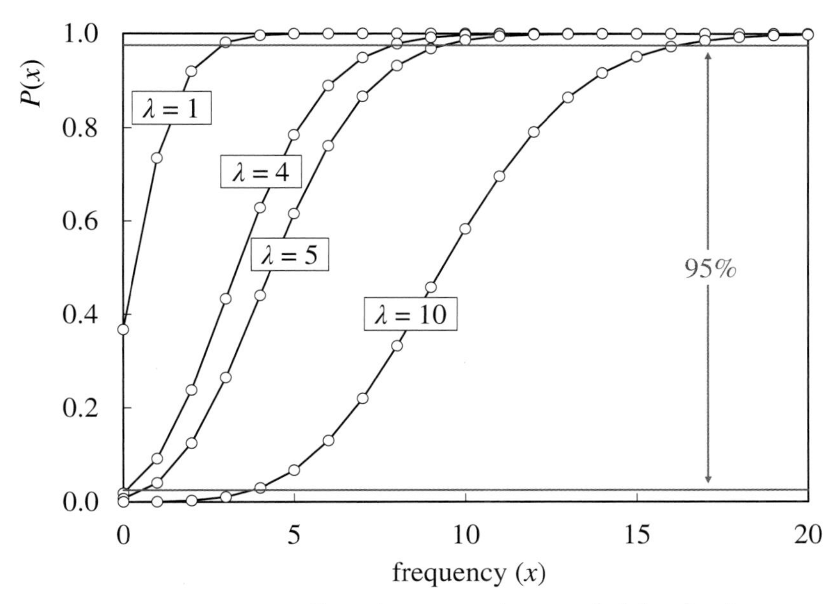

Figure 12.56 *Poisson: Effect of λ on cumulative probability distribution*

$$P(x>k)=1-F(x)=1-e^{-\lambda x}\sum_{i=0}^{k-1}\frac{(\lambda x)^i}{i!} \tag{12.127}$$

To obtain the PDF we differentiate the CDF described by Equation (12.127).

$$\begin{aligned} f(x)=\frac{dF(x)}{dx}&=\lambda e^{-\lambda x}\sum_{i=0}^{k-1}\frac{(\lambda x)^i}{i!}-e^{-\lambda x}\sum_{i=0}^{k-1}\frac{i\lambda(\lambda x)^{i-1}}{i!} \\ &=\lambda e^{-\lambda x}\sum_{i=0}^{k-1}\left[\frac{(\lambda x)^i}{i!}-\frac{i(\lambda x)^{i-1}}{i!}\right] \end{aligned} \tag{12.128}$$

$$=\lambda e^{-\lambda x}\left[(1-0)+(\lambda x-1)+\left(\frac{(\lambda x)^2}{2}-\lambda x\right)+\left(\frac{(\lambda x)^3}{6}-\frac{(\lambda x)^2}{2}\right)\ldots+\left(\frac{(\lambda x)^{k-1}}{(k-1)!}-\frac{(\lambda x)^{k-2}}{(k-2)!}\right)\right] \tag{12.129}$$

Examination of Equation (12.129) reveals that all but one of the terms cancel out to leave the PDF for the gamma distribution.

$$f(x)=\lambda e^{-\lambda x}\left[\frac{(\lambda x)^{k-1}}{(k-1)!}\right]=\frac{\lambda^k x^{k-1}e^{-\lambda x}}{(k-1)!} \tag{12.130}$$

As with the binomial distribution, the factorial can cause an overflow error if k is large. This can be avoided by changing the calculation to a product of fractions.

$$\frac{x^{k-1}}{(k-1)!}=\frac{x}{k-1}\times\frac{x}{k-2}\ldots\times\frac{x}{1}=\prod_{i=1}^{k-1}\frac{x}{k-i} \tag{12.131}$$

This then restricts k to integer values and hence the distribution becomes the Erlang distribution. To further reduce the risk of overflow, taking the logarithm of Equation (12.131) and substituting into Equation (12.130) gives

$$f(x) = \exp\left[\sum_{i=1}^{k-1} \ln\left(\frac{x}{k-i}\right) + k\ln(\lambda) - \lambda x\right] \tag{12.132}$$

13

Extreme Value Analysis

When fitting a distribution to process data, by definition, there are relatively few values that lie in the tail(s). As a result there is less confidence that the fitted distribution truly represents behaviour in these regions. The overall fit may appear good but, if we wish to determine the probability of the process operating at the extreme(s), a more reliable approach should be adopted. This technique is known as *extreme value analysis (EVA)* which was developed from *extreme value theory (EVT)*. It allows assessment of the probability of events being more extreme than previously observed. It is used when events have a low frequency but high severity.

We will use as an example the variation in the LPG splitter reflux flow. On any column reflux is a key manipulated variable, often automatically adjusted to maintain product composition or, in some cases, reflux drum level. Its value will vary in response to any type of disturbance to the column. The data comprise 5,000 measurements collected hourly. We saw in Section 10.2 that they appear to be normally distributed. We want to assess the probability of a very high reflux flow – perhaps because it is known to cause column flooding.

There are two methods of identifying which of the values should be classed as extreme. The first is to choose the highest value in a defined time period. Known as the *period maxima* (or sometimes the *block maxima*), in this example, we have chosen the highest value that occurs in each 24 hour period. Our set of extreme data now therefore comprises 208 values. Figure 13.1 shows how they are distributed. As usual we choose a prior distribution to fit to these data. Many texts suggest the use of the generalised extreme value (GEV) distribution that we covered in Section 12.4. Repeating Equation (12.38), its CDF is

$$F(x) = \exp\left[-\left(1 + \frac{\delta(x-\alpha)}{\beta}\right)^{\frac{-1}{\delta}}\right] \qquad \delta x \geq \delta\alpha - \beta;\ \beta > 0 \tag{13.1}$$

In this case, as shown by Figure 13.2, it fits reasonably well with δ set to –0.0332, α to 67.7 and β to 4.19. If we were concerned that the reflux might exceed, for example, 85 m^3/hr then this distribution would indicate the probability of this occurring is 0.0118. Remembering that, since this is the probability of it occurring in a 24 hour period, this is equivalent to expecting it

Statistics for Process Control Engineers: A Practical Approach, First Edition. Myke King.

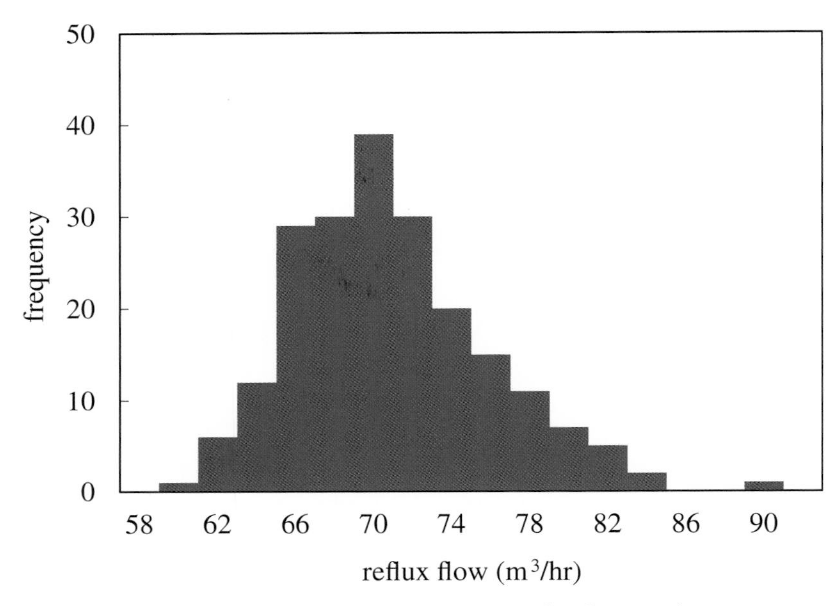

Figure 13.1 Frequency distribution of extreme reflux flow (24-hour maxima)

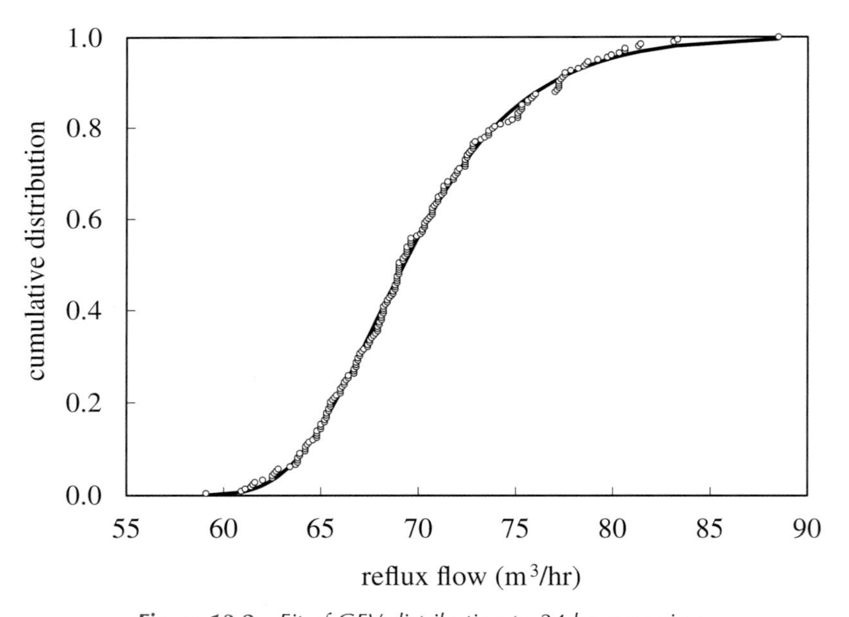

Figure 13.2 Fit of GEV distribution to 24-hour maxima

to occur once in 85 days. Figure 13.1 shows that of the 5,000 measurements, one exceeds 85 m^3/hr. This is equivalent to one event every 208 days – significantly less frequent than indicated by the fitted distribution.

To explore the sensitivity of the result, we can repeat the above with the period chosen at 48 hours. This reduces the number of maxima to 104, with their frequency distribution shown

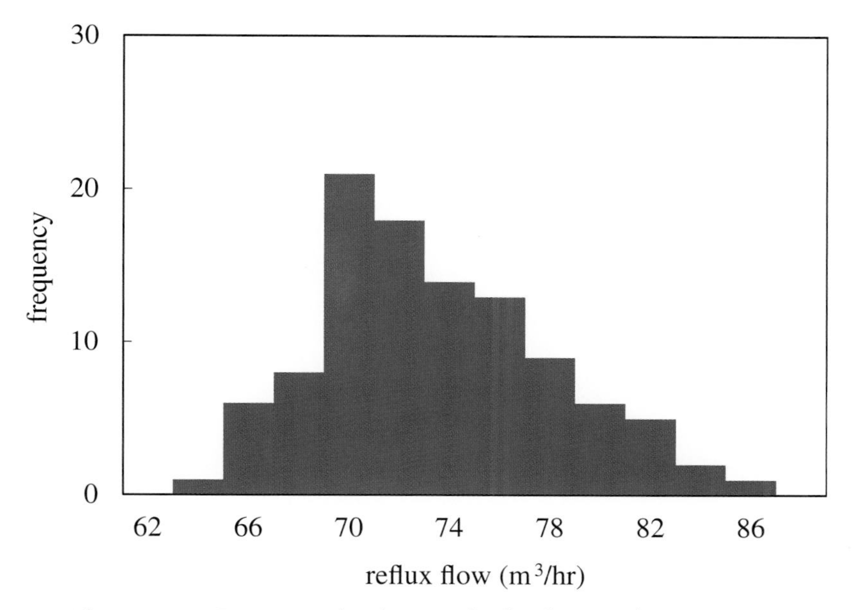

Figure 13.3 Frequency distribution of reflux flow (48-hour maxima)

as Figure 13.3. Fitting the GEV distribution (Figure 13.4) sets δ to −0.0169, α to 70.4 and β to 4.10. *RSS* is 0.0379. The probability of the reflux exceeding 85 m^3/hr is now 0.0248, equivalent to one event every 40.4 periods. Remembering the period is now 2 days, we expect an event every 81 days. This is close to that determined using the 24 hour maxima, confirming that we might anticipate a high reflux event to occur 2.5 times more frequently than the collected data suggests.

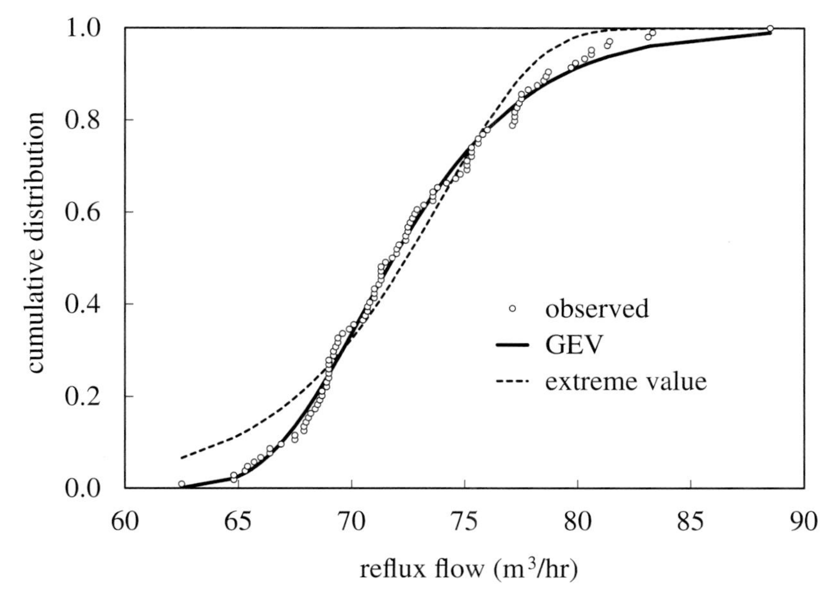

Figure 13.4 Fit of GEV and extreme value distributions to 48-hour maxima

There are alternatives to the GEV distribution. These include the Gumbel, Fréchet and reverse Weibull distributions. In Section 12.4 we showed that, like the GEV distribution, the Gumbel distribution is a special case of the Hosking distribution. It can be obtained by setting δ to 0 to give

$$F(x)=\exp\left[-\exp\left(-\frac{x-\alpha}{\beta}\right)\right] \tag{13.2}$$

This is also known as the *extreme value max distribution*, intended for use when the extreme values are maxima. When they are minima the distribution it is modified to the *extreme value min distribution*.

$$F(x)=\exp\left[-\exp\left(\frac{x+\alpha}{\beta}\right)\right] \tag{13.3}$$

However, when fitting the distribution, permitting both α and β to be either negative or positive avoids the need to select either the max or min version.

The Fréchet distribution, covered later in Section 33.6, is also a special case the GEV distribution.

$$F(x)=\exp\left[-\left(\frac{x-\alpha_f}{\beta_f}\right)^{-\delta_f}\right] \tag{13.4}$$

$$\delta_f=\frac{1}{\delta} \qquad \alpha_f=\alpha-\frac{\beta}{\delta} \qquad \beta_f=\frac{\beta}{\delta} \tag{13.5}$$

The *reverse Weibull distribution* is another special case.

$$F(x)=\exp\left[-\left(\frac{x-\alpha_w}{\beta_w}\right)^{\delta_w}\right] \tag{13.6}$$

$$\delta_w=-\frac{1}{\delta} \qquad \alpha_w=\alpha-\frac{\beta}{\delta} \qquad \beta_w=\frac{\beta}{\delta} \tag{13.7}$$

So, if fitting the GEV distribution, there is no need to consider as alternatives the Gumbel, Fréchet or reverse Weibull distributions.

A distribution bearing a similar name, the *extreme value distribution*, is described by

$$f(x)=\frac{\delta}{\beta}\exp\left[\delta(x-\alpha)-\frac{\exp[\delta(x-\alpha)]}{\beta}\right] \tag{13.8}$$

$$F(x)=1-\exp\left[-\frac{\exp[\delta(x-\alpha)]}{\beta}\right] \tag{13.9}$$

$$x(F)=\alpha+\frac{\ln[-\beta\ln(1-F)]}{\delta} \tag{13.10}$$

Fitting this to the same 48 hour maxima gives δ to 0.233, α to 75.7 and β to 0.67. The fit is poor with *RSS* much larger at 0.2630. The dashed line in Figure 13.4 further illustrates this.

The period maxima approach has now been generally superseded by the *threshold exceedance* method. Values are classed as extreme if they exceed a chosen threshold (u). Before fitting a distribution the values are converted to *exceedances* (y), where

$$y_i = x_i - u \qquad x_i \geq u \tag{13.11}$$

There is no mathematical technique for choosing u. Clearly it must be large enough to exclude values we would consider as not extreme but small enough to ensure that there are sufficient extreme values so that we can reliably fit a distribution. In this example we will first determine u as the 95% centile, i.e. reflux flows in the top 5% are considered as extreme. From Table A1.8, for a normal distribution, this is at $\mu + 1.645\sigma$. The mean of the 5,000 reflux flows is 52.1 and the standard deviation (calculated using Equation (4.35)) is 8.73. These values give u as 66.5 m^3/hr. Of the 5,000 values, 284 exceed this value. The frequency distribution of these exceedances is shown in Figure 13.5. Had the distribution been exactly normal, we would have expected 250 (i.e. 5% of 5,000) exceedances. Substantially exceeding this number suggests that the distribution is not normal for the upper tail and adds justification for the use of EVA.

Again we should choose the distribution that best fits the data but the convention is to use the generalised Pareto distribution (second version) that we will cover in Section 23.8. Its PDF is

$$f(y) = \frac{1}{\beta}\left[1 + \frac{\delta(y-\alpha)}{\beta}\right]^{-\frac{1}{\delta}-1} \qquad y > \alpha - \frac{\beta}{\delta};\ \beta, \delta > 0 \tag{13.12}$$

The CDF is an example of a *conditional excess distribution function*. In the case of the generalised Pareto distribution it is

$$F(y) = 1 - \left[1 + \frac{\delta(y-\alpha)}{\beta}\right]^{\frac{-1}{\delta}} \qquad \delta > 0 \tag{13.13}$$

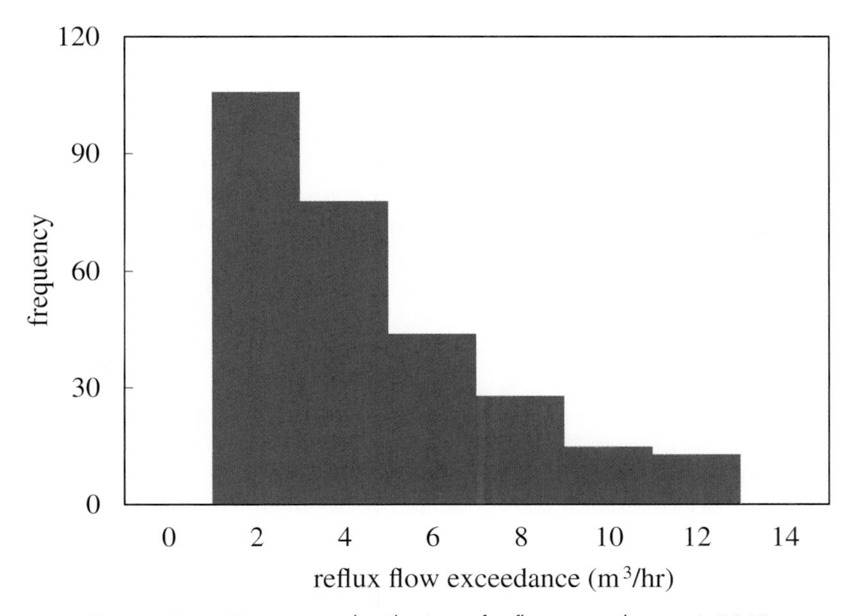

Figure 13.5 *Frequency distribution of reflux exceedances (>66.5)*

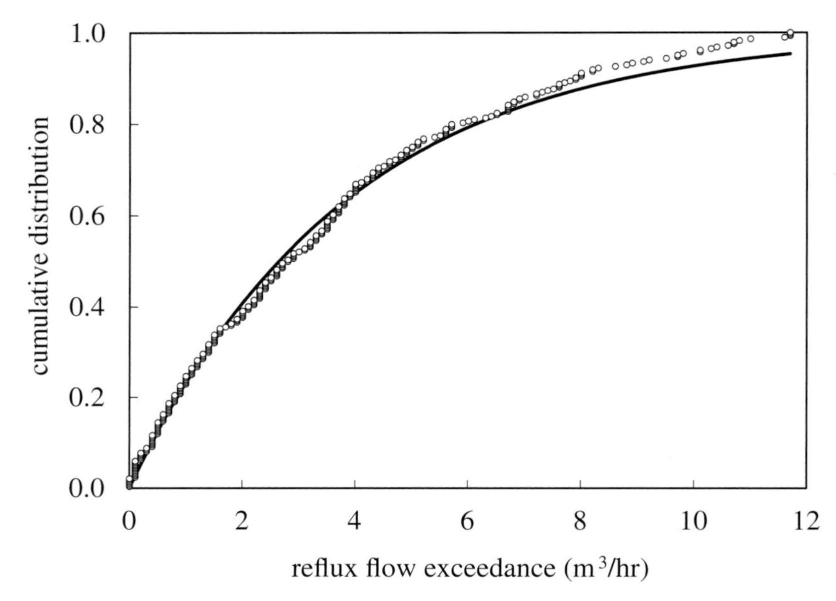

Figure 13.6 *Fit of generalised Pareto distribution to exceedances (>66.5)*

$$F(y) = 1 - \exp\left(-\frac{y-\alpha}{\beta}\right) \qquad \delta = 0 \tag{13.14}$$

The term δ can be described as the *tail index*. As Figure 13.6 shows, in this case, the distribution fits well with δ set to 0, α set to 0 and β set to 3.81. If, as before, we are concerned about the reflux exceeding 85 m^3/hr (an exceedance of 18.5 m^3/hr) then this distribution tells us that the probability is 0.00776. Remembering that there are 284 exceedances out of 5,000 data points, the observed probability of any exceedance is 0.0568. The probability of an exceedance violating the maximum flow limit is therefore the product of these two probabilities, i.e. 0.000441. This is equivalent to expecting the event once in 2,270 hours, or 95 days.

Choosing a higher threshold excludes less extreme values and should give greater confidence that the distribution is a better fit to the truly extreme values. As an alternative approach, we can set the threshold so that there are a chosen number of exceedances. For example, if we choose 30, we should set u at 75.8 m^3/hr. The frequency distribution is shown as Figure 13.7. Figure 13.8 shows the result of fitting the same distribution function with δ still set to 0, but with α now set to 0.131 and β to 3.02. The probability of an exceedance being greater than 9.2 m^3/hr (i.e. a reflux of 85 m^3/hr) now increases to 0.0498. The probability of any exceedance falls to 0.006 (30/5000) and so the overall probability is 0.000299. This is equivalent to expecting the event once in 3,346 hours, or 139 days.

The same approach can be applied to determining the probability of very low values. There were two occasions when the reflux fell below 27 m^3/hr, so the expected interval between such events is 104 days.

The frequency distribution of the 72 hour period minima is shown in Figure 13.9. Figure 13.10 shows the GEV distribution with δ set to −0.405, α to 32.390 and β to 2.60. The probability of the reflux falling below 27 m^3/hr is 0.0109, equivalent to one event every 91 days.

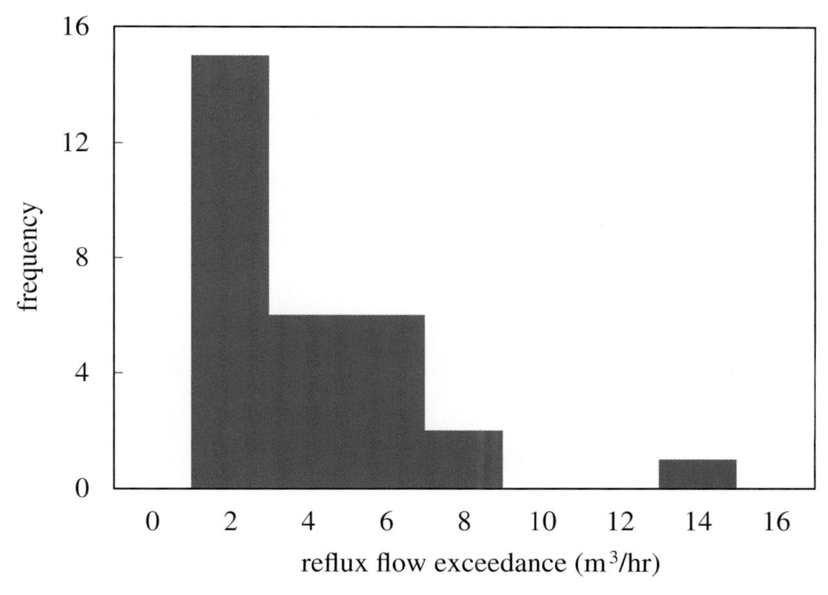

Figure 13.7 *Frequency distribution of reflux exceedances (>75.8)*

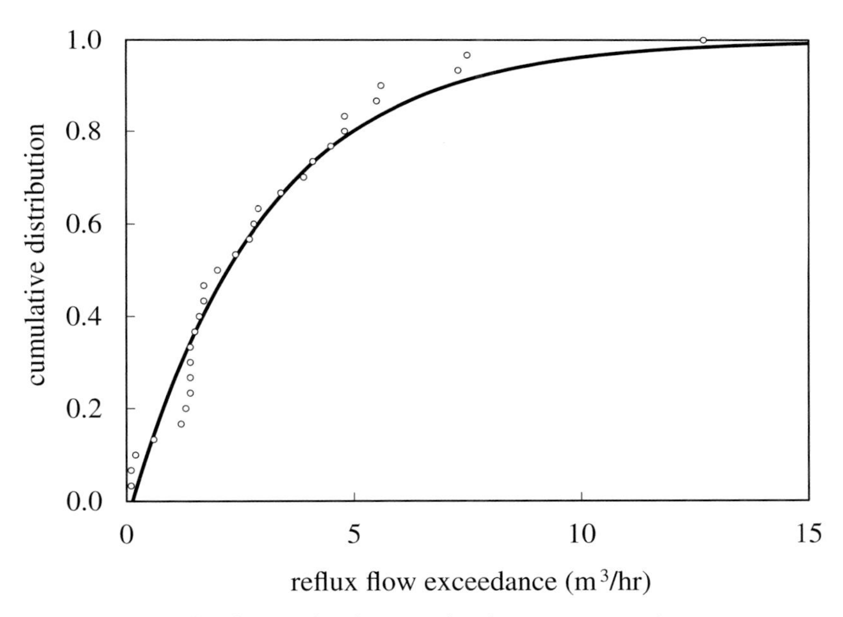

Figure 13.8 *Fit of generalised Pareto distribution to exceedances (>75.8)*

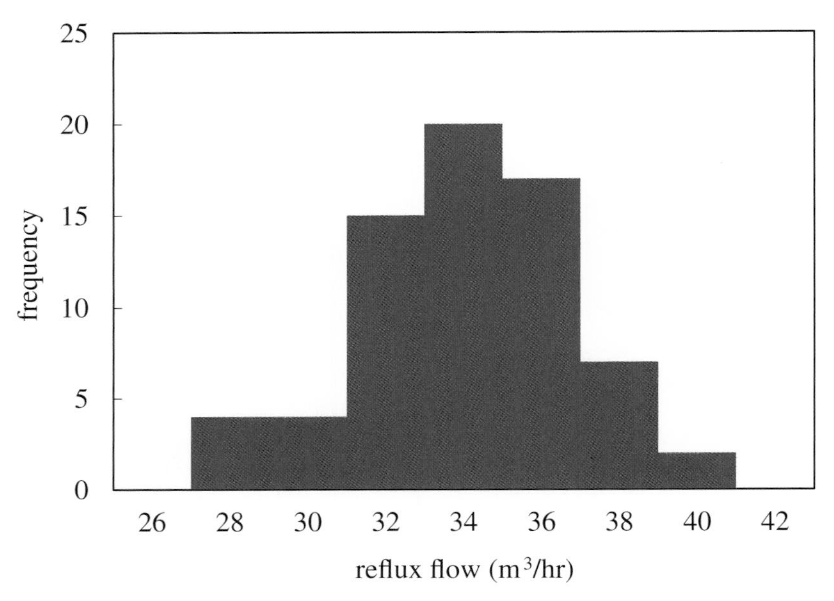

Figure 13.9 Frequency distribution of reflux flow (72-hour minima)

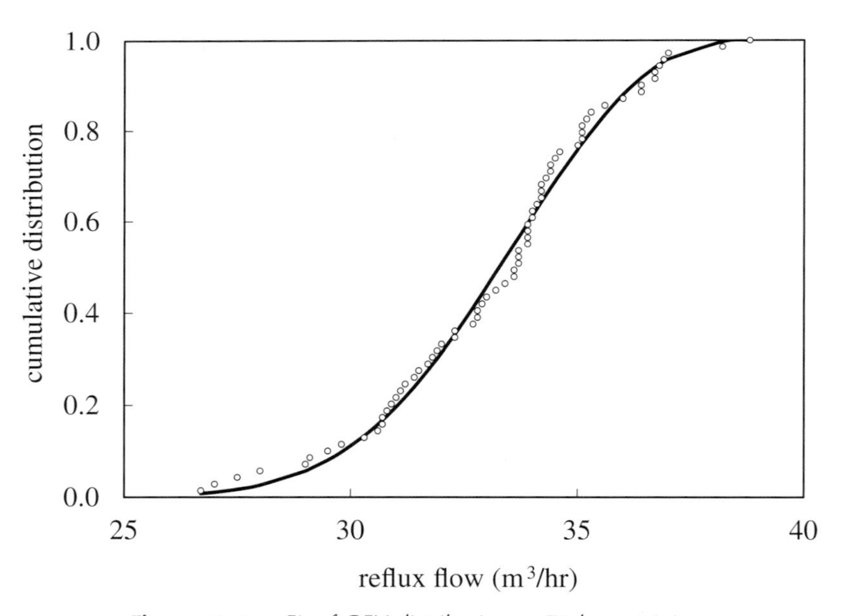

Figure 13.10 Fit of GEV distribution to 72-hour minima

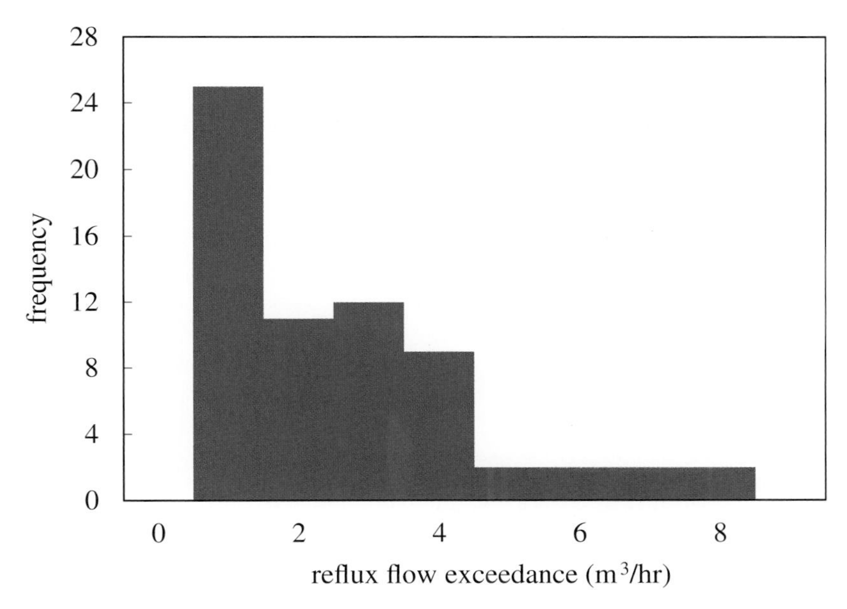

Figure 13.11 *Frequency distribution of reflux exceedances (<34.2)*

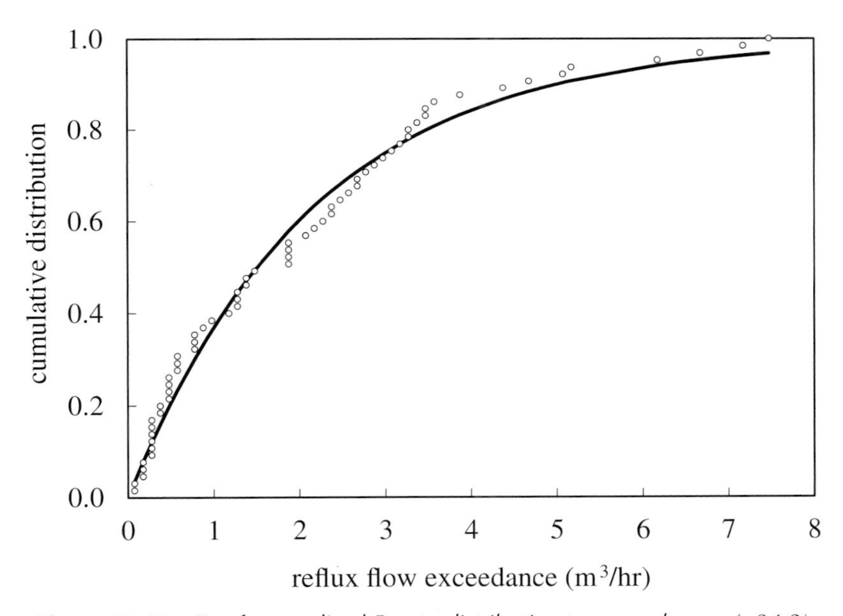

Figure 13.12 *Fit of generalised Pareto distribution to exceedances (<34.2)*

To apply the threshold approach, we modify the calculation of exceedances to

$$y_i = u - x_i \qquad x_i \leq u \tag{13.15}$$

Using the 2% centile, the threshold is given by $\mu - 2.054\sigma$ or 34.2 m^3/hr. The reflux was below this on 65 occasions, with their frequency distribution shown as Figure 13.11. We would have expected 100 occasions if the distribution were truly normal. Figure 13.12

shows the result of fitting the generalised Pareto distribution with δ set to 0, α to 0 and β to 2.17. The probability of an exceedance being greater than 7.2 m^3/hr (i.e. a reflux of less than 27 m^3/hr) is 0.0367. The probability of any exceedance is 0.0130 (65/5000) and so the overall probability is 0.000576. This is equivalent to expecting the event once in 2,098 hours, or 87 days.

14

Hazard Function

The *hazard function*, sometimes described as the *conditional failure density function* or, more commonly, the *failure rate* is defined as

$$h(x) = \frac{f(x)}{1 - F(x)} \tag{14.1}$$

The reciprocal of the hazard function is known as the *conditional failure density function.*

Applying Equation (14.1) to the exponential distribution, described by Equations (12.61) and (12.62), gives

$$h(x) = \lambda \tag{14.2}$$

This tells us that the failure rate is constant at λ; it does not vary with time. The process has no memory.

Commonly the EL (exponential-logarithmic) distribution is used. We will cover this in detail in Section 28.9. From Equations (28.33) and (28.34)

$$h(x) = -\frac{\lambda(1-\delta)e^{-\lambda x}}{[1-(1-\delta)e^{-\lambda x}]\ln[1-(1-\delta)e^{-\lambda x}]} \tag{14.3}$$

This function is plotted as Figure 14.1, with λ adjusted so that $h(x)$ is 1 when x is zero. As $\delta \rightarrow 1$, Equation (14.3) becomes Equation (14.2). As δ is reduced below 1, the failure rate reduces over time. This might apply to the number of occasions a control scheme is disabled because of technical issues that are resolved over time. Or it may be because the operators initially did not properly understand its operation and this resolved by ongoing training.

The cumulative function, as usual, is derived by integrating $h(x)$ from $-\infty$ to x and so, from Equation (14.1), is

$$H(x) = -\ln[1 - F(x)] \tag{14.4}$$

This is plotted for the EL distribution as Figure 14.2.

Taking the LPG splitter reflux flow as an example, we might be interested in whether commissioning improved control has affected the frequency at which the flow exceeded the problematic value of 70 m^3/hr. This occurred 129 times in the 5,000 hourly measurements, i.e. an

Statistics for Process Control Engineers: A Practical Approach, First Edition. Myke King.

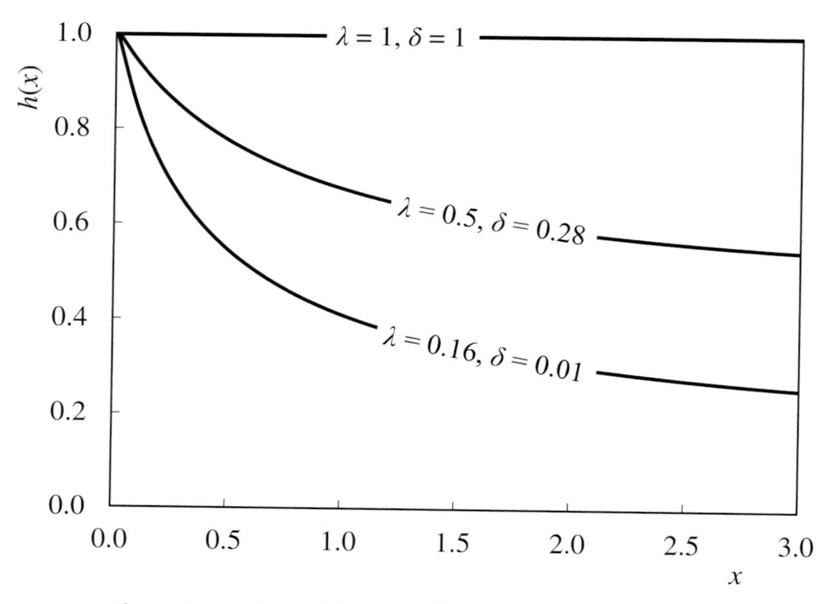

Figure 14.1 Hazard function derived from EL distribution

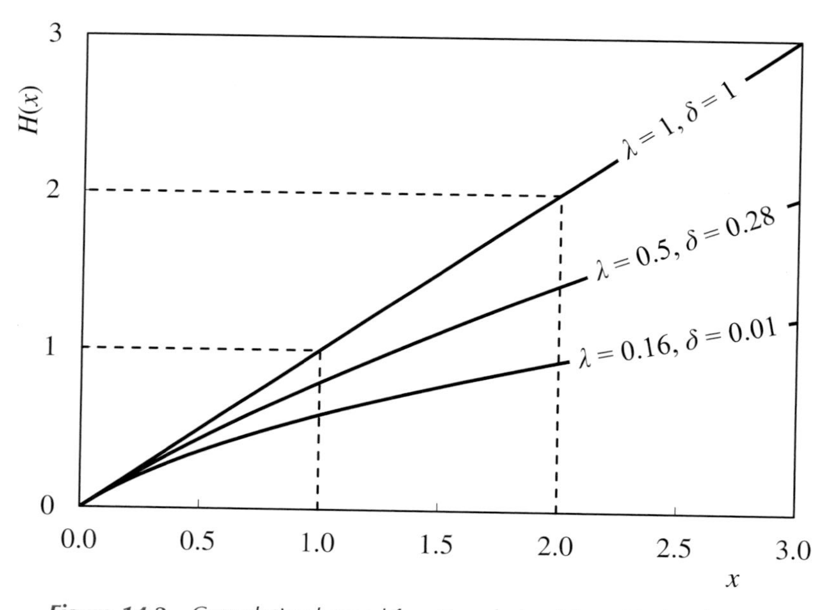

Figure 14.2 Cumulative hazard function derived from EL distribution

average rate of 0.0258 hr^{-1} or 33.8 hours between events. Figure 14.3 shows the frequency distribution of the interval between events.

Fitting the EL distribution, Equation (28.34) gives a value for λ of 0.0266 hr^{-1} and a value for δ of 0.815. Figure 14.4 shows that the distribution fits the data well. Using these parameters, Equation (14.3) is plotted as the black line in Figure 14.5. It shows that, at the beginning, the frequency of events is 0.0295 hr^{-1}, reducing to 0.0266 hr^{-1} over the 10 days since the improved

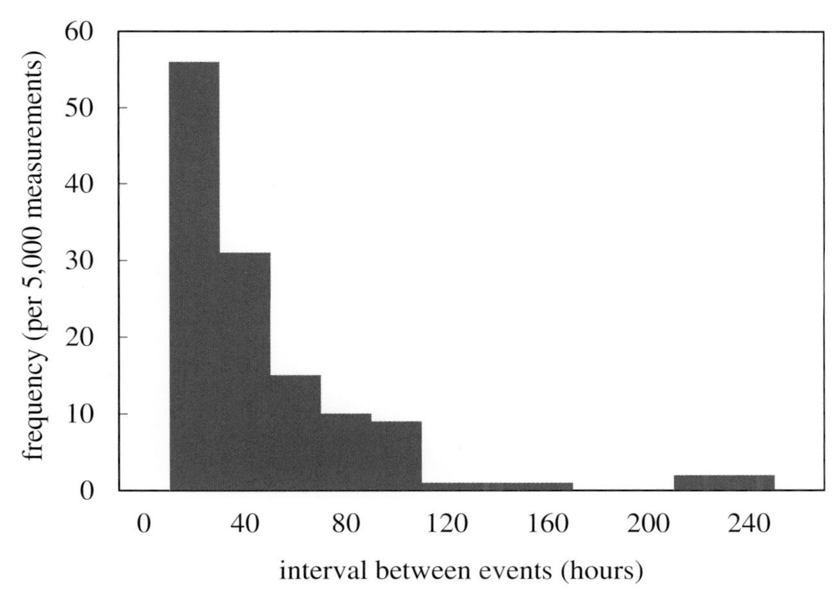

Figure 14.3 Distribution of intervals between high reflux flow events

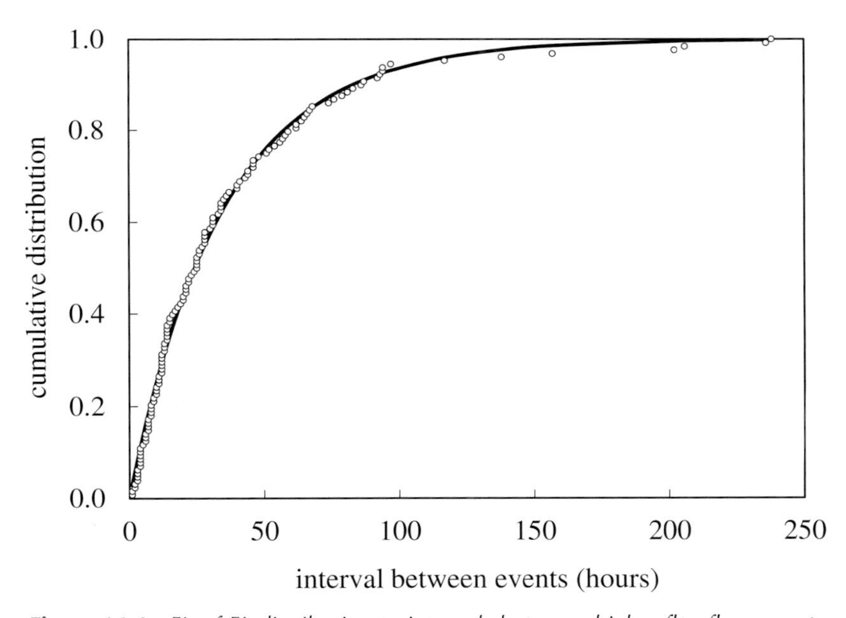

Figure 14.4 Fit of EL distribution to intervals between high reflux flow events

control was commissioned. This is equivalent to the period between events, plotted as the coloured line, increasing from 33.9 to 37.6 hours.

One of the benefits of improved process control can be an increase in the *mean time between failure (MTBF)* of critical process equipment, where

$$MTBF = \frac{1}{\lambda} \tag{14.5}$$

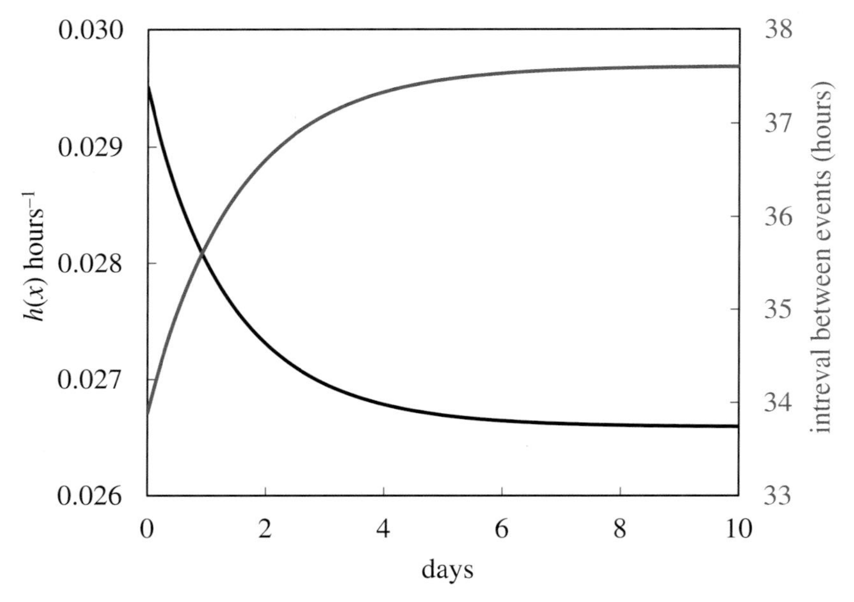

Figure 14.5 *Reduction in frequency of events assuming EL distribution*

There are examples of the reduction in process disturbances resulting in a threefold increase in the MTBF of pumps and compressors that operate in a difficult service. While such a benefit is unlikely to be quantifiable before implementing improved control, later proving its existence does much to foster a positive attitude to the technology. The relatively modest 11% increase in the mean time between high reflux events is because δ is close to 1. To achieve a threefold increase would require the fitted value of δ to be around 0.15.

Because δ cannot exceed 1, the EL distribution cannot represent a process where failure rate increases over time. Among those distributions, which can cover a decrease or an increase, is the gamma distribution. From Equation (12.92)

$$f(x) = \frac{\lambda^k x^{k-1} e^{-\lambda x}}{(k-1)!} \tag{14.6}$$

Figure 14.6 was developed by determining $F(x)$ from $f(x)$, using the trapezium rule, and then applying Equation (14.1). It shows the impact of changing k, with λ adjusted so that $h(x)$ is 1 when x is 0. Like the EL distribution, values less than 1 give a declining failure rate. Values greater than 1 might represent the failure rate of process equipment that becomes less reliable with age. But it might also be helpful in identifying if problems detected in a MPC application are becoming more frequent and whether the controller should be re-engineered. It can similarly apply to key instrumentation, such as on-stream analysers, where a proven reduction in reliability can justify replacement.

Other distributions that can be fitted to failure rate data include the Weibull-II distribution. From Equations (12.71) and (12.72)

$$h(x) = \frac{\delta x^{\delta-1}}{\beta^{\delta}} \tag{14.7}$$

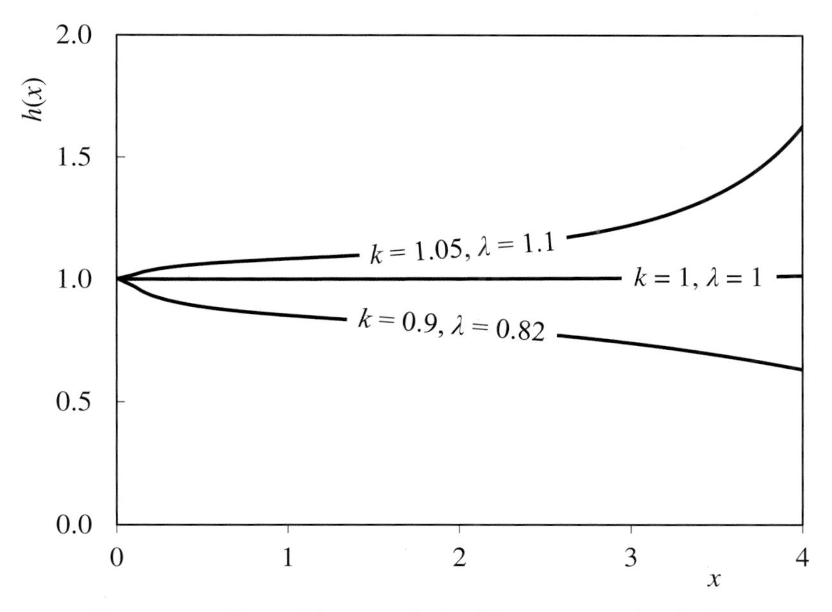

Figure 14.6 *Hazard function derived from gamma distribution*

If δ is 1 then the failure rate remains constant. Figure 14.7 shows the effect that adjusting δ has on failure rate over time. Again, β has been adjusted so that $h(x)$ is 1 when x is 0.

Fitting Equation (12.72) to the high reflux event data gives values for δ and β as 0.991 and 35.2 respectively. As the black line in Figure 14.8 shows, improved control reduced the frequency from 0.0291 to 0.0270 hr^{-1}, increasing the interval (coloured line) from 34.4 to

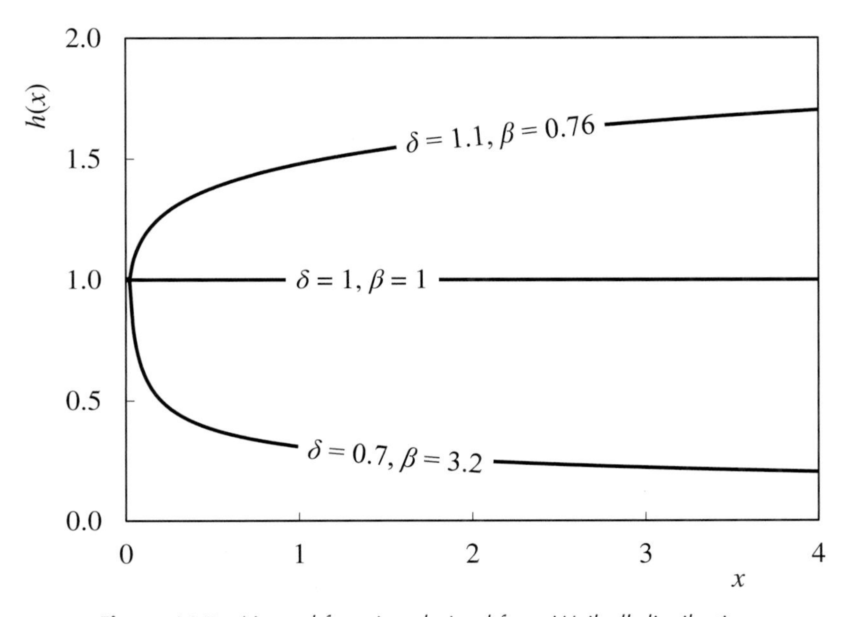

Figure 14.7 *Hazard function derived from Weibull distribution*

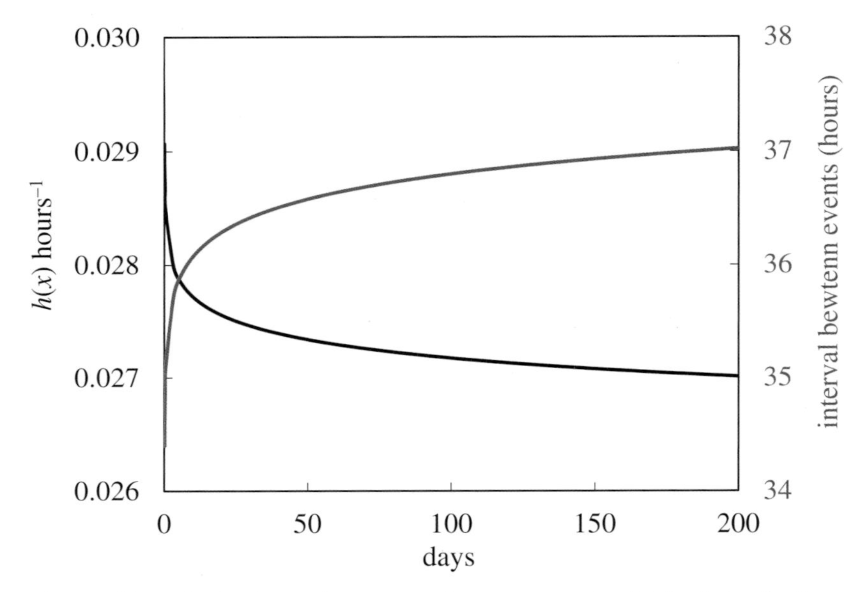

Figure 14.8 *Reduction in frequency of events assuming Weibull distribution*

37.0 hours. This is 30% less improvement than that estimated by fitting the EL distribution and over a much longer period of 200 days. If due to a control improvement, this would appear excessive. Figure 14.9 shows that the Weibull distribution is only a marginally worse fit to the data. Indeed, *RSS* is 0.0412 compared to 0.0393 for the EL distribution.

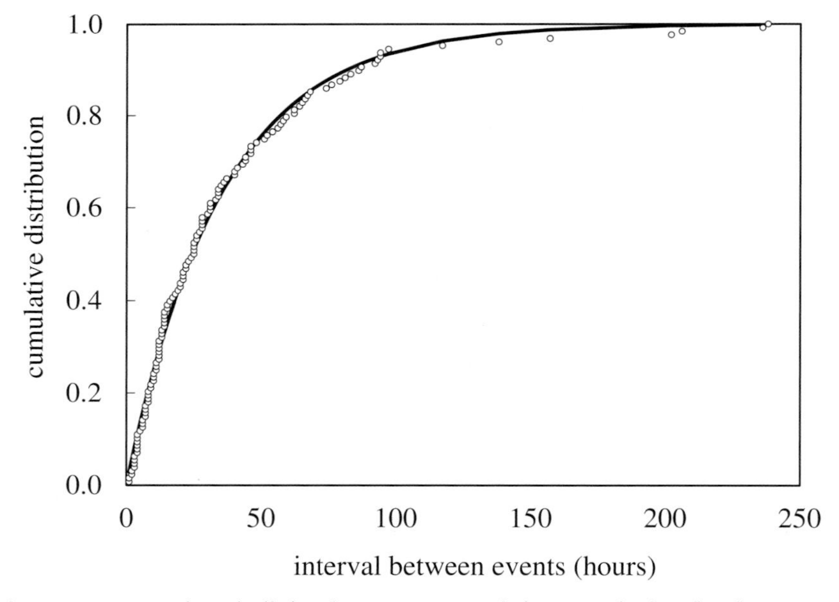

Figure 14.9 *Fit of Weibull distribution to intervals between high reflux flow events*

Many aspects of process operation can follow *bathtub* failure curves. For example, a new advanced control scheme might initially prove unreliable until all the teething troubles are resolved. In the longer term, lack of technical support might make it fall back slowly into disuse. This is illustrated by Figure 14.10 which shows the variation in *downtime* – the percentage of time that the scheme was out of service.

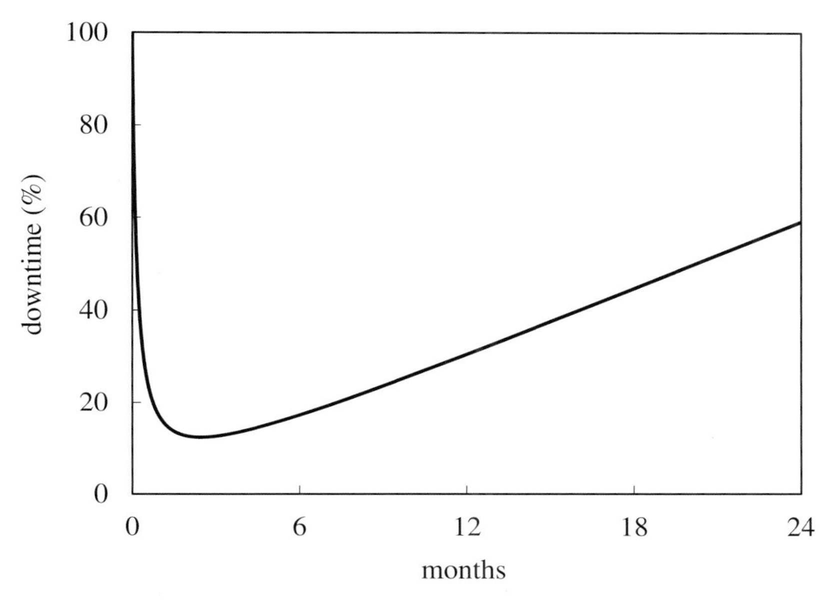

Figure 14.10 *Bathtub curve for advanced control application*

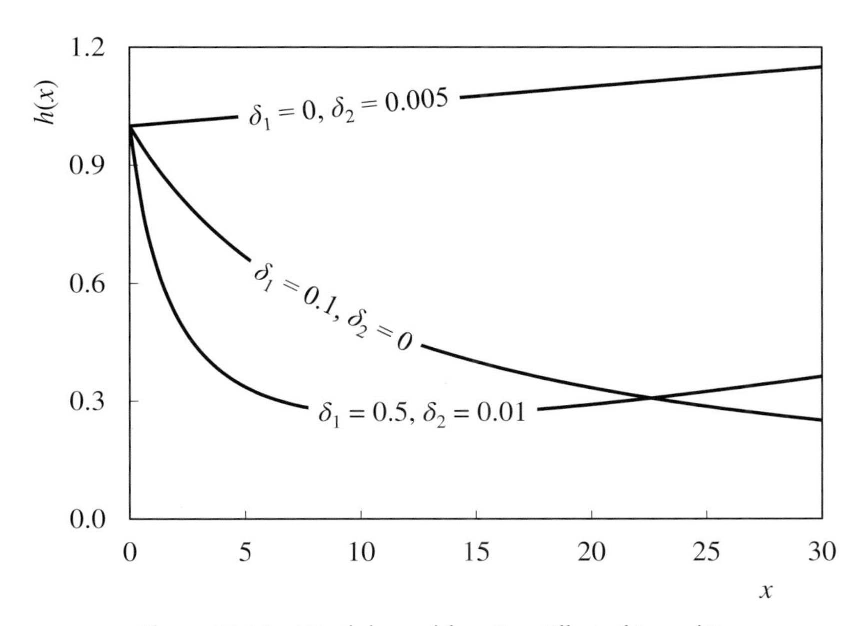

Figure 14.11 *Hjorth hazard function: Effect of δ_1 and δ_2*

Such curves can be generated by the *Hjorth distribution*, described by

$$f(x) = \frac{(1+\delta_1 x)\delta_2 x + \beta}{(1+\delta_1 x)^{\beta/\delta_1 + 1}} \exp\left(-\frac{\delta_2 x^2}{2}\right) \qquad x > 0;\ \beta, \delta_1, \delta_2 > 0 \tag{14.8}$$

$$F(x) = 1 - \frac{1}{(1+\delta_1 x)^{\beta/\delta_1}} \exp\left(-\frac{\delta_2 x^2}{2}\right) \tag{14.9}$$

$$\therefore h(x) = \frac{\beta}{1+\delta_1 x} + \delta_2 x \tag{14.10}$$

Figure 14.11 shows the effect of δ_1 and δ_2, with β fixed at 1. If δ_1 is set to zero, the failure rate increases. If δ_2 is set to zero, it decreases. If both are greater than 0, we obtain the bathtub curve. The initial letters of these three behaviours give the distribution its alternative name of the *IDB distribution*. Figure 14.10 was produced by setting β to 1, δ_1 to 0.2 and δ_2 to 0.0008. Time (x), although plotted in months, was defined in days in Equation (14.10).

15

CUSUM

CUSUM, or *cumulative sum*, is a simple mathematical technique that separates any underlying trend in data from random behaviour. For the process control engineer its applications include the detection of bias error in a measurement and determining whether a process has memory.

Measurements are subject to two forms of error – random and bias. A common requirement is to determine into which category falls an error that has been detected in an inferential property (or on-stream analyser) when it is compared to a laboratory sample. A random error may result from instrument repeatability, poor time-stamping and mistakes. A bias error may result from change in feed composition or degradation of catalyst activity. In practice, a recorded error is likely to comprise both. Applying a correction term to the inferential in response to a random error will reduce the accuracy of the inferential. Overlooking a bias error will result in off-grade production.

CUSUM, in this case, is calculated as the cumulative sum of disagreements between two measurements of the same property. Using the example of C_3 in butane, we have 1,152 sequential measurements from an on-stream analyser collected every 30 minutes. We also have the corresponding measurements from the inferential. We define the error as the inferential measurement minus that of the analyser. Figure 15.1 plots this error over time. Visually it would appear to be randomly distributed either side of zero. However, Figure 15.2 plots the cumulative sum of the errors and shows distinct downward and upward trends. If the error was indeed entirely random, we would expect the trend to be horizontal – albeit noisy. In fact, for the first 400 samples, CUSUM reduced steadily by about 20. Its slope is therefore –0.05. This is the bias error in the inferential. For the initial period of 200 hours, it consistently under-read by 0.05. By monitoring CUSUM, this error could have been detected within a few hours and dealt with by a simple change to the bias term in the inferential.

We showed earlier that most distributions, such as the Weibull distribution, assume that processes have memory, whereas a few, like the exponential distribution, do not. A bias error is a symptom of the process having memory. There are several ways of detecting this. For example, if we define an event as the inferential error exceeding 0.5, we can plot the distribution of times between events. This is shown as Figure 15.3; the solid line is the result of fitting the Weibull-II distribution. It shows that around one third of the events have an interval of 30 minutes. Since

Statistics for Process Control Engineers: A Practical Approach, First Edition. Myke King.

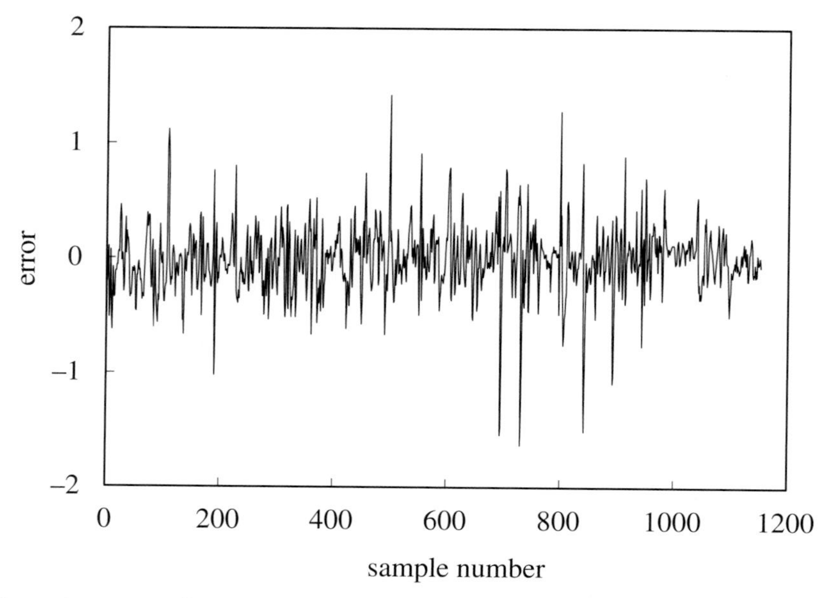

Figure 15.1 *Error between inferential for C_3 in butane and the on-stream analyser*

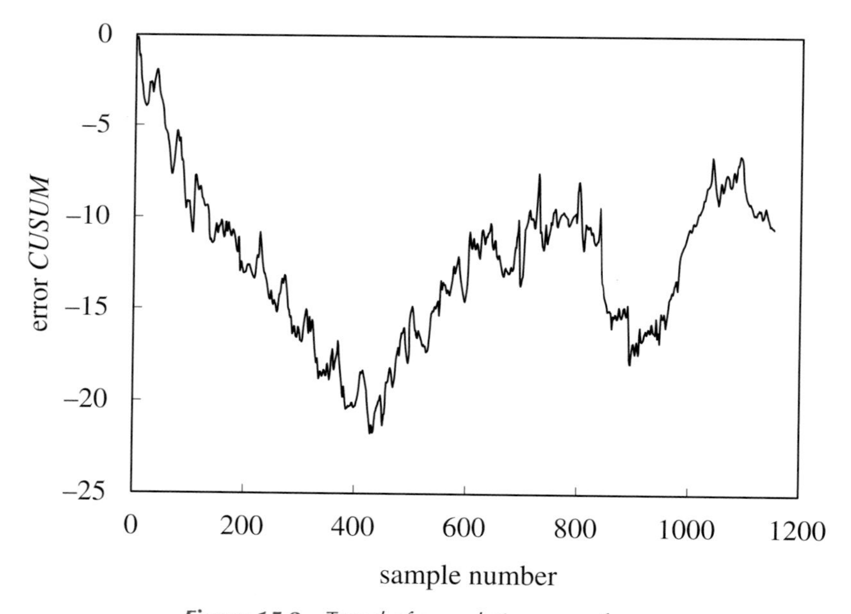

Figure 15.2 *Trend of cumulative sum of error*

this is also the data collection interval it shows that once an event occurs, it is likely to be immediately followed by another. Errors do not occur as a purely random event but tend to persist for a longer period.

We can also fit the Weibull-II distribution, using the method described in Section 12.8, as shown by Figure 15.4. The slope of the line (k) is 1.13. If the process had no memory, we would expect k to be 1 – equivalent to the exponential distribution.

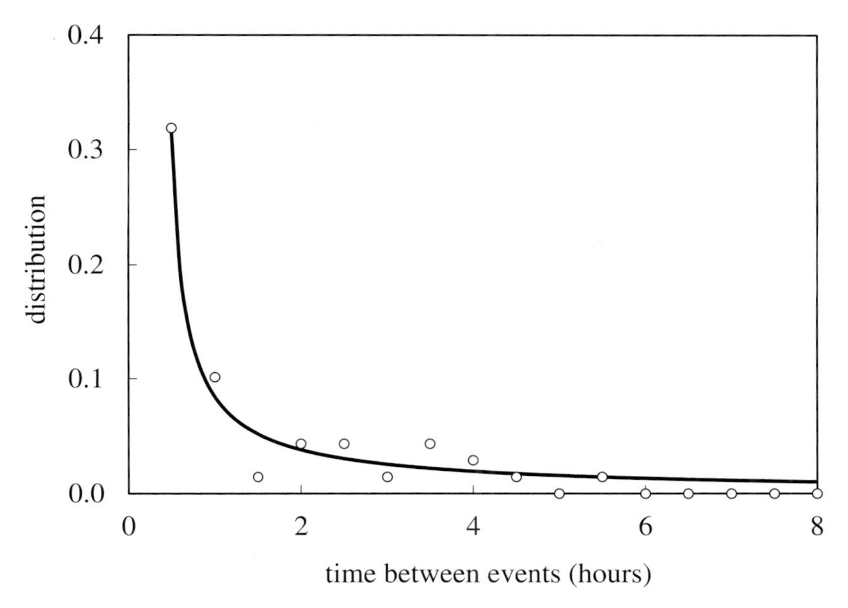

Figure 15.3 Fit of Weibull-II distribution to interval between errors exceeding 0.5 vol%

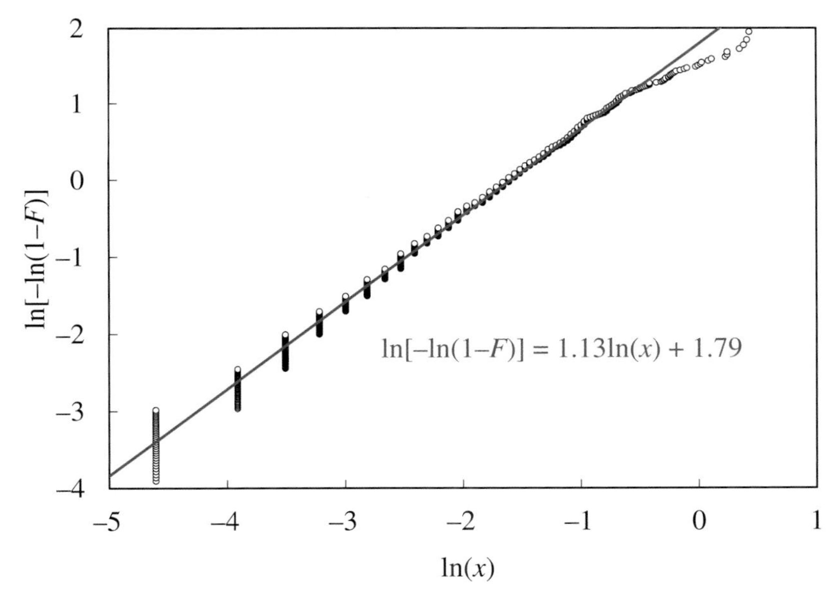

Figure 15.4 Demonstrating that the process has memory

Of course, if we have a reliable on-stream analyser installed, then we can use this to continuously update the bias term in the inferential. A different approach is required if we have only laboratory results. These will be far less frequent. If a result is incorrect and used to update the inferential, it could be many hours before this is undone by the next result. Like any measurement, the laboratory is subject to random error. We need to distinguish between this and a genuine deviation from the inferential. We can use the slope of the CUSUM to determine what correction is necessary. We need only to decide on the number (n) of previous measurements used in determining the slope.

We define the most recent error (E_n) as the inferred measurement minus the laboratory result, E_{n-1} as the previous error, etc. Without delving into the mathematics involved, correction based on the CUSUM slope reduces to a simple formula. In general, the bias correction is determined by

$$bias_{n+1} = bias_n - \frac{6}{n(n+1)(n+2)} \sum_{i=1}^{n} i(n-i+1)E_{n-i+1} \tag{15.1}$$

So, as a trivial example, choosing n as 1 gives

$$bias_{n+1} = bias_n - E_n \tag{15.2}$$

This effectively assumes that the laboratory is correct and the inferential is immediately adjusted to agree with it. Choosing n as 2 gives

$$bias_{n+1} = bias_n - \frac{E_n + E_{n-1}}{2} \tag{15.3}$$

This applies the average of the last two discrepancies as the correction. Higher values of n apply a weighted average of recent discrepancies. For example, increasing n to 3 gives

$$bias_{n+1} = bias_n - \frac{3E_n + 4E_{n-1} + 3E_{n-2}}{10} \tag{15.4}$$

The alternative, more traditional, approach is to apply a partial correction determined by a filter parameter (K) – typically set to around 0.3.

$$bias_{n+1} = bias_n - KE_n \tag{15.5}$$

Figure 15.5 compares the two techniques. The CUSUM technique implements the correction fully, although not until after three samples. The more commonly used filtering method takes six results to eliminate 90% of the error. As it is the size of error multiplied by duration, the shaded area is representative of the total product giveaway this approach causes. It can be shown that this is determined from $-1/\ln(1 - K)$. So if K is 0.3, the area is 280%. The equivalent area, if the CUSUM method is applied, is comparable at 300%. So, while the CUSUM method implements the bias update more quickly, it does not necessarily have much impact on overall production.

Large values of n will cause the updating to become unstable. The best solution is a combination of the two methods, but with K set to a value slightly less than 1. For example, if n is 4

$$bias_{n+1} = bias_n - K\left(\frac{2E_n + 3E_{n-1} + 3E_{n-2} + 2E_{n-4}}{10}\right) \tag{15.6}$$

The choice of n and K are best made by including them among the coefficients that are regressed when developing the inferential. Note that the correction is applied no more frequently than every n results. Indeed it is better to make the correction only if it is larger than some set limit. This limit might be based on the standard deviation of the error (σ_{error}) determined when the inferential was developed. For example, setting the limit to $1.96\sigma_{error}$ would mean that the bias would be updated only when we are more than 95% certain that the error is genuine.

CUSUM can be used to identify whether a dataset has memory. To do so, we cumulatively sum the deviations from mean.

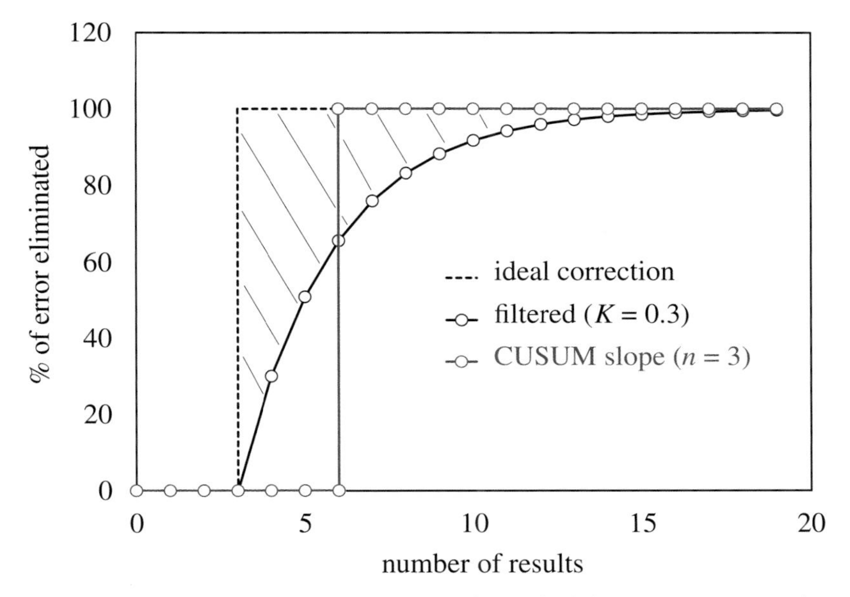

Figure 15.5 Comparison between speeds at which bias error is removed

$$CUSUM = \sum_{i=1}^{n} (x_i - \bar{x}) \tag{15.7}$$

For example, using the inventory data from Table A1.6, we can determine the mean as 1292. The first value in the table (x_1) is 422 and the deviation from mean is therefore −870. So, when n is 1, *CUSUM* is −870. The second value in the table (x_2) is 520, the deviation from mean is −772 and, when n is 2, *CUSUM* is −1642. The coloured line in Figure 15.6 is the continuation of this

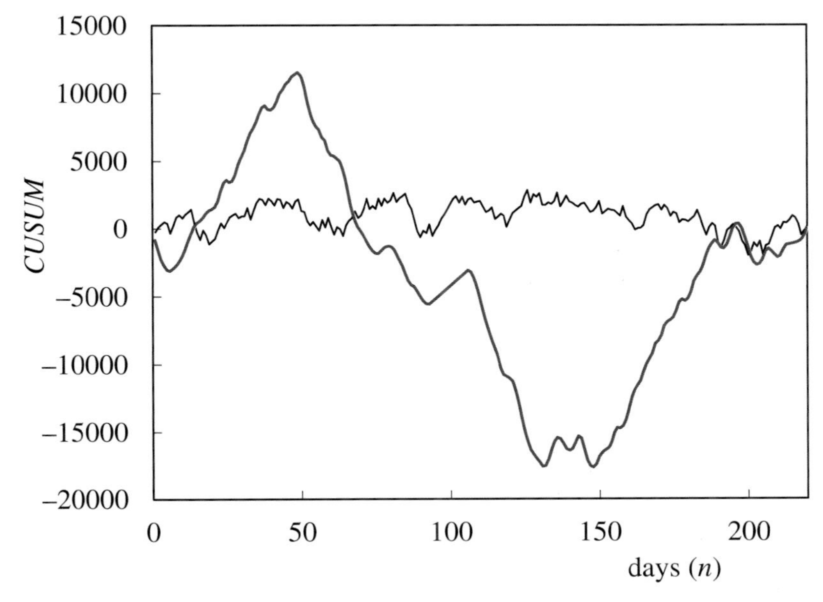

Figure 15.6 Trend of cumulative sum of deviation of stock level from the mean

calculation up to 220 for n. It shows sustained periods when the inventory is above or below the average. In other words, if the inventory were above (or below) the mean yesterday it is highly likely that it will be again today. The process has memory.

We also confirmed this by fitting the Weibull-III distribution, as shown in Figure 12.38. Without memory, we would expect the CUSUM trend to resemble the black line. This was produced by randomly shuffling the data in Table A1.6 before plotting CUSUM. The data clearly have the same mean and standard deviation but are no longer related to time.

16

Regression Analysis

To perform regression we first need to decide which is the independent variable (x) and which is the dependent variable (y). Used by control engineers mainly to develop inferential properties, the result is intended to predict the dependent property (e.g. product composition) from independent variables such as flow, temperature and pressure.

To illustrate the principle of regression analysis we will fit a straight line to three points with (x,y) coordinates (2,3), (3,9) and (7,12). To do so, we first calculate the mean values of x and y. These are respectively 4 and 8. We then calculate the total shaded area shown in Figure 16.1. This is known as the *total sum of squares*. As is usual, the side of each square is the vertical distance between the point and in the line, i.e. in the y direction.

$$\sum_{i=1}^{3}(y_i-\bar{y})^2=(3-8)^2+(9-8)^2+(12-8)^2=42 \tag{16.1}$$

Regression analysis rotates the coloured line around the point ($\bar{x}$,$\bar{y}$) to minimise the shaded area. In effect this minimises the difference between the predicted and actual values of the dependent variable. We will show later how to derive the equation of the line of best fit. For this example we would predict y from x as

$$\hat{y}=1.5x+2 \tag{16.2}$$

And so

$$\hat{y}_{1'}=5 \qquad \hat{y}_{2'}=6.5 \qquad \hat{y}_{3'}=12.5 \tag{16.3}$$

The minimised shaded area is known as the *residual sum of squares (RSS)*.

$$\sum_{i=1}^{3}(y_i-\hat{y}_i)^2=(3-5)^2+(9-6.5)^2+(12-12.5)^2=10.5 \tag{16.4}$$

We can also calculate the *explained sum of squares (ESS)*.

$$\sum_{i=1}^{3}(\hat{y}_i-\bar{y})^2=(5-8)^2+(6.5-8)^2+(12.5-8)^2=31.5 \tag{16.5}$$

Statistics for Process Control Engineers: A Practical Approach, First Edition. Myke King.

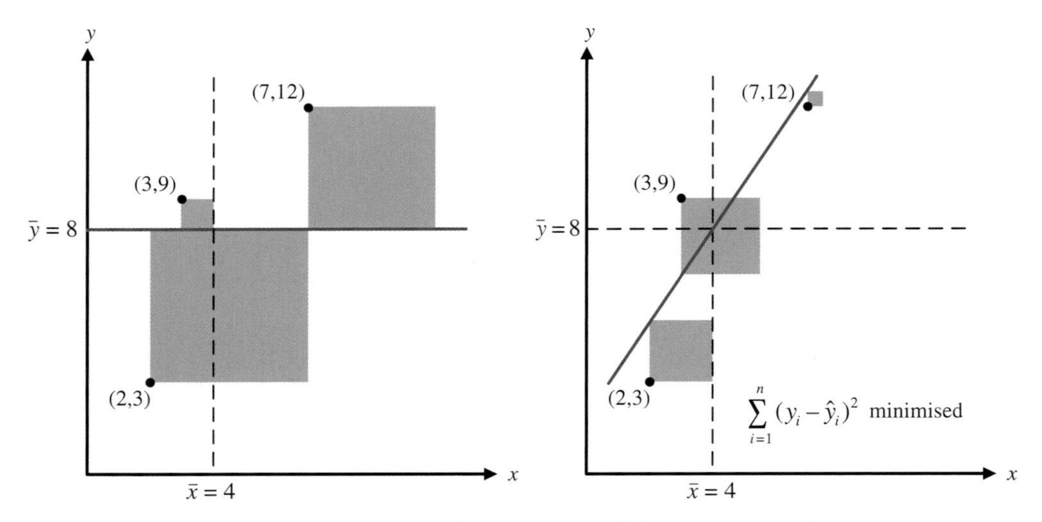

Figure 16.1 *Minimisation of the total sum of the squares in y direction*

Equation (16.1) gives the variance of y multiplied by $(n - 1)$. Similarly Equation (16.5) gives the variance of $\hat{y}$, also multiplied by $(n - 1)$. The ratio of these results is the *coefficient of determination (R^2)* – the proportion of the variance of y that is explained by the variance of x.

$$R^2 = \frac{\sum_{i=1}^{3} (\hat{y}_i - \bar{y})^2}{\sum_{i=1}^{3} (y_i - \bar{y})^2} = \frac{31.5}{42} = 0.75 \tag{16.6}$$

We know that variances are additive; the explained variance and the residual variance will sum to the total variance. R^2 can therefore also be calculated from the residual sum of squares.

$$R^2 = 1 - \frac{\sum_{i=1}^{3} (y_i - \hat{y}_i)^2}{\sum_{i=1}^{3} (y_i - \bar{y})^2} = 1 - \frac{10.5}{42} = 0.75 \tag{16.7}$$

This is not to be confused with the Pearson R^2. If we square Equation (4.75) and replace x with $\hat{y}$, we obtain

$$R^2 = \frac{\left[\sum_{i=1}^{n} (\hat{y}_i - \bar{y})(y_i - \bar{y})\right]^2}{\sum_{i=1}^{n} (\hat{y}_i - \bar{y})^2 \sum_{i=1}^{n} (y_i - \bar{y})^2} \tag{16.8}$$

This will also have a value of 0.75 but behaves differently if the sequence of the y values is shuffled while retaining the sequence of the $\hat{y}$ values. The value determined by Equation (16.6) would remain unchanged while that from Equation (16.8) will vary. If we want to assess whether a product quality correlates well with measurements of flow, temperature and pressure

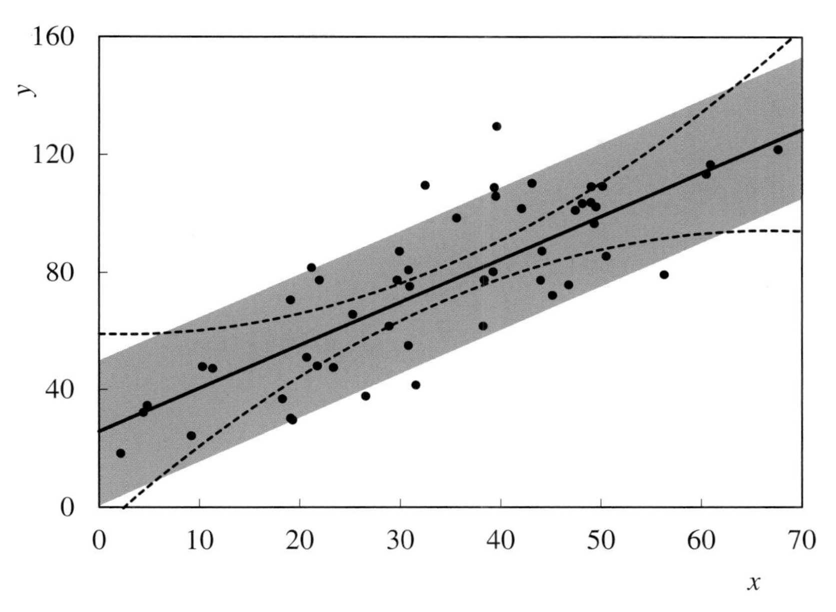

Figure 16.2 Linear regression analysis

we would clearly expect that correlation to be poor if the sequence of one of the measurements is randomly changed. Pearson R^2 would reflect this. The alternative coefficient of determination has very limited practical application.

Figure 16.2 shows a more typical example with a larger number (n) of data points (x,y). Again we wish to identify the equation of the (straight) line of best fit, as shown, where a_0 is the intercept on the y axis and a_1 is the slope of the line, i.e.

$$\hat{y} = a_0 + a_1 x \tag{16.9}$$

We have seen that the equation of the line is developed to minimise the residual sum of the squares (RSS) between the predicted value of y and the actual value, i.e.

$$\sum_{i=1}^{n} (\hat{y}_i - y_i)^2 = \sum_{i=1}^{n} (a_0 + a_1 x_i - y_i)^2 \tag{16.10}$$

Partially differentiating with respect to each of a_1 and a_0 and setting the derivative to 0 will identify the best choice of these values. Firstly with respect to a_0

$$\frac{\partial \sum_{i=1}^{n} (\hat{y}_i - y_i)^2}{\partial a_0} = \sum_{i=1}^{n} 2(a_0 + a_1 x_i - y_i) = 0 \tag{16.11}$$

$$\therefore \; n a_0 + a_1 \sum_{i=1}^{n} x_i - \sum_{i=1}^{n} y_i = 0 \quad \text{or} \quad a_0 + a_1 \bar{x} - \bar{y} = 0 \tag{16.12}$$

As we expect from the principles of the method, calculating the mean of the predicted values from Equation (16.9) and comparing the result with Equation (16.12) shows that it is identical to the mean of the actual values.

Partially differentiating with respect to a_1

$$\frac{\partial \sum_{i=1}^{n} (\hat{y}_i - y_i)^2}{\partial a_1} = \sum_{i=1}^{n} 2x_i(a_0 + a_1 x_i - y_i) = 0 \tag{16.13}$$

$$\therefore \; a_0 \sum_{i=1}^{n} x_i + a_1 \sum_{i=1}^{n} x_i^2 - \sum_{i=1}^{n} x_i y_i = 0 \quad \text{or} \quad n a_0 \bar{x} + a_1 \sum_{i=1}^{n} x_i^2 - \sum_{i=1}^{n} x_i y_i = 0 \tag{16.14}$$

Solving Equations (16.12) and (16.14) gives

$$a_0 = \frac{\bar{y} \sum_{i=1}^{n} x_i^2 - \bar{x} \sum_{i=1}^{n} x_i y_i}{\sum_{i=1}^{n} x_i^2 - n\bar{x}^2} \quad \text{and} \quad a_1 = \frac{\sum_{i=1}^{n} x_i y_i - n\bar{x}.\bar{y}}{\sum_{i=1}^{n} x_i^2 - n\bar{x}^2} \tag{16.15}$$

This line of best fit is plotted in Figure 16.2. The minimised *RSS* is

$$\sum_{i=1}^{n} (\hat{y}_i - y_i)^2 = \sum_{i=1}^{n} (y_i - \bar{y})^2 - a_1^2 \sum_{i=1}^{n} (x_i - \bar{x})^2 \tag{16.16}$$

The last term in Equation (16.16) is the variability of y that is due to the variability of x. Subtracting this from the variability of y gives the variability of the prediction error. We will show later how this variability can be converted to standard deviation.

This same approach is theoretically possible for more than one independent variable but manually solving the resulting equations would be impractical. Rewriting Equations (16.12) and (16.14) for the case of a single independent variable x_1

$$a_0 n + a_1 \sum x_1 = \sum y \tag{16.17}$$

$$a_0 \sum x_1 + a_1 \sum x_1^2 = \sum x_1 y \tag{16.18}$$

Applying the same approach we can derive similar equations for two independent variables, x_1 and x_2.

$$a_0 n + a_1 \sum x_1 + a_2 \sum x_2 = \sum y \tag{16.19}$$

$$a_0 \sum x_1 + a_1 \sum x_1^2 + a_2 \sum x_1 x_2 = \sum x_1 y \tag{16.20}$$

$$a_0 \sum x_2 + a_1 \sum x_1 x_2 + a_2 \sum x_2^2 = \sum x_2 y \tag{16.21}$$

And for three independent variables, x_1, x_2 and x_3

$$a_0 n + a_1 \sum x_1 + a_2 \sum x_2 + a_3 \sum x_3 = \sum y \tag{16.22}$$

$$a_0 \sum x_1 + a_1 \sum x_1^2 + a_2 \sum x_1 x_2 + a_3 \sum x_1 x_3 = \sum x_1 y \tag{16.23}$$

$$a_0 \sum x_2 + a_1 \sum x_1 x_2 + a_2 \sum x_2^2 + a_3 \sum x_2 x_3 = \sum x_2 y \tag{16.24}$$

$$a_0 \sum x_3 + a_1 \sum x_1 x_3 + a_2 \sum x_2 x_3 + a_3 \sum x_3^2 = \sum x_3 y \tag{16.25}$$

We can also arrange these last four equations into matrix form.

$$\begin{pmatrix} n & \sum x_1 & \sum x_2 & \sum x_3 \\ \sum x_1 & \sum x_1^2 & \sum x_1x_2 & \sum x_1x_3 \\ \sum x_2 & \sum x_1x_2 & \sum x_2^2 & \sum x_2x_3 \\ \sum x_3 & \sum x_1x_3 & \sum x_2x_3 & \sum x_3^2 \end{pmatrix} \begin{pmatrix} a_0 \\ a_1 \\ a_2 \\ a_3 \end{pmatrix} = \begin{pmatrix} \sum y \\ \sum x_1y \\ \sum x_2y \\ \sum x_3y \end{pmatrix} \tag{16.26}$$

The reader should now see the pattern by which higher numbers of independent variables would be incorporated. The coefficients (a) would then be determined by solving these equations.

The partial differentiation performed in developing the formulae above assumes that the variables are truly independent. If this is not the case then the resulting correlation may be suspect. Indeed, if there is a perfect correlation between two variables, the formulae will fail. Consider the case where there is a linear relationship between x_2 and x_1 (known as *collinearity*) such that

$$x_2 = p + qx_1 \tag{16.27}$$

Then, by substituting for x_2 in Equation (16.24), we obtain

$$a_0\sum(px_1+q) + a_1\sum x_1(px_1+q) + a_2\sum(px_1+q)x_2 + a_3\sum(px_1+q)x_3 = \sum(px_1+q)y \tag{16.28}$$

This can be rewritten as

$$p\left(a_0\sum x_1 + a_1\sum x_1^2 + a_2\sum x_1x_2 + a_3\sum x_1x_3\right) + q\left(a_0n + a_1\sum x_1 + a_2\sum x_2 + a_3\sum x_3\right) = p\sum x_1y + q\sum y \tag{16.29}$$

But this could also have been derived by multiplying Equation (16.23) by p, multiplying Equation (16.22) by q and adding the results together. In other words we no longer have four independent equations with four unknown coefficients. In order to solve the equations we have to set either a_1 or a_2 to zero – effectively removing either x_1 or x_2 from the analysis. One of these variables is *redundant*; the inclusion of both would result in one row of the $\boldsymbol{\Sigma}$ matrix in Equation (16.26) being derivable from others. The matrix would be *singular*.

The same problem will arise if there is *multi-collinearity*, where an independent variable may not show a strong correlation with another but does correlate perfectly with a linear combination of others. As an example, consider the development of an inferential property on a binary distillation column. Inputs that might be considered include the flows of feed, distillate and bottoms. While there will be some correlation between any chosen two of these, it would be unwise to exclude either one. Indeed variation in the distillate-to-feed (or bottoms-to-feed) ratio is likely to be a key input. However, because the process must mass balance, there will be a strong correlation involving all three inputs. They will fit the general equation

$$x_1 = b_0 + b_2x_2 + b_3x_3 + \ldots \tag{16.30}$$

To detect multi-collinearity Equation (16.30), along with those for x_1, x_3 etc., is derived by regression. The coefficient of determination (R^2) is then calculated for each predicted independent variable.

$$R_j^2 = \frac{\sum_{i=1}^{n} \left(\hat{x}_{ji} - \bar{x}_j\right)^2}{\sum_{i}^{n} \left(x_{ji} - \bar{x}_j\right)^2} \tag{16.31}$$

Consideration would then be given to excluding the jth independent variable if R_j^2 is greater than 0.9. Of course, if a strong correlation is identified such that, for example, x_1 can be predicted from x_2 and x_3, then it follows that x_2 can be predicted from x_1 and x_3; similarly x_3 can be predicted from x_1 and x_2. We would therefore have to decide which of the three variables should be excluded. Pragmatically this would probably be the one we might feel is least reliably measured or might exhibit dynamic behaviour quite different from the others.

In practice most cases of collinearity will be detected by including only one input in Equation (16.30). Indeed there is no need to perform the regression; R^2 can be determined directly from Equation (4.75). There are however situations that would justify including more than one input. Including all n flows into and out of a process would require correlations with up to $n - 1$ inputs to be regressed. While this would be reasonable on a binary distillation column, including a large number of inputs all in Equation (16.30) can generate results that can be virtually impossible to interpret properly. As always, judgement is required in including only those measurements that make engineering sense.

Occasionally referred to is the *variance inflation factor (VIF)*. It is defined as

$$VIF_j = \frac{1}{1 - R_j^2} \tag{16.32}$$

This is the factor by which the variance of the estimate of the coefficient a_j is increased due to collinearity. For example, if R_j^2 was 0.9 then VIF_j would be 10. The standard deviation is thus multiplied by $\sqrt{10}$. We saw in Section 8.2 that confidence interval is expressed as a multiple of the standard deviation, so (for example) the 95% confidence interval in our estimate of a_j is 3.16 times larger than it would be if there were no collinearity. The reciprocal of VIF is known as the *tolerance*.

Process data, used to develop inferential properties, will usually show some level of cross-correlation. However because of random errors in the measurements, even if the correlation is perfect, the matrix used to derive Equation (16.30) will not be exactly singular. While the level of cross-correlation in well-chosen process data is usually not sufficient to undermine the value of the resulting inferential, it is wise to check for any strong collinearity and remove any offending variable(s). Otherwise the coefficients (a) will be unreliable and would likely show large variation if derived from different subsets of the data.

But great care should be taken before rejecting any measurements. For example, many reactors have a large number of measurements of bed temperature. Many of the intermediate temperatures are likely to be predictable from others around it. It is common in reactors for product composition to correlate strongly with the rise (or fall) in temperature across the bed. Two temperatures may show a strong correlation but the difference between them varies sufficiently to account for the variation in product composition. Rejecting either temperature would compromise the accuracy of the inferential. The same issue can arise with distillation tray temperatures. These are usually highly correlated but the difference between two temperatures is an effective measure of separation and a valuable input to the inferential.

We can define a confidence interval for the resulting correlation. The standard deviation of the prediction error (σ_{error}) is

$$\sigma_{error} = \sqrt{\frac{\sum_{i=1}^{n}(\hat{y}-y)^2}{n-2}} \tag{16.33}$$

The denominator is $n - 2$ rather than $n - 1$ because we use two degrees of freedom to calculate intermediate values (a_0 and a_1). In the more conventional definition of standard deviation we calculate only one – the mean.

Without presenting the complex derivation, the confidence interval for the line of regression is

$$\sigma_y^2 = t.\sigma_{error}^2\left[\frac{1}{n} + \frac{(x-\bar{x})^2}{\sum_{i=1}^{n}(x_i-\bar{x})^2}\right] \tag{16.34}$$

Strictly t is from the Student t distribution, determined by the number of degrees of freedom and the required confidence interval. In our example we have 51 data points, giving 50 degrees of freedom. The 95% confidence interval would require a value of 2.0086 for t. The value approaches that from the normal distribution of 1.9600 shown in Table A1.8. Since regression analysis usually involves a large number of values, little accuracy would be lost by using this estimate. The confidence interval has been included in Figure 16.2 as the dashed lines. We might expect 95% of the points to lie within the dashed lines but only about 40% do so. This is because the lines are the boundaries of all possible straight lines, not the confidence interval of the predicted value. (To illustrate the difference, the 95% confidence interval for the predicted value is shown as the shaded band.) Rapidly diverging boundaries are an indication that there is poor *scatter* and that points should be collected over a wider range.

These *confidence boundaries* for the line of regression can only be drawn based on a single independent variable. A correlation can of course be based on several independents. We can plot the boundaries for each independent by compensating its values to remove the effect of the other independents. Imagine that we have developed a correlation based on three independent variables.

$$\hat{y} = a_0 + a_1x_1 + a_2x_2 + a_3x_3 \tag{16.35}$$

From Equation (16.12) we can see that the same correlation will apply to the means.

$$\bar{y} = a_0 + a_1\bar{x}_1 + a_2\bar{x}_2 + a_3\bar{x}_3 \tag{16.36}$$

Subtracting gives

$$\hat{y} - \bar{y} = a_1(x_1-\bar{x}_1) + a_2(x_2-\bar{x}_2) + a_3(x_3-\bar{x}_3) \tag{16.37}$$

Suppose we want to determine the confidence boundaries for the line of regression defined by a_2. Rearranging we get

$$\hat{y} = a_2\left[x_2 + \frac{a_1}{a_2}(x_1-\bar{x}_1) + \frac{a_3}{a_2}(x_3-\bar{x}_3)\right] + \bar{y} - a_2\bar{x}_2 \tag{16.38}$$

We can therefore rewrite the correlation using the compensated independent variable (x_2').

$$x_2' = x_2 + \frac{a_1}{a_2}(x_1 - \bar{x}_1) + \frac{a_3}{a_2}(x_3 - \bar{x}_3) \tag{16.39}$$

We have adjusted each value of x_2 to reflect what they would need to be for the correlation to give the same prediction if the other independent values were fixed at their means. In fact we can choose any other reference value. This technique is applied to define a linear *PCT (pressure compensated temperature)* used in the control of distillation columns. Removing the effect of pressure means that product composition will more closely correlate with the temperature. In other words, controlling PCT will more closely control product quality. The technique is similarly applied to derive the weighting coefficients in the *WABT (weighted average bed temperature)* of reactors so that it correlates well with conversion.

We can now write a correlation based on a single input and use this to determine the confidence interval for the line of regression.

$$\hat{y} = a_0' + a_2 x_2' \quad \text{where} \quad a_0' = \bar{y} - a_2 \bar{x}_2 \tag{16.40}$$

Further, if the plot of $\hat{y}$ against x_2' shows a nonlinear relationship, this can be used to help select a suitable nonlinear function that might be applied to x_2. Figure 16.3 shows an example where y is the concentration of heavy key component in the overhead product from a distillation column and x_2 the temperature measured on a tray close to the top. The black points show that there appears to be no correlation. This is because of variations in x_1 – the column pressure. Applying Equation (16.39) removes the effect of this variation. The coloured points now show a clear correlation that suggests the PCT would be better included as a quadratic, rather than linear, function.

It should be noted that regression does not tell us which is the independent variable and which is the dependent. It merely enables us to show that they are correlated. Indeed both could be dependent variables changing in response to changes in an unmeasured independent. The formulae above were developed assuming that we would wish to minimise, in the y direction,

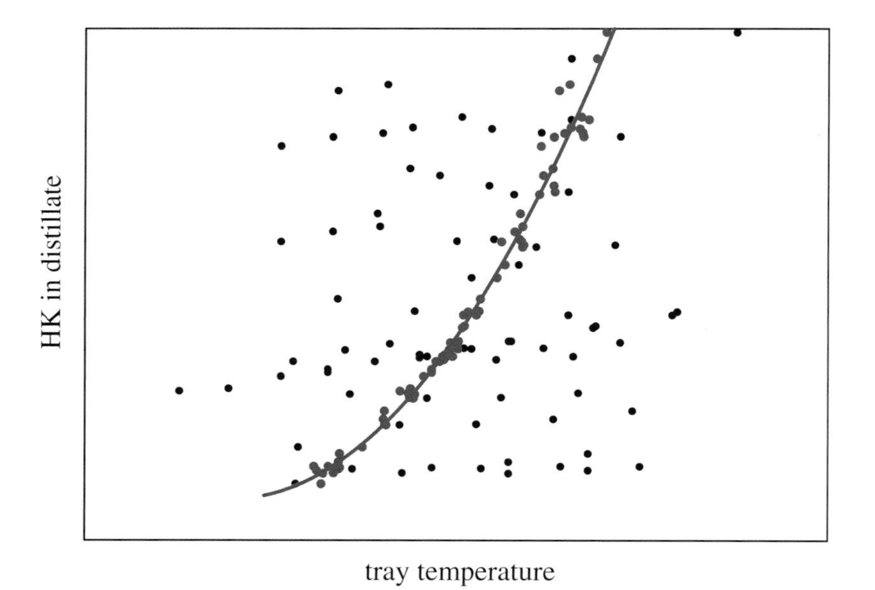

Figure 16.3 *Compensation of one independent variable for variation in others*

the sum of the squares of the distances from each point to the regressed line. If ε is the unaccounted error then each value of y is

$$y_i = a_0 + a_1 x_i + \varepsilon_i \tag{16.41}$$

This assumes that the error arises in y, for example because there are unmeasured variables affecting its value. However the error may be present in x, for example because it is inaccurately measured. In this case we should develop a correlation of the form

$$y_i = a_0 + a_1(x_i + \varepsilon_i) \tag{16.42}$$

In other words, we would wish to minimise

$$\sum_{i=1}^{n} (\hat{x}_i - x_i)^2 = \frac{1}{a_1^2} \sum_{i=1}^{n} (\hat{y}_i - y_i)^2 \tag{16.43}$$

Using the simple example from the beginning of this chapter, this is illustrated as Figure 16.4.

Alternatively x can be regressed against y to give the relationship

$$\hat{x} = b_0 + b_1 y \tag{16.44}$$

Inverting this equation gives

$$\hat{y} = a_0 + a_1 x \quad \text{where} \quad a_0 = -\frac{b_0}{b_1} \quad \text{and} \quad a_1 = \frac{1}{b_1} \tag{16.45}$$

We should adopt this approach if the error in x is large compared to the range of values of x. However in practice there are likely to be measurement errors in both x and y. It is possible to

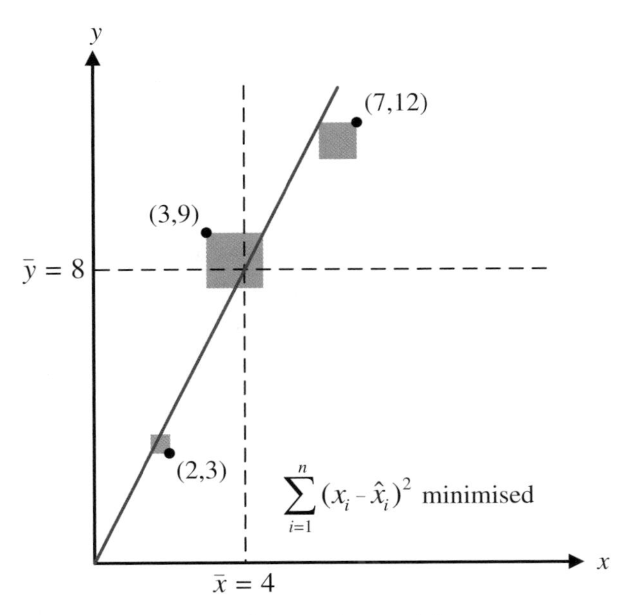

Figure 16.4 Minimisation of the total sum of the squares in x direction

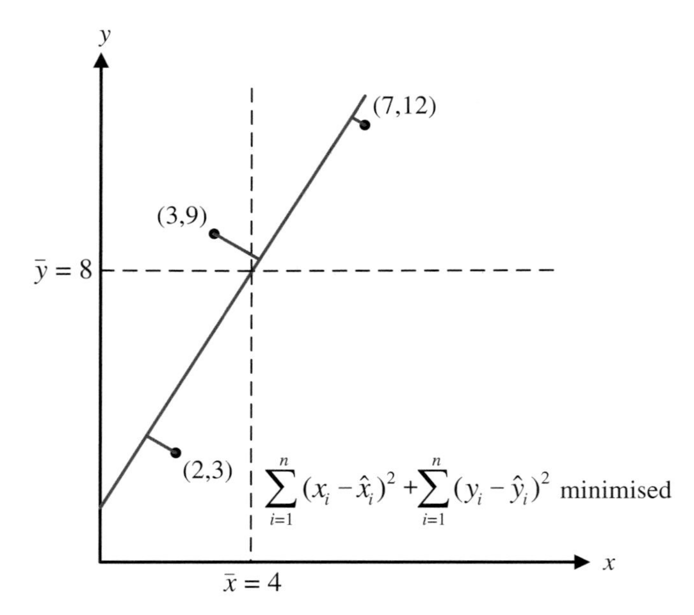

Figure 16.5 *Minimisation of the total sum of the squares of the perpendicular distance*

give equal weighting to both by instead using the perpendicular distance from each point to the regressed line. This is illustrated as Figure 16.5. The function we would then minimise is

$$\sum_{i=1}^{n}(\hat{x}_i - x_i)^2 + \sum_{i=1}^{n}(\hat{y}_i - y_i)^2 = \left(\frac{1}{a_1^2} + 1\right)\sum_{i=1}^{n}(\hat{y}_i - y_i)^2 \qquad (16.46)$$

It can be shown that the value of a_1 derived using this penalty is the geometric mean of the values derived using the penalties described by Equations (16.10) and (16.43).

An alternative method of giving equal weight to potential errors in both x and y is, for each point, to multiply the distances from the line in both the horizontal and vertical directions. This is the area of the rectangle with one vertex at the actual value and the adjacent vertices on the line, as illustrated in Figure 16.6. The penalty function is then the total area of all the rectangles and becomes

$$\sum_{i=1}^{n}|\hat{x}_i - x_i|.|\hat{y}_i - y_i| = \frac{1}{|a_1|}\sum_{i=1}^{n}(\hat{y}_i - y_i)^2 \qquad (16.47)$$

Using the data included in Table 16.1, the impact on the resulting correlation between $\%C_4$ and tray 17 temperature of using each of the penalty functions is given in Table 16.2. Because there is only one independent, each method gives the same result for Pearson R^2 – no matter what coefficients are chosen. This is one reason why R^2 has only limited value in assessing the accuracy of inferentials. The preferred performance index (ϕ), explained at the end of this chapter, measures the accuracy of predicting y. So, not surprisingly, it is worse for the correlation derived to accurately predict x. Similarly the two methods giving equal weight to errors in both x and y, as might be expected, give intermediate values for ϕ. In this case, all the methods give similar coefficients; the difference in their accuracy is unlikely to be noticeable in view of other potential sources of error. The high value of Pearson R^2 reflects that there appears to be little error in the measurement of both x and y.

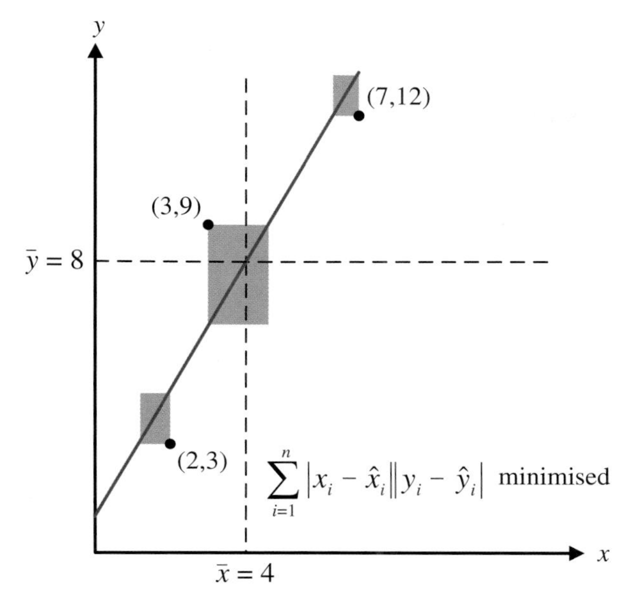

Figure 16.6 *Minimisation of the total sum of the rectangles in* x *and* y *direction*

Table 16.1 *LPG splitter data for development of inferential*

tray 15 (x_1)	tray 17 (x_2)	%C_4 (y)
66.5	60.8	5.34
64.9	58.0	3.79
66.4	59.4	4.28
67.4	62.5	6.46
66.2	57.6	3.41
66.3	60.5	5.26
67.2	61.9	5.91
67.2	62.1	5.86
66.8	61.1	5.43
65.1	58.2	3.90
65.6	59.0	4.41
65.4	58.7	4.13
66.2	60.1	4.80
64.3	57.0	3.48
66.4	60.4	5.08
67.7	62.8	6.61
66.9	61.5	5.70
66.4	60.5	5.08
65.9	59.5	4.37
64.6	57.5	3.67

Table 16.2 Impact of choice of penalty function

penalty function	a_0	a_1	R^2	ϕ
$\sum_{i=1}^{n}(a_0+a_1x_i-y_i)^2$	−28.09	0.5494	0.978	0.978
$\frac{1}{a_1^2}\sum_{i=1}^{n}(a_0+a_1x_i-y_i)^2$	−28.65	0.5587	0.978	0.977
$\left(\frac{1}{a_1^2}+1\right)\sum_{i=1}^{n}(a_0+a_1x_i-y_i)^2$	−28.46	0.5555	0.978	0.960
$\frac{1}{\lvert a_1\rvert}\sum_{i=1}^{n}(a_0+a_1x_i-y_i)^2$	−28.83	0.5617	0.978	0.939

Applying the same techniques to less accurate data would result in significantly different estimates for a_1. For example, Figure 16.7 shows the correlations developed using data with a Pearson R^2 of 0.722. The value of a_1 varies substantially. For the conventional penalty function, the slope of the coloured line is −0.564. For that based on Equation (16.43), the slope of the black line is −0.781. Remembering this is likely to be a process gain, the variation of around ±16% is substantial. As expected, the functions described by Equations (16.46) and (16.47), shown as dashed lines, generate intermediate values.

The formulae presented above can each be expanded to develop inferentials involving more than one independent variable. Additionally, weighting factors can be included that reflect the relative accuracy of each variable. For example, we apply the weighting w_0 to the dependent variable, w_1 to the independent variable x_1, w_2 to x_2, etc. The penalty function then becomes

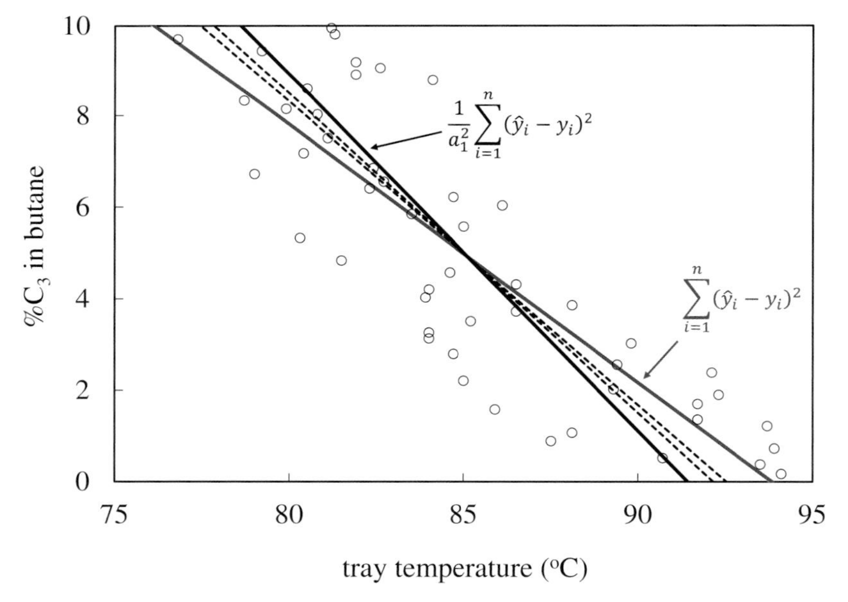

Figure 16.7 Impact of choice of penalty function

$$\sum_{i=1}^{n}\left[w_0(\hat{y}_i-y_i)^2+w_1(\hat{x}_{1i}-x_{ii})^2+w_2(\hat{x}_{2i}-x_{2i})^2+\ldots\right]$$
$$=\left(w_0+\frac{w_1}{a_1^2}+\frac{w_2}{a_2^2}+\ldots\right)\sum_{i=1}^{n}(\hat{y}_i-y_i)^2 \tag{16.48}$$

The penalty functions described by Equations (16.16), (16.43) and (16.46) can all be developed from this equation by choosing appropriate weighting factors. It should be noted that selection of the best weighting factors cannot be achieved by choosing those that give the highest value of Pearson R^2 or ϕ. Both R^2 and ϕ are based on only the prediction error in y and so will always be maximised by setting w_0 to 1 and all the other weighting factors to 0. Weighting coefficients can only be selected using judgement as to which measurements the engineer believes are more reliable. Avoiding having to make this judgement provides additional motivation to ensure all the data used are as reliable as possible. The more complex penalty functions would then offer no advantage over using the simplest conventional approach.

Equations could be identified for regressing nonlinear functions, for example

$$\hat{y}=a_0+a_1x+a_2x^2 \tag{16.49}$$

However a simpler approach is to treat x and x^2 as independent variables; x_1 is defined as x and x_2 as x^2. While x_1 and x_2 will now be correlated, the relationship between them is nonlinear and will not therefore cause problems in solving the equations. It is generally not advisable to exclude the linear term when adding the quadratic term. By differentiating Equation (16.49) we can determine at what value of x the minimum (or maximum) occurs for the predicted value.

$$\frac{d\hat{y}}{dx}=a_1+2a_2x=0 \quad \text{hence} \quad x=-\frac{a_1}{2a_2} \tag{16.50}$$

By excluding the linear term we force the minimum (or maximum) to occur at $x=0$. This is similar, when fitting a linear correlation, to omitting the constant term (a_0) and so force the predicted value to be 0 when $x=0$.

An alternative is to apply some transformation to the predicted value, for example

$$\sqrt{\hat{y}}=a_0+a_1x \tag{16.51}$$

This is effectively the same as Equation (16.49) – although the values of a_0 and a_1 will be different. Other transformations such as the logarithm of the predicted value, or its square, might also be considered. Indeed, predicting the logarithm or the square root of a property guarantees that the property cannot be negative. This is quite useful for properties such as compositions.

It may be beneficial to include *compound* inputs – those derived from a calculation involving several inputs. For example, the coefficients in an inferential may need to be adjusted if the type of feed being processed is changed. So, for Type 1 feed, the correlation might be

$$\hat{y}=a_0+a_1x_1 \tag{16.52}$$

For Type 2 it might be

$$\hat{y}=b_0+b_1x_1 \tag{16.53}$$

To ensure the same inferential can be used all the time we can incorporate the feed type by adding a second input (x_2) that is set to 1 when the feed is Type 1 and set to 0 when it is Type 2. So

$$\hat{y}=x_2(a_0+a_1x_1)+(1-x_2)(b_0+b_1x_1) \tag{16.54}$$

Rewriting

$$\hat{y} = c_0 + c_1x_1 + c_2x_2 + c_3x_1x_2 \quad (16.55)$$

where

$$c_0 = b_0 \quad c_1 = b_1 \quad c_2 = a_0 - b_0 \quad c_3 = a_1 - b_1 \quad (16.56)$$

Now imagine that the unit has never processed 100% of either feed type. We therefore do not have any data to regress to determine coefficients in Equations (16.52) and (16.53). But, provided we know x_2 (the proportion of Type 1 in the total feed), we can regress to determine directly the coefficients in Equation (16.55). However for this to be successful, if a_1 is significantly different from b_1, we must include the compound input x_1x_2. The coefficient of x_1 is now effectively $(c_1 + c_3x_2)$, or we can think of the coefficient of x_2 being effectively $(c_2 + c_3x_1)$. Such functions, in which the coefficient of one input changes as another input varies, are known as *heterogenic*. Those in which this is not the case are *homogenic*.

The regression technique so far described is known as *OLS (ordinary least squares)*. OLS assumes that the data is *homoscedastic*, i.e. each data point is equally reliable. A refinement is *WLS (weighted least squares)* where weights are assigned to individual data points. For example, a lower weighting might be applied to an input when its value falls within a range over which its accuracy is known to be suspect. An indication of this might be a correlation between the variance of the prediction error and one of the inputs. For example, using that data presented in Table 16.3, applying OLS gives the inferential for %C_4 in distillate based on PCT.

$$\hat{y} = -12.09 + 0.3031x \quad (16.57)$$

For convenience the table has been sorted by increasing PCT. By separately calculating variance of the prediction error for each of the 5°C intervals we can see that the reliability of the prediction varies greatly depending on the value of the PCT. We thus choose a weight (w) for each interval that, in this case, is the reciprocal of the variance of the error. We then modify Equation (16.10) to convert it to a weighted penalty function

$$\sum_{i=1}^{n} w_i(\hat{y}_i - y_i)^2 = \sum_{i=1}^{n} w_i(a_0 + a_1x_i - y_i)^2 \quad (16.58)$$

Minimising this results in a modified inferential.

$$\hat{y} = -11.58 + 0.2930x \quad (16.59)$$

The weighted error variances for each 5°C interval will now be approximately the same. Whether the resulting inferential is any more reliable is debatable. In practice choosing the weights, without a strong engineering basis for their inclusion, will add nothing to the accuracy of the resulting correlation.

A better approach would be to explore what is causing the variation in accuracy. For example, in this case, it might be that the PCT calculation is not reliable over the whole range. If so, then correcting this would be a far better solution. It does however demonstrate that checking how the prediction error varies can highlight an opportunity to improve an inferential that may not be obvious otherwise. Figure 16.8, for example, shows the same correlation as Figure 16.3 but with prediction error plotted against the actual dependent value. The prediction is based on a linear function of pressure and temperature. The curve crossing the zero error line in more than one place would suggest that the relationship is nonlinear. Indeed Figure 16.3 reveals that a nonlinear PCT should be used. Other plots of prediction error can be helpful. For example, plotting it against an independent not included in the correlation would, if a trend appears, suggest the independent should be included.

Table 16.3 *Use of weighted least squares regression*

PCT (x_i)	measured %C_4 (y_i)	predicted %C_4 ($\hat{y}_i$)	$(\hat{y}_i-y_i)^2$	$\frac{\sum(\hat{y}_i-y_i)^2}{n}$	w	weighted %C4 prediction
40.3	0.39	0.05	0.1140	0.4346	2.301	0.23
42.3	1.89	0.65	1.5252			0.82
43.9	1.42	1.14	0.0800			1.29
44.3	1.12	1.26	0.0189			1.41
45.9	1.27	1.74	0.2206	1.1766	0.850	1.87
46.3	3.32	1.86	2.1310			1.99
47.0	1.84	2.07	0.0534			2.20
47.9	1.73	2.34	0.3749			2.46
48.0	1.73	2.37	0.4127			2.49
48.3	2.74	2.46	0.0768			2.58
49.9	1.53	2.94	2.0019			3.05
50.0	0.94	2.98	4.1412			3.08
50.3	3.57	3.07	0.2546	1.4968	0.668	3.16
51.9	2.59	3.55	0.9167			3.63
52.0	3.02	3.58	0.3109			3.66
52.3	5.27	3.67	2.5664			3.75
53.9	3.28	4.15	0.7570			4.22
54.0	3.22	4.18	0.9220			4.25
54.3	6.45	4.27	4.7498			4.34
55.9	4.64	4.75	0.0127	3.5368	0.283	4.80
56.0	4.09	4.78	0.4800			4.83
56.3	6.92	4.87	4.1894			4.92
57.9	5.42	5.36	0.0042			5.39
58.0	4.78	5.39	0.3665			5.42
58.1	2.23	5.42	10.1476			5.45
58.3	8.37	5.48	8.3765			5.51
59.3	9.11	5.78	11.1083			5.80
59.9	6.33	5.96	0.1385			5.98
60.0	5.25	5.99	0.5446			6.01
60.1	3.87	6.02	4.6145	5.0110	0.200	6.04
61.9	9.83	6.56	10.6898			6.56
62.0	6.04	6.59	0.3032			6.59
62.1	4.13	6.62	6.2037			6.62
64.0	7.43	7.19	0.0561			7.18
64.1	4.36	7.22	8.1987			7.21
66.0	9.98	7.80	4.7707	1.9080	0.524	7.76
66.1	6.52	7.83	1.7055			7.79
68.1	8.70	8.43	0.0737			8.38
69.1	9.77	8.73	1.0820			8.67

The *PLS (partial least squares)* method is applied in situations where the number of inputs exceeds the number of datasets. If the number of inputs is equal to the number of datasets then, in general, an exact fit to the data will be possible. Increasing the number of inputs beyond this will cause the technique covered in this section to fail. This is unlikely to arise for inferentials that the control engineer is likely to develop. But it does arise, for example, with NIR (near

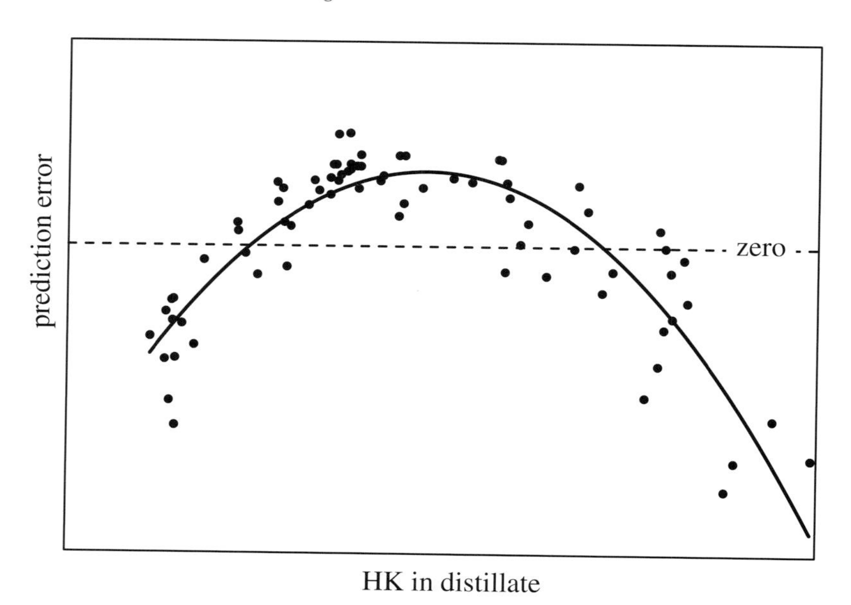

Figure 16.8 *Detection that a nonlinear correlation should be used*

infra-red) analysers. For each sample, these generate a frequency spectrum that will contain the amplitudes of around 1,000 frequencies. Only a few properties are predicted from this result. It would be impractical to collect the many thousands of samples that would be needed to apply OLS. Instead, PLS is used to first search for *latent vectors* that explain as much as possible of the relationship between the dependent properties and the independent amplitudes. These vectors are then used in developing the correlation. The mathematics are quite complex, well beyond the scope of this book, and in any case would be built into a software tool.

Like any statistical technique, regression analysis is open to abuse if applied without an understanding of the process. With modern spreadsheets and statistical packages it is relatively easy to extract large quantities of data from the process information database and search for all possible correlations. By including a large number of process variables and a wide range of arithmetical transformations (such as powers, logarithms, ratios, cross-products, etc.) it will certainly be possible to apparently improve the accuracy of the inferential. However, this is likely to be only a mathematical coincidence. Consider the example where we attempt to predict a value (y) from one input (x_1) such that

$$y = a_0 + a_1 x_1 \tag{16.60}$$

If we have three datasets (or records) from which to derive this correlation then a_0 and a_1 would be chosen to give the best fit to the available data. Imagine that this does not give the required accuracy, and so we introduce a second input (x_2). Since we have only three records, the coefficients (a_0, a_1 and a_2) could be determined by solving the following equations simultaneously.

$$(y)_1 = a_0 + a_1 (x_1)_1 + a_2 (x_2)_1 \tag{16.61}$$

$$(y)_2 = a_0 + a_1 (x_1)_2 + a_2 (x_2)_2 \tag{16.62}$$

$$(y)_3 = a_0 + a_1 (x_1)_3 + a_2 (x_2)_3 \tag{16.63}$$

This would then give us a perfect fit. However this 'perfection' would be achieved even if x_2 was a random number completely unrelated to y. Of course, we generally have far more datasets than independent variables but this illustrates the point that the inclusion of any additional input will appear to improve the correlation. The relationship between the degree of correlation and the number of inputs is approximately linear. So, for example, if the number of inputs is half the number of records we would achieve 50% of perfection even if the inputs are random numbers.

There are several techniques, described in the next few sections, which can help us decide whether the addition of an input to an inferential property is truly beneficial.

16.1 F Test

The main application of the Fisher distribution, described in Section 12.6, is in support of multi-variable regression analysis. Using the measurements shown in Table 16.1 for the LPG splitter, the aim is to build an inferential for the C_4 content of the overhead propane product (y). In this example we have measurements (x_1 and x_2) of the temperature of two trays close to the top of the 20-tray column – on trays 15 and 17. As a measure of the accuracy of the predicted value ($\hat{y}$) we will use the variance of the error (σ^2_{error}), defined as

$$\sigma^2_{error} = \frac{\sum_{i=1}^{n} (\hat{y}_i - y_i)^2}{n-p} \tag{16.64}$$

The denominator in Equation (16.64) takes account of the p degrees of freedom used in estimating the p coefficients in the inferential.

Depending on which temperature is used, we obtain two possible inferentials.

$$\hat{y} = -56.38 + 0.9253x_1 \tag{16.65}$$

$$\hat{y} = -28.09 + 0.5494x_2 \tag{16.66}$$

The first of these has a σ^2_{error} of 0.1972; the second, 0.0220. Of these we would therefore clearly choose the second. The performance of both is shown in Figure 16.9. As expected the temperature of the tray nearer the top of the column gives the better prediction. However we could use both temperatures in the inferential.

$$\hat{y} = -20.54 - 0.2092x_1 + 0.6544x_2 \tag{16.67}$$

This has a σ^2_{error} of 0.0177. As expected it is better than either of the single-input possibilities. Its performance is shown in Figure 16.10. However we would expect x_1 to be highly correlated with x_2 and so add little value. Indeed, as Figure 16.11 shows, this is the case for most of the measurements with the exception of two – which might be considered outliers. This explains the relatively small improvement in σ^2_{error}. We need to check whether including the second temperature gives a significantly better prediction. To do so, we first calculate the *variance ratio (F)* of the improvement in error variance to the variance beforehand. This is

$$F = \frac{\left[\dfrac{\sigma_1^2 - \sigma_2^2}{f_1}\right]}{\left[\dfrac{\sigma_2^2}{f_2}\right]} \tag{16.68}$$

In our example σ_1^2 is that for the better of the two single-input inferentials and σ_2^2 is that for the two-input version. σ_1^2 will always be greater than σ_2^2. The degrees of freedom are derived from the number of coefficients (p) used in the inferential. In this example p_1 is 2 and p_2 is 3. The total number of measurements (n) in our example is 20.

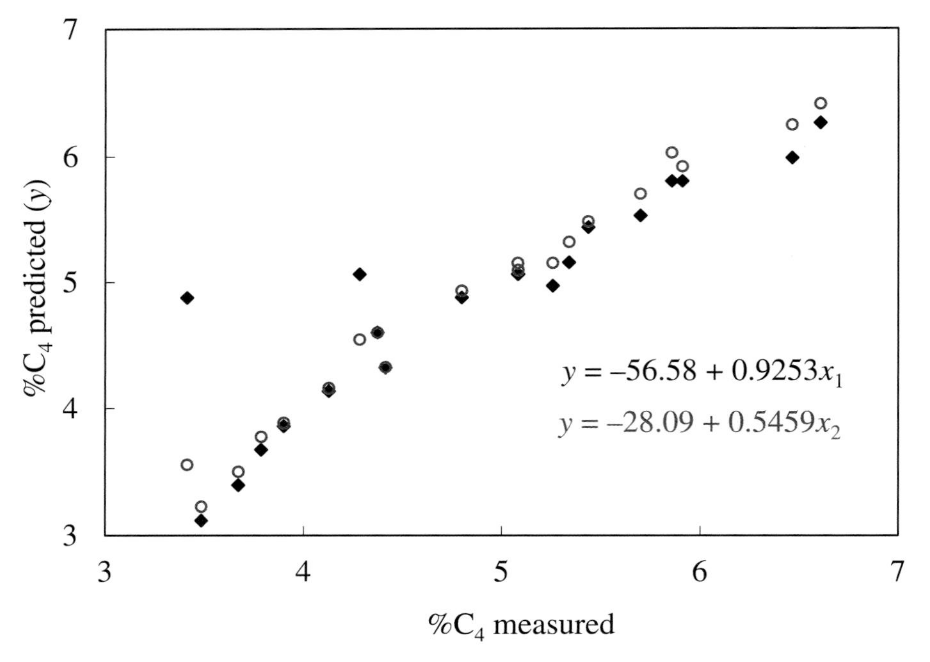

Figure 16.9 *Single input inferentials*

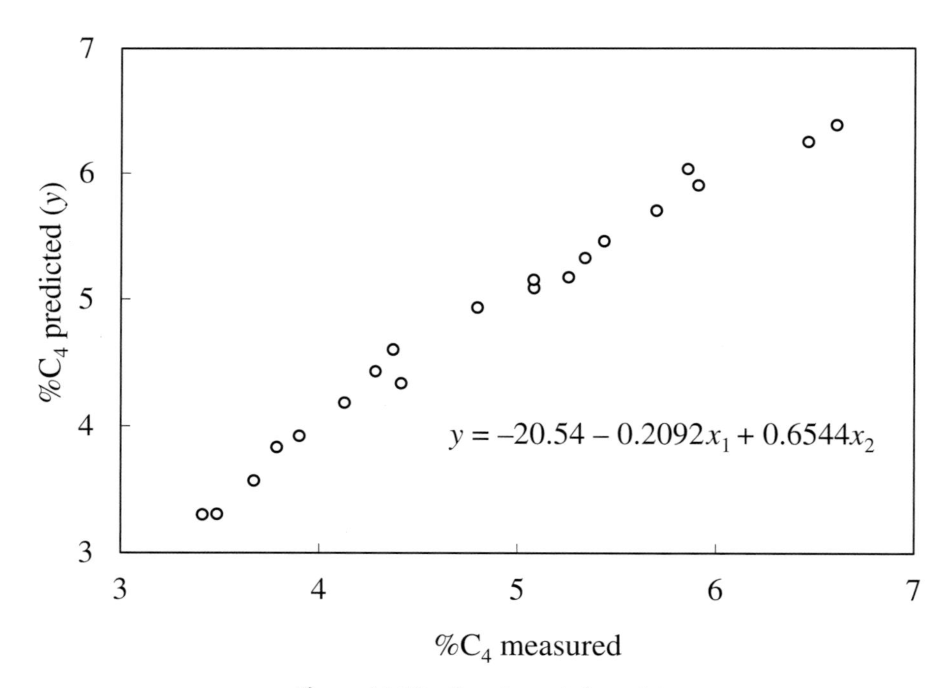

Figure 16.10 *Two-input inferential*

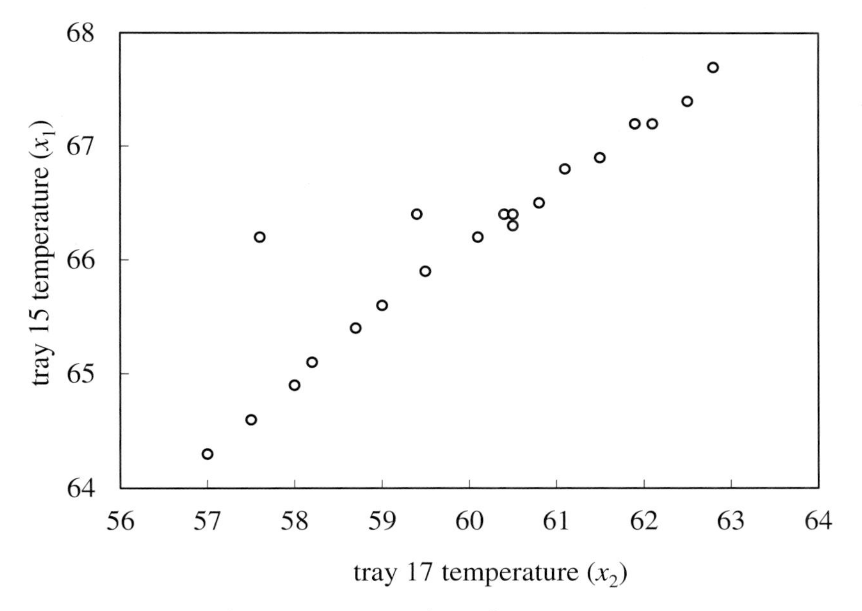

Figure 16.11 *Correlation between inputs*

$$f_1 = p_2 - p_1 = 1 \tag{16.69}$$

$$f_2 = n - p_2 = 17 \tag{16.70}$$

$$\therefore \; F = \frac{\left[\dfrac{0.0220 - 0.0177}{3-2}\right]}{\left[\dfrac{0.0177}{20-3}\right]} = 4.19 \tag{16.71}$$

To perform the F test, we make the null hypothesis that the two-input correlation is not significantly different from the single-input alternative. The curve in Figure 16.12 is the cumulative probability determined by applying the trapezium rule to Equation (12.55). It is plotted to give the probability that the null hypothesis is false. In our example they show a probability of around 94% that this is the case – only marginally below the value of 95% at which we should adopt the two-input inferential.

The alert engineer might at first be suspicious of the end result. We would expect $\%C_4$ to increase as the tray temperatures increase and so would consider suspect the negative coefficient for x_1 in Equation (16.67). But we can rewrite this equation.

$$\hat{y} = -20.54 + 0.4452x_2 - 0.2092(x_1 - x_2) \tag{16.72}$$

This is consistent with Equation (16.66) which shows that the tray 17 temperature (x_2) is the better choice. Since x_1 is a temperature lower down the column it will be greater than x_2 and so $(x_1 - x_2)$ will always be positive. This difference is a measure of the separation taking place in this section of the column. Increasing separation will reduce $\%C_4$ and so the negative coefficient for this parameter is correct.

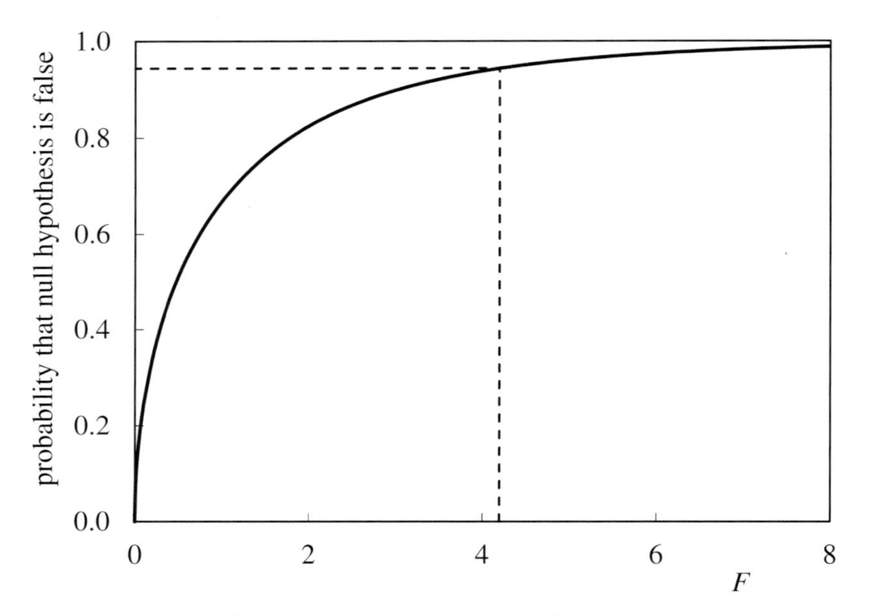

Figure 16.12 Probability that two-input inferential is better than single input version

16.2 Adjusted R^2

Squaring Equation (4.75) gives

$$R^2 = \frac{\left(\sum_{i=1}^{n}(x_i-\bar{x})(y_i-\bar{y})\right)^2}{\sum_{i=1}^{n}(x_i-\bar{x})^2\sum_{i}^{n}(y_i-\bar{y})^2} \tag{16.73}$$

Since the addition of an input will always improve a correlation, R^2 cannot be used to determine whether the addition is truly beneficial. Instead we should use the *adjusted* R^2 – usually written as $\bar{R}^2$. The principle behind the adjusted version is to include a penalty as additional inputs are used, where n is the number of data points and p the number of inputs.

$$\bar{R}^2 = 1-(1-R^2)\frac{n-1}{n-p-1} \tag{16.74}$$

If n is 2 we have two data points (x_1,y_1) and (x_2,y_2) and so

$$\bar{x} = \frac{x_1+x_2}{2} \quad \text{and} \quad \bar{y} = \frac{y_1+y_2}{2} \tag{16.75}$$

Equation (16.73) will then give a result of 1 for R^2. If there is only one input ($p = 1$) then $\bar{R}^2$ will be indeterminate – although usually then assumed to be zero. In general n must be greater than $p + 1$.

Unlike R^2, $\bar{R}^2$ can be less than zero. For example, if we retain only the one input, but this time with 3 data points, and if R^2 is at its lowest value of zero, then from Equation (16.74), $\bar{R}^2$ will be at its lowest possible value of −1. In general, if R^2 is zero and n is at is at its lowest possible value of $p+2$ then $\bar{R}^2$ will be $-p$.

If we increase the number of inputs from p to $p+1$, then for the correlation to have genuinely improved

$$1-\left(1-R_{p+1}^2\right)\frac{n-1}{n-(p+1)-1}>1-\left(1-R_p^2\right)\frac{n-1}{n-p-1} \tag{16.76}$$

or

$$R_{p+1}^2>\frac{1+(n-p-2)\,R_p^2}{n-p-1} \tag{16.77}$$

Using the inferential developed as Equation (16.66), the left hand side evaluates to 0.991, while the right, from Equation (16.67), is 0.979, showing that the improvement justifies the use of an additional input.

Adjusted R^2 can also be applied to dynamic model identification. Model order is determined by the number of historical values of PV and MV included. Increasing the order will always increase the accuracy of the model. Determining the adjusted R^2 as the order is incremented will ensure that the order is not increased to the level where only noise is being modelled and hence results in the best choice of order for the dynamic model.

The adjusted R^2 can similarly be used to determine whether using a larger number of data points to develop the correlation has produced a significantly more reliable version. If the original was developed from n points and the revised version developed from m points then, for it to be significantly better

$$R_m^2>\frac{(m-p-1)(n-1)\,R_n^2-p(m-n)}{(n-p-1)(m-1)} \tag{16.78}$$

16.3 Akaike Information Criterion

The *Akaike information criterion (AIC)* offers another method of deciding whether the inclusion of additional parameters in a correlation gives a statistically significant improvement in its accuracy. There are several published definitions but if the correlation is developed using the conventional least squares regression and the prediction error is normally distributed then

$$AIC=n\ln\left(\sigma_{error}^2\right)+2p \tag{16.79}$$

As usual n is the number of data points; p is the number of independent variables used in the correlation. If n is small compared to p it is advisable to use the *second order AIC* to avoid overfitting.

$$AIC=n\ln\left(\sigma_{error}^2\right)+\frac{2pn}{n-p-1} \tag{16.80}$$

The correlation described by Equation (16.66) has an AIC given by

$$AIC_1=20\ln(0.0220)+\frac{20\times2\times1}{20-1-1}=-74.08 \tag{16.81}$$

That described by Equation (16.67) has an AIC given by

$$AIC_2=20\ln(0.0177)+\frac{20\times2\times2}{20-2-1}=-76.01 \tag{16.82}$$

The better correlation is the one with the lower AIC. Whether the difference is significant is determined from ΔAIC. A value less than 2 for ΔAIC indicates that increasing the number of independent variables is probably not justified by the improvement in accuracy. In this case

$$\Delta AIC = AIC_1 - AIC_2 = 1.93 \tag{16.83}$$

This agrees with the conclusion of the F test in Section 16.1. The two-input inferential does not quite meet the criterion for selection over the one-input version. Very large values of ΔAIC, those greater than 10, indicate that there are variables that are better excluded from the original correlation.

An alternative comparison is to determine the relative probability.

$$\exp\left[\frac{-\Delta AIC}{2}\right] = 0.38 \tag{16.84}$$

This indicates that there is a probability of 38% that the inclusion of the additional parameter will minimise the loss of information. Alternatively we are only 62% certain that the inferential is not improved by the addition of the second variable. On this basis we might elect to include it.

If n is significantly larger than p then we can apply Equation (16.79) to determine the condition necessary to justify the inclusion of one additional parameter. We require

$$AIC_p - AIC_{p+1} > 2 \tag{16.85}$$

Therefore

$$\left(n\ln\left(\sigma_p^2\right) + 2p\right) - \left(n\ln\left(\sigma_{p+1}^2\right) + 2(p+1)\right) > 2 \tag{16.86}$$

Rearranging

$$\ln\left(\frac{\sigma_p^2}{\sigma_{p+1}^2}\right) - \frac{2}{n} > 0 \quad \text{or} \quad \ln\left(\frac{\sigma_p}{\sigma_{p+1}}\right) - \frac{1}{n} > 0 \tag{16.87}$$

In our example the left hand side of the second inequality evaluates to 0.060, marginally confirming that including the additional parameter is justified. If required, a similar but more complex inequality can be derived from Equation (16.80).

While normally the number of inputs to an inferential would be increased incrementally with the AIC test applied at each stage, this may not always be possible. For example, we might wish to compare an inferential that uses three inputs with one that uses only one (which is not one of the three). Equation (16.87) can be extended to handle this. For an inferential using p_1 inputs to be preferable to one using p_2 inputs

$$\ln\left(\frac{\sigma_{p_1}}{\sigma_{p_2}}\right) - \frac{p_2 - p_1}{n} > 0 \tag{16.88}$$

Like adjusted R^2, AIC can also be applied to dynamic model identification. It can similarly be used to determine whether a correlation has become more reliable by developing it from a larger number of data points. If the original was developed from n points and the revised version developed from m points, then, for it to be significantly better,

$$\ln\left(\frac{\sigma_n}{\sigma_m}\right) - \frac{m}{2n} > 0 \tag{16.89}$$

16.4 Artificial Neural Networks

Artificial neural networks are developed using regression. Figure 16.13 shows the network as usually drawn. It comprises the *input layer* of *neurons*, each of which scales each of the n inputs using *weighting coefficients (w)* and a *bias (b)*.

$$x_j' = b + \sum_{i=1}^{n} w_{ij} x_i \tag{16.90}$$

The next *hidden layer* comprises the j neurons, in which each scaled input passes through a *transfer function* (also known as the *activation function*). This is chosen to be highly nonlinear with the output bounded within a narrow range. Most commonly used is the *sigmoid* (S-shaped) *function*, an example of which (shown as the coloured curve in Figure 16.14) is

$$y_j = \frac{1}{1 + \exp\left(-x_j'\right)} \tag{16.91}$$

where

$$-\infty < x_j' < \infty \quad \text{and} \quad 0 < y_j < 1 \tag{16.92}$$

There are other examples (shown as the black curves in Figure 16.14) where

$$-\infty < x_j' < \infty \quad \text{and} \quad -1 < y_j < 1 \tag{16.93}$$

such as

$$y_j = \frac{1 - \exp\left(-x_j'\right)}{1 + \exp\left(-x_j'\right)} \tag{16.94}$$

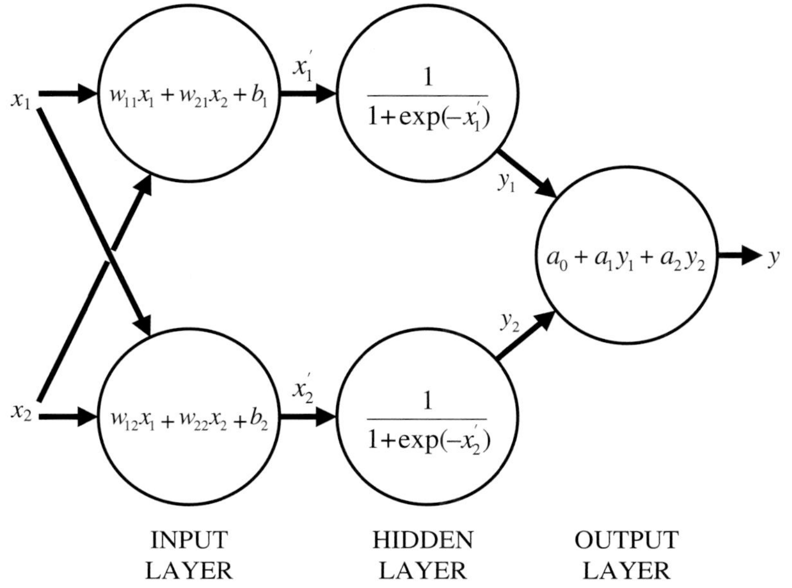

Figure 16.13 *Structure of artificial neural network*

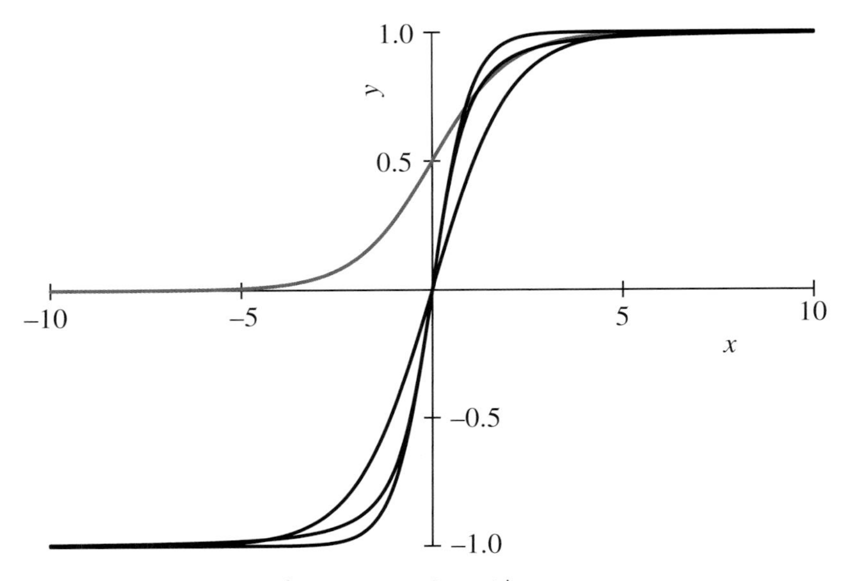

Figure 16.14 Sigmoid curves

$$y_j = \tanh\left(x'_j\right) = \frac{\exp\left(x'_j\right) - \exp\left(-x'_j\right)}{\exp\left(x'_j\right) + \exp\left(-x'_j\right)} \tag{16.95}$$

$$y_j = \frac{x'_j}{\sqrt{1 + x'^2_j}} \tag{16.96}$$

Others include the *radial basis function (RBF)*; one example of which uses a Gaussian activation function based on the mean (μ) and standard deviation (σ) of x'_j.

$$y_j = \exp\left[-\frac{\left(x'_j - \mu_j\right)^2}{\sigma_j^2}\right] \tag{16.97}$$

There is also a *triangular basis function (TBF)*.

$$y_j = 1 - \left|x'_j\right| \quad \text{if } -1 < x'_j < 1 \quad ; \quad y_j = 0 \quad \text{otherwise} \tag{16.98}$$

In the *output layer* of neurons the outputs from each transfer function are combined on a linear basis as a weighted sum

$$y = a_0 + \sum_{j=1}^{n} a_j y_j \tag{16.99}$$

Since neurons in one layer only pass information to all the neurons in the next layer, the network is of the *feedforward* type. In a *recurrent* network neurons also pass information to the previous layer. The coefficients (b, w and a) are chosen, as with linear regression, to minimise the residual sum of squares (RSS). With neural networks this process is iterative and is known as *training*.

Table 16.4 LPG splitter data collected over wide operating range

tray 15 (x_1)	tray 17 (x_2)	%C_4 (y)
63.7	52.2	1.58
57.3	48.9	1.36
62.2	54.1	2.57
66.8	61.1	5.38
67.2	61.9	5.84
65.8	54.4	1.99
59.6	51.1	1.82
64.6	57.5	3.68
68.9	58.4	2.89
68.5	64.3	7.63
45.2	40.9	0.20
67.4	56.4	2.40
60.6	52.2	2.08
48.6	42.6	0.40
51.3	44.2	0.61
66.4	60.5	5.11
52.9	45.3	0.76
66.4	60.4	4.99
60.1	49.0	1.10
54.8	46.7	0.99
69.0	65.3	8.50
66.2	57.6	3.21
40.8	39.2	0.04
66.4	59.4	4.24
69.2	65.8	9.05
65.1	58.2	3.96
69.6	66.5	9.87
67.4	62.5	6.36
62.8	55.0	2.84
61.5	53.2	2.33

Table 16.4 shows process data collected from the same distillation column as that which generated the data in Table 16.1, except that it covers a wider range of operating conditions. Revising, by linear regression, the coefficients in Equation (16.67) will make if fit the process data reasonably well with *RSS* of 8.0.

$$y = -10.68 - 0.3619x_1 + 0.6668x_2 \tag{16.100}$$

It is possible to develop a neural network with a single neuron in the hidden layer. The formula for this can be derived by combining Equations (16.90), (16.91) and (16.99).

$$y = a_0 + \frac{a_1}{1 - \exp[-b_1 - w_{11}x_1 - w_{21}x_2]} \tag{16.101}$$

Training results in

$$b_1 = -8.8896578 \quad w_{11} = -0.0409135 \quad w_{21} = 0.1101842 \tag{16.102}$$

$$a_0 = -1.3965854 \quad a_1 = 919.7932 \tag{16.103}$$

This gives *RSS* a value of 0.8, considerably better than that from the linear regression. Figure 16.15 confirms that the predicted values now match the measured values very closely. While it would be possible to include nonlinear functions in conventional regression it is unlikely that the use of simple transformations, such as logarithms and powers, will match the *RSS* achieved by the neural network.

As usual, it is important that the inferential makes good engineering sense. With simple functions the meaning of unusual terms (for example x_1x_2) can be checked and the signs of their coefficients can be validated. With a neural network this is impractical. Indeed, if using a proprietary package, the values of the coefficients are not always made available to the engineer. However we can explore how the inferential behaves by varying each of its inputs. Figure 16.16 shows the result of this exercise; the dashed lines compare the performance to the linear inferential. While analysis does not show that the inferential is accurate it demonstrates that it behaves as expected. The $\%C_4$ increases both as the tray 17 temperature increases and as the temperature difference between the trays falls. While this gives a strong indication that the inferential is reliable we would need to be cautious in applying it if any input falls outside the range over which training was performed.

It would be possible, in this example, to increase the number of neurons. While this would reduce the RSS it would do so by modelling more of the noise in the measurements. This *overfitting* would likely result in a reduction in accuracy when the inferential is put into use. One check is to look at the relative values of the weighting coefficients. If some are very much smaller than others then this may indicate that a neuron is adding little to the accuracy.

A great deal is published about the development and use of neural networks. Much of this is applicable to any form of regression. For example, it is good discipline to randomly split the data into two groups – one for training and one for testing. However, failing the test will result in the engineer reconfiguring the inferential. In effect the test data is also used for training. Under these circumstances a third independent dataset should then be used for testing. This

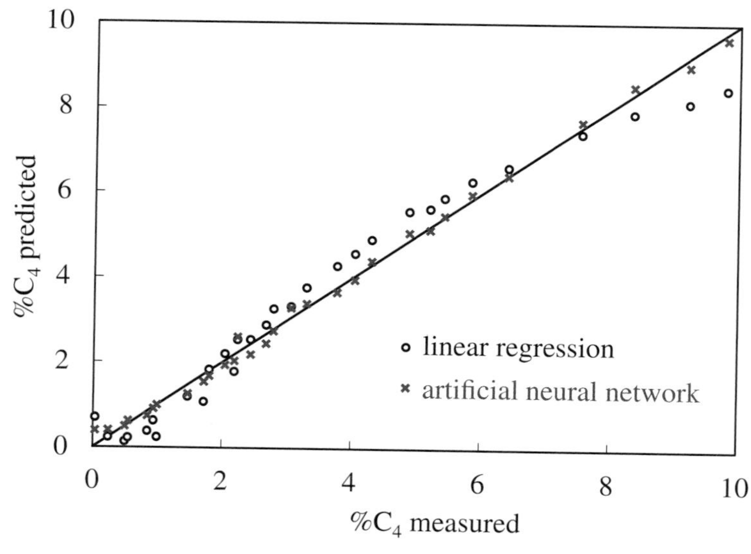

Figure 16.15 *Artificial neural network outperforming linear regression*

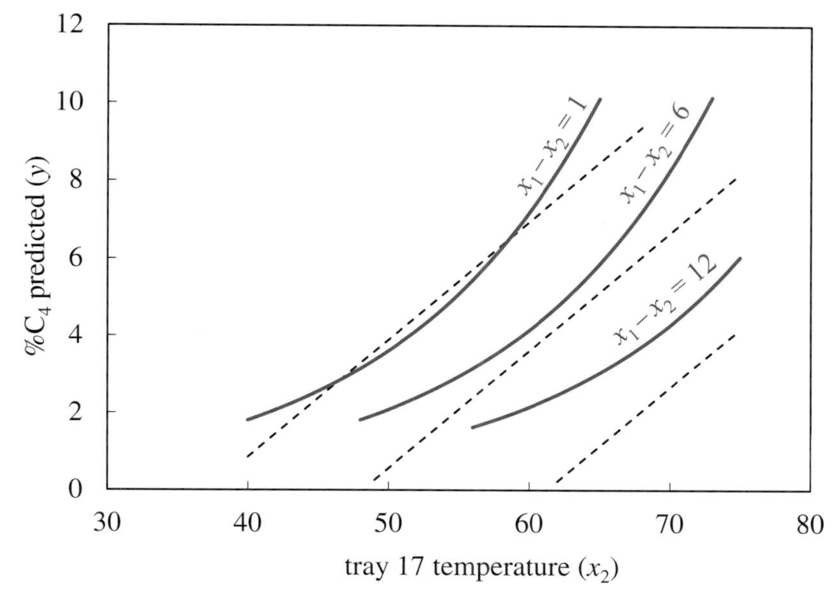

Figure 16.16 *Validation of inferential based on artificial neural network*

is particularly important with neural networks because they can extrapolate poorly outside the region in which they have been trained. Indeed it should be common practice to check whether the current inputs fall within the training envelope before using the output. It would be wise to temporarily disable any control action under these circumstances until its result has been validated and the inferential retrained as necessary.

There is a tendency to 'throw data' at neural networks. Data should be pre-processed, for example to remove outliers and to identify any cross-correlations. Consideration should be given to including derived measurements, such as ratios and cross-products. Including sufficient neurons will permit any data to be fitted, in much the same way that including additional terms in conventional regression does. However the lack of transparency in a neural network will hide any nonsensical relationship.

It is possible to design a neural network to generate more than one output. In our example we could include additional measurements taken around the bottom of the column and also infer the %C_3 in bottoms butane product. This might seem advantageous if we were including measurements that affect both product compositions – such as column pressure. However maintenance is likely to be more difficult than supporting two separate networks.

The neural network that we have described is known as the *back propagation (BP)* type. Its name comes from the method used to train it and is the one most commonly used for inferentials. There are others and it is often worth trying each of them and comparing the results.

So far we have presented neural networks as a steady-state technique. Indeed their most common application in the process industry is inferential properties. These are usually developed on the assumption that the input data were collected at steady state. However neural networks can also be applied to dynamic problems. They can be set up as a dynamic model of a process, where historical values are used to predict future values. They can also be used as controllers, where historical values are used to determine future control moves.

16.5 Performance Index

Once developed, an inferential property calculation has to be assessed as to whether it is sufficiently accurate to capture the expected benefits. Many engineers will use the Pearson R^2 coefficient, setting some minimum value to decide whether sufficient accuracy has been achieved. This is described in more detail in Section 4.9. If there are n datasets, where x is the measured property and $\hat{x}$ the prediction, this is defined as

$$R^2 = \frac{\left(\sum_{i=1}^{n}(x_i-\bar{x})(\hat{x}_i-\bar{x})\right)^2}{\sum_{i=1}^{n}(x_i-\bar{x})^2\sum_{i}^{n}(\hat{x}_i-\bar{x})^2} \qquad (16.104)$$

A perfect correlation would have a value of 1 for R^2. However a value close to 1 does not necessarily indicate that an inferential is useful. As an illustration, consider the graph shown in Figure 16.17 for the stock price of a process control vendor. Figure 16.18 shows the performance of an inferential developed by the author. With R^2 of 0.99 one would question why the developer is not a multibillionaire. The reason is that it failed to predict the large falls in the value of the stock. The three occasions circled undermine completely the usefulness of the prediction. The same is true of an inferential property. If there is no change in the property then, no matter how accurate, the inferential has no value. If it then fails to respond to any significant change, it may as well be abandoned.

A further limitation of the use of R^2 is that, if there is a perfect relationship between inferential ($PV_{inferential}$) and laboratory result ($PV_{laboratory}$), the value of R^2 will also be 1 for any linear function, i.e.

$$PV_{inferential} = a_1 PV_{laboratory} + a_0 \qquad (16.105)$$

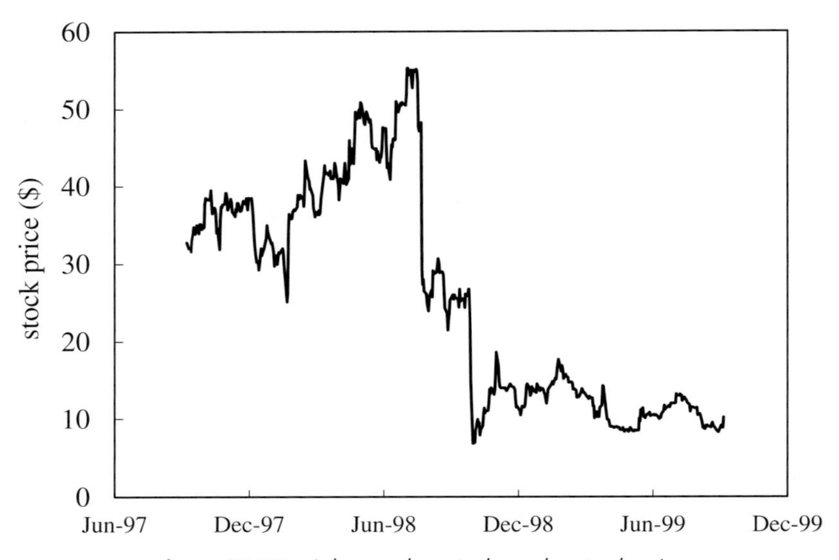

Figure 16.17 Advanced control vendor stock price

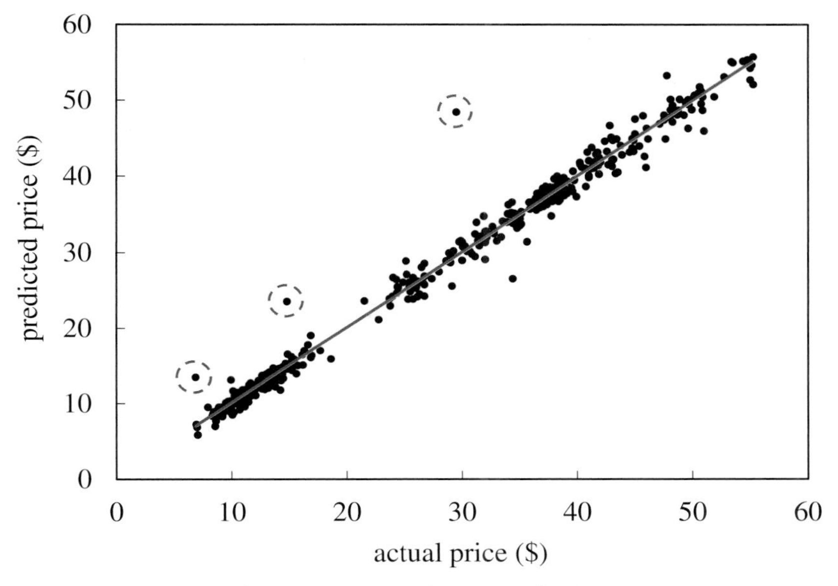

Figure 16.18 *Predicting stock price*

So, for example, if a_1 had a value of 3 and a_0 a value of 0, then the inferential would be treble the laboratory result but, according to the statistical test, be working perfectly. The same would apply if a_1 were negative – even though this reverses the sign of the process gain. We have to distinguish between correlation and accuracy. A high value for R^2 tells us that two variables correlate. It does not tell us they are equal. At best, R^2 might be used to compare inferentials, for example, to select the best of several possible. It does not provide an absolute measure of usefulness.

A better approach is to compare the standard deviation of the prediction error (σ_{error}) against the variation in the actual property (σ_{actual}). Benefit calculations are usually based on the assumption that the standard deviation of the actual property is halved. If we assume that our control scheme is perfect and the only disturbance comes from the random error in the prediction then, to capture the benefits

$$\sigma_{error} \leq 0.5\sigma_{actual} \tag{16.106}$$

This can be written in the form of a performance parameter (ϕ)

$$\phi = 1 - \frac{\sigma_{error}^2}{\sigma_{actual}^2} \geq 0.75 \tag{16.107}$$

This parameter clearly has a value of 1 when the inferential is perfect. To understand how it works over its full range, consider the inferential $Q = a_0$ where a_0 is the mean of all the property measurements used to build the inferential. The inferential will have a bias error of zero. But, since the inferential always generates the same value, the standard deviation of the prediction error will be the same as the standard deviation of the actual property. The inferential clearly has no value and the value of ϕ will be zero.

Next consider the case when the true property does not change. Any error in the prediction will cause the controller to wrongly take corrective action. In general, if the standard deviation of the prediction error is greater than that of the true property, ϕ will be negative – indicating

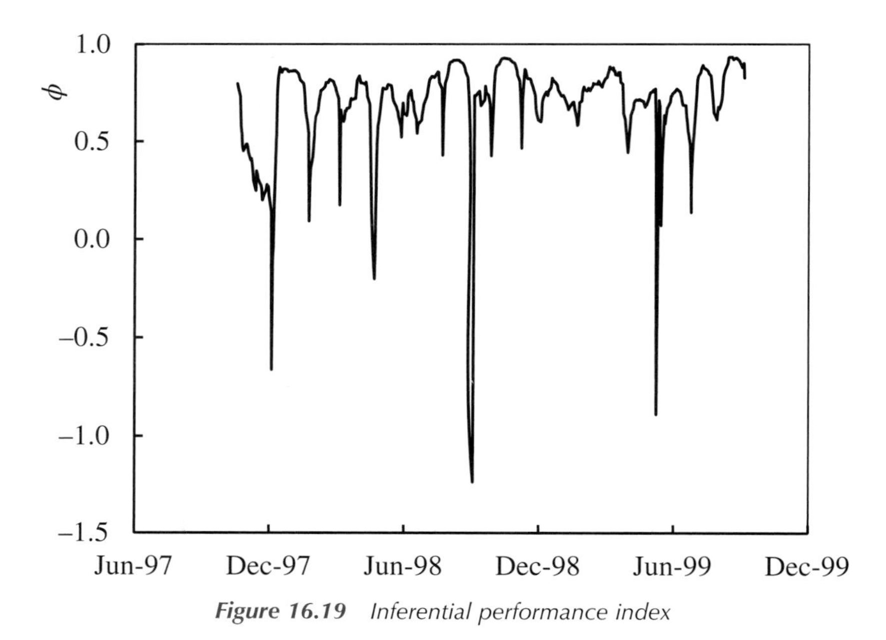

Figure 16.19 *Inferential performance index*

that the inferential is so bad that process performance would be improved by switching off the controller. Figure 16.19 trends this parameter for the stock price example. It confirms what we know, that the prediction will lose us money on several occasions.

The parameter (ϕ) can be used both in the development of an inferential and (with modification) its monitoring. At the development stage we clearly need its value to be greater than 0.75 but, given that this assumes perfect control, in reality it needs to be higher if we are to capture the benefits claimed. A more realistic target is 0.9.

It is common to attempt to improve the accuracy of an existing inferential by collecting more recent process data. However, if the existing inferential has been at least partially successful in improving control, this will have caused a reduction in σ_{actual} and result in a lower value of ϕ. If the history is available, σ_{actual} would be better calculated from data collected before the original inferential was commissioned. This would also permit ϕ to be determined for the existing inferential and therefore any improvement quantified. It would also permit the economic impact of the new inferential to be assessed. If ϕ_1 and ϕ_2 are the before and after values then the potential percentage increase in benefits captured (Δ) is

$$\Delta = 100\left(1 - \sqrt{\frac{1-\phi_2}{1-\phi_1}}\right) \qquad (16.108)$$

This formula might also be used to justify enhancing (or replacing) a poorly performing inferential with an on-stream analyser. It might also be used to justify simplifying an inferential to make it more robust, accepting that the inferential might capture fewer benefits per hour, but more than make up for this by remaining in service for longer.

As a monitoring tool ϕ can be very valuable in the early detection of degradation in the accuracy of an inferential and disabling it before its poor performance does any real harm. However it needs to be used with care.

- If our controller is successful, it will reduce σ_{actual}. Our performance parameter will then fall, misleadingly indicating that the performance of the inferential has degraded. For example if, at the design stage, σ_{error} was half of σ_{actual} then ϕ would have a value of 0.75. If the controller successfully halves σ_{actual}, then ϕ will drop to zero – suggesting the inferential is of no value. To avoid this we choose a constant value for σ_{actual}, equal to the variation before the controller was commissioned.
- The observant reader may have noticed that the sudden drops in ϕ in Figure 16.19 do not occur at the same time as the unforeseen drops in stock price. So while the technique may be effectively used to assess accuracy at the design stage, it has little value in this form as a monitoring tool. The problem arises because, on the day that the stock price drops, there is a large increase not only in the variance of the error but also in the variance of the actual value. Their ratio therefore changes little. This too would be resolved by using a constant value for σ_{actual}. The large error is not therefore associated with a large change in the actual value and the value of ϕ will show a spike at the same time as the error occurs.
- We have to use a number of historical values to calculate σ_{actual} – usually 30. Thus, even if a problem with the inferential is resolved, the performance index will indicate a problem until 30 more laboratory results are taken. While we can reduce the number of historical values used, a better approach would be to treat as outliers the occasion(s) where the inferential is now known to have failed and remove them from the calculation of the index.
- If ϕ is calculated at a high frequency, e.g. by the use of on-stream analyser measurements, then care must be taken to ensure that the process is at steady state. Because the dynamics of the analyser will be longer than those of the inferential, any change in the inferential will be reflected some time later in the analyser measurement. There will therefore appear to be a transient error, even if both the inferential and analyser are accurate. Alternatively, dynamic compensation can be applied.
- Finally we should recognise that a failure may not be due to a problem with the inferential but a problem with the on-stream analyser or laboratory result.

The use of the performance index is not restricted to inferential properties. It can be used in any situation where there are two methods of determining a measurement. For example, MPC includes a prediction of every CV, based on changes in MV or DV. Any discrepancy in the prediction is automatically corrected by adjusting a bias term in the prediction. But we can compare the unbiased CV against that measured, monitoring the performance index to ensure that the level of correction remains within an acceptable limit.

We cover data reconciliation in Chapter 18 as a means of resolving conflicting process measurements. The performance index provides a method of detecting whether any discrepancy is getting worse. For example, we might have difficulty closing a mass balance around a unit because of measurement errors. The standard deviation of the actual flow (σ_{actual}) might be determined from the feed flow meter, while σ_{error} would be the standard deviation of the mass imbalance.

17

Autocorrelation

Autocorrelation is a correlation of a set of values with itself, sometimes also called *serial correlation*. External inputs are not considered. We apply Equation (4.75) to check for a correlation between a value measured now and the same measurement taken earlier, where k is the number of collection intervals between the two.

$$R_k = \frac{\sum_{i=k+1}^{n} (x_i - \bar{x})(x_{i-k} - \bar{x})}{\sqrt{\sum_{i=1}^{n-k} (x_i - \bar{x})^2 \sum_{i=k+1}^{n} (x_{i-k} - \bar{x})^2}} \tag{17.1}$$

One use is to identify if there is a repetitive pattern. For example, Figure 17.1 shows a noisy measurement taken every second for 10 minutes. At first glance it might appear that the underlying measurement is constant. However, by determining the Pearson correlation coefficient (R) at different values of k, we can see from the *correlogram* (shown as Figure 17.2) that the correlation varies cyclically. The strongest positive correlation occurs at regular intervals of about 180 seconds, as does the strongest negative correlation. This tells us that the apparently constant measurement has an underlying oscillation with a period of about three minutes. This might highlight a previously undetected problem with controller tuning or an issue with the control valve. Indeed this technique forms the basis of some controller oscillation detection techniques. The correlogram has the same frequency of oscillation as the raw measurement but with a much improved signal-to-noise ratio – making automatic detection of oscillation more reliable.

While not strictly autocorrelation (because it involves another measure of the same parameter), it can be used to explore whether a correlation between two measurements requires timing to be taken into account. For example, the engineer may wish to validate an inferential property by comparison with an on-stream analyser. The inferential property is likely to respond more quickly to process changes than the analyser. This is not an issue if comparison data are only collected when the process is at steady state. If this is not the case then the two measurements will not agree, potentially wrongly leading to the conclusion that one is suspect. Again, plotting

Statistics for Process Control Engineers: A Practical Approach, First Edition. Myke King.

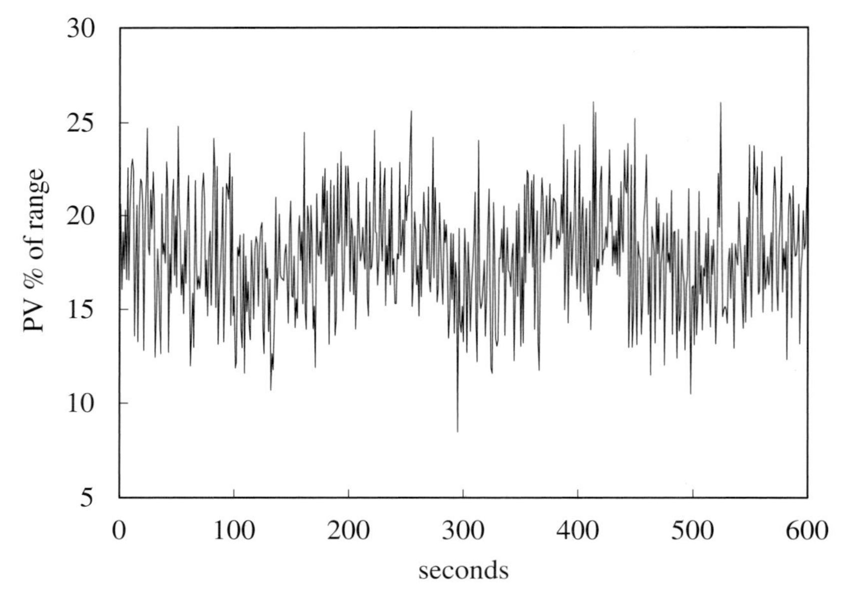

Figure 17.1 *Noisy measurement*

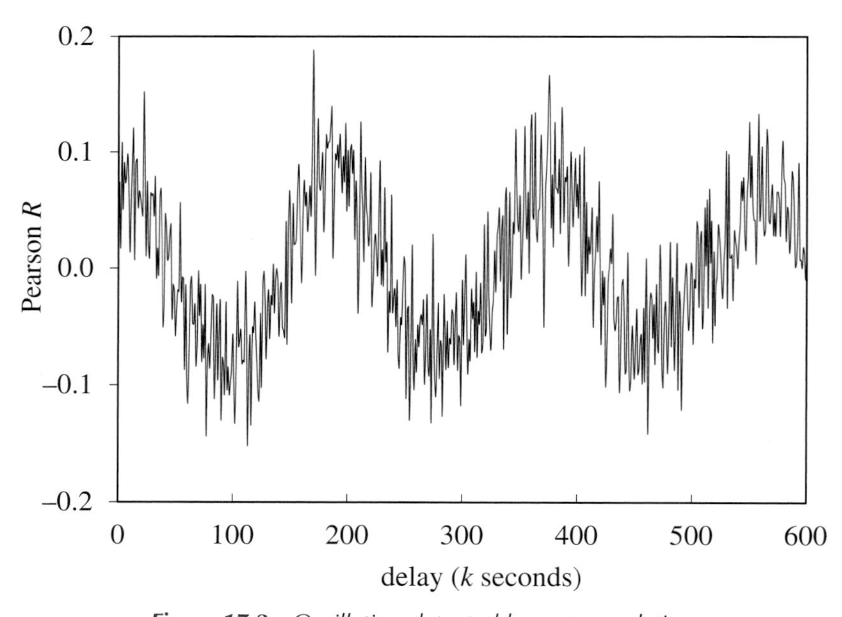

Figure 17.2 *Oscillation detected by autocorrelation*

R_k against k will identify what delay to apply to the inferential measurement before comparing it to the analyser. Data collected under changing conditions can then be combined with data collected at steady state. This is illustrated by Figure 17.3. In response to a series of step-tests the inferential (coloured line) shows a deadtime of 0.5 minutes and a lag of 2 minutes. The analyser (black line) has a deadtime of 10 minutes and a lag of 0.5 minute. We could identify these

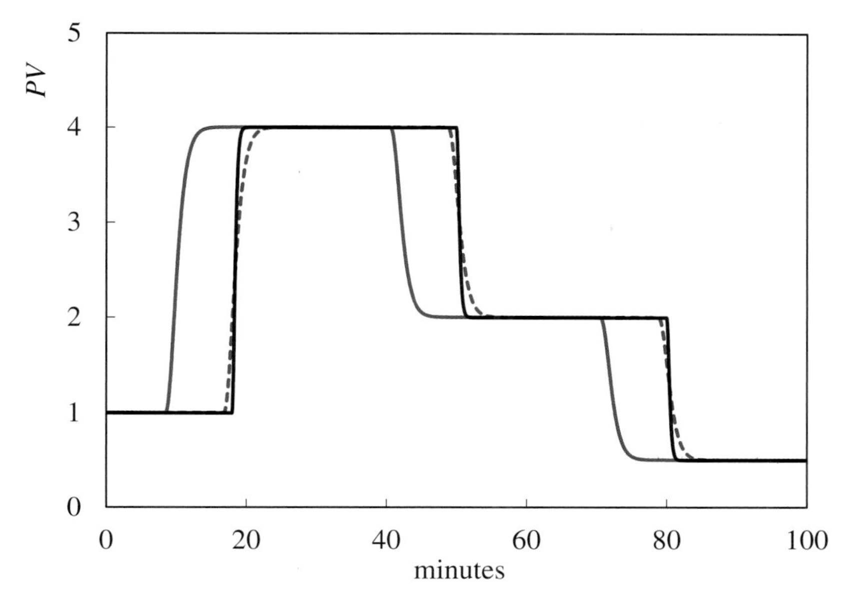

Figure 17.3 Comparison between on-stream analyser and inferential

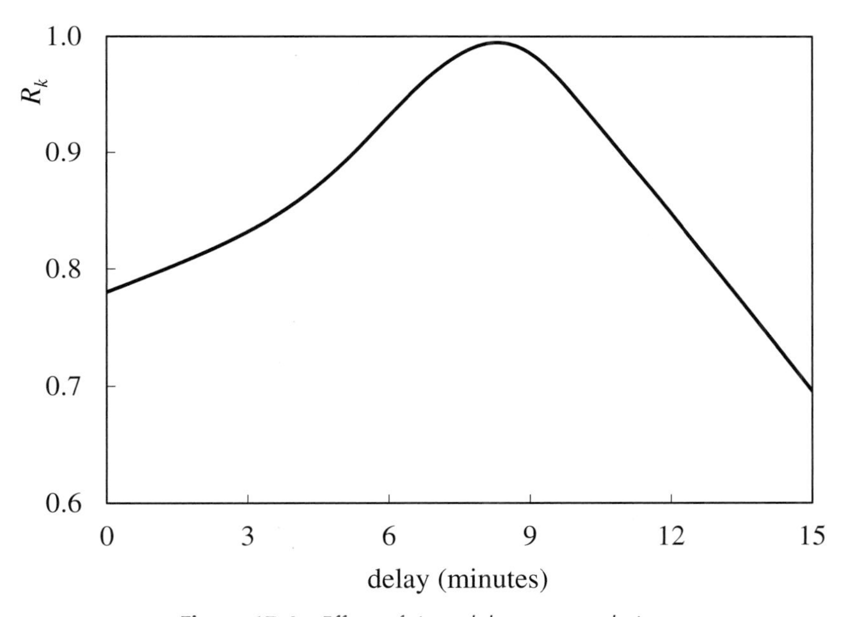

Figure 17.4 Effect of time delay on correlation

models and apply full dynamic compensation to the inferential. But for simpler comparison, Figure 17.4 shows the optimum delay to apply would be 8.3 minutes. The dashed line in Figure 17.3 shows this applied to the inferential. Without compensation for the delay, the correlation coefficient (R) between analyser and inferential would be around 0.78 but, with it, it approaches a value of 1.

The same approach can be applied to the development of inferentials, checking whether a delay should be applied to any of the independent variables being considered as inputs to the inferential calculation. Similarly it can be applied to dynamic model identification to determine a good starting estimate for process deadtime that is then improved by iteration. It can also be used to validate the dynamic models in MPC.

Also used is *autocovariance*, derived from Equation (4.72).

$$g_k = \frac{\sum_{i=k+1}^{n} (x_i - \bar{x})(x_{i-k} - \bar{x})}{n-k-1} \tag{17.2}$$

While similarly effective, like covariance, its magnitude depends not only on whether there is a correlation but also on the magnitude of the values x. While a nonzero value would indicate a correlation, it is not immediately obvious how strong it is.

An important use is in modelling process behaviour. Consider the inventory of liquid in a product storage tank. The inventory is clearly a function of the product flow into the tank but is also a function of liquid being pumped from the tank to another process or intermittently in batches for delivery to a customer. While not necessarily of obvious value to the control engineer, modelling stock levels can be an important part of feasibility studies. For example, gasoline is a blend of naphtha type components and small amounts of butane. It is usually economic to maximise the butane content up to the RVP specification of the product. In industry there are many examples where the economic advantage of doing so has been overestimated because of periods when insufficient butane was available in storage. Being able to predict such occasions would be an important part of estimating the benefits before embarking on a blend optimisation project.

We are likely to have stock level data collected regularly, usually at the end of each day. Analysing these data is likely to show that the level follows a normal distribution with mean and standard deviation that we can readily determine. However we cannot use this distribution to generate stock levels for our simulation. If a tank is almost full one day, it is very unlikely to be nearly empty the next. Today's level is a function of yesterday's – as we saw in Section 12.8, the process has memory. As an example, Table A1.6 and Figure 3.15 show the stock level recorded daily over 220 days.

Our data is a *time series* and needs to be modelled as such. If x is the inventory, a possible approach would be to develop a model such as

$$x_i = x_{i-1} + z_i \tag{17.3}$$

In this case z is a normally distributed random variable with a mean and standard deviation that we have identified by analysing the difference between x_{i-1} and x_i. Thus z is the change in inventory each day; this however, if selected from a normal distribution, could result in the predicted inventory being negative or greater than the tank capacity (in this case 2258). Further, the changes in inventory may also be a time series. As we saw previously in Figure 15.6, showing the cumulative sum of deviations from the mean inventory, if changes in inventory were truly random then we would expect the trend to be very noisy around zero, not show long trends upward or downward. We might also expect cyclic patterns, for example product withdrawal may be different at weekends and we would then expect x_i to be correlated with x_{i-7}.

Figure 17.5 shows how the autocorrelation coefficient (R^2) varies with the age of the previous inventory measurement. As expected, there is a strong correlation with yesterday's level; the correlation then declines rapidly for older values. But, as we anticipated, there are peaks at one and two weeks. The peak at six weeks might be due to some other cyclic behaviour.

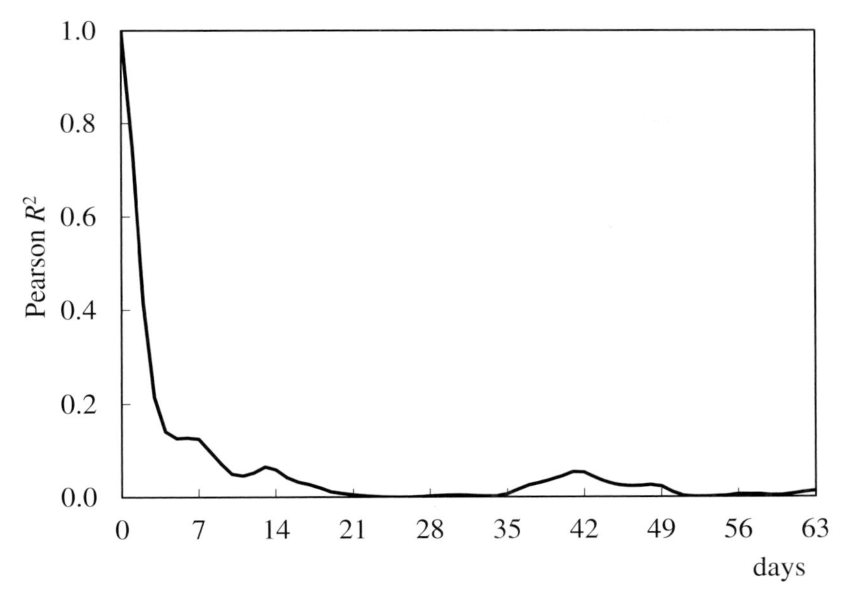

Figure 17.5 Autocorrelation of stock levels

It coincides with the changes in direction of the CUSUM trend and might result from a regular change in operating mode. Otherwise it would indicate quite an unlikely correlation. It should be included with care.

The general approach is to develop a model using the last p deviations from the mean to predict the next deviation.

$$x_i = \bar{x} - \sum_{j=1}^{p} b_j \left(x_{i-j} - \bar{x}\right) + z_i \tag{17.4}$$

The coefficients (b) are chosen to give the best fit. The difference between the actual deviation and that predicted will be a random number (z_i) with a mean of zero and a known standard deviation. This variable is sometimes described as *white noise*.

To determine the coefficients (b) we choose an initial value for the number of days (p) used in the model. Using Equation (17.2) we then calculate p autocovariances (g_0 to g_p). Note that g_0 is the variance of x (σ^2). Using the method developed by Anderson,[8] we can find the coefficients by solving the following simultaneous equations.

$$\begin{pmatrix} g_0 & g_1 & g_2 & \cdot & \cdot & g_{p-1} \\ g_1 & g_0 & g_1 & \cdot & \cdot & g_{p-2} \\ g_2 & g_1 & g_0 & \cdot & \cdot & \cdot \\ \cdot & \cdot & \cdot & \cdot & \cdot & g_2 \\ g_{p-2} & \cdot & \cdot & \cdot & g_0 & g_1 \\ g_{p-1} & g_{p-2} & \cdot & g_2 & g_1 & g_0 \end{pmatrix} \begin{pmatrix} b_1 \\ b_2 \\ b_3 \\ \cdot \\ b_{p-1} \\ b_p \end{pmatrix} = \begin{pmatrix} -g_1 \\ -g_2 \\ -g_3 \\ \cdot \\ -g_{p-1} \\ -g_p \end{pmatrix} \tag{17.5}$$

The solution is given by multiplying this equation by the inverse of the autocovariance matrix. This is a relatively simple technique in most spreadsheet packages.

$$\begin{pmatrix} b_1 \\ b_2 \\ b_3 \\ \cdot \\ b_{p-1} \\ b_p \end{pmatrix} = \begin{pmatrix} g_0 & g_1 & g_2 & \cdot & \cdot & g_{p-1} \\ g_1 & g_0 & g_1 & \cdot & \cdot & g_{p-2} \\ g_2 & g_1 & g_0 & \cdot & \cdot & \cdot \\ \cdot & \cdot & \cdot & \cdot & \cdot & g_2 \\ g_{p-2} & \cdot & \cdot & \cdot & g_0 & g_1 \\ g_{p-1} & g_{p-2} & \cdot & g_2 & g_1 & g_0 \end{pmatrix}^{-1} \begin{pmatrix} -g_1 \\ -g_2 \\ -g_3 \\ \cdot \\ -g_{p-1} \\ -g_p \end{pmatrix} \tag{17.6}$$

The mean of the random variable (z) will be zero; its variance is

$$\sigma_z^2 = \frac{1}{n-p} \sum_{i=p+1}^{n} \left((x_i - \bar{x}) + \sum_{j=1}^{p} b_j (x_{i-1} - \bar{x}) \right)^2 \tag{17.7}$$

To avoid over-fitting the model we check whether each coefficient is significantly different from zero. So we make the null hypothesis that $b_i = 0$. Using a_{ii}, the appropriate diagonal element from the inverted autocovariance matrix, we determine ε_i.

$$\varepsilon_i = \frac{b_i}{\sigma_z} \sqrt{\frac{n-p}{a_{ii}}} \tag{17.8}$$

The value ε_i will be normally distributed with a mean of 0 and a standard deviation of 1. From Table A1.9, if the absolute value of ε_i is greater than 1.64, then we are 95% sure that b_i should not be zero. If this test shows that b_p should be nonzero then it is likely that p should be increased before proceeding further. Once we are satisfied that we have included sufficient historical values, other coefficients failing this test should be set to zero. For the model to be stable

$$-1 < \sum_{i=1}^{p} b_i < 0 \tag{17.9}$$

The calculated values of b_i will always obey this; however setting to zero any coefficients found to be insignificant can cause the total of the remainder to lie outside this range. In any case, because they are not independent, the remaining coefficients should be recalculated. This is done by removing the corresponding rows and columns from the autocovariance matrix and repeating the calculations above. There is no guarantee that this will result in the remaining coefficients being significantly different from zero; so the process may need repeating.

The first estimate of the coefficients gave the results in Table 17.1. The analysis was initially performed by including the last 30 values in the model but the coefficients applied to values more than 14 days old were shown to be insignificant ($\varepsilon < 1.64$). The remainder were recalculated with p set at 14. The table shows that the coefficients b_3 to b_6 and b_9 to b_{12} should all be set to zero. Table 17.2 shows the impact of doing so and reveals that b_8 should also be zero. Table 17.3 shows the impact of removing this coefficient. All the remaining coefficients are significantly different from zero.

The mean of the inventories is 1292 and the variance 218712. From Equation (17.7) we find that the variance of the prediction error (σ_z) is 45085. Our model therefore accounts for 173627 of the variance, or about 79% of the variation in inventory.

To generate the remaining behaviour we randomly select, for each predicted value, v_1 to v_{12} from the range 0 to 1 and use these to generate z_i.

$$z_i = \sqrt{45085} \left[\sum_{j=1}^{12} v_j - 6 \right]_i \tag{17.10}$$

Table 17.1 Initial estimate of coefficients

	b	ε
1	−1.231	18.10
2	0.508	4.73
3	−0.030	0.27
4	−0.049	0.44
5	−0.071	0.64
6	0.096	0.87
7	−0.197	1.79
8	0.198	1.79
9	−0.174	1.57
10	0.172	1.54
11	−0.146	1.30
12	0.132	1.17
13	−0.218	2.03
14	0.127	1.87
total	−0.833	

Table 17.2 Intermediate estimate of coefficients

	b	ε
1	−1.178	18.88
2	0.397	6.40
7	−0.147	2.35
8	0.099	1.59
13	−0.149	2.40
14	0.111	1.78
total	−0.866	

Table 17.3 Final estimate of coefficients

	b	ε
1	−1.177	18.82
2	0.402	6.46
7	−0.066	1.84
13	−0.147	2.37
14	0.115	1.84
total	−0.873	

$$x_i = 1292 + 1.177x_{i-1} - 0.402x_{i-2} + 0.066x_{i-7} + 0.147x_{i-13} - 0.115x_{i-14} + z_i \tag{17.11}$$

Unlike process dynamic models, this model will not closely reproduce previous behaviour. In this example the daily inventory is only partially predictable; a significant component is noise caused by the semi-random nature of the process. Figure 17.6 shows its performance. While this might appear a poor model it does generate a time series with the same statistics as the actual data and is therefore quite suitable for simulation studies.

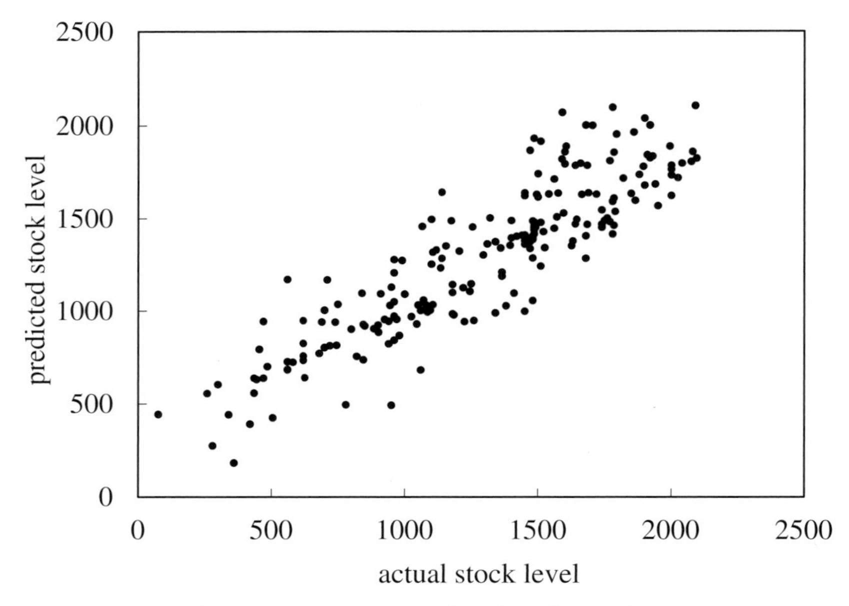

Figure 17.6 *Accuracy of predicted stock level*

18

Data Reconciliation

No process measurement can be considered perfect. The instrumentation itself is subject to error and data collection is subject to inaccuracies in time-stamping. It is for this reason that we never expect heat and mass balances to close perfectly. However performing such a balance is effectively a comparison between two 'opinions' of the true value – one measured directly and the other derived from the other measurements involved in the balance. Data reconciliation is a technique that uses these multiple estimates to produce an estimate that is more reliable than any of them.

Consider, as a simple example, that we have two measurements of the same property – both subject to error. The first has a standard deviation of σ_1, the second σ_2. The values of each of these measurements can be considered to have come from two distributions with different means, i.e. μ_1 and μ_2. Our aim is to choose the most likely estimate. This will be a weighted average of the two measurements, where a and $(1 - a)$ are the weighting coefficients. This estimate will therefore have the mean

$$\mu = a.\mu_1 + (1-a)\mu_2 \tag{18.1}$$

Provided the errors in the two measurements are not correlated then the standard deviation is

$$\sigma = \sqrt{(a.\sigma_1)^2 + ((1-a)\sigma_2)^2} = \sqrt{(\sigma_1^2 + \sigma_2^2)a^2 - 2\sigma_2^2 a + \sigma_2^2} \tag{18.2}$$

The best estimate will have the smallest standard deviation. This will occur when

$$\frac{d\sigma}{da} = 0 \quad \text{or} \quad \frac{(\sigma_1^2 + \sigma_2^2)a - \sigma_2^2}{\sqrt{(\sigma_1^2 + \sigma_2^2)a^2 - 2\sigma_2^2 a + \sigma_2^2}} = 0 \tag{18.3}$$

Thus the best choice of a is

$$a = \frac{\sigma_2^2}{\sigma_1^2 + \sigma_2^2} \tag{18.4}$$

Statistics for Process Control Engineers: A Practical Approach, First Edition. Myke King.

Substituting this value into Equation (18.1) gives the best estimate as

$$\mu = \frac{\sigma_2^2\mu_1 + \sigma_1^2\mu_2}{\sigma_1^2 + \sigma_2^2} \tag{18.5}$$

And this estimate will have a standard deviation given by substituting Equation (18.4) into Equation (18.2)

$$\sigma = \frac{\sigma_1\sigma_2}{\sqrt{\sigma_1^2 + \sigma_2^2}} \tag{18.6}$$

Rearranging Equation (18.6) shows that σ must be less than both σ_1 and σ_2.

$$\sigma = \frac{\sigma_1}{\sqrt{1 + \left(\frac{\sigma_2}{\sigma_1}\right)^2}} = \frac{\sigma_2}{\sqrt{1 + \left(\frac{\sigma_1}{\sigma_2}\right)^2}} \tag{18.7}$$

As an example, consider a distillation column that has been operating at a feed rate of 100 for the last 24 hours, as determined by a flow controller on the feed. This is the value for μ_1. The flow meter has a quoted reproducibility (1.96σ) of 4% of the instrument range that, in this case, is set at 125. Assuming no other sources of error we can assume σ_1 has a value of 2.5.

The feed rate is also estimated from the change in level in the feed tank. Let us assume that this gives an estimate of 103. This is the value for μ_2. Tank gauging typically has a reproducibility of 2 mm. Since the calculation uses the difference between two measurements, the estimate of the change in level will have a variance of $2\sigma^2$ and so the standard deviation will be 1.41 mm. The effect this has on the estimate of flow depends on the magnitude of the change in tank level which, in turn, depends on the interval over which it is measured and the tank's cross-sectional area. To simplify the arithmetic let us assume that the tank level changed by 100 mm. The value of σ_2 is $103 \times 1.41/100$ or around 1.5.

Applying Equations (18.5) and (18.6) gives a best estimate of 102.2 with a standard deviation of 1.3. As expected, this standard deviation is less than that of either measurement. Not only have we reconciled the difference between the two measurements but we have greater confidence in the resulting estimate than in either of the measurements. In other words, taking into account other measurements, even if inaccurate, can improve the accuracy of the estimate. This is illustrated as Figure 18.1.

We can add a third measurement by first replacing, in Equation (18.5), μ_2 with μ_3 and σ_2 with σ_3. We then replace μ_1 with the expression used to determine μ from the same equation and replace σ_1 with the expression used to determine σ, i.e. Equation (18.6). This gives

$$\mu = \frac{\sigma_2^2\sigma_3^2\mu_1 + \sigma_1^2\sigma_3^2\mu_2 + \sigma_1^2\sigma_2^2\mu_3}{\sigma_2^2\sigma_3^2 + \sigma_1^2\sigma_3^2 + \sigma_1^2\sigma_2^2} \tag{18.8}$$

Making similar substitutions in Equation (18.6) we get

$$\sigma = \frac{\sigma_1\sigma_2\sigma_3}{\sqrt{\sigma_2^2\sigma_3^2 + \sigma_1^2\sigma_3^2 + \sigma_1^2\sigma_2^2}} \tag{18.9}$$

In practice we often have many more than two estimates of flows. In addition to those above we can derive another from the overall unit mass balance – from summing all the product flows. And we can assign a standard deviation to this measurement by considering the contribution made by the reproducibility of each instrument used in the calculation. For example, our column might simply produce two products – distillate and bottoms. Let us assume that these two

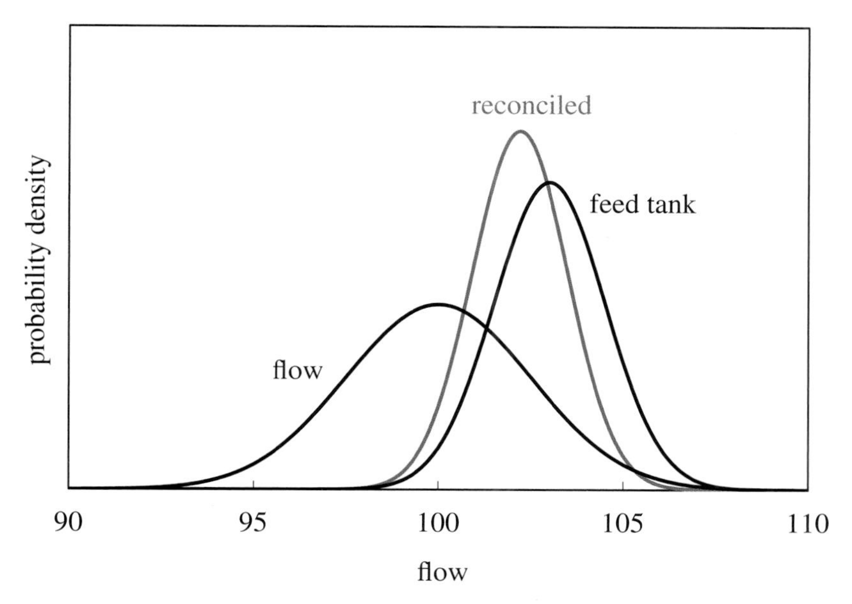

Figure 18.1 *Reconciliation of two measurements*

flows, measured over the same period, are respectively averaged at 42 and 53. Giving a total of 95, this gives us another measurement of feed rate. Like the feed meter, both product meters have a reproducibility of 4% of range. The range of the distillate meter is 90 and so the standard deviation of its measurement is 1.8. The range of the bottoms meter is 110, giving a standard deviation of 2.2. The standard deviation of the derived feed flow measurement is the square root of the sum of the squares of these two values, i.e. 2.8.

Figure 18.2 shows the effect of including this measurement. From Equation (18.8) we now find that the best estimate is 101.0. Equation (18.9) shows that the standard deviation has improved to 1.2.

The difference between this best estimate and the measurement derived by mass balance is 6.0. This is more than double the standard deviation and therefore outside the quoted reproducibility, suggesting that this measurement is suspect. Whether this is caused by a problem with the distillate meter or bottoms meter would require reconciliation of those flows using other measurements. For example, we may also be able to include changes in product tank inventories or a mass balance around a downstream process. If we have analysis of the composition of the streams we can perform component balances, each of which generates another estimate. And, with sufficient temperature measurements, we might also be able to derive another value by energy balance.

Formulae using four measurements can be developed by extending Equations (18.8) and (18.9) and so on. However, for complex problems, proprietary *DRS* (*data reconciliation system*) software would be used. These products typically minimise the penalty function (C), where μ is the reconciled value derived from the n measurements of μ_1 to μ_n.

$$C = \sum_{i=1}^{n} \left(\frac{\mu - \mu_i}{\sigma_i} \right)^2 \tag{18.10}$$

Figure 18.3 shows the effect of applying this technique to our example process. Again it demonstrates that the best estimate is 101.

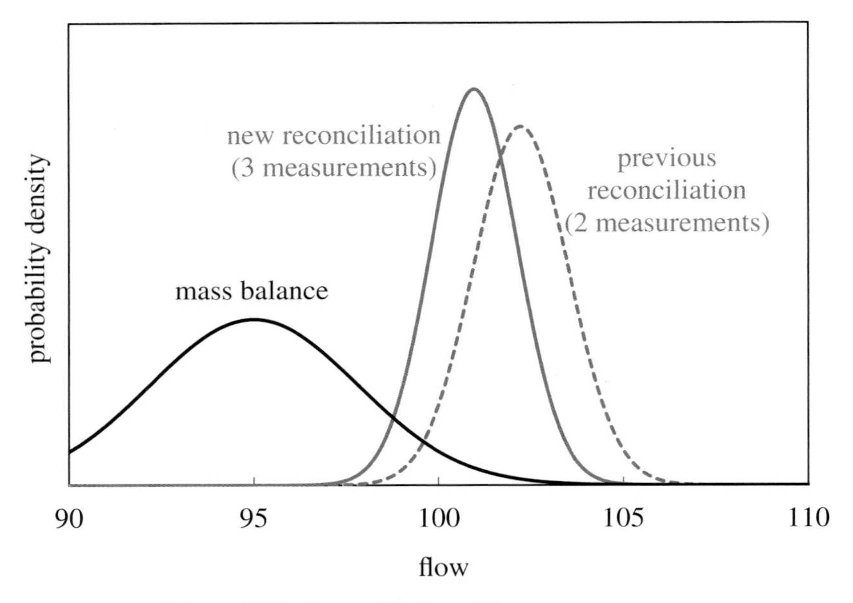

Figure 18.2 Reconciliation of three measurements

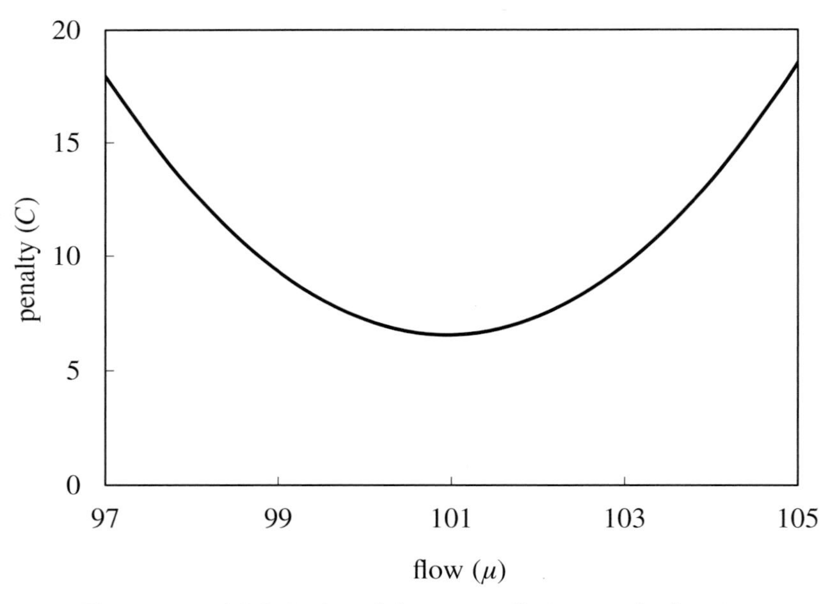

Figure 18.3 Minimisation of data reconciliation penalty function

DRS software permits the user to define complex process flowsheets and so, for example, reconcile all the measurements across a whole manufacturing site. Such information is valuable both for accounting purposes but also for process modelling. Indeed a similar approach must be taken with the data provided to CLRTO (closed-loop real-time optimisation) – particularly if based on rigorous process simulation.

The software also allows the user to identify suspect measurements by considering how different the best estimate is from each measurement. In our example the best estimate is 2.2 higher than that measured by the flow meter, i.e. within the quoted reproducibility. Had it not been then this would indicate that the meter was in need of attention. Similarly it is 0.8 lower than that calculated from tank levels – again within the expected range. Indeed it is this systematic method of identifying, and so rectifying, measurement problems that is perhaps of the greatest benefit since it improves the quality of the raw process measurements used by all. Indeed applying DRS, without rectifying such problems, risks generating yet another set of process measurements with no confidence that they are any more reliable than the raw data.

Many DRS packages also support 'what if?' analysis. The inclusion of additional measurements and improving the accuracy of existing measurements can be assessed in terms of the impact they would have on overall accuracy. There are many examples where the accuracy of the estimate (derived by mass balance) of an unmeasured flow may not be significantly improved by the installation of a direct measurement. It might however be improved by replacing poor instrumentation elsewhere in the process. What-if analysis permits definition of the scope of the most cost effective measurement upgrade project.

The technique is not limited to the measurement of flows. It can be applied wherever there are multiple measurements of the same parameter. For example, by comparing inferential (or on-stream analyser) against the laboratory result, we can decide whether the difference between the two measurements is significant enough to merit attention.

19

Fourier Transform

Strictly, the Fourier transform is not a statistical technique. It is basically a solution of a large number of simultaneous equations. However, its application is strongly linked to both regression analysis and autocorrelation that we covered in previous chapters. As we will show, it offers the engineer another valuable diagnostic tool.

Most texts covering the Fourier transform do so in a highly mathematical way, making it difficult for the control engineer to identify where its application might be beneficial. Here we restrict its use to identifying cyclic disturbances to process measurements. Often such disturbances are not immediately obvious, presenting themselves as random noise. This is common problem with averaging level control. Applying this technique has identified many controllers that appear to be working well, using the available surge capacity, but the variation in level is in fact a very slow oscillation (disguised by process disturbances) caused by excessive integral action. The technique can also help diagnose control valve problems, such as stiction and hysteresis.

Since control engineers deal largely with process data collected at a fixed time interval we will focus on the *discrete Fourier transform (DFT)*. Strictly it is the *real DFT* where the input is restricted to real data. There is also an *imaginary DFT* but this has no application here.

Fourier showed that any signal can be decomposed into a number of sinusoidal signals. Each of these signals will have a different *frequency* and *amplitude*; further they will not necessarily be *in phase*. We need to be careful with the definitions and units of measurement of these terms. If we consider the continuous function for the kth frequency

$$x_k = a.\sin(2\pi f_k t + \phi) \tag{19.1}$$

The term a is known as the *peak amplitude*. Many engineers will use the term 'amplitude' to describe the *peak-to-peak* distance $2a$. Amplitude can also be referred to as *energy* or *power*. If time (t) is measured in seconds then the frequency (f_k) will be in Hz (sec^{-1}). The factor 2π is required to convert f_k from cycles per unit time to radians per unit time. Similarly the phase shift (ϕ) is expressed in radians.

With discrete Fourier transforms the frequency (k) is expressed as the number of cycles that occur within the period over which the input data are collected. The term $f_k t$ is replaced by ki/N.

Statistics for Process Control Engineers: A Practical Approach, First Edition. Myke King.

The sine wave at this frequency has a peak amplitude of a_k and now comprises a series of discrete values of x_k. N is the total number of values collected in the period and i is the index (ranging from 0 to $N-1$) of each value of x_k.

$$(x_k)_i = a_k.\sin\left(\frac{2\pi ki}{N} + \phi_k\right) \tag{19.2}$$

This equation can be rewritten as

$$(x_k)_i = a_k\left[\sin(\phi_k)\cos\left(\frac{2\pi ki}{N}\right) + \cos(\phi_k)\sin\left(\frac{2\pi ki}{N}\right)\right] \tag{19.3}$$

The result of the Fourier transform is, for each frequency, a combination of in-phase sine and cosine waves of peak amplitude p_k and q_k respectively. For example, at frequency k, the series of x_k values is

$$(x_k)_i = p_k.\cos\left(\frac{2\pi ki}{N}\right) + q_k.\sin\left(\frac{2\pi ki}{N}\right) \tag{19.4}$$

Comparing this with Equation (19.3) gives

$$p_k = a_k.\sin(\phi_k) \tag{19.5}$$

$$q_k = a_k.\cos(\phi_k) \tag{19.6}$$

Squaring and adding Equations (19.5) and (19.6) gives the peak amplitude.

$$a_k = \sqrt{p_k^2 + q_k^2} \tag{19.7}$$

One might reasonably deduce that dividing Equation (19.5) by Equation (19.6) gives the phase angle.

$$\phi_k = \tan^{-1}\left(\frac{p_k}{q_k}\right) \tag{19.8}$$

However, this will only generate a value for ϕ_k in the range $-\pi/2$ to $\pi/2$. Further it is indeterminate when q_k is zero. To obtain values from the full range of $-\pi$ to π requires the use of the *arctangent2 (atan2)* function. This is shown in Figure 19.1 and is defined as

$$\text{atan2}\left(\frac{p_k}{q_k}\right) = \tan^{-1}\left(\frac{p_k}{q_k}\right) \quad \text{when } q_k > 0 \tag{19.9}$$

$$\text{atan2}\left(\frac{p_k}{q_k}\right) = \tan^{-1}\left(\frac{p_k}{q_k}\right) + \pi \quad \text{when } p_k \geq 0 \text{ and } q_k < 0 \tag{19.10}$$

$$\text{atan2}\left(\frac{p_k}{q_k}\right) = \tan^{-1}\left(\frac{p_k}{q_k}\right) - \pi \quad \text{when } p_k < 0 \text{ and } q_k < 0 \tag{19.11}$$

$$\text{atan2}\left(\frac{p_k}{q_k}\right) = \frac{\pi}{2} \quad \text{when } p_k \geq 0 \text{ and } q_k = 0 \tag{19.12}$$

$$\text{atan2}\left(\frac{p_k}{q_k}\right) = -\frac{\pi}{2} \quad \text{when } p_k < 0 \text{ and } q_k = 0 \tag{19.13}$$

We need to consider the conditions under which the transformation can be applied. Firstly, the data must be a repeating waveform. If an insufficient number of values are analysed then the

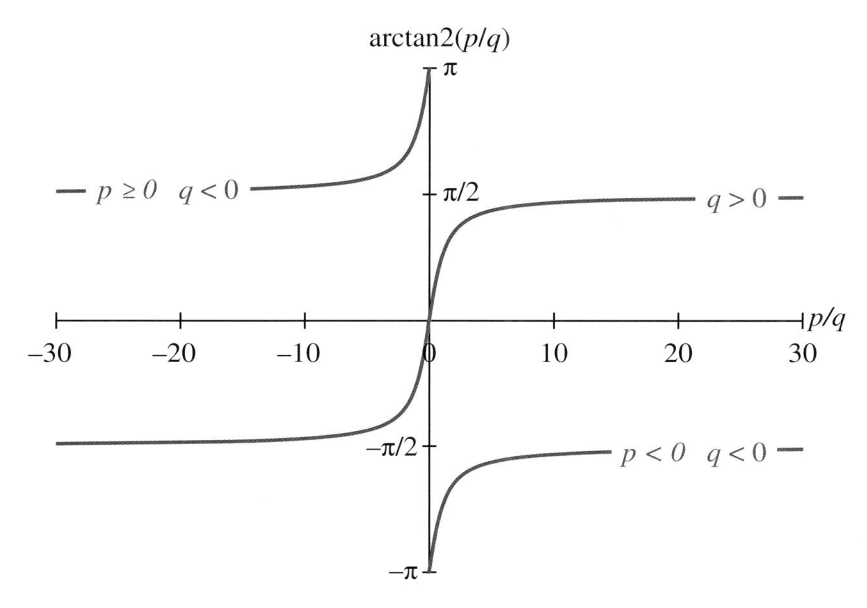

Figure 19.1 *Arctangent2 function*

resulting distortion is known as *leakage*. Secondly, if an analog signal contains no frequency higher than or equal to f_{max} then it can be reconstructed from the digital signal provided the sampling exceeds $2f_{max}$ samples per second. This is known as the *Nyquist–Shannon sampling theorem*. For example, many process historians collect data at one minute intervals, in which case the maximum frequency detectable will be 30 hr^{-1}, i.e. 0.5 min^{-1} or 1/120 Hz. Collecting data at too low a frequency causes distortion known as *aliasing*.

The maximum value of k is therefore $N/2$; k has a minimum value of zero. Thus the Fourier transform will comprise $N/2 + 1$ sine waves and $N/2 + 1$ cosine waves. It should be noted therefore that N should be an even number and that only discrete frequencies that are an exact multiple of the data collection interval are included. Some texts suggest that N should be a power of 2 (e.g. 512, 1024 …). This permits the use of the *fast Fourier transform* technique, substantially reducing the calculations required and hence the execution time of any analysis tool. However this only becomes an issue when analysing very high frequencies and is of less advantage for process data.

The coefficients (p) for each frequency (k) are determined from the actual sampled data (x) using the formulae

$$p_k = \frac{2}{N}\sum_{i=0}^{N-1} x_i \cos\left(\frac{2\pi ki}{N}\right) \tag{19.14}$$

except

$$p_0 = \frac{1}{N}\sum_{i=0}^{N-1} x_i \quad \text{and} \quad p_{N/2} = \frac{1}{N}\sum_{i=0}^{N-1} x_i \cos(2\pi i) \tag{19.15}$$

The reason that the calculations for p_0 and $p_{N/2}$ are slightly different has to do with *bandwidth*. Bandwidth is the difference between the highest and lowest frequencies in the band. The total bandwidth is split into $N/2$ bands, but there are $N/2 + 1$ frequencies. The width of each band, for the frequencies from $k = 1$ to $k = N/2 - 1$, is $2/N$. However the frequencies at each end, i.e. $k = 0$ and $k = N/2$, have a bandwidth of half this value, or $1/N$.

The coefficient p_0 is applied to the cosine of zero, which is unity. It is therefore the mean of all the values collected and represents the *offset* of the signal from zero.

The coefficients (q) for each frequency (k) are also determined from the actual sampled data (x) but using the formulae

$$q_k = \frac{2}{N}\sum_{i=0}^{N-1} x_i \sin\left(\frac{2\pi ki}{N}\right) \qquad (19.16)$$

Since q_0 is derived from the sine of zero, it will also be zero. Similarly $q_{N/2}$ will be zero since it is derived from the sine of integer multiples of π, each of which will be zero.

The analog signal can be reconstructed from the *inverse discrete Fourier transform (IDFT)*.

$$x_i = \sum_{k=0}^{N/2} p_k \cos\left(\frac{2\pi ki}{N}\right) + \sum_{k=0}^{N/2} q_k \sin\left(\frac{2\pi ki}{N}\right) \qquad (19.17)$$

Figure 19.2 shows 60 values collected at one minute intervals for an hour – typically what might be readily available from a process historian. The coefficients p_0 to p_{30} were calculated using Equations (19.14) and (19.15); q_0 to q_{30} using Equation (19.16). These were then used in Equation (19.17) to reconstruct the analog signal shown as the coloured line. This simply confirms that the method works and is a useful check that the calculations have been performed correctly. It should be noted that for digital signals the reconstructed analog signal will pass through all the discrete values; it is not a curve of best fit but an exact solution. Neglecting q_0 and $q_{N/2}$, which are both zero, we have calculated N coefficients to fit a curve through N points.

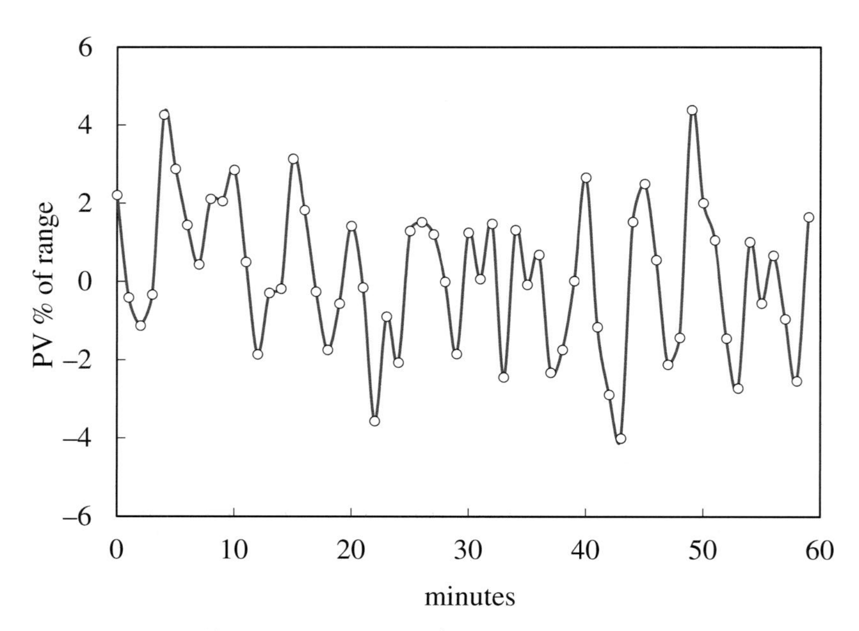

Figure 19.2 *Apparently noisy measurement*

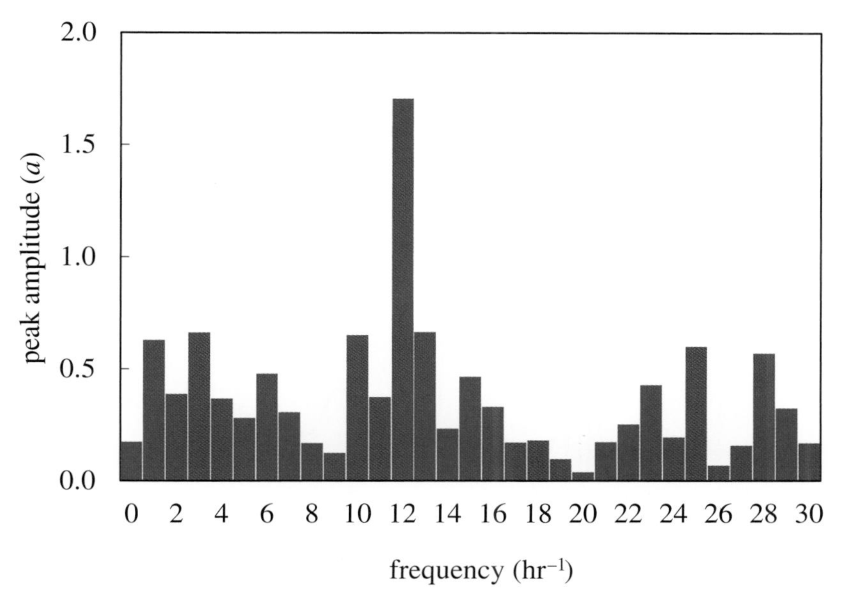

Figure 19.3 *Power (or frequency) spectrum*

The value of the technique is that, by applying Equation (19.7), we can determine the peak amplitude of every frequency and present this as the *power spectrum* shown as Figure 19.3. We have converted a signal that is in the *time domain* to one in the *frequency domain*. In our example the average signal is approximately zero and so a_0 is small. If this is not the case then the zero frequency should be omitted from the chart so that it does not dwarf the others. Alternatively the mean (μ) can be subtracted from each value of x before calculating the amplitudes. This affects only p_0, reducing it (and hence a_0) to zero.

Figure 19.3 is, of course, also a histogram. We can therefore apply many of the techniques we covered in earlier chapters. For example, we can plot a kernel density estimate. If it were unimodal, we could fit a distribution and so determine statistical parameters such as mean and variance. These might be useful in determining whether the spectrum changes over time.

Figure 19.4 shows the same power spectrum but based on the period (wavelength) of oscillation and converted to a kernel density estimate.

The power spectrum shows that the apparently random noise in the process measurement is dominated by a waveform that completes 12 cycles within the period covered by the data, i.e. it has a period of 5 minutes. This might give some clue to the source of the oscillation – particularly if the technique is also applied to values collected over the same period from potential sources. The peak amplitude at this frequency (a_{12}) is about 1.7, giving a peak-to-peak amplitude of 3.4.

If required, we can also obtain the *phase spectrum* by applying the appropriate choice from Equations (19.9) to (19.13). This is shown in Figure 19.5 and gives a value for ϕ_{12} of 1.63 radians.

Using the values a_{12} and ϕ_{12} in Equation (19.1) we can superimpose the waveform on the original signal, as shown in Figure 19.6. There is a strong correlation between the two curves with a value for Pearson R of 0.64. While not accounting for all the noise, it does explain a large part of the variation. Very similar values for a_{12} and ϕ_{12} would be obtained by regressing the coefficients to give a wave of best fit.

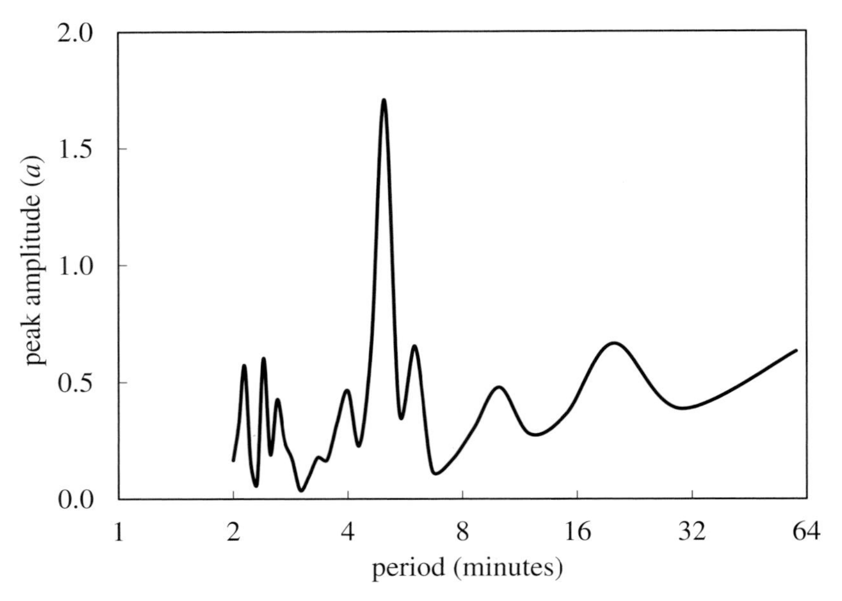

Figure 19.4 *Period spectrum*

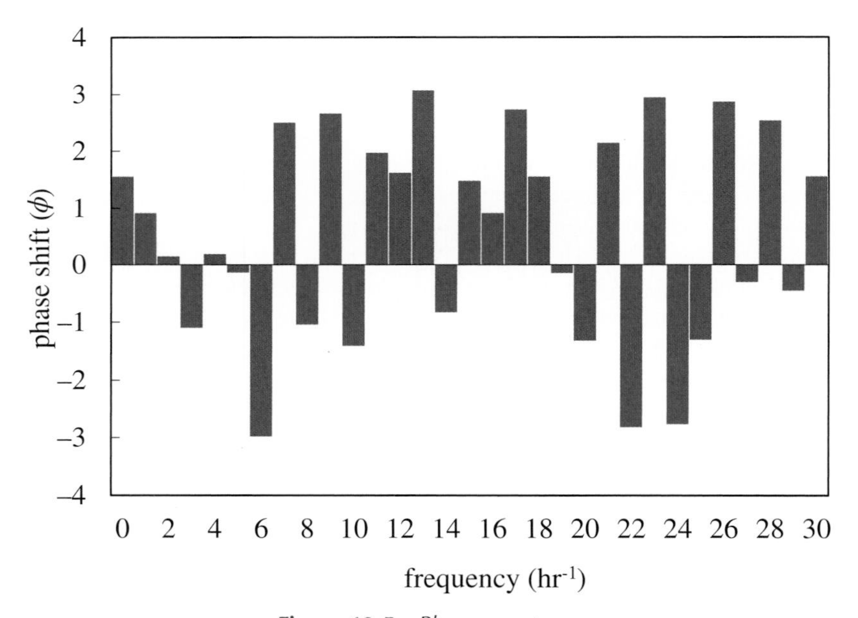

Figure 19.5 *Phase spectrum*

A limitation of this discrete method is that the frequency spectrum can only contain frequencies equivalent to integer values of k. So, for example, if we increase the number of values collected from 60 to 62 the wavelength corresponding to k having a value of 12 will increase to 5.17 minutes (62/12). If there is truly dominant waveform, with a wavelength of 5 minutes, its dominance will be masked by adjacent frequencies. Figure 19.7 shows how the spectrum

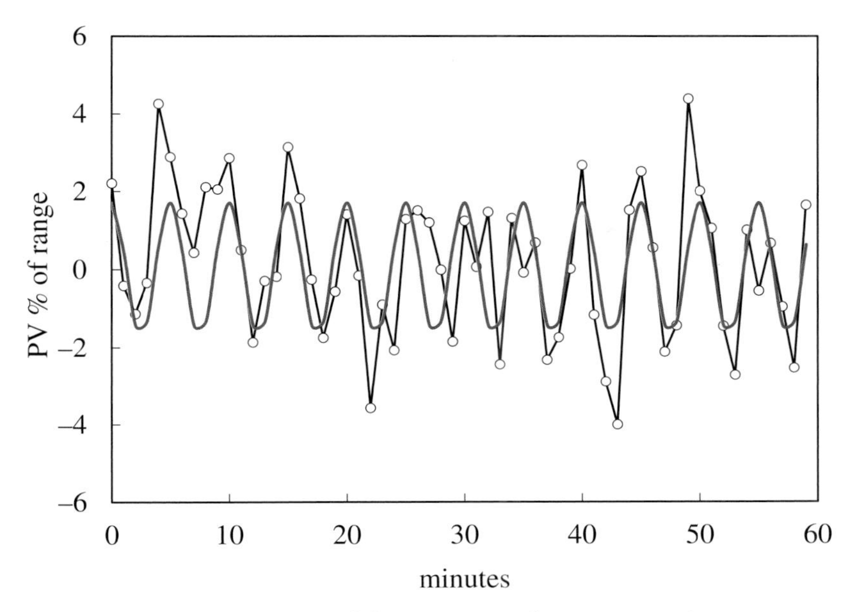

Figure 19.6 *Superimposition of dominant wave form on original measurement*

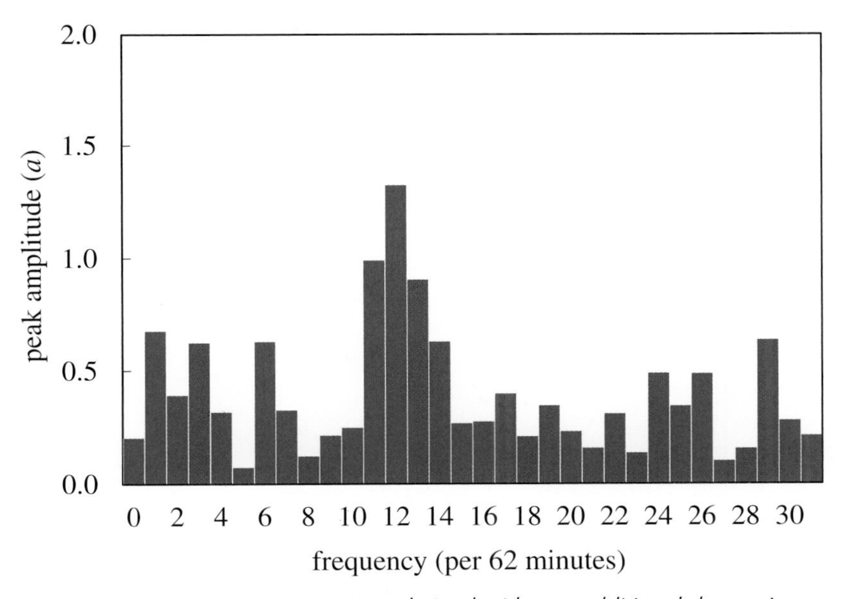

Figure 19.7 *Frequency spectrum derived with two additional data points*

will now appear. Superimposing the dominant waveform on the original signal will show a poorer match; indeed in this case Pearson R reduces to 0.48. As might be expected, synthesising a signal by also including the two frequencies either side of the dominant one increases R to 0.67 – close to that derived from the original dataset. It is therefore good practice to analyse different datasets, making small changes to the number of values included in each set.

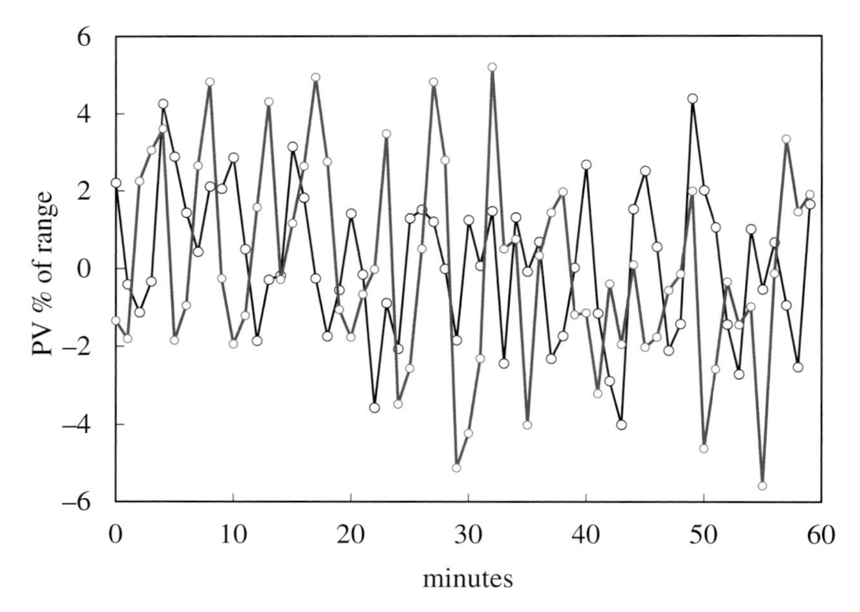

Figure 19.8 Addition of second noisy measurement

While DFT will highlight a problem with a controller, it does not necessarily mean that it is caused by the controller. The problem may persist even if the controller is switched to manual – so demonstrating that the cause may be elsewhere in the process. To illustrate this, Figure 19.8 shows data collected from another controller (PV_2) superimposed on the trend for the controller analysed previously (PV_1). It would be difficult to conclude from these trends whether changes in PV_2 are causing the variation in PV_1. Indeed the scatter plot show as Figure 19.9 shows no

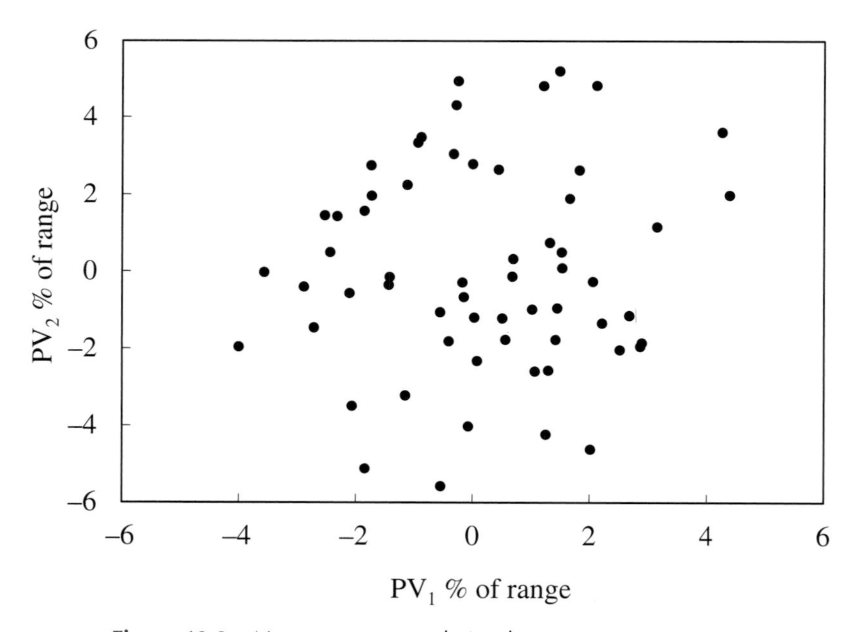

Figure 19.9 No apparent correlation between measurements

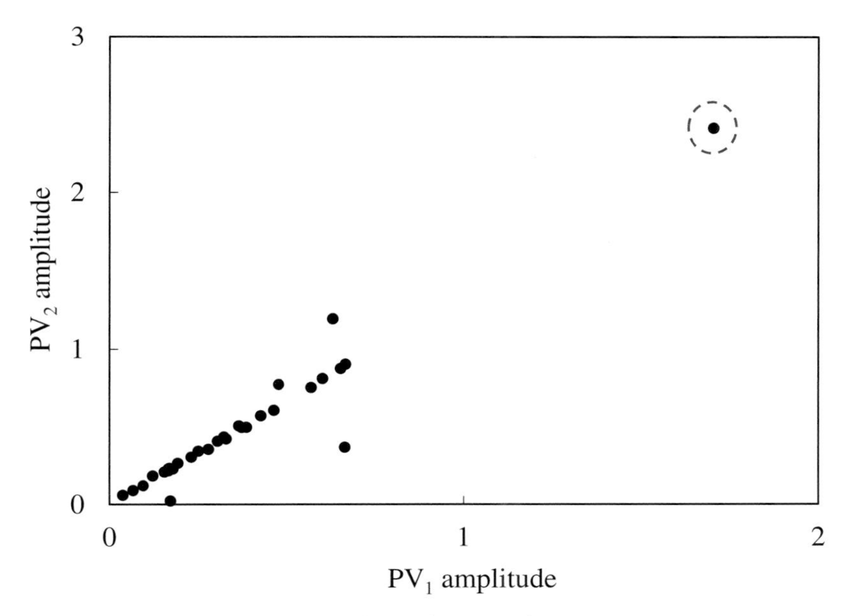

Figure 19.10 *Correlation between frequency spectra*

obvious relationship. However, we can apply DFT to generate the power spectrum for PV_2 and plot the amplitude of each frequency against the corresponding amplitude from the PV_1 power spectrum. Figure 19.10 shows not only are both process values dominated by the frequency of 12 cycles/minute (highlighted) but that they contain many other frequencies in common. The very strong correlation (Pearson R^2 of 0.92) suggests that oscillation in PV_2 may be the cause of that in PV_1 – or vice versa. This technique is used by several controller diagnostics products. Indeed these will often analyse all the process values and generate a *power spectrum correlation map* – a grid using colour to show where correlations are strong. While valuable, this method does not distinguish cause from effect. To confirm the cause, it is still necessary to switch to manual any controller thought to be the source of the problem.

Part 2

Catalogue of Distributions

20

Normal Distribution

In Chapter 5 we covered the classic normal distribution along with its bivariate version. Chapter 12 addressed the lognormal distribution – one of the common variations. Here we cover the wide range of other variations, many of which better fit our examples.

20.1 Skew-Normal

As its name suggests the *skew-normal distribution* is a modified form of the normal distribution that can take account of skewness. It is described by the PDF

$$f(x) = \frac{1}{\beta\sqrt{2\pi}} \exp\left[\frac{-(x-\alpha)^2}{2\beta^2}\right]\left[1 + \operatorname{erf}\left(\frac{\delta(x-\alpha)}{\beta\sqrt{2(1-\delta^2)}}\right)\right] \quad \beta > 0;\ -1 \le \delta \le 1 \tag{20.1}$$

The parameter δ determines skewness. Figure 20.1 shows its effect, with α set to 0 and β to 1. If δ is less than zero the distribution is skewed to the left, if greater than zero it is skewed to the right. If set to -1 or $+1$ we obtain the *half-normal distribution* that we will describe later.

The mean, variance, skewness and kurtosis are

$$\mu = \alpha + \beta\delta\sqrt{\frac{2}{\pi}} \tag{20.2}$$

$$\sigma^2 = \beta^2\left(1 - \frac{2\delta^2}{\pi}\right) \tag{20.3}$$

$$\gamma = \left(\frac{4-\pi}{2}\right)\delta^3\left(\frac{2}{\pi - 2\delta^2}\right)^{\frac{3}{2}} \tag{20.4}$$

$$\kappa = 3 + \frac{8\delta^4(\pi-3)}{\left(\pi - 2\delta^2\right)^2} \tag{20.5}$$

Statistics for Process Control Engineers: A Practical Approach, First Edition. Myke King.

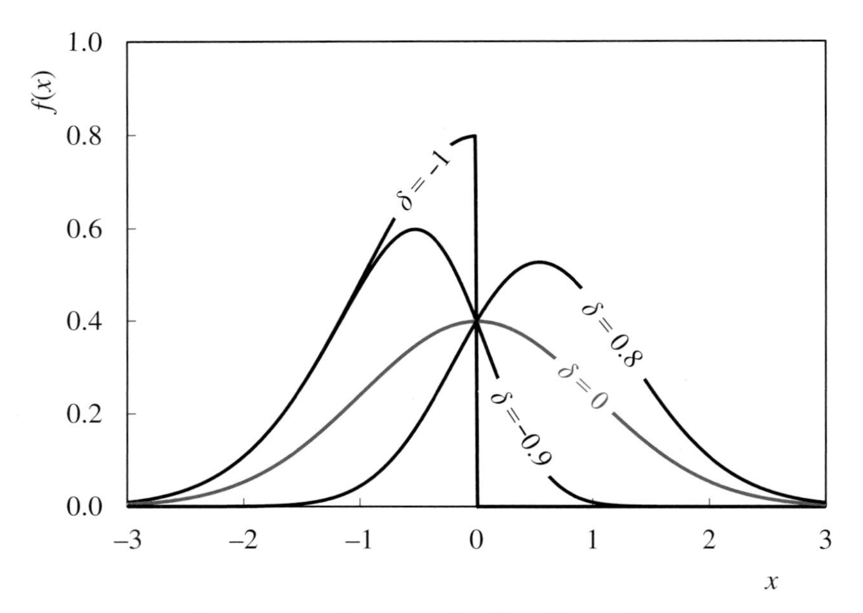

Figure 20.1 Skew-normal: Effect of δ on shape

Figure 20.2 plots Equations (20.4) and (20.5) as δ varies. It shows two limitations of the distribution. The first is that skewness is limited to the range −1 to 1 and kurtosis limited to the range 3 to 3.87. Secondly, skewness and kurtosis are highly correlated; adjusting δ affects both. Fitting the distribution to data will usually be a compromise between matching skewness and matching kurtosis.

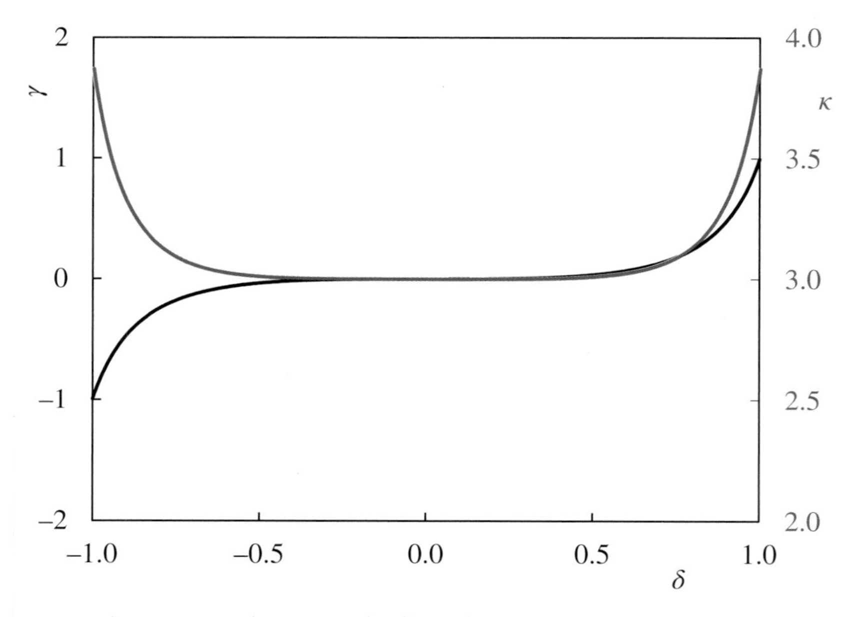

Figure 20.2 Skew-normal: Effect of δ on skewness and kurtosis

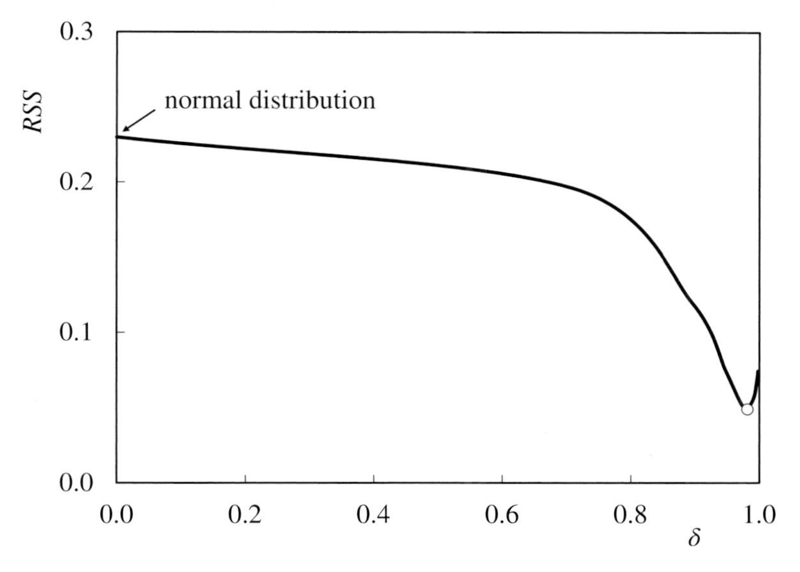

Figure 20.3 *Skew-normal: Impact δ has on fitting to C_4 in propane data*

Nevertheless the skew-normal distribution fits the C_4 in propane data better than the normal distribution. Figure 20.3 shows the impact that including δ has on RSS. It passes through minimum of 0.0488 – considerably better than 0.2303 given by the normal distribution. Fitting Equation (20.1) gives δ as 0.982, α as 1.88 and β as 3.39. Equation (20.2) then gives the mean as 4.54 and Equation (20.3) the standard deviation as 2.11.

The fitted skewness and kurtosis must lie on the line plotted in Figure 20.4. As shown, the calculated values are far from the line; indeed skewness exceeds the maximum of 1 and kurtosis

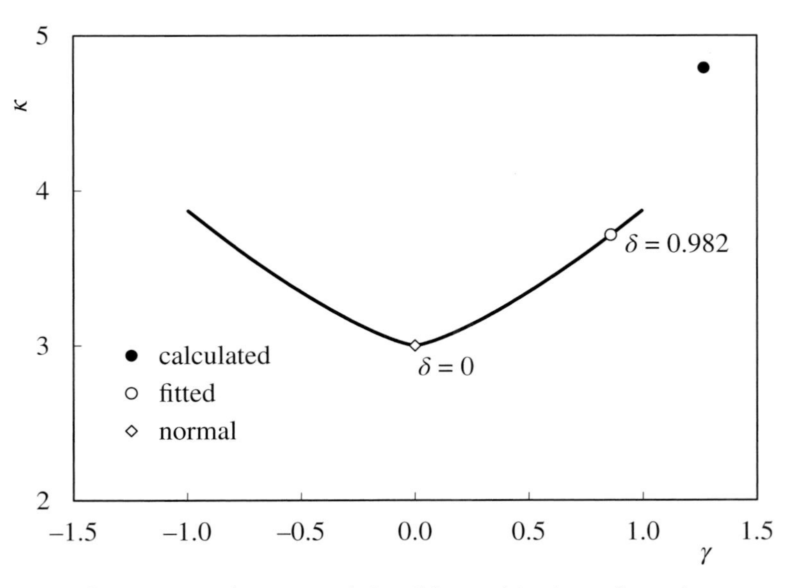

Figure 20.4 *Skew-normal: Feasible combinations of γ and κ*

exceeds the maximum of 3.66. As we have seen with mean and standard deviation, simply calculating their value gives a less accurate result than curve fitting. The same is true of skewness and kurtosis. In this case, from Equations (20.4) and (20.5), the fitted skewness is 0.86 and kurtosis is 3.57.

An alternative approach described in some texts is to determine the skewness by calculation and use the result to solve Equation (20.4) to determine the maximum likelihood value for δ. This would fail here because the calculated skewness is larger than that supported by the distribution; the value for δ would exceed 1. In any case such an approach does not ensure that best fit is achieved.

20.2 Gibrat

The *standard lognormal distribution* is described by Equation (12.1), with α is set to 0 and β to 1. This is also known as the *Gibrat distribution*. The key parameters are then reduced to constants. This distribution clearly has no practical value to the control engineer.

$$\mu = \sqrt{e} \approx 1.649 \tag{20.6}$$

$$\sigma = e(e-1) \approx 4.671 \tag{20.7}$$

$$\gamma = (e+2)\sqrt{e-1} \approx 6.185 \tag{20.8}$$

$$\kappa = e^4 + 2e^3 + 3e^2 - 3 \approx 113.9 \tag{20.9}$$

20.3 Power Lognormal

Also known as the *Marshall–Olkin distribution*, as the name suggests, the *power lognormal distribution* is derived from the lognormal distribution. Its PDF is

$$f(x) = \frac{p}{\beta x\sqrt{2\pi}} \exp\left[-\frac{(\ln(x)-\alpha)^2}{2\beta^2}\right]\left\{\frac{1}{2}-\frac{1}{2}\operatorname{erf}\left[\frac{\ln(x)-\alpha}{\beta\sqrt{2}}\right]\right\}^{p-1} \quad x>0;\ p,\beta>0 \tag{20.10}$$

If the *power parameter (p)* is set to 1, the distribution reverts to the classic lognormal distribution described by Equation (12.1). The effect of changing p, with α fixed at 0 and β at 1, is shown in Figure 20.5.

The CDF is

$$F(x) = 1 - \left\{\frac{1}{2}-\frac{1}{2}\operatorname{erf}\left[\frac{\ln(x)-\alpha}{\beta\sqrt{2}}\right]\right\}^{p} \tag{20.11}$$

Fitting to the C_4 in propane data gives a value of 0.267 for p, 1.001 for α and 0.312 for β. As might be expected, the additional parameter (p) results in a fit better than that achieved with the lognormal distribution – reducing *RSS* from 0.0477 to 0.0322. However there are no simple formulae for calculation of mean, variance, skewness or kurtosis – so undermining its advantage in terms of accuracy.

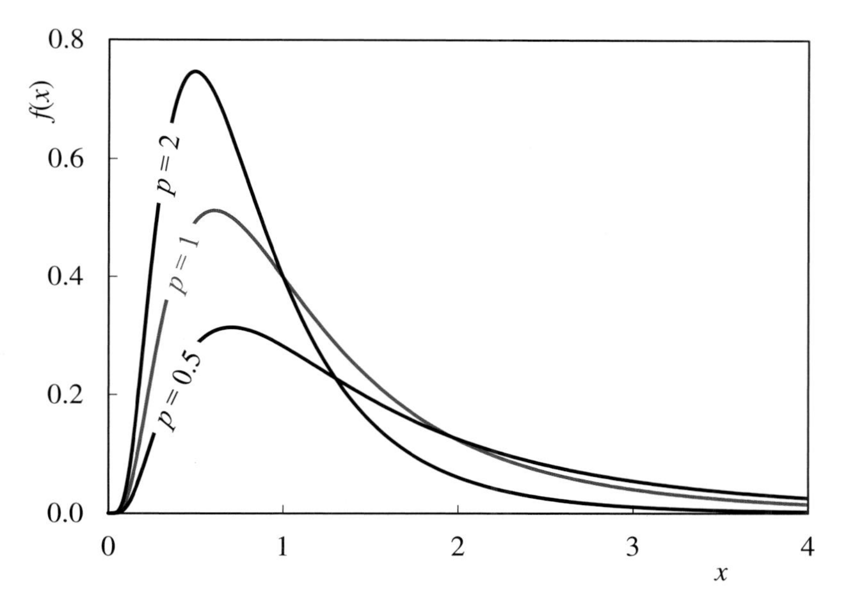

Figure 20.5 Power lognormal: Effect of p on shape

20.4 Logit-Normal

The *logit-normal distribution* uses the logit function, as described in Section 11.3. Its PDF and CDF are very similar to the lognormal distribution.

$$f(x) = \frac{1}{\beta x(1-x)\sqrt{2\pi}} \exp\left[\frac{-(\text{logit}(x)-\alpha)^2}{2\beta^2}\right] \quad 0<x<1;\ \beta>0 \qquad (20.12)$$

$$F(x) = \frac{1}{2} + \frac{1}{2}\text{erf}\left[\frac{\text{logit}(x)-\alpha}{\beta\sqrt{2}}\right] \qquad (20.13)$$

Figure 20.6 shows the effect of varying β with α fixed at zero. Values of β greater than $\sqrt{2}$ result in a bimodal distribution and so are unlikely to be applicable to process data. Figure 20.7 shows the effect of varying α with β fixed at 1. The distribution can be fitted to highly skewed data; positive values of α give negative skewness.

Because the logit function requires x to be in the range 0 to 1 the data need to be scaled accordingly before fitting. Scaling the C_4 in propane data, over the range 1 to 16 vol%, and fitting gives α as -1.26 and β as 0.852. *RSS* is 0.0607 – much better than the normal distribution but not quite as good as the lognormal distribution.

The parameters α and β are the mean and standard deviation of logit(x). Unfortunately there are no formulae to convert them to μ and σ for x. Nor are there formulae for skewness or kurtosis.

20.5 Folded Normal

If data follow the normal distribution then so will their deviation from any fixed value, e.g. $(x - \alpha)$. The absolute value of the deviation will then follow the *folded normal distribution*. Its PDF is

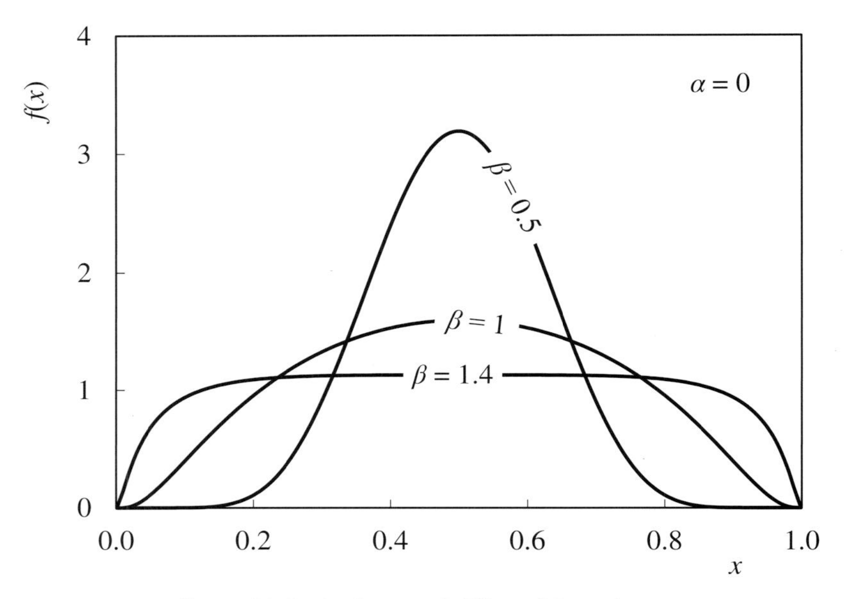

Figure 20.6 Logit-normal: Effect of β on shape

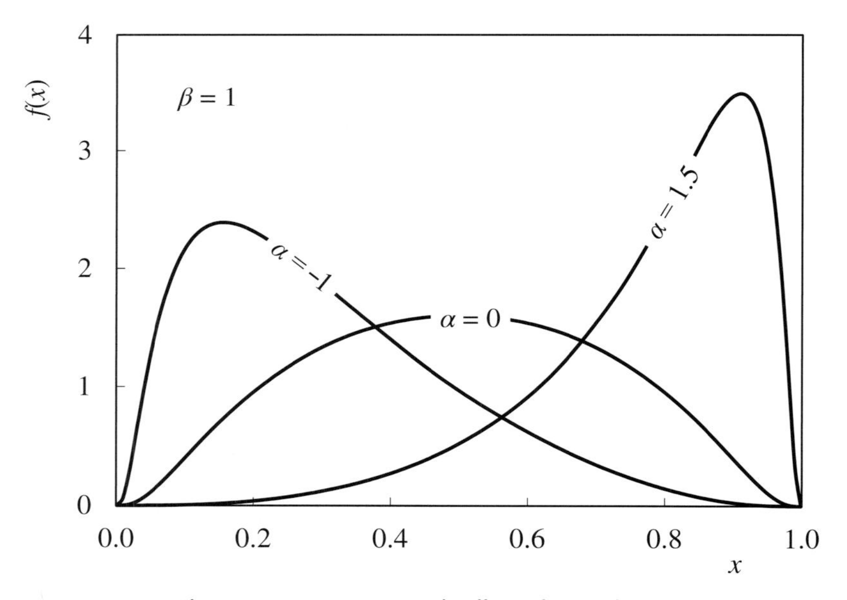

Figure 20.7 Logit-normal: Effect of α on shape

$$f(x) = \frac{1}{\beta\sqrt{2\pi}}\left\{\exp\left[-\frac{(x-\alpha)^2}{2\beta^2}\right] + \exp\left[-\frac{(x+\alpha)^2}{2\beta^2}\right]\right\} \quad \beta > 0 \tag{20.14}$$

Figure 20.8 shows the effect of varying α with β fixed at 1. If α exceeds β, the distribution is bimodal and therefore unlikely to be applicable to process data. The formulae for mean and standard deviation are too complex to justify inclusion here. However, if α is zero, the

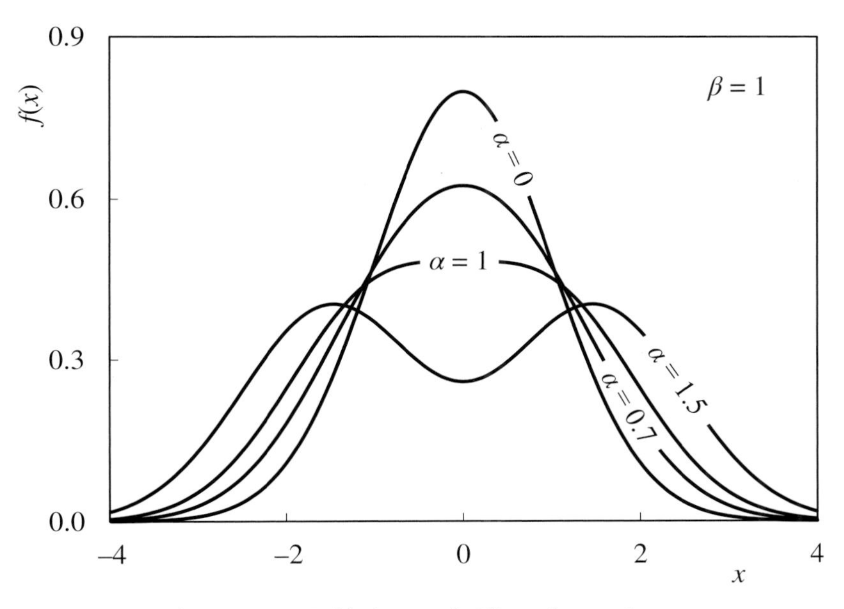

Figure 20.8 *Folded normal: Effect of α on shape*

distribution becomes the *half-normal distribution*, which does have applications in the process industry.

In general, the CDF of a half-distribution can be derived from that of the full distribution by

$$F_{\text{half}}(x) = 2F_{\text{full}}(x) - 1 \tag{20.15}$$

So, from Equation (5.43), for the half-normal distribution

$$F(x) = 2\Phi\left(\frac{x-\mu}{\sigma}\right) - 1 = \text{erf}\left(\frac{x-\mu}{\sigma\sqrt{2}}\right) \tag{20.16}$$

This is also a special case of the chi distribution that we will cover, with a worked example, in Section 30.1.

20.6 Lévy

The *Lévy distribution* can also be described as the *reciprocal Gaussian distribution*. Its PDF is

$$f(x) = \sqrt{\frac{\beta}{2\pi(x-\alpha)^3}} \exp\left[-\frac{\beta}{2(x-\alpha)}\right] \qquad x > \alpha;\ \beta > 0 \tag{20.17}$$

Figure 20.9 shows the effect of varying β with α fixed at zero. Its CDF is

$$F(x) = 1 - \text{erf}\left[\sqrt{\frac{\beta}{2(x-\alpha)}}\right] \tag{20.18}$$

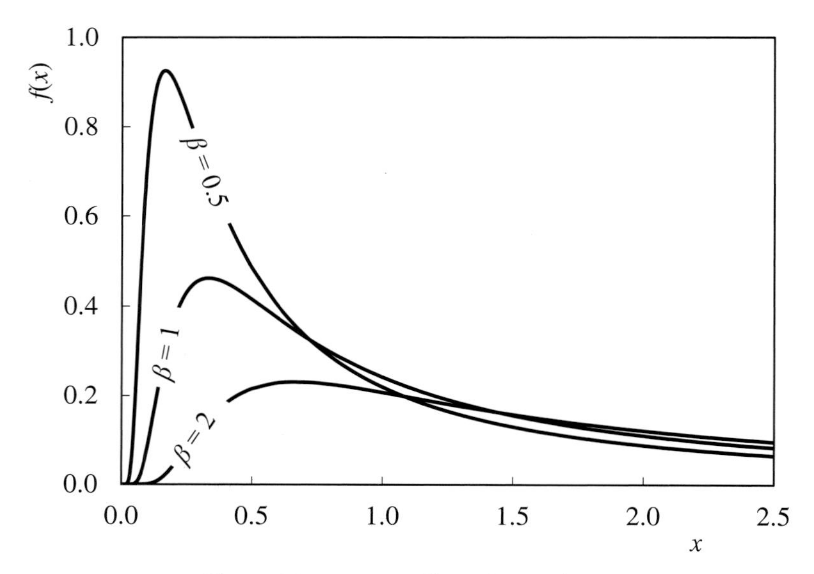

Figure 20.9 *Levy: Effect of β on shape*

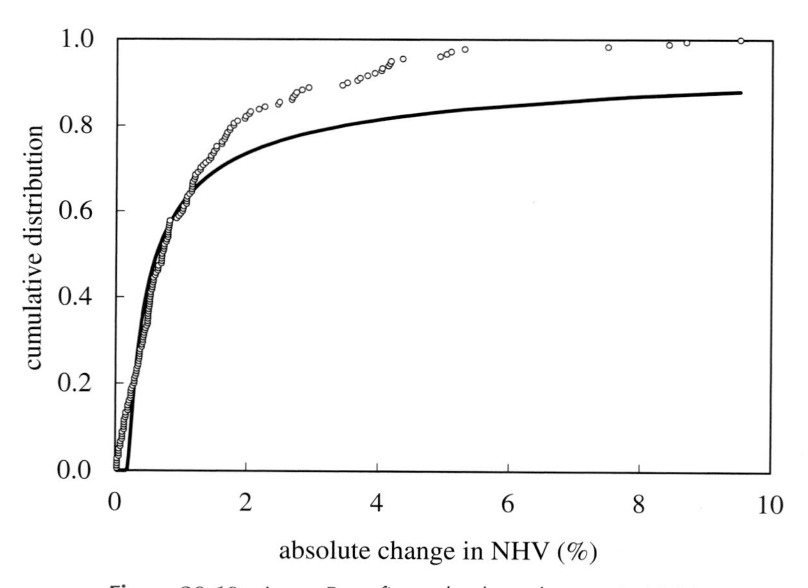

Figure 20.10 *Levy: Poor fit to absolute changes in NHV*

Fitting to the NHV disturbance data gives α as 0.160, β as 0.211 with *RSS* as a very poor 0.9225. This is confirmed by Figure 20.10 showing that the Lévy distribution is very highly skewed with a tail inherently much larger than that of the data.

The mean is

$$\mu = \alpha - \beta \tag{20.19}$$

There are no formulae for standard deviation, skewness or kurtosis.

20.7 Inverse Gaussian

The *inverse Gaussian distribution* is commonly described using the mean (μ) and a measure of dispersion (β).

$$f(x)=\sqrt{\frac{\beta}{2\pi x^3}}\exp\left(\frac{-\beta(x-\mu)^2}{2\mu^2 x}\right) \qquad x>0;\ \beta>0 \tag{20.20}$$

The variance is

$$\sigma^2=\frac{\mu^3}{\beta} \qquad \therefore \beta=\frac{\mu^3}{\sigma^2} \tag{20.21}$$

Substituting into Equation (20.20) avoids the use of β and instead uses σ directly.

$$f(x)=\sqrt{\frac{\mu^3}{2\pi\sigma^2 x^3}}\exp\left(\frac{-\mu(x-\mu)^2}{2\sigma^2 x}\right) \qquad x>0;\ \mu>0 \tag{20.22}$$

Skewness and kurtosis are

$$\gamma=\frac{3\sigma}{\mu} \tag{20.23}$$

$$\kappa=\frac{15\sigma^2}{\mu^2}+3 \tag{20.24}$$

If, in Equation (20.20) μ is set to 1, then it becomes the PDF of the *Wald distribution*.
Figure 20.11 shows the effect of changing σ, keeping μ constant at 5. Figures 20.12 and 20.13 show the effect that μ and σ have on skewness and kurtosis. Equations (20.23) and (20.24) show that, as σ approaches 0, the distribution approaches the normal distribution.

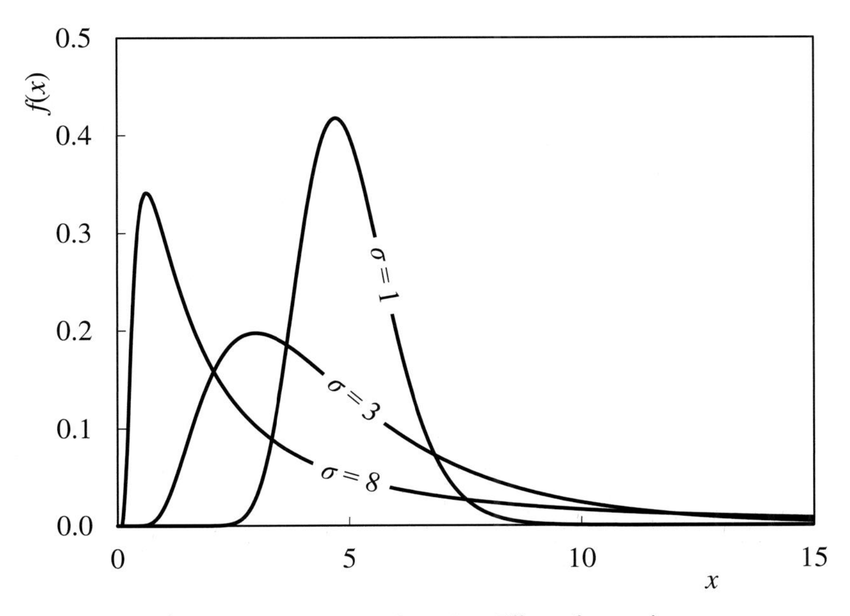

Figure 20.11 Inverse Gaussian: Effect of σ on shape

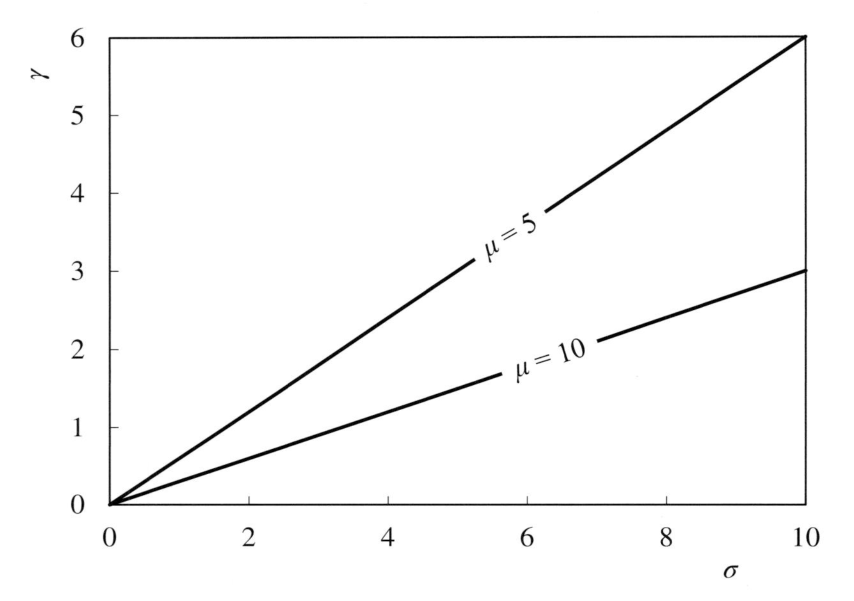

Figure 20.12 Inverse Gaussian: Effect of μ and σ on skewness

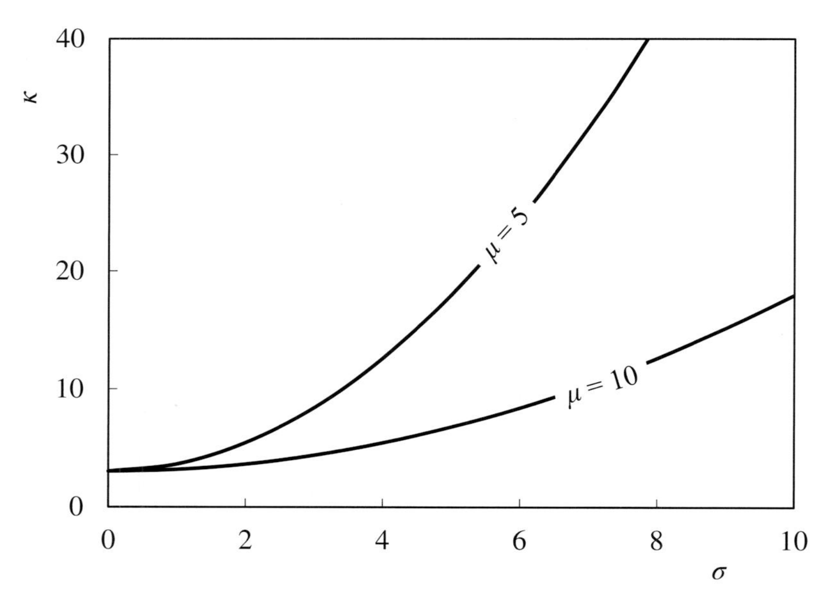

Figure 20.13 Inverse Gaussian: Effect of μ and σ on kurtosis

The CDF is

$$F(x)=\frac{1}{2}\left\{1+\text{erf}\left[\sqrt{\frac{\mu}{2\sigma^2 x}}(x-\mu)\right]\right\}+\frac{1}{2}\exp\left(\frac{2\mu^2}{\sigma^2}\right)\left\{1-\text{erf}\left[\sqrt{\frac{\mu}{2\sigma^2 x}}(x+\mu)\right]\right\} \tag{20.25}$$

Using the C_4 in propane data, minimising *RSS* to 0.0402 is achieved by adjusting μ to 4.85 and σ to 2.48. While not as good a fit as the (less practical) power lognormal it is marginally better

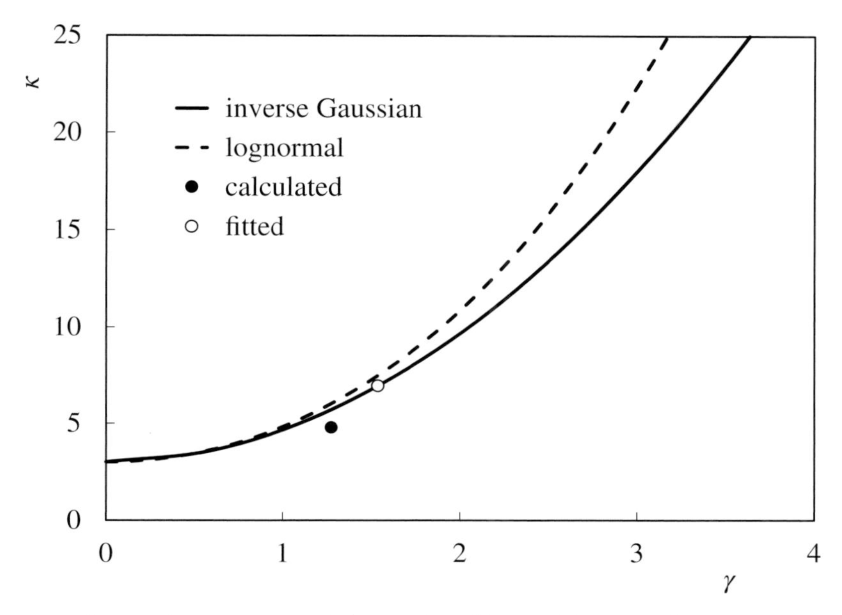

Figure 20.14 *Inverse Gaussian: Feasible combinations of γ and κ*

than that achieved with the lognormal distribution. Figure 20.14 shows, as the solid line, the feasible combinations of skewness and kurtosis. The calculated values lie close to the line and so fitting works well. Shown for comparison, as the dashed line, is the relationship for the lognormal distribution. The line is further from the calculated skewness and kurtosis – explaining why the inverse Gaussian distribution, in this case, would be the better choice.

The standard deviation is over 25% larger than that estimated using the normal distribution. If this was being used as the basis of a calculation to determine the benefits of an improved control strategy, the conventional approach would have substantially underestimated the benefits. This does not imply that the use of the normal distribution always gives a conservative estimate. If the data had been skewed in the opposite direction the assumption that the data are normally distributed would have overstated the benefits.

As we will see, of the distributions described in this chapter, the inverse Gaussian distribution provides the best fit to the C_4 in propane data. This is illustrated by Figure 20.15 which compares it (the solid line) to that of the normal distribution (dashed line) and by the P–P plot in Figure 20.16.

Figure 20.17 shows the derived PDF. To show the actual distribution, the original data has been split into 32 bands each 0.5% wide. The scatter around the derived PDF is not a reflection on its accuracy but is mainly caused by splitting the data into bands. It can only be reduced by increasing the number of measurements in each band – either by having fewer wider bands or by increasing the total number of measurements. Nevertheless it shows clearly that the skewness has been well represented.

It can be a requirement, for example for use in Monte Carlo simulation, to generate data that has an asymmetric distribution. A feature of the inverse Gaussian distribution is that this is straightforward. If the number to be generated is y we first require a number (x) selected at random from a normal distribution that has a mean of 0 and a standard deviation of 1. We described, in Chapter 5, how this can be done.

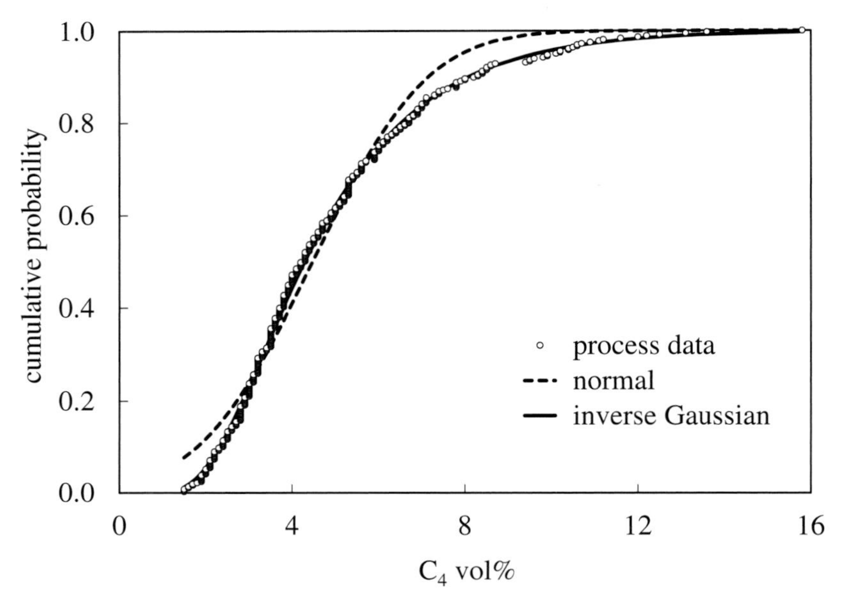

Figure 20.15 *Inverse Gaussian: Fitted to the C_4 in propane data*

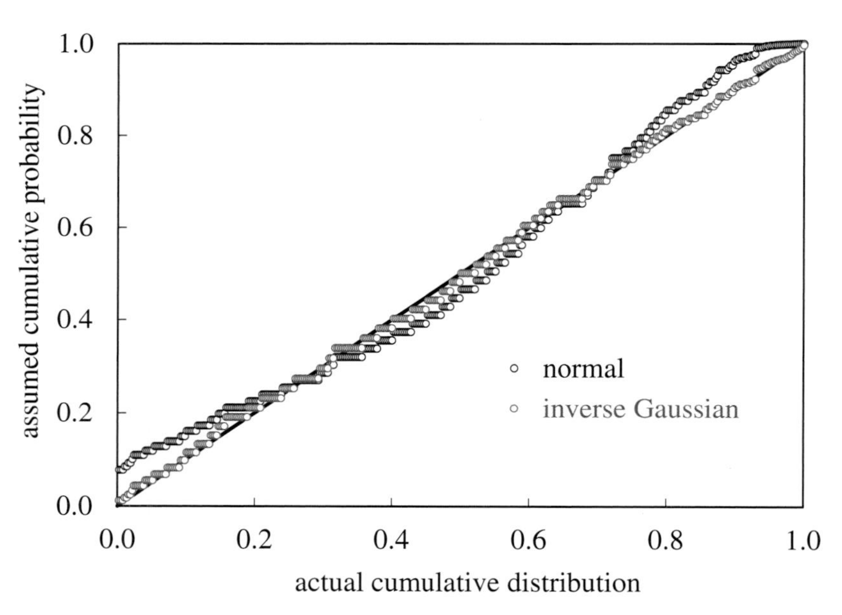

Figure 20.16 *Inverse Gaussian: P–P plot showing improved fit to C_4 data*

$$y = \mu + \frac{\sigma^2}{2\mu}\left[x^2 - \sqrt{\frac{4\mu^2 x^2}{\sigma^2} + x^4}\right] \tag{20.26}$$

However, before using this value, we next choose another number between 0 and 1, selected at random from a uniform distribution. If this number is greater than $\mu/(\mu + y)$ then we replace y with μ^2/y.

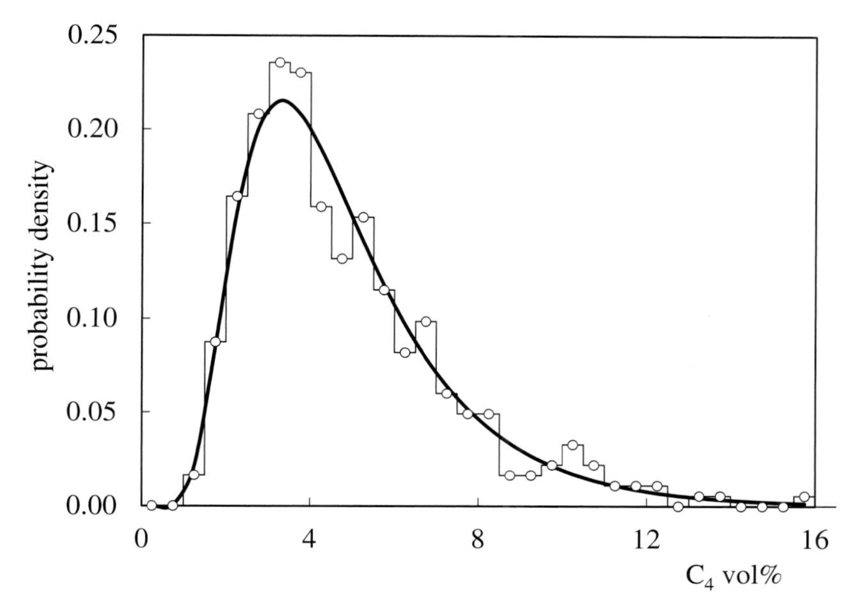

Figure 20.17 *Inverse Gaussian: Fit to histogram of C_4 in propane data*

The inclusion of a location parameter (α) in Equation (20.22) gives the *three-parameter inverse Gaussian distribution*.

$$f(x) = \sqrt{\frac{\mu^3}{2\pi\sigma^2(x-\alpha)^3}} \exp\left(\frac{-\mu(x-\alpha-\mu)^2}{2\sigma^2(x-\alpha)}\right) \quad x > \alpha;\ \mu, \sigma > 0 \tag{20.27}$$

Fitting to the C_4 in propane data gives α as 0.468 and μ as 4.44; adding these gives the mean $(\mu + \alpha)$ slightly increased at 4.90. Standard deviation (σ) is slightly increased to 2.59. As might be expected, the inclusion of another fitting parameter reduces *RSS* – to 0.0305.

20.8 Generalised Inverse Gaussian

The *generalised inverse Gaussian* (*GIG*) *distribution*[9] is also known as the *Sichel distribution*. Its PDF is

$$f(x) = \left(\frac{\alpha}{\beta}\right)^{\frac{\delta}{2}} \frac{x^{\delta-1}}{2K_\delta(\sqrt{\alpha\beta})} \exp\left(-\frac{\alpha x^2 + \beta}{2x}\right) \quad x > 0;\ \alpha, \beta > 0 \tag{20.28}$$

K_δ is the modified Bessel function of the second kind, as described in Section 11.8.

Figure 20.18 shows the effect of varying β and δ, with α fixed at 2. Raw moments are given by

$$m_n = \left(\frac{\beta}{\alpha}\right)^{\frac{n}{2}} \frac{K_{\delta+n}(\sqrt{\alpha\beta})}{K_\delta(\sqrt{\alpha\beta})} \tag{20.29}$$

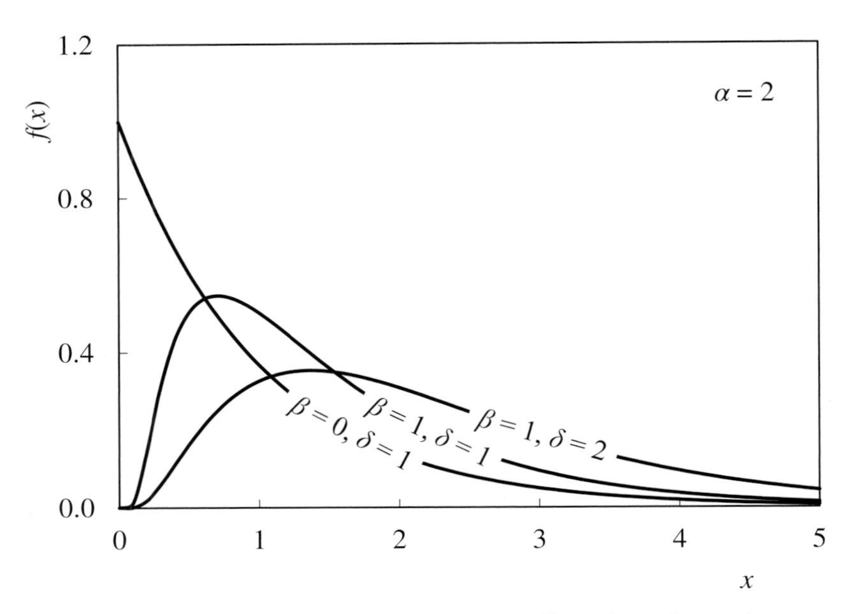

Figure 20.18 Generalised inverse Gaussian: Effect of β and δ on shape

Hence

$$\mu = \sqrt{\frac{\beta}{\alpha}} \frac{K_{\delta+1}(\sqrt{\alpha\beta})}{K_{\delta}(\sqrt{\alpha\beta})} \tag{20.30}$$

$$\sigma^2 = \frac{\beta}{\alpha}\left[\frac{K_{\delta+2}(\sqrt{\alpha\beta})}{K_{\delta}(\sqrt{\alpha\beta})} - \left(\frac{K_{\delta+1}(\sqrt{\alpha\beta})}{K_{\delta}(\sqrt{\alpha\beta})}\right)^2\right] \tag{20.31}$$

Most spreadsheet packages do not support non-integer values of δ in Bessel functions. This distribution therefore can only be properly fitted using custom software.

20.9 Normal Inverse Gaussian

The *normal inverse Gaussian distribution* has the PDF

$$f(x) = \frac{\lambda\beta K_1\left(\lambda\sqrt{\beta^2 + (x-\alpha)^2}\right)}{\pi\sqrt{\beta^2 + (x-\alpha)^2}} \exp\left(\beta\sqrt{\lambda^2 - \delta^2} + \delta(x-\alpha)\right) \quad -1 \le \delta \le 1;\ \beta, \lambda > 0 \tag{20.32}$$

K_1 is the first order modified Bessel function of the second kind, as described in Section 11.8. The effect of the *asymmetry parameter* (δ) is shown in Figure 20.19. Negative values give negative skewness; positive values give positive skewness. The effect of the *tail heaviness parameter* (λ) is shown in Figure 20.20. Increasing λ increases kurtosis. The effect of the dispersion parameter (β) is shown in Figure 20.21. As usual, variation in dispersion is difficult to distinguish from variation in kurtosis. This can cause problems when fitting the distribution to data; there may be many combinations of λ and β that give a very similar fit.

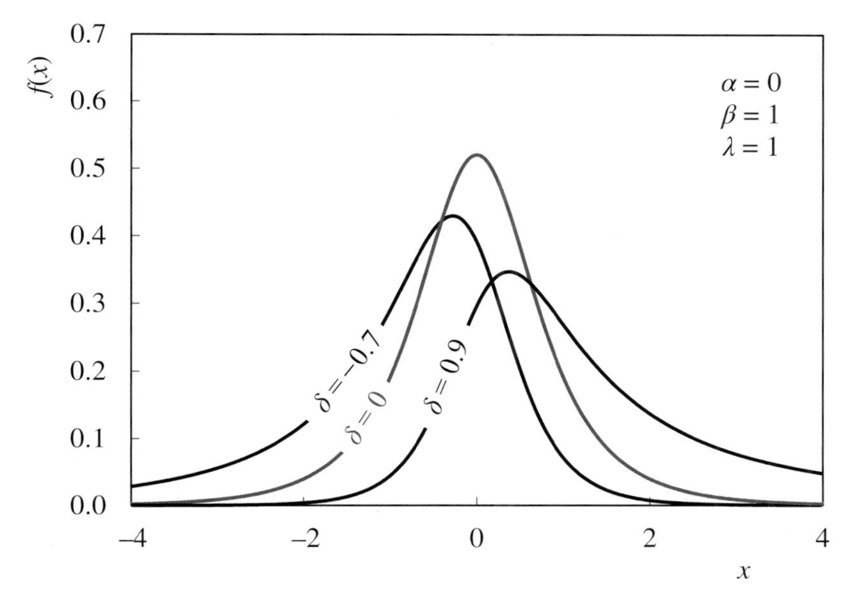

Figure 20.19 *Normal inverse Gaussian: Effect of δ on shape*

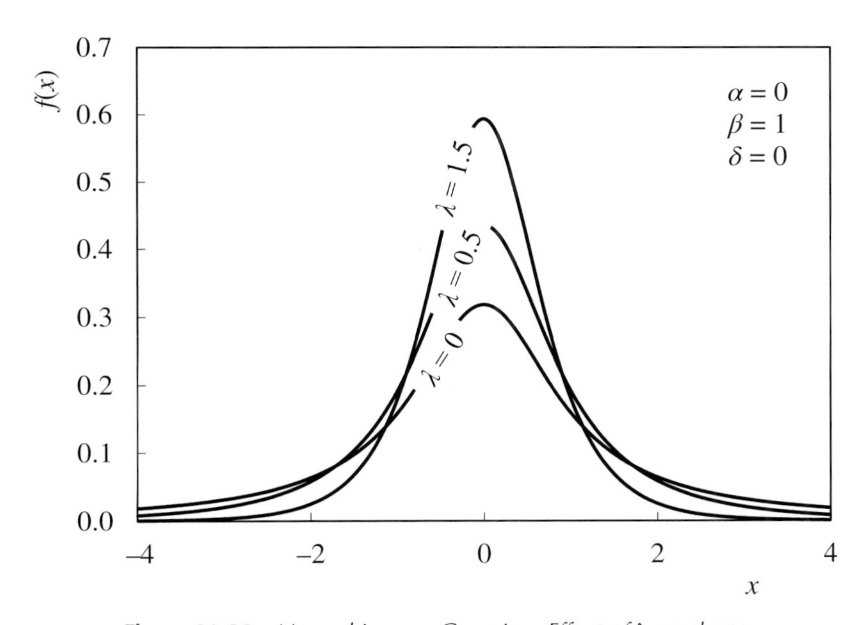

Figure 20.20 *Normal inverse Gaussian: Effect of λ on shape*

Fitting to the C_4 in propane data gives α as 2.27, β as 1.88, λ as 1.22, δ as 1.00 and $F(x_1)$ as 0.0223. With δ at its upper limit, the skewness of the data cannot be fully represented. With *RSS* at 0.0442 the fit is much closer than that of the normal distribution but bettered by other distributions in this chapter.

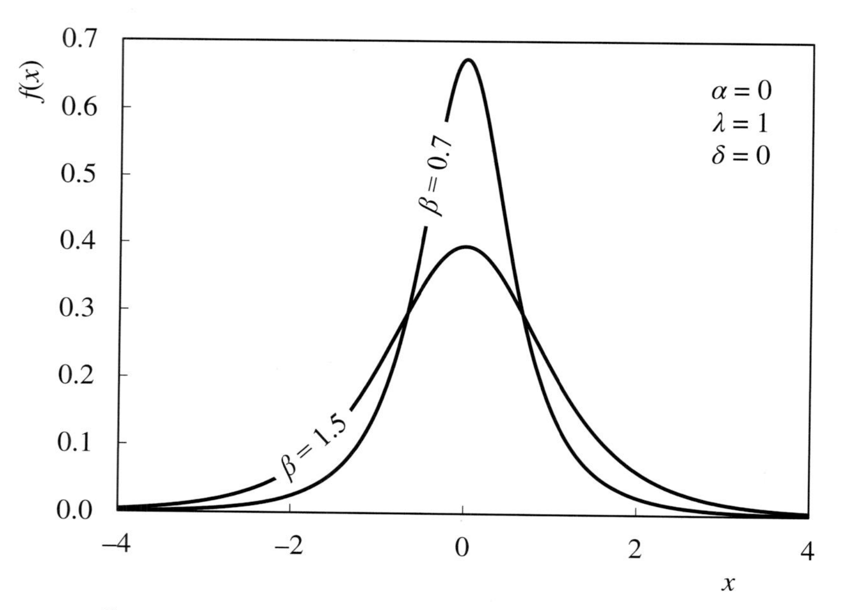

Figure 20.21 *Normal inverse Gaussian: Effect of β on shape*

Mean and variance are

$$\mu = \alpha + \frac{\delta\beta}{\sqrt{\lambda^2 - \delta^2}} \tag{20.33}$$

$$\sigma^2 = \frac{\lambda^2\beta}{\left(\lambda^2 - \delta^2\right)^{3/2}} \tag{20.34}$$

These give μ as 4.94 and σ as 2.83.

The formulae for skewness and kurtosis are unusual in that they include the dispersion parameter (β).

$$\gamma = \frac{3\delta}{\lambda\sqrt{\beta}\left(\lambda^2 - \delta^2\right)^{1/4}} \tag{20.35}$$

$$\kappa = \frac{12\delta^2 + 3\lambda^2\left(\beta\sqrt{\lambda^2 - \delta^2} + 1\right)}{\beta\lambda^2\sqrt{\lambda^2 - \delta^2}} \tag{20.36}$$

These give γ as 2.13 and κ as 9.24. Both values are considerably larger than those calculated from the data and from those determined from other well-fitting distributions. This, and the unusual formulation, place some suspicion on the reliability of this distribution.

20.10 Reciprocal Inverse Gaussian

The *reciprocal inverse Gaussian distribution* has the PDF

$$f(x) = \sqrt{\frac{\lambda}{2\pi x}} \exp\left(\frac{-\lambda(1 - \alpha x)^2}{2\alpha^2 x}\right) \quad x > 0;\ \alpha, \lambda > 0 \tag{20.37}$$

Figure 20.22 shows the effect of varying α and λ. Fitting the distribution to the C_4 in propane data gives α as 0.269 and λ as 1.03. *RSS* is 0.0387. As illustrated by Figure 20.23, while the fit is numerically better than that achieved with the inverse Gaussian distribution, the upper tail is not so well matched. This would be important if we were primarily interested in assessing the probability of such extreme behaviour.

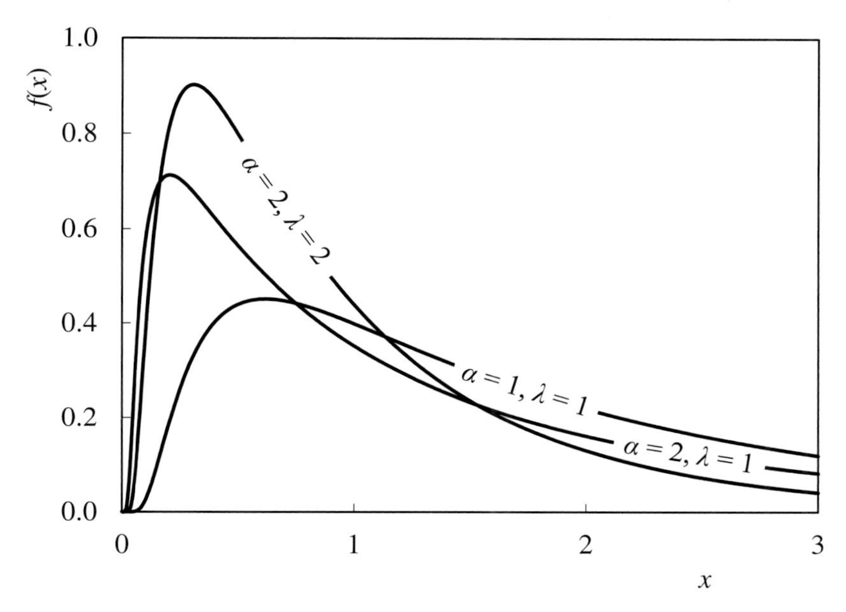

Figure 20.22 Reciprocal inverse Gaussian: Effect of α and λ on shape

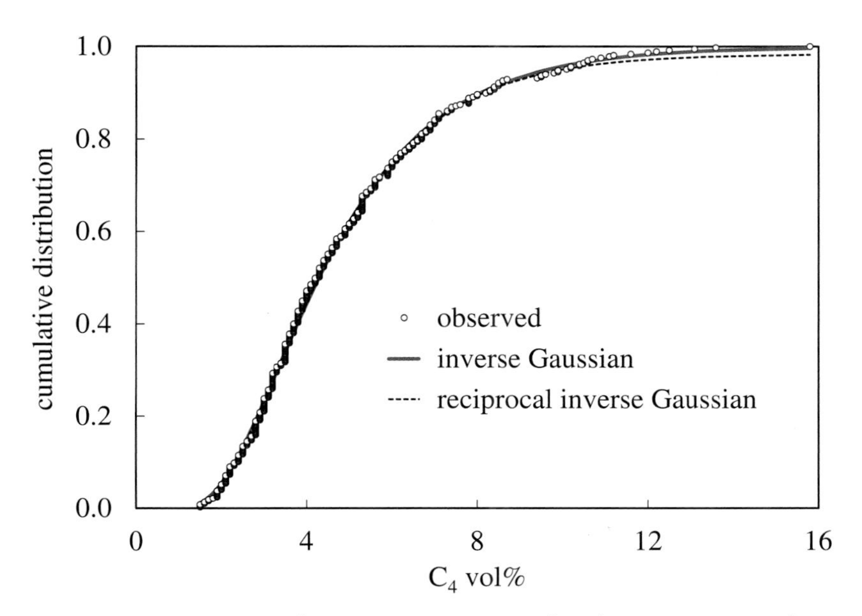

Figure 20.23 Reciprocal inverse Gaussian: Fitted to the C_4 in propane data

The mean and variance are

$$\mu = \frac{\lambda + \alpha}{\lambda \alpha} \tag{20.38}$$

$$\sigma^2 = \frac{\lambda + 2\alpha}{\lambda^2 \alpha} \tag{20.39}$$

These give μ as 4.69 and σ as 2.35 – close to the values calculated from the data and generated by other well-fitting distributions.

Skewness and kurtosis are

$$\gamma = \frac{\frac{3\alpha}{\lambda} + 8}{\left(\frac{\alpha}{\lambda} + 2\right)^{\frac{3}{2}}} \tag{20.40}$$

$$\kappa = \frac{3\left(\frac{\alpha^2}{\lambda^2} + 9\frac{\alpha}{\lambda} + 12\right)}{\left(\frac{\alpha}{\lambda} + 2\right)^2} \tag{20.41}$$

20.11 Q-Gaussian

The *q-Gaussian distribution* is a member of the family of *Tsallis distributions*.[10] Like the normal distribution it is symmetrical about the mean and hence its skewness is zero. It can, however, represent highly leptokurtic data. Its PDF depends on the choice of the parameter q.

For $q < 1$

$$f(x) = \sqrt{\frac{1-q}{2\pi}} \frac{\Gamma\left(\frac{5-3q}{2(1-q)}\right)}{\beta\Gamma\left(\frac{2-q}{1-q}\right)} \left[1 + (q-1)\frac{(x-\mu)^2}{2\beta^2}\right]^{\frac{1}{1-q}} \quad \beta > 0 \tag{20.42}$$

The distribution is then bounded by

$$\mu - \beta\sqrt{\frac{2}{1-q}} \leq x \leq \mu + \beta\sqrt{\frac{2}{1-q}} \tag{20.43}$$

As $q \to 1$ the q-Gaussian distribution approaches the normal distribution

$$f(x) \to \frac{1}{\beta\sqrt{2\pi}} \exp\left[-\frac{(x-\mu)^2}{2\beta^2}\right] \tag{20.44}$$

For $1 < q < 3$ the distribution is unbounded. The PDF becomes

$$f(x) = \sqrt{\frac{q-1}{2\pi}} \frac{\Gamma\left(\frac{1}{q-1}\right)}{\beta\Gamma\left(\frac{3-q}{2(q-1)}\right)} \left[1 + (q-1)\frac{(x-\mu)^2}{2\beta^2}\right]^{\frac{1}{1-q}} \quad \beta > 0 \tag{20.45}$$

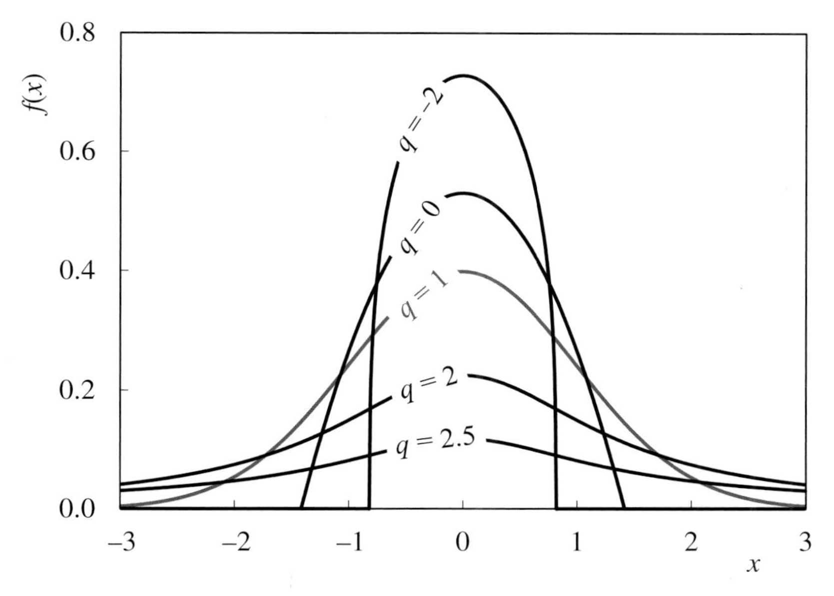

Figure 20.24 *Q-Gaussian: Effect of* q *on shape*

Figure 20.24 shows the effect of changing q, with the mean (μ) is fixed at 0 and β fixed at 1. It would appear that increasing q above 1 makes the curve flatter and therefore reduces kurtosis. However this is not the case because q also changes the variance. This is given by

$$\sigma^2 = \frac{2\beta^2}{5-3q} \qquad q < \frac{5}{3} \tag{20.46}$$

The coloured curve in Figure 20.25 is the normal distribution ($q = 1$) with β set to 1. Equation (20.46) confirms that σ is 1. Increasing q to 1.5 requires that β be reduced to 0.5 to maintain σ at 1. The black curve in Figure 20.25 shows the resulting increase in kurtosis. Kurtosis is

$$\kappa = \frac{15-9q}{7-5q} \qquad q < \frac{7}{5} \tag{20.47}$$

As Figure 20.26 shows, values of q just less than 1.4 give a highly leptokurtic distribution. Values greater than 1.4 appear to give a highly platykurtic distribution – although, strictly, Equation (20.47) should not be used in this region.

Fitting the PDF to the C_4 in propane data gives values for q, μ, β and $F(x_1)$ of 1.14, 3.95, 1.82 and 0.0264 respectively. This gives a value of *RSS* of 0.2269, representing only a marginal improvement over the classic normal distribution. The q-Gaussian distribution has a skewness of zero and cannot therefore match well the highly skewed C_4 data.

Fitting the PDF to the NHV disturbance data (with q restricted to the maximum of 1.4) sets β to 0.198 – reducing *RSS* to 0.0898. Relaxing the restriction gives a value for 1.81 for q and 0.641 for β – reducing *RSS* to 0.0364. Figure 20.27 shows the fit is considerably better than the normal distribution. Similarly the P–P plot in Figure 20.28 shows that virtually every point on the q-Gaussian distribution is closer to the ideal line than those points on the normal distribution. Since q exceeds 1.67, we cannot determine σ (or κ). However, we could sacrifice the ability to calculate σ in order to achieve the significantly more accurate fit. We can treat

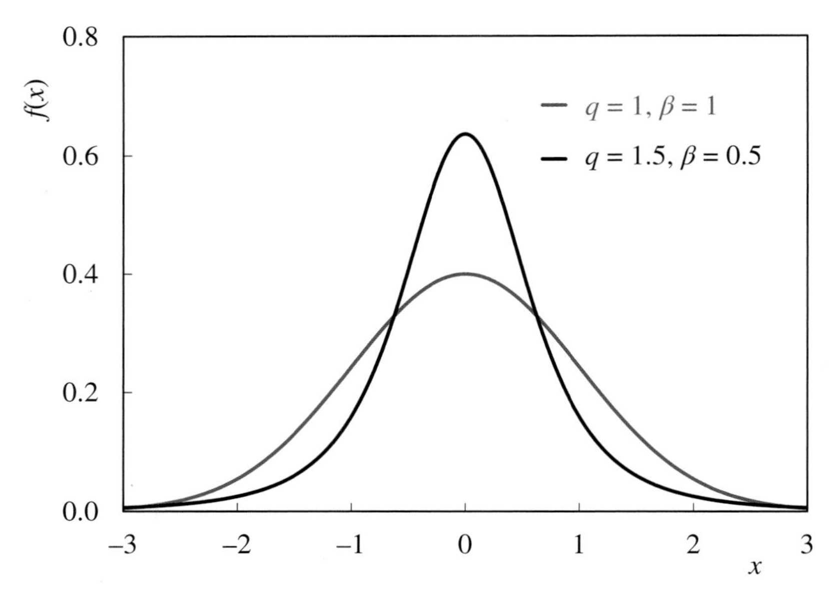

Figure 20.25 *Q-Gaussian: Effect of q on kurtosis, keeping σ fixed*

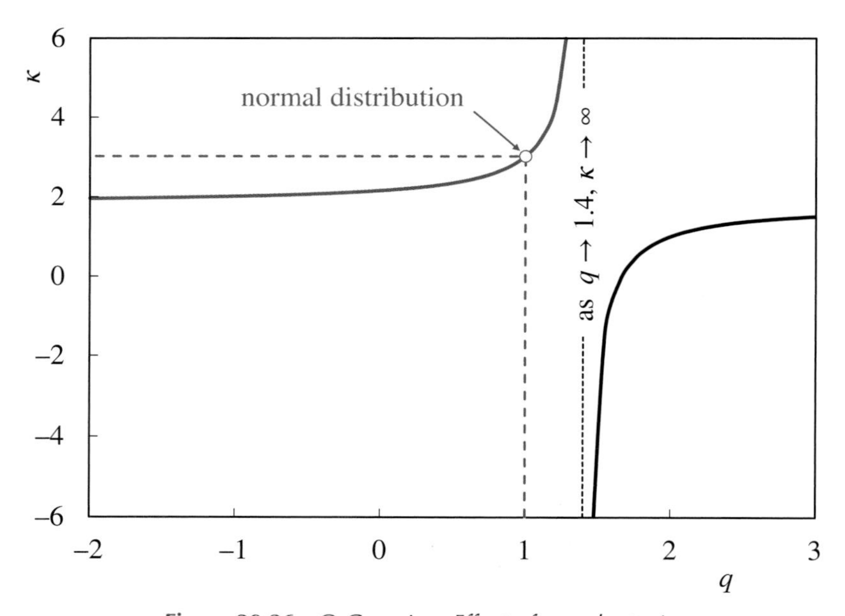

Figure 20.26 *Q-Gaussian: Effect of q on kurtosis*

β as if it is the standard deviation to explore, for example, the impact of halving the variation in NHV.

Rearranging Equation (20.47)

$$q = \frac{7\kappa - 15}{5\kappa - 9} \tag{20.48}$$

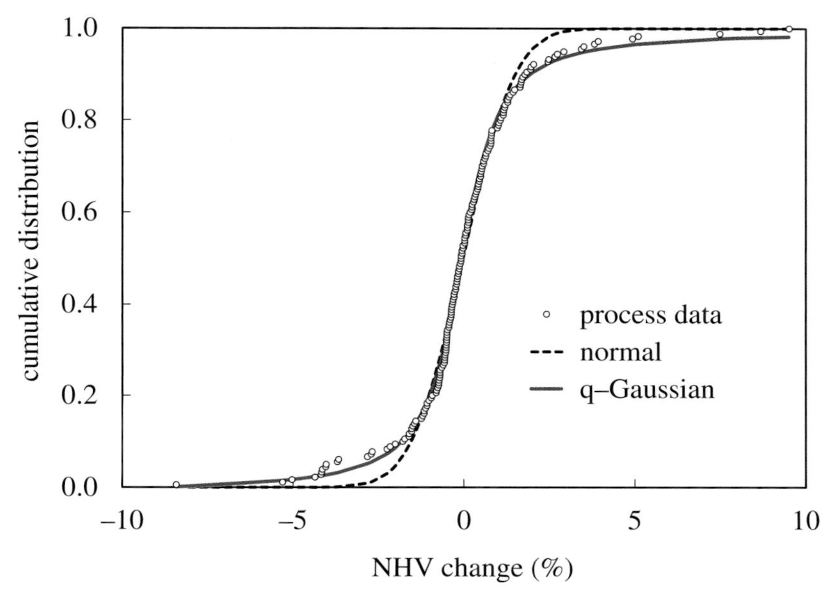

Figure 20.27 Q-Gaussian: Fitted to the NHV disturbance data

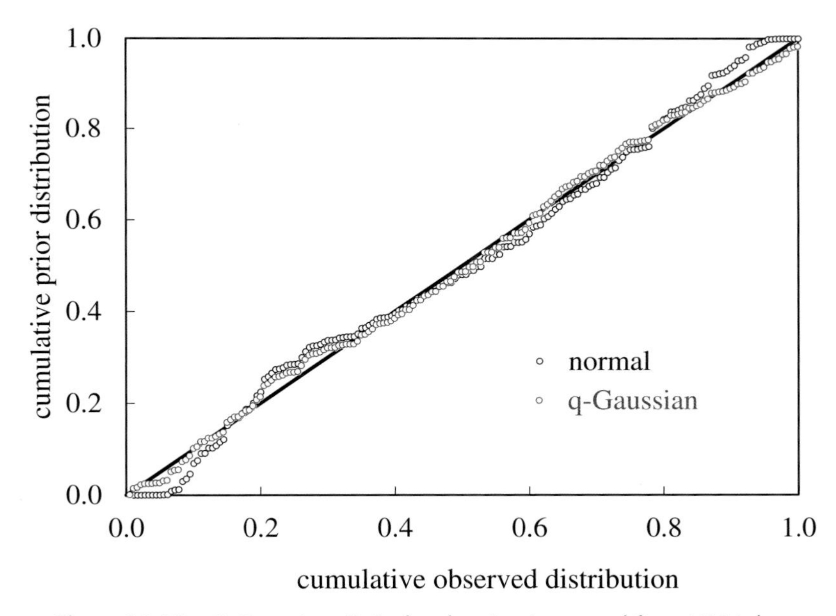

Figure 20.28 Q-Gaussian: P–P plot showing improved fit to NHV data

Rearranging Equation (20.46) and substituting Equation (20.48) gives

$$\beta = \sigma\sqrt{\frac{5-3q}{2}} = \sigma\sqrt{\frac{2\kappa}{5\kappa-9}} \tag{20.49}$$

This would suggest, since κ is calculated from the data as 8.87 and σ as 1.19, that q should be 1.33 and β should be 0.843. These maximum likelihood estimates can be used as initial values for improvement by fitting. As Figure 20.26 shows a small adjustment in q, when close to 1.4, can result in very large changes in kurtosis. This can give problems with the fitting algorithm.

20.12 Generalised Normal

The title *generalised normal distribution* or *generalised Gaussian distribution* is applied to at least three forms of the distribution. While each is developed from the normal distribution, by the addition of shape parameters, the way in which this is done is quite different.

The first version can also be entitled the *exponential power distribution* although, as we will see later, is also the name of a very different distribution. It is better known as the *generalised error distribution (GED)*. Its PDF is

$$f(x) = \frac{\delta}{2\beta\Gamma\left(\frac{1}{\delta}\right)}\exp\left[-\left(\frac{|x-\mu|}{\beta}\right)^{\delta}\right] \qquad \delta,\beta>0 \tag{20.50}$$

The term μ is the mean and β is directly proportional to the standard deviation (σ).

$$\beta = \sigma\sqrt{\frac{\Gamma\left(\frac{1}{\delta}\right)}{\Gamma\left(\frac{3}{\delta}\right)}} \quad \text{or} \quad \sigma^2 = \frac{\Gamma\left(\frac{3}{\delta}\right)}{\Gamma\left(\frac{1}{\delta}\right)}\beta^2 \tag{20.51}$$

The term δ determines the type of distribution. For example, if δ is set to 2, then Equation (20.50) reduces to Equation (5.35) and so becomes the normal distribution. If δ is set to 1, then Equation (20.50) becomes the Laplace distribution that we will cover in Section 32.7. However δ does not have to be an integer and so can be fitted along with μ and σ. As $\delta \to \infty$, the distribution becomes $U(\mu-\beta, \mu+\beta)$. Figure 20.29 shows the effect of changing δ – keeping μ fixed at 0 and σ at 1.

The skewness of the GED is zero. Kurtosis is

$$\kappa = \frac{\Gamma\left(\frac{5}{\delta}\right)\Gamma\left(\frac{1}{\delta}\right)}{\Gamma^2\left(\frac{3}{\delta}\right)} \tag{20.52}$$

While a very large kurtosis can be represented, as Figure 20.30 shows, the smallest possible is 1.8. This value is approached as δ approaches ∞.

The dashed line in Figure 20.31 is the result of fixing δ at 2, i.e. the normal distribution, with the resulting standard deviation of 1.19.The solid line is the result of fitting Equation (20.50) to the NHV disturbance data, resulting with a value for δ of 0.83, a mean of –0.03 and a standard deviation of 1.76. As Figure 20.32 shows, the fit is slightly better than that achieved with the Laplace distribution. The minimum *RSS* is 0.0553. Changing the prior distribution from normal to GED, in this case, reduces *RSS* by over 60%. Figure 20.33 shows the impact on the estimate of standard deviation; the more reliable estimate is about 50% higher.

The second version includes a parameter that adds skewness to the distribution. Its PDF is

$$f(x) = \frac{\frac{1}{\sqrt{2\pi}}\exp\left(\frac{-z^2}{2}\right)}{\beta-\delta(x-m)} \qquad x<m+\frac{\beta}{\delta} \tag{20.53}$$

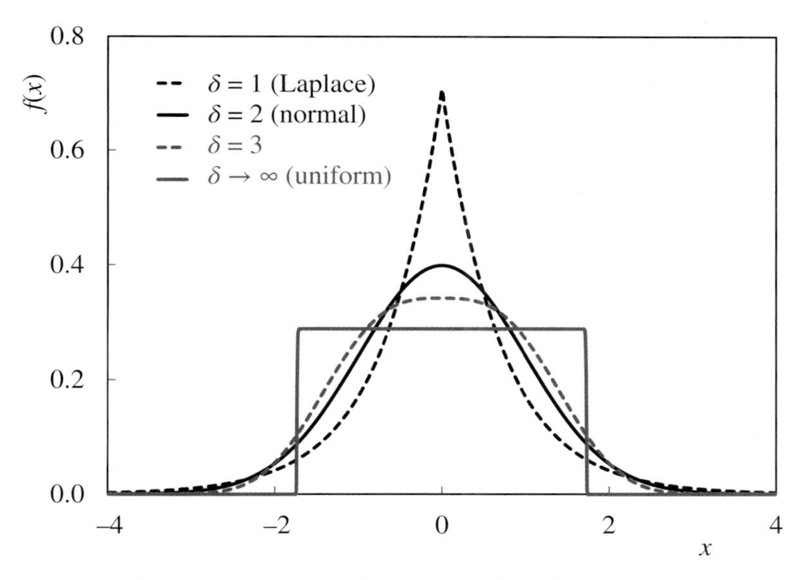

Figure 20.29 Generalised error: Effect of δ on shape

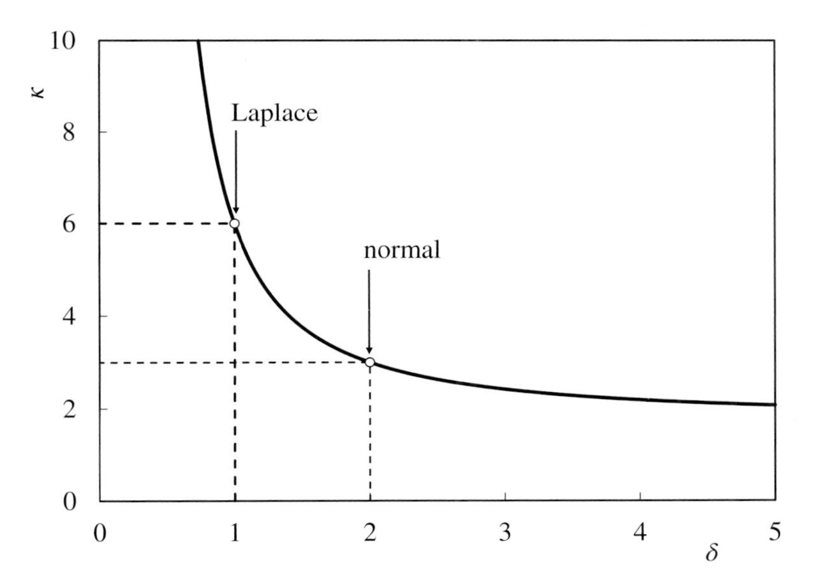

Figure 20.30 Generalised error: Effect of δ on kurtosis

$$z=-\frac{1}{\delta}\log\left[1-\frac{\delta(x-m)}{\beta}\right] \quad \text{for} \quad \delta\neq 0 \tag{20.54}$$

$$z=\frac{x-m}{\beta} \quad \text{for} \quad \delta=0 \tag{20.55}$$

This PDF is unusual in that it uses the median (m) rather than the mean to locate the distribution. The scale factor (β) has the same effect as changing the standard deviation. The additional

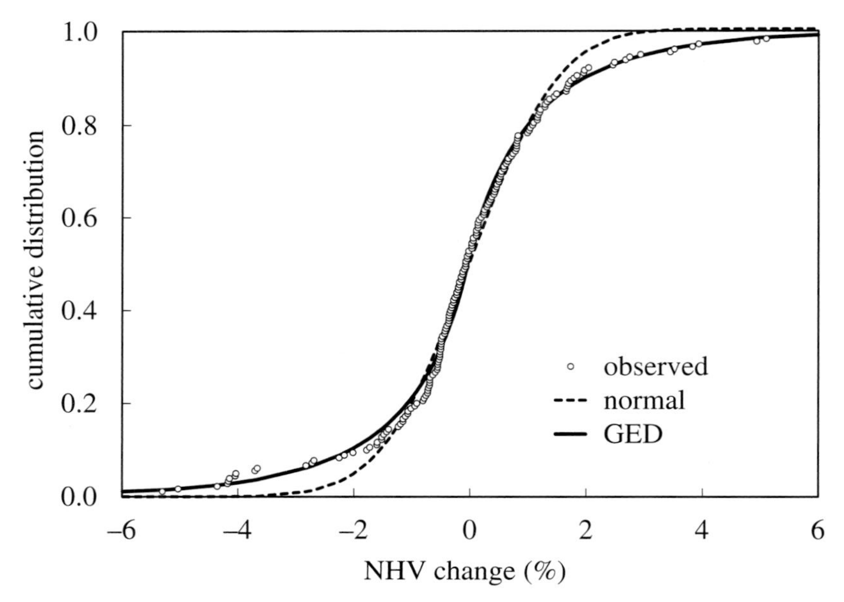

Figure 20.31 Generalised error: Fitted to the NHV disturbance data

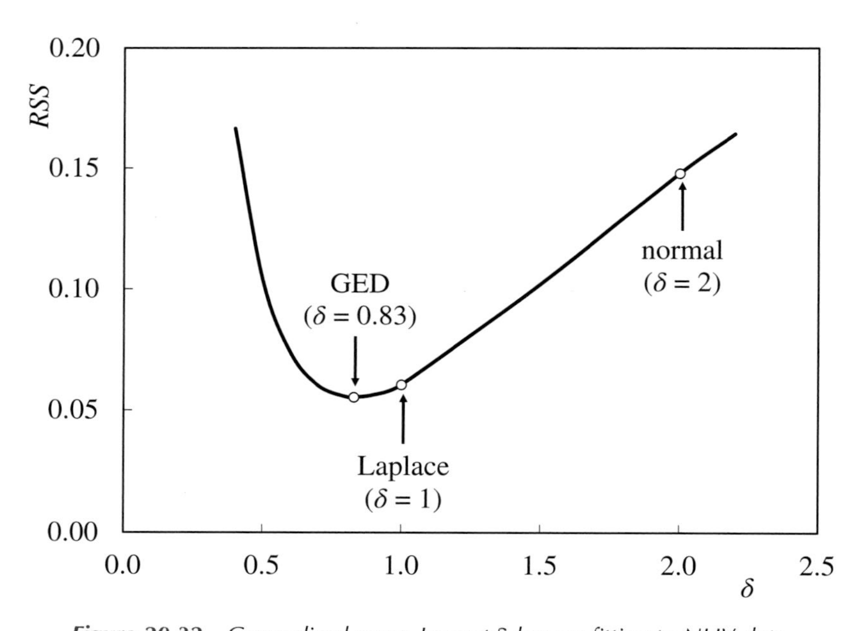

Figure 20.32 Generalised error: Impact δ has on fitting to NHV data

shape factor (δ) determines skewness. Figure 20.34 shows the effect of changing β from 1 (coloured curve) with m and k both fixed at 0. Figure 20.35 shows the effect of changing δ.

The CDF is

$$F(x) = \frac{1}{2}\left[1 + \operatorname{erf}\left(\frac{z}{\sqrt{2}}\right)\right] \tag{20.56}$$

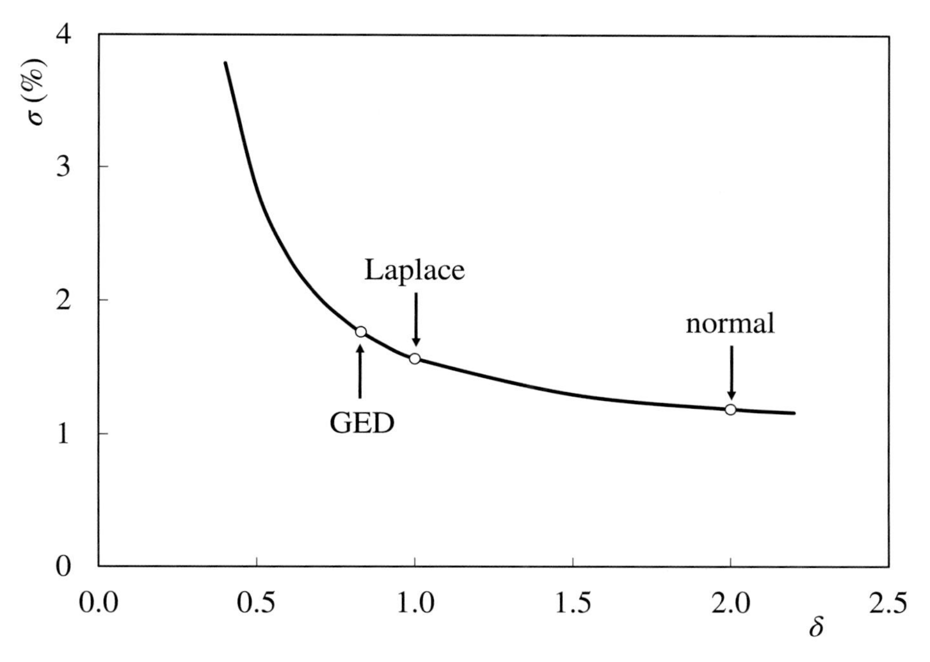

Figure 20.33 *Generalised error: Impact δ has on estimating σ*

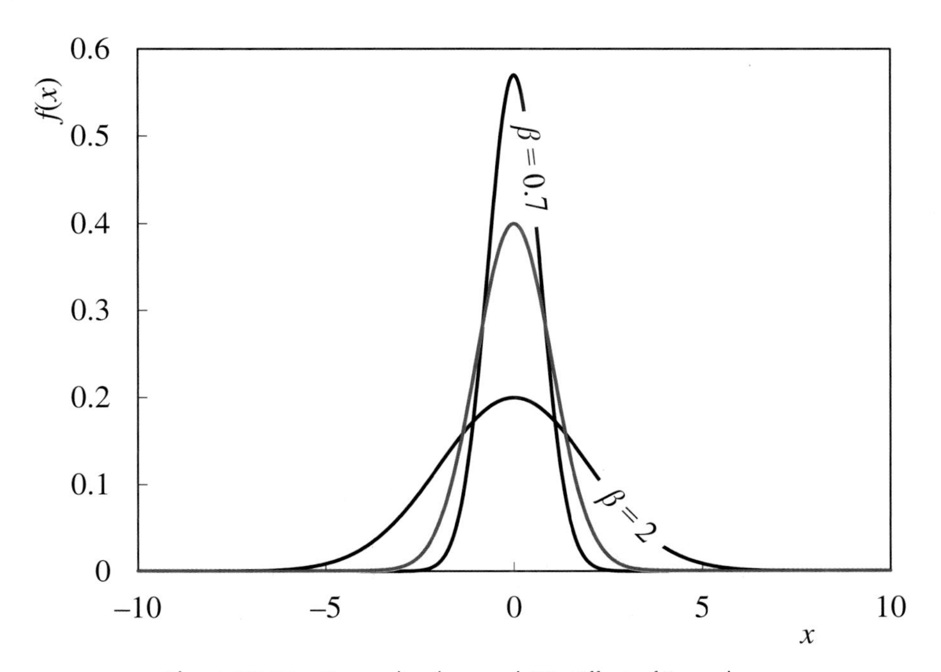

Figure 20.34 *Generalised normal (2): Effect of β on shape*

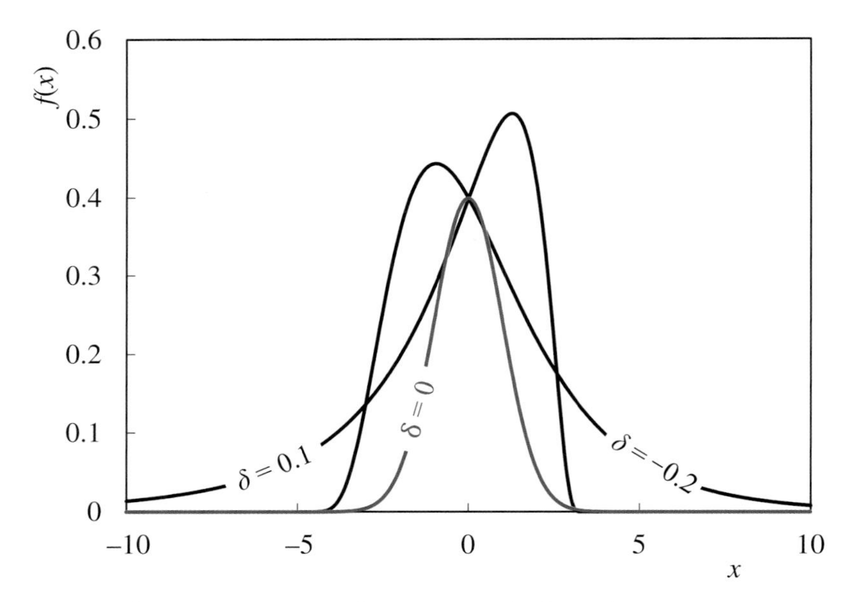

Figure 20.35 Generalised normal (2): Effect of δ on shape

Key parameters are

$$\mu = m - \frac{\beta}{\delta}\left[\exp\left(\frac{\delta^2}{2}\right) - 1\right] \tag{20.57}$$

$$\sigma^2 = \left(\frac{\beta}{\delta}\right)^2 \exp(\delta^2)\left[\exp(\delta^2) - 1\right] \tag{20.58}$$

$$\gamma = \frac{3\exp(\delta^2) - \exp(3\delta^2) - 2}{\left[\exp(\delta^2) - 1\right]^{1.5}} \times \frac{\delta}{|\delta|} \tag{20.59}$$

$$\kappa = \exp(4\delta^2) + 2\exp(3\delta^2) + 3\exp(2\delta^2) - 3 \tag{20.60}$$

Fitting Equation (20.56) to the NHV disturbance data gives values of −0.04 for m, 0.51 for β and −0.063 for δ. Equation (20.57) then gives μ as −0.02 and Equation (20.58) gives σ as 0.51. *RSS* is minimised to 0.1309 – not significantly better than that achieved by the benchmark normal distribution.

Figure 20.36 plots Equations (20.59) and (20.60). While changing δ varies skewness as expected, it also changes kurtosis. Fitting the distribution is likely therefore to be a compromise in matching both these to the actual distribution. This is shown more clearly by Figure 20.37 which shows what combinations of skewness and kurtosis are supported by the distribution. As expected from the poor fit, the calculated values lie well away from the line.

The PDF of the third version includes two additional shape parameters.

$$f(x) = \frac{\delta_1 x^{\delta_1\delta_2 - 1}}{\beta^{\delta_1\delta_2}\Gamma(\delta_2)} \exp\left[-\left(\frac{x-\alpha}{\beta}\right)^{\delta_1}\right] \quad x \ge \alpha;\ \delta_1, \delta_2, \beta > 0 \tag{20.61}$$

Figure 20.38 shows the effect of varying δ_1 and δ_2 with α fixed at 0 and β at 1.

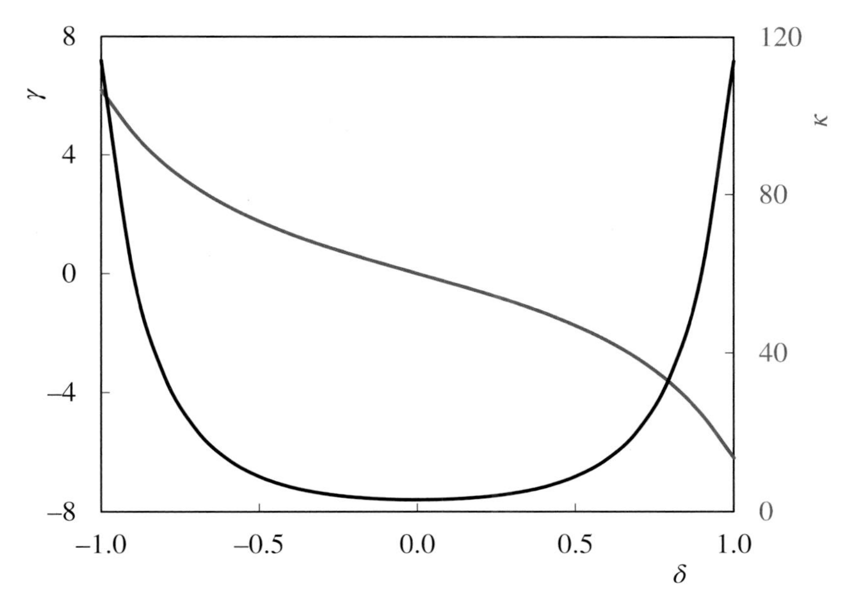

Figure 20.36 *Generalised normal (2): Effect of δ on γ and κ*

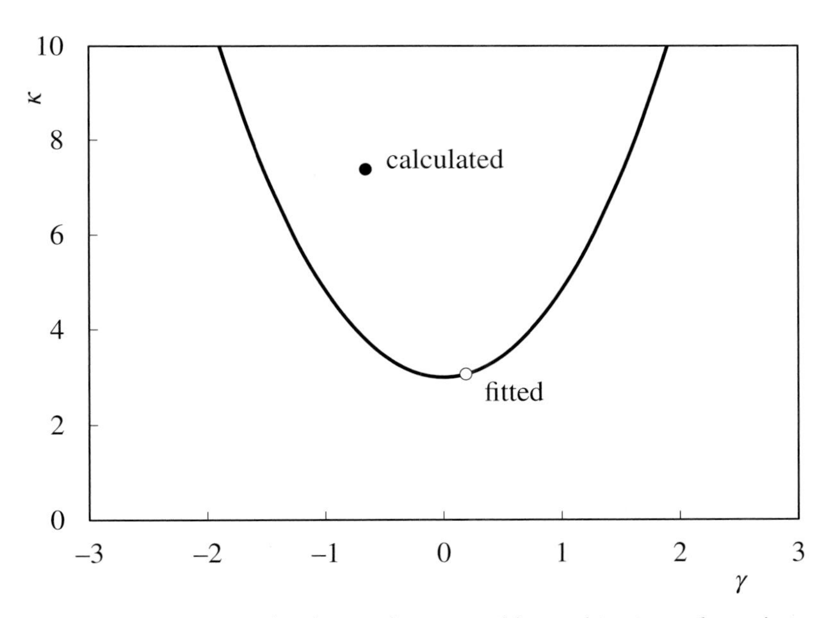

Figure 20.37 *Generalised normal (2): Feasible combinations of γ and κ*

Fitting to the absolute value of the NHV disturbance data gives α as –0.0113, β as 0.00284, δ_1 as 0.348, δ_2 as 7.13 and $F(x_1)$ as 0.0282. With *RSS* at 0.0328, it is one of the better fits.

Raw moments are given by

$$m_n = \frac{\Gamma\left(\frac{n}{\delta_1} + \delta_2\right)}{\Gamma(\delta_2)} \beta^n \tag{20.62}$$

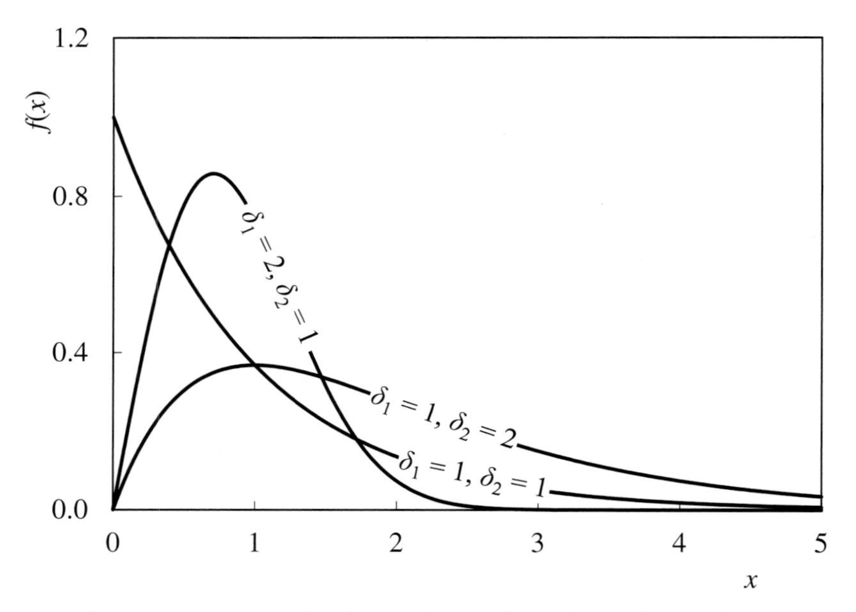

Figure 20.38 *Generalised normal (3): Effect of δ_1 and δ_2 on shape*

This gives the mean as 1.12 and the standard deviation as 1.35, both close to the values calculated from the data.

Feasible combinations of skewness and kurtosis are shown in Figure 20.39. Also shown are the values calculated from the data and those from fitting the distribution.

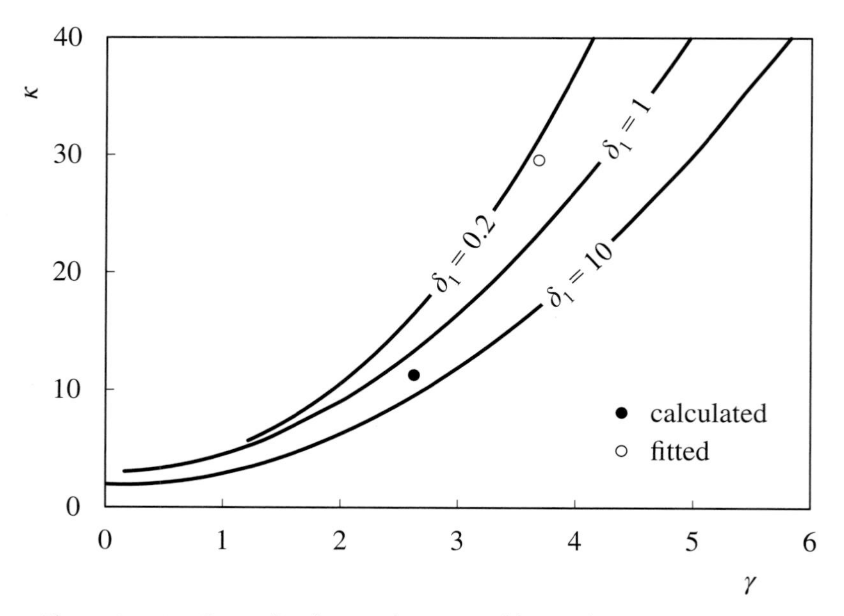

Figure 20.39 *Generalised normal (3): Feasible combinations of γ and κ*

20.13 Exponentially Modified Gaussian

Instead of the mean of the normal distribution being a fixed value, it is assumed to follow the exponential distribution (which we covered in Section 12.7). The result is the *exponentially modified Gaussian (EMG) distribution*. It is an example of a compound distribution. Its PDF is

$$f(x)=\frac{\lambda}{2}\exp\left[\frac{\lambda}{2}\left(\lambda\beta^2+2\alpha-2x\right)\right]\left[1-\operatorname{erf}\left(\frac{\lambda\beta^2+\alpha-x}{\sqrt{2}\beta}\right)\right]\quad \beta,\lambda>0 \tag{20.63}$$

Figure 20.40 shows the effect of varying the rate parameter (λ), with α fixed at 0 and β at 1. As λ is increased, its effect reduces and the distribution approaches the normal distribution. Its CDF is

$$F(x)=\frac{1}{2}\left\{1-\exp\left[\frac{\lambda}{2}\left(\lambda\beta^2+2\alpha-2x\right)\right]\left[1-\operatorname{erf}\left(\frac{\lambda\beta^2+\alpha-x}{\sqrt{2}\beta}\right)\right]+\operatorname{erf}\left(\frac{x-\alpha}{\sqrt{2}\beta}\right)\right\} \tag{20.64}$$

Fitting to the C_4 in propane data sets α as 2.35, β as 0.736 and λ as 0.388. With *RSS* at 0.0329, the fit is reasonably good.

The mean and variance are

$$\mu=\alpha+\frac{1}{\lambda} \tag{20.65}$$

$$\sigma^2=\beta^2+\frac{1}{\lambda^2} \tag{20.66}$$

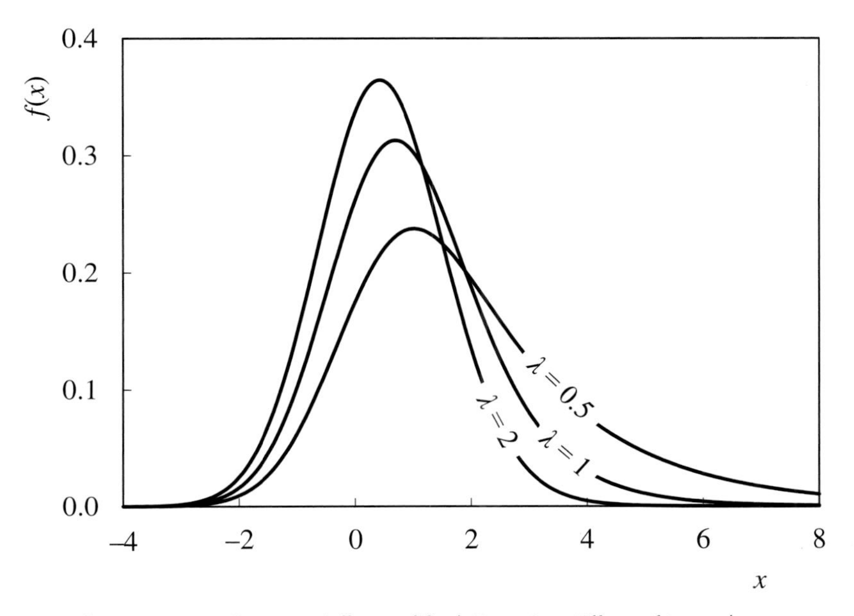

Figure 20.40 Exponentially modified Gaussian: Effect of λ on shape

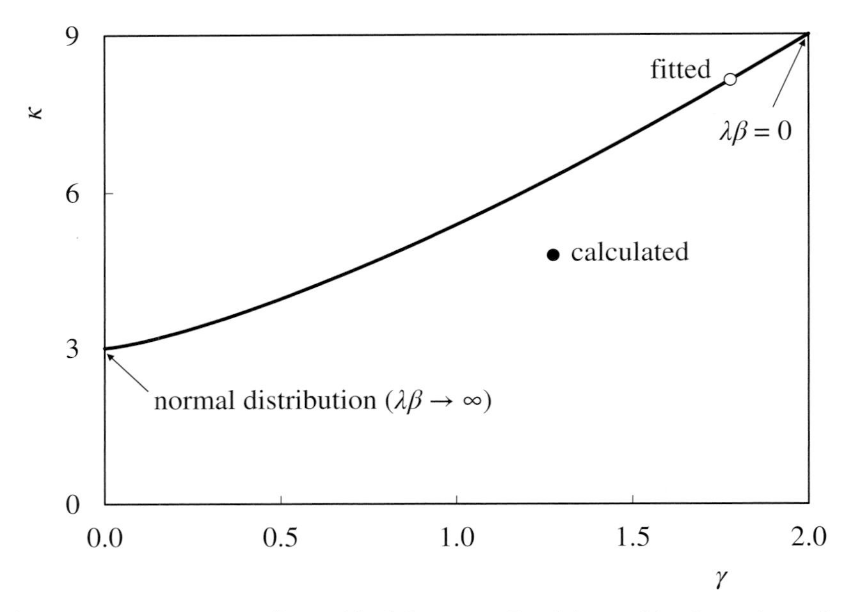

Figure 20.41 *Exponentially modified Gaussian: Feasible combinations of γ and κ*

These give μ as 4.93 and σ as 2.68 – both quite close to the values calculated from the data. Skewness and kurtosis are

$$\gamma = \frac{2}{\left(\lambda^2\beta^2 + 1\right)^{3/2}} \tag{20.67}$$

$$\kappa = \frac{3\left(\lambda^4\beta^4 + 2\lambda^2\beta^2 + 3\right)}{\left(\lambda^2\beta^2 + 1\right)^2} \tag{20.68}$$

While both γ and κ clearly depend on both λ and β, the feasible combinations of skewness and kurtosis do not. This is because the parameters always appear as $\lambda\beta$. This is illustrated by Figure 20.41. The line is plotted by choosing a value for β and then calculating γ and κ for a range of values for λ. No matter what value is chosen for β, the resulting curve remains the same. From Equation (20.66), we see that λ must have units that are the reciprocal of those of β and so $\lambda\beta$ (and hence γ and κ) are dimensionless.

Figure 20.41 shows the limitation on skewness, which can only be positive. Equation (20.67) gives it as 1.78 – a little larger than that calculated from the data. Similarly the distribution can only represent leptokurtic data. Equation (20.68) gives kurtosis as 8.13, which is considerably larger than that calculated. Nevertheless, the distribution would be a reasonably practical choice.

We can similarly develop, from the normal distribution, a compound distribution in which the mean is constant but the standard deviation follows a distribution. For example, if we assume it follows the inverse gamma distribution, we obtain the Student t distribution that we covered this in Section 12.5.

Compounding distributions can be taken to extremes. For example, there is the *normal exponential gamma (NEG) distribution*. Starting as the normal distribution, the variance is assumed to follow the exponential distribution. Then the rate parameter of the exponential distribution is assumed to follow the gamma distribution. Such complexity is rarely justified.

20.14 Moyal

The *Moyal distribution* is described by

$$f(x)=\frac{1}{\beta\sqrt{2\pi}}\exp\left[-\frac{x-\alpha}{2\beta}-\frac{1}{2}\exp\left(-\frac{x-\alpha}{\beta}\right)\right]\quad \beta>0 \tag{20.69}$$

This PDF is also an approximation to the *Landau distribution*. Figure 20.42 shows the effect of changing β, keeping α at zero. The CDF is

$$F(x)=1-\text{erf}\left[\frac{1}{\sqrt{2}}\exp\left(-\frac{x-\alpha}{2\beta}\right)\right] \tag{20.70}$$

Fitting to the C_4 in propane data gives α as 3.41 and β as 1.12. *RSS* is 0.0481. The equations below give the mean as 4.84 and the standard deviation as 2.49.

$$\mu\approx\alpha+1.2704\beta \tag{20.71}$$

$$\sigma^2=\frac{\pi^2\beta^2}{2} \tag{20.72}$$

The *beta-Moyal distribution* involves the use of incomplete gamma functions and so puts its complexity beyond the scope of this book.

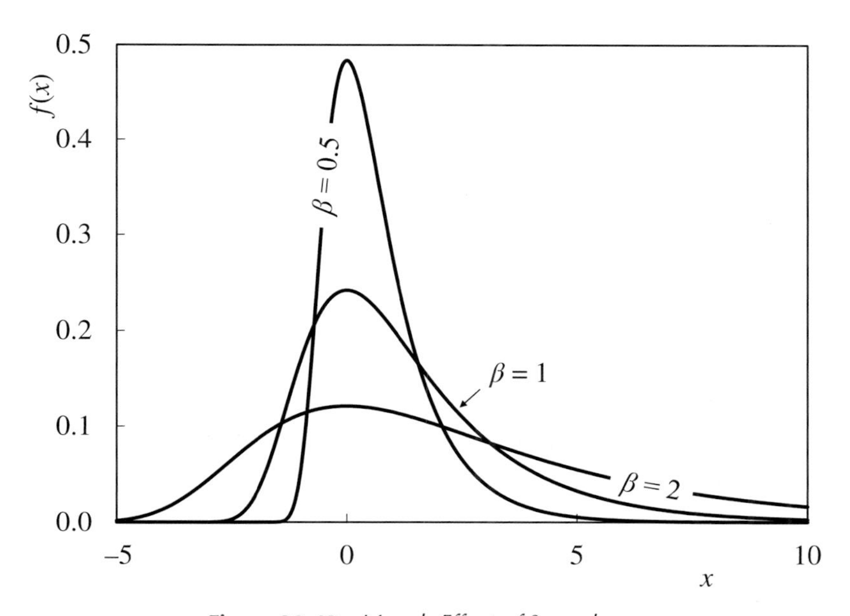

Figure 20.42 Moyal: Effect of β on shape

21

Burr Distribution

There are 12 Burr distributions.[11] In many cases there are no published formulae for mean, variance, skewness or kurtosis. Although these might be developed, the majority of the distributions would offer a poor fit to process data that would almost certainly be bettered by other distributions for which the formulae exist. Indeed, they are largely neglected in the literature. The exception is Type XII, which we will show gives one of the best fits to the C_4 in propane data.

21.1 Type I

The *Burr-I distribution* is more commonly known as the uniform distribution, as described in Section 5.1.

21.2 Type II

The *Burr-II distribution* is also known as the generalised logistic distribution, as described in the next chapter.

21.3 Type III

The *Burr-III distribution* is described by the PDF

$$f(x) = \frac{\delta_1 \delta_2}{\beta}\left(\frac{x-\alpha}{\beta}\right)^{-\delta_1 - 1}\left[1 + \left(\frac{x-\alpha}{\beta}\right)^{-\delta_1}\right]^{-\delta_2 - 1} \quad x \geq \alpha,\ \delta_1, \delta_2, \beta > 0 \tag{21.1}$$

Figure 21.1 shows the effect of varying δ_1 and δ_2 with α fixed at 0 and β at 1. The CDF is

$$F(x) = \left[1 + \left(\frac{x-\alpha}{\beta}\right)^{-\delta_1}\right]^{-\delta_2} \tag{21.2}$$

Statistics for Process Control Engineers: A Practical Approach, First Edition. Myke King.

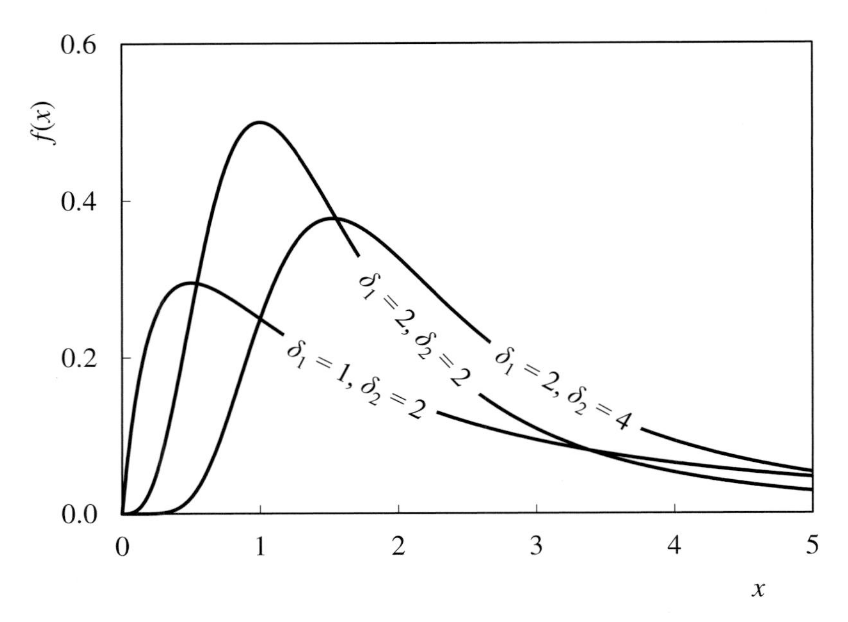

Figure 21.1 Burr-III: Effect of δ_1 and δ_2 on shape

Fitting to the C_4 in propane data gives α as −4.35, β as 3.84, δ_1 as 4.86 and δ_2 as 35.3. *RSS* is 0.0399.

The CDF can be inverted to give the QF.

$$x(F) = \alpha + \beta\left(F^{-1/\delta_2} - 1\right)^{-1/\delta_1} \qquad 0 \le F \le 1 \tag{21.3}$$

21.4 Type IV

The *Burr-IV distribution* is described by the PDF

$$f(x) = \frac{\delta_2\beta}{(x-\alpha)^2}\left(\frac{\delta_1\beta - x + \alpha}{x-\alpha}\right)^{\frac{1}{\delta_1}-1}\left[1 + \left(\frac{\delta_1\beta - x + \alpha}{x-\alpha}\right)^{\frac{1}{\delta_1}}\right]^{-\delta_2-1} \qquad -\alpha < x < \alpha + \beta\delta_1;\ \delta_1, \delta_2, \beta > 0 \tag{21.4}$$

As Figure 21.2 shows, with α set to 0 and β set to 1, the width of the distribution is determined by δ_1. With this set to 10, the effect of varying δ_2 is shown. Due to its shape, the distribution is unlikely to be applicable to process data.

The CDF is

$$F(x) = \left[1 + \left(\frac{\delta_1\beta - x + \alpha}{x-\alpha}\right)^{\frac{1}{\delta_1}}\right]^{-\delta_2} \tag{21.5}$$

The QF is

$$x(F) = \alpha + \frac{\delta_1\beta}{1 + \left(F^{-1/\delta_2} - 1\right)^{\delta_1}} \qquad 0 \le F \le 1 \tag{21.6}$$

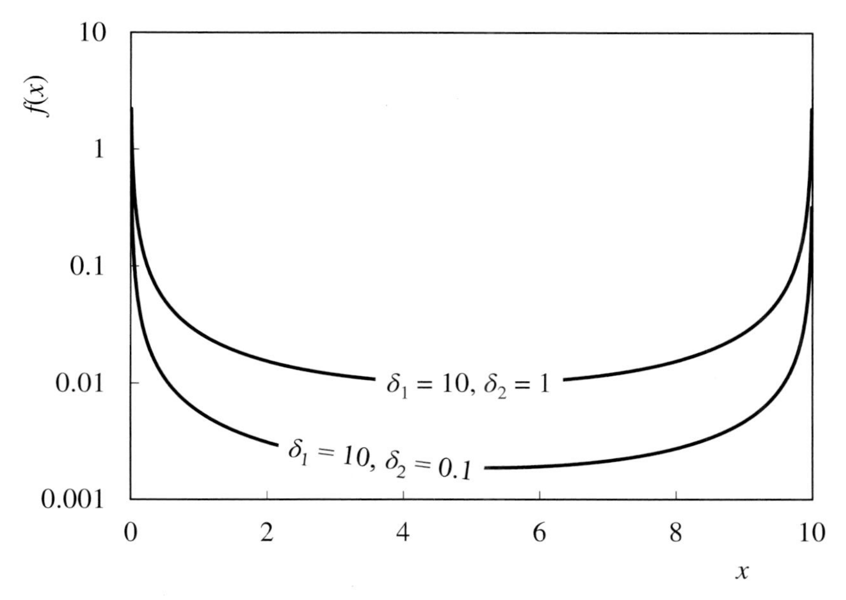

Figure 21.2 Burr-IV: Effect of δ_1 and δ_2 on shape

21.5 Type V

The *Burr-V distribution* is described by the PDF

$$f(x)=\delta_1\delta_2\sec^2\left(\frac{x-\alpha}{\beta}\right)\exp\left[-\tan\left(\frac{x-\alpha}{\beta}\right)\right]\left\{1+\delta_1\exp\left[-\tan\left(\frac{x-\alpha}{\beta}\right)\right]\right\}^{-\delta_2-1} \tag{21.7}$$

$$\alpha-\frac{\pi}{2}\leq x\leq\alpha+\frac{\pi}{2};\ \delta_1,\delta_2,\beta>0$$

Figure 21.3 shows the effect of varying δ_1 and δ_2 with α fixed at 0 and β at 1. The distribution is bimodal unless δ_1 is significantly different from δ_2. This and the inconvenient bounds on x make the distribution unsuitable for process data.

The CDF is

$$F(x)=\left\{1+\delta_1\exp\left[-\tan\left(\frac{x-\alpha}{\beta}\right)\right]\right\}^{-\delta_2} \tag{21.8}$$

The QF is

$$x(F)=\alpha+\beta\tan^{-1}\left[-\ln\left(\frac{F^{-1/\delta_2}-1}{\delta_1}\right)\right]\quad 0\leq F\leq 1 \tag{21.9}$$

21.6 Type VI

The *Burr-VI distribution* is described by the PDF

$$f(x)=\frac{\delta_1\delta_2}{\beta}\cosh\left(\frac{x-\alpha}{\beta}\right)\exp\left[-\delta_1\sinh\left(\frac{x-\alpha}{\beta}\right)\right]\left\{1+\exp\left[-\delta_1\sinh\left(\frac{x-\alpha}{\beta}\right)\right]\right\}^{-\delta_2-1}\quad \delta_1,\delta_2,\beta>0 \tag{21.10}$$

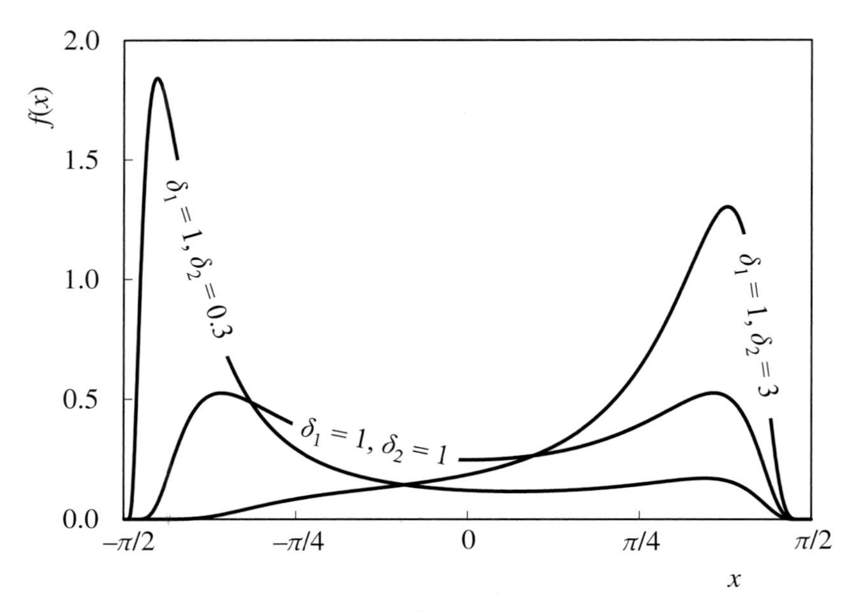

Figure 21.3 Burr-V: Effect of δ_1 and δ_2 on shape

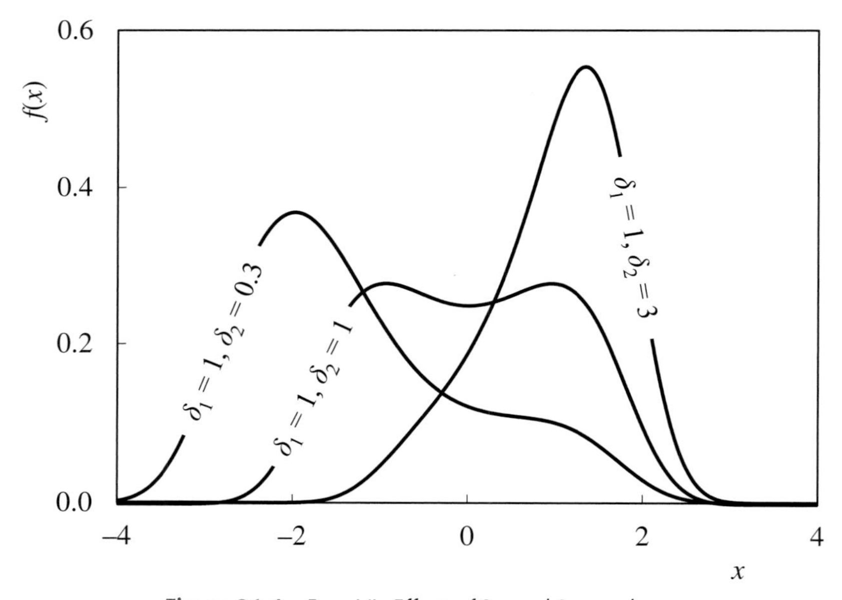

Figure 21.4 Burr-VI: Effect of δ_1 and δ_2 on shape

Figure 21.4 shows the effect of varying δ_1 and δ_2 with α fixed at 0 and β at 1. The distribution is bimodal unless δ_1 is significantly different from δ_2.

The CDF is

$$F(x) = \left\{1 + \exp\left[-\delta_1 \sinh\left(\frac{x-\alpha}{\beta}\right)\right]\right\}^{-\delta_2} \tag{21.11}$$

Fitting to the C_4 in propane data gives α as −6.09, β as 52.0, δ_1 as 28.6 and δ_2 as 224. *RSS* is, not surprisingly, poor at 0.1424.

The QF is

$$x(F) = \alpha + \beta \sinh^{-1}\left[-\frac{\ln\left(F^{-1/\delta_2} - 1\right)}{\delta_1}\right] \quad 0 \le F \le 1 \tag{21.12}$$

21.7 Type VII

The *Burr-VII distribution* is described by the PDF

$$f(x) = \frac{\delta}{2^{\delta}\beta}\left[1 - \tanh^2\left(\frac{x-\alpha}{\beta}\right)\right]\left[1 + \tanh\left(\frac{x-\alpha}{\beta}\right)\right]^{\delta-1} \quad \delta, \beta > 0 \tag{21.13}$$

Figure 21.5 shows the effect of varying δ with α fixed at 0 and β at 1.

The CDF is

$$F(x) = \left\{\frac{1}{2}\left[1 + \tanh\left(\frac{x-\alpha}{\beta}\right)\right]\right\}^{\delta} \tag{21.14}$$

Fitting to the C_4 in propane data gives α as −2.27, β as 3.56 and δ as 28.8. With just three shape parameters, *RSS* is poor at 0.1463.

The QF is

$$x(F) = \alpha + \beta \tanh^{-1}\left\{(2F)^{1/\delta} - 1\right\} \quad 0 \le F \le 1 \tag{21.15}$$

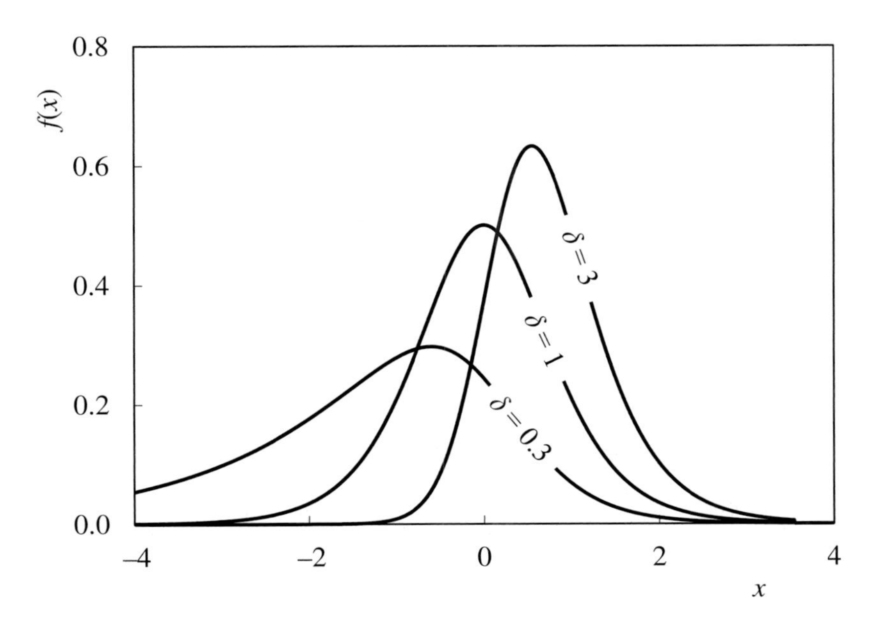

Figure 21.5 *Burr-VII: Effect of δ on shape*

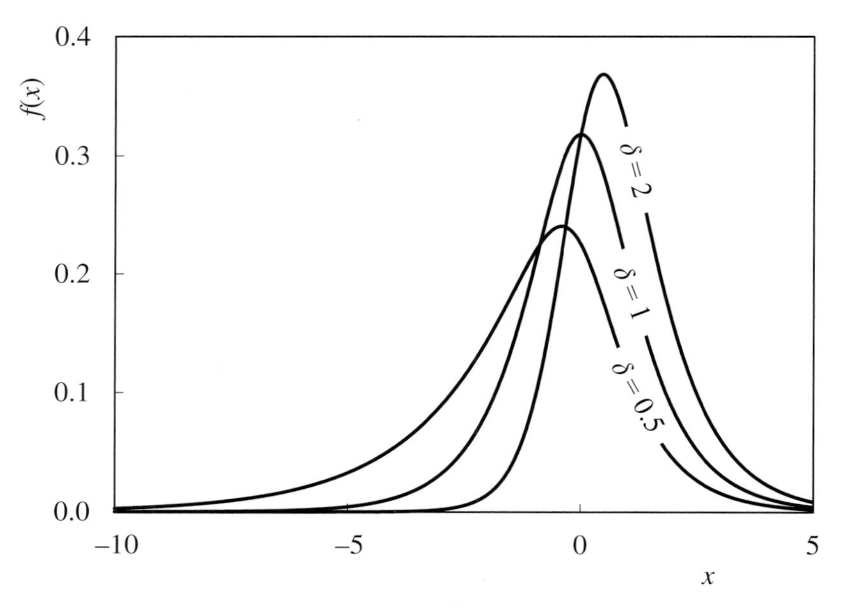

Figure 21.6 Burr-VIII: Effect of δ on shape

21.8 Type VIII

The *Burr-VIII distribution* is described by the PDF

$$f(x) = \frac{2\delta}{\pi\beta} \frac{\exp\left(\frac{x-\alpha}{\beta}\right)}{1 + \left[\exp\left(\frac{x-\alpha}{\beta}\right)\right]^2} \left\{\frac{2}{\pi}\tan^{-1}\left[\exp\left(\frac{x-\alpha}{\beta}\right)\right]\right\}^{\delta-1} \quad \delta, \beta > 0 \tag{21.16}$$

Figure 21.6 shows the effect of varying δ with α fixed at 0 and β at 1.

The CDF is

$$F(x) = \left\{\frac{2}{\pi}\tan^{-1}\left[\exp\left(\frac{x-\alpha}{\beta}\right)\right]\right\}^{\delta} \tag{21.17}$$

Because the distribution is skewed in the wrong direction, fitting it to the C_4 in propane data will give a poor result. However, fitting it to (100 − C_4) gives α as 97.3, β as 0.631 and δ as 0.239. With *RSS* at 0.0362, the fit is one of the best.

The QF is

$$x(F) = \alpha + \beta \ln\left[\tan\left(\frac{\pi F^{1/\delta}}{2}\right)\right] \quad 0 \leq F \leq 1 \tag{21.18}$$

21.9 Type IX

The *Burr-IX distribution* is described by the PDF

$$f(x) = \frac{2\delta_1\delta_2}{\beta} \frac{\exp\left(\frac{x-\alpha}{\beta}\right)\left[1 + \exp\left(\frac{x-\alpha}{\beta}\right)\right]^{\delta_2 - 1}}{\left(2 + \delta_1\left\{\left[1 + \exp\left(\frac{x-\alpha}{\beta}\right)\right]^{\delta_2} - 1\right\}\right)^2} \quad \delta_1, \delta_2, \beta > 0 \tag{21.19}$$

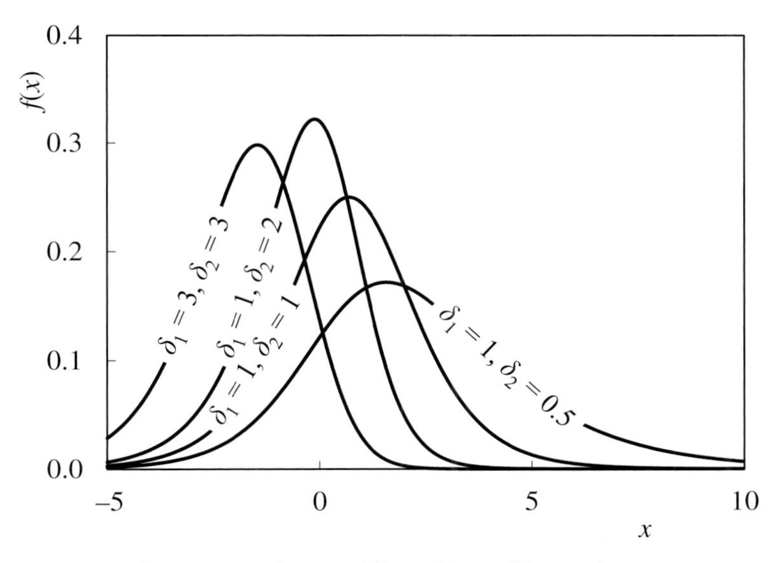

Figure 21.7 *Burr-IX: Effect of δ_1 and δ_2 on shape*

Figure 21.7 shows the effect of varying δ_1 and δ_2 with α fixed at 0 and β at 1.
The CDF is

$$F(x) = 1 - \frac{2}{2 + \delta_1 \left\{ \left[1 + \exp\left(\frac{x-\alpha}{\beta} \right) \right]^{\delta_2} - 1 \right\}} \tag{21.20}$$

Fitting to the C_4 in propane data gives α as 0.188, β as 5.00, δ_1 as 0.0787 and δ_2 as 0.139. *RSS* is very poor at 0.4716.
The QF is

$$x(F) = \alpha + \beta \ln \left\{ \left[1 + \frac{F}{\delta_1 (1-F)} \right]^{\frac{1}{\delta_2}} \right\} \qquad 0 \le F \le 1 \tag{21.21}$$

21.10 Type X

The *Burr-X distribution* is described by the PDF

$$f(x) = 2\delta \left(\frac{x-\alpha}{\beta} \right) \exp\left[-\left(\frac{x-\alpha}{\beta} \right)^2 \right] \left\{ 1 - \exp\left[-\left(\frac{x-\alpha}{\beta} \right)^2 \right] \right\}^{\delta-1} \qquad x \ge 0; \delta, \beta > 0 \tag{21.22}$$

Figure 21.8 shows the effect of varying δ with α fixed at 0 and β at 1.
The CDF is

$$F(x) = \left\{ 1 - \exp\left[-\left(\frac{x-\alpha}{\beta} \right)^2 \right] \right\}^{\delta} \tag{21.23}$$

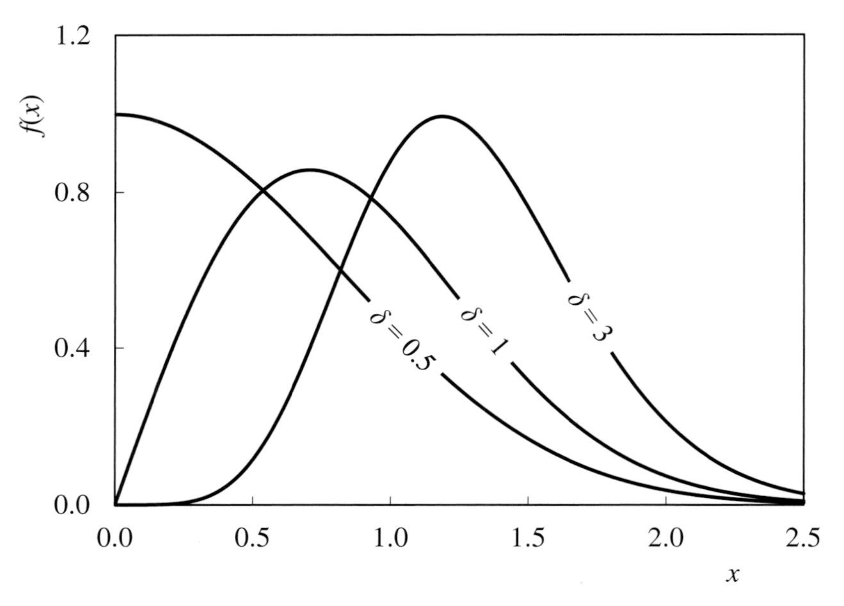

Figure 21.8 Burr-X: Effect of δ on shape

Fitting to the C_4 in propane data gives α as −35.5, β as 12.0 and δ as 45100. With the rather suspect result for δ, it is not surprising that *RSS* is poor at 0.1811.

The QF is

$$x(F) = \alpha + \beta\sqrt{-\ln(1-F^{1/\delta})} \quad 0 \le F \le 1 \tag{21.24}$$

21.11 Type XI

The *Burr-XI distribution* is described by the PDF

$$f(x) = \frac{\delta}{\beta}\left\{1-\cos\left[\frac{2\pi(x-\alpha)}{\beta}\right]\right\}\left\{\frac{x-\alpha}{\beta}-\frac{1}{2\pi}\sin\left[\frac{2\pi(x-\alpha)}{\beta}\right]\right\}^{\delta-1} \quad \alpha \le x \le \alpha+\beta;\delta,\beta>0 \tag{21.25}$$

Figure 21.9 shows the effect of varying δ with α fixed at 0 and β at 1.

The CDF is

$$F(x) = \left\{\frac{x-\alpha}{\beta}-\frac{1}{2\pi}\sin\left[\frac{2\pi(x-\alpha)}{\beta}\right]\right\}^{\delta} \tag{21.26}$$

Fitting to the C_4 in propane data gives α as 1.97, β as 9.47 and δ as 0.290. *RSS* is poor at 0.1552.

The CDF cannot be inverted.

21.12 Type XII

The *Burr-XII distribution* was covered in Section 12.2.

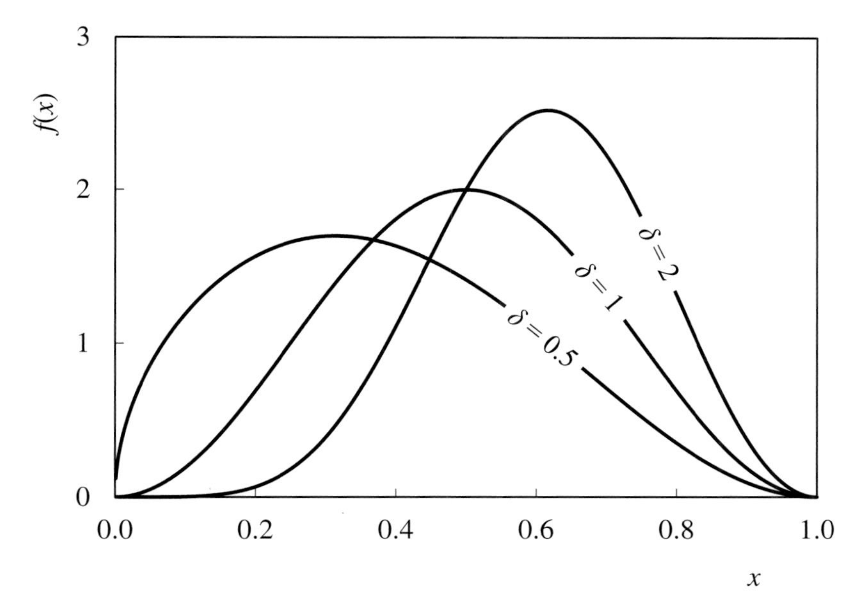

Figure 21.9 *Burr-XI: Effect of δ on shape*

21.13 Inverse

The *inverse Burr distribution* is the inverse of the *Burr-XII distribution*. It is also known as the *Dagum-I distribution*.[12] Its PDF is

$$f(x) = \frac{\frac{\delta_1\delta_2}{\beta}}{\left(\frac{x-\alpha}{\beta}\right)^{\delta_1+1}\left[1+\left(\frac{x-\alpha}{\beta}\right)^{-\delta_1}\right]^{\delta_2+1}} = \frac{\frac{\delta_1\delta_2}{\beta}\left(\frac{x-\alpha}{\beta}\right)^{\delta_1\delta_2-1}}{\left[1+\left(\frac{x-\alpha}{\beta}\right)^{\delta_1}\right]^{\delta_2+1}} \quad x \geq \alpha;\ \beta, \delta_1 > 1;\ \delta_2 > 0 \qquad (21.27)$$

The effect that δ_1 and δ_2 have on the shape of the distribution is shown in Figures 21.10 and 21.11. Setting δ_1 to δ_2 gives the inverse paralogistic distribution that we cover in the next chapter.

The CDF is commonly presented in one of two different ways.

$$F(x) = \left[1+\left(\frac{x-\alpha}{\beta}\right)^{-\delta_1}\right]^{-\delta_2} = \frac{(x-\alpha)^{\delta_1\delta_2}}{\left(\beta^{\delta_1}+(x-\alpha)^{\delta_1}\right)^{\delta_2}} \qquad (21.28)$$

The CDF can be inverted to

$$x(F) = \alpha + \beta\left(F^{-1/\delta_2}-1\right)^{-1/\delta_1} \quad 0 \leq F \leq 1 \qquad (21.29)$$

The raw moments are given by

$$m_n = \frac{\Gamma\left(1-\frac{n}{\delta_1}\right)\Gamma\left(\delta_2+\frac{n}{\delta_1}\right)}{\Gamma(\delta_2)}\beta^n \quad \delta_1 > n \qquad (21.30)$$

From this we can derive formulae for the mean and standard deviation.

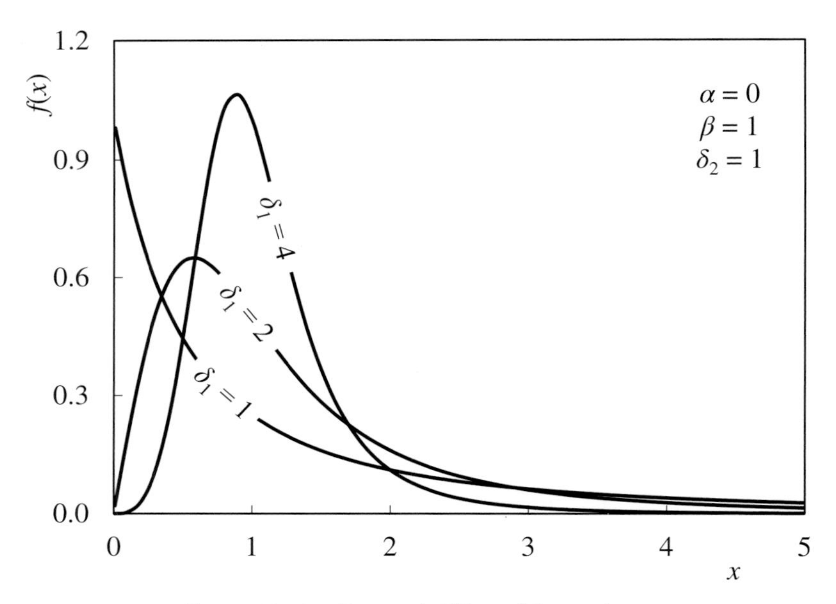

Figure 21.10 Dagum-I: Effect of δ_1 on shape

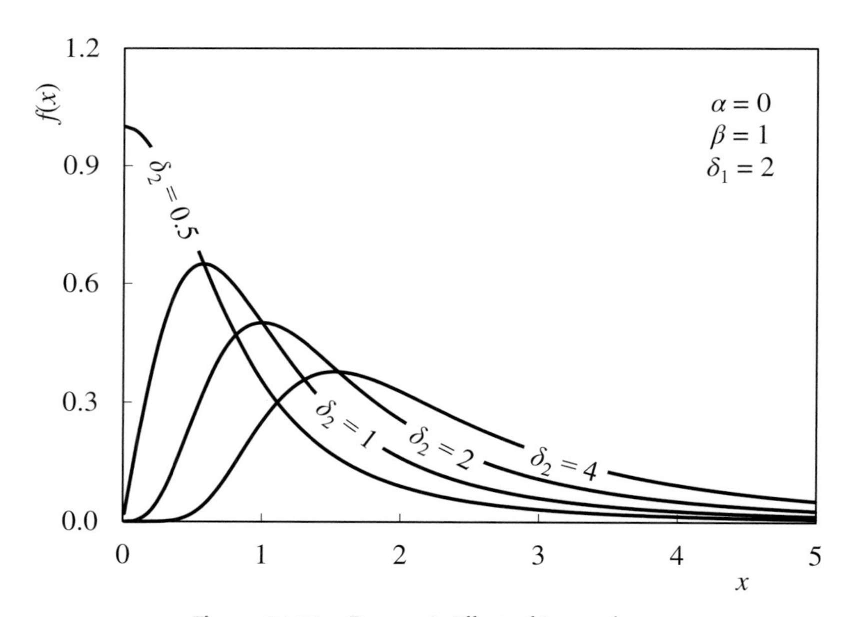

Figure 21.11 Dagum-I: Effect of δ_2 on shape

$$\mu = \alpha + \frac{\Gamma\left(1-\frac{1}{\delta_1}\right)\Gamma\left(\delta_2+\frac{1}{\delta_1}\right)}{\Gamma(\delta_2)}\beta \quad \delta_1 > 1 \tag{21.31}$$

$$\sigma^2 = -\frac{\beta^2}{\Gamma^2(\delta_2)}\left[2\delta_1\Gamma\left(1-\frac{2}{\delta_1}\right)\Gamma\left(\delta_2+\frac{2}{\delta_1}\right)\Gamma(\delta_2) + \Gamma^2\left(1-\frac{1}{\delta_1}\right)\Gamma^2\left(\delta_2+\frac{1}{\delta_1}\right)\right] \quad \delta_1 > 2 \tag{21.32}$$

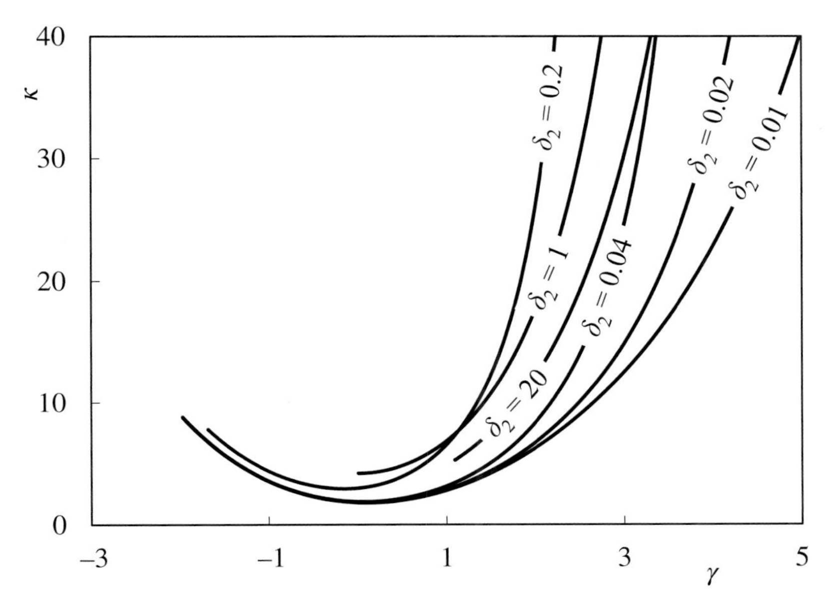

Figure 21.12 *Dagum-I: Feasible combinations of γ and κ*

While formulae for skewness and kurtosis can similarly be derived, the result is too complex to present here. They are better determined by first calculating the moments and applying Equations (4.52) and (4.55). By varying δ_1, for each of a range of values for δ_2, Figure 21.12 shows what combinations are possible.

Fitting Equation (21.28) to the C_4 in propane data gives 3.23 for δ_1, 2.07 for δ_2, −0.573 for α and 3.63 for β – minimising *RSS* to 0.0510. This gives a skewness of 17.2 compared to the 1.27 calculated from the data. To calculate kurtosis, δ_1 must be greater than 4 and so is not possible in this case. The poor match in skewness and the inability to check kurtosis might lead to this distribution being deemed unsuitable in this case. Also, in terms of *RSS*, the fit is poorer than many of the distributions considered so far.

The *Dagum-II distribution* is described by

$$f(x) = \delta_3 + (1-\delta_3)\frac{\delta_1\delta_2}{\beta}\frac{\left(\frac{x-\alpha}{\beta}\right)^{\delta_1\delta_2-1}}{\left[1+\left(\frac{x-\alpha}{\beta}\right)^{\delta_1}\right]^{\delta_2+1}} \qquad x>\alpha;\ \beta,\delta_1>1;\ \delta_2>0;\ 0\le\delta_3\le 1 \qquad (21.33)$$

$$F(x) = \delta_3 + (1-\delta_3)\left[1+\left(\frac{x-\alpha}{\beta}\right)^{-\delta_1}\right]^{-\delta_2} \qquad (21.34)$$

The *Dagum-III distribution* has the same PDF and CDF but with different bounds on δ_3 and x.

$$\delta_3<0,\ x>\alpha+\left\{\beta\left[\left(1-\frac{1}{\delta_1}\right)^{\frac{1}{\delta_2}}-1\right]\right\}^{-\frac{1}{\delta_1}} \qquad (21.35)$$

Fitting to the C_4 data gives 3.25 for δ_1, 1.83 for δ_2, −0.0168 for δ_3, −0.606 for α and 3.80 for β – minimising *RSS* to 0.0473. As usual, the introduction of an additional shape parameter has improved the fit. The parameter δ_3 and all values of x are within the bounds of the Dagum-III distribution.

The QF is

$$x(F) = \alpha + \beta \left[\left(\frac{F - \delta_3}{1 - \delta_3} \right)^{-\frac{1}{\delta_2}} - 1 \right]^{-\frac{1}{\delta_1}} \qquad 0 \le F \le 1 \tag{21.36}$$

22

Logistic Distribution

22.1 Logistic

A generalised form of the *logistic distribution* is described by

$$f(x) = \frac{\left(1 + \frac{\delta(x-\alpha)}{\beta}\right)^{-\frac{1}{\delta}-1}}{\beta\left[1 + \left(1 + \frac{\delta(x-\alpha)}{\beta}\right)^{-\frac{1}{\delta}}\right]^2} \qquad \beta > 0 \tag{22.1}$$

Depending on the sign of δ, the distribution is either lower-bounded or upper-bounded.

$$\delta < 0 \text{ then } x \le \alpha - \frac{\beta}{\delta} \tag{22.2}$$

$$\delta > 0 \text{ then } x \ge \alpha - \frac{\beta}{\delta} \tag{22.3}$$

The CDF is

$$F(x) = \frac{1}{1 + \left(1 + \frac{\delta(x-\alpha)}{\beta}\right)^{-\frac{1}{\delta}}} \tag{22.4}$$

The CDF can be inverted

$$x(F) = \alpha + \frac{\beta}{\delta}\left[\left(\frac{F}{1-F}\right)^{\delta} - 1\right] \tag{22.5}$$

Figure 22.1 shows the effect of changing δ with α fixed at 0 and β at 1. Note that all the curves pass through the point (α, 0.25β). Fitting Equation (22.4) to the C_4 in propane data gives α as 2.30, β as 1.27 and δ as 0.360. *RSS* is 0.0529, m\aking the distribution a relative poor choice.

Statistics for Process Control Engineers: A Practical Approach, First Edition. Myke King.

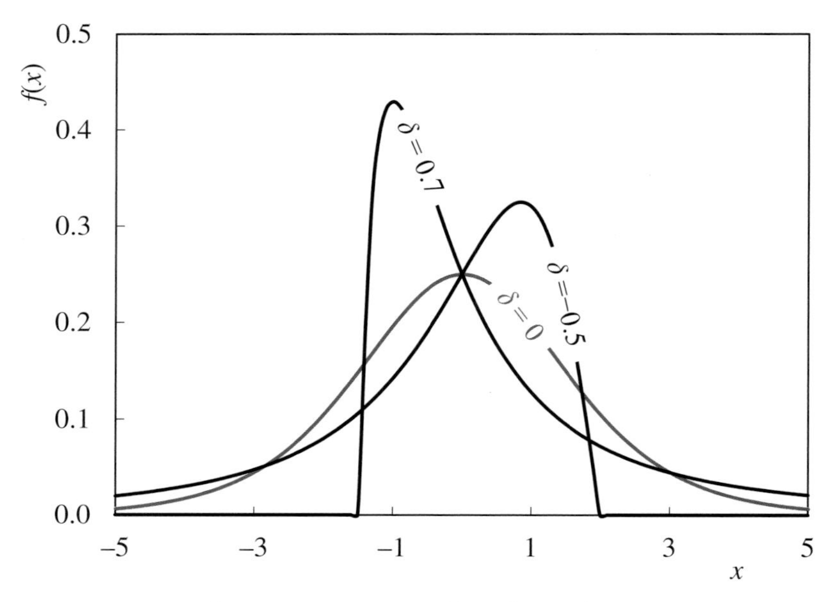

Figure 22.1 *Logistic: Effect of δ on shape*

The mean and variance are

$$\mu = \alpha + \frac{\pi\beta}{\sin(\pi\delta)} - \frac{\beta}{\delta} \tag{22.6}$$

$$\sigma^2 = \frac{2\pi\beta^2}{\delta\sin(2\pi\delta)} - \left(\frac{\pi\beta}{\sin(\pi\delta)}\right)^2 \tag{22.7}$$

As $\delta \to 0$, from Equation (11.4), we obtain the density functions

$$f(x) = \frac{\exp\left(-\frac{(x-\mu)}{\beta}\right)}{\beta\left[1+\exp\left(-\frac{(x-\mu)}{\beta}\right)\right]^2} \qquad \beta > 0 \tag{22.8}$$

$$F(x) = \frac{1}{1+\exp\left(-\frac{(x-\mu)}{\beta}\right)} \tag{22.9}$$

Equation (22.8) is represented by the coloured line in Figure 22.1. This form is what many might consider the classic logistic distribution. It is similar to the normal distribution with σ set to about 1.63β. Its skewness is zero but it is slightly leptokurtic with kurtosis fixed at 4.2.

Equation (22.9) can be written using the logistic function, described in Section 11.4, giving the distribution its name.

$$F(x) = \text{logistic}\left(\frac{x-\mu}{\beta}\right) \tag{22.10}$$

Perhaps less usefully, it is occasionally presented using the *hyperbolic secant (sech)* and *hyperbolic tangent (tanh)* functions.

$$f(x) = \frac{1}{4\beta}\text{sech}^2\left(\frac{x-\mu}{2\beta}\right) \tag{22.11}$$

$$F(x) = \frac{1}{2}\left[1 + \tanh\left(\frac{x-\mu}{2\beta}\right)\right] \tag{22.12}$$

The mean is μ as usual. The variance is

$$\sigma^2 = \frac{\pi^2}{3}\beta^2 \tag{22.13}$$

The CDF can be inverted to

$$x(F) = \mu + \beta\ln\left(\frac{F}{1-F}\right) \tag{22.14}$$

Figure 22.2 shows, as the coloured line, the normal distribution N(5,1). Theoretically, from Equation (22.13), setting β set to 0.551 would give the equivalent logistic distribution – shown as the solid line. In practice, setting β to 0.615σ gives a match that is closer visually, as shown by the dashed line.

Fitting Equation (22.9) to the C_4 in propane data gives μ as 4.49, β as 1.26 and hence σ as 2.29. Not surprisingly forcing δ to be 0 increases *RSS*. At 0.6051, it is much higher than that of the normal distribution.

Fitting to the NHV disturbance data gives μ as –0.02, β as 0.715 and hence σ as 1.30. *RSS* is then 0.1061. Calculating the kurtosis from the process data gives a value of 8.87. The logistic distribution, which has a kurtosis of 4.2, will (for this example) fit better than the normal distribution that has a kurtosis of 3. Of course, this may not be the case for other datasets. For this reason, a number of modifications to the distribution have been published to address its limitations. This chapter describes those of interest.

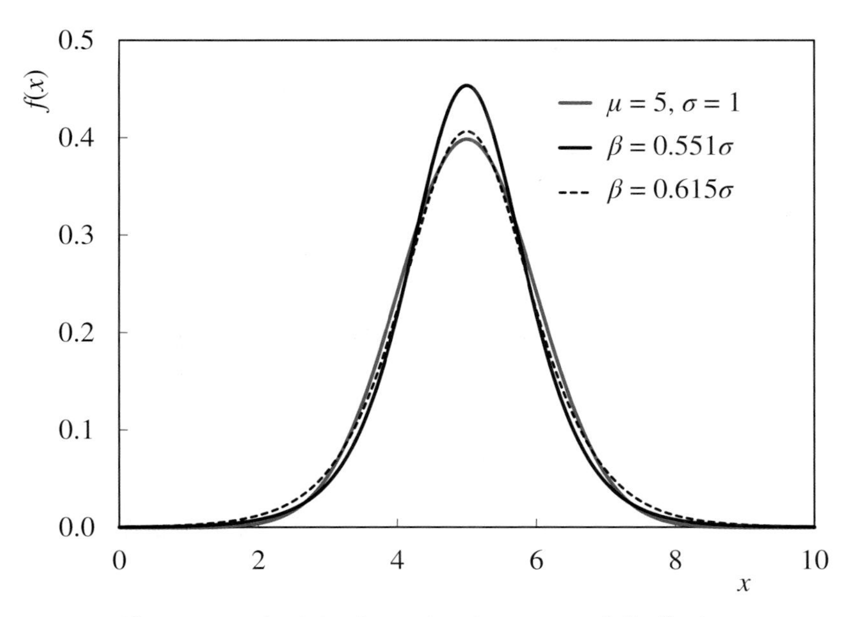

Figure 22.2 *Logistic: Approximation to normal distribution*

22.2 Half-Logistic

The *half-logistic distribution* can be applied to data where it is realistic to use the absolute value of each measurement. For example, process disturbances and inferential errors can be in either direction but we are often concerned only with their size.

Its PDF is almost identical to that of the classic logistic distribution, given by Equation (22.8), replacing μ with α. It is also common to replace β with $1/\lambda$. We need to add the factor 2 to ensure the area under the PDF curve is unity.

$$f(x)=\frac{2\lambda\exp\{-\lambda(x-\alpha)\}}{[1+\exp\{-\lambda(x-\alpha)\}]^2} \qquad x\geq\alpha;\ \lambda>0 \tag{22.15}$$

The CDF is

$$F(x)=\frac{1-\exp\{-\lambda(x-\alpha)\}}{1+\exp\{-\lambda(x-\alpha)\}} \tag{22.16}$$

Figure 22.3 shows the effect of changing λ.

The mean and variance are

$$\mu=\alpha+\frac{\ln(4)}{\lambda} \tag{22.17}$$

$$\sigma^2=\frac{1}{\lambda^2}\left(\frac{\pi^2}{3}-\ln^2(4)\right) \tag{22.18}$$

Fitting to the absolute values of the NHV disturbance data gives α as –0.06 and λ as 1.30. Hence μ is 1.33 and σ is 1.52 – both close to the values calculated from the data. But, with *RSS* at 0.2206, there are several other distributions that fit better.

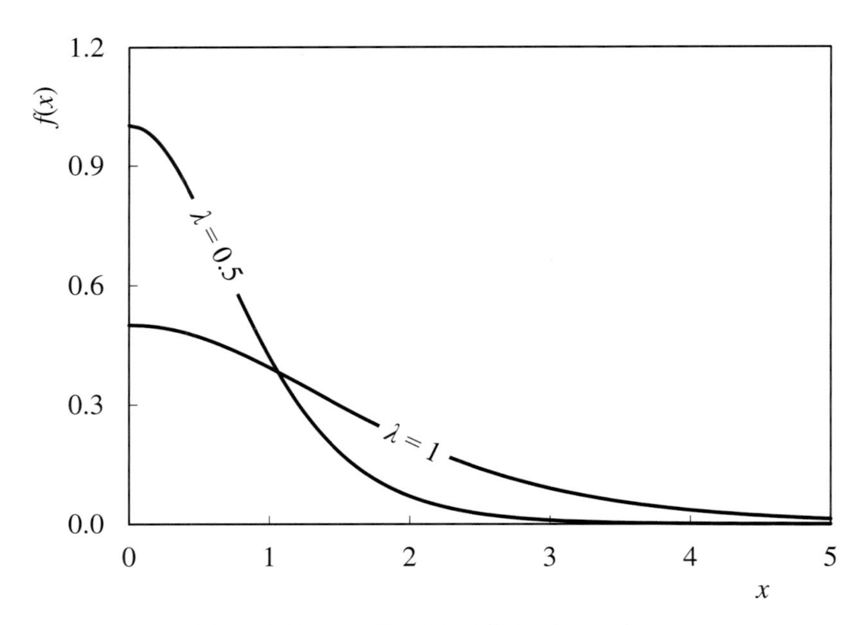

Figure 22.3 *Half-logistic: Effect of λ on shape*

22.3 Skew-Logistic

The *skew-logistic distribution*, as its name suggests, permits skewness to be added to the logistic distribution. There are at least three distributions published with this title. Two are described in this section. The third is also known as the *generalised logistic distribution* and is described in Section 22.7.

Its PDF of the first is

$$f(x) = \frac{\frac{2}{\beta}\exp\left(-\frac{x-\alpha}{\beta}\right)}{\left[1+\exp\left(-\frac{x-\alpha}{\beta}\right)\right]^2\left[1+\exp\left(-\frac{\delta(x-\alpha)}{\beta}\right)\right]} \qquad \beta>0 \tag{22.19}$$

Figure 22.4 shows the effect of varying the skewness parameter (δ); if it is set to zero, the distribution becomes the classic logistic distribution (as shown by the coloured curve).

Fitting to the C_4 in propane data, with α set to 2.04, β to 2.11 and δ far from zero at 10.63, results in *RSS* falling to 0.0419 – comparable to that achieved by the inverse Gaussian distribution. However, while α is representative of the mean and β of the standard deviation, there are no formulae published that convert one to the other. Similarly there are none for skewness or kurtosis.

Fitting to the NHV disturbance data sets α to –0.291, β to 0.727 and δ much closer to zero at 0.281. *RSS* is 0.1031 which is very close to that of the classic logistic distribution. Clearly, because the data is only slightly skewed, there is little advantage in choosing a distribution that can represent skewness.

The second form of the skew-logistic distribution[13] is defined in two parts.

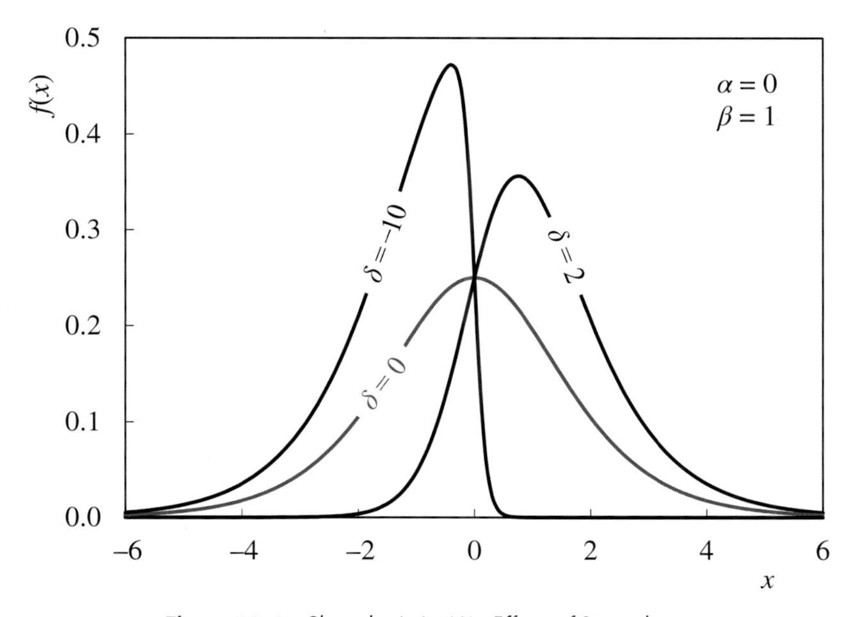

Figure 22.4 *Skew-logistic (1): Effect of δ on shape*

For $x<\alpha$

$$f(x)=\frac{\delta}{1+\delta^2}\frac{\frac{2}{\beta}\exp\left(-\frac{x-\alpha}{\delta\beta}\right)}{\left[1+\exp\left(-\frac{x-\alpha}{\delta\beta}\right)\right]^2}\qquad \beta>0,\ \delta>0 \tag{22.20}$$

$$F(x)=\frac{\left(\frac{2\delta}{1+\delta^2}\right)}{1+\exp\left(-\frac{x-\alpha}{\delta\beta}\right)} \tag{22.21}$$

For $x\geq\alpha$

$$f(x)=\frac{\delta}{1+\delta^2}\frac{\frac{2}{\beta}\exp\left(-\frac{\delta(x-\alpha)}{\beta}\right)}{\left[1+\exp\left(-\frac{\delta(x-\alpha)}{\beta}\right)\right]^2}\qquad \beta>0;\ \delta>0 \tag{22.22}$$

$$F(x)=\frac{\delta^2}{1+\delta^2}\left[1+\frac{2}{\delta^2}\frac{1-\exp\left(-\frac{\delta(x-\alpha)}{\beta}\right)}{1+\exp\left(-\frac{\delta(x-\alpha)}{\beta}\right)}\right] \tag{22.23}$$

As Figure 22.5 shows, the discontinuity at $x=\alpha$ is unlikely to closely fit process data. Fitting to the C_4 in propane data gives values of 4.18 for α, 1.21 for β and 1.05 for δ. *RSS* then becomes 0.3461 – considerably larger than that for the normal distribution. Again there are no formulae published to enable calculation of any of the key statistical parameters.

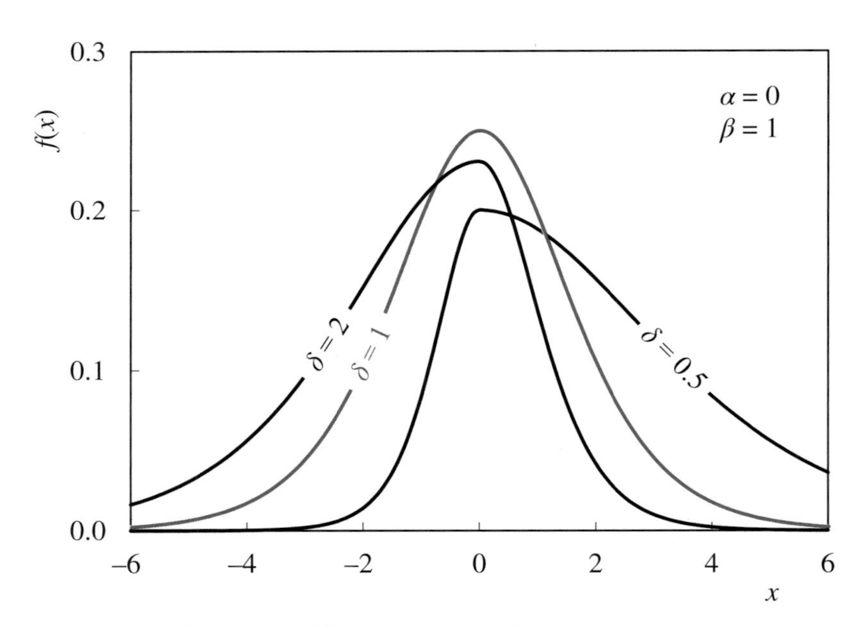

Figure 22.5 *Skew-logistic (2): Effect of δ on shape*

22.4 Log-Logistic

The *log-logistic distribution* is related to the logistic distribution in the same way that the log-normal distribution is related to the normal distribution. It similarly allows it to represent skewness. It is also known as the *Fisk distribution*. Its PDF is

$$f(x)=\frac{\left(\frac{\delta}{\beta}\right)\left(\frac{x-\alpha}{\beta}\right)^{\delta-1}}{\left[1+\left(\frac{x-\alpha}{\beta}\right)^{\delta}\right]^{2}} \qquad x\geq\alpha;\ \beta,\delta>0 \tag{22.24}$$

Its CDF is

$$F(x)=\frac{\left(\frac{x-\alpha}{\beta}\right)^{\delta}}{1+\left(\frac{x-\alpha}{\beta}\right)^{\delta}}=\frac{(x-\alpha)^{\delta}}{\beta^{\delta}+(x-\alpha)^{\delta}} \tag{22.25}$$

Its QF is

$$x(F)=\alpha+\beta\left(\frac{F}{1-F}\right)^{1/\delta} \qquad 0\leq F\leq 1 \tag{22.26}$$

If δ is set to 1, it becomes the *standard log-logistic distribution*. (According to our definition, this is a misuse of the word 'standard' which is more generally used for the case when α is 0 and β is 1.) Generally, δ need not be an integer and so the distribution must be lower-bounded at α.

Fitting this distribution to the C_4 in propane data gives α as 0.743, β as 3.52 and δ as 2.78. Figure 22.6 shows this as the coloured line and also the impact of changing these values. *RSS* is 0.0529 which, although considerably better than that using the normal distribution, is slightly bettered by the inverse Gaussian distribution.

The moments are given by

$$m_n=\frac{n\pi}{\delta\sin\left(\frac{n\pi}{\delta}\right)}\beta^n=\Gamma\left(1+\frac{n}{\delta}\right)\Gamma\left(1-\frac{n}{\delta}\right)\beta^n \qquad \delta>n \tag{22.27}$$

This leads to

$$\mu=\alpha+\frac{\pi\beta}{\delta\sin\left(\frac{\pi}{\delta}\right)}=\Gamma\left(1+\frac{n}{\delta}\right)\Gamma\left(1-\frac{n}{\delta}\right)\beta \qquad \delta>1 \tag{22.28}$$

$$\sigma^2=\pi\left(\frac{\beta}{\delta}\right)^2\left(\frac{2\delta}{\sin\left(\frac{2\pi}{\delta}\right)}-\frac{\pi}{\sin^2\left(\frac{\pi}{\delta}\right)}\right)=\left\{\Gamma\left(1+\frac{2}{\delta}\right)\Gamma\left(1-\frac{2}{\delta}\right)-\left[\Gamma\left(1+\frac{n}{\delta}\right)\Gamma\left(1-\frac{n}{\delta}\right)\right]^2\right\}\beta^2 \qquad \delta>2 \tag{22.29}$$

These give μ as 5.14 and σ as 4.13. Figure 22.7 shows the extended tail of the fitted distribution, explaining why the standard deviation is so much larger than other estimates.

It is possible, although complicated, to derive formulae for skewness and kurtosis. Skewness approaches zero and kurtosis approaches 4.2 as $\delta\rightarrow\infty$. Reducing δ increases both skewness and kurtosis, as shown by Figure 22.8. Skewness can only be defined if δ is greater than 3 and kurtosis only if it is greater than 4. Indeed this is the problem encountered in this case.

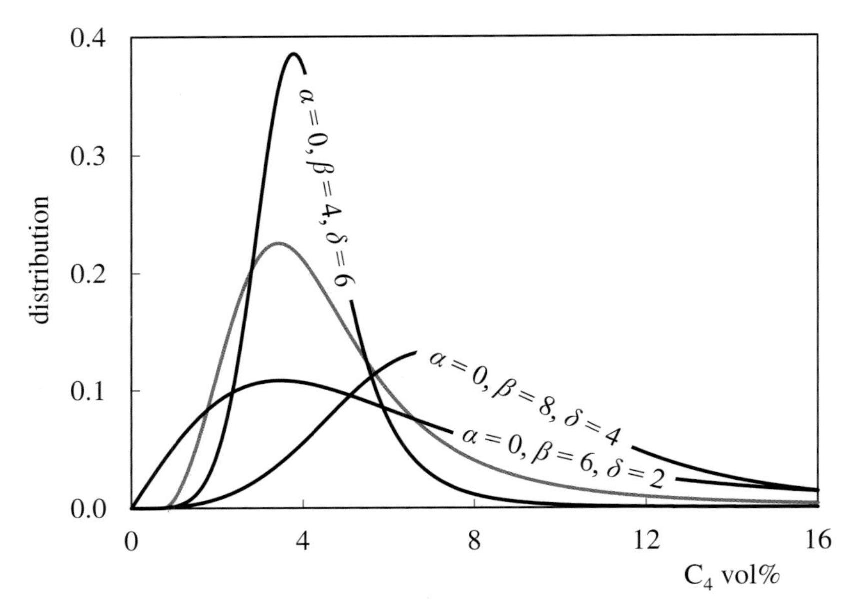

Figure 22.6 Log-logistic: Effect of α, β and δ on shape

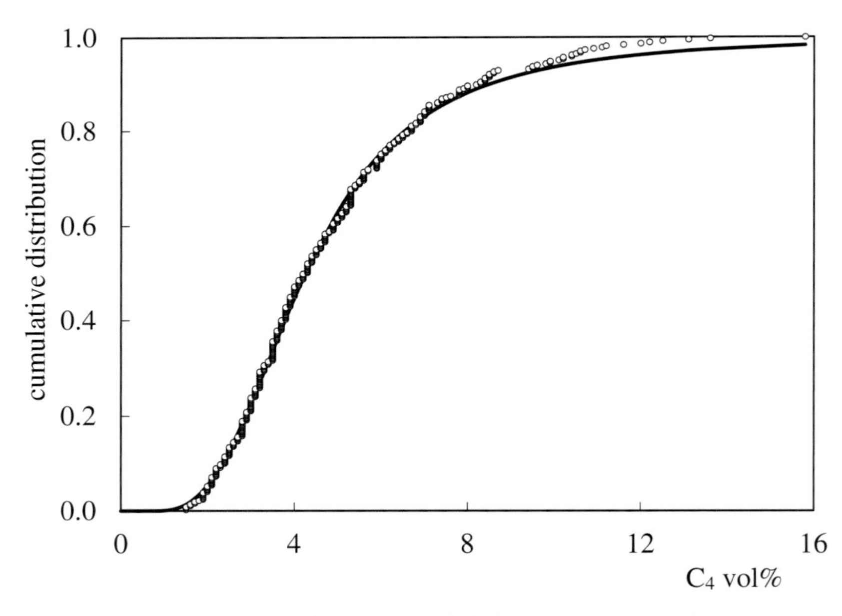

Figure 22.7 Log-logistic: Fitted to the C_4 in propane data

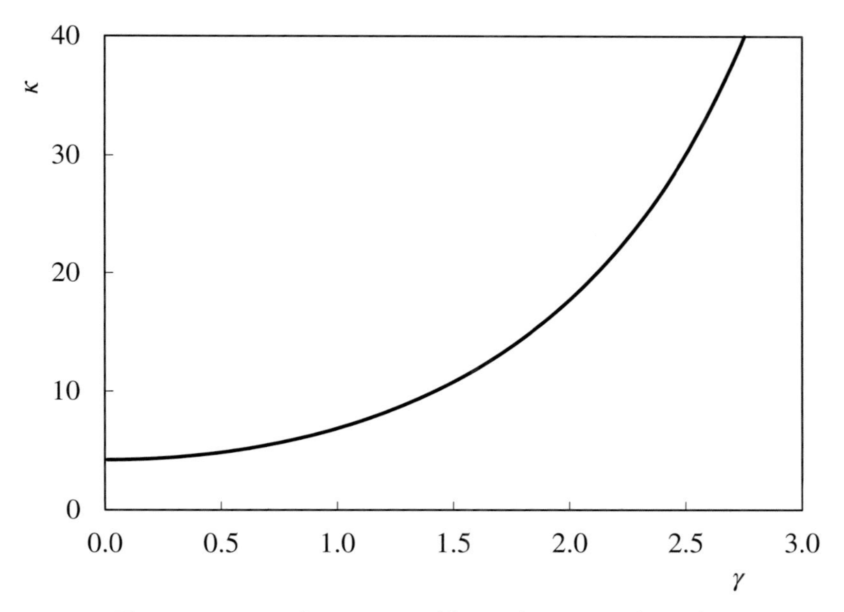

Figure 22.8 Log-logistic: Feasible combinations of γ and κ

Fitting to the NHV disturbance data gives α as −8.98, β as 8.94 and δ as 22.6. Equation (22.28) gives μ as 0.05. Equation (22.29) gives σ as 1.31. *RSS* is 0.0961, again better than that of the normal distribution.

22.5 Paralogistic

The PDF of the *paralogistic distribution* is

$$f(x) = \frac{\frac{\delta^2}{\beta}\left(\frac{x-\alpha}{\beta}\right)^{\delta-1}}{\left[1+\left(\frac{x-\alpha}{\beta}\right)^{\delta}\right]^{\delta+1}} \quad x \geq \alpha;\ \beta,\delta > 0 \tag{22.30}$$

Figure 22.9 shows the effect of varying δ with α fixed at 0 and β at 1. Fitting to the C_4 in propane data sets α to 1.56, β to 4.22 and δ to 1.78 – resulting in *RSS* of 0.0327. The fit is comparable to the best of the distributions.

The CDF is

$$F(x) = 1 - \left[1+\left(\frac{x-\alpha}{\beta}\right)^{\delta}\right]^{-\delta} \tag{22.31}$$

The QF is

$$x(F) = \alpha + \beta\left[(1-F)^{-1/\delta} - 1\right]^{1/\delta} \quad 0 \leq F \leq 1 \tag{22.32}$$

Moments are given by

$$m_n = \frac{\Gamma\left(1+\frac{n}{\delta}\right)\Gamma\left(\delta-\frac{n}{\delta}\right)}{\Gamma(\delta)}\beta^n \quad \delta > \sqrt{n} \tag{22.33}$$

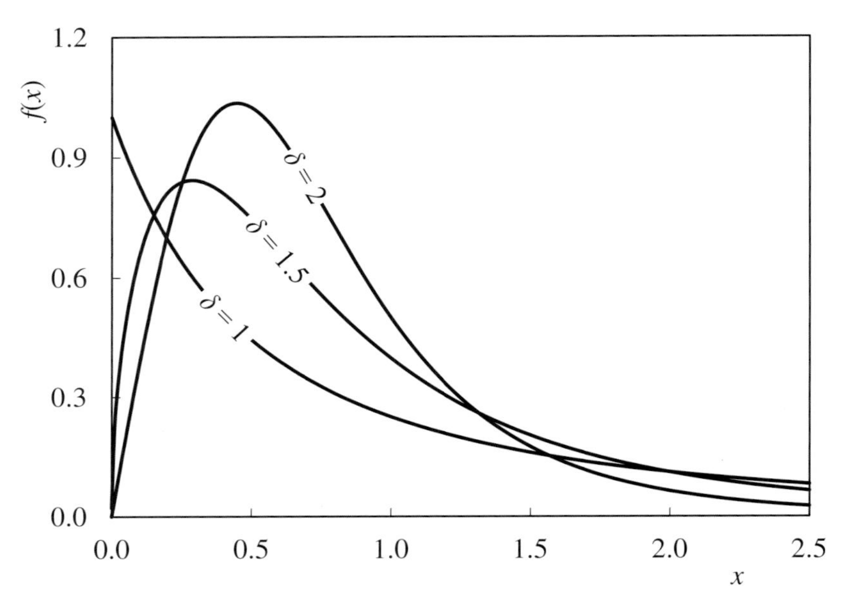

Figure 22.9 *Paralogistic: Effect of δ on shape*

Thus the mean and variance are

$$\mu = \alpha + \frac{\beta}{\Gamma(\delta)}\Gamma\left(1+\frac{1}{\delta}\right)\Gamma\left(\delta-\frac{1}{\delta}\right) \tag{22.34}$$

$$\sigma^2 = \left\{\frac{1}{\Gamma(\delta)}\Gamma\left(1+\frac{2}{\delta}\right)\Gamma\left(\delta-\frac{2}{\delta}\right) - \left[\frac{1}{\Gamma(\delta)}\Gamma\left(1+\frac{1}{\delta}\right)\Gamma\left(\delta-\frac{1}{\delta}\right)\right]^2\right\}\beta^2 \tag{22.35}$$

This gives μ as 5.29 and σ as 3.75.

Figure 22.10 shows the feasible combinations of skew and kurtosis. In this example δ is below the limit for kurtosis to be calculated.

22.6 Inverse Paralogistic

The *inverse paralogistic distribution* is described by the PDF

$$f(x) = \frac{\dfrac{\delta^2}{\beta}\left(\dfrac{x-\alpha}{\beta}\right)^{\delta^2-1}}{\left[1+\left(\dfrac{x-\alpha}{\beta}\right)^{\delta}\right]^{\delta+1}} \qquad x \geq \alpha;\ \beta, \delta > 0 \tag{22.36}$$

Fitting this to the C_4 in propane data gives values for α, β and δ as 2.70, 2.48 and 0.942 respectively. At 1.171, *RSS* is far worse than that of the normal distribution. As Figure 22.11 confirms, this is a poor fit.

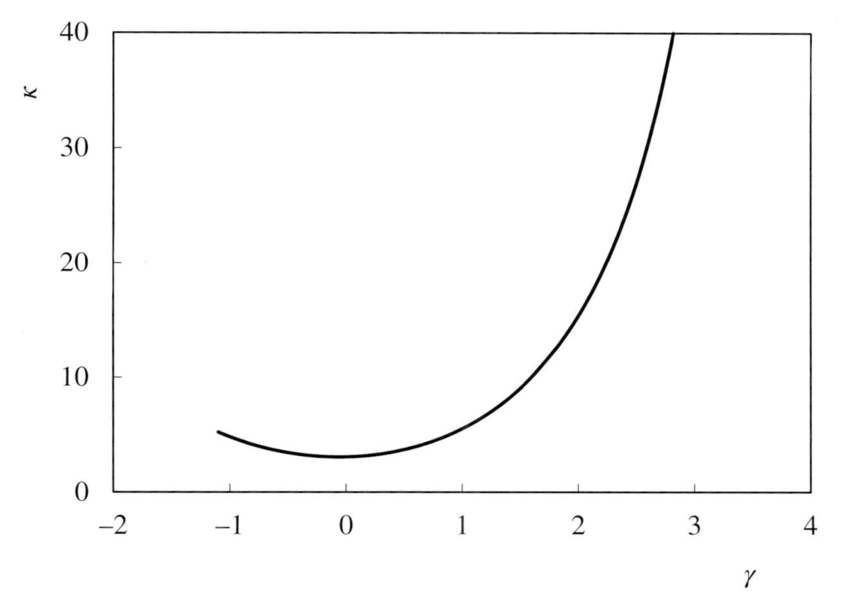

Figure 22.10 Paralogistic: Feasible combinations of γ and κ

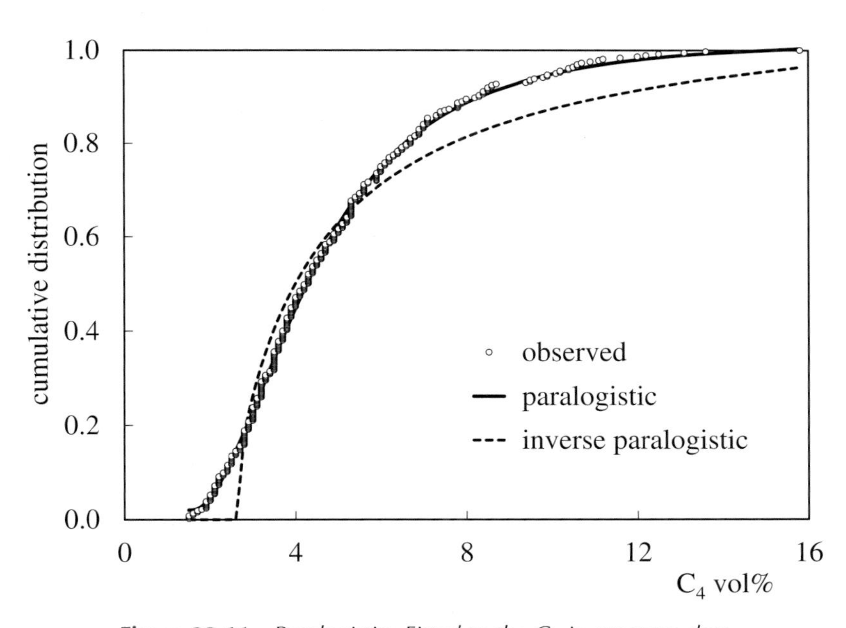

Figure 22.11 Paralogistic: Fitted to the C_4 in propane data

22.7 Generalised Logistic

The *generalised logistic distribution* has four types. All three include the additional shape parameter (δ). Although used in different ways, if set to 1, the generalised version will revert to the classic logistic distribution.

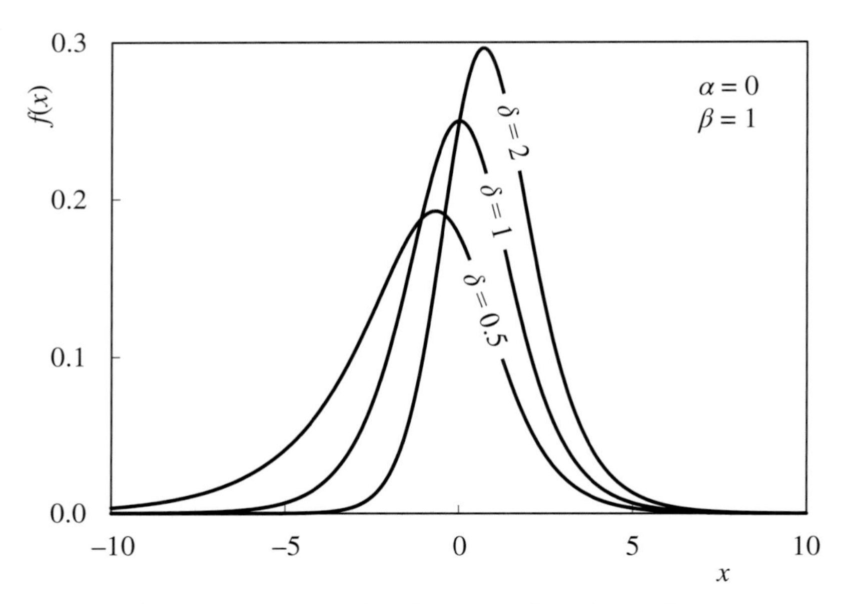

Figure 22.12 Generalised logistic-I: Effect of δ on shape

The PDF of Type I is

$$f(x)=\frac{\frac{\delta}{\beta}\exp\left(-\frac{x-\alpha}{\beta}\right)}{\left[1+\exp\left(-\frac{x-\alpha}{\beta}\right)\right]^{\delta+1}} \qquad x\geq\alpha;\ \beta,\delta>0 \tag{22.37}$$

Figure 22.12 shows the impact of changing δ. Skewness is zero when δ is 1 and increases as δ is increased. The CDF is

$$F(x)=\frac{1}{\left[1+\exp\left(-\frac{x-\alpha}{\beta}\right)\right]^{\delta}} \tag{22.38}$$

The CDF can be inverted

$$x(F)=\alpha-\beta\ln\left[\frac{F^{1/\delta}}{F^{1/\delta}-1}\right] \qquad 0\leq F\leq 1 \tag{22.39}$$

The PDF of Type II is

$$f(x)=\frac{\frac{\delta}{\beta}\exp\left(-\frac{\delta(x-\alpha)}{\beta}\right)}{\left[1+\exp\left(-\frac{x-\alpha}{\beta}\right)\right]^{\delta+1}} \qquad \beta,\delta>0 \tag{22.40}$$

Figure 22.13 shows the impact of changing δ. Again, skewness is zero when δ is 1 but now decreases as δ is increased. Figure 22.13 is the mirror image of Figure 22.12. The CDF is

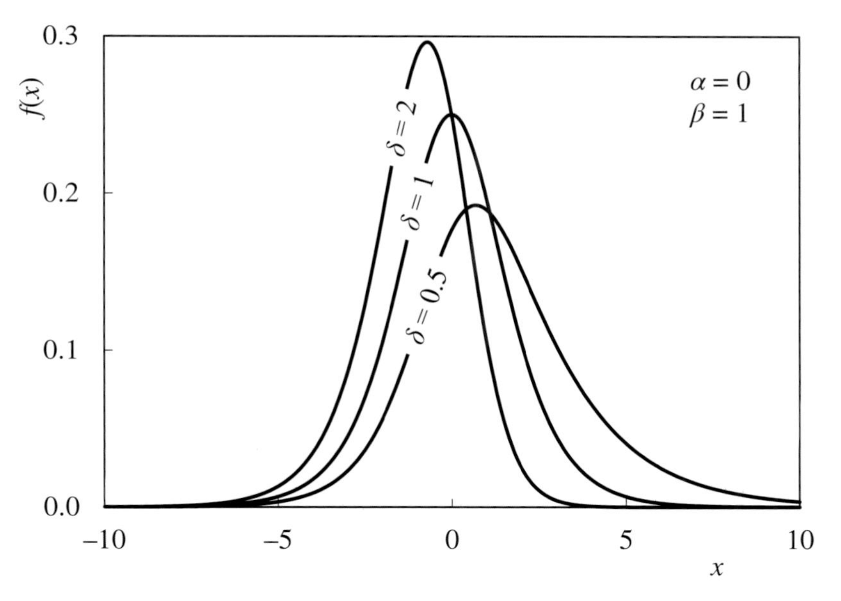

Figure 22.13 *Generalised logistic-II: Effect of δ on shape*

$$F(x) = 1 - \frac{\exp\left(-\frac{\delta(x-\alpha)}{\beta}\right)}{\left[1+\exp\left(-\frac{x-\alpha}{\beta}\right)\right]^{\delta}} \tag{22.41}$$

The CDF can be inverted

$$x(F) = \alpha - \beta \ln\left[\frac{1-(1-F)^{1/\alpha}}{(1-F)^{1/\alpha}}\right] \quad 0 \le F \le 1 \tag{22.42}$$

The PDF of Type III is

$$f(x) = \frac{\frac{1}{\beta}\exp\left(-\frac{\delta(x-\alpha)}{\beta}\right)}{\mathrm{B}(\delta,\delta)\left[1+\exp\left(-\frac{x-\alpha}{\beta}\right)\right]^{2\delta}} \quad x \ge \alpha;\ \beta,\delta > 0 \tag{22.43}$$

Figure 22.14 shows the impact of changing δ. Skewness stays at zero; δ now changes kurtosis.

Type IV also known as the *exponential generalised beta-II distribution*. Its PDF is

$$f(x) = \frac{\frac{1}{\beta}\exp\left(-\frac{\delta_2(x-\alpha)}{\beta}\right)}{\mathrm{B}(\delta_1,\delta_2)\left[1+\exp\left(-\frac{x-\alpha}{\beta}\right)\right]^{\delta_1+\delta_2}} \quad x \ge \alpha;\ \beta,\delta_1,\delta_2 > 0 \tag{22.44}$$

The CDF for both Types III and IV include a mathematical function that is beyond the scope of this book. They are more easily derived from the trapezium rule.

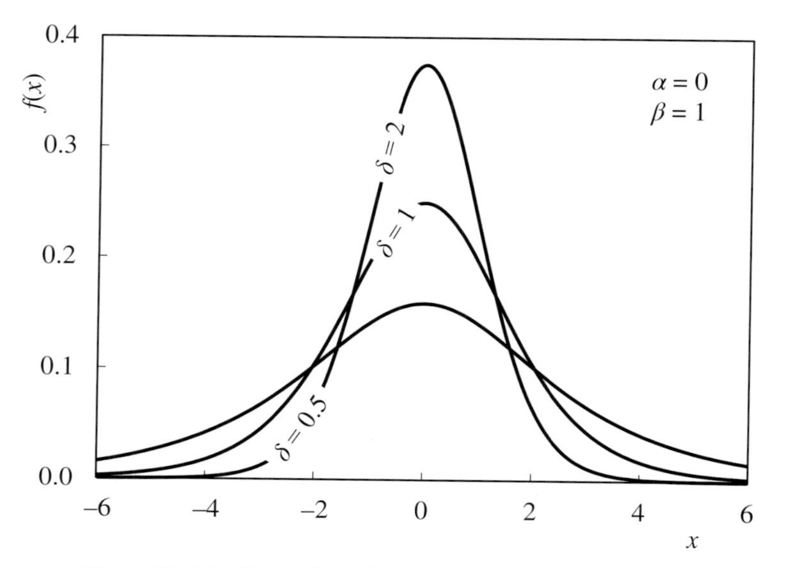

Figure 22.14 Generalised logistic-III: Effect of δ on shape

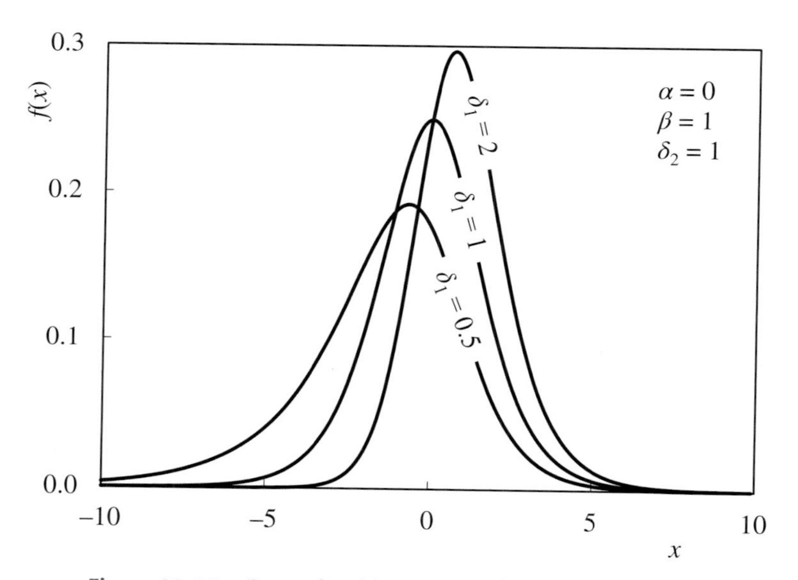

Figure 22.15 Generalised logistic-IV: Effect of δ_1 on shape

Figure 22.15 shows the impact of changing δ_1 with δ_2 fixed at 1. The curves are the same as those for Type I. If δ_1 is set to 1, Type IV becomes Type II – as shown by Figure 22.16, which shows the impact of changing δ_2. If δ_2 is kept equal to δ_1, Type IV becomes Type III. We therefore need explore fitting only Type IV.

Fitting Equation (22.44) to the C_4 in propane data gives values of 0.0443 for α, 0.0122 for β, 0.0129 for δ_1 and 0.01 for δ_2 – giving a value of 0.0288 for *RSS*. Not surprisingly the inclusion of a fourth parameter results in one of the best fits.

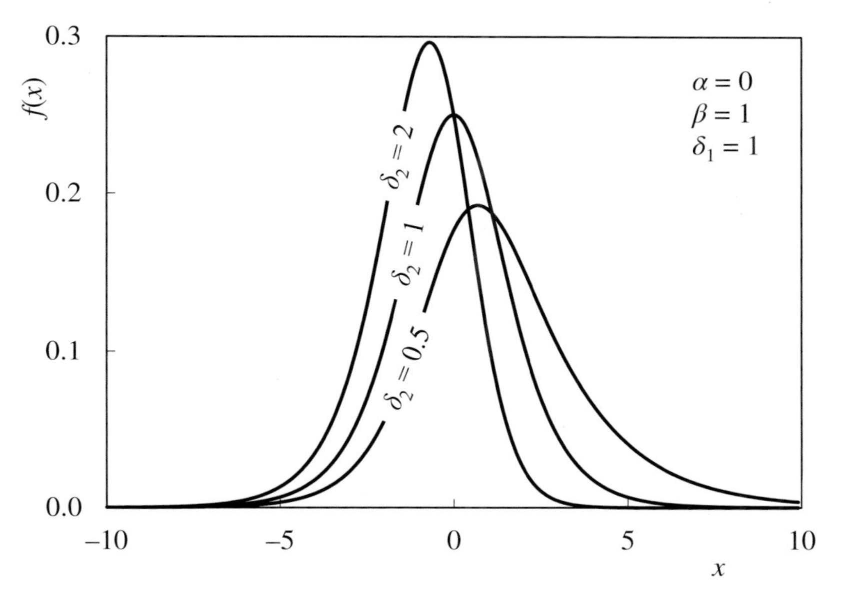

Figure 22.16 *Generalised logistic-IV: Effect of δ_2 on shape*

Fitting to the NHV disturbance data gives values of −0.212 for α, 0.7174 for β, 23.2 for δ_1 and 0.282 for δ_2. Also adjusting $F(x_1)$ to 0.0122 gives a value of 0.0374 for *RSS*. Again this is the one of the best fits.

This is one of very few distributions that give the best fit to both datasets. However, none of the fitted parameters have any obvious relationship to mean or standard deviation. In principle they can be derived, using the techniques described in Section 4.7, from the moment generating function

$$M(t) = \frac{\Gamma(\delta_2 - t)\Gamma(\delta_1 + t)}{\Gamma(\delta_1)\Gamma(\delta_2)} \tag{22.45}$$

However, differentiating Equation (22.45) requires us to differentiate the gamma function. Equation (11.40) allows us to do this but only for the gamma function of integers. Since δ_1 and δ_2 are almost certainly not integers, we would have to adopt a far more complex approach. This illustrates well the compromise we often have to make in selecting the prior distribution between the accuracy of fit achievable, versatility and practicality.

22.8 Generalised Log-Logistic

The *generalised log-logistic distribution* is described by the PDF

$$f(x) = \frac{1}{\mathrm{B}(\delta_1, \delta_2)} \frac{\dfrac{1}{\beta}\left(\dfrac{x-\alpha}{\beta}\right)^{\delta_1 - 1}}{\left(1 + \dfrac{x-\alpha}{\beta}\right)^{\delta_1 + \delta_2}} \qquad x \geq \alpha;\ \beta, \delta_1, \delta_2 > 0 \tag{22.46}$$

The effect of δ_1 and δ_2 is shown in Figure 22.17. Fitting to the C_4 in propane data, with α set to 0, gives 1.15 for β, 19.8 for δ_1, 5.59 for δ_2 and 0.0354 for *RSS*. There are no published formulae for mean, standard deviation, skewness and kurtosis.

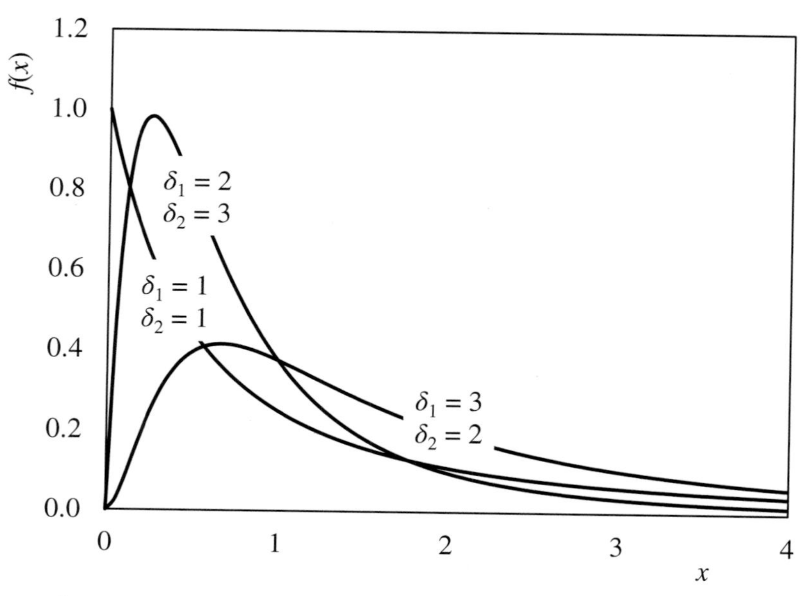

Figure 22.17 Generalised log-logistic: Effect of δ_1 and δ_2 on shape

22.9 Exponentiated Kumaraswamy–Dagum

The *exponentiated Kumaraswamy–Dagum (EKD) distribution*[14] is an extension of those described above. It was developed to accommodate a long tail of high values. Its CDF is

$$F(x) = \left\{1 - \left[1 - \left(1 + \lambda(x-\alpha)^{-\delta_1}\right)^{-\delta_2}\right]^{\delta_3}\right\}^{\delta_4} \qquad x \geq 0;\ \lambda, \delta_1, \delta_2, \delta_3, \delta_4 > 0 \qquad (22.47)$$

Its PDF can be obtained by differentiating Equation (22.47) but the result is too complex to merit inclusion here. The QF can be obtained by inversion.

$$x(F) = \alpha + \left(\frac{1}{\lambda}\left\{\left[1 - \left(1 - F^{1/\delta_4}\right)^{1/\delta_3}\right]^{-1/\delta_2} - 1\right\}\right)^{-1/\delta_1} \qquad 0 \leq F \leq 1 \qquad (22.48)$$

By selectively setting coefficients to 1, Equation (22.47) can in principle describe many of the distributions covered previously. This would remove the need to explore each distribution separately. However some of these distributions have additional parameters not present in the equation. Considering only the EKD distribution would not therefore fully explore the potential of the related distributions.

Fitting to the C_4 in propane data gives one solution as α as 0.00, λ as 14.9, δ_1 as 0.761, δ_2 as 0.852, δ_3 as 15.4 and δ_4 as 30.8 – resulting in a value of 0.0350 for *RSS*. However there are multiple sets of parameters that give very similar values. This suggests that some of the parameters influence the shape of the distribution in very similar ways and perhaps could be omitted.

No simple formula can be developed to determine the moments and hence there is no easy way of determining the mean or standard deviation. While it might well give the best fit, this problem and the lack of robustness mean the EKD distribution is unlikely to be of practical use in the process industry.

23

Pareto Distribution

There are four types of Pareto distribution. Originally developed for exploring the distribution of wealth, they lead to Pareto analysis and to the Pareto principle, or '80/20 rule', familiar to many engineers.

23.1 Pareto Type I

The *Pareto-I distribution* is described by the PDF

$$f(x) = \frac{\delta}{x}\left(\frac{\beta}{x}\right)^{\delta} \quad x \geq \beta;\ \beta, \delta > 0 \tag{23.1}$$

The shape parameter δ is sometimes described as the *tail index*; β is the minimum nonzero value that x can have. Figure 23.1 shows the effect of changing these parameters.

The CDF is

$$F(x) = 1 - \left(\frac{\beta}{x}\right)^{\delta} \tag{23.2}$$

The QF is

$$x(F) = \beta(1-F)^{-1/\delta} \quad 0 \leq F \leq 1 \tag{23.3}$$

The moments are given by

$$m_n = \frac{\delta}{\delta - n}\beta^n \quad \delta > n \tag{23.4}$$

From this can be derived

$$\mu = \frac{\delta}{\delta - 1}\beta \quad \delta > 1 \tag{23.5}$$

Statistics for Process Control Engineers: A Practical Approach, First Edition. Myke King.

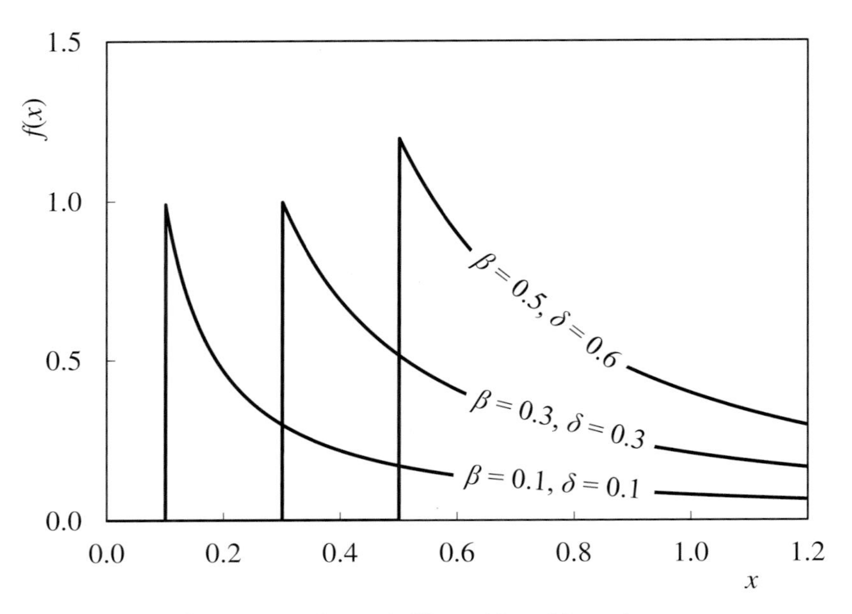

Figure 23.1 *Pareto-I: Effect of β and δ on shape*

$$\sigma^2 = \frac{\delta}{(\delta-1)^2(\delta-2)}\beta^2 \qquad \delta>2 \tag{23.6}$$

$$\gamma = \frac{2(\delta+1)}{\delta-3}\sqrt{\frac{\delta-2}{\delta}} \qquad \delta>3 \tag{23.7}$$

$$\kappa = \frac{3(\delta-2)\left(3\delta^2+\delta+2\right)}{\delta(\delta-3)(\delta-4)} \qquad \delta>4 \tag{23.8}$$

One application is to assess the probability of the size of process disturbances. We can explore this using the NHV disturbance data. However, as Figure 23.1 shows, the Pareto distribution is non-symmetrical. Generally, we are not interested whether the disturbance is an increase or a decrease so we can use its absolute value. Figure 23.2 shows the resulting observed distribution. As might be expected, there are a large number of very small changes with the very occasional large change. However, the likelihood of a change being exactly zero is remote. Rather than attempt to count these, we effectively treat as zero any change that is less than β.

Fitting gives β as 0.22 and δ as 0.67. At 0.6831 *RSS* is much larger than that for the distributions tested so far. In this case, δ is less than 1 and so Equations (23.5) to (23.8) cannot be applied.

23.2 Bounded Pareto Type I

The *bounded Pareto-I distribution* is described by

$$f(x) = \delta\frac{x_{min}^{\delta}\,x_{max}^{\delta}}{\left(x_{max}^{\delta}-x_{min}^{\delta}\right)x^{\delta+1}} \qquad \delta>0 \tag{23.9}$$

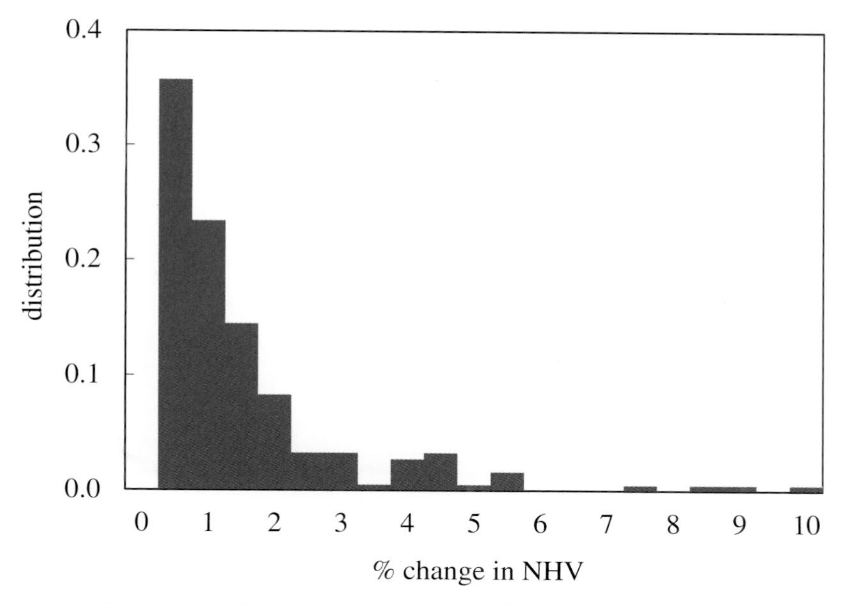

Figure 23.2 *Observed distribution of absolute change in NHV*

$$F(x) = \frac{x^{\delta} - x_{\min}^{\delta}}{x_{\max}^{\delta} - x_{\min}^{\delta}} \tag{23.10}$$

$$x(F) = \left[F\left(x_{\max}^{\delta} - x_{\min}^{\delta}\right) + x_{\min}^{\delta}\right]^{1/\delta} \quad 0 \le F \le 1 \tag{23.11}$$

$$m_n = \frac{\delta}{\delta - n} \frac{x_{\min}^{n} x_{\max}^{\delta} - x_{\min}^{\delta} x_{\max}^{n}}{x_{\max}^{\delta} - x_{\min}^{\delta}} \quad \delta > n \tag{23.12}$$

Hence

$$\mu = \frac{\delta}{\delta - 1} \frac{x_{\min} x_{\max}^{\delta} - x_{\min}^{\delta} x_{\max}}{x_{\max}^{\delta} - x_{\min}^{\delta}} \quad \delta > 1 \tag{23.13}$$

$$\sigma^2 = \frac{\delta}{\delta - 2} \frac{x_{\min}^{2} x_{\max}^{\delta} - x_{\min}^{\delta} x_{\max}^{2}}{x_{\max}^{\delta} - x_{\min}^{\delta}} \quad \delta > 2 \tag{23.14}$$

Fitting to the NHV disturbance data sets $x_{\min}$ to 0.09, $x_{\max}$ to 3.23 and δ to 0.20. Again, although *RSS* is improved at 0.2343, the fit is still poor. Also of concern are the 20 values that exceed $x_{\max}$ and so have been excluded from the fit. Again the value of δ prevents the calculation of any of the raw moments.

23.3 Pareto Type II

The *Pareto-II distribution* is a special case of the *Burr-XII distribution*. It is described by the PDF

$$f(x) = \frac{\delta}{\beta}\left(1 + \frac{x - \alpha}{\beta}\right)^{-(\delta+1)} \quad x \ge \alpha;\ \beta > 0 \tag{23.15}$$

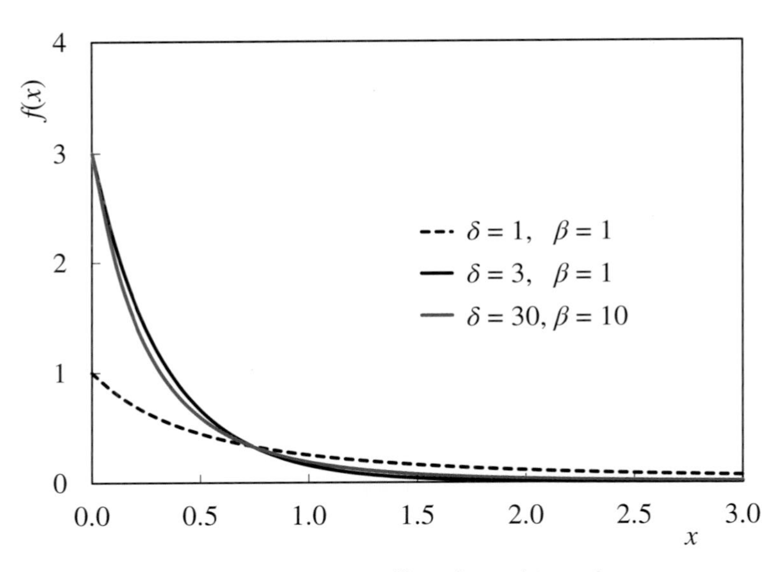

Figure 23.3 Pareto-II: Effect of β and δ on shape

Figure 23.3 shows the distribution plotted for a range of values of δ and β with α fixed at 0. It is notable that the shape is mainly governed by the δ/β ratio rather than the individual values. For example, the two solid curves are almost identical despite the change, by a factor of 10, in the shape parameters. Curve fitting can therefore generate vary large values for δ and β with negligible improvement in accuracy.

The CDF is

$$F(x) = 1 - \left(1 + \frac{x-\alpha}{\beta}\right)^{-\delta} \tag{23.16}$$

Raw moments are given by

$$m_n = \frac{\Gamma(\delta-n)\Gamma(n+1)}{\Gamma(\delta)}\beta^n \quad \delta > n \tag{23.17}$$

Hence

$$\mu = \alpha + \frac{\beta}{\delta-1} \quad \delta > 1 \tag{23.18}$$

$$\sigma^2 = \frac{\delta\beta^2}{(\delta-1)^2(\delta-2)} \quad \delta > 2 \tag{23.19}$$

Fitting Equation (23.16) to the NHV disturbance data gives values of 0.0385 for α, 3.96 for δ, 3.64 for β and 0.0501 for *RSS*. This is the best fit so far with μ now determined as 1.27 and σ as 1.75.

Inverting Equation (23.16) gives the QF

$$x(F) = \alpha + \beta\left\{(1-F)^{-1/\delta} - 1\right\} \quad 0 \le F \le 1 \tag{23.20}$$

So for example, if F is 0.95, x is 4.15. This means that 95% of the disturbances can be expected to be below 4.15. In other words, there is 5% probability of a disturbance being greater than this.

23.4 Lomax

The *Lomax distribution* is a special form of the Pareto-II distribution in which α is set to zero. In the case of process disturbances and inferential errors, we would expect the mean value to be zero. We saw this confirmed by fitting the Pareto-II distribution that set α very close to zero. For this example, the Lomax distribution might be expected to give an almost identical fit.

It is described by

$$f(x) = \frac{\delta}{\beta}\left(1 + \frac{x}{\beta}\right)^{-(\delta+1)} \qquad x \geq 0;\ \beta > 0 \tag{23.21}$$

$$F(x) = 1 - \left(1 + \frac{x}{\beta}\right)^{-\delta} \tag{23.22}$$

$$x(F) = \beta\left\{(1-F)^{-1/\delta} - 1\right\} \qquad 0 \leq F \leq 1 \tag{23.23}$$

The raw moments are the same as those for the Pareto-II distribution. Only the mean is affected, i.e.

$$\mu = \frac{\beta}{\delta - 1} \qquad \delta > 1 \tag{23.24}$$

To fit the distribution, we would normally calculate μ and σ from the process data and, by rearranging Equations (23.24) and (23.19), obtain maximum likelihood estimates for δ and β.

$$\delta = \frac{2\mu^2 + \sigma^2 + \sigma\sqrt{\sigma^2 - 4\mu^2}}{2\mu} \qquad \sigma \geq 2\mu \tag{23.25}$$

$$\beta = \frac{2\mu^2 + \sigma^2 - 2\mu + \sigma\sqrt{\sigma^2 - 4\mu^2}}{2} \qquad \sigma \geq 2\mu \tag{23.26}$$

Calculated from the NHV disturbance data, μ is 1.30 and σ is 1.60. However, in this example, these values violate the criterion for applying Equations (23.25) and (23.26). Instead, we must guess a starting point; fitting then gives values of 6.93 for δ and 7.12 for β. *RSS* is 0.0721. From Equations (23.24) and (23.19), μ is now determined as 1.20 and σ as 1.42. Forcing α to zero has a surprisingly large impact, particularly on the estimate of σ.

23.5 Inverse Pareto

The *inverse Pareto distribution* is described by

$$f(x) = \frac{\dfrac{\delta}{\beta}\left(\dfrac{x-\alpha}{\beta}\right)^{\delta-1}}{\left[1 + \dfrac{x-\alpha}{\beta}\right]^{\delta+1}} \qquad x \geq \alpha;\ \beta > 0 \tag{23.27}$$

Figure 23.4 shows the effect of changing δ, with α fixed at 0 and β at 1.

The CDF is

$$F(x) = \left(\frac{x-\alpha}{x-\alpha+\beta}\right)^{\delta} = \left[\frac{\left(\dfrac{x-\alpha}{\beta}\right)}{1 + \left(\dfrac{x-\alpha}{\beta}\right)}\right]^{\delta} \tag{23.28}$$

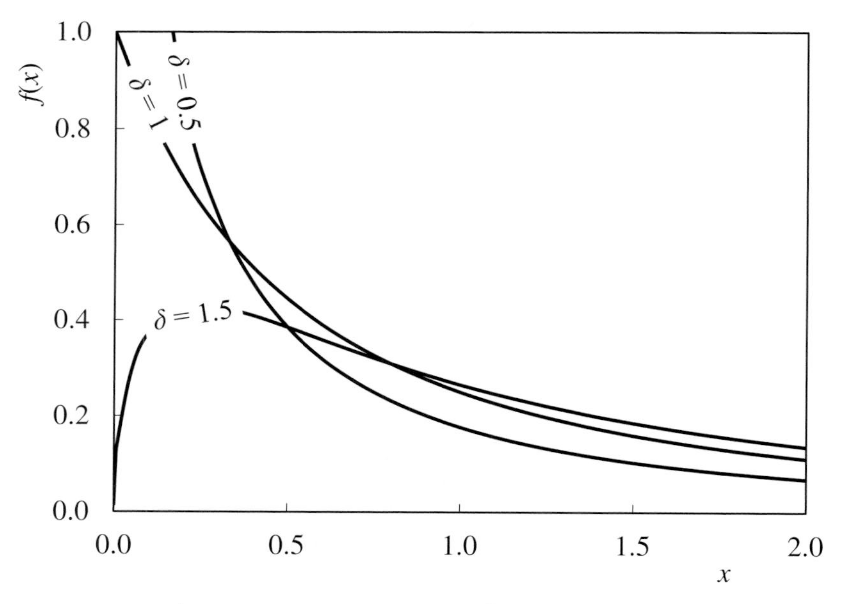

Figure 23.4 *Inverse Pareto: Effect of δ on shape*

Fitting to the C_4 in propane data gives α as 2.18, β as 0.0333 and δ as 38.4. With *RSS* at 1.1500, the fit is very poor.

The QF is

$$x(F) = \alpha + \frac{\beta}{(1 - F^{1/\delta})} \qquad 0 \le F \le 1 \tag{23.29}$$

23.6 Pareto Type III

The *Pareto-III distribution* is described by the PDF

$$f(x) = \frac{\frac{\delta}{\beta}\left(\frac{x-\alpha}{\beta}\right)^{\delta-1}}{\left[1 + \left(\frac{x-\alpha}{\beta}\right)^{\delta}\right]^2} \qquad x \ge \alpha;\ \beta, \delta > 0 \tag{23.30}$$

Figure 23.5 shows the effect of changing δ, with α fixed at 0 and β at 1. The CDF is

$$F(x) = 1 - \left[1 + \left(\frac{x-\alpha}{\beta}\right)^{\delta}\right]^{-1} \tag{23.31}$$

The QF is

$$x(F) = \alpha + \beta\left(\frac{F}{1-F}\right)^{\frac{1}{\delta}} \qquad 0 \le F \le 1 \tag{23.32}$$

Raw moments are given by

$$m_n = \Gamma\left(1 - \frac{n}{\delta}\right)\Gamma\left(1 + \frac{n}{\delta}\right)\beta^n \qquad \delta > n \tag{23.33}$$

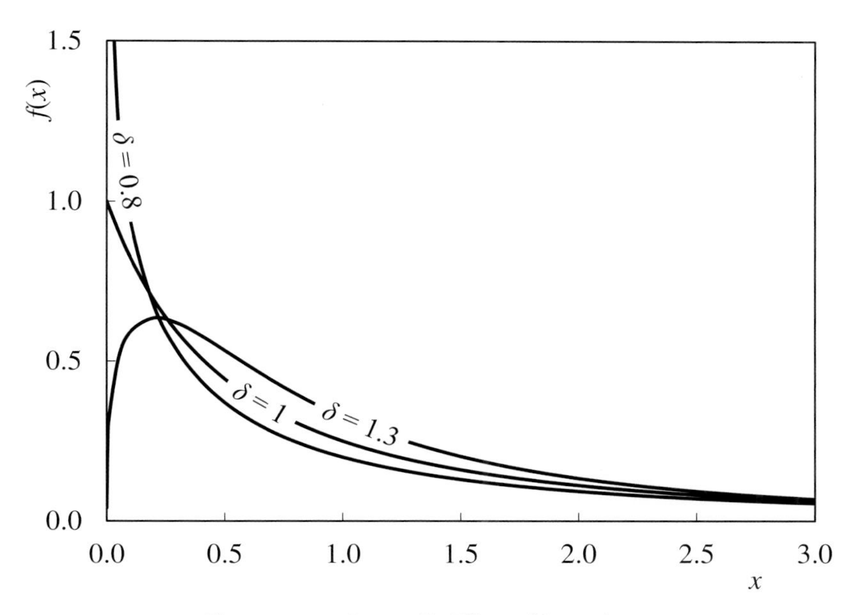

Figure 23.5 *Pareto-III: Effect of δ on shape*

Hence

$$\mu = \alpha + \Gamma\left(1-\frac{1}{\delta}\right)\Gamma\left(1+\frac{1}{\delta}\right)\beta \quad \delta > 1 \tag{23.34}$$

$$\sigma^2 = \left[\Gamma\left(1-\frac{2}{\delta}\right)\Gamma\left(1+\frac{2}{\delta}\right) - \Gamma^2\left(1-\frac{1}{\delta}\right)\Gamma^2\left(1+\frac{1}{\delta}\right)\right]\beta^2 \quad \delta > 2 \tag{23.35}$$

23.7 Pareto Type IV

The *Pareto-IV distribution* is the same as the Burr-XII distribution – as described in Section 12.2.

23.8 Generalised Pareto

The *generalised Pareto distribution* is the title of three different distributions. The most straightforward is developed by adding a power term to the Pareto-I distribution. It is also known as the *Stoppa distribution* as covered in the next chapter.

The second version is described by

$$f(x) = \frac{1}{\beta}\left[1 + \frac{\delta(x-\alpha)}{\beta}\right]^{-\frac{1}{\delta}-1} \quad x \geq \alpha;\ \beta, \delta > 0 \tag{23.36}$$

$$F(x) = 1 - \left[1 + \frac{\delta(x-\alpha)}{\beta}\right]^{-\frac{1}{\delta}} \tag{23.37}$$

$$x(F)=\alpha+\frac{\beta}{\delta}\left[(1-F)^{-\delta}-1\right] \tag{23.38}$$

$$\mu=\alpha+\frac{\beta}{1-\delta} \qquad \delta<1 \tag{23.39}$$

$$\sigma^2=\frac{\beta^2}{(1-\delta)^2(1-2\delta)} \qquad \delta<\frac{1}{2} \tag{23.40}$$

$$\gamma=\frac{2(1+\delta)\sqrt{1-2\delta}}{(1-3\delta)} \qquad \delta<\frac{1}{3} \tag{23.41}$$

$$\kappa=\frac{3(1-2\delta)\left(2\delta^2+\delta+3\right)}{(1-3\delta)(1-4\delta)} \qquad \delta<\frac{1}{4} \tag{23.42}$$

The effect of changing δ, with α fixed at 0 and β at 1, is shown in Figure 23.6. Figure 23.7 shows the feasible combinations of skewness and kurtosis. Fitting to the NHV disturbance data gives α as 0.0385, β as 0.919, δ as 0.252 and *RSS* as 0.0501. The fit is as good as the Pareto-II distribution but, because δ is greater than 0.25, κ can be determined.

The third form of the generalised Pareto distribution is described by

$$f(x)=\left(\alpha+\frac{\delta}{x+\beta}\right)\left(1+\frac{x}{\beta}\right)^{-\delta}\exp(-\alpha x) \qquad \alpha,\beta,\delta>0 \tag{23.43}$$

$$F(x)=1-\left(1+\frac{x}{\beta}\right)^{-\delta}\exp(-\alpha x) \tag{23.44}$$

Fitting the CDF to the NHV disturbance data results in α being set to its minimum of 0. This reduces the distribution to the Lomax distribution described previously. Removing the limit

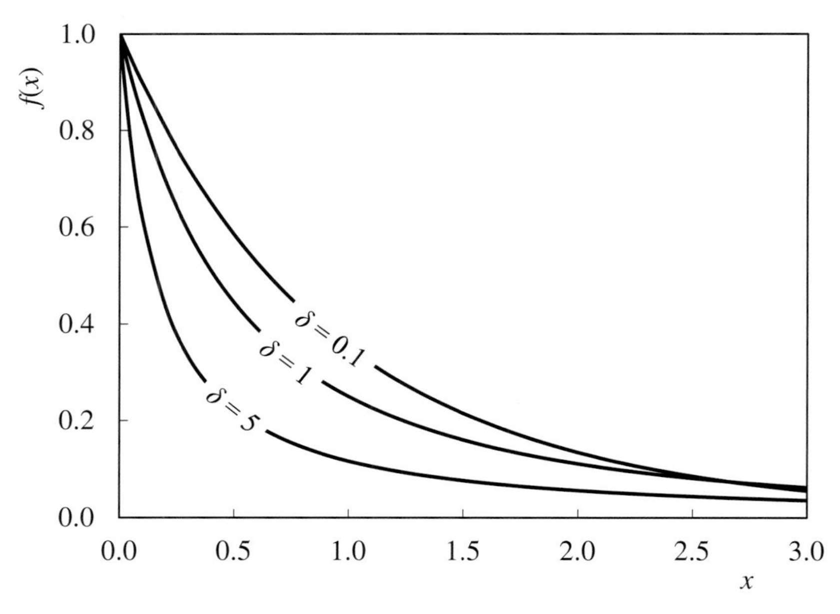

Figure 23.6 *Generalised Pareto: Effect of δ on shape*

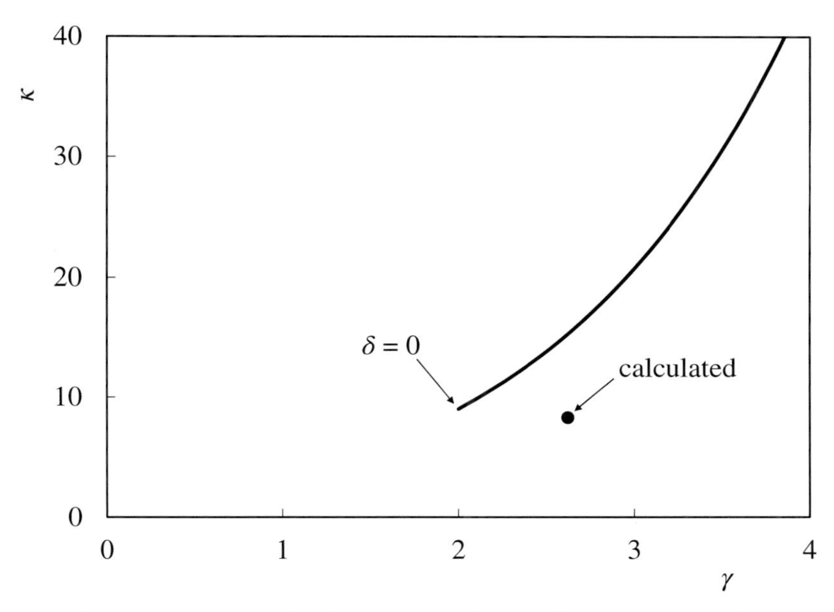

Figure 23.7 *Generalised Pareto: Feasible combinations of γ and κ*

gives α as −3.67, δ as 172 and β as 37.1. The latter two parameters are suspiciously large and result in a small improvement to *RSS* of 0.0677. This, and the lack of any formulae to calculate mean or variance, makes this version of the distribution impractical.

23.9 Pareto Principle

The *Pareto principle* is also known colloquially as the *80/20 rule*. It is often quoted by control engineers, for example, 'capture 80% of the benefits for 20% of the cost'. The values 20% and 80% will only be approximate; nor are they likely to sum to 100%. The principle is that the best technical solution may not be the best economically. For example, equation-based CLRTO is costly to install and difficult to maintain. While, in theory, it should capture all the available benefits, in practice it will likely fall into disuse and capture none. A much less complex solution may not capture all the benefits but will continue to perform well in the future. The same principle is enshrined in the *KIS* (*keep it simple*) doctrine of control design.

The principle was first used in describing the distribution of wealth in Italy as '20% of the population owns 80% the land'. It also goes further, for example, it would state that 'of the remaining 80% of the population, 20% own 80% of the remaining 20% of the land'. In the world of process control this effectively states that, if a less complex solution has been successfully implemented and we wish to consider capturing the remaining benefits, the 80/20 rule can be applied repetitively to these.

The principle is illustrated by the *Lorenz curve* which is normally plotted using real economic data but can be represented by

$$f(x) = 1 - (1 - x)^{1 - 1/\alpha} \tag{23.45}$$

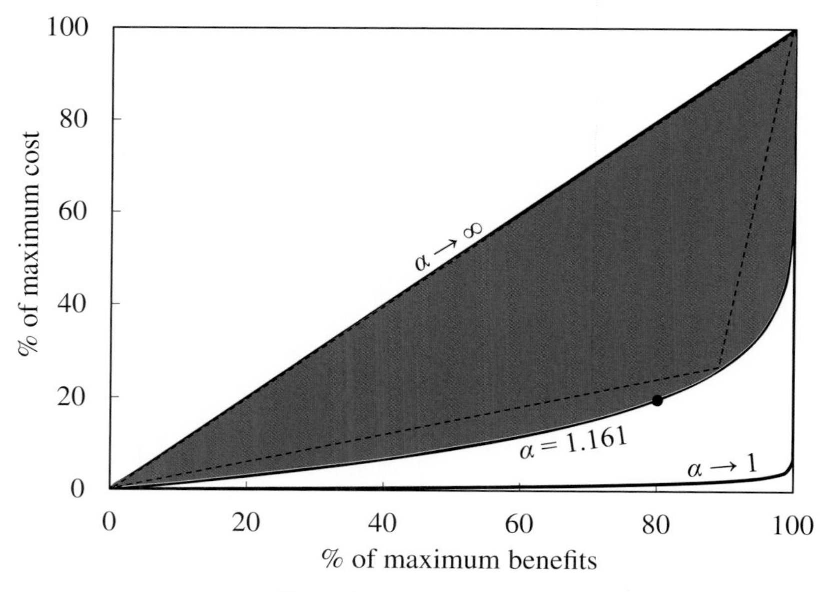

Figure 23.8 *Lorenz curve*

For our purposes we can consider x as the fraction of benefits captured and $f(x)$ as the fraction of the maximum cost. Solving Equation (23.45) for $x = 0.8$ and $f(x) = 0.2$ gives α as 1.161. This is shown in Figure 23.8. The *Gini coefficient* or *Gini index (G)* is the area (shown shaded) between the curve and the *equidistribution line* ($\alpha \to \infty$) expressed as a fraction of the total area under the equidistribution line. Since the total area is 0.5, the Gini coefficient is double the shaded area. It can be calculated from

$$G = \frac{1}{2\alpha - 1} \qquad \alpha > 1 \tag{23.46}$$

As α varies from 1 to ∞, G varies from 1 to 0. In the 80/20 case G is 0.756. It can be viewed as a measure of how economic it is to capture the remaining benefits from improved control. For example, if 90% of the remaining benefits can be captured for 10% of the cost, G is 0.912.

The *Pietra ratio* (P) provides a similar measure but instead is based on the area of the largest triangle that can be fitted in the shaded area. This is shown by the dashed line in Figure 23.8. In the 80/20 case it has the value 0.627, determined from

$$P = \left(\frac{\alpha - 1}{\alpha}\right)^{\alpha - 1} - \left(\frac{\alpha - 1}{\alpha}\right)^{\alpha} \tag{23.47}$$

The Pareto principle forms the basis of *Pareto analysis* – a technique for prioritising actions, as illustrated in Figure 23.9. In this example the number of hours that a MPC application has been out of service has been broken down by cause. The causes are plotted as bars in decreasing order of the number of hours lost. Also plotted as the solid line is the cumulative percentage of total hours lost. To the left of the dashed line are the causes that must be addressed to achieve an 80% reduction in lost hours. In this example instrument maintenance and operator training are the clear priorities. A competent control engineer will already be aware of such problems

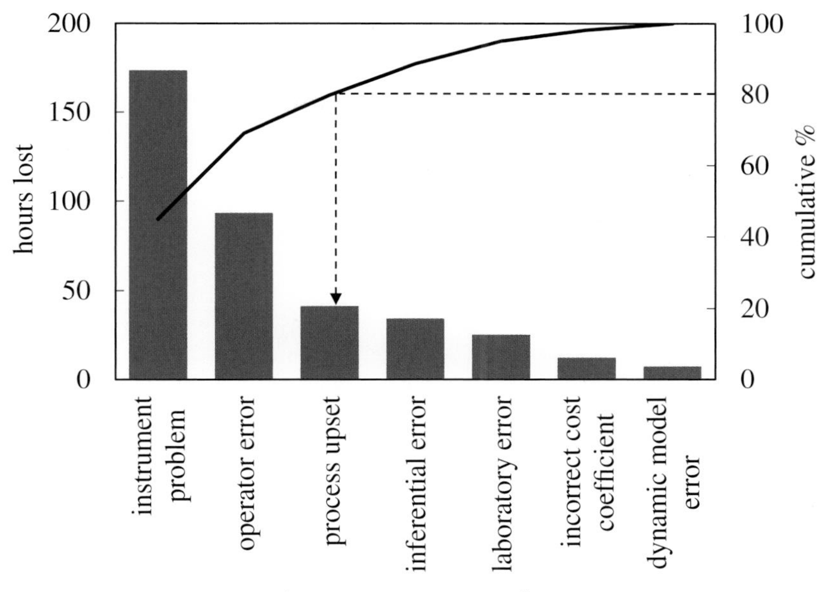

Figure 23.9 *Pareto analysis*

and the technique would do little other than to confirm what is already known. It does however offer a way of presenting the problem to others, for example as part of reporting MPC performance to senior management.

24

Stoppa Distribution

What is generally described as the *Stoppa distribution*[15] is the second of five types. Three of the five were also published by others, are better known under different titles and so described elsewhere in this book.

24.1 Type I

The *Stoppa-I distribution* is commonly known as the *power distribution*, described later in Section 33.20.

24.2 Type II

The *Stoppa-II distribution* is one of three generalisations of the Pareto-I distribution, defined by adding the shape parameter (δ_2). Its PDF is

$$f(x) = \frac{\delta_1 \delta_2}{x}\left(\frac{\beta}{x}\right)^{\delta_1}\left[1 - \left(\frac{\beta}{x}\right)^{\delta_1}\right]^{\delta_2 - 1} \qquad x \geq \beta;\ \beta, \delta_1, \delta_2 > 0 \tag{24.1}$$

Setting δ_2 to 1 (and δ_1 to δ) gives the classic Pareto distribution described by Equation (23.1). Figure 24.1 shows this, as the coloured line, and the effect of increasing it above this value.

The CDF is

$$F(x) = \left[1 - \left(\frac{\beta}{x}\right)^{\delta_1}\right]^{\delta_2} \tag{24.2}$$

Fitting to the NHV disturbance data gives values for β, δ_1 and δ_2 are respectively 0, 1.01 and 379. With *RSS* at 0.2108 the fit is an improvement on the standard Pareto distribution but is still poor – as shown by Figure 24.2.

Statistics for Process Control Engineers: A Practical Approach, First Edition. Myke King.

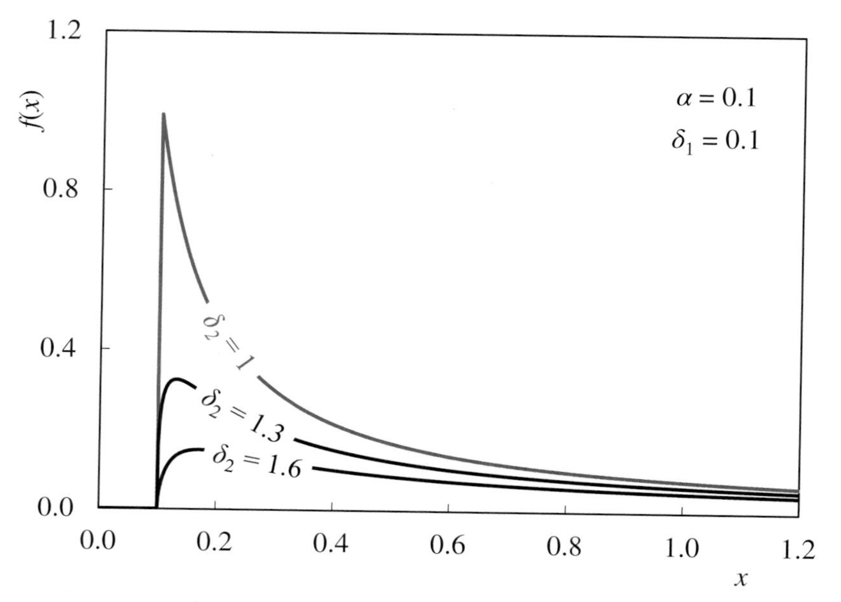

Figure 24.1 *Stoppa-II: Effect of δ_2 on shape*

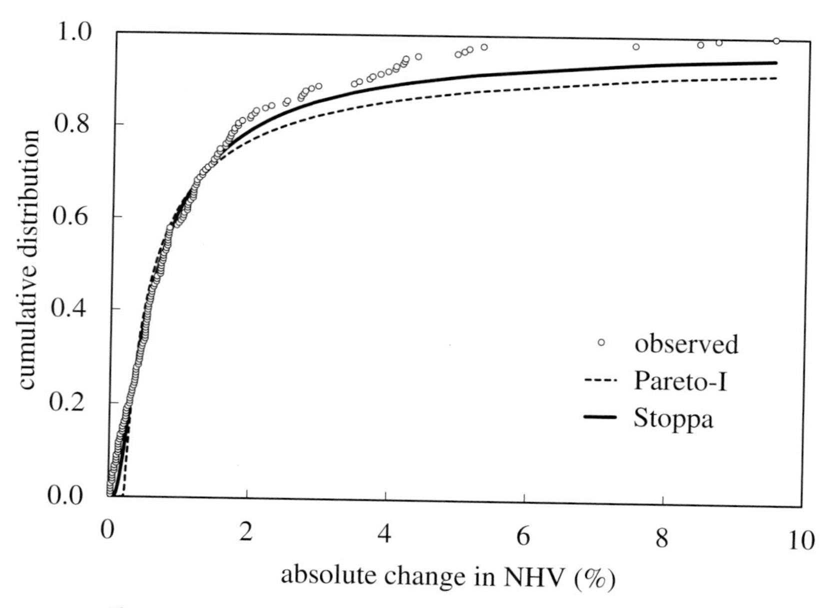

Figure 24.2 *Stoppa-II: Fitted to the NHV disturbance data*

Inverting gives the QF

$$x(F) = \beta\left(1 - F^{1/\delta_2}\right)^{-1/\delta_1} \quad 0 \le F \le 1 \tag{24.3}$$

The moments are given by

$$m_n = \delta_2 \mathrm{B}\left(1 - \frac{n}{\delta_1}, \delta_2\right)\beta^n \quad \delta_1 > n \tag{24.4}$$

However, in this case, the value of δ_2 does not permit calculation of the beta function (B). This requires the factorial of ($\delta_2 - 1$), which is too large for most software. This result is common to most datasets. In theory, the best fit is often obtained as δ_2 approaches infinity. However, the resulting improvement in *RSS* is usually negligible. Constraining δ_2 to some sensible upper limit usually has an indistinguishable impact on the accuracy of fit.

Fitting the Stoppa distribution to the C_4 in propane data gives a similarly poor result, with β, δ_1 and δ_2 respectively set to 1.03, 2.30 and 16.6. While it is possible now to calculate the moments, the long tail inherent to the distribution makes the estimates of mean and variance unrealistically high.

24.3 Type III

The *Stoppa-III distribution* is better known as the *generalised exponential distribution*, described later in Section 28.1.

24.4 Type IV

The *Stoppa-IV distribution* is described by the PDF

$$f(x)=\frac{1}{(x-\alpha)^2}\left(\frac{\beta}{x-\alpha}-1\right)^{\frac{1}{\beta\delta}-1}\left[1-\left(\frac{\beta}{x-\alpha}-1\right)^{\frac{1}{\beta\delta}}\right]^{\delta-1} \qquad \alpha+\frac{\beta}{2}\leq x\leq\alpha+\beta;\ \beta>0;\ \delta\geq 1 \tag{24.5}$$

Figure 24.3 shows the effect of varying δ, with α set to 0 and β to 1. It is unlikely that the distribution would fit any source of process data. The CDF is

$$F(x)=\left[1-\left(\frac{\beta}{x-\alpha}-1\right)^{\frac{1}{\beta\delta}}\right]^{\delta} \tag{24.6}$$

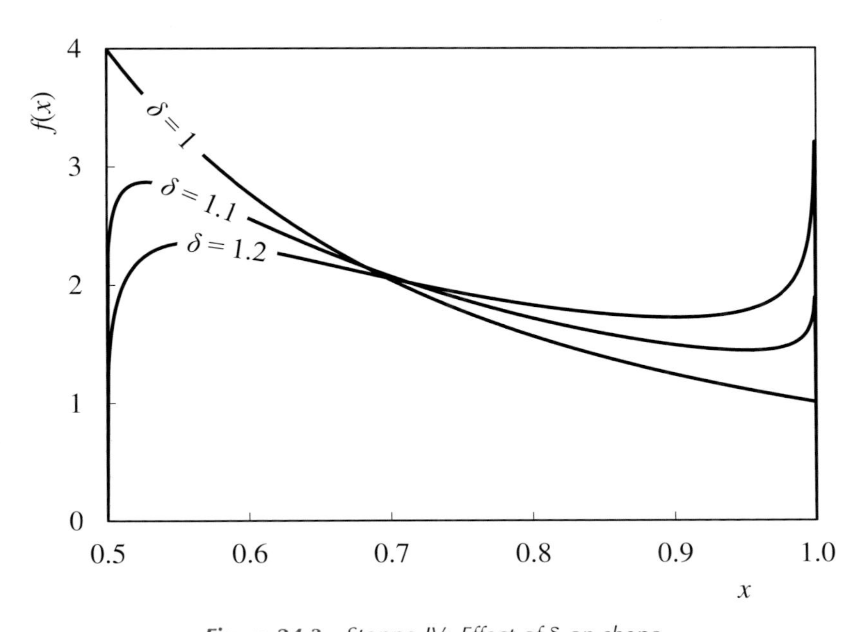

Figure 24.3 *Stoppa-IV: Effect of δ on shape*

This can be inverted to give the QF

$$x(F) = \alpha + \frac{\beta}{1 + \left(1 - F^{1/\delta}\right)^{\beta\delta}} \qquad 0 \leq F \leq 1 \tag{24.7}$$

24.5 Type V

The *Stoppa-V distribution* is also known as the *Burr-V distribution*, described in Section 21.5.

25

Beta Distribution

As one of the more commonly used distributions, the classic beta distribution was described in Section 12.3. Here we cover special cases and enhancements.

25.1 Arcsine

Setting, in Equation (12.13), both δ_1 and δ_2 to 0.5 gives the special case known as the *arcsine distribution*

$$f(x) = \frac{1}{\pi\sqrt{x(1-x)}} \qquad 0<x<1 \tag{25.1}$$

It is the CDF that gives the distribution its name

$$F(x) = \frac{2\sin^{-1}(\sqrt{x})}{\pi} \tag{25.2}$$

Inverting give the QF

$$x(F) = \sin^2\left(\frac{F}{2}\pi\right) \qquad 0 \leq F \leq 1 \tag{25.3}$$

Statistical parameters are

$$\mu = \frac{1}{2} \tag{25.4}$$

$$\sigma^2 = \frac{1}{8} \tag{25.5}$$

$$\gamma = 0 \tag{25.6}$$

$$\kappa = -\frac{3}{2} \tag{25.7}$$

Statistics for Process Control Engineers: A Practical Approach, First Edition. Myke King.

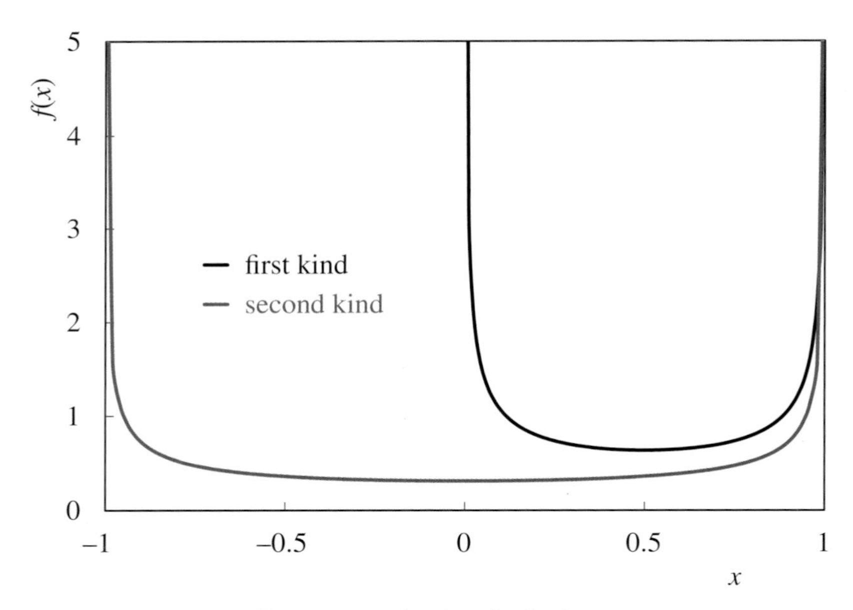

Figure 25.1 Arcsine distributions

There is also the *second kind* of the arc-sine distribution; its PDF is

$$f(x) = \frac{1}{\pi\sqrt{(1-x^2)}} \qquad -1 < x < 1 \tag{25.8}$$

Figure 25.1 shows both kinds. Without a shape factor they cannot be fitted to data but, in any case, it is difficult to see what application either of the anti-modal distributions might have in the process industry.

25.2 Wigner Semicircle

Setting, in Equation (12.13), both δ_1 and δ_2 to 1.5 and replacing x with $(x + r)/r$ gives another special case of the beta distribution, known as the *Wigner semicircle distribution*. It is described by

$$f(x) = \frac{2\sqrt{r^2-x^2}}{\pi r^2} \qquad -r < x < r \tag{25.9}$$

$$F(x) = \frac{1}{\pi}\left[\frac{x\sqrt{r^2-x^2}}{2r^2} + \sin^{-1}\left(\frac{x}{r}\right)\right] \tag{25.10}$$

$$\mu = 0 \tag{25.11}$$

$$\sigma^2 = \frac{r^2}{4} \tag{25.12}$$

$$\gamma = 0 \tag{25.13}$$

$$\kappa = 2 \tag{25.14}$$

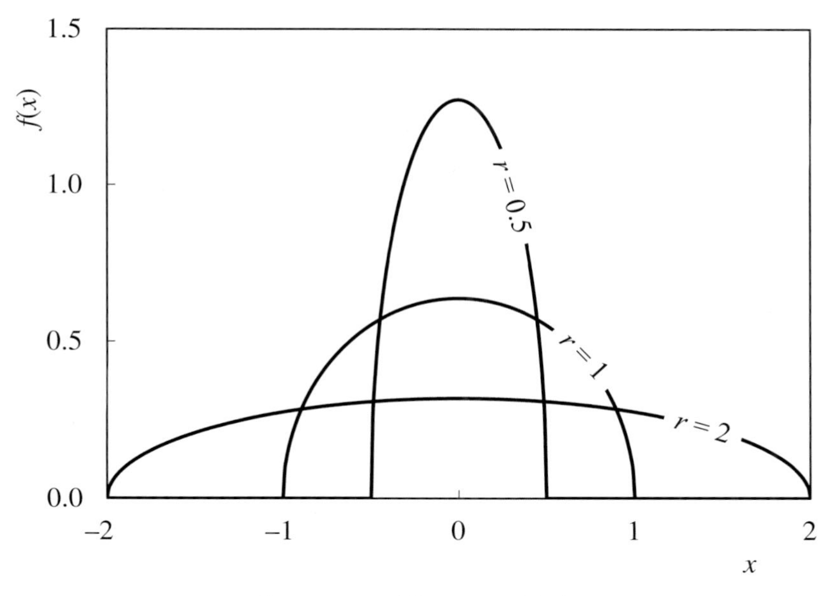

Figure 25.2 *Wigner semicircle*

Figure 25.2 shows the effect of r. More strictly it shows semi-ellipses. It is difficult to see that it has any application to the process industry.

25.3 Balding–Nichols

The *Balding–Nichols distribution* is the beta-I distribution with

$$\delta_1 = \frac{1-p}{p}\mu \tag{25.15}$$

$$\delta_2 = \frac{1-p}{p}(1-\mu) \tag{25.16}$$

Rearranging

$$\mu = \frac{\delta_1}{\delta_1 + \delta_2} \qquad 0 < \mu < 1 \tag{25.17}$$

$$p = \frac{1}{\delta_1 + \delta_2 + 1} \qquad 0 < p < 1 \tag{25.18}$$

Using the values of δ_1 and δ_2 obtained from fitting the beta-I distribution, p is 0.105 and μ is 0.235. These values could also have been obtained from directly fitting the Balding–Nichols distribution.

The variance is

$$\sigma^2 = p\mu(1-\mu) \tag{25.19}$$

To calculate skewness and kurtosis, we can use Equations (12.16) and (12.17) with δ_1 and δ_2 as calculated above.

25.4 Generalised Beta

The PDF of the *generalised beta distribution* is

$$f(x) = \frac{\delta_3}{B(\delta_1,\delta_2)x}\left[\left(\frac{x}{\beta}\right)^{\delta_3}\right]^{\delta_1}\left[1-\left(\frac{x}{\beta}\right)^{\delta_3}\right]^{\delta_2-1} \qquad 0 \le x \le \beta;\ \delta_1,\delta_2,\delta_3,\beta > 0 \tag{25.20}$$

If β and δ_3 are both set to 1, the distribution reverts to the beta-I distribution. While the addition of these coefficients theoretically will improve the fit, their identification can prove problematic. Using the example of the NHV data, the optimum values for δ_1 and δ_2 are probably infinite. As they are permitted to increase the changes have a diminishing effect on *RSS*. For example limiting β to 2 and δ_3 to 0.6, results in δ_1 increasing to 58 and δ_2 to 61. *RSS* is then reduced to 0.0808. Relaxing the limits soon results in the calculation of the beta function failing because of numerical overflow. It is unlikely that the distribution is robust enough for use with process data.

Raw moments are given by

$$m_n = \frac{\Gamma(\delta_1+\delta_2)\Gamma\left(\delta_1+\frac{n}{\delta_3}\right)}{\Gamma(\delta_1)\Gamma\left(\delta_1+\delta_2+\frac{n}{\delta_3}\right)}\beta^n \tag{25.21}$$

25.5 Beta Type II

The *beta-II distribution* is also known as the *beta prime distribution* or the *inverted beta distribution*. Its PDF is

$$f(x) = \frac{x^{\delta_1-1}}{B(\delta_1,\delta_2)(1+x)^{\delta_1+\delta_2}} \qquad x > 0;\ \delta_1,\delta_2 > 0 \tag{25.22}$$

The main difference between it and beta-I is that there is no upper limit on x. Fitting to the (scaled) C_4 in propane data gives values for δ_1 and δ_2 of 3.16 and 13.5 respectively. Figure 25.3 shows the effect of changing from beta-I to beta-II. The curve no longer terminates when x is 1.

The mean and variance are

$$\mu = \frac{\delta_1}{\delta_2-1} \qquad \delta_2 > 1 \tag{25.23}$$

$$\sigma^2 = \frac{\delta_1(\delta_1+\delta_2-1)}{(\delta_2-1)^2(\delta_2-2)} \qquad \delta_2 > 2 \tag{25.24}$$

The skewness and kurtosis are

$$\gamma = \frac{2(2\delta_1+\delta_2-1)}{\delta_2-3}\sqrt{\frac{\delta_2-2}{\delta_1(\delta_1+\delta_2-1)}} \qquad \delta_2 > 3 \tag{25.25}$$

$$\kappa = \frac{3\delta_1(\delta_1+\delta_2-1)(\delta_2+5)(\delta_2-2)+6(\delta_2-1)^2(\delta_2-2)}{\delta_1(\delta_1+\delta_2-1)(\delta_2-3)(\delta_2-4)} \qquad \delta_2 > 4 \tag{25.26}$$

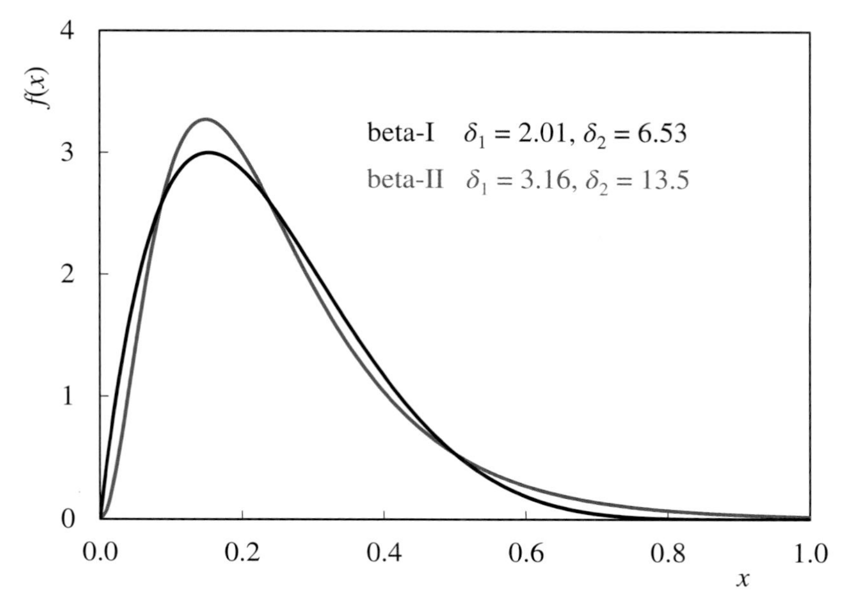

Figure 25.3 *Beta-II: Comparison to the beta-I distribution fitted to the C_4 data*

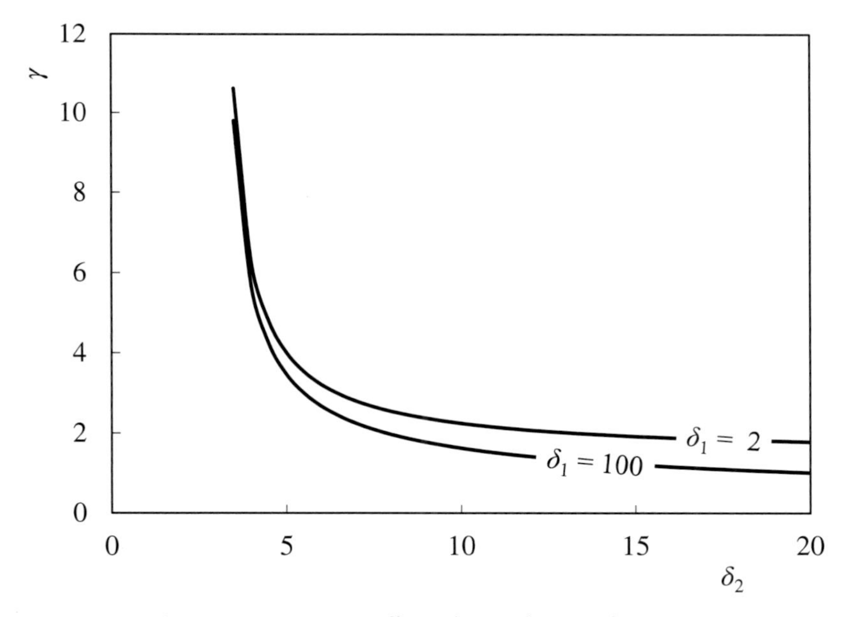

Figure 25.4 *Beta-II: Effect of δ_1 and δ_2 on skewness*

From Equation (25.23) we get an estimate of the mean as 0.254 (4.81 vol%) and, from Equation (25.24), a standard deviation of 0.177 (2.65 vol%). *RSS* is 0.0355 which compares closely to the Burr distribution – one of the best of the more practical distributions.

Unlike beta-I, the distribution cannot model skewness in both directions. Figures 25.4 and 25.5 show the effect that δ_1 and δ_2 have on both this and kurtosis. It can be seen that large

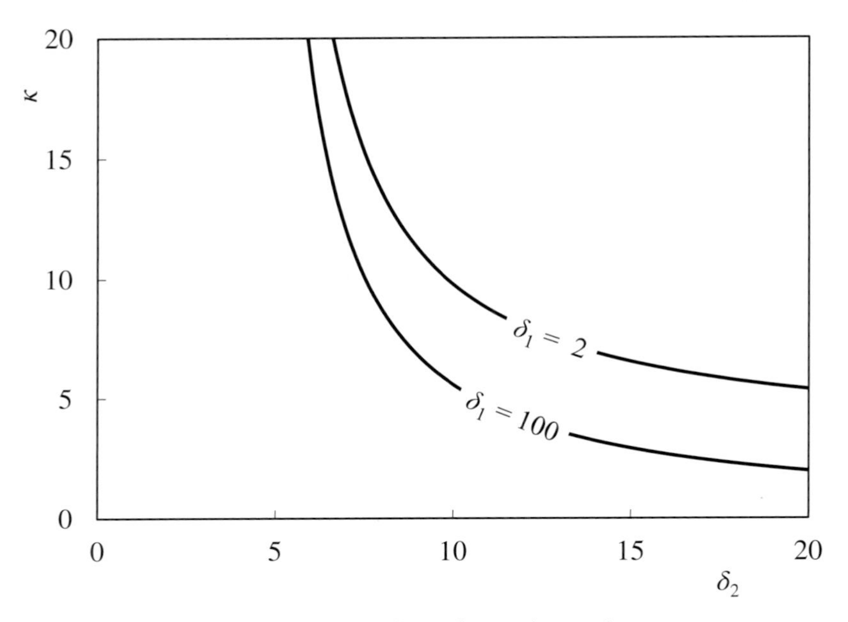

Figure 25.5 *Beta-II: Effect of δ_1 and δ_2 on kurtosis*

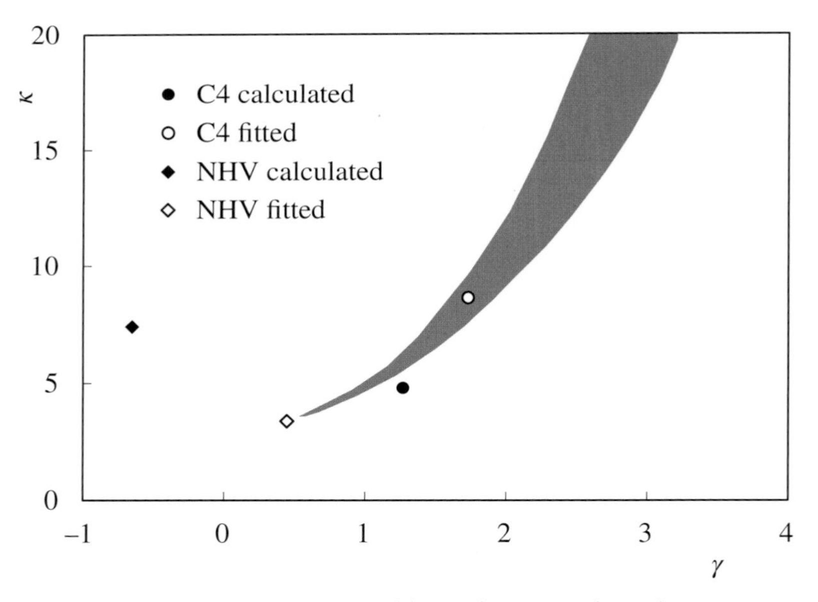

Figure 25.6 *Beta-II: Feasible combinations of γ and κ*

changes in δ_1 are required to have significant influence on either. Should the ratio of skewness to kurtosis be very different from that calculated from the C_4 in propane data then fitting a beta-II distribution would give a poor result.

To demonstrate this we can fit Equation (25.22) to the NHV disturbance data. This gives values for δ_1 and δ_2 of 63.5 and 105, with *RSS* being minimised to 0.0650. While a reasonable fit, it is outperformed by the several other distributions. Figure 25.6 shows the range of skewness and kurtosis of the distribution, constructed by plotting a number of combinations of δ_1 and δ_2. The calculated combination of skewness and kurtosis for the C_4 data lie close to this region, while that for the NHV data are well outside.

25.6 Generalised Beta Prime

The *generalised beta prime distribution* is also known as the *transformed beta distribution*. It adds a location parameter (α), a scale parameter (β) and an additional shape parameter (δ_3). If α is set to 0 and the other two parameters to 1, it reverts to the beta prime distribution described in the previous section. It is actually a compound distribution derived from two gamma distributions. We covered the gamma distribution in Section 12.10, but the mathematics of combining two of them do not merit inclusion in this book. The resulting PDF is

$$f(x) = \frac{\delta_3 \left(\frac{x-\alpha}{\beta}\right)^{\delta_1\delta_3 - 1}}{\beta B(\delta_1,\delta_2)\left[1 + \left(\frac{x-\alpha}{\beta}\right)^{\delta_3}\right]^{\delta_1+\delta_2}} \qquad x \geq \alpha;\ \beta,\delta_1,\delta_2,\delta_3 > 0 \qquad (25.27)$$

Figure 25.7 shows the effect of varying δ_1, δ_2 and δ_3 – with α fixed at 0 and β at 1. Other distributions that are also covered include:

Burr-XII	$\delta_1 = 1$
Dagum-I	$\delta_2 = 1$
Pearson-VI	$\delta_3 = 1$
inverse Pareto	$\delta_1 = 1$ and $\delta_2 = 1$
inverse paralogistic	$\delta_1 = 1$ and $\delta_2 = \delta_3$
log-logistic	$\delta_1 = 1$ and $\delta_3 = 1$
paralogistic	$\delta_1 = \delta_2$ and $\delta_3 = 1$
Pareto-II	$\delta_2 = 1$ and $\delta_3 = 1$

If none of these conditions are met, the CDF and QF do not exist.

Fitting to the C_4 in propane data gives α as 1.52, β as 3.58, δ_1 as 1.31, δ_2 as 1.89, δ_3 as 0.970 and $F(x_1)$ as 0.0119. With *RSS* at 0.0305, not surprisingly the fit is better than any of the distributions that can be represented. However, with δ_3 very close to 1 and neither δ_1 nor δ_2 meeting any of the conditions above, the fitted distribution is effectively the Pearson-VI distribution.

Raw moments are given by

$$m_n = \frac{\Gamma\left(\delta_1 + \frac{n}{\delta_3}\right)\Gamma\left(\delta_2 - \frac{n}{\delta_3}\right)}{\Gamma(\delta_1)\Gamma(\delta_2)}\beta^n \qquad -\delta_1\delta_3 < n < \delta_2\delta_3 \qquad (25.28)$$

For this example, the bounds on n are therefore −1.27 to 1.84. Only the first moment can therefore be determined. This gives μ as 6.99 – much higher than expected. Fitting six parameters might appear to give a good fit but the result is inconsistent with our understanding of the process. This is explained further in Section 27.6, where we cover the Pearson-VI distribution.

Another version includes an additional parameter (δ_4). Its PDF is

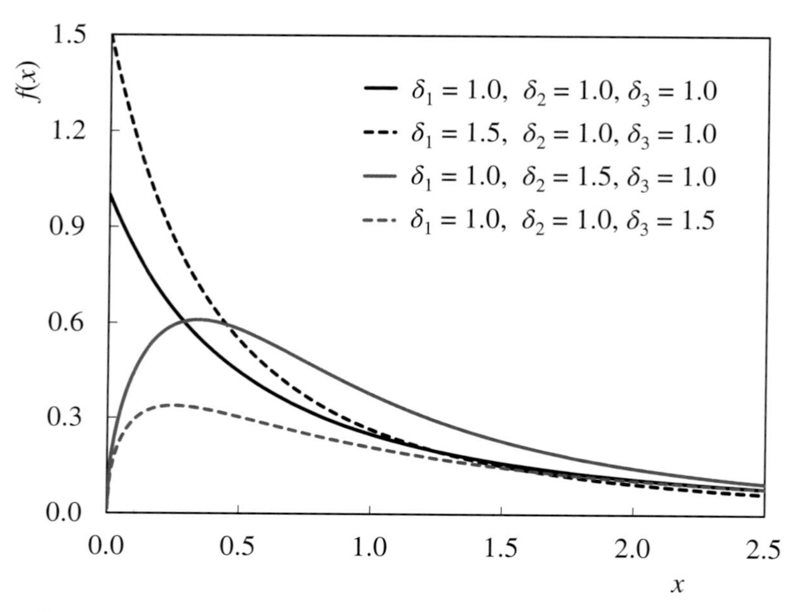

Figure 25.7 Generalised beta prime: Effect of δ_1, δ_2 and δ_3 on shape

$$f(x)=\frac{\delta_3\left(\frac{x}{\beta}\right)^{\delta_1\delta_3-1}\left[1-(1-\delta_4)\left(\frac{x}{\beta}\right)^{\delta_3}\right]^{\delta_2-1}}{\beta\mathrm{B}(\delta_1,\delta_2)\left[1+\delta_4\left(\frac{x}{\beta}\right)^{\delta_3}\right]^{\delta_1+\delta_2}} \qquad 0<x<1;\ \beta,\delta_1,\delta_2,\delta_3>0;\ 0\le\delta_4\le1 \tag{25.29}$$

The purpose of δ_4 is to mix the classic beta distribution with the beta prime distribution. To help understand this we first set β and δ_3 to 1. If δ_4 is then set to 0, Equation (25.29) reverts to Equation (12.13). If set to 1, it becomes Equation (25.22).

The added complexity of these distributions brings little benefit to their application to the process industry. While they will theoretically fit more closely to the data they are unlikely to be robust.

25.7 Beta Type IV

The *beta-IV distribution* or *beta subjective distribution* is identical to the beta-I distribution except that it incorporates the range of x in the PDF.

$$f(x)=\frac{(x-x_{min})^{\delta_1-1}(x_{max}-x)^{\delta_2-1}}{\mathrm{B}(\delta_1,\delta_2)(x_{max}-x_{min})^{\delta_1+\delta_2-1}} \qquad x_{min}<x<x_{max} \tag{25.30}$$

There is no need therefore to scale x over the range 0 to 1. The formulae for mean and variance also include the range of x. Since skewness and kurtosis are dimensionless, their formulae remain unchanged.

$$\mu=x_{min}+\frac{\delta_1}{\delta_1+\delta_2}(x_{max}-x_{min}) \tag{25.31}$$

$$\sigma^2 = \frac{\delta_1 \delta_2}{(\delta_1 + \delta_2)^2 (\delta_1 + \delta_2 + 1)} (x_{max} - x_{min})^2 \tag{25.32}$$

In addition to adjusting δ_1 and δ_2 to obtain the best fit, Equation (25.30) permits the range to be optimised. Doing so results in the range being selected as 1.6 to 85. The corresponding values for δ_1 and δ_2 are then 1.72 and 42.6, minimising *RSS* to 0.0377 – close to that achieved by the beta-II distribution. Equations (25.31) and (25.32) give the mean and standard deviation as 4.84 and 2.45.

A similar approach is to apply the transformation to the raw data (x).

$$z = \frac{x - \alpha}{\beta} \tag{25.33}$$

Initially, α is chosen as the lowest value of x and β as the highest value less α. The transformed variable (z) will then range from 0 to 1.

Remembering, from Equations (5.61) to (5.63)

$$f(x) = f(z)\frac{dz}{dx} = \frac{f(z)}{\beta} \tag{25.34}$$

The PDF, $f(z)$ is the beta-I distribution described by Equation (12.13).

$$f(z) = \frac{z^{\delta_1 - 1}(1 - z)^{\delta_2 - 1}}{\mathrm{B}(\delta_1, \delta_2)} \tag{25.35}$$

Combining Equations (25.33), (25.34) and (25.35)

$$f(x) = \frac{1}{\mathrm{B}(\delta_1, \delta_2)} \frac{1}{\beta} \left(\frac{x - \alpha}{\beta}\right)^{\delta_1 - 1} \left(1 - \frac{x - \alpha}{\beta}\right)^{\delta_2 - 1} = \frac{(x - \alpha)^{\delta_1 - 1}(\alpha + \beta - x)^{\delta_2 - 1}}{\mathrm{B}(\delta_1, \delta_2)\beta^{\delta_1 + \delta_2 - 1}} \tag{25.36}$$

What, at first glance, might appear to be a very different distribution, with additional shape parameters, remains the beta-I distribution. It simply provides an alternative method by which the range parameters can be included in those fitted. Comparison with Equation (25.30) shows that α is x_{min} and $\alpha + \beta$ is x_{max}.

From Equations (12.14) and (12.15), the mean and variance are now

$$\mu = \alpha + \frac{\delta_1}{\delta_1 + \delta_2}\beta \tag{25.37}$$

$$\sigma^2 = \frac{\delta_1 \delta_2}{(\delta_1 + \delta_2)^2 (\delta_1 + \delta_2 + 1)} \beta^2 \tag{25.38}$$

Skewness and kurtosis remain as described by Equations (12.16) and (12.17).

25.8 PERT

The beta-IV can also be known as the *PERT distribution*, arising from its use with PERT project planning charts. The distribution relies on the *three point estimation technique*. In this case it requires three estimates of the time taken to complete each task – the 'best' (x_{min}), the 'worst' (x_{max}) and the 'most likely' (x_{mode}). The triangular distribution, described in Section 5.2, is replaced with a beta-I distribution passing through the three points. The resulting PDF for each task are then combined to produce a PDF for the whole project schedule.

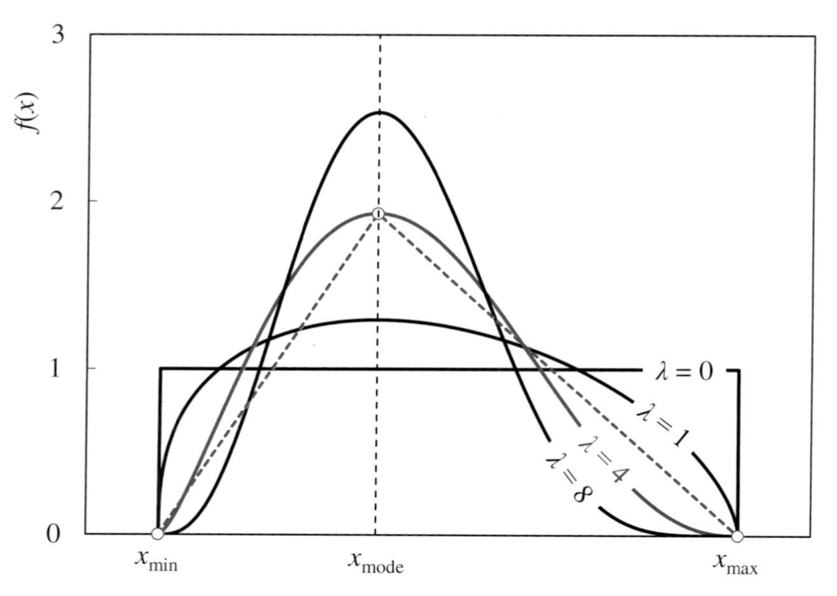

Figure 25.8 *PERT: Effect of λ on shape*

To apply Equation (12.13), the mean (μ) is first defined as the weighted average of these three values.

$$\mu = \frac{x_{min} + \lambda x_{mode} + x_{max}}{\lambda + 2} \tag{25.39}$$

The shape factor (λ) affects the height of the peak of the PDF. Typically a value of 4 is chosen.

From this we can derive δ_1 and δ_2.

$$\delta_1 = 1 + \lambda \frac{x_{mode} - x_{min}}{x_{max} - x_{min}} = \frac{(2x_{mode} - x_{min} - x_{max})}{(x_{mode} - \mu)(x_{max} - x_{min})}(\mu - x_{min}) \tag{25.40}$$

$$\delta_2 = 1 + \lambda \frac{x_{max} - x_{mode}}{x_{max} - x_{min}} = \frac{(2x_{mode} - x_{min} - x_{max})}{(x_{mode} - \mu)(x_{max} - x_{min})}(x_{max} - \mu) \tag{25.41}$$

The coloured lines in Figure 25.8 show how the triangular distribution is converted to the beta-I distribution with the same range and mode. The black lines show the effect of the choice of λ.

However, after some spectacularly underestimated project schedules, this *three-point estimation technique* is now largely discredited. It is only as good as the assumed best, worst and most likely estimates. Further, selecting the form of distribution based only on three points is unlikely to be reliable.

Fitting to the C_4 in propane data gives x_{min} as 1.60, x_{mode} as 3.03, x_{max} as 98.1, λ as 49.5 and $F(x_1)$ as 0.0010. From Equations (25.40) and (25.41), this is equivalent to setting δ_1 at 1.73 and δ_2 at 49.7. Although these values are slightly different from the result obtained in Section 12.3, at 0.0374, *RSS* is the same. In other words, we could achieve the same result by directly fitting the beta-I distribution.

One might argue that fitting the PERT distribution also gives us the mode. Indeed, from Figure 3.6, we can see that the true mode is 3.5 – close to the fitted mode. However, the same result can be derived directly from the fitted beta-I distribution.

$$x_{mode} = \frac{\delta_1 - 1}{\delta_1 + \delta_2 - 2}(x_{max} - x_{min}) + x_{min} \tag{25.42}$$

Fitting the PERT distribution therefore offers no advantage over fitting the beta-I distribution.

25.9 Beta Rectangular

The *beta rectangular distribution* is a mixture of the beta-I and uniform distributions. Its PDF is

$$f(x) = \frac{\delta x^{\delta_1 - 1}(1-x)^{\delta_2 - 1}}{B(\delta_1, \delta_2)} + 1 - \delta \qquad \delta, \delta_1, \delta_2 > 0 \tag{25.43}$$

The term δ determines the relative contribution. If set to 0, Equation (25.43) gives the uniform distribution U(0,1). If set to 1, it becomes the beta-I distribution. As written, x must be scaled over the range 0 to 1. The true range can be included as

$$f(x) = \frac{\delta (x - x_{min})^{\delta_1 - 1}(x_{max} - x)^{\delta_2 - 1}}{B(\delta_1, \delta_2)(x_{max} - x_{min})^{\delta_1 + \delta_2 - 1}} + \frac{1 - \delta}{x_{max} - x_{min}} \tag{25.44}$$

As Figure 25.9 shows, since $f(x)$ steps down to zero at the extreme values of x, it is unlikely to fit process data. Formulae for mean and variance exist but do not merit inclusion here.

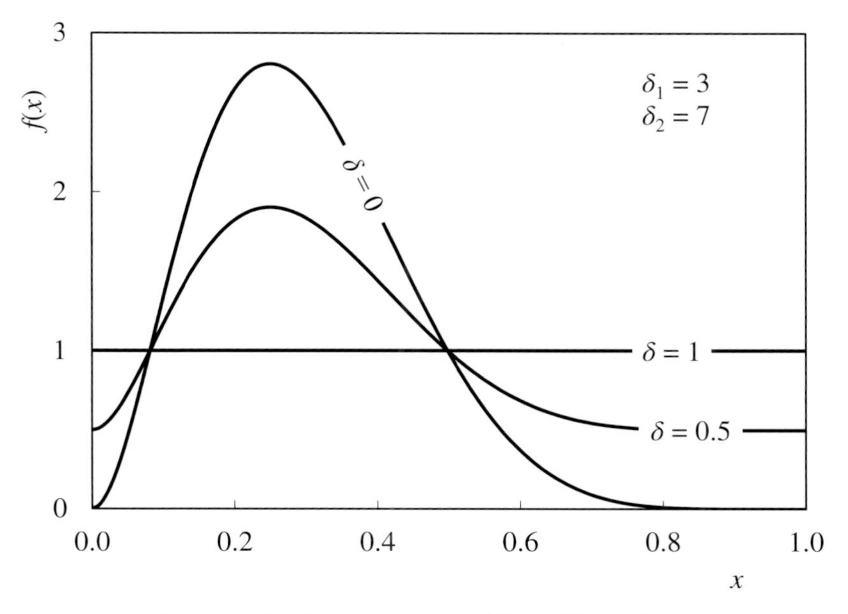

Figure 25.9 *Beta rectangular: Effect of δ on shape*

While the distribution is unlikely to be applicable, it does demonstrate the principle of combining distributions. In the event that a well-fitting distribution cannot be identified, the problem might be resolved by a combination. However it is likely that determining parameters, such as mean and variance, would then be very complex.

25.10 Kumaraswamy

Related to the beta-I distribution is the *Kumaraswamy distribution*. It is also known as the *minimax distribution*.[16] It is described by the PDF

$$f(x) = \delta_1 \delta_2 x^{\delta_1 - 1}\left(1 - x^{\delta_1}\right)^{\delta_2 - 1} \qquad 0 \le x \le 1;\ \delta_1, \delta_2 > 0 \tag{25.45}$$

Like the beta-I distribution, some choices of the shape parameters δ_1 and δ_2 can produce anti-modal distributions that are unlikely to be found in the process industry. More useful values are used as examples in Figures 25.10 and 25.11.

Its CDF is

$$F(x) = 1 - \left(1 - x^{\delta_1}\right)^{\delta_2} \tag{25.46}$$

Applying Equation (11.4) we can see that, as δ_2 is increased, $1/\delta_2 \to 0$ and this distribution approaches

$$F(x) = 1 - \exp\left(x^{\delta_1}\right) \tag{25.47}$$

This is the Weibull-I distribution that we covered in Section 12.8.

Again using the C_4 in propane data, as with the beta-I distribution, we need to select a range. Unlike the beta-I distribution, for which the range should be optimised, in order to achieve the

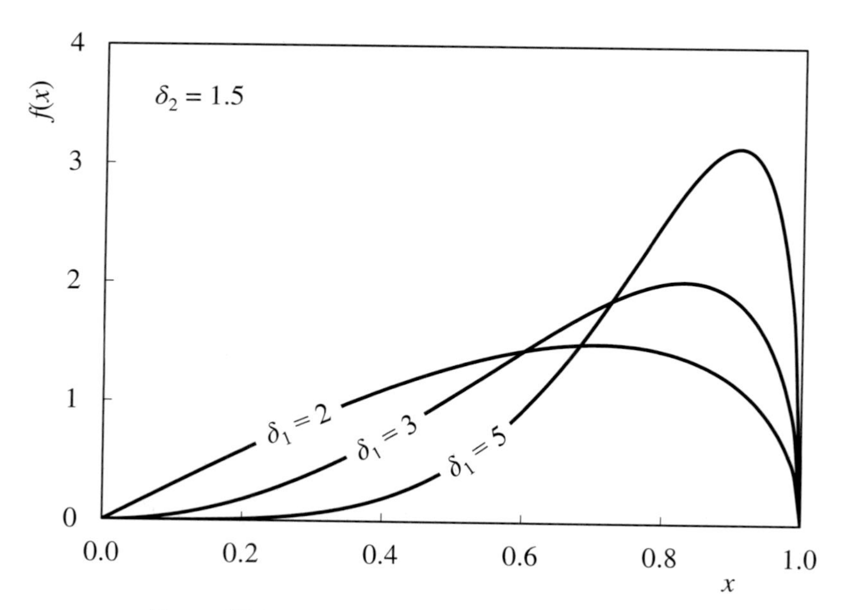

Figure 25.10 Kumaraswamy: Effect of δ_1 on shape

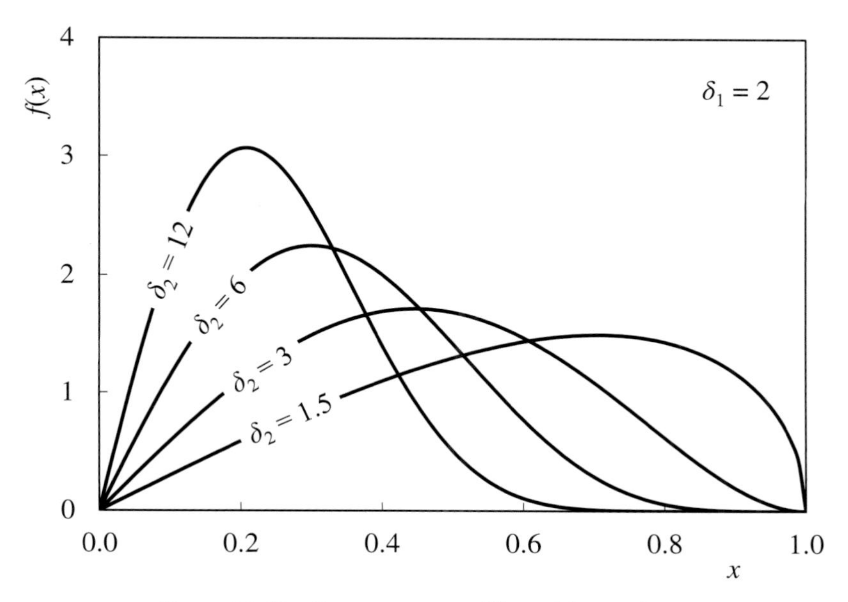

Figure 25.11 *Kumaraswamy: Effect of* δ_2 *on shape*

best fit the range for the Kumaraswamy distribution should be as small as possible. The narrowest it can be, without losing data, is 1.5 to 15.8. Fitting gives values of 1.38 and 6.04 for δ_1 and δ_2 – minimising *RSS* to 0.1118. While better than the normal distribution, compared to most other distributions, the fit is poor.

However it has two advantages over the beta-I distribution. The first is that the calculations of skewness and kurtosis is more straightforward, although the formulae are too complex to present sensibly here. Instead, they are derived from the raw moments, given by

$$m_n = \frac{\delta_2 \Gamma\left(1 + \frac{n}{\delta_1}\right)\Gamma(\delta_2)}{\Gamma\left(1 + \delta_2 + \frac{n}{\delta_1}\right)} \tag{25.48}$$

Calculating the moments gives

$$m_1 = 0.225 \qquad m_2 = 0.0723 \qquad m_3 = 0.0285 \qquad m_4 = 0.0129 \tag{25.49}$$

From Equations (4.46), (4.49), (4.52) and (4.55) the key parameters can be calculated as

$$\mu = 0.225 \qquad \sigma = 0.148 \qquad \gamma = 0.775 \qquad \kappa = 3.19 \tag{25.50}$$

Remembering that x is ranged 0 to 1, converting to engineering units (in the same way as we did for the beta-I distribution) gives the mean as 4.71 and the standard deviation as 2.11.

Figure 25.12 shows feasible values for skewness and kurtosis. Like beta-I, the Kumaraswamy distribution is one of the few that can exhibit both positive and negative skewness. The skewness calculated from the data is 1.27 and the kurtosis 4.79. This combination cannot quite be covered by the distribution.

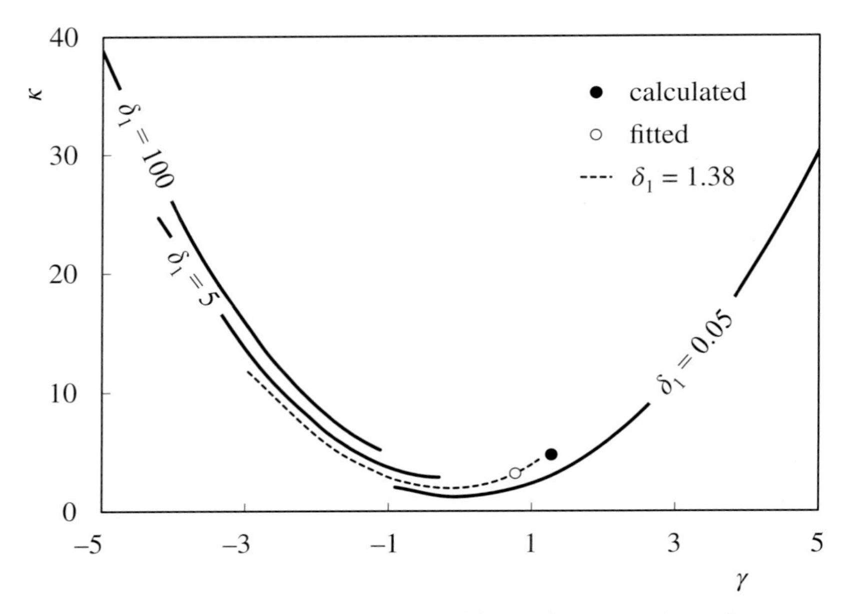

Figure 25.12 Kumaraswamy: Feasible combinations of γ and κ

The second advantage is that, unlike the beta-I distribution, not only is there a simple formula for $F(x)$ but it can also be inverted. Rearranging Equation (25.46)

$$x(F) = \left\{1 - [1-F]^{\frac{1}{\delta_2}}\right\}^{\frac{1}{\delta_1}} \quad 0 \le F \le 1 \tag{25.51}$$

For example, choosing F as 0.95 gives x as 0.506 which, when converted to engineering units, is 8.73. In other words we expect 5% of the results to exceed this value.

An extension of the minimax distribution is the *minimax odds distribution*. Applying Equation (5.87) to convert odds to probability, Equation (25.45) becomes

$$f(x) = \frac{\delta_1 \delta_2}{(x+1)^{\delta_1 \delta_2 + 1}} \left[(x+1)^{\delta_1} - 1\right]^{\delta_2 - 1} \quad x \ge 0 \tag{25.52}$$

Figure 25.13 shows the effect of varying δ_1 and δ_2. In comparison with the minimax, it exhibits a much fatter tail.

The CDF is

$$F(x) = \left[1 - \frac{1}{(x+1)^{\delta_1}}\right]^{\delta_2} \tag{25.53}$$

Fitting to C_4 in propane data gives δ_1 as 2.91 and δ_2 as 83.9. *RSS*, although slightly improved, is still poor at 0.0919. This is primarily because the fatter tail is not an advantage in this case. Formulae for the moments can be developed by applying the inverse transformation, given by Equation (5.87), but the added complexity is not justified by the improvement in accuracy of fit.

The QF is

$$x(F) = \left(1 - F^{1/\delta_2}\right)^{1-/\delta_1} - 1 \tag{25.54}$$

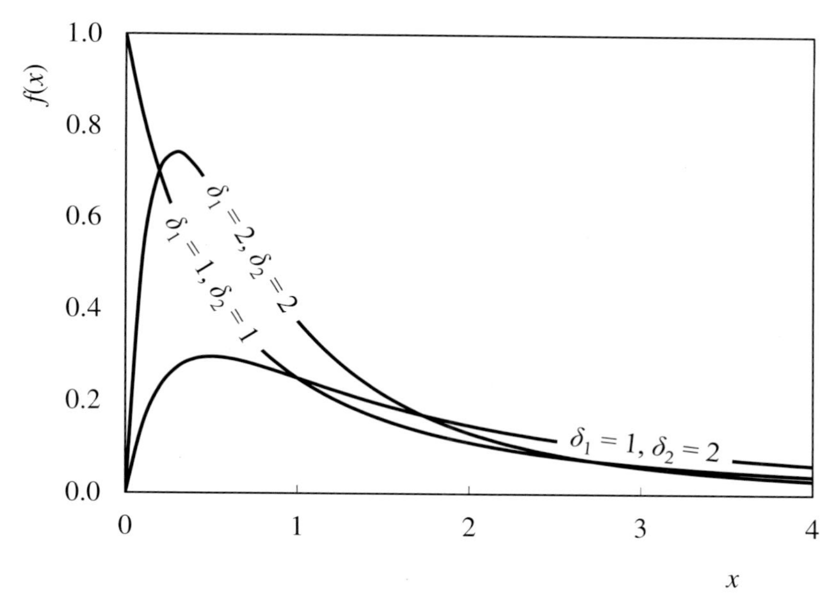

Figure 25.13 *Minimax odds: Effect of δ_1 and δ_2 on shape*

25.11 Noncentral Beta

The *noncentral beta distribution* is described by the PDF

$$f(x) = \sum_{j=0}^{\infty} \left(\frac{\delta}{2}\right)^j \frac{x^{\alpha+j-1}(1-x)^{\beta-1}}{j!\mathrm{B}(\alpha+j,\beta)} \exp\left(-\frac{\delta}{2}\right) \quad 0 \le x \le 1;\ \alpha,\beta,\delta > 0 \tag{25.55}$$

The parameters α and β, as usual, define to location and scale of the distribution. The term δ determines the non-centrality. The CDF is

$$F(x) = \sum_{j=0}^{\infty} \left(\frac{\delta}{2}\right)^j \frac{\mathrm{I}_x(\alpha+j,\beta)}{j!} \exp\left(-\frac{\delta}{2}\right) \tag{25.56}$$

Generally, spreadsheet packages require x to be integer in Bessel functions – including the modified version of the first kind (I_x). If this is not the case then the trapezium rule can be used to fit the PDF to data. While possible in the spreadsheet environment, the summation to infinity in the PDF adds complexity that probably is not justified by better accuracy of fit. Further the calculation of mean and variance involve the use of mathematics far beyond the scope of this book.

26

Johnson Distribution

There are four types of *Johnson distribution*.[17] The most commonly referred to is the version known as S_U – probably because it is unbounded. Not included here, for any of the distributions, are the very complex formulae for the calculation of skewness and kurtosis. If required, the calculations can realistically only be performed by a software product.

26.1 S_N

The first, known as *Johnson* S_N, is the normal distribution but with additional shape parameters α and β. It is described by

$$f(x) = \frac{\beta}{\sigma_J\sqrt{2\pi}}\exp\left[-\frac{1}{2}\left(\alpha + \frac{\beta(x-\mu_J)}{\sigma_J}\right)^2\right] \quad \beta, \sigma_J > 0 \tag{26.1}$$

$$F(x) = \frac{1}{2} + \frac{1}{2}\text{erf}\left[\frac{1}{\sqrt{2}}\left(\alpha + \frac{\beta(x-\mu_J)}{\sigma_J}\right)\right] \tag{26.2}$$

$$\mu = \mu_J - \frac{\alpha}{\beta}\sigma_J \tag{26.3}$$

$$\sigma^2 = \left(\frac{\sigma_J}{\beta}\right)^2 \tag{26.4}$$

It has no advantage over the classic normal distribution. As Figure 26.1 illustrates, as usual, α determines the location and β the dispersion. The modification that α and β make to the shape of the distribution can be exactly replicated by adjusting μ and σ in the classic normal distribution. Indeed, fitting all four parameters in Equation (26.1) will be problematic because there is not a unique set of values that minimises *RSS*.

Statistics for Process Control Engineers: A Practical Approach, First Edition. Myke King.

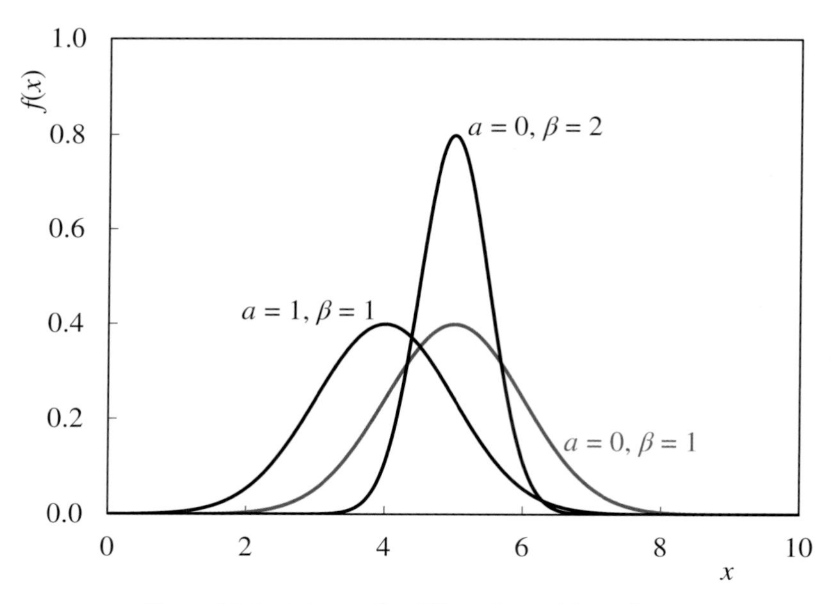

Figure 26.1 *Johnson S_N: Effect of α and β on shape*

26.2 S_U

The second, known as *Johnson S_U*, is described by the PDF

$$f(x)=\frac{\beta}{\sqrt{2\pi\left((x-\mu_J)^2+\sigma_J^2\right)}}\exp\left\{-\frac{1}{2}\left[\alpha+\beta\sinh^{-1}\left(\frac{x-\mu_J}{\sigma_J}\right)\right]^2\right\}\quad \beta,\sigma_J>0 \tag{26.5}$$

Alternatively

$$f(x)=\frac{\beta}{\sqrt{2\pi\left((x-\mu_J)^2+\sigma_J^2\right)}}\exp\left\{-\frac{1}{2}\left[\alpha+\beta\ln\left(\frac{x-\mu_J+\sqrt{(x-\mu_J)^2+\sigma_J^2}}{\sigma_J}\right)\right]^2\right\} \tag{26.6}$$

With α set to 0 and β set to 1, Figure 26.2 shows that, compared to the normal (S_N) distribution, the S_U distribution remains symmetrical but leptokurtic. Keeping μ_J at 5 and σ_J at 1, Figure 26.3 shows that increasing β increases the kurtosis and adjusting α permits either a positive or negative skewness to be introduced.

Its CDF is

$$F(x)=\frac{1}{2}+\frac{1}{2}\text{erf}\left\{\frac{1}{\sqrt{2}}\left[\alpha+\beta\sinh^{-1}\left(\frac{x-\mu_J}{\sigma_J}\right)\right]\right\} \tag{26.7}$$

Alternatively

$$F(x)=\frac{1}{2}+\frac{1}{2}\text{erf}\left\{\frac{1}{\sqrt{2}}\left[\alpha+\beta\ln\left(\frac{x-\mu_J+\sqrt{(x-\mu_J)^2+\sigma_J^2}}{\sigma_J}\right)\right]\right\} \tag{26.8}$$

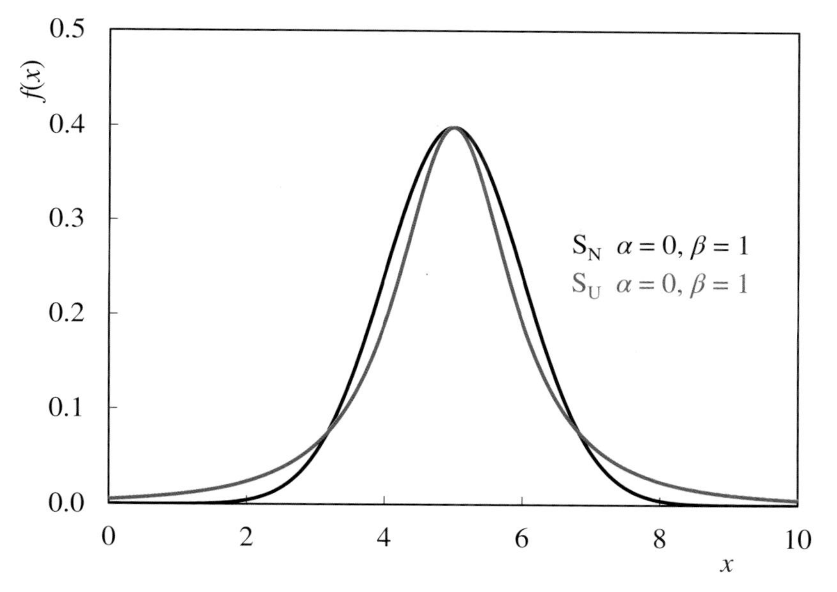

Figure 26.2 *Johnson S_U: Comparison with Johnson S_N distribution*

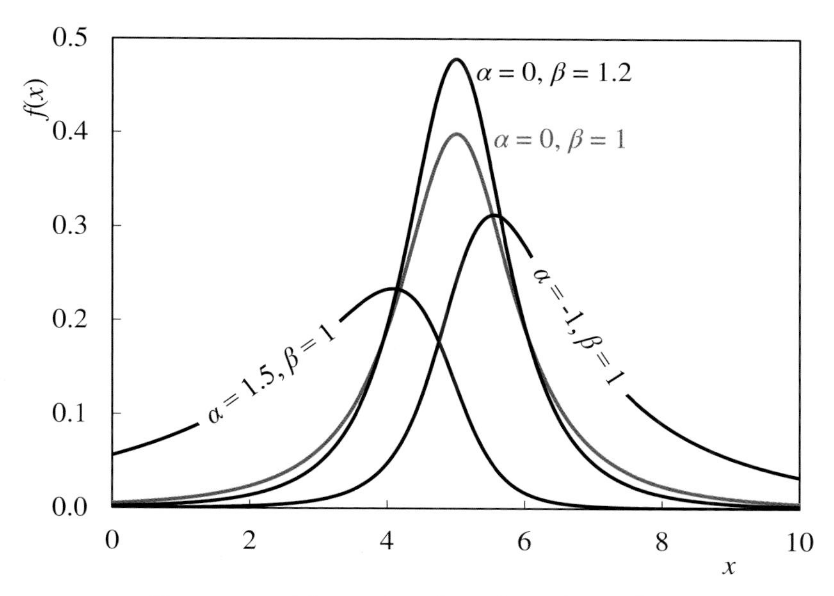

Figure 26.3 *Johnson S_U: Effect of α and β on shape*

Mean and variance are

$$\mu = \mu_J - \sigma_J \sinh\left(\frac{\alpha}{\beta}\right)\exp\left(\frac{1}{2\beta^2}\right) \tag{26.9}$$

$$\sigma^2 = \left(\frac{\sigma_J}{2}\right)^2 \exp\left(-\frac{2\alpha}{\beta}\right)\left[\exp\left(\frac{1}{\beta^2}\right)-1\right]\left[\exp\left(\frac{1}{\beta^2}\right)+2\exp\left(\frac{2\alpha}{\beta}\right)+\exp\left(\frac{4\alpha\beta+1}{\beta^2}\right)\right] \tag{26.10}$$

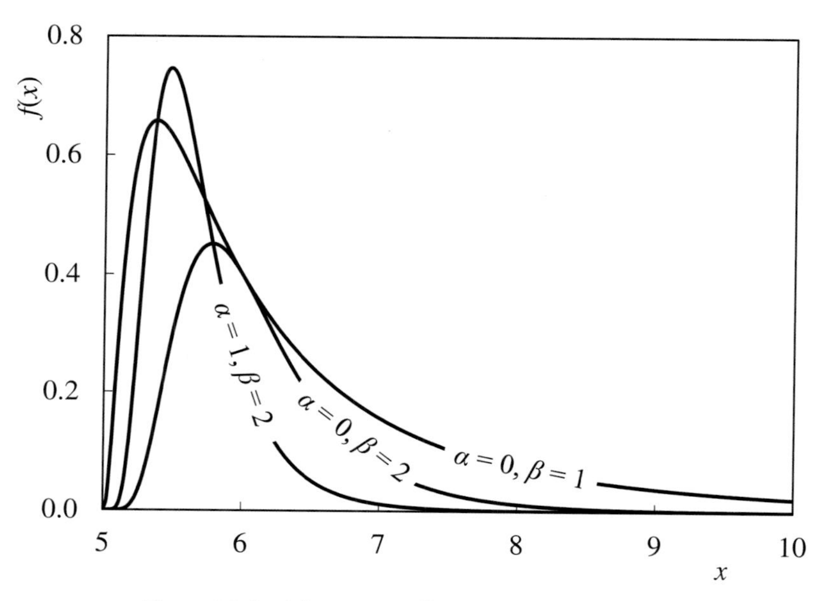

Figure 26.4 Johnson S_L: Effect of α and β on shape

26.3 S_L

The third, known as *Johnson* S_L, is described by the PDF

$$f(x)=\frac{\beta}{\sqrt{2\pi}(x-\mu_J)}\exp\left\{-\frac{1}{2}\left[\alpha+\beta\ln\left(\frac{x-\mu_J}{\sigma_J}\right)\right]^2\right\} \quad x>\mu_J;\ \beta,\sigma_J>0 \tag{26.11}$$

Keeping μ_J at 5 and σ_J at 1, Figure 26.4 shows the effect of α and β.

Its CDF is

$$F(x)=\frac{1}{2}+\frac{1}{2}\text{erf}\left\{\frac{1}{\sqrt{2}}\left[\alpha+\beta\ln\left(\frac{x-\mu_J}{\sigma_J}\right)\right]\right\} \tag{26.12}$$

Mean and variance are

$$\mu=\mu_J+\sigma_J\exp\left(\frac{1-2\alpha\beta}{2\beta^2}\right) \tag{26.13}$$

$$\sigma^2=\sigma_J^2\exp\left(\frac{1-2\alpha\beta}{\beta^2}\right)\left[\exp\left(\frac{1}{\beta^2}\right)-1\right] \tag{26.14}$$

26.4 S_B

The fourth, known as *Johnson* S_B, has the PDF

$$f(x)=\frac{\alpha\beta}{\sqrt{2\pi}(x-\mu_J)(\sigma_J-x+\mu_J)}\exp\left\{-\frac{1}{2}\left[\alpha+\beta\ln\left(\frac{x-\mu_J}{\sigma_J-x+\mu_J}\right)\right]^2\right\} \quad \mu_J\le x\le\mu_J+\sigma_J;\ \alpha,\beta>0 \tag{26.15}$$

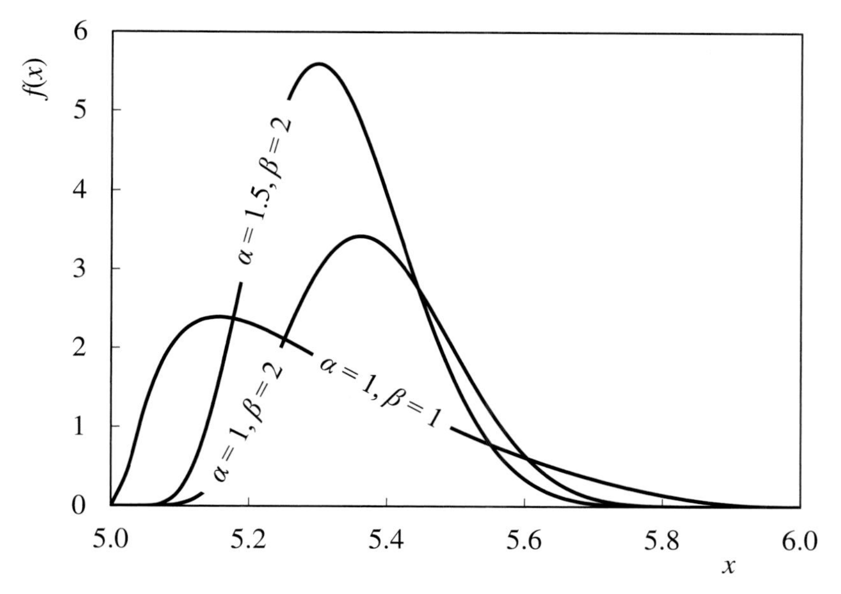

Figure 26.5 *Johnson S_B: Effect of α and β on shape*

Keeping μ_J at 5 and σ_J at 1, Figure 26.5 shows the effect of α and β.
Its CDF is

$$F(x) = \frac{1}{2} + \frac{1}{2}\text{erf}\left\{\frac{1}{\sqrt{2}}\left[\alpha + \beta \ln\left(\frac{x - \mu_J}{\sigma_J - x + \mu_J}\right)\right]\right\} \tag{26.16}$$

26.5 Summary

Table 26.1 shows the results of fitting each of the Johnson distributions to the C_4 in propane data. As seen previously, the highly skewed data cannot be represented well by the normal distribution (S_N). The three others are all a substantial improvement with, in this case, the S_B distribution providing the best fit. However, for this distribution, the calculation of mean and variance are extremely complicated. So, unless implemented in a software product, this limits its use. Further, with the best fit, the results are bounded between 1.08 (μ_J) and 27.7 ($\mu_J + \sigma_J$). The range over the year was 1.5 to 15.8 so, while it is unlikely that the lower or upper bound will be violated in the future, the lower bound has already been approached closely. This might also

Table 26.1 *Results of fitting Johnson distributions to C_4 data*

distribution	μ_J	σ_J	α	β	*RSS*	μ	σ
S_N	4.47	2.09	0.00	1.00	0.6138	4.47	2.09
S_U	0.896	0.841	−3.47	1.655	0.0328	4.94	2.76
S_L	0.634	3.66	0.010	1.724	0.0322	4.94	2.72
S_B	1.08	26.6	2.60	1.31	0.0308		

be of concern, especially if there is interest in exploring the statistics of the higher purity product.

The next best fit is provided by the S_L distribution in which the lower bound is less likely to be approached and there is no upper bound. The calculations of mean and standard deviation are also considerably easier.

27

Pearson Distribution

Karl Pearson, perhaps most known for the R^2 coefficient used to assess correlations, developed a family of 12 distributions.[18–21] His work predates the publication of almost all others, bar the normal distribution. Indeed, some well-used distributions such as the F distribution, gamma distribution and Student t distribution are special cases of Pearson's work. Pearson's objective was that a distribution be fitted not only to the mean and variance of the data, but also to the skewness and kurtosis. Indeed, it was Pearson who originally defined skewness and kurtosis as additional measures of the character of a distribution. In principle, maximum likelihood estimates for each of the shape parameters can be then determined from the distribution's moments. In practice this is mathematically complex. Instead, as we have done throughout this book, we apply the numerical curve fitting approach described in Chapter 9.

All the distributions are described only in PDF form. In most cases the CDF exists but is too complex to be used other than by specialist software products. Calculation of skewness and kurtosis is, for some of the distributions, similarly complex, in which case they have been omitted.

Figure 27.1 shows the range of skewness and kurtosis that can be represented by each of the distribution types. Skewness is plotted as γ^2 only so that the plots are linear or approximately so. This does not imply that, by taking the negative root, the distribution can represent data that are negatively skewed. Types II, III, V and VII are restricted to lines. Types I, IV and VI cover regions. Types IV and VI overlap, with Type V forming the boundary of the overlap. Type III forms the boundary between Types I and VI. Initial indications are that Type I would be a good choice for the C_4 data and either Type IV or Type VI for the NHV disturbance data.

Types VIII, IX and XII appear to be documented more as completion of a mathematical exercise and none have attracted a practical application. Type X we now know as the *exponential distribution* and Type XI as the *Pareto-I distribution*.

Statistics for Process Control Engineers: A Practical Approach, First Edition. Myke King.

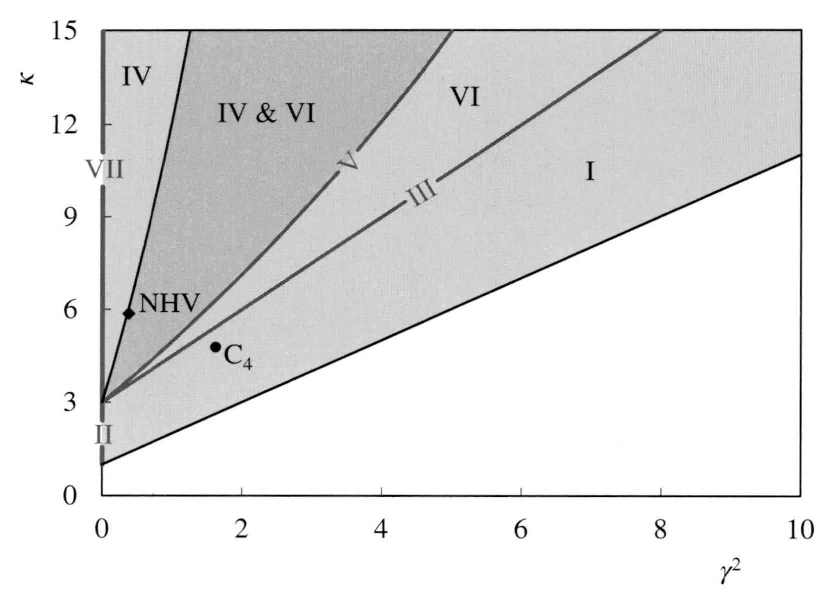

Figure 27.1 Pearson: Range of γ and κ covered

27.1 Type I

The *Pearson-I distribution* can also be described as the *beta-I distribution*. We covered the classic version in Section 12.3, where we showed it fitted the C_4 data well. Including location and scaling parameters, its PDF is

$$f(x) = \frac{1}{B(\delta_1, \delta_2)} \left(\frac{x-\alpha}{\beta} \right)^{\delta_1 - 1} \left(1 - \frac{x-\alpha}{\beta} \right)^{\delta_2 - 1} \quad \alpha \le x \le \alpha + \beta;\ \beta, \delta_1, \delta_2 > 0 \tag{27.1}$$

Although δ_1 and δ_2 can be less than 1, the resulting distribution is unlikely to fit process data.

The PDF is occasionally written in a less useful form that changes the bounds on x and shifts the value of δ_1 and δ_2 by -1.

$$f(x) = \frac{1}{2^{\delta_1 + \delta_2} B(\delta_1 + 1, \delta_2 + 1)} \left(1 + \frac{x-\alpha}{\beta} \right)^{\delta_1} \left(1 - \frac{x-\alpha}{\beta} \right)^{\delta_2} \quad \alpha - \beta \le x \le \alpha + \beta;\ \beta, \delta_1, \delta_2 > 1 \tag{27.2}$$

27.2 Type II

The *Pearson-II distribution* is also known as the *symmetric-beta distribution*. It is obtained by setting both δ_1 and δ_2, in the Type I distribution, equal to the same value (δ). It is therefore described by the PDF

$$f(x) = \frac{\Gamma(2\delta)}{\Gamma^2(\delta)} \left(\frac{x-\alpha}{\beta} \right)^{\delta - 1} \left(1 - \frac{x-\alpha}{\beta} \right)^{\delta - 1} \quad \alpha \le x \le \alpha + \beta;\ \beta, \delta > 0 \tag{27.3}$$

Key parameters are

$$\mu = 0.5 \tag{27.4}$$

$$\sigma^2 = \frac{1}{4(2\delta+1)} \tag{27.5}$$

$$\gamma = 0 \tag{27.6}$$

$$\kappa = \frac{3(2\delta+1)}{(2\delta+3)} \tag{27.7}$$

As $\delta \to \infty$ the distribution approaches the normal distribution. The feasible range of kurtosis is from 1 to 3. This very restrictive range is shown as the coloured line in Figure 27.1.

27.3 Type III

The *Pearson-III distribution* is described by the PDF

$$f(x) = \frac{1}{\Gamma(\delta)\beta}\left|\frac{x-\alpha}{\beta}\right|^{\delta-1} \exp\left(-\frac{x-\alpha}{\beta}\right) \qquad \beta, \delta > 0 \tag{27.8}$$

Figure 27.2 shows the effect of changing δ. As $\delta \to \infty$ the distribution approaches the normal distribution.

Fitting to the C_4 in propane data gives values of 1.60 for α, 1.84 for β, 1.79 for δ and 0.0001 for $F(x_1)$. With *RSS* at 0.0358, the fit is one of the best evaluated.

The mean and variance are

$$\mu = \alpha + \delta\beta \tag{27.9}$$

$$\sigma^2 = \delta\beta^2 \tag{27.10}$$

These give μ as 4.89 and σ as 2.46.

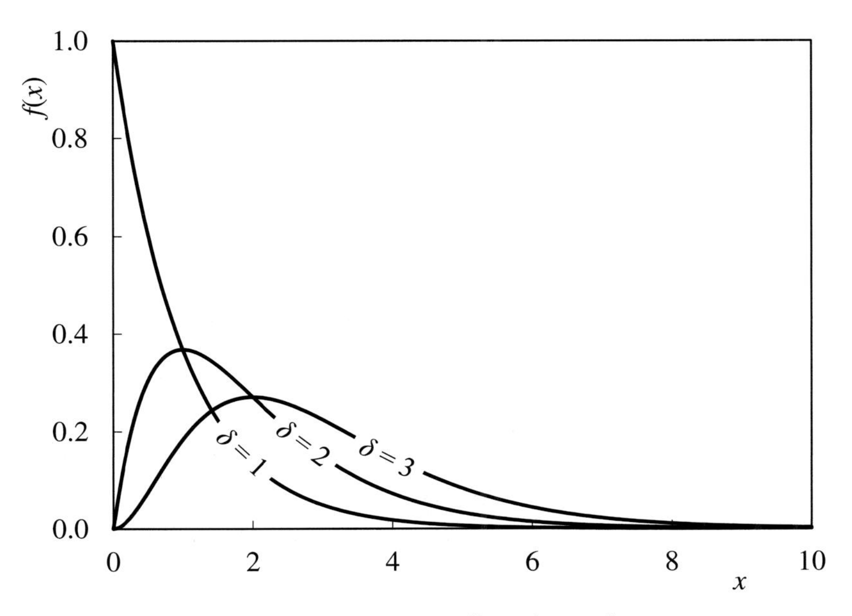

Figure 27.2 Pearson-III: Effect of δ on shape

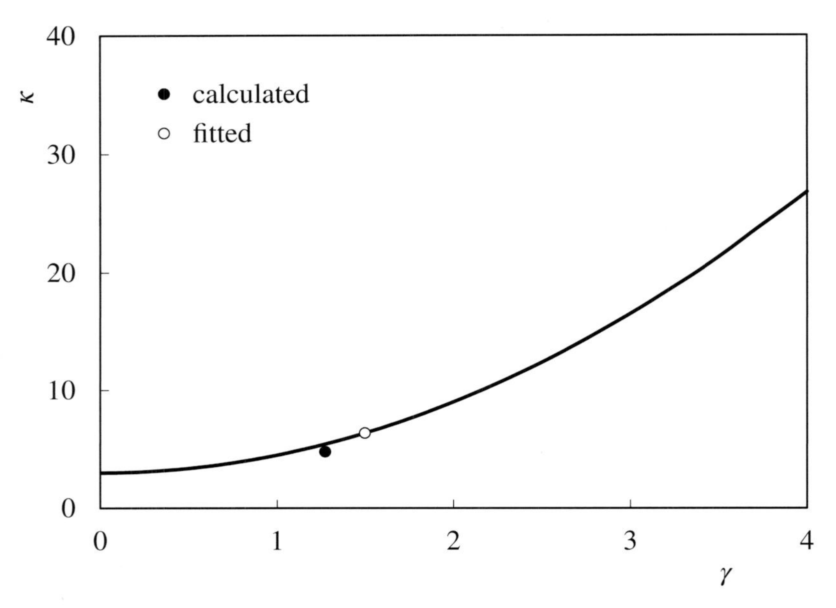

Figure 27.3 Pearson-III: Feasible combinations of γ and κ

Skewness and kurtosis are

$$\gamma^2 = \frac{4}{\delta} \tag{27.11}$$

$$\kappa = \frac{6}{\delta} + 3 \tag{27.12}$$

The feasible combinations of skewness and kurtosis are shown as a line in Figure 27.1. The equations above give γ as 1.50 and κ as 6.36. Showing the same line, Figure 27.3 also shows that these fitted values are close to those calculated from the data. This confirms that, in this case, the Pearson-III distribution would be a good choice.

A special case of the Type III distribution is obtained by setting

$$\alpha = 0 \quad \beta = \frac{1}{\lambda} \quad \delta = k \tag{27.13}$$

This gives the *gamma distribution* that we covered in Section 12.10.

27.4 Type IV

The *Pearson-IV distribution* is described by the PDF

$$f(x) = K\left[1 + \left(\frac{x-\alpha}{\beta}\right)^2\right]^{-\delta_1} \exp\left[-\delta_2 \tan^{-1}\left(\frac{x-\alpha}{\beta}\right)\right] \quad \beta, \delta_1, \delta_2 > 0 \tag{27.14}$$

K is the normalisation constant.

$$K = \frac{\Gamma^2\left(\delta_1 + \frac{\delta_2\sqrt{-1}}{2}\right)}{\Gamma(\delta_1)\Gamma\left(\delta_1 - \frac{1}{2}\right)\beta\sqrt{\pi}} \tag{27.15}$$

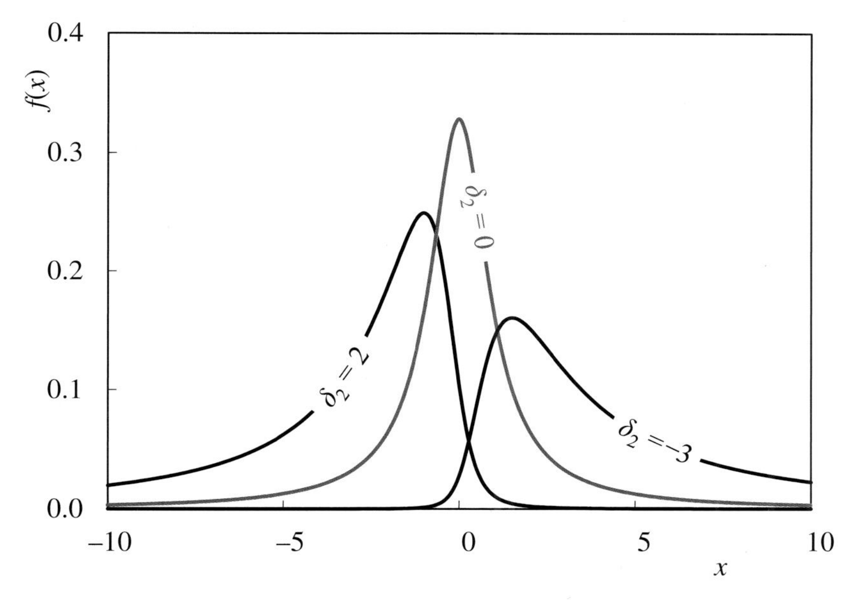

Figure 27.4 *Pearson-IV: Effect of δ_2 on shape*

The problem with this definition is that it includes $\sqrt{-1}$. While the gamma function of a complex number can be a real number, its calculation is beyond the scope of this book and generally not included in spreadsheet packages. Without specialist software, the alternative, when fitting the distribution, is to adjust K to make $F(x_\infty)$ equal to 1. This is achieved by extending Table 9.2 – adding some large values for x in column 1 and adjusting K until the last element in column 8 is 1. If the distribution were upper-bounded (which this one is not) then K is adjusted to make $F(x_{max})$ equal to 1.

With α set to 0, β to 1 and δ_1 to 1, Figure 27.4 shows that varying δ_2 changes skewness – positive values giving positive skewness.

The mean and variance are based on δ.

$$\delta = 2(\delta_1 - 1) \tag{27.16}$$

$$\mu = \alpha - \frac{\delta_2}{\delta}\beta \qquad \delta_1 > 1 \tag{27.17}$$

$$\sigma^2 = \frac{\delta^2 + \delta_2^2}{\delta^2(\delta-1)}\beta^2 \qquad \delta_1 > \frac{3}{2} \tag{27.18}$$

Raw moments are given by the recurrence formula

$$m_0 = 1 \tag{27.19}$$

$$m_1 = 0 \tag{27.20}$$

$$m_n = \frac{(n-1)}{\delta^2(\delta-n+1)}\left[\beta\left(\delta^2+\delta_2^2\right)m_{n-2} - 2\delta\delta_2 m_{n-1}\right]\beta \qquad n \geq 2 \tag{27.21}$$

From these we can derive

$$\gamma = \frac{4\delta_2}{\delta-2}\sqrt{\frac{\delta-1}{\delta^2+\delta_2^2}} \qquad \delta_1 > 2 \tag{27.22}$$

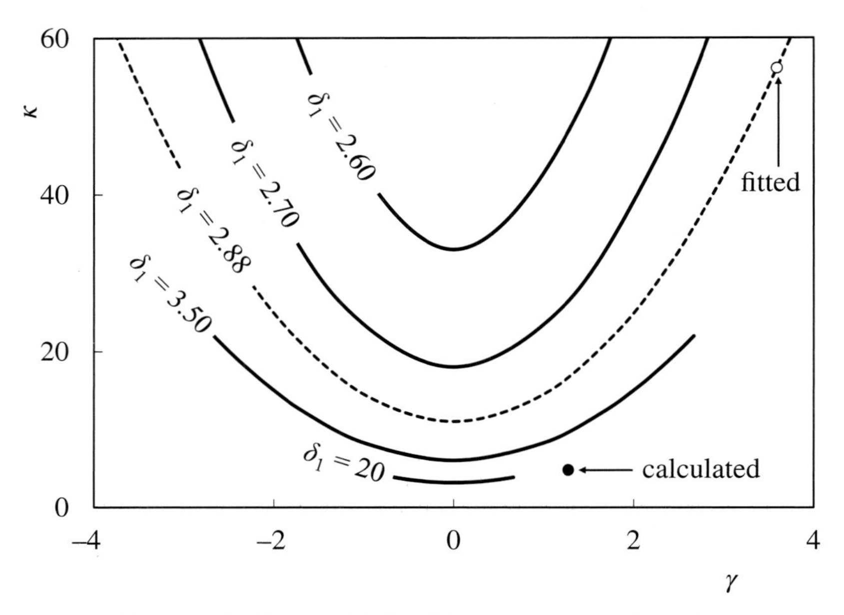

Figure 27.5 Pearson-IV: Feasible combinations of γ and κ

$$\kappa = \frac{3(\delta-1)\left[(\delta+6)\left(\delta^2+\delta_2^2\right)-8\delta^2\right]}{(\delta-2)(\delta-3)\left(\delta^2+\delta_2^2\right)} \qquad \delta_1 > \frac{5}{2} \tag{27.23}$$

Fitting to the C_4 in propane data gives values of 0.000103 for K, 0.03 for α, 1.68 for β, 2.88 for δ_1 and −11.1 for δ_2. With *RSS* at 0.0326, it is one of the best fits. From Equation (27.17) the mean is then 5.01; from Equation (27.18) the standard deviation is 3.17.

From Equation (27.22) skewness is 3.59. From Equation (27.23), kurtosis is 56.1 – substantially higher than that calculated from the data as 4.79. This places some suspicion on whether the distribution would be a good choice. The result is plotted in Figure 27.5, along with the feasible combinations of skewness and kurtosis.

Fitting to the NHV disturbance data gives values of 0.382 for K, 0.00 for α, 1.76 for β, 2.40 for δ_1 and 0.104 for δ_2. With *RSS* at 0.0513, the fit is close to the best achieved by others. The mean is –0.07 and the standard deviation 1.31. Skewness is –0.248. Because δ_1 is less than 2.5, kurtosis cannot be calculated.

By setting δ_2 to 0, the distribution becomes symmetrical and is then known as the *Pearson-VII distribution*. Because this removes the complex term, K can then be calculated.

By also setting δ_1 to 1 we obtain the *Cauchy distribution*. This is also known as the *McCullagh distribution*, the *Breit–Wigner distribution*, the *Lorenz distribution* or the *Cauchy–Lorenz distribution*. While kurtosis cannot be determined, the distribution can represent data for which this is very large. It is described by the PDF

$$f(x) = \frac{1}{\pi\beta}\left[1+\left(\frac{x-\alpha}{\beta}\right)^2\right]^{-1} \qquad \beta > 0 \tag{27.24}$$

Figure 27.6 shows the effect of changing α and β.

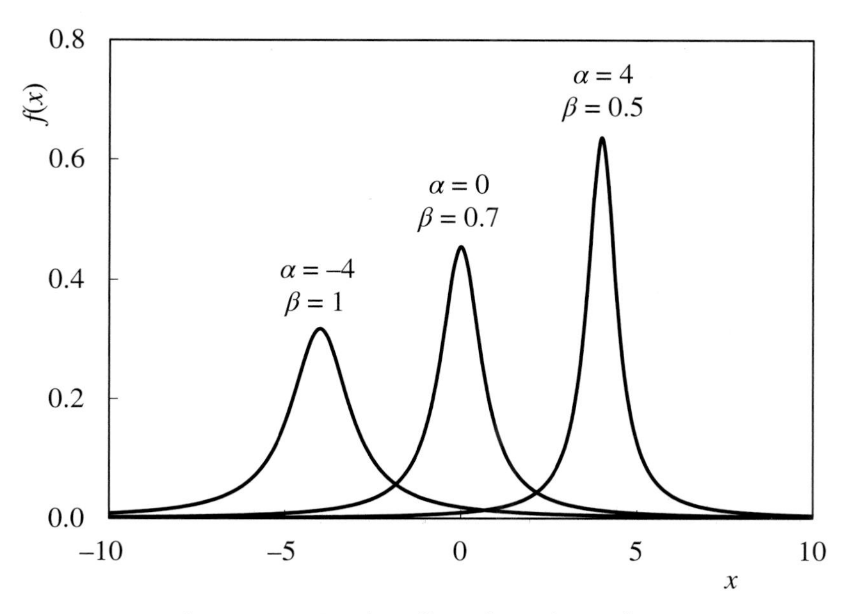

Figure 27.6 *Cauchy: Effect of α and β on shape*

The CDF is

$$F(x) = \frac{1}{\pi}\tan^{-1}\left(\frac{x-\alpha}{\beta}\right) + \frac{1}{2} \tag{27.25}$$

If α is set to 0 and β to 1, as usual, the distribution becomes the *standard Cauchy distribution*.

Fitting Equation (27.25) to the NHV disturbance data gives values of –0.04 for α, 0.71 for β and 0.0605 for *RSS*. Kurtosis, calculated from the data, is 5.87 – considerably higher than that of a normal distribution. In this case, Cauchy would therefore be the better choice.

By inverting Equation (27.25), the QF is

$$x(F) = \alpha + \beta\tan\left[\pi\left(F - \frac{1}{2}\right)\right] \quad 0 \le F \le 1 \tag{27.26}$$

Figure 27.7 plots Equation (27.25) against the process data. The fit to the tails is poor but it can be used elsewhere. So, if we choose values for F as 0.1 and 0.9, x is –2.24 and 2.15 respectively. In other words we can be 80% sure that a disturbance will be within ±2.2 (or 20% sure that the disturbance will exceed this). This detail is shown in colour in Figure 27.7.

While β is not the standard deviation, it can be used in much the same way. For example, we might expect improved control to halve it. Halving it in Equation (27.25), and using the ±1.1 values of x, gives values for F of 0.051 and 0.949. In other words, with improved control, the probability of a disturbance being larger than ±2.2 falls to 10.2%. As might be expected, if we halve the size of disturbances, we will violate any limit half as often.

The PDF of the *log Cauchy distribution* is derived from Equation (27.24).

$$f(x) = \frac{1}{\pi\beta x}\left[1 + \left(\frac{\ln(x)-\alpha}{\beta}\right)^2\right]^{-1} \quad x > 0;\ \beta > 0 \tag{27.27}$$

Figure 27.8 shows the effect of varying α, with β fixed at 2. Normally only affecting position, here α also affects shape. Figure 27.9 shows the effect of varying β with α fixed at 0. Normally

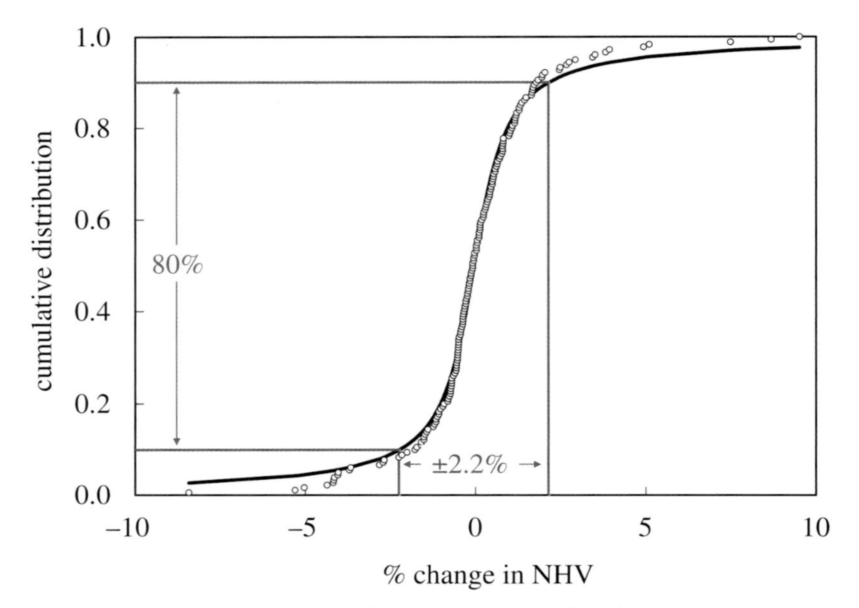

Figure 27.7 Cauchy: Fitted to NHV disturbances

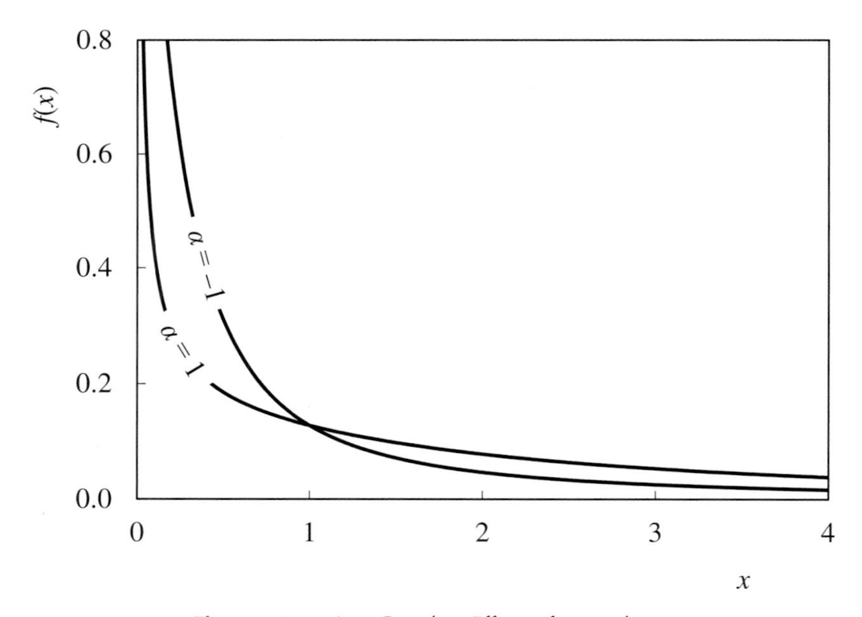

Figure 27.8 Log-Cauchy: Effect of α on shape

only affecting scale (or dispersion), at values less than 1, it introduces an inflexion into the curve. Process data are unlikely to follow such a distribution.

The CDF is

$$F(x) = \frac{1}{\pi}\tan^{-1}\left(\frac{\ln(x)-\alpha}{\beta}\right) + \frac{1}{2} \tag{27.28}$$

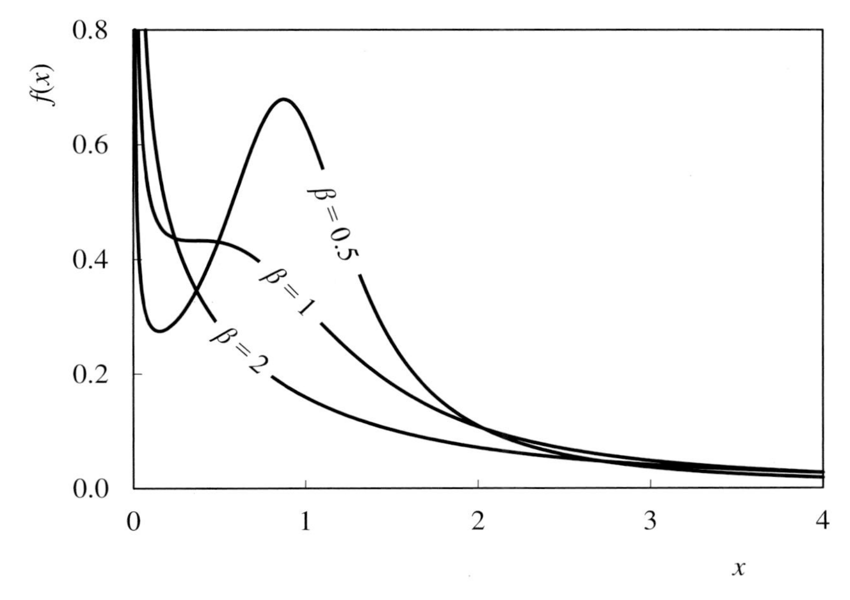

Figure 27.9 Log-Cauchy: Effect of β on shape

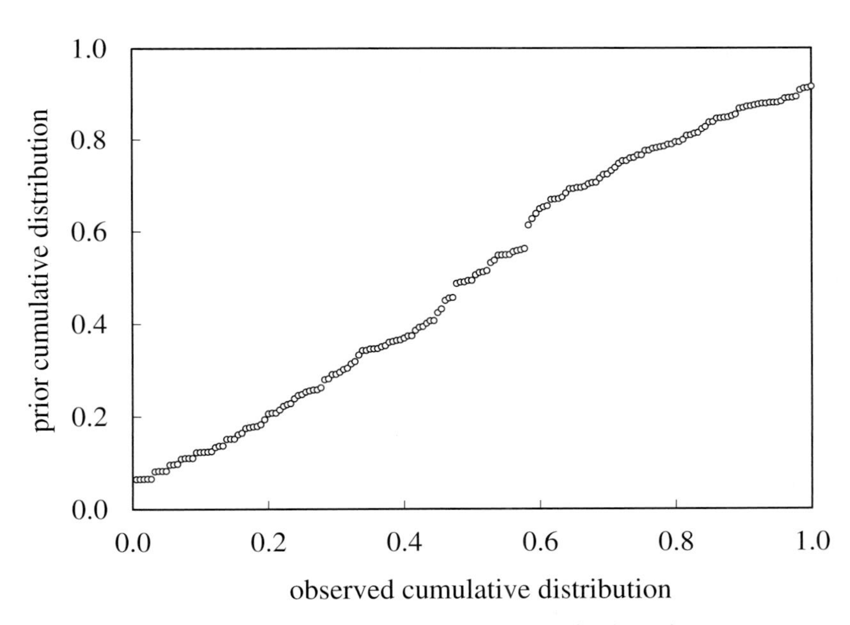

Figure 27.10 Log-Cauchy: P–P plot showing fit to absolute changes in NHV

Since x must be greater than zero, we will evaluate the distribution by fitting it to the absolute value of the NHV disturbances. This gives α as –0.34 and β as 0.69. *RSS* is 0.1713, which is better than that achieved by several other distributions. Figure 27.10 is a P–P plot confirming that the fit appears reasonable, although perhaps of concern that it does not pass through (0,0) and (1,1). However, this hides a more significant problem that can only be seen by plotting the PDF.

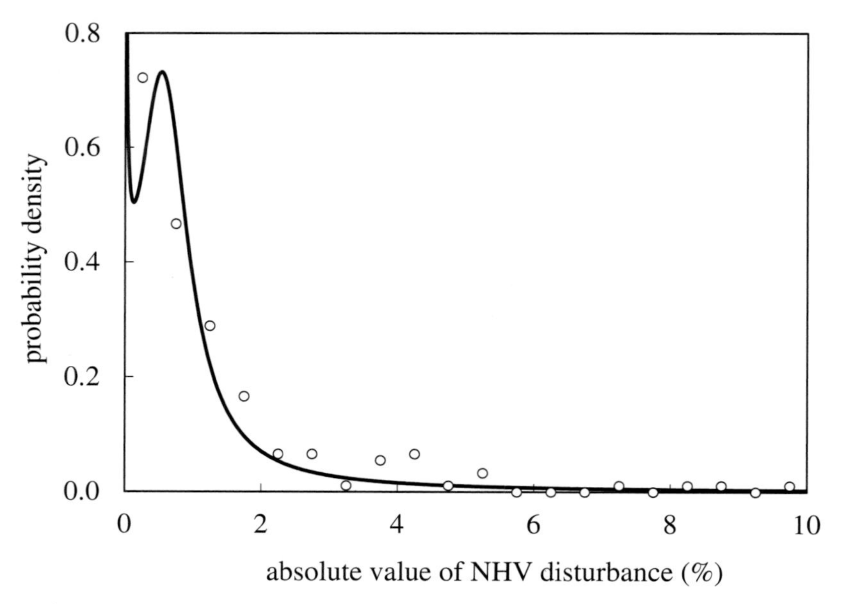

Figure 27.11 Log-Cauchy: Plot of PDF fitted to absolute changes in NHV

As Figure 27.11 shows, mathematically the fit might be reasonable but, as engineers, we should be highly suspicious that it does not realistically represent the behaviour of the process.

The CDF can be inverted to produce the QF

$$x(F) = \exp\left\{\alpha + \beta \tan\left[\pi\left(F - \frac{1}{2}\right)\right]\right\} \quad 0 \leq F \leq 1 \tag{27.29}$$

There are no formulae for mean, variance, skewness or kurtosis. Given the problem illustrated above, the log-Cauchy distribution has little practical value to the process industry.

27.5 Type V

When first published, the *Pearson-V distribution* was the normal distribution. Pearson subsequently replaced it with this version and referred to the normal distribution as the *Pearson-0 distribution*. The replacement Pearson-V distribution, also now known as the *inverse gamma distribution*, is described by the PDF

$$f(x) = \frac{1}{\beta\Gamma(\delta)}\left|\frac{\beta}{x-\alpha}\right|^{\delta+1} \exp\left(-\frac{\beta}{x-\alpha}\right) \quad \beta, \delta > 0 \tag{27.30}$$

The effect of δ is shown in Figure 27.12.

Moments are given by

$$m_n = \frac{\Gamma(\delta)}{\Gamma(\delta - n)}\beta^n \quad \delta > n \tag{27.31}$$

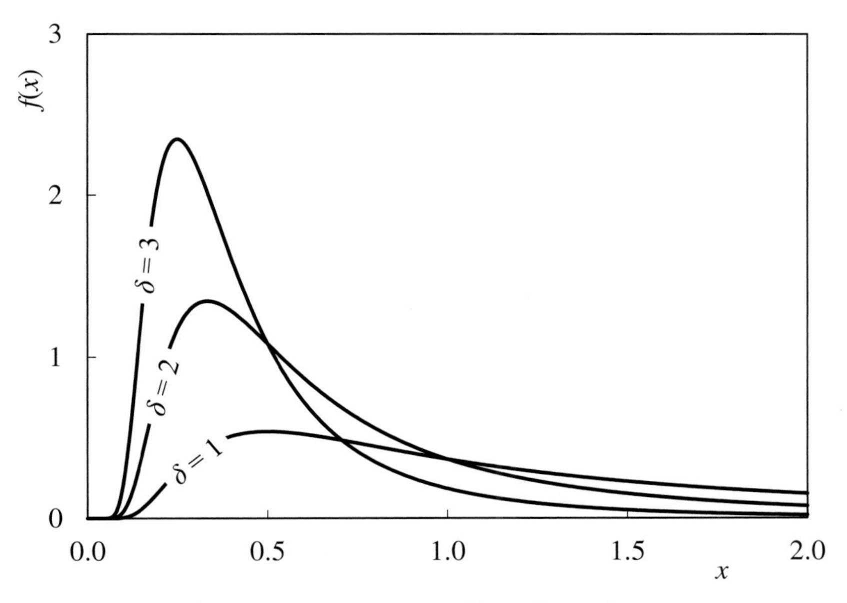

Figure 27.12 *Pearson-V: Effect of δ on shape*

Hence

$$\mu = \alpha + \frac{\beta}{\delta - 1} \qquad \delta > 1 \tag{27.32}$$

$$\sigma^2 = \frac{\beta^2}{(\delta-1)^2(\delta-2)} \qquad \delta > 2 \tag{27.33}$$

$$\gamma = \frac{4\sqrt{\delta-2}}{\delta-3} \qquad \delta > 3 \tag{27.34}$$

$$\kappa = \frac{3(\delta+5)(\delta-2)}{(\delta-3)(\delta-4)} \qquad \delta > 4 \tag{27.35}$$

From Equation (27.35), as $\delta \to \infty$, $\kappa \to 3$ and so Equation (27.30) cannot be fitted accurately to platykurtic data.

Fitting to the C_4 in propane data gives values of 0.01 for α, 17.6 for β, 4.38 for δ and 0.0204 for $F(x_1)$. The resulting *RSS* is 0.0336. Mean (μ) is therefore 5.20 and standard deviation (σ) 3.36. Skewness is 4.46 and kurtosis an unbelievable 126. So while the fit appears to be good, some suspicion must be placed on the result.

27.6 Type VI

The *Pearson-VI distribution* is described by

$$f(x) = \frac{\left(\frac{x-\alpha}{\beta}\right)^{\delta_1 - 1}}{\mathrm{B}(\delta_1,\delta_2)\beta\left(1 + \frac{x-\alpha}{\beta}\right)^{\delta_1+\delta_2}} \qquad \beta, \delta_1, \delta_2 > 0 \tag{27.36}$$

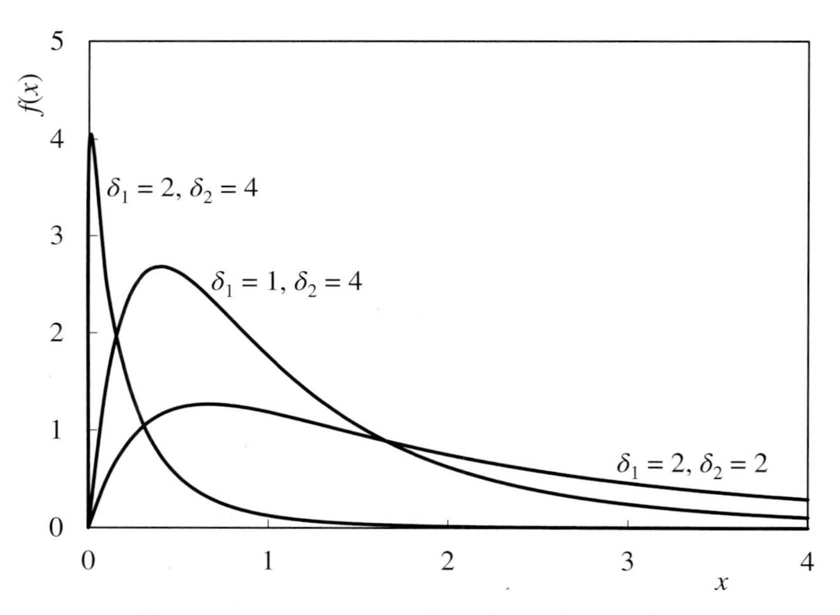

Figure 27.13 *Pearson-VI: Effect of δ_1 and δ_2 on shape*

Figure 27.13 shows the effect of δ_1 and δ_2.

Key parameters are

$$\mu = \alpha + \frac{\beta\delta_1}{\delta_2 - 1} \quad \delta_2 > 1 \tag{27.37}$$

$$\sigma^2 = \frac{\beta^2\delta_1(\delta_1 + \delta_2 - 1)}{(\delta_2 - 1)^2(\delta_2 - 2)} \quad \delta_2 > 2 \tag{27.38}$$

$$\gamma = 2\frac{2\delta_1 + \delta_2 - 1}{\delta_2 - 3}\sqrt{\frac{\delta_2 - 2}{\delta_1(\delta_1 + \delta_2 - 1)}} \quad \delta_2 > 3 \tag{27.39}$$

$$\kappa = \frac{3(\delta_2 - 2)}{(\delta_2 - 3)(\delta_2 - 4)}\left[\frac{2(\delta_2 - 1)^2}{\delta_1(\delta_1 + \delta_2 - 1)} + \delta_2 + 5\right] \quad \delta_2 > 4 \tag{27.40}$$

Fitting to the C_4 in propane data gives values of 1.46 for α, 8.00 for β, 2.66 for δ_1, 6.78 for δ_2 and 0.0172 for $F(x_1)$. The resulting *RSS*, at 0.0298, appears to be good. However, while the mean (μ) of 5.14 seems reasonable, the standard deviation (σ) is a suspiciously large 10.4. This is explained by Figure 27.14. This was drawn by fixing β at different values in the range 1 to 40 and then fitting the remaining shape parameters. Plotting *RSS* against σ shows that the curve is extremely flat. While mathematically *RSS* is minimised when σ is 10.4, large changes in σ would have little effect on *RSS*.

Skewness is 2.71 and kurtosis 20.16. These values, along with feasible range of γ and κ, are included in Figure 27.15. However, plotting *RSS* against these parameters would show a similar insensitivity and so undermines the confidence we have in the result.

Although Figure 27.1 would indicate that this distribution would be a possible choice for the NHV disturbance data, the calculated values of skewness and kurtosis are very close to the limit.

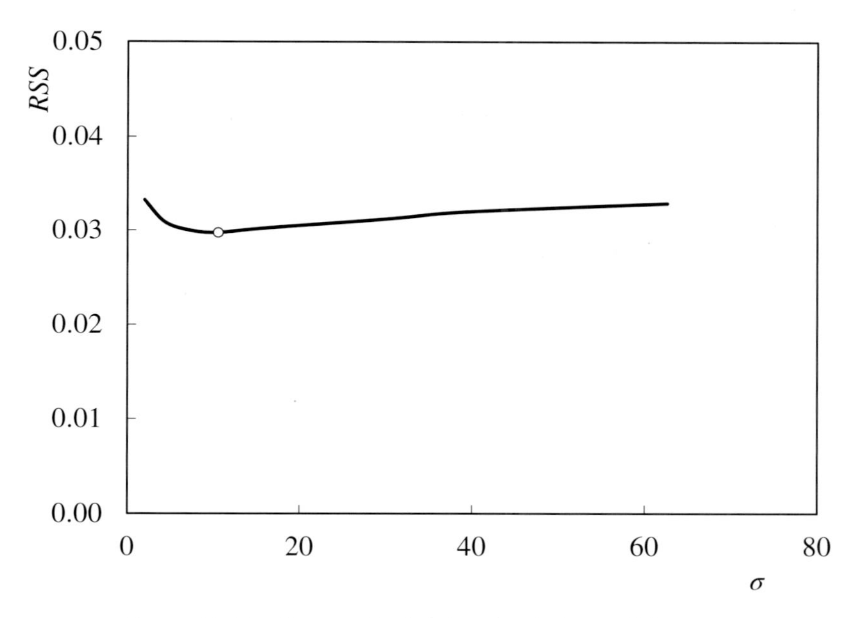

Figure 27.14 Pearson-VI: Relationship between RSS and σ

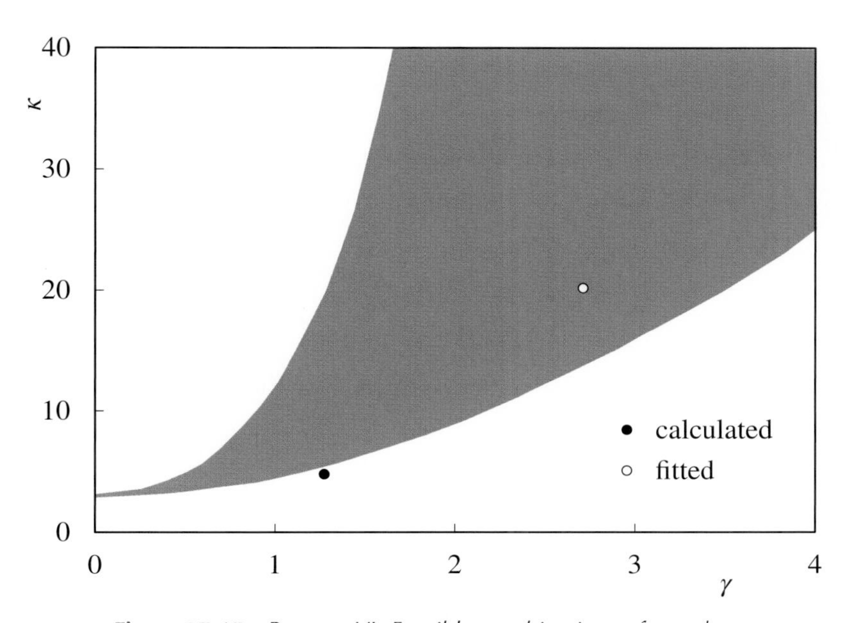

Figure 27.15 Pearson-VI: Feasible combinations of γ and κ

Fitting requires impractically high values gives values for δ_1 and δ_2. Type IV would be a much better choice.

A special case of the Type VI distribution is obtained by setting

$$\delta_1 = \frac{f_1}{2} \quad \delta_2 = \frac{f_2}{2} \quad \alpha = 0 \quad \beta = \frac{f_2}{f_1} \quad F = x \tag{27.41}$$

This gives the *F distribution* described in Section 12.6.

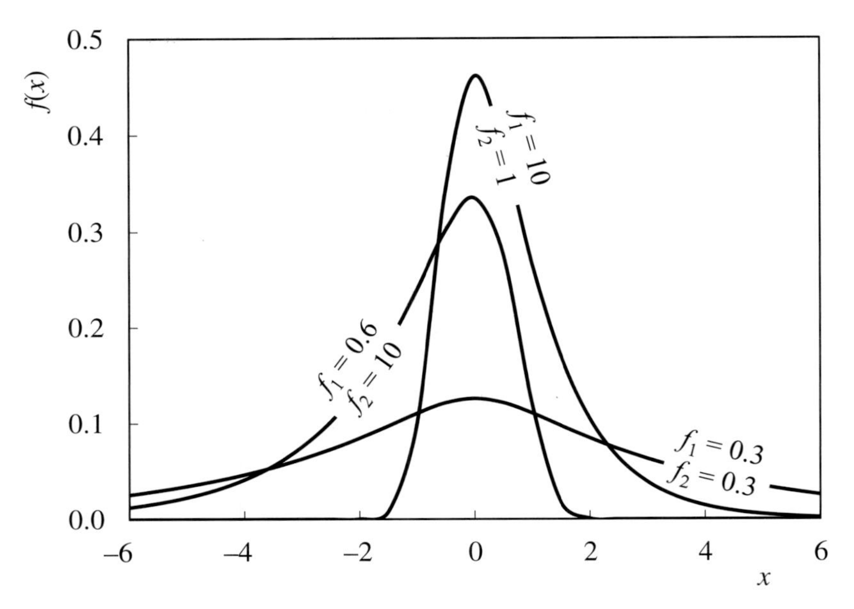

Figure 27.16 *Log-F: Effect of f_1 and f_2 on shape*

The *log-F distribution*, an extension of the F distribution, is also known as *Fisher's z distribution*. It can also be derived from the Type IV generalised logistic distribution and has the PDF

$$f(x)=\frac{2f_1^{\frac{f_1}{2}}f_2^{\frac{f_2}{2}}\exp\left(\frac{f_1(x-\alpha)}{\beta}\right)}{\mathrm{B}\left(\frac{f_1}{2},\frac{f_2}{2}\right)\beta\left[f_2+f_1\exp\left(\frac{2(x-\alpha)}{\beta}\right)\right]^{\frac{f_1+f_2}{2}}}\qquad \beta,f_1,f_2>0 \tag{27.42}$$

Figure 27.16, drawn with α set to 0 and β set to 1, shows the effect of changing f_1 and f_2. The formulae for mean and variance employ mathematics beyond the scope of this book but can be approximated by

$$\mu\approx\alpha+\frac{f_1-f_2}{2f_1f_2}\beta \tag{27.43}$$

$$\sigma^2\approx\frac{f_1+f_2}{2f_1f_2}\beta^2 \tag{27.44}$$

Fitting to the C_4 in propane data gives values of 3.18 for α, 1.00 for β, 5.57 for f_1, 0.371 for f_2 and 0.0231 for $F(x_1)$. The resulting value of 0.0287 for *RSS* is the lowest of all the distributions. From Equations (27.43) and (27.44), the mean is approximately 4.44 and the standard deviation approximately 1.20. There are no simple formulae for skewness or kurtosis.

Fitting to the NHV disturbance data gives values of –0.209 for α, 0.0274 for β, 0.0289 for f_1, 0.0224 for f_2 and 0.0001 for $F(x_1)$. The resulting value of 0.0374 for *RSS* is one of the lowest. A better fit could be obtained by reducing β but, at lower values $e^{2(x-\alpha)/\beta}$ exceeds 2^{1024} – the maximum value permitted by many software products. The mean, at –0.07, is close to 0 as expected but the standard deviation suspiciously low at 0.17.

The PDF of the *noncentral F distribution* is

$$f(x)=\sum_{i=0}^{\infty}\frac{\left(\frac{f_1}{f_2}\right)^{\frac{f_1}{2}+i}\left(\frac{\delta}{2}\right)^{i}\exp\left(-\frac{\delta}{2}\right)x^{\frac{f_1}{2}+i-1}}{\mathrm{B}\left(\frac{f_1}{2}+i,\frac{f_2}{2}\right)i!\left(1+\frac{f_1}{f_2}x\right)^{\frac{f_1+f_2}{2}+i}}\qquad f_1,f_2,\delta>0 \tag{27.45}$$

Fitting to the C_4 in propane data gives f_1 as 19.0, f_2 as 10.9 and δ as 58.3 with *RSS* as 0.0355. Strictly, f_1 and f_2 are degrees of freedom and so should theoretically be integers. Already very close, fixing them at 19 and 11 and refitting gives δ as 58.4 with *RSS* as 0.0356.

While the sum to infinity may appear impractical, it converges fairly quickly. The smaller the number of degrees of freedom, the more quickly is the convergence. In this example summing up to $i = 100$ was more than adequate. This is well within the capabilities of any spreadsheet package.

The mean and variance are

$$\mu=\frac{f_2(f_1+\delta)}{f_1(f_2-2)}\qquad f_2>2 \tag{27.46}$$

$$\sigma^2=2\left(\frac{f_2}{f_1}\right)^2\frac{(f_1+\delta)^2+(f_1+2\delta)(f_2-2)}{(f_2-2)^2(f_2-4)}\qquad f_2>4 \tag{27.47}$$

These give μ as 4.98 and σ as 2.92. These are reasonably close to the values calculated from the data. The higher estimate for the standard deviation reflects the fact that the upper tail of the distribution is a little too long.

There are no formulae for skewness or kurtosis.

As might be predicted, the *doubly noncentral F distribution* is too complex to be included here. It includes an additional shape parameter and so likely to better fit the data. However, it has no formulae for mean, variance, skewness and kurtosis. It is of little practical value in the process industry.

27.7 Type VII

The *Pearson-VII distribution* is a special (symmetrical) case of the Type IV distribution. It is described by the PDF

$$f(x)=\frac{\left[1+\left(\frac{x-\alpha}{\beta}\right)^2\right]^{-\delta}}{\beta\mathrm{B}\left(\delta-\frac{1}{2},\frac{1}{2}\right)}\qquad \beta,\delta>0 \tag{27.48}$$

By definition

$$\mathrm{B}\left(\delta-\frac{1}{2},\frac{1}{2}\right)=\frac{\Gamma\left(\delta-\frac{1}{2}\right)\Gamma\left(\frac{1}{2}\right)}{\Gamma(\delta)}=\frac{\Gamma\left(\delta-\frac{1}{2}\right)\sqrt{\pi}}{\Gamma(\delta)} \tag{27.49}$$

So the PDF can alternatively be written as

$$f(x)=\frac{\Gamma(\delta)\left[1+\left(\frac{x-\alpha}{\beta}\right)^2\right]^{-\delta}}{\beta\sqrt{\pi}\Gamma\left(\delta-\frac{1}{2}\right)} \tag{27.50}$$

The mean and variance are

$$\mu=\alpha \tag{27.51}$$

$$\sigma^2=\frac{\beta^2}{2\delta-3} \qquad \delta>\frac{3}{2} \tag{27.52}$$

Skewness and kurtosis are

$$\gamma=0 \tag{27.53}$$

$$\kappa=\frac{3(2\delta-3)}{2\delta-5} \qquad \delta>\frac{5}{2} \tag{27.54}$$

Fitting to the C_4 in propane data gives values of 3.95 for α, 6.76 for β, 6.92 for δ and 0.0264 for $F(x_1)$. But, at 0.2269, *RSS* shows that the fit is very poor. Not surprisingly, because the distribution is symmetrical, it cannot properly represent the highly skewed data.

Fitting to the almost unskewed NHV disturbance data gives values of 0 for α, 1.05 for β, 1.26 for δ and 0.0146 for $F(x_1)$. At 0.0485, *RSS* shows that the fit is reasonable – although improved upon by other distributions. Because δ is less than 1.5, it is not possible to determine the variance.

Examination of Equation (27.54) shows that κ reaches a minimum of 3 as $\delta \rightarrow \infty$, so the distribution cannot be fitted accurately to platykurtic data.

Figure 27.17 shows the effect of varying δ. In each case α is 0 and β calculated from Equation (27.52) to fix σ at 1. There are two special cases. Firstly, as δ approaches infinity,

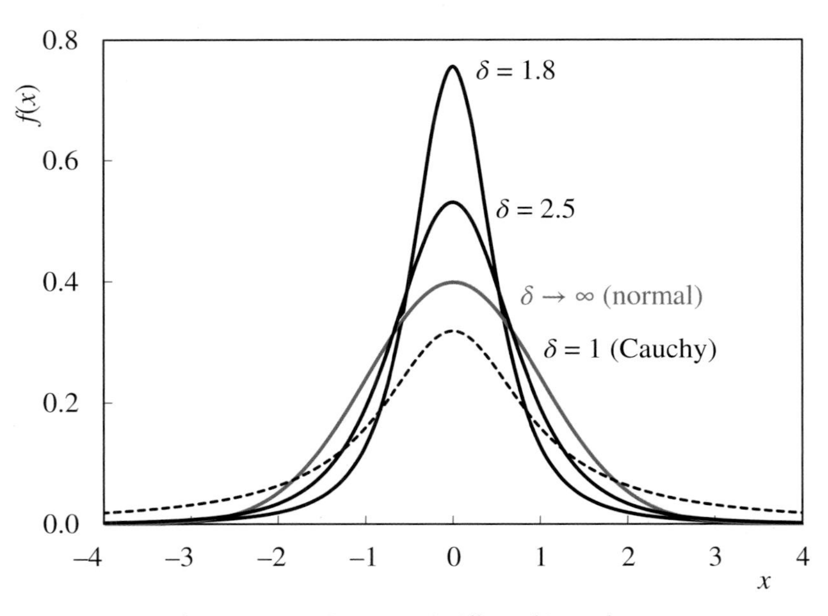

Figure 27.17 *Pearson-VII: Effect of δ on shape*

the distribution approaches the normal distribution – as shown by the coloured line. Secondly, setting δ to 1 gives the Cauchy distribution. Equation (27.52) does not permit the calculation of β to fix σ at 1. Instead the distribution is shown, as the dashed line, in its standard form with β set to 1. This is also a special case of the *standard Student t distribution*, which we covered in Section 12.5, obtained by setting

$$\delta = \frac{f+1}{2} \quad \beta = \sqrt{f} \quad \alpha = 0 \quad x = t \tag{27.55}$$

The PDF of the *log Student t distribution* is

$$f(x) = \frac{\Gamma\left(\frac{f+1}{2}\right)}{x\beta\Gamma\left(\frac{f}{2}\right)\sqrt{\pi f}}\left[1+\frac{1}{f}\left(\frac{\ln x-\alpha}{\beta}\right)^2\right]^{-\frac{f+1}{2}} \quad x>0;\ \beta, f>0 \tag{27.56}$$

In much the same way as the lognormal distribution is derived from the normal distribution, the log-Student t distribution is derived from the Student t distribution by defining t as

$$t = \frac{\ln x - \alpha}{\beta} \tag{27.57}$$

Figure 27.18 shows the effect of varying f, with α fixed at 0 and β at 1. The bimodal shape is unlikely to be applicable to process data. In the same way that the t distribution approaches the normal distribution as f in increased, the log Student t distribution approaches the lognormal distribution – shown as the coloured curve.

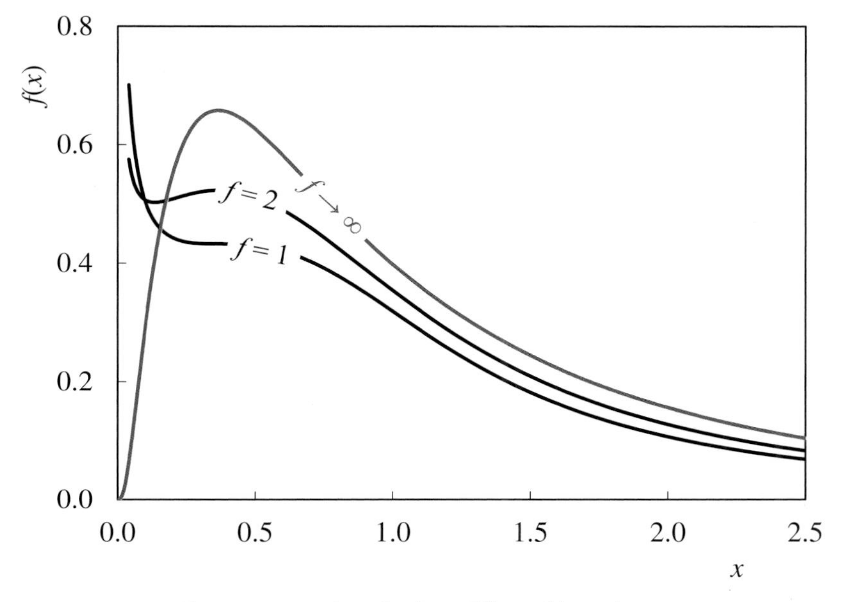

Figure 27.18 *Log-Student: Effect of f on shape*

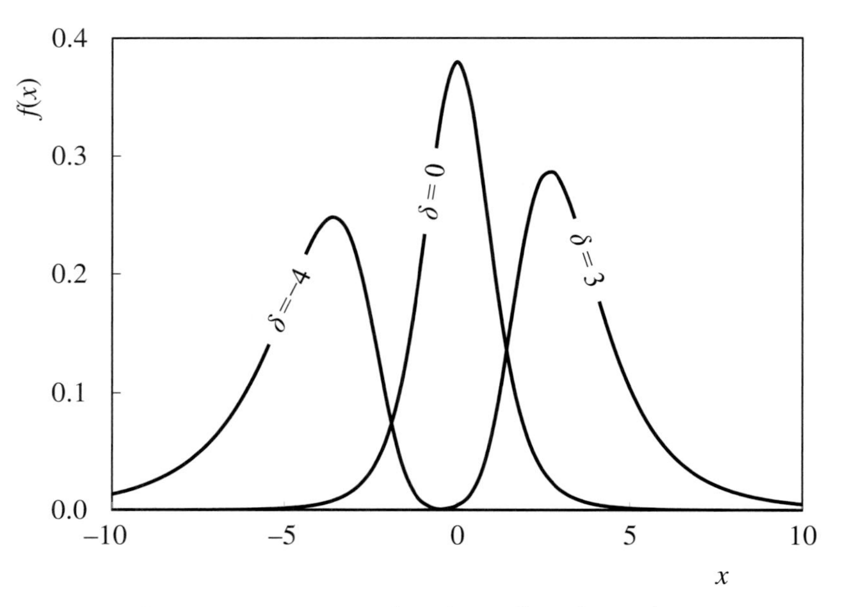

Figure 27.19 Noncentral Student: Effect of δ on shape

The PDF of the *noncentral Student t distribution* is

$$f(t) = \frac{f^{\frac{f}{2}} \exp\left(-\frac{\delta^2}{2}\right)}{\Gamma\left(\frac{f}{2}\right)\sqrt{\pi}(t^2+f)^{\frac{f+1}{2}}} \sum_{j=0}^{\infty} \frac{1}{j!} \Gamma\left(\frac{f+1+j}{2}\right)\left(\frac{\delta\sqrt{2}t}{\sqrt{f+t^2}}\right)^j \quad f>0 \tag{27.58}$$

The effect of δ, with f fixed at 5, is shown in Figure 27.19.

Fitting to the C_4 in propane data gives f as 3.59 and δ as 3.87. With *RSS* at 0.0471, the fit is reasonable but bettered by other distributions that are easier to apply. Further, as Figure 27.20 shows, the distribution's upper tail is excessive.

The mean is

$$\mu = \delta \frac{\Gamma^2\left(\frac{f-1}{2}\right)}{\Gamma^2\left(\frac{f}{2}\right)} \sqrt{\frac{f}{2}} \tag{27.59}$$

This gives μ as 4.84, close to that calculated from the data.

The variance is

$$\sigma^2 = \frac{f\left(1+\delta^2\right)}{f-2} - \delta^2 \frac{\Gamma^2\left(\frac{f-1}{2}\right)}{\Gamma^2\left(\frac{f}{2}\right)} \frac{f}{2} \tag{27.60}$$

This gives σ as 3.32. This is substantially larger than the calculated value, illustrating the problem caused by the excessive tail.

While the *doubly noncentral t distribution* is likely to better fit the data, its complexity undermines its practical value.

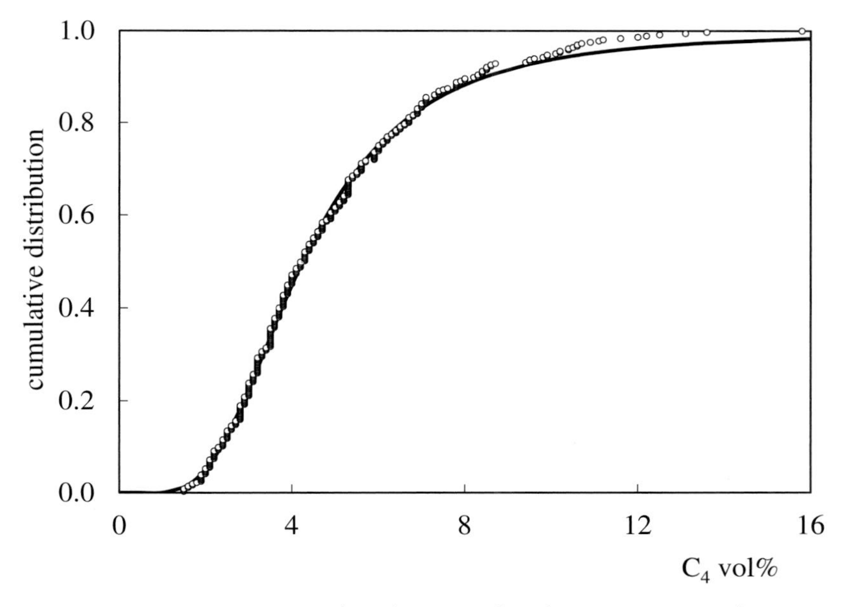

Figure 27.20 Noncentral Student: Fitted to the C_4 in propane data

27.8 Type VIII

The PDF of the *Pearson-VIII distribution* is

$$f(x) = \frac{K}{\beta}\left(1 + \frac{x-\alpha}{\beta}\right)^{-\delta} \qquad \alpha-\beta < x \leq \alpha;\ \beta, \delta > 0 \tag{27.61}$$

It has little application to process data.

27.9 Type IX

The PDF of the *Pearson-IX distribution* is

$$f(x) = \frac{K}{\beta}\left(1 + \frac{x-\alpha}{\beta}\right)^{\delta} \qquad \alpha-\beta < x \leq \alpha;\ \beta, \delta > -1 \tag{27.62}$$

It too has little application to process data.

27.10 Type X

The *Pearson-X distribution* is now more commonly known as the exponential distribution that we covered in Section 12.7.

27.11 Type XI

The *Pearson-XI distribution* is more commonly known as the *Pareto-I distribution* and was covered in Section 23.1.

27.12 Type XII

The PDF of the *Pearson-XII distribution* is

$$f(x) = K\left(1 + \frac{x-\alpha}{\beta_1}\right)^{\delta}\left(1 - \frac{x-\alpha}{\beta_2}\right)^{-\delta} \quad -\beta_1 < x < \beta_2;\ \beta_1, \beta_2 > 0;\ -1 < \delta < 1 \tag{27.63}$$

It has little application to process data.

28

Exponential Distribution

We covered the classic exponential distribution in Section 12.7. Here we describe a range of extensions.

28.1 Generalised Exponential

The *generalised exponential distribution*, confusingly, can be referred to by using the same *GED* acronym as the *generalised error distribution* that we covered in Section 20.12. Less ambiguously known as the *exponentiated exponential distribution*, it extends the exponential distribution by adding a shape parameter (δ). It is described by

$$f(x) = \frac{\delta e^{-\lambda x}}{\beta}\left[1 - \exp\left(-\frac{x-\alpha}{\beta}\right)\right]^{\delta-1} \qquad x \geq \alpha;\ \lambda, \delta > 0 \tag{28.1}$$

$$F(x) = \left[1 - \exp\left(-\frac{x-\alpha}{\beta}\right)\right]^{\delta} \tag{28.2}$$

$$x(F) = \alpha - \beta \ln\left(1 - F^{1/\delta}\right) \qquad 0 \leq F \leq 1 \tag{28.3}$$

Fitting to the intervals between events of the LPG splitter reflux exceeding 65 m^3/hr gives α as 0.0626, β as 12.5 and δ as 0.950. With δ so close to 1, in this case *RSS* is only slightly improved, from 0.0058 for the classic exponential distribution, to 0.0053.

28.2 Gompertz–Verhulst

The *Gompertz–Verhulst distribution*[22] includes another shape parameter. It is described by

$$f(x) = \frac{\delta_1 \delta_2}{\beta} \exp\left(-\frac{x-\alpha}{\beta}\right)\left[1 - \delta_1 \exp\left(-\frac{x-\alpha}{\beta}\right)\right]^{\delta_2 - 1} \qquad x \geq \alpha - \beta \ln \delta_1;\ \delta_1, \delta_2 > 0 \tag{28.4}$$

Statistics for Process Control Engineers: A Practical Approach, First Edition. Myke King.

$$F(x) = \left[1 - \delta_1 \exp\left(-\frac{x-\alpha}{\beta}\right)\right]^{\delta_2} \tag{28.5}$$

If δ_1 is set to 1, the distribution reverts to the generalised exponential distribution. However its inclusion does not offer a better fit. This can be seen by examination of the QF

$$x(F) = \alpha + \beta \ln \delta_1 - \ln\left(1 - F^{1/\delta_2}\right) \qquad 0 \le F \le 1 \tag{28.6}$$

The effect of δ_1 is to translate x by $\beta \ln \delta_1$. The same could be achieved simply by adjusting α.

28.3 Hyperexponential

The *hyperexponential distribution* can be applied when the rate of events varies, depending on operating mode. It is described by

$$f(x) = \sum_{i=1}^{n} p_i \lambda_i e^{-\lambda_i x} \qquad x \ge 0;\ p, \lambda > 0 \tag{28.7}$$

$$F(x) = \sum_{i=1}^{n} p_i \left(1 - e^{-\lambda_i x}\right) \tag{28.8}$$

$$\mu = \sum_{i=1}^{n} \frac{p_i}{\lambda_i} \tag{28.9}$$

$$\sigma^2 = \sum_{i=1}^{n} \frac{2p_i - p_i^2}{\lambda_i^2} \tag{28.10}$$

$$\gamma = \sum_{i=1}^{n} \frac{2\left(3p_i - 2p_i^2 + p_i^3\right)}{\left(p_i(2-p_i)\right)^{3/2}} \tag{28.11}$$

$$\kappa = \sum_{i=1}^{n} \frac{3\left(8p_i - 8p_i^2 + 4p_i^3 - p_i^4\right)}{p_i^2 (2-p_i)^2} \tag{28.12}$$

The probability that the current mode is mode 1 is p_1, that it is mode 2 is p_2, etc. The probabilities must clearly sum to 1.

As an example, imagine a process has three operating modes ($n = 3$). In mode 1 the event in questions occurs, on average, every 50 days ($\lambda_1 = 0.02$). It operates in this mode for 20% of the time ($p_1 = 0.2$). The process operates in mode 2 for 30% of the time, during which the average interval is 100 days. In mode 3 it is 125 days. Using Equation (28.7), these values were used to develop Figure 28.1.

From Equation (28.9), μ is 103 days and, from Equation (28.10), σ is 133 days. The standard deviation being larger than the mean might be somewhat surprising – particularly as the number of days cannot be negative. This is an inherent property of the distribution. If n is 1 (and hence p_1 is 1), the distribution becomes the classic exponential distribution. Examination of Equations (28.9) and (28.10) shows that the coefficient of variation (σ/μ) will then be unity.

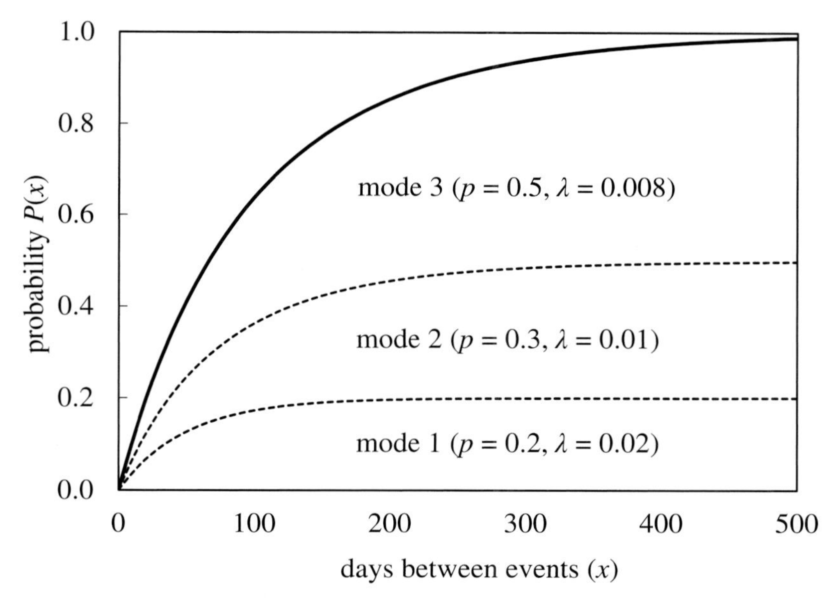

Figure 28.1 *Hyperexponential: Cumulative probability of interval between events*

28.4 Hypoexponential

The *hypoexponential distribution* is also known as the *generalised Erlang distribution*. It is a series of exponential distributions – each with a different value for λ. Its coefficient of variation (σ/μ) will always be less than 1. The mathematics of the full distribution are too complex to merit inclusion here but are considerably simpler if the series comprises just two exponential distributions. The PDF is then

$$f(x)=\frac{\lambda_1\lambda_2}{\lambda_1-\lambda_2}[\exp(-\lambda_2 x)-\exp(-\lambda_1 x)] \quad x\geq 0;\ \lambda_1,\lambda_2>0 \tag{28.13}$$

The parameters λ_1 and λ_2 are interchangeable; their effect is shown in Figure 28.2. The CDF is

$$F(x)=1-\frac{\lambda_2}{\lambda_2-\lambda_1}\exp(-\lambda_1 x)+\frac{\lambda_1}{\lambda_2-\lambda_1}\exp(-\lambda_2 x) \tag{28.14}$$

While not designed for the purpose, it is possible to fit it to the C_4 in propane data – to give λ_1 as 0.323 and λ_2 as 0.735. At 0.2682 *RSS*, not surprisingly, shows the fit is poor.

The mean and variance are

$$\mu=\frac{1}{\lambda_1}+\frac{1}{\lambda_2} \tag{28.15}$$

$$\sigma^2=\frac{1}{\lambda_1^2}+\frac{1}{\lambda_2^2} \tag{28.16}$$

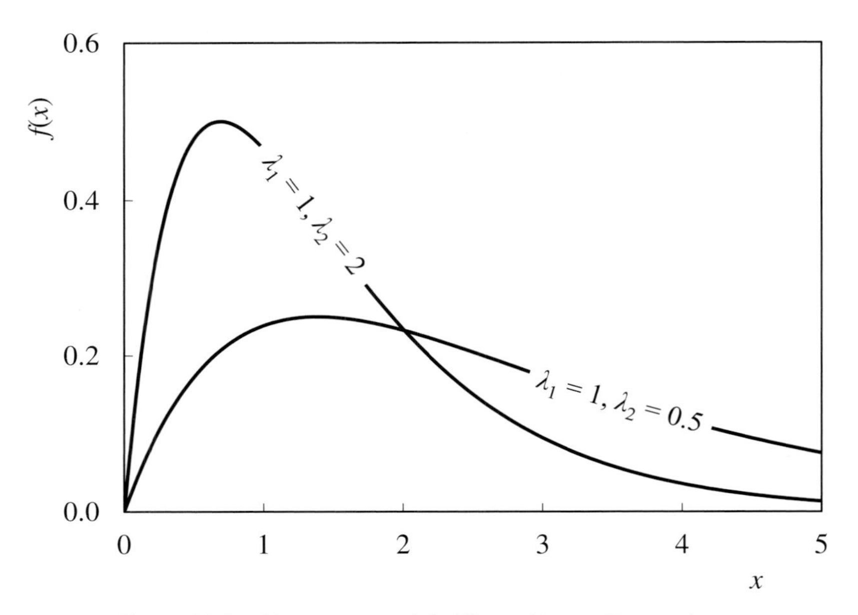

Figure 28.2 *Hypoexponential: Effect of λ_1 and λ_2 on shape*

These give values of 4.46 for μ and 3.38 for σ. There is no formula for kurtosis; skewness is 1.67, calculated from

$$\gamma = \frac{2\left(\frac{1}{\lambda_1^3} + \frac{1}{\lambda_2^3}\right)}{\left(\frac{1}{\lambda_1^2} + \frac{1}{\lambda_2^2}\right)^{3/2}} \tag{28.17}$$

28.5 Double Exponential

The *double exponential distribution* allows x to be less than α by using the absolute value of their difference. It is described by

$$f(x) = \frac{1}{2\beta}\exp\left(\frac{-|x-\alpha|}{\beta}\right) \qquad \beta > 0 \tag{28.18}$$

$$F(x) = \frac{1}{2} + \frac{x-\alpha}{2|x-\alpha|}\left[1-\exp\left(\frac{-|x-\alpha|}{\beta}\right)\right] \tag{28.19}$$

$$\mu = \alpha \tag{28.20}$$

$$\sigma^2 = 2\beta^2 \tag{28.21}$$

Also known as the *Laplace distribution*, we will cover it in detail in Section 32.7.

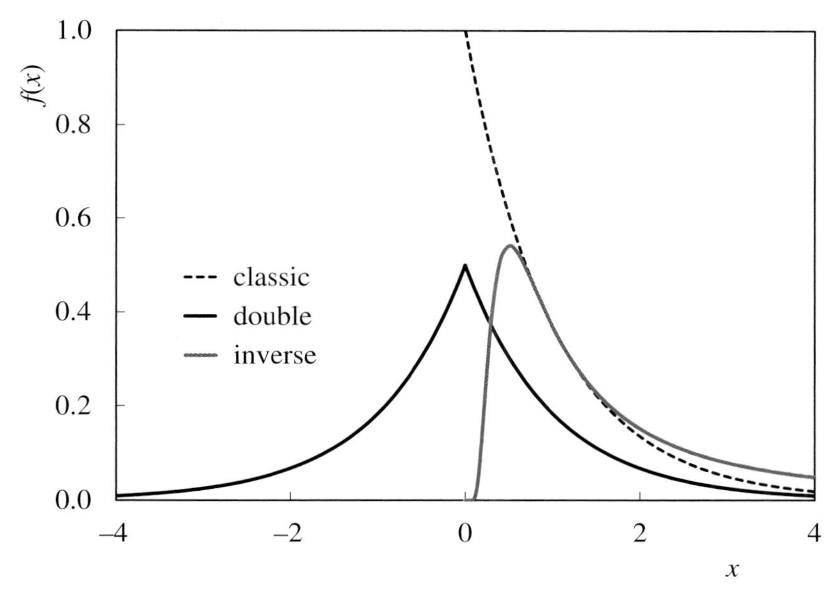

Figure 28.3 Exponential: Comparison between versions

28.6 Inverse Exponential

The *inverse exponential distribution* is described by

$$f(x) = \frac{\beta}{(x-\alpha)^2} \exp\left(-\frac{\beta}{x-\alpha}\right) \quad x \geq \alpha, \beta > 0 \tag{28.22}$$

$$F(x) = \exp\left[-\frac{\beta}{x-\alpha}\right] \tag{28.23}$$

$$x(F) = \alpha - \frac{\beta}{\ln(F)} \quad 0 \leq F \leq 1 \tag{28.24}$$

Figure 28.3 compares both this and the double exponential distribution to the standard exponential distribution (shown as the dashed line).

The raw moments of the distribution are all infinite. Thus the mean, variance, etc. are also infinite.

28.7 Maxwell–Jüttner

The *Maxwell–Jüttner distribution* is an extension of the Maxwell–Boltzmann distribution we will cover in Section 30.4. It is included here because of its similarity to the exponential distribution. It was developed to describe the distribution of speeds of gas molecules and atoms. Its PDF contains several constants that are defined theoretically and would not be applicable to process data. Here they have been combined into shape parameters that would be determined empirically.

$$f(x) = \frac{\lambda^3 (x-\alpha)^2}{2} \exp[-\lambda(x-\alpha)] \quad x > 0;\ \lambda > 0 \tag{28.25}$$

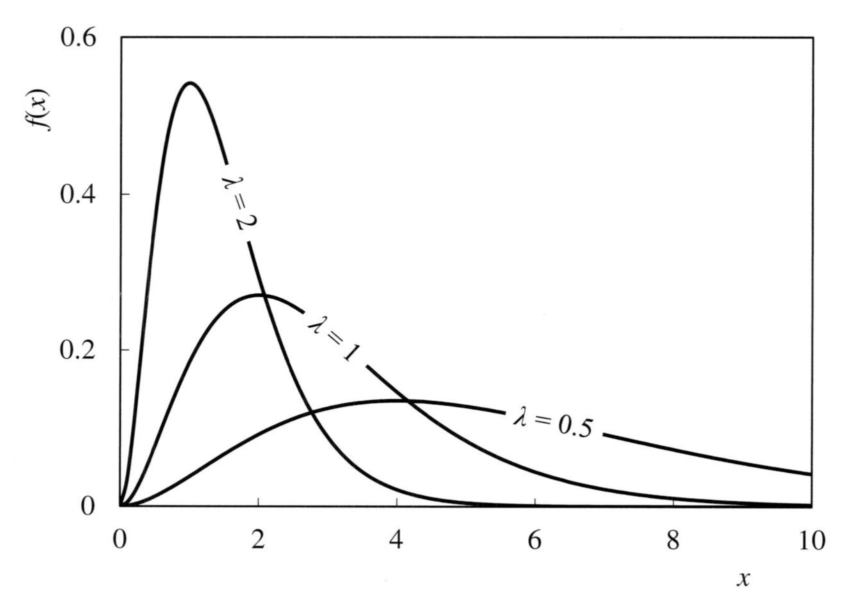

Figure 28.4 Maxwell–Jüttner: Effect of λ on shape

Figure 28.4 shows the effect of λ with α fixed at zero. The PDF can be integrated to give the CDF

$$F(x) = 1 - \left[\frac{\lambda^2(x-\alpha)^2}{2} + \lambda(x-\alpha) + 1\right]\exp[-\lambda(x-\alpha)] \tag{28.26}$$

Fitting to the C_4 in propane data gives α as 0.818 and λ as 0.761. *RSS* is 0.0807, which is considerably better than many of the more commonly used distributions. But there are no formulae published for mean, variance, etc.

28.8 Stretched Exponential

The *stretched exponential distribution* is described by

$$f(x) = \delta\lambda^{\delta}x^{\delta-1}\exp\left\{-(\lambda x)^{\delta}\right\} \quad x>0;\ \delta,\lambda>0 \tag{28.27}$$

$$F(x) = 1 - \exp\left\{-(\lambda x)^{\delta}\right\} \tag{28.28}$$

$$x(F) = \frac{[-\ln(1-F)]^{1/\delta}}{\lambda} \quad 0 \le F \le 1 \tag{28.29}$$

If δ is set to 1, it becomes the exponential distribution, as shown by the coloured line in Figure 28.5. Increasing δ reduces the tails.

Moments are given by

$$m_n = \frac{n}{\lambda\delta}\Gamma\left(\frac{n}{\delta}\right) \tag{28.30}$$

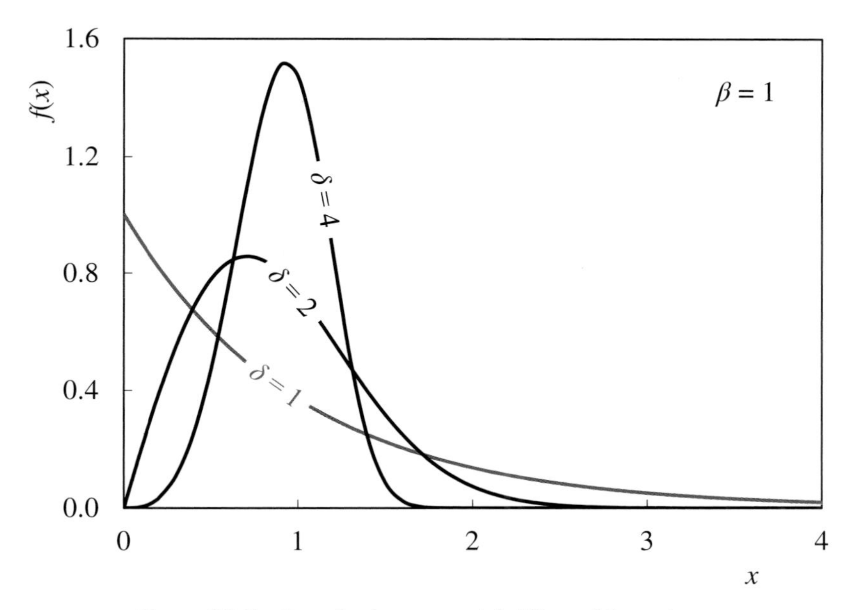

Figure 28.5 Stretched exponential: Effect of δ on shape

Hence

$$\mu = \frac{1}{\lambda\delta}\Gamma\left(\frac{1}{\delta}\right) \tag{28.31}$$

$$\sigma^2 = \frac{1}{\lambda^2\delta^2}\left[2\lambda\delta\Gamma\left(\frac{2}{\delta}\right) - \Gamma^2\left(\frac{1}{\delta}\right)\right] \tag{28.32}$$

28.9 Exponential Logarithmic

The *exponential-logarithmic (EL) distribution* includes an additional shape parameter (δ). It is described by

$$f(x) = -\frac{1}{\ln\delta}\frac{\lambda(1-\delta)e^{-\lambda x}}{1-(1-\delta)e^{-\lambda x}} \qquad x > 0;\ \lambda > 0;\ 0 \leq \delta \leq 1 \tag{28.33}$$

$$F(x) = 1 - \frac{\ln\left[1-(1-\delta)e^{-\lambda x}\right]}{\ln\delta} \tag{28.34}$$

$$x(F) = -\frac{1}{\lambda}\ln\left\{\frac{1-\exp[(1-F)\ln\delta]}{1-\delta}\right\} \qquad 0 \leq F \leq 1 \tag{28.35}$$

As δ approaches 1, the distribution becomes the classic exponential distribution – as shown as the coloured curve in Figure 28.6.

The calculation of the mean and variance of this distribution involves the use of *polylogarithms* – a topic beyond the scope of this book and a complexity probably not justified by any better fit that the distribution might offer.

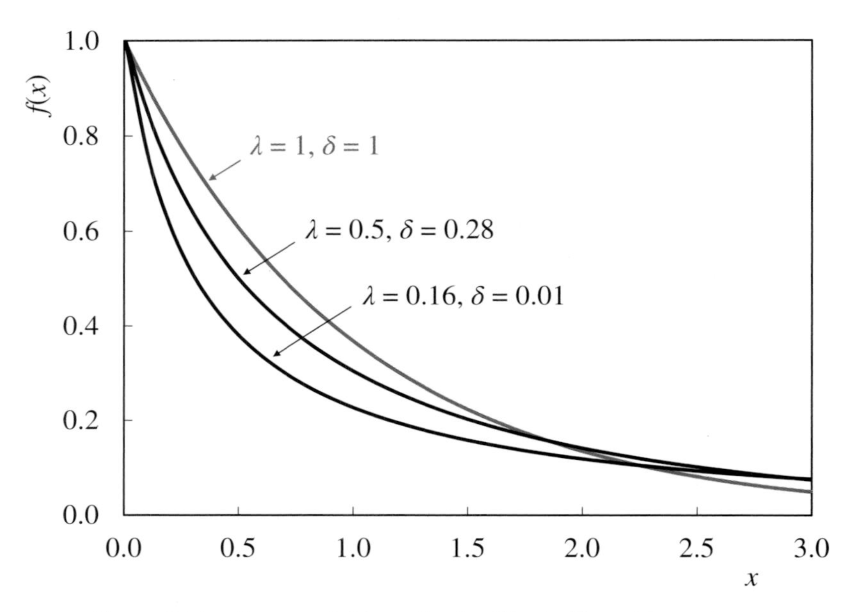

Figure 28.6 *Exponential logarithmic: Effect of λ and δ on shape*

28.10 Logistic Exponential

The *logistic exponential* is described by the PDF

$$f(x) = \frac{\lambda\delta[\exp(\lambda x) - 1]^{\delta-1}\exp(\lambda x)}{\left[1 + \{\exp(\lambda x) - 1\}^{\delta}\right]^2} \quad x > 0;\ \lambda, \delta > 0 \tag{28.36}$$

Figure 28.7 shows the effect of changing λ and δ. The CDF is

$$F(x) = \frac{[\exp(\lambda x) - 1]^{\delta}}{1 + [\exp(\lambda x) - 1]^{\delta}} \tag{28.37}$$

Fitting to the C_4 in propane data gives λ as 0.516 and δ as 6.12. *RSS* is a mediocre 0.0802, which with no formulae for mean or variance, means the distribution would not be the first choice.

The CDF can be inverted to give the QF.

$$x(F) = \frac{1}{\lambda}\ln\left[\left(\frac{F}{1-F}\right)^{\frac{1}{\delta}} + 1\right] \quad 0 \le F \le 1 \tag{28.38}$$

28.11 Q-Exponential

Employing a method similar to that used to develop the q-Gaussian distribution from the normal distribution, the *q-exponential distribution* is developed from the exponential distribution. It is another in the family of Tsallis distributions. Its PDF is

$$f(x) = (2-q)\lambda[1 + (q-1)\lambda x]^{\frac{1}{1-q}} \quad x > 0;\ \lambda, 0 < q < 2 \tag{28.39}$$

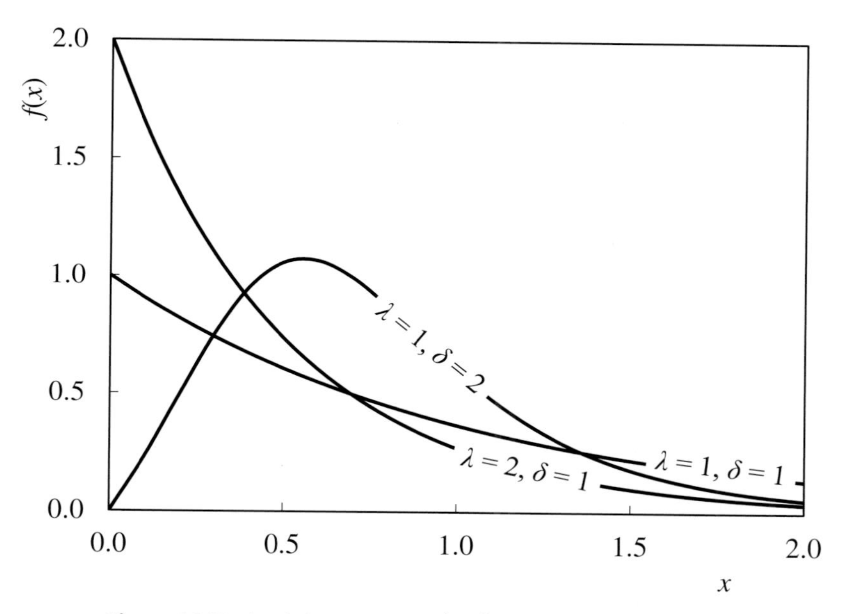

Figure 28.7 *Logistic exponential: Effect of λ and δ on shape*

If $q < 1$ then the distribution becomes double-bounded.

$$0 < x < \frac{1}{\lambda(1-q)} \tag{28.40}$$

If q is set to 1, the q-exponential becomes the exponential distribution. Figure 28.8 shows the effect of varying q. For the curves to all start at the same point, λ has been set accordingly. The CDF is

$$F(x) = 1 - [1 + (q-1)\lambda x]^{\frac{2-q}{1-q}} \tag{28.41}$$

The CDF can be inverted to give the QF

$$x(F) = \frac{(1-F)^{\frac{1-q}{2-q}} - 1}{\lambda(q-1)} \qquad 0 \le F \le 1 \tag{28.42}$$

The mean and variance are

$$\mu = \frac{1}{\lambda(3-2q)} \qquad q < \frac{3}{2} \tag{28.43}$$

$$\sigma^2 = \frac{q-2}{\lambda^2(2q-3)^2(3q-4)} \qquad q < \frac{4}{3} \tag{28.44}$$

The skewness and kurtosis are

$$\gamma = \frac{2}{(5-4q)}\sqrt{\frac{3q-4}{q-2}} \qquad q < \frac{5}{4} \tag{28.45}$$

$$\kappa = \frac{3(6q^3 - 55q^2 + 88q - 48)}{(q-2)(4q-5)(5q-6)} \qquad q < \frac{6}{5} \tag{28.46}$$

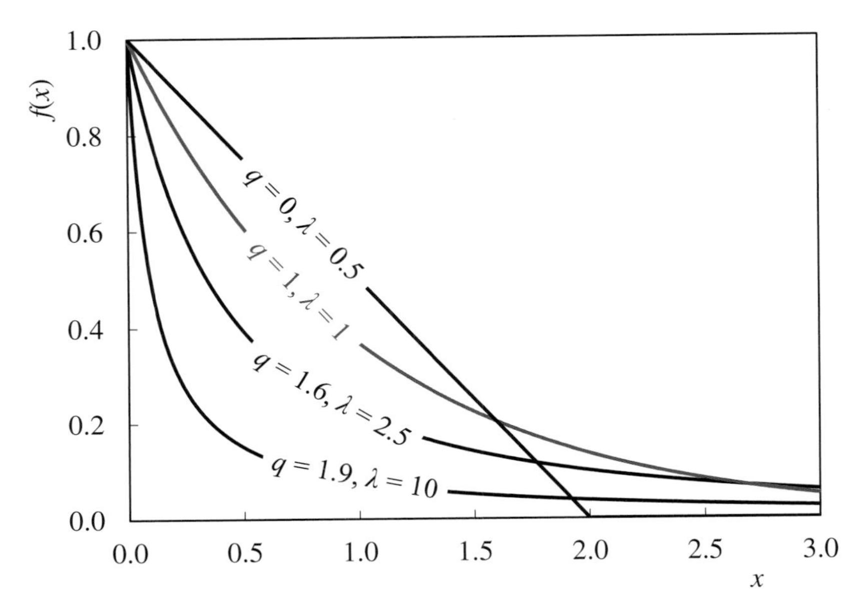

Figure 28.8 *Q-exponential: Effect of q and λ on shape*

Figure 28.9 shows the best fit to the NHV disturbances, with q set to 1.13 and λ set to 1.11 resulting in *RSS* of 0.0721. The fit is noticeably better than the classic exponential with λ fitted at 0.912 with *RSS* at 0.0935. For example, from Equation (28.42), the probability of a disturbance being no more than 5 is 0.975 and so there is a 2.5% chance that it will be exceeded. The

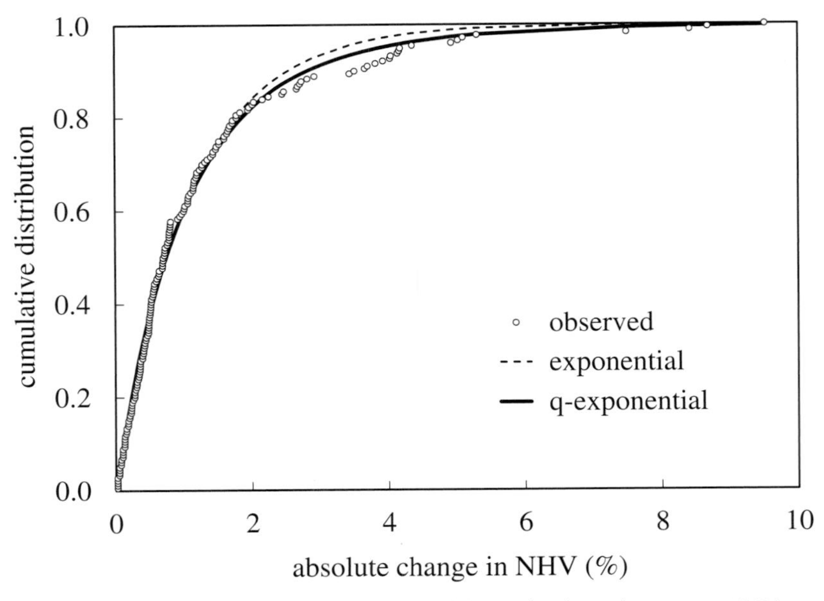

Figure 28.9 *Q-exponential: Improved fit to absolute changes in NHV*

NHV data was collected on a daily basis, so this is equivalent to expecting nine problems per year. Equation (12.60) puts the chance at 1.0%, badly underestimating the number of problems per year as four.

28.12 Benktander

The *Benktander distribution* is related to the exponential distribution. There are two types; Type I is described by the PDF

$$f(x) = \left\{ \left[1 + \frac{2\delta \ln(x+1)}{\lambda}\right] [1 + \lambda + 2\delta \ln(x+1)] - \frac{2\delta}{\lambda} \right\} (x+1)^{-[2+\lambda+\delta \ln(x+1)]} \tag{28.47}$$

$x > 0,\ \lambda, \delta > 0$

Figure 28.10 shows the effect of λ and δ. The CDF is

$$F(x) = 1 - \left[1 + \frac{2\delta \ln(x+1)}{\lambda}\right] (x+1)^{-[1+\lambda+\delta \ln(x+1)]} \tag{28.48}$$

Type II is described by

$$f(x) = \left[\lambda (x+1)^{\delta} - \delta + 1\right] (x+1)^{\delta-2} \exp\left\{ \frac{\lambda \left[1 - (x+1)^{\delta}\right]}{\delta} \right\} \quad x > 0;\ \lambda > 0;\ 0 < \delta \le 1 \tag{28.49}$$

$$F(x) = 1 - (x+1)^{\delta-1} \exp\left\{ \frac{\lambda \left[1 - (x+1)^{\delta}\right]}{\delta} \right\} \tag{28.50}$$

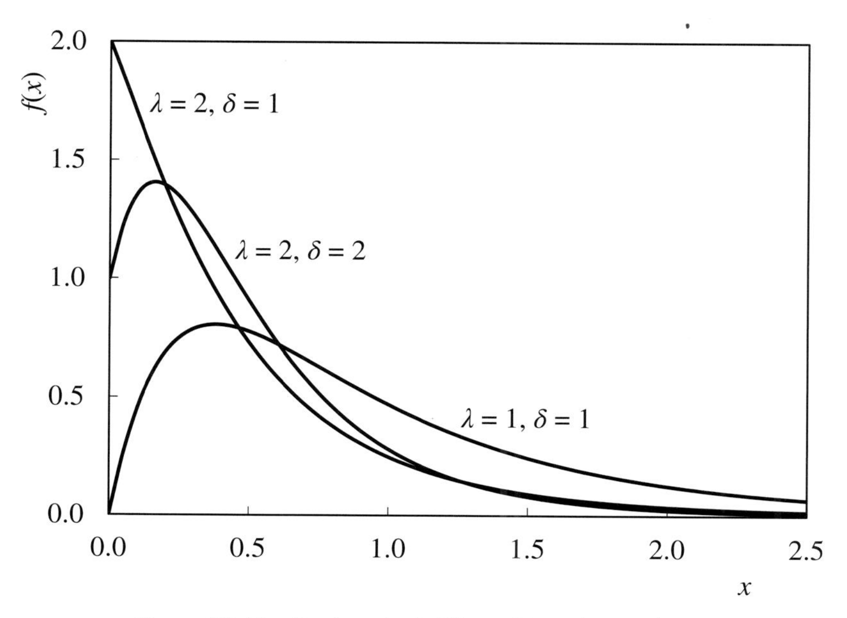

Figure 28.10 *Benktander-I: Effect of λ and δ on shape*

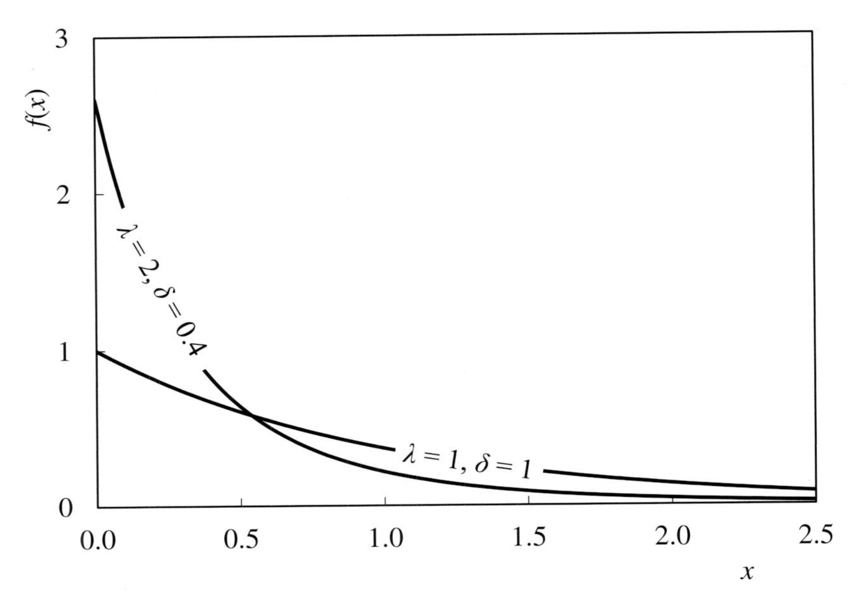

Figure 28.11 *Benktander-II: Effect of λ and δ on shape*

Figure 28.11 shows the effect of λ and δ. Setting δ to 1 gives the exponential distribution. Fitting Equation (28.50) to the LPG splitter reflux 65 m^3/hr event data gives λ as 0.0822 and δ as 0.991 – very close to the exponential distribution. *RSS*, as might be expected, is very slightly improved at 0.0053.

While the mean for both types is $1/\lambda$, as it is for the exponential distribution, the formula for variance is too complex to be included here.

29

Weibull Distribution

We covered a number of the Weibull distributions in Section 12.8. Here we describe two enhanced versions.

29.1 Nukiyama–Tanasawa

The *Nukiyama–Tanasawa distribution* is an extension of the Weibull-II distribution.

$$f(x) = \frac{\delta_2 x^{\delta_1} \exp\left[-\left(\frac{x}{\beta}\right)^{\delta_2}\right]}{\beta^{\delta_1+1}\Gamma\left(\frac{\delta_1+1}{\delta_2}\right)} \qquad x>0;\ \beta,\delta_1,\delta_2>0 \tag{29.1}$$

If δ_1 is set to $\delta_2 - 1$, Equation (29.1) reverts to the Weibull-II distribution, as shown by the coloured line in Figure 29.1. Varying δ_1 has an effect similar to that of δ_2. However, together they make the distribution a little more flexible.

Fitting to the C_4 in propane data, in this case, requires that $F(x_n)$ is forced to 1; otherwise the fit in this area is very poor. Doing so sets β, δ_1 and δ_2 to 1.01, 2.954 and 0.948 respectively. With *RSS* at 0.0779 the fit is poor compared to several alternative distributions.

29.2 Q-Weibull

Employing a method similar to that used to develop the q-Gaussian distribution from the normal distribution, the *q-Weibull distribution* is developed from the Weibull distribution. It is another in the family of Tsallis distributions.

Statistics for Process Control Engineers: A Practical Approach, First Edition. Myke King.

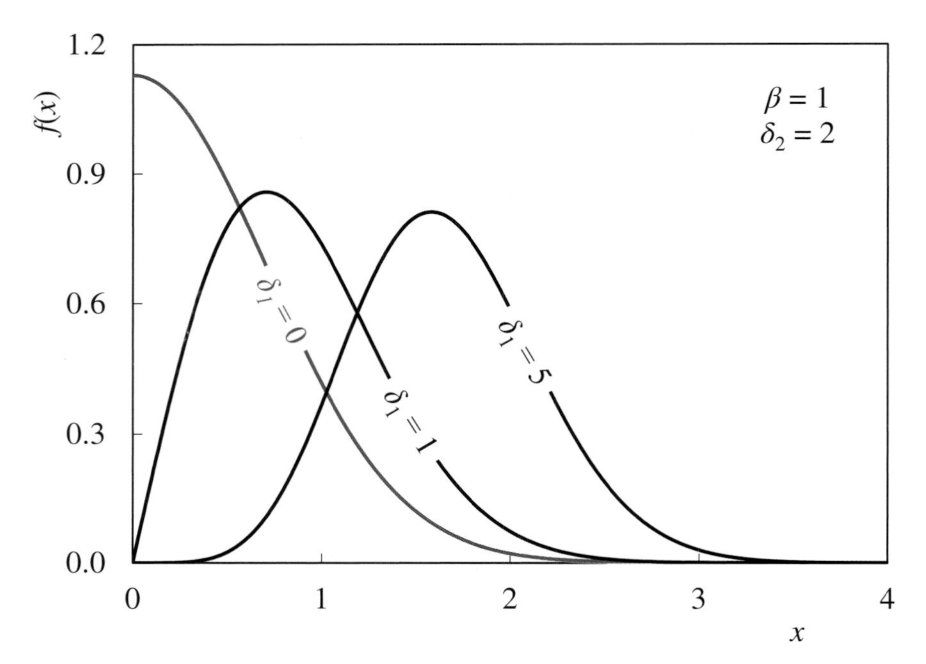

Figure 29.1 *Nukiyama–Tanasawa: Effect of δ_1 on shape*

For $q \geq 1$ the PDF is

$$f(x) = (2-q)\frac{\delta x^{\delta-1}}{\beta^{\delta}}\left[1+(q-1)\left(\frac{x}{\beta}\right)^{\delta}\right]^{\frac{1}{1-q}} \quad x>0;\ \beta,\delta>0;\ q<2 \tag{29.2}$$

If $q < 1$ Equation (29.2) applies only if

$$0 < x < \beta(1-q)^{-\frac{1}{\delta}} \tag{29.3}$$

If q is set to 1, the distribution reverts to the Weibull-II distribution. Figures 29.2 and 29.3 show the effect of adjusting q.

The CDF is

$$F(x) = 1 - \left[1+(q-1)\left(\frac{x}{\beta}\right)^{\delta}\right]^{\frac{2-q}{1-q}} \tag{29.4}$$

The CDF can be inverted to

$$x = \beta\left\{\frac{[1-F(x)]^{(1-q)/(2-q)}-1}{q-1}\right\}^{\frac{1}{\delta}} \tag{29.5}$$

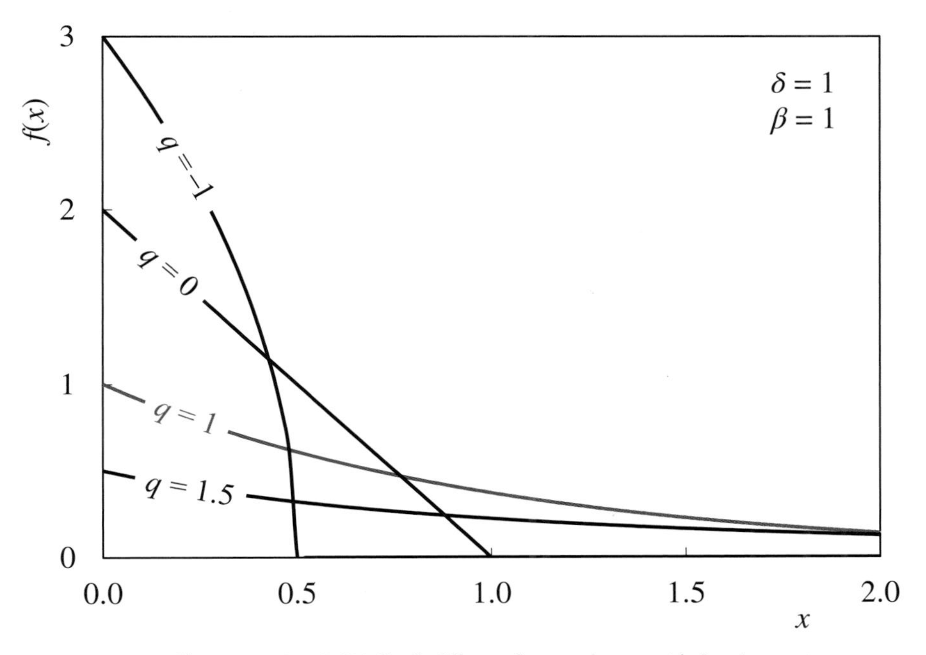

Figure 29.2 *Q-Weibull: Effect of q on shape with δ = 1*

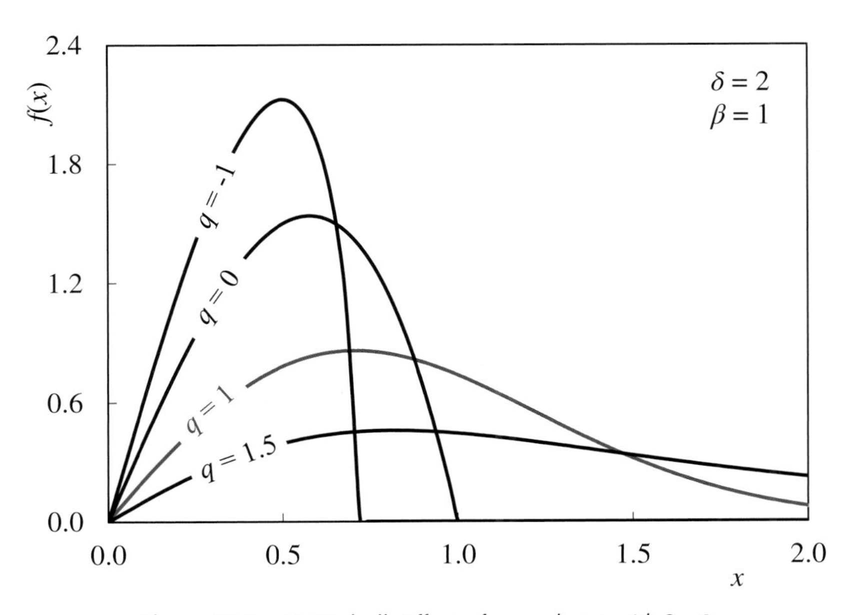

Figure 29.3 *Q-Weibull: Effect of q on shape with δ = 2*

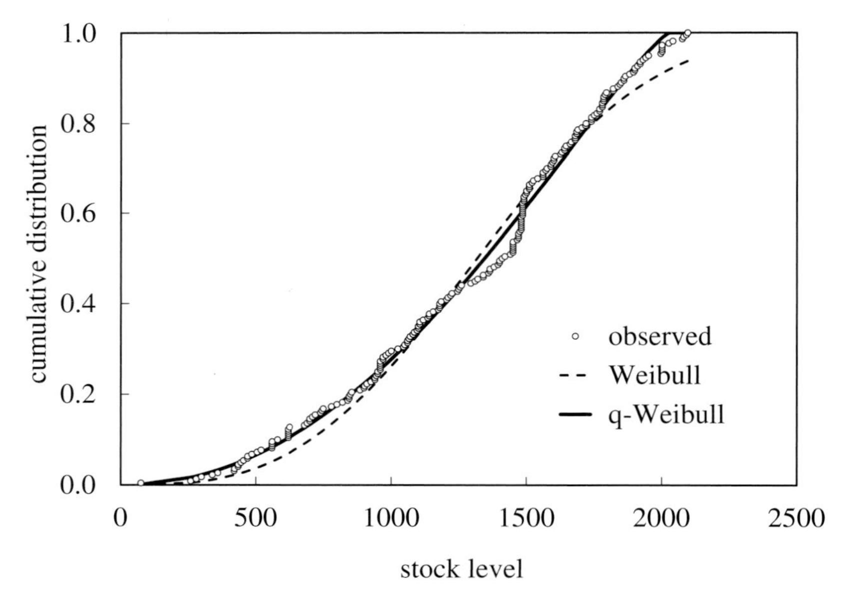

Figure 29.4 *Q-Weibull: Improved fit to stock level data*

Using the stock level example, Figure 29.4 shows that q-Weibull fits the data much better than Weibull. The parameter q is fitted as −2.75, with δ as 2.10 and β as 3802. As an example, the Weibull distribution predicts the probability of the stock level falling below 300 tonnes is 0.8%, equivalent to three occasions per year. The q-Weibull puts it much higher at 2.3%, or eight occasions per year. It actually occurred three times in 220 days – equivalent to five occasions per year. Close inspection of Figure 29.4, at the stock level of 300 tonnes, shows the actual number of occurrences is between the two predictions.

30

Chi Distribution

The *chi-distribution* is described by the PDF

$$f(x) = \frac{2(x-\alpha)^{f-1}\exp\left(\frac{-(x-\alpha)^2}{2\beta^2}\right)}{\beta^f 2^{\frac{f}{2}}\Gamma\left(\frac{f}{2}\right)} \quad \beta, f \geq 1 \tag{30.1}$$

The parameter f is often referred to as the *degrees of freedom*, in which case it should be restricted to integers. Figure 30.1 shows the effect of changing f with α fixed at 0 and β at 1.

30.1 Half-Normal

With f set to 1, Equation (30.1) becomes almost the same as Equation (5.35), describing the normal distribution.

$$f(x) = \sqrt{\frac{2}{\pi}}\frac{1}{\beta}\exp\left[\frac{-(x-\alpha)^2}{2\beta^2}\right] \quad \beta > 0 \tag{30.2}$$

The only difference is a factor of 2 – required because it only applies to half the values of x, i.e. those greater than the mean. It is therefore known as the *half-normal distribution*. Process disturbances and inferential errors, which are likely to have a mean close to zero, might follow a normal distribution. Their absolute value will then follow the half-normal distribution.

Fitting to the absolute values of the NHV disturbances gives α as –0.08 and β as 1.31. As expected, because of the high kurtosis of the data, the fit is poor with *RSS* at 0.3248.

Mean and variance are

$$\mu = \alpha + \beta\sqrt{\frac{2}{\pi}} \tag{30.3}$$

Statistics for Process Control Engineers: A Practical Approach, First Edition. Myke King.

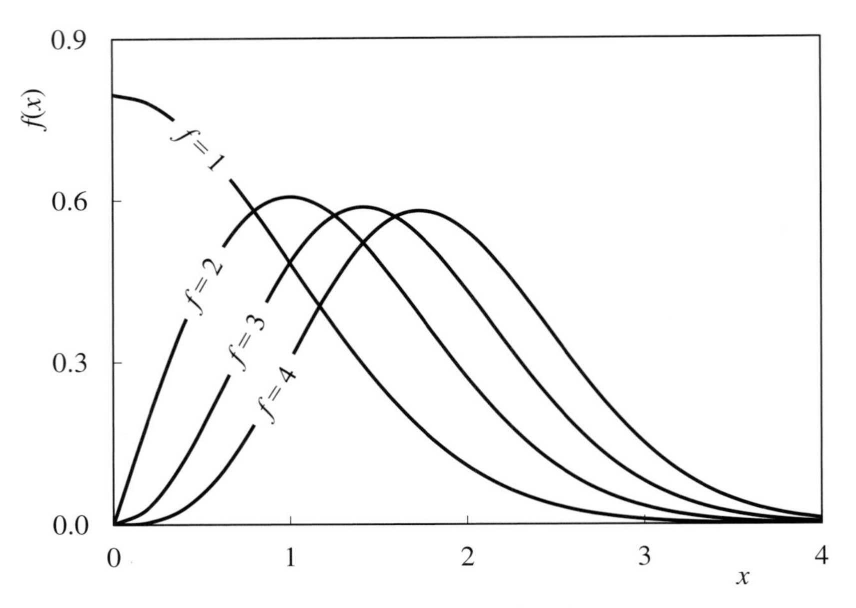

Figure 30.1 *Chi: Effect of f on shape*

$$\sigma^2 = \beta^2\left(1 - \frac{2}{\pi}\right) \tag{30.4}$$

These give μ as 0.96 and σ as 0.52 – somewhat away from the values calculated from the data.

Unfortunately, skewness and kurtosis are fixed and so application is therefore restricted to situations where these are close to the values calculated from the data.

$$\gamma = \frac{\sqrt{2}(4-\pi)}{(\pi-2)^{1.5}} \approx 0.995 \tag{30.5}$$

$$\kappa = \frac{8(\pi-3)}{(\pi-2)^2} \approx 0.869 \tag{30.6}$$

30.2 Rayleigh

Setting f to 2 gives the *Rayleigh distribution*. It is also a special case of the Weibull-III distribution that arises from setting, in Equation (12.74), δ to 2 and multiplying β by $\sqrt{2}$.

It is described by the PDF

$$f(x) = \frac{x-\alpha}{\beta^2}\exp\left(\frac{-(x-\alpha)^2}{2\beta^2}\right) \qquad \beta > 0 \tag{30.7}$$

In theory a variable (x) will follow this distribution if it is the geometric mean of two values (x_1 and x_2), both sampled from the same normal distribution that has a mean of zero.

$$x = \sqrt{x_1^2 + x_2^2} \tag{30.8}$$

While one might conceive, for example, an inferential property in which two variables are combined in this way, they are unlikely to meet the condition that their means are both zero. In practice the resulting distribution is likely to be close to normal. However, as usual, we do not let the theory restrict us in considering the Rayleigh distribution as a contender for the prior distribution.

Keeping α fixed at zero, Figure 30.2 plots Equation (30.7) for a range of values of β.

The CDF and its inverse are

$$F(x) = 1 - \exp\left(\frac{-(x-\alpha)^2}{2\beta^2}\right) \tag{30.9}$$

$$x(F) = \alpha + \sqrt{-2\beta^2 \ln(1-F)} \quad 0 \le F \le 1 \tag{30.10}$$

The mean and variance are

$$\mu = \alpha + \beta\sqrt{\frac{\pi}{2}} \tag{30.11}$$

$$\sigma^2 = \beta^2\left(2 - \frac{\pi}{2}\right) \tag{30.12}$$

Again skewness and kurtosis are fixed.

$$\gamma = \frac{2\sqrt{\pi}(\pi-3)}{(4-\pi)^{1.5}} \approx 0.631 \tag{30.13}$$

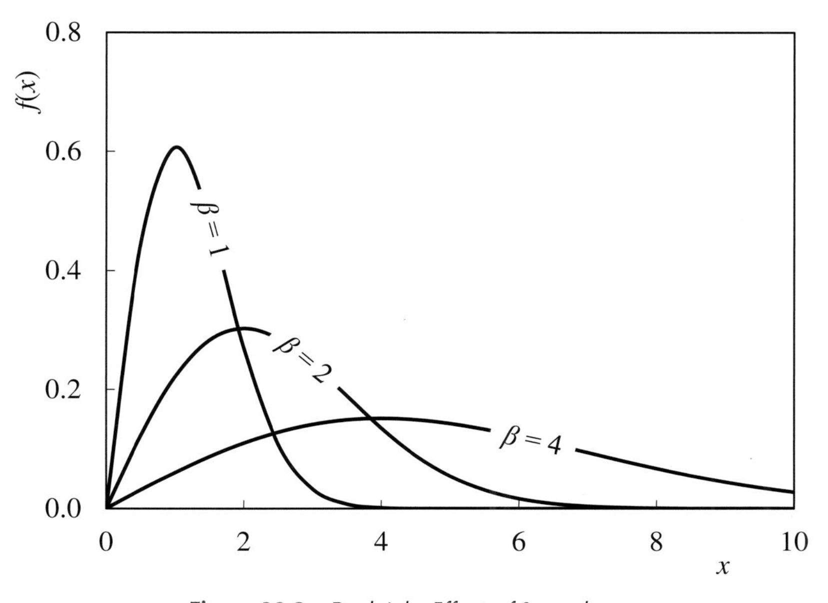

Figure 30.2 *Rayleigh: Effect of β on shape*

$$\kappa = -\frac{6\pi^2 - 24\pi + 16}{(4-\pi)^2} \approx 0.245 \quad (30.14)$$

30.3 Inverse Rayleigh

The *inverse Rayleigh distribution* is described by the PDF

$$f(x) = \frac{2}{\beta}\left(\frac{\beta}{x-\alpha}\right)^3 \exp\left[-\left(\frac{\beta}{x-\alpha}\right)^2\right] \quad \beta > 0 \quad (30.15)$$

Figure 30.3 shows the effect of varying β.

Raw moments are given by

$$m_n = \Gamma\left(1 - \frac{n}{2}\right)\beta^n \quad (30.16)$$

Therefore

$$\mu = \alpha + \sqrt{\pi}\beta \quad (30.17)$$

The second moment (m_2) includes $\Gamma(0)$, which is infinite. Variance, skewness and kurtosis cannot therefore be defined.

30.4 Maxwell

Setting f to 3 gives the *Maxwell distribution* or, more fully, the *Maxwell–Boltzmann distribution*. This was developed for the very specific purpose of estimating the probability of the speed of atoms (or molecules) in an ideal gas. With some similarity to the Rayleigh distribution, it

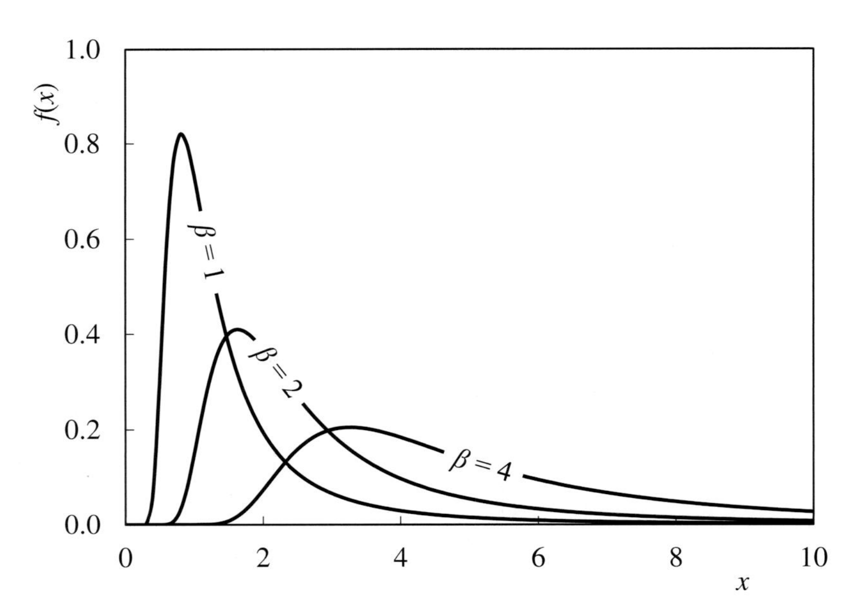

Figure 30.3 Inverse Rayleigh: Effect of β on shape

describes the distribution resulting from the square root of the sum of the squares of three values. It is described by the PDF

$$f(x)=\sqrt{\frac{2}{\pi}}\frac{(x-\alpha)^2}{\beta^3}\exp\left(\frac{-(x-\alpha)^2}{2\beta^2}\right) \quad \beta>0 \tag{30.18}$$

When used for its original purpose, the dispersion parameter (β) has a theoretical value based on particle size and temperature. The PDF is plotted for a range of values of β as Figure 30.4.

The CDF is

$$F(x)=\text{erf}\left(\frac{x-\alpha}{\beta\sqrt{2}}\right)-\sqrt{\frac{2}{\pi}}\frac{x}{\beta}\exp\left(\frac{-(x-\alpha)^2}{2\beta^2}\right) \tag{30.19}$$

Mean and variance are

$$\mu=\alpha+2\beta\sqrt{\frac{2}{\pi}} \tag{30.20}$$

$$\sigma^2=\beta^2\left(3-\frac{8}{\pi}\right) \tag{30.21}$$

Again skewness and kurtosis are fixed.

$$\gamma=\frac{2\sqrt{2}(16-5\pi)}{(3\pi-8)^{1.5}}\approx 0.486 \tag{30.22}$$

$$\kappa=-\frac{4(3\pi^2-40\pi+96)}{(3\pi-8)^2}\approx 0.108 \tag{30.23}$$

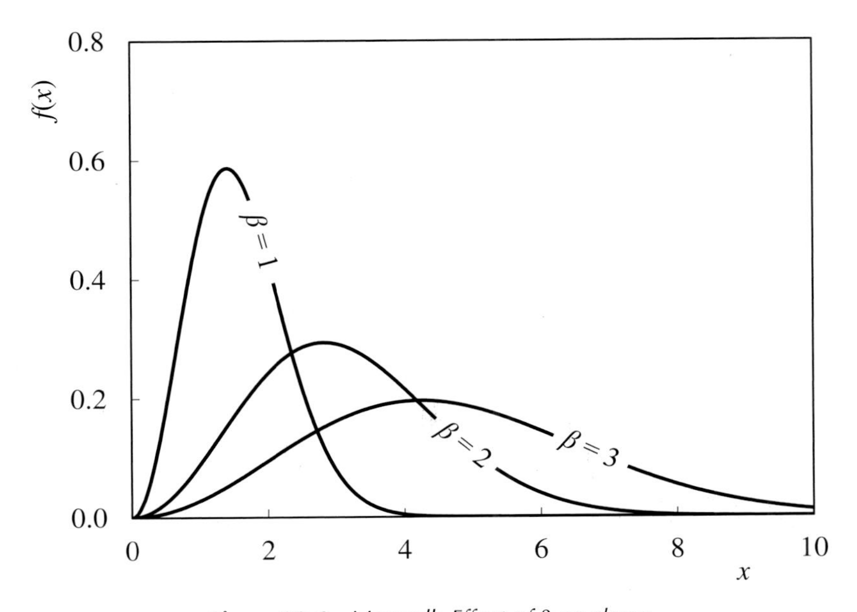

Figure 30.4 *Maxwell: Effect of β on shape*

In general, for the chi distribution, the mean and standard deviation can be calculated from

$$\mu = \alpha + \beta\sqrt{2}\frac{\Gamma\left(\frac{f+1}{2}\right)}{\Gamma\left(\frac{f}{2}\right)} \tag{30.24}$$

$$\sigma^2 = \left[f - \frac{2\Gamma^2\left(\frac{f+1}{2}\right)}{\Gamma^2\left(\frac{f}{2}\right)}\right]\beta^2 \tag{30.25}$$

And the skewness and kurtosis from

$$\gamma = \frac{\left(\frac{\mu-\alpha}{\beta}\right)\left(1-2\frac{\sigma^2}{\beta^2}\right)}{\left(\frac{\sigma^2}{\beta^2}\right)^{\frac{3}{2}}} \tag{30.26}$$

$$\kappa = \frac{\frac{1}{2}-(2f-1)\left(1-2\frac{\sigma^2}{\beta^2}\right)-\frac{3}{2}\left(1-2\frac{\sigma^2}{\beta^2}\right)^2}{\left(\frac{\sigma^2}{\beta^2}\right)^2} \tag{30.27}$$

While it would appear that γ and κ are affected by the choice of α and β, some simple manipulation of the equations shows that this is not the case. Rearranging Equation (30.24)

$$\frac{\mu-\alpha}{\beta} = \frac{\Gamma\left(\frac{f+1}{2}\right)\sqrt{2}}{\Gamma\left(\frac{f}{2}\right)} \tag{30.28}$$

Rearranging Equation (30.25)

$$\frac{\sigma^2}{\beta^2} = f - \frac{2\Gamma^2\left(\frac{f+1}{2}\right)}{\Gamma^2\left(\frac{f}{2}\right)} \tag{30.29}$$

Substituting Equations (30.28) and (30.29) into Equation (30.26) eliminates α and β; so γ is independent of both. Similarly we can substitute Equation (30.29) into Equation (30.27) to show the same for κ. The skewness and kurtosis depend only on the choice of f. The numerical values for skewness and kurtosis for the half-normal, Rayleigh and Maxwell distributions were derived from these equations.

Because a negative number cannot be raised to a fractional power then, if any value of x is less than α, f in the chi distribution must be an integer. As special cases of the chi distribution, the half-normal, Rayleigh and Maxwell distributions need not be considered separately. They exist, however, because a CDF can only be defined for values of f of

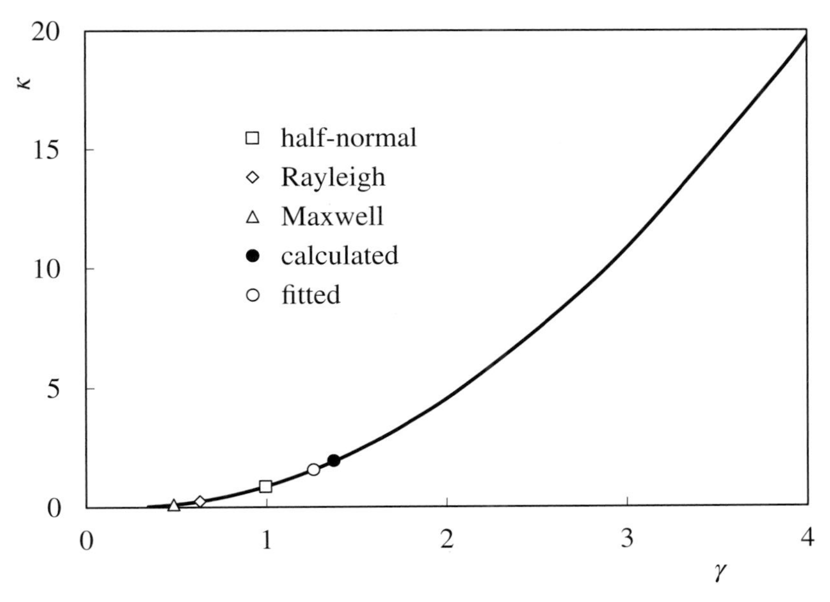

Figure 30.5 *Chi: Feasible combinations of γ and κ*

1, 2 or 3. While a non-integer value of f has no engineering meaning, provided all values of x are greater than α, there is no reason why f should not be adjusted, along with α and β, to fit real data.

Fitting to the absolute value of the changes in the C_4 content of the propane rundown gives 0.066 for α, 3.65 for β and 0.691 for f. Equation (30.24) gives the mean as 2.30 and Equation (30.25) gives the standard deviation as 2.05.

Equations (30.26) and (30.27) give the skewness as 1.26 and kurtosis as 1.57. These are very close to the values of 1.37 and 1.96, calculated from the original data. Figure 30.5 shows the possible combinations of these parameters, with the calculated values happening to fall on the line.

Figure 30.6 shows (as the black curve) the resulting cumulative distribution. Since we have no CDF, we cannot invert it to obtain the QF. However, we can use the distribution as plotted. For example, if the maximum tolerable change is 10, then we can be 97.8% sure that this will not be exceeded. In other words we could expect to have disturbances larger than this eight times per year. However, in 357 days only two disturbances actually exceed 10. Close examination of the fit shows that it does not match well as $F(x)$ approaches unity.

One approach is, in the calculation of RSS, to more highly weight mismatching in the region of interest. By applying the technique described by Equations (9.9) and (9.10), revised values are 0.059 for α, 3.67 for β and 0.710 for f. The result, shown as the coloured curve in Figure 30.6, more closely approaches the extreme values with an imperceptible change elsewhere. Indeed, key parameters such as μ, σ, γ and κ are very little changed. But, importantly, the probability of a disturbance being less than 10 increases to 98.6%. We would now expect five disturbances per year to exceed this value. Still large compared to historical performance, a better choice of prior distribution might be an option. Alternatively, as described in Chapter 13, extreme value analysis would be applicable.

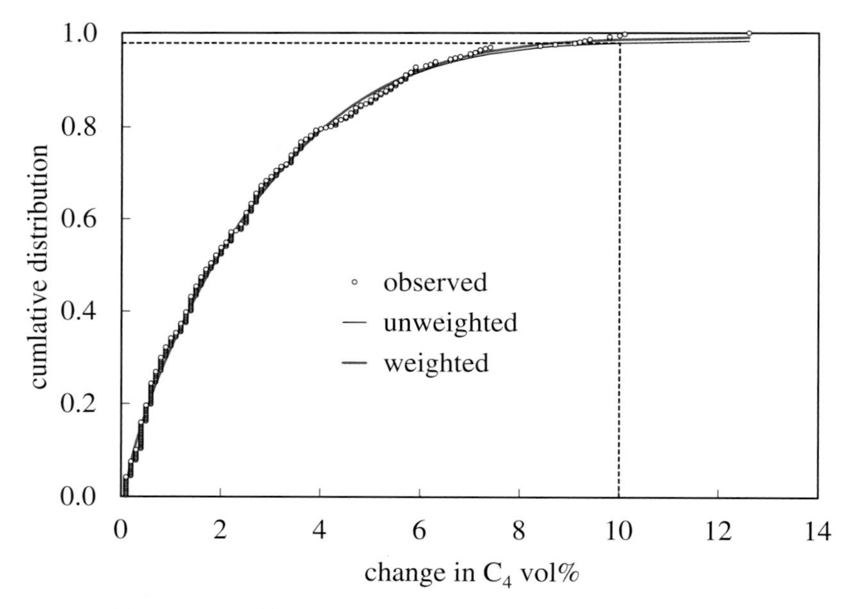

Figure 30.6 Chi: Fitted to changes in C_4 content of propane

30.5 Inverse Chi

The *inverse chi distribution* is described by the PDF

$$f(x) = \frac{2\beta 2^{\frac{f}{2}}}{(x-\alpha)^{f+1}\Gamma\left(\frac{f}{2}\right)} \exp\left(-\frac{2\beta^2}{(x-\alpha)^2}\right) \quad \beta, f > 0 \tag{30.30}$$

Figure 30.7 shows the effect of changing *f*, with α fixed at 0 and β at 1.

Fitting to the C_4 in propane data gives α as 1.67, β as 1.11, *f* as 1.13 and $F(x_1)$ as 0.0939. $F(x_1)$ being so far from zero indicates that the fit is poor. This confirmed by the *RSS* of 0.2725.

The fit to the NHV disturbance data is similarly poor with α at 11.79, β at 1.05, *f* at 1.83 and $F(x_1)$ at 0.0978.

Raw moments are given by

$$m_n = \frac{\Gamma\left(\frac{f-n}{2}\right)}{\Gamma\left(\frac{f}{2}\right)} 2^{\frac{n}{2}} \beta^n \tag{30.31}$$

So, the mean and variance are

$$\mu = \alpha + \frac{\Gamma\left(\frac{f-1}{2}\right)}{\Gamma\left(\frac{f}{2}\right)} \sqrt{2}\beta \tag{30.32}$$

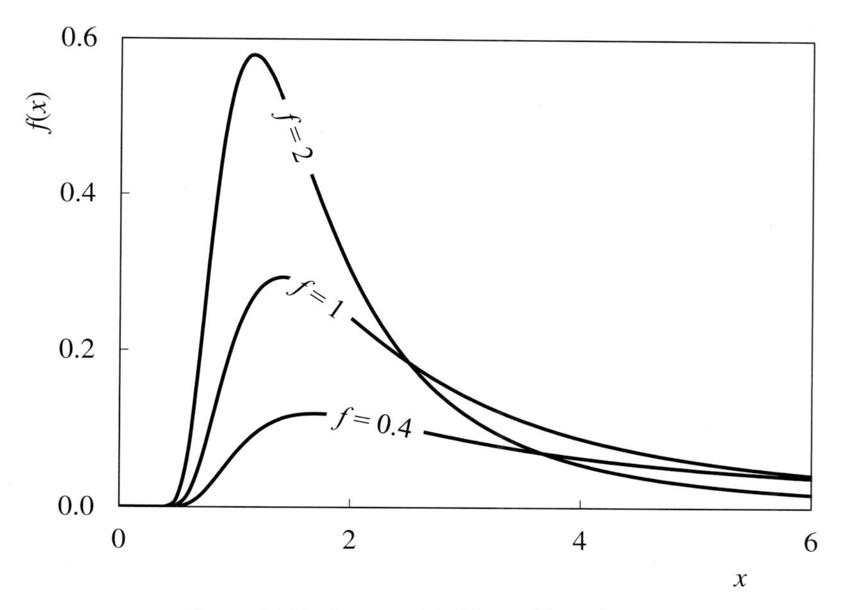

Figure 30.7 *Inverse chi: Effect of f on shape*

$$\sigma^2 = 2\left[\frac{\Gamma\left(\frac{f}{2}\right)\Gamma\left(\frac{f-2}{2}\right) - \Gamma^2\left(\frac{f-1}{2}\right)}{\Gamma^2\left(\frac{f}{2}\right)}\right]\beta^2 \tag{30.33}$$

30.6 Inverse Chi-Squared

The chi-squared distribution was covered in Section 12.9. The *inverse chi-squared distribution* is defined by the PDF

$$f(x) = \frac{\left(\frac{\beta}{2}\right)^{\frac{f}{2}} \exp\left(-\frac{\beta}{2(x-\alpha)}\right)}{\Gamma\left(\frac{f}{2}\right)(x-\alpha)^{\frac{f}{2}+1}} \qquad \beta, f > 0 \tag{30.34}$$

Figure 30.8 shows the effect of varying f, with α fixed at 0 and β at 1. As $f \to \infty$, the distribution becomes the normal distribution.

Fitting to the C_4 in propane example gives α as 0, β as 35.4, f as 8.81 and $F(x_1)$ as 0.0202. *RSS* is 0.0336, which is comparable to that achieved by the best fitting distributions.

The mean and variance are

$$\mu = \alpha + \frac{\beta}{f-2} \quad \text{for} \quad f > 2 \tag{30.35}$$

$$\sigma^2 = \frac{2\beta^2}{(f-2)^2(f-4)} \quad \text{for} \quad f > 4 \tag{30.36}$$

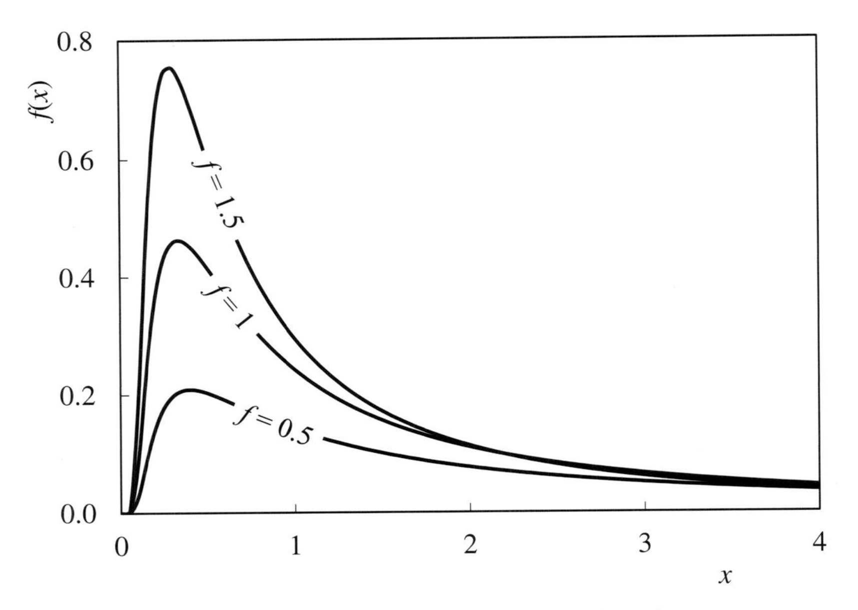

Figure 30.8 Inverse chi-squared: Effect of f on shape

These give μ as 5.20 and σ as 3.35, both of which are somewhat different from those calculated from the data.

Skewness and kurtosis are

$$\lambda = \frac{4\sqrt{2(f-4)}}{f-4} \quad \text{for} \quad f > 6 \tag{30.37}$$

$$\kappa = \frac{3(f+10)(f-4)}{(f-6)(f-8)} \quad \text{for} \quad f > 8 \tag{30.38}$$

These give γ as 4.41 and κ as an unbelievable 119. Feasible combinations are shown in Figure 30.9.

30.7 Noncentral Chi-Squared

The *noncentral chi-squared distribution* is defined by the PDF

$$f(x) = \frac{1}{2}\left(\frac{x-\alpha}{\beta}\right)^{\frac{\delta-2}{4}} \mathrm{I}_{\frac{\delta-2}{2}}\left(\sqrt{\beta(x-\alpha)}\right)\exp\left(-\frac{x-\alpha+\beta}{2}\right) \quad \beta > 0 \tag{30.39}$$

The effect of varying δ and β, with α set to zero, is shown in Figure 30.10. The CDF is

$$F(x) = 1 - \mathrm{Q}_{\delta/2}\left(\sqrt{\beta}, \sqrt{x}\right) \tag{30.40}$$

The CDF uses the Marcum Q-function, as described in Section 11.9. For the reasons given, fitting is simpler if instead we apply the trapezium rule to the PDF. This still involves the use of the modified Bessel function of the first kind (I). Most spreadsheet packages that support this only do so for integer values of $(\delta - 2)/2$. This restricts δ to even integers. Fixing δ as 2,

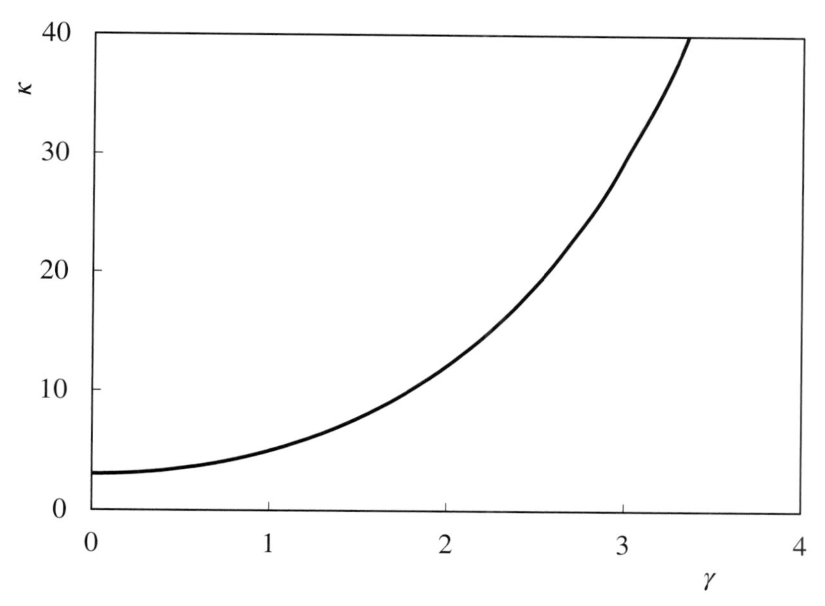

Figure 30.9 Inverse chi-squared: Feasible combinations of γ and κ

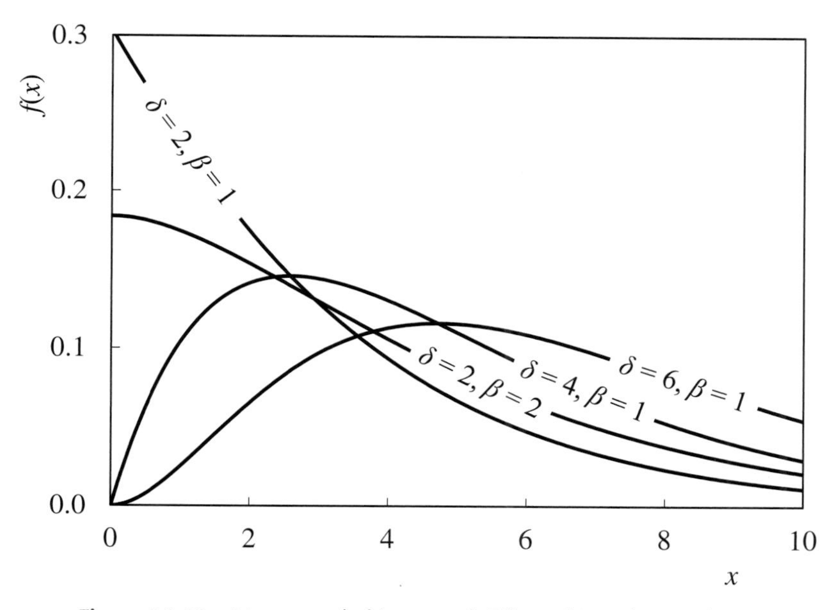

Figure 30.10 Noncentral chi-squared: Effect of δ and β on shape

fitting to the C_4 in propane data gives α as 1.70 and β as 1.65, with *RSS* as 0.6580. This is the worst that has been achieved by any distribution evaluated. Fixing δ as 4 gives only a marginally better fit with *RSS* at 0.6276. Fixing it at 6 gives a much poorer result. It is likely therefore that the best choice of δ is somewhere between 2 and 4, but it would require specialist software to determine this.

Key parameters are

$$\mu = \alpha + \delta + \beta \tag{30.41}$$

$$\sigma^2 = 2(\delta + \beta) \tag{30.42}$$

$$\gamma = \left(\frac{2}{\delta + 2\beta}\right)^{\frac{3}{2}} (\delta + 3\beta) \tag{30.43}$$

$$\kappa = \frac{12(\delta + 4\beta)}{(\delta + 2\beta)^2} + 3 \tag{30.44}$$

Given the complexity involved, and the likelihood of a very poor fit, the distribution is unlikely to be of value to the control engineer.

31

Gamma Distribution

As a commonly used distribution, we covered the gamma distribution in detail in Section 12.10. This chapter describes the range of possible modifications.

31.1 Inverse Gamma

The *inverse gamma distribution* is also known as the *Pearson-V distribution* and was covered in Section 27.5.

31.2 Log-Gamma

There are at least five distributions that are described as the *log-gamma distribution*. In no particular order, the PDF of the first is

$$f(x)=\frac{(x+1-\alpha)^{-\frac{\delta_2+1}{\delta_2}}}{\delta_2^{\delta_1}\Gamma(\delta_1)}[\ln(x+1-\alpha)]^{\delta_1-1} \quad x>\alpha-1;\ \delta_1,\delta_2>0 \tag{31.1}$$

Raw moments are given by

$$m_n=\frac{1}{(1-n\delta_2)^{\delta_1}} \tag{31.2}$$

This leads to

$$\mu=\alpha-1+\frac{1}{(1-\delta_2)^{\delta_1}} \tag{31.3}$$

$$\sigma^2=\frac{1}{(1-2\delta_2)^{\delta_1}}-\frac{1}{(1-\delta_2)^{2\delta_1}} \tag{31.4}$$

Statistics for Process Control Engineers: A Practical Approach, First Edition. Myke King.

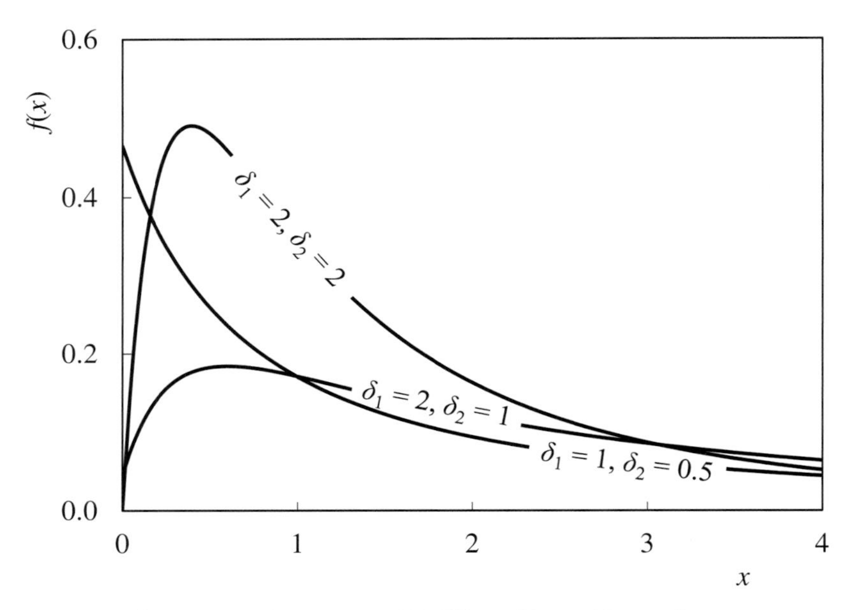

Figure 31.1 *Log-gamma (1): Effect of δ_1 and δ_2 on shape*

By replacing δ_2 with $1/\delta_2$, we obtain the same distribution that might, at first glance, appear different

$$f(x) = \frac{\delta_2^{\delta_1}}{\Gamma(\delta_1)(x+1-\alpha)^{\delta_2+1}}[\ln(x+1-\alpha)]^{\delta_1-1} \quad x>\alpha-1;\ \delta_1,\delta_2>0 \tag{31.5}$$

Figure 31.1 plots Equation (31.5) and shows, with α fixed at zero, the effect of varying δ_1 and δ_2.

Raw moments are given by

$$m_n = \left(\frac{\delta_2}{\delta_2-n}\right)^{\delta_1} \tag{31.6}$$

This leads to

$$\mu = \alpha - 1 + \left(\frac{\delta_2}{\delta_2-1}\right)^{\delta_1} \tag{31.7}$$

$$\sigma^2 = \left(\frac{\delta_2}{\delta_2-2}\right)^{\delta_1} - \left(\frac{\delta_2}{\delta_2-1}\right)^{2\delta_1} \tag{31.8}$$

Fitting to the C_4 in propane data, with α set to zero, gives δ_1 as 17.9, δ_2 as 10.5 and $F(x_1)$ as 0.0100. *RSS* is a respectable 0.0355. The mean is 4.92 and the standard deviation 2.76.

The second form of the log-gamma distribution is very similar but x is restricted to the range 0 to 1. Its PDF is

$$f(x) = \frac{\delta_2^{\delta_1}}{\Gamma(\delta_1)x^{1-\delta_2}}[-\ln(x)]^{\delta_1-1} \quad 0<x<1;\ \delta_1,\delta_2>0 \tag{31.9}$$

Figure 31.2 shows the effect of varying δ_1 and δ_2.

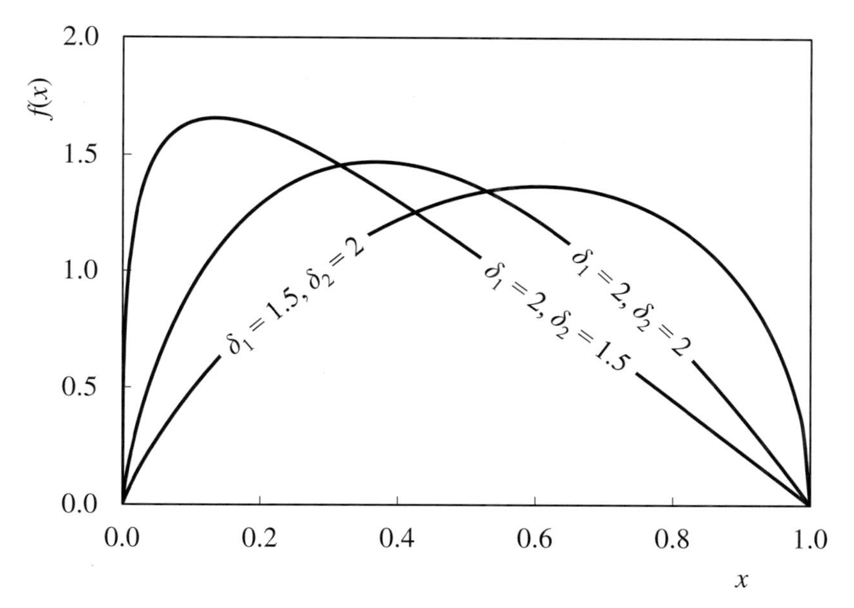

Figure 31.2 *Log-gamma (2): Effect of δ_1 and δ_2 on shape*

Raw moments are given by

$$m_n = \left(\frac{\delta_2}{\delta_2 + n}\right)^{\delta_1} \tag{31.10}$$

Therefore

$$\mu = \left(\frac{\delta_2}{\delta_2 + 1}\right)^{\delta_1} \tag{31.11}$$

$$\sigma^2 = \left(\frac{\delta_2}{\delta_2 + 2}\right)^{\delta_1} - \left(\frac{\delta_2}{\delta_2 + 1}\right)^{2\delta_1} \tag{31.12}$$

Before fitting, the C_4 in propane data is scaled over their range, i.e. 1.5 to 15.8 vol%, so that x covers the range 0 to 1. Fitting then gives δ_1 as 4.50 and δ_2 as 2.56. *RSS* is not as good at 0.0634. The mean is 0.227, or 4.75 vol%. The standard deviation is 0.152, or a rather low 2.17 vol%.

The third version is described by the PDF

$$f(x) = \frac{[\ln(x+1-\alpha)]^{\delta_1 - 1}}{\delta_2^{\delta_1}\Gamma(\delta_1)(x+1-\alpha)}\exp\left[-\frac{1}{\delta_2}\ln(x+1-\alpha)\right] \quad x > \alpha - 1;\ \delta_1, \delta_2 > 0 \tag{31.13}$$

Figure 31.3 shows the effect, with α set to zero, of varying δ_1 and δ_2.

Fitting to the C_4 in propane data gives α as 1.31, δ_1 as 11.6, δ_2 as 0.137 and $F(x_1)$ as 0.0200. *RSS* is a respectable 0.0321.

The PDF of the fourth version is substantially different.

$$f(x) = \frac{1}{\delta_2^{\delta_1}\Gamma(\delta_2)}\exp[\delta_1(x-\alpha)]\exp\left[-\frac{1}{\delta_2}\exp(x-\alpha)\right] \quad \delta_1, \delta_2 > 0 \tag{31.14}$$

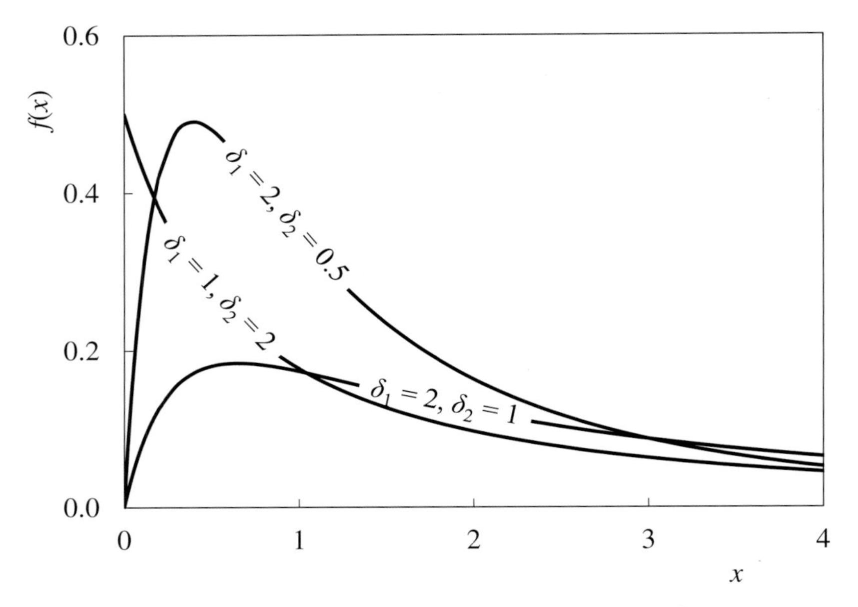

Figure 31.3 *Log-gamma (3): Effect of δ_1 and δ_2 on shape*

Figure 31.4 shows the effect of varying δ_1 and δ_2, with α fixed at zero. Of note is, that unlike those described previously, it is unbounded.

Fitting to the C_4 in propane data gives α as 0.967, δ_1 as 0.441, δ_2 as 97.6 and $F(x_1)$ as 0.0601. As Figure 31.4 shows, the distribution is skewed in the opposite direction to the data and so, not

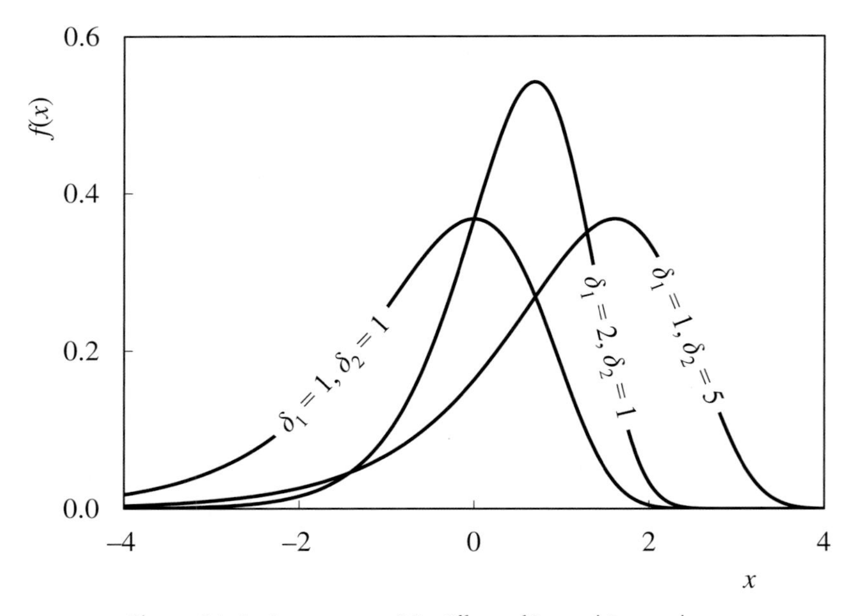

Figure 31.4 *Log-gamma (4): Effect of δ_1 and δ_2 on shape*

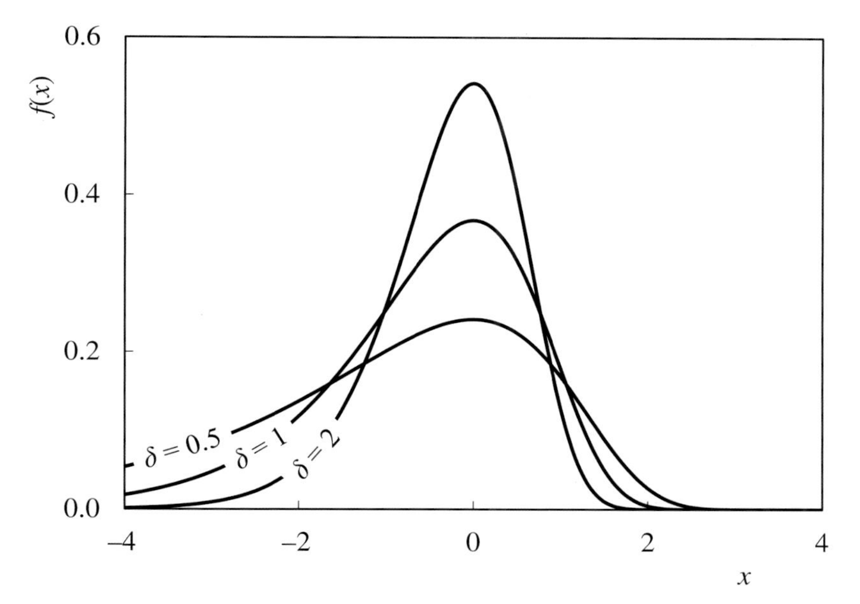

Figure 31.5 *Log-gamma (5): Effect of δ on shape*

surprisingly, *RSS* is very poor at 0.7886. However, fitting to (100 – C_4) gives α as 92.9, δ_1 as 0.436, δ_2 as 97.1 and $F(x_1)$ as 0.0140. The resulting *RSS* is a respectable 0.0351.

The PDF of the fifth version is

$$f(x) = \frac{\delta^\delta}{\delta\Gamma(\delta)}\exp\left[\delta\left(\frac{x-\alpha}{\beta}\right) - \delta\exp\left(\frac{x-\alpha}{\beta}\right)\right] \qquad \delta,\beta > 0 \tag{31.15}$$

Figure 31.5 shows the effect of varying δ, with α fixed at 0 and β at 1. Again the skew is in the wrong direction to fit the C_4 in propane data. Fitting to (100 – C_4) gives α as 96.8, β as 0.711, δ as 0.276 and $F(x_1)$ as 0. *RSS* is an excellent 0.0283.

31.3 Generalised Gamma

The *generalised gamma distribution* is also known, the *transformed gamma distribution*[23] and the *Stacy–Mihram distribution*. It is the un-shifted version of the Amoroso distribution that we will cover in Chapter 34. Its PDF is

$$f(x) = \frac{\delta_2}{\beta\Gamma(\delta_1)}\left(\frac{x}{\beta}\right)^{\delta_1\delta_2-1}\exp\left[-\left(\frac{x}{\beta}\right)^{\delta_2}\right] \tag{31.16}$$

It includes the gamma, Weibull-II and exponential distributions as special cases, but these are also covered by the Amoroso distribution.

Fitting to the C_4 in propane data is problematic. Figure 31.6 shows the result of fixing β and fitting δ_1 and δ_2. *RSS* does not pass through a minimum, suggesting that β should be 0.

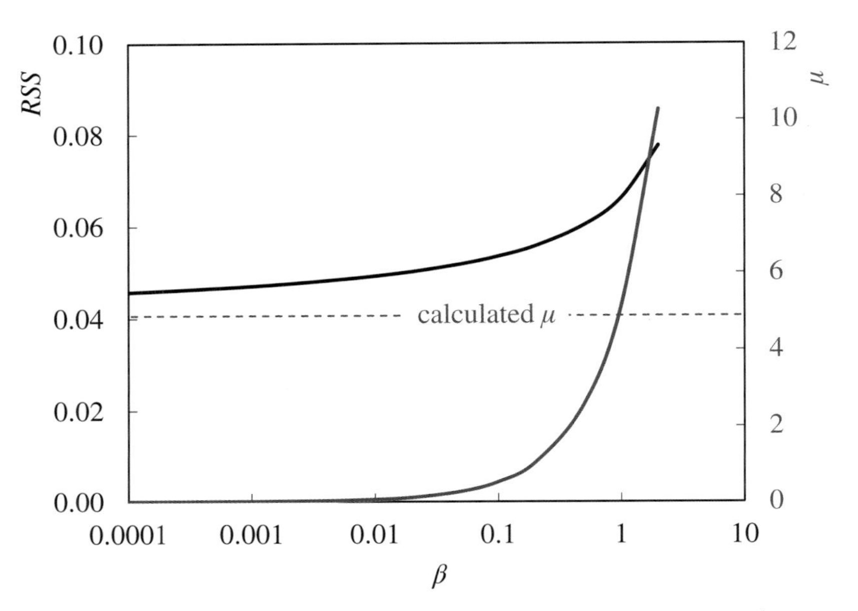

Figure 31.6 Generalised gamma: Problem fitting to C_4 in propane data

We will show in Chapter 34 that the mean is

$$\mu = \frac{\Gamma\left(\delta_1 + \frac{1}{\delta_2}\right)}{\Gamma(\delta_1)}\beta \tag{31.17}$$

Also shown in Figure 31.6, as the coloured curve, is the resulting estimate of μ. By calculation from the data, this is 4.87 – suggesting that β should be close to 1.

31.4 Q-Gamma

The *q-gamma distribution* is another in the family of Tsallis distributions. Depending on the value of q, the PDF is

$$f(x) = \frac{(1-q)^{\delta} x^{\delta-1}[1+(q-1)x]^{-\frac{\lambda}{q-1}}}{\mathrm{B}\left(\delta, \frac{\lambda}{1-q}+1\right)} \qquad x>0;\ \lambda,\delta>0;\ 0<q<1 \tag{31.18}$$

$$f(x) = \frac{(q-1)^{\delta} x^{\delta-1}[1+(q-1)x]^{-\frac{\lambda}{q-1}}}{\mathrm{B}\left(\delta, \frac{\lambda}{q-1}-\delta\right)} \qquad x>0;\ \lambda,\delta>0;\ 1<q<\frac{\lambda}{\delta}+1 \tag{31.19}$$

If q is set to 1, the distribution reverts to the classic gamma distribution described by Equation (12.92). Figure 31.7 shows this as the coloured line and the effect of varying q.

Fitting to the C_4 in propane data, with q fixed at 1, gives values of 4.38 for δ and 0.988 for λ, with *RSS* as 0.0660. Fully fitting gives q as 1.40, δ as 11.5 and λ as 7.41. *RSS* reduces

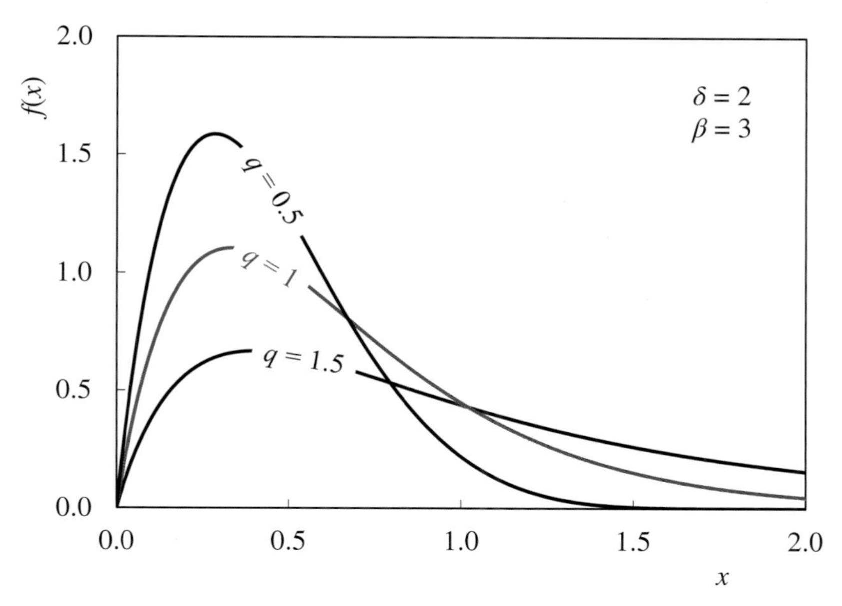

Figure 31.7 *Q-gamma: Effect of q on shape*

significantly to 0.0418. Not surprisingly, the addition of a shape parameter will usually mean that the q-gamma distribution will be a better fit than the gamma distribution.

Raw moments are given by

$$m_n = \frac{\Gamma(\delta+n)\Gamma\left(\frac{\lambda}{1-q}+\delta+1\right)}{\Gamma(\delta)\Gamma\left(\frac{\lambda}{1-q}+\delta+n+1\right)(1-q)^n} \qquad 0<q<1 \tag{31.20}$$

$$m_n = \frac{\Gamma(\delta+n)}{\Gamma(\delta)\lambda^n} \qquad q=1 \tag{31.21}$$

$$m_n = \frac{\Gamma(\delta+n)\Gamma\left(\frac{\lambda}{q-1}-\delta-n\right)}{\Gamma(\delta)\Gamma\left(\frac{\lambda}{q-1}-\delta\right)(1-q)^n} \qquad 1<q<\frac{\lambda}{\delta+n}+1 \tag{31.22}$$

These give μ as 4.79 and σ as 2.63, close to the values calculated from the data. Somewhat different are γ at 2.27 and κ at 15.2.

32

Symmetrical Distributions

There remain a large number of distributions that are not extensions to those covered so far in this book. Further they bear little relationship to each other. This chapter includes those that are symmetrical, i.e. their skewness is zero. Asymmetric distributions follow in the next chapter.

The simplest example of a symmetrical distribution is the uniform distribution that we covered in Section 5.1. The most used is the normal distribution (Section 5.3). We have also covered others such as the Student t distribution (Section 12.5), logistic distribution (Section 22.1) and the Cauchy distribution (Section 27.4). This chapter groups together others that fall into this category. In the absence of any mathematical argument for arranging them in a particular sequence, they are presented largely alphabetically.

There are several situations where we might reasonably expect the skewness of process data to be zero. For example, when we check an inferential property against a laboratory result, it is equally probable that error will be positive or negative. The distribution of errors will have a skewness close to zero. Perhaps the most common application is in the analysis of process disturbances. Disturbances in one direction are likely to be equally probable as those in the opposite direction. Indeed, we will use the NHV data as the example to evaluate each of the distributions. Of the symmetrical distributions considered so far, the best choice is logistic distribution – giving a value of 0.1061 for *RSS*. This provides the benchmark against which the distributions in this chapter can be assessed.

32.1 Anglit

The *anglit distribution* is described by the PDF

$$f(x) = \frac{1}{\beta}\sin\left(\frac{2(x-\mu)}{\beta} + \frac{\pi}{2}\right) = \frac{1}{\beta}\cos\left(\frac{2(x-\mu)}{\beta}\right) \quad \mu - \frac{\pi\beta}{4} < x < \mu + \frac{\pi\beta}{4}, \; \beta > 0 \qquad (32.1)$$

Figure 32.1 shows the effect of varying μ and β. The CDF is

$$F(x) = \sin^2\left(\frac{x-\mu}{\beta} + \frac{\pi}{4}\right) \qquad (32.2)$$

Statistics for Process Control Engineers: A Practical Approach, First Edition. Myke King.

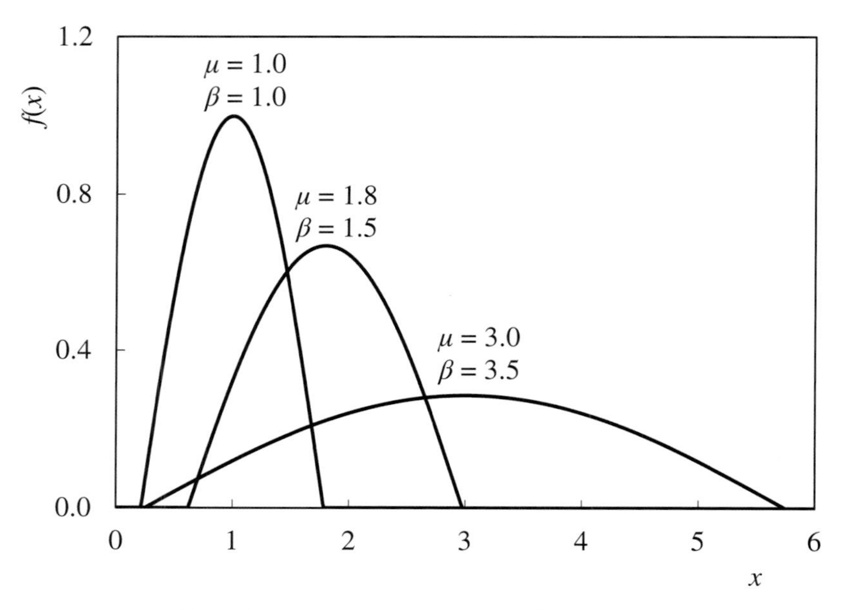

Figure 32.1 *Anglit: Effect of μ and β on shape*

The QF is

$$x(F) = \left[\sin^{-1}\left(\sqrt{F}\right) - \frac{\pi}{4}\right]\beta + \mu \quad 0 \le F \le 1 \tag{32.3}$$

The mean is μ. The variance is

$$\sigma^2 = \frac{\pi^2 - 8}{16}\beta^2 \approx 0.117\beta^2 \tag{32.4}$$

Fitting to the NHV disturbance data gives μ as −0.02 and β as 3.16, from which the standard deviation is derived as 1.08. At 0.2165 *RSS* is considerably worse than that from fitting the normal distribution and much worse than that from the benchmark logistic distribution.

Kurtosis is fixed.

$$\kappa = 3 - \frac{2(\pi^4 - 96)}{(\pi^2 - 8)^2} \approx 2.194 \tag{32.5}$$

32.2 Bates

The *Bates distribution* is that of the mean of n values taken from the uniform distribution U (0,1). It is described by the PDF

$$f(x) = \frac{n}{2(n-1)!}\sum_{i=0}^{n}\frac{(-1)^i n!(nx-i)^{n-1}}{i!(n-i)!}\frac{|nx-i|}{nx-i} \quad 0 \le x \le 1;\ n > 1 \tag{32.6}$$

Because of its definition, values (x) must first be scaled from 0 to 1 by applying the formula

$$x_{\text{scaled}} = \frac{x - x_{\min}}{x_{\max} - x_{\min}} \tag{32.7}$$

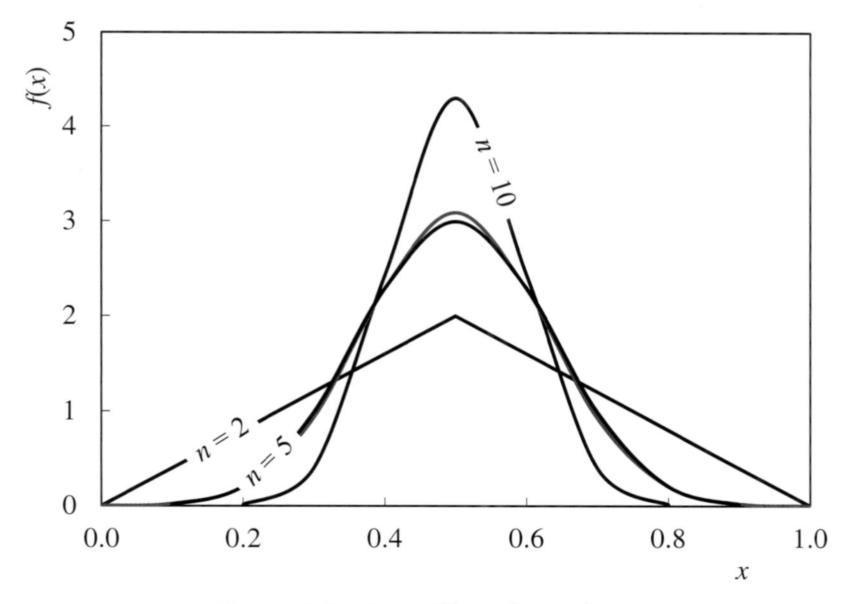

Figure 32.2 Bates: Effect of n on shape

Key statistical parameters are

$$\mu = \frac{x_{\min} + x_{\max}}{2} \tag{32.8}$$

$$\sigma^2 = \frac{(x_{\max} - x_{\min})^2}{12n} \tag{32.9}$$

$$\kappa = 3 - \frac{6}{5n} \tag{32.10}$$

Figure 32.2 shows the effect of changing n. As $n \to \infty$, the distribution approaches the normal distribution. In fact the approach occurs at very low values of n. Shown as the coloured line is the plot of the normal distribution based the mean and standard deviation calculated from Equations (32.8) and (32.9) for $n = 5$. Examination of Equation (32.10) confirms this behaviour; kurtosis approaches 3 as n is increased. As n reaches 3, kurtosis is already well within the range 3 ± 0.5, over which we might treat the distribution as normal. Since the distribution is triangular when $n = 2$ and n must be an integer, it is unlikely that the distribution offers any improvement on the fit achievable with the normal distribution. It is also considerably more complex to use, for example, in a spreadsheet.

32.3 Irwin–Hall

The *Irwin–Hall distribution* is closely related to the Bates distribution. But, rather the distribution of the mean of n values taken from the uniform distribution U(0,1), it is the distribution of their sum. It can also therefore be known as the *uniform sum distribution*. It is described by

$$f(x)=\frac{1}{2(n-1)!}\sum_{i=0}^{n}\frac{(-1)^{i}n!(x-i)^{n-1}}{i!(n-i)!}\frac{|x-i|}{x-i} \quad x>0;\ n>1 \tag{32.11}$$

Because the distribution is symmetrical, remembering (from Section 5.1) that $\lfloor x \rfloor$ is the floor of x, i.e. the largest integer that is less than or equal to x, the PDF can also be written as

$$f(x)=\frac{1}{(n-1)!}\sum_{i=0}^{\lfloor x \rfloor}\frac{(-1)^{i}n!(x-i)^{n-1}}{i!(n-i)!} \tag{32.12}$$

Key parameters are

$$\mu=\frac{n}{2} \tag{32.13}$$

$$\sigma^{2}=\frac{n}{12} \tag{32.14}$$

$$\kappa=3-\frac{6}{5n} \tag{32.15}$$

Figure 32.3 shows the effect of varying n. As n is increased the distribution approaches the normal distribution. Indeed, using $n=12$ is the basis of the technique, described in Section 5.6, for generating a normal distribution.

For the same reasons as those for the Bates distribution, fitting the Irwin–Hall distribution to process data offers no advantage over choosing the normal distribution.

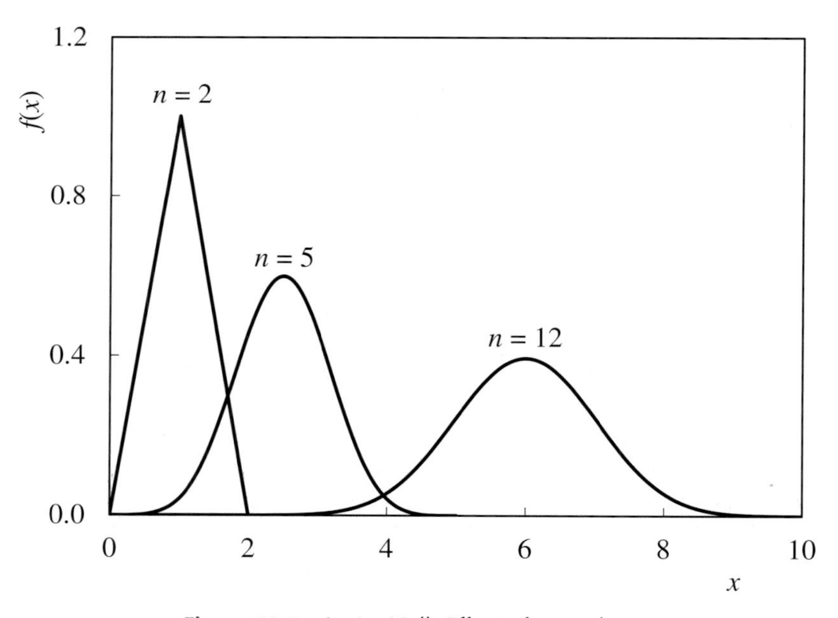

Figure 32.3 *Irwin–Hall: Effect of* n *on shape*

32.4 Hyperbolic Secant

The *hyperbolic secant distribution* is also known as the *sech distribution*. Its PDF is

$$f(x)=\frac{1}{2\sigma}\operatorname{sech}\left[\frac{\pi(x-\mu)}{2\sigma}\right] \qquad \sigma>0 \tag{32.16}$$

Remembering that

$$\operatorname{sech}(x)=\frac{2}{e^{x}+e^{-x}} \tag{32.17}$$

An alternative formulation of the PDF is therefore

$$f(x)=\frac{1}{\sigma}\left[\exp\left(\frac{\pi(x-\mu)}{2\sigma}\right)+\exp\left(-\frac{\pi(x-\mu)}{2\sigma}\right)\right]^{-1} \tag{32.18}$$

Figure 32.4 shows the effect of varying μ and σ.

The CDF can similarly be expressed in two ways.

$$F(x)=\frac{1}{2}+\frac{1}{\pi}\tan^{-1}\left[\sinh\left(\frac{\pi(x-\mu)}{2\sigma}\right)\right] \tag{32.19}$$

$$F(x)=\frac{2}{\pi}\tan^{-1}\left\{\exp\left[\frac{\pi(x-\mu)}{2\sigma}\right]\right\} \tag{32.20}$$

Fitting to the NHV disturbance data gives μ as –0.03, σ as 1.38 and *RSS* as 0.0837. Kurtosis, calculated from the data, is 5.87 which happens to be close to the distribution's fixed value of 5. So, for example, it fits much better than the normal distribution with its kurtosis of 3.

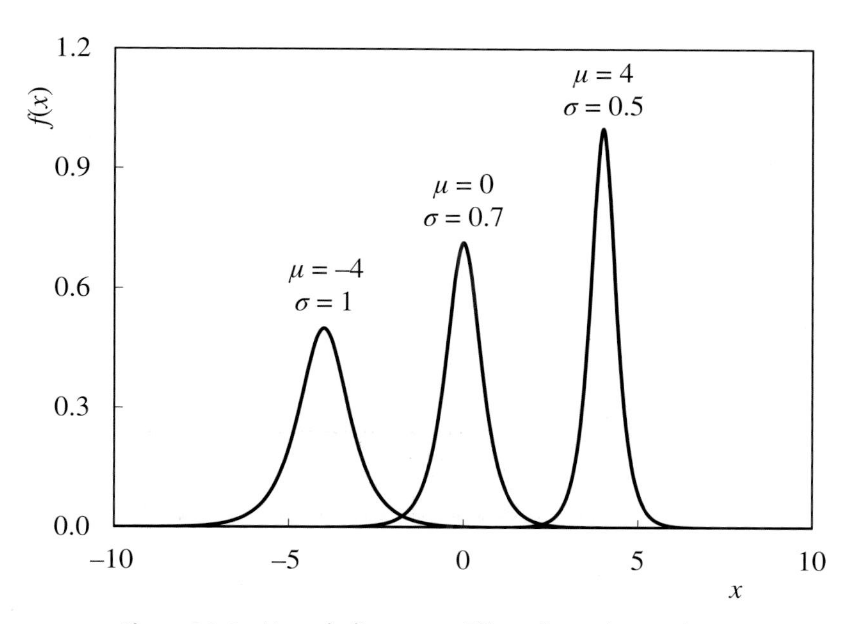

Figure 32.4 *Hyperbolic secant: Effect of μ and σ on shape*

The QF can similarly be expressed in several ways.

$$x(F) = \mu + \frac{2\sigma}{\pi}\sinh^{-1}\left[\tan\left(\pi F - \frac{\pi}{2}\right)\right]$$

$$= \mu - \frac{2\sigma}{\pi}\sinh^{-1}[\cot(\pi F)] \qquad (32.21)$$

$$= \mu + \frac{2\sigma}{\pi}\ln\left[\tan\left(\frac{\pi F}{2}\right)\right] \qquad 0 \le F \le 1$$

32.5 Arctangent

The PDF of the *arctangent distribution* is

$$f(x) = \frac{\beta}{\left[\tan^{-1}(\delta\beta) + \frac{\pi}{2}\right]\left[1 + \beta^2(x - \alpha - \delta)^2\right]} \qquad x \ge \alpha;\ \beta > 0 \qquad (32.22)$$

Figure 32.5 shows the effect of varying δ and β, with α fixed at 0. Strictly, because the distribution has a lower bound and no upper bound, it is not completely symmetrical.

The CDF is

$$F(x) = \frac{\tan^{-1}(\delta\beta) - \tan^{-1}[\beta(\alpha + \delta - x)]}{\tan^{-1}(\delta\beta) + \frac{\pi}{2}} \qquad (32.23)$$

Fitting to the NHV disturbance data gives α as −8.20, δ as 8.11 and β as 1.34. *RSS* is 0.0445. There are no formulae for mean or variance, but $(\alpha + \delta)$ is a measure of location and β of dispersion. There are similarly no formulae for skewness or kurtosis.

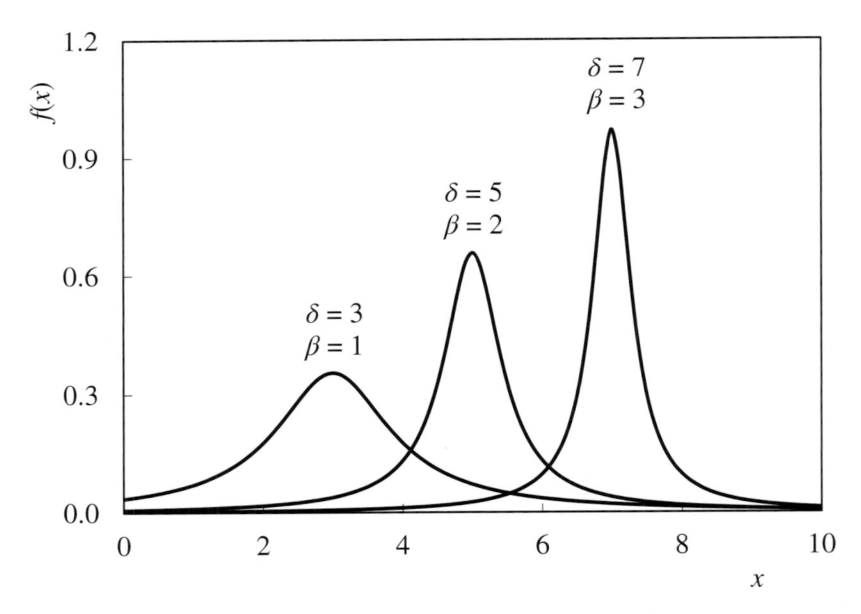

Figure 32.5 *Arctangent: Effect of δ and β on shape*

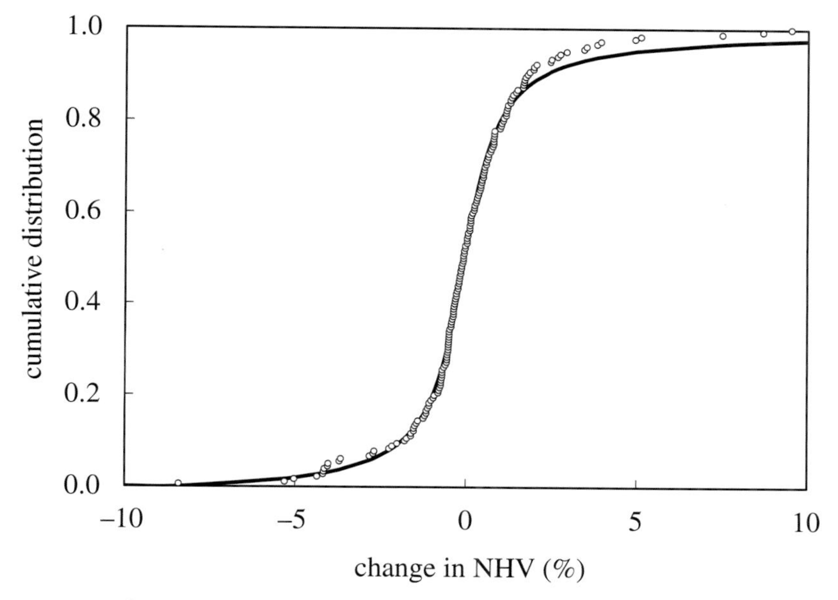

Figure 32.6 Arctangent: Poor fit to NHV disturbance data

The QF is

$$x(F) = \alpha + \delta + \frac{1}{\beta}\tan\left\{F\left[\tan^{-1}(\alpha\beta) + \frac{\pi}{2}\right] - \tan^{-1}(\alpha\beta)\right\} \quad 0 \le F \le 1 \tag{32.24}$$

As Figure 32.6 shows, the best fit has an extremely long upper tail that is entirely unsuitable for this application. To illustrate this, consider an event that occurs, on average, once per year. F in Equation (32.24) would be 364/365 and so x would be 89.5 – ten times larger than the largest disturbance recorded.

32.6 Kappa

We previously covered the Hosking four parameter kappa distribution. This *kappa distribution* is, however, very different. Its PDF is

$$f(x) = \frac{K}{(\pi\delta\beta^2)^{\frac{3}{2}}}\frac{\Gamma(\delta+1)}{\Gamma\left(\delta-\frac{1}{2}\right)}\left[1+\frac{(x-\alpha)^2}{\delta\beta^2}\right]^{-(\delta+1)} \quad \delta > 0.5 \tag{32.25}$$

When applied in the area of physics for which the distribution was developed, K is assigned a numerical value. When fitting to process data K is adjusted to ensure $F(x_\infty)$, or at least $F(x_n)$, is unity.

Figure 32.7 shows, with α set to zero and β to 1, the effect of changing δ. Fitting to the NHV disturbance data gives α as 0, β as 1.81, δ as 0.502, K as 2235 and $F(x_1)$ as 0.0142. *RSS*, at 0.0526, shows the fit is reasonable. There are however no simple formulae to calculate mean, variance or kurtosis.

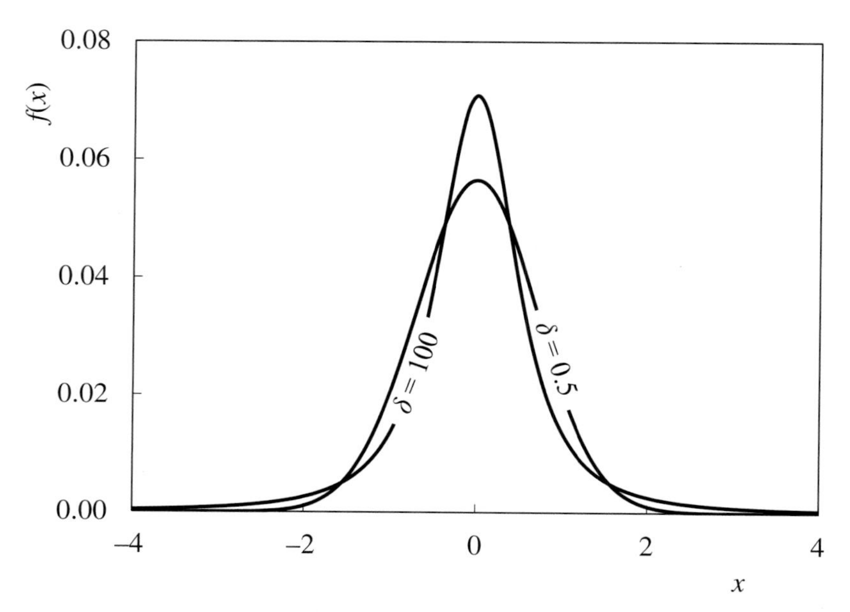

Figure 32.7 *Kappa: Effect of δ on shape*

While of limited practical application to the process industry, features of the kappa distribution are incorporated into several other distributions – some of which are described in this book.

32.7 Laplace

Devised by Pierre-Simon Laplace, the same mathematician that developed the transforms found daunting by many control engineers, the *Laplace distribution* might be considered as an alternative to the Cauchy distribution. It is special case of the generalised normal distribution described in Section 20.12. Although its kurtosis is fixed at 6, by adjusting σ it can be fitted to represent a wide range of leptokurtic distributions. It is effectively back-to-back exponential distributions, reflected in the line $x=\mu$, and also therefore known as the double exponential distribution (as mentioned in Section 28.5). Its PDF is

$$f(x)=\frac{1}{\sigma\sqrt{2}}\exp\left(\frac{-|x-\mu|\sqrt{2}}{\sigma}\right) \qquad \sigma>0 \tag{32.26}$$

Figure 32.8 shows the effect of varying its parameters. Its CDF is

$$F(x)=\frac{1}{2}+\frac{x-\mu}{2|x-\mu|}\left[1-\exp\left(\frac{-|x-\mu|\sqrt{2}}{\sigma}\right)\right] \tag{32.27}$$

The QF is slightly complicated by the inclusion of the absolute value. We can split Equation (32.27) into its two cases and then invert these. We then determine $x(F)$ using both Equations (32.29) and (32.31) and pick the solution that is within its bound.

$$F(x)=\frac{1}{2}\exp\left[\frac{(x-\mu)\sqrt{2}}{\sigma}\right] \qquad x>\mu \tag{32.28}$$

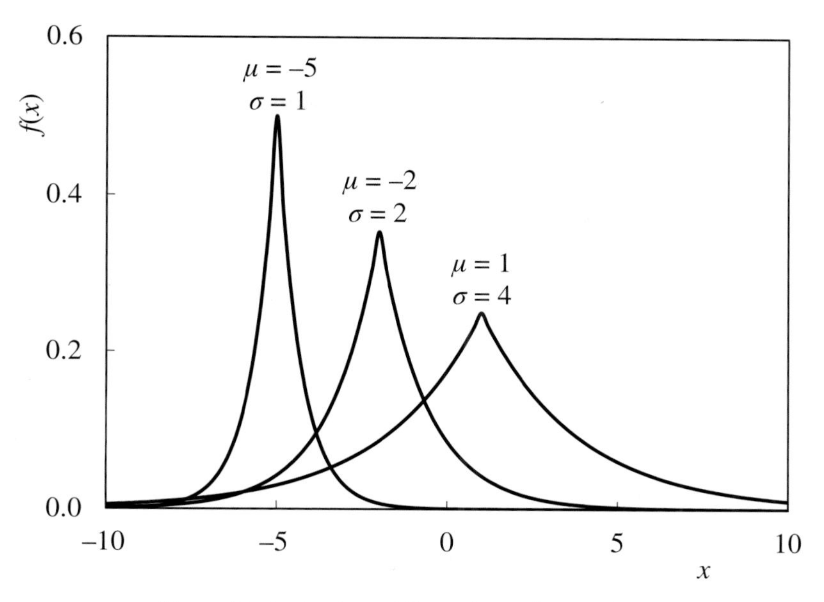

Figure 32.8 *Laplace: Effect of* μ *and* σ *on shape*

$$x(F)=\mu+\frac{\sigma\ln(2F)}{\sqrt{2}} \qquad 0\le F\le 1 \tag{32.29}$$

$$F(x)=1-\frac{1}{2}\exp\left[-\frac{(x-\mu)\sqrt{2}}{\sigma}\right] \qquad x\le\mu \tag{32.30}$$

$$x(F)=\mu-\frac{\sigma\ln(2-2F)}{\sqrt{2}} \qquad 0\le F\le 1 \tag{32.31}$$

Fitting to the NHV disturbance data gives μ as −0.04, σ as 1.56 and *RSS* as 0.0556. So, while a slightly better fit than the Cauchy distribution, the QF make it slightly more cumbersome to apply.

32.8 Raised Cosine

The *cosine distribution* is defined by the PDF, CDF and QF

$$f(x)=\frac{1}{2\beta}\cos\left(\frac{x-\mu}{2\beta}\right) \qquad \mu-\frac{\pi\beta}{2}<x<\mu+\frac{\pi\beta}{2};\ \beta>0 \tag{32.32}$$

$$F(x)=\frac{1}{2}\left[1+\sin\left(\frac{x-\mu}{\beta}\right)\right] \tag{32.33}$$

$$x(F)=\alpha+\beta\sin^{-1}(2F-1) \qquad 0\le F\le 1 \tag{32.34}$$

Its mean is μ; its variance is

$$\sigma^2=\beta^2\left(\pi^2-8\right) \tag{32.35}$$

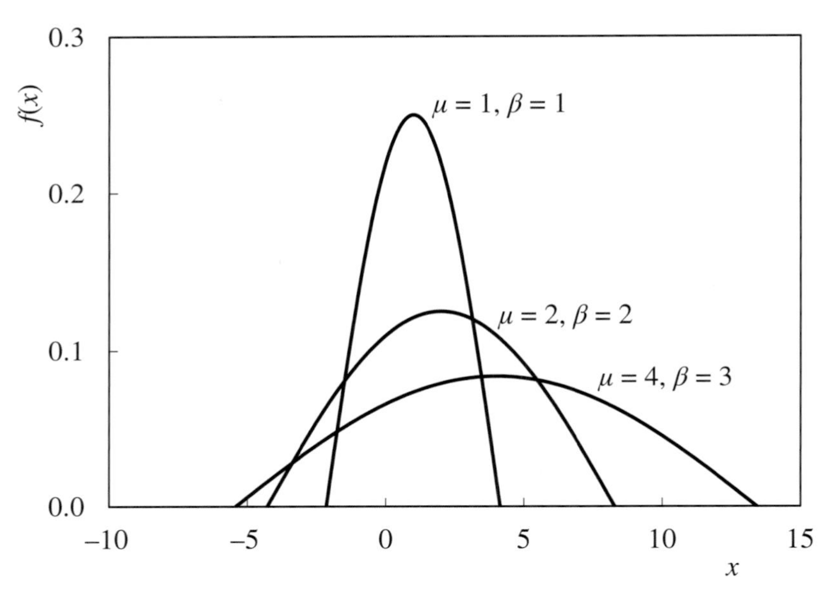

Figure 32.9 *Cosine: Effect of μ and β on shape*

Figure 32.9 shows the effect of varying μ and β. It is unlikely to fit any form of process data. However, the *raised cosine distribution* may be of more use. Its PDF is

$$f(x) = \frac{1}{2\beta}\left[1 + \cos\left(\frac{x-\mu}{\beta}\pi\right)\right] \quad \mu-\beta < x < \mu+\beta \tag{32.36}$$

Figure 32.10 shows the effect of varying μ and β. Its CDF is

$$F(x) = \frac{1}{2}\left[1 + \frac{x-\mu}{\beta} + \frac{1}{\pi}\sin\left(\frac{x-\mu}{\beta}\pi\right)\right] \tag{32.37}$$

Its mean is μ; its variance is

$$\sigma^2 = \beta^2\left(\frac{1}{3} - \frac{2}{\pi^2}\right) \tag{32.38}$$

Fitting Equation (32.37) to the NHV disturbance data gives values of 0.00 for α and 3.09 for β. With *RSS* at 0.1887, the fit is poor. Furthermore, 20 of the 180 measurements fall outside the limits of Equation (32.36) and are therefore ignored. Since we are likely to be most interested in the probability of such extreme behaviour, this distribution would be a most unsuitable choice.

The kurtosis calculated from the data is 5.87. The kurtosis for the distribution is fixed, as below, giving another reason for rejecting the distribution.

$$\kappa = \frac{9(\pi^4 - 20\pi^2 + 120)}{5(\pi^2-6)^2} \approx 2.41 \tag{32.39}$$

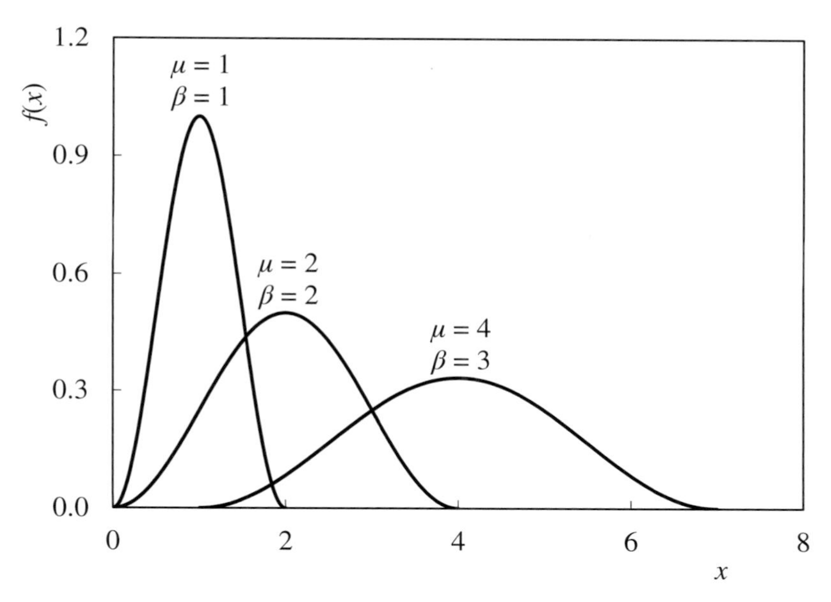

Figure 32.10 *Raised cosine: Effect of μ and β on shape*

32.9 Cardioid

Related to the raised cosine distribution is the *cardioid distribution*, defined by the PDF

$$f(x) = \frac{1}{2\pi}[1 - 2\beta\cos(x-\mu)] \quad 0 \le x < 2\pi,\ 0 < \beta < 0.5 \tag{32.40}$$

It is an example of a *circular distribution*, in that x can be an angle representing a direction – such as a compass bearing. The PDF might then be the distribution of wind strength. Plotting $f(x)$ as a radar plot conventionally places the origin at the bottom. Adding π to the wind direction (in radians) rotates the plot to put North at the top. Figure 32.11 shows the cardioid shape that gives the distribution its name. It shows that the mean wind direction is south-east (a bearing of 135°), defined by

$$\mu = 135 \times \frac{2\pi}{360} + \pi \tag{32.41}$$

Figure 32.12 is a more conventional plot, showing the effect of β (with μ fixed at 0). If β is 0, we obtain the uniform distribution U(0,2π). If β is 0.5 we obtain the raised cosine distribution. Note that all curves pass through the points ($\mu + \pi/2$, $1/2\pi$) and ($\mu + 3\pi/2$, $1/2\pi$).

The CDF is

$$F(x) = \frac{x}{2\pi} - \frac{\beta}{\pi}\sin(x-\mu) \tag{32.42}$$

32.10 Slash

While there is no definition of kurtosis for the *slash distribution*, it can represent the distribution of data for which this is very large. Theoretically it is the distribution of the ratio of two numbers with the numerator selected from N(0,1) and the denominator from U(0,1). This need not

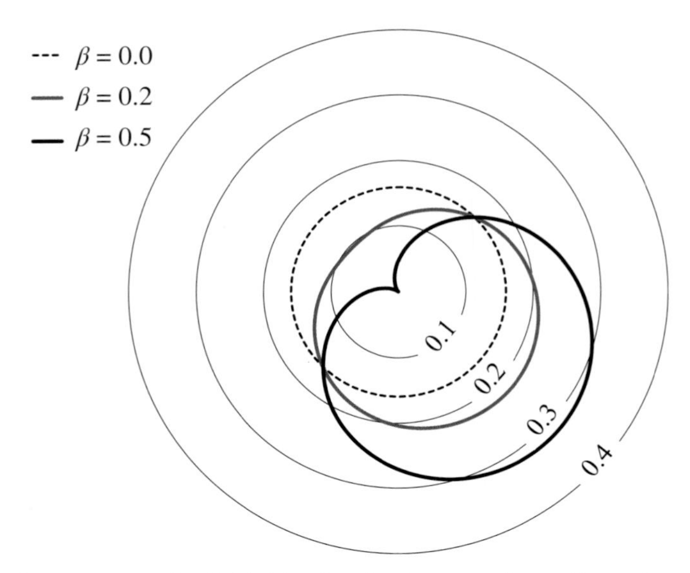

Figure 32.11 Cardioid: Circular plot showing effect of β on shape

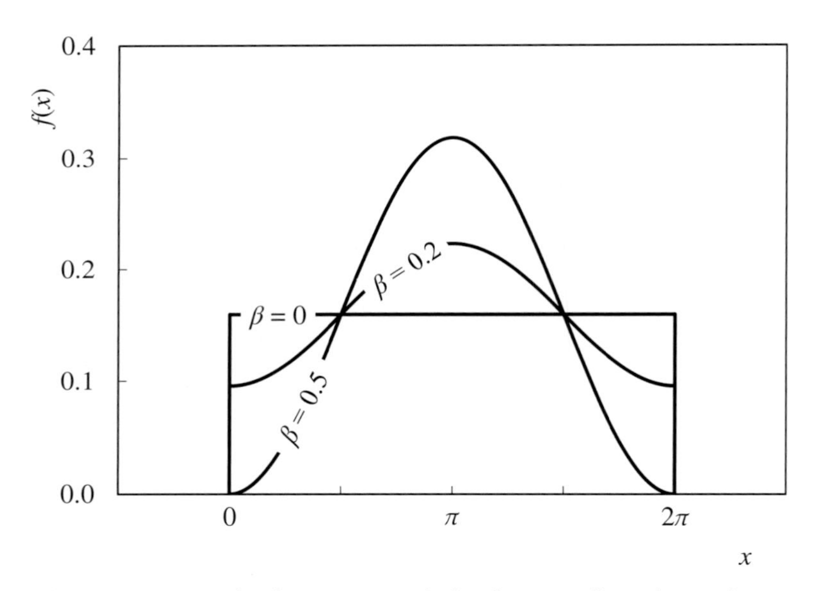

Figure 32.12 Cardioid: Conventional plot showing effect of β on shape

concern us; as usual we choose simply it on the basis that it provides the best fit to the data. It is described by the PDF

$$f(x) = \frac{\sigma}{\sqrt{2\pi}(x-\mu)^2}\left\{1-\exp\left[-\frac{1}{2}\left(\frac{x-\mu}{\sigma}\right)^2\right]\right\} \quad x \neq \mu;\ \sigma > 0 \tag{32.43}$$

$$f(x) = \frac{1}{2\sigma\sqrt{2\pi}} \quad x = \mu;\ \sigma > 0 \tag{32.44}$$

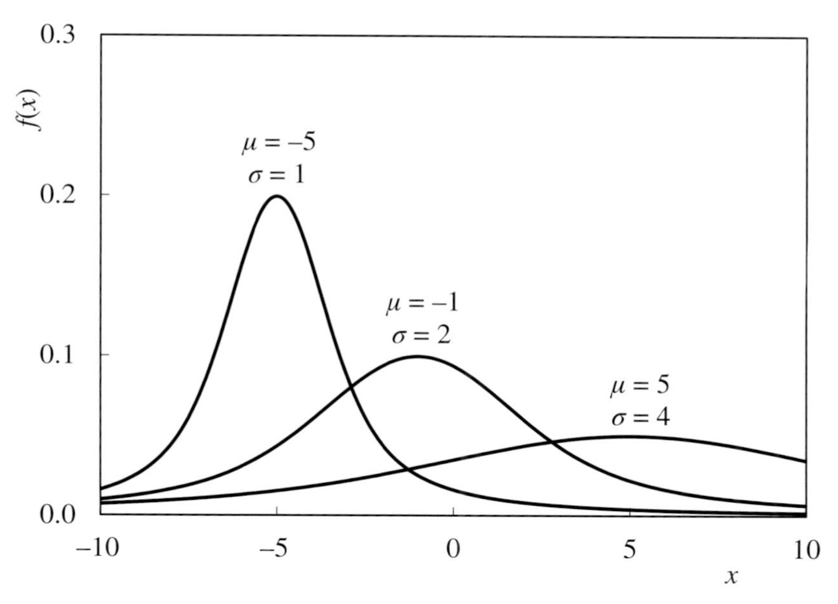

Figure 32.13 *Slash: Effect of μ and σ on shape*

Figure 32.13 shows the effect of varying μ and σ. The distribution gives the fattest tails and so is suitable if a highly platykurtic distribution is required. Its CDF is

$$F(x)=\frac{1}{2}\left[1+\operatorname{erf}\left(\frac{x-\mu}{\sigma\sqrt{2}}\right)\right]-\frac{\sigma}{(x-\mu)\sqrt{2\pi}}\left\{1-\exp\left[-\frac{1}{2}\left(\frac{x-\mu}{\sigma}\right)^2\right]\right\} \tag{32.45}$$

Fitting to the NHV disturbance data gives μ as –0.04, σ as 0.49 and *RSS* as 0.0440. This provides the best fit of all the distributions considered so far. By more properly identifying what is kurtosis, it also gives the lowest estimate for the standard deviation.

32.11 Tukey Lambda

The *Tukey lambda distribution* is unusual in that, generally, it can only be described as a QF.

$$x(F)=\frac{1}{\lambda}\left[F^{\lambda}-(1-F)^{\lambda}\right] \quad 0\le F\le 1;\ \lambda\neq 0 \tag{32.46}$$

$$x(F)=\ln\left(\frac{F}{1-F}\right) \quad 0\le F\le 1;\ \lambda=0 \tag{32.47}$$

Figure 32.14 shows the effect of changing λ. Values of 1 or greater produce a uniform distribution. Values between 0.5 and 1 produce a bounded inverted U-shaped distribution. A value of 0.135 gives approximately the normal distribution with a standard deviation of about 1.46. A value of –1 gives approximately the Cauchy distribution. If λ is 0, the distribution becomes the standard logistic distribution. It can then be expressed as a PDF and CDF.

$$f(x)=\frac{\exp(-x)}{(1+\exp(-x))^2} \tag{32.48}$$

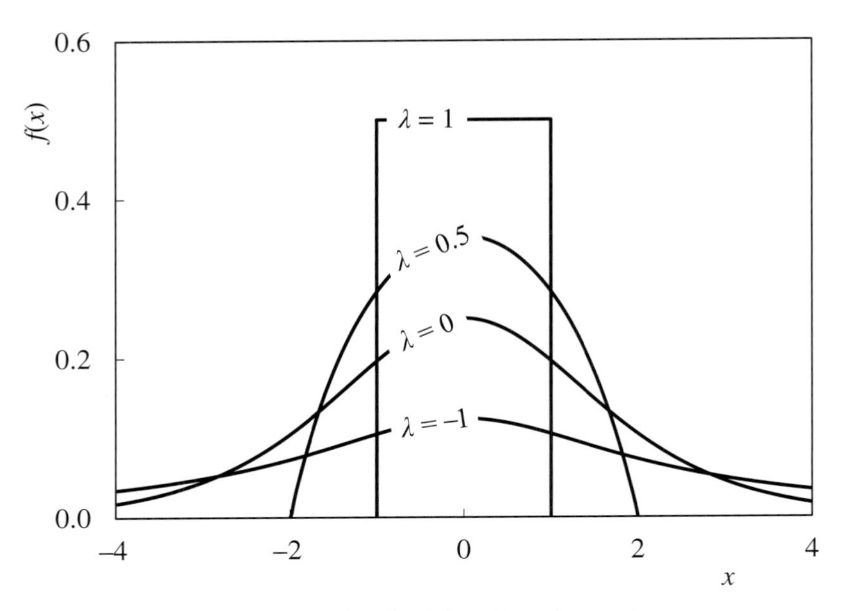

Figure 32.14 *Tukey lambda: Effect of λ on shape*

$$F(x)=\frac{1}{1+\exp(-x)} \tag{32.49}$$

The variance is

$$\sigma^2=\frac{2\{\Gamma(2\lambda+1)-\Gamma^2(\lambda+1)\}}{\lambda^2\Gamma(2\lambda+2)} \qquad \lambda>-\frac{1}{2} \tag{32.50}$$

$$\sigma^2=\frac{\pi^2}{3} \qquad \lambda=0 \tag{32.51}$$

Fitting Equation (32.46) to the NHV disturbance data, assuming a mean of 0, gives a value of −0.0961 for λ. From Equation (32.50) we obtain a value for σ of 2.17. *RSS* is 41.0 but, since it is based on predicting x rather than F, cannot be compared to the values achieved by other distributions. However we can make the comparison provided the alternative distribution can be expressed as a QF. For example, the best fit so far was achieved with the slash distribution but its CDF cannot readily be inverted. Next best was the Cauchy distribution. Its QF is given by Equation (27.26). Fitting this to the NHV data, with α fixed at 0, gives a value for β as 0.219. *RSS* is 167, showing the Cauchy distribution is substantially outperformed by the Tukey distribution.

The kurtosis is

$$\kappa=\frac{(2\lambda+1)^2\Gamma^2(2\lambda+1)\{3\Gamma^2(2\lambda+1)-4\Gamma(\lambda+1)\Gamma(3\lambda+1)+\Gamma(4\lambda+1)\}}{2(4\lambda+1)\Gamma(4\lambda+1)\{\Gamma^2(\lambda+1)-\Gamma(2\lambda+1)\}^2} \tag{32.52}$$

This is plotted as Figure 32.15, showing that the distribution can represent both platykurtic and leptokurtic data. It confirms that the normal distribution ($\kappa=3$) is approximated by setting λ to 0.135. For the NHV data κ is 6.61, as shown. This is close to the value calculated from the data as 5.87, giving further evidence that the Tukey distribution would be a good choice.

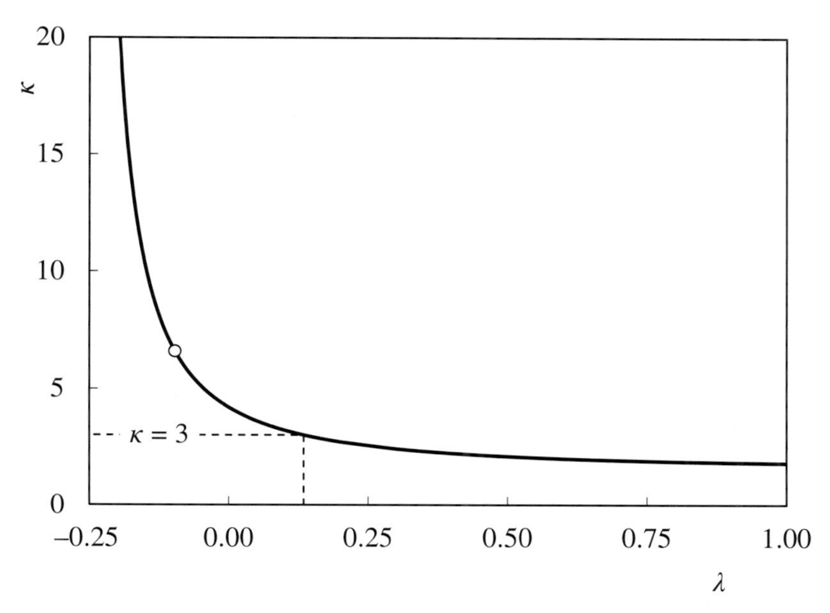

Figure 32.15 *Tukey lambda: Effect of λ on kurtosis*

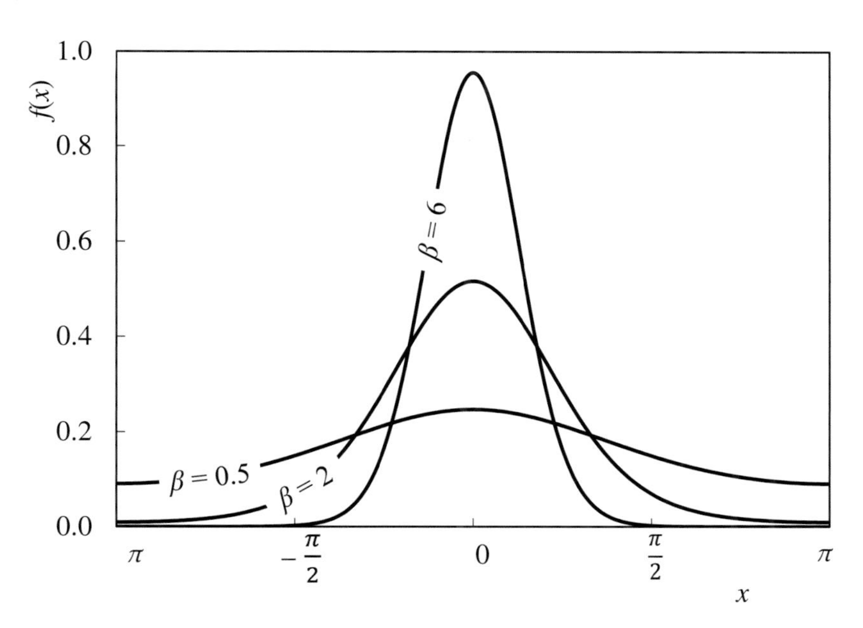

Figure 32.16 *Von Mises: Effect of β on shape*

32.12 Von Mises

The *von Mises distribution*, also known as the *Tikhonov distribution*, is described by the PDF

$$f(x)=\frac{\exp\{\beta\cos(x-\mu)\}}{2\pi I_0(\beta)} \qquad \mu-\pi\leq x\leq\mu+\pi;\ \beta>0 \tag{32.53}$$

I_0 is the zero order modified Bessel function, as described in Section 11.8. Figure 32.16 shows the effect of adjusting β, with α fixed at 0.

The inclusion of the trigonometric function would make the function cyclic if it were not for the bounds on x. For smaller values of β, $f(x)$ does not approach zero at the bounds and so is unlikely to match process behaviour. Indeed, as β approaches 0, the distribution approaches the uniform distribution.

Fitting to the NHV disturbance data gives μ as −0.01 and β as 1.08. The lower and upper bounds are therefore set at −3.15 and 3.13 respectively. Of the 180 data points, 20 fall outside this range and are therefore ignored in fitting the data. Since the whole point of the study is to explore the probability of large disturbances, the von Mises distribution would, in this example, be an entirely unsuitable choice. This is confirmed by the value of 0.1149 for *RSS* – making it one of the worse distributions to select for this dataset.

While there is no formula to calculate standard deviation, at higher values of β the distribution closely approaches the normal distribution. Its standard deviation (σ) can be obtained from Figure 32.17. Applying this to the NHV data, as shown, gives σ as 1.17 – comparable to that obtained by choosing the ill-fitting normal distribution.

There are no formulae for skewness and kurtosis.

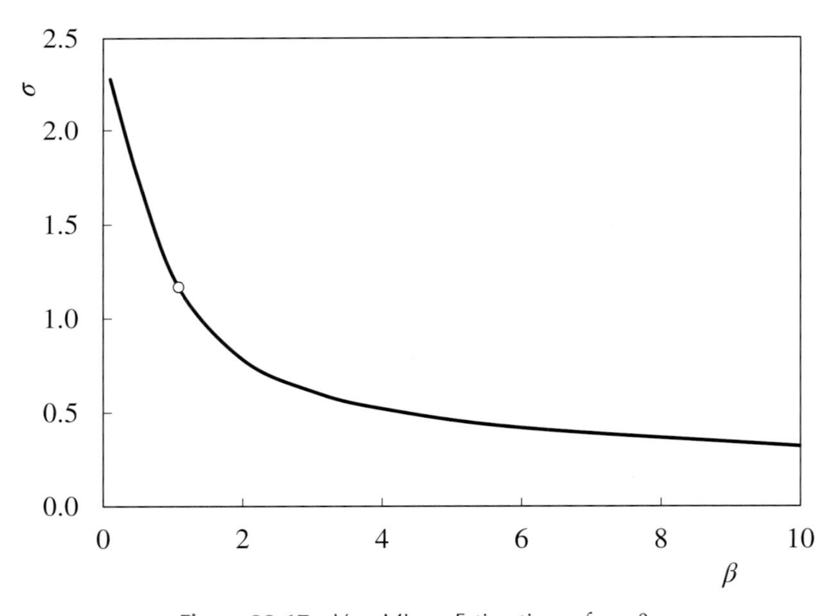

Figure 32.17 Von Mises: Estimating σ from β

33
Asymmetrical Distributions

There are a wide range of continuous distributions that accommodate nonzero skewness. This chapter includes those that are not obvious members of the groups covered in the previous chapters. For the lack of any better way of sorting them they are largely in alphabetical order.

33.1 Benini

The *three-parameter Benini distribution* is described by

$$f(x)=\left[\frac{\alpha}{x}+\frac{2\beta}{x}\ln\left(\frac{x}{x_{\min}}\right)\right]\exp\left\{-\alpha\ln\left(\frac{x}{x_{\min}}\right)-\beta\left[\ln\left(\frac{x}{x_{\min}}\right)\right]^2\right\}\qquad x>x_{\min};\ x_{\min},\alpha,\beta>0 \tag{33.1}$$

$$F(x)=1-\exp\left\{-\alpha\ln\left(\frac{x}{x_{\min}}\right)-\beta\left[\ln\left(\frac{x}{x_{\min}}\right)\right]^2\right\}=1-\left(\frac{x}{x_{\min}}\right)^{-\alpha-\beta(\ln x-\ln x_{\min})} \tag{33.2}$$

The *two-parameter Benini distribution* is obtained by setting α to 0.

Figure 33.1 shows the effect of varying α and β with $x_{\min}$ fixed at 0.1. Fitting Equation (33.2) to the NHV disturbance data gives $x_{\min}$ as 0.088, α very close to 0 and β as 0.164. With *RSS* at 0.1678 the fit is poor and the three-parameter version offers no advantage over the two-parameter one.

Fitting to the C_4 in propane data gives 1.75 for $x_{\min}$, 0 for α and 0.900 for β. *RSS* is 0.0865, making it one of the poorer choices.

For both versions, the calculations for mean and variance require mathematics beyond the scope of this book.

Statistics for Process Control Engineers: A Practical Approach, First Edition. Myke King.

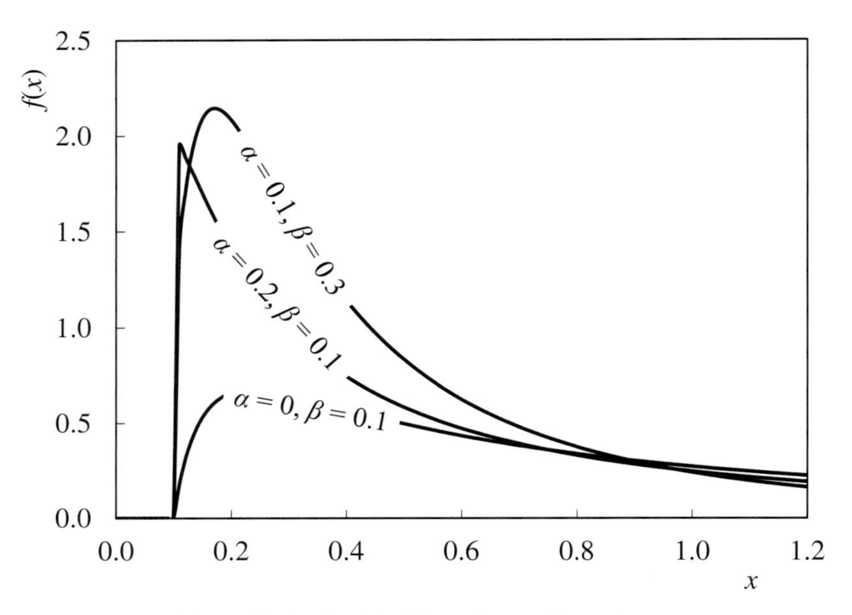

Figure 33.1 *Benini: Effect of α and β on shape*

33.2 Birnbaum–Saunders

The *Birnbaum–Saunders distribution* is also known as the *fatigue life distribution*, indicating the purpose for which it was originally developed. Its PDF is

$$f(x)=\frac{\left(\sqrt{\frac{x-\alpha}{\beta}}+\sqrt{\frac{\beta}{x-\alpha}}\right)\exp\left[\frac{1}{2\delta^2}\left(\sqrt{\frac{x-\alpha}{\beta}}-\sqrt{\frac{\beta}{x-\alpha}}\right)^2\right]}{2\delta(x-\alpha)\sqrt{2\pi}} \qquad x>\alpha;\ \beta,\delta>0 \tag{33.3}$$

Figure 33.2 shows the effect of varying δ (with α fixed at 0 and β at 1). The CDF is

$$F(x)=\frac{1}{2}+\frac{1}{2}\text{erf}\left[\frac{1}{\delta\sqrt{2}}\left(\sqrt{\frac{x-\alpha}{\beta}}-\sqrt{\frac{\beta}{x-\alpha}}\right)\right] \tag{33.4}$$

Fitting to the C_4 in propane data gives values for α, β and k as 0.603, 3.67 and 0.58 respectively. *RSS* is 0.0304 – a very close fit. The equations below give the mean as 4.89 and the standard deviation as 2.55.

$$\mu=\alpha+\beta\left(1+\frac{\delta^2}{2}\right) \tag{33.5}$$

$$\sigma^2=\delta^2\beta^2\left(1+\frac{5\delta^2}{4}\right) \tag{33.6}$$

Applying the Equations (33.7) and (33.8), Figure 33.3 shows the distribution can only represent positive skewness and leptokurtosis (with $\kappa>6$).

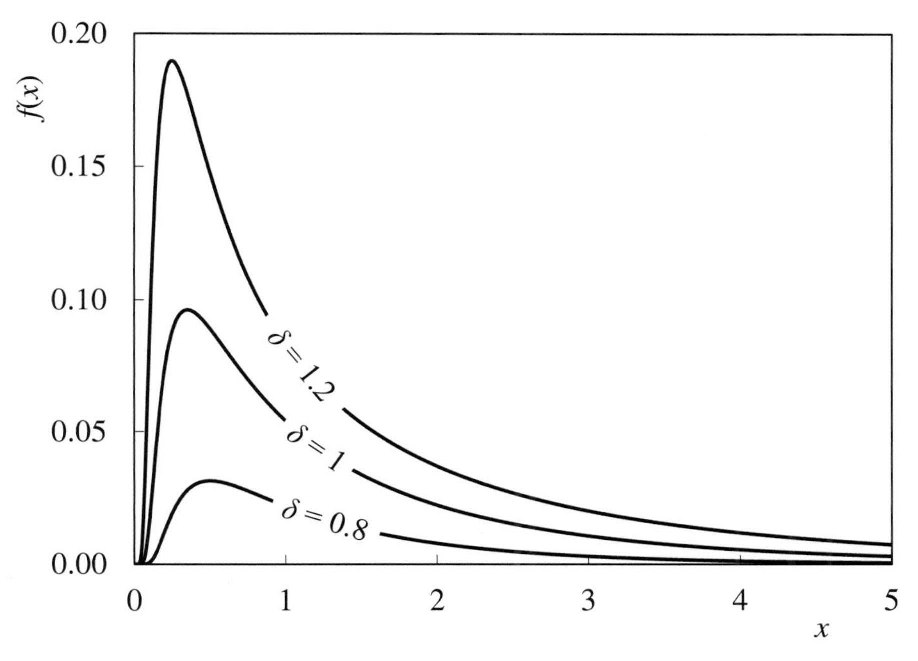

Figure 33.2 *Birnbaum–Saunders: Effect of δ on shape*

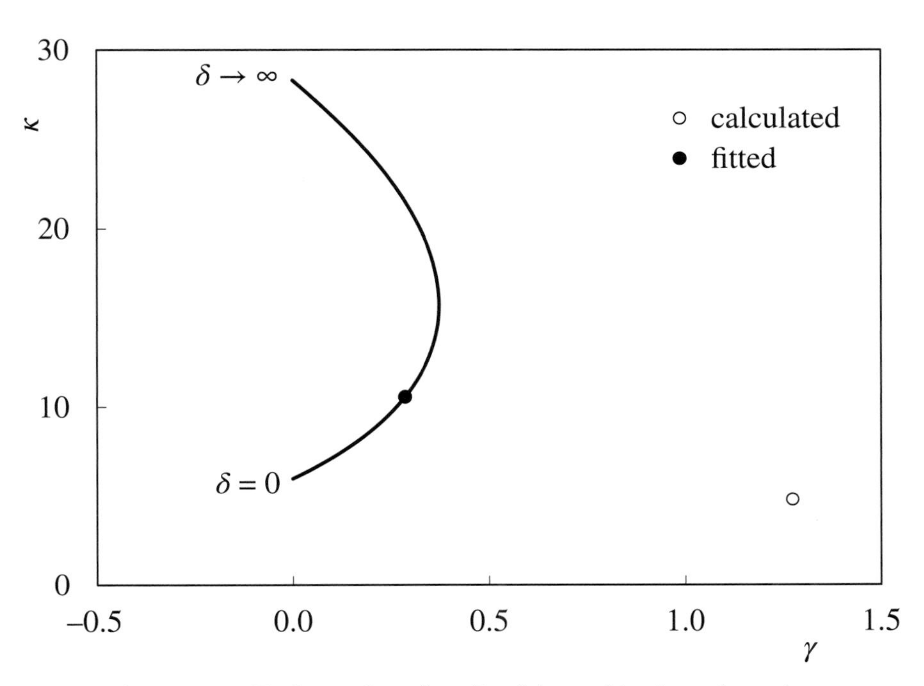

Figure 33.3 *Birnbaum–Saunders: Feasible combinations of γ and κ*

$$\gamma = \frac{16\delta^2\left(11\delta^2+6\right)}{\left(5\delta^2+4\right)^3} \tag{33.7}$$

$$\kappa = \frac{6\left(118\delta^4+81\delta^2+16\right)}{\left(5\delta^2+4\right)^2} \tag{33.8}$$

33.3 Bradford

The PDF of the *Bradford distribution* is

$$f(x) = \frac{\left(\frac{\lambda}{x_{max}-x_{min}}\right)}{\left[\lambda\left(\frac{x-x_{min}}{x_{max}-x_{min}}\right)+1\right]\ln(\lambda+1)} \qquad x_{min} \leq x \leq x_{max};\ \lambda > -1 \tag{33.9}$$

With x_{min} set to 0 and x_{max} to 1, Figure 33.4 shows the effect of changing λ. Setting it to zero gives the uniform distribution. The CDF is

$$F(x) = \frac{\ln\left[\frac{\lambda(x-x_{min})}{x_{max}-x_{min}}+1\right]}{\ln(\lambda+1)} \tag{33.10}$$

Choosing x_{min} as 0 and x_{max} as 10, fitting to the NHV disturbance data gives a value of 131 for λ. With *RSS* at 1.55, the Bradford distribution cannot be adjusted to fit the data well.

The mean and variance are

$$\mu = \frac{\lambda(x_{max}-x_{min})+\ln(\lambda+1)\{x_{min}(\lambda+1)-x_{max}\}}{\lambda\ln(\lambda+1)} \tag{33.11}$$

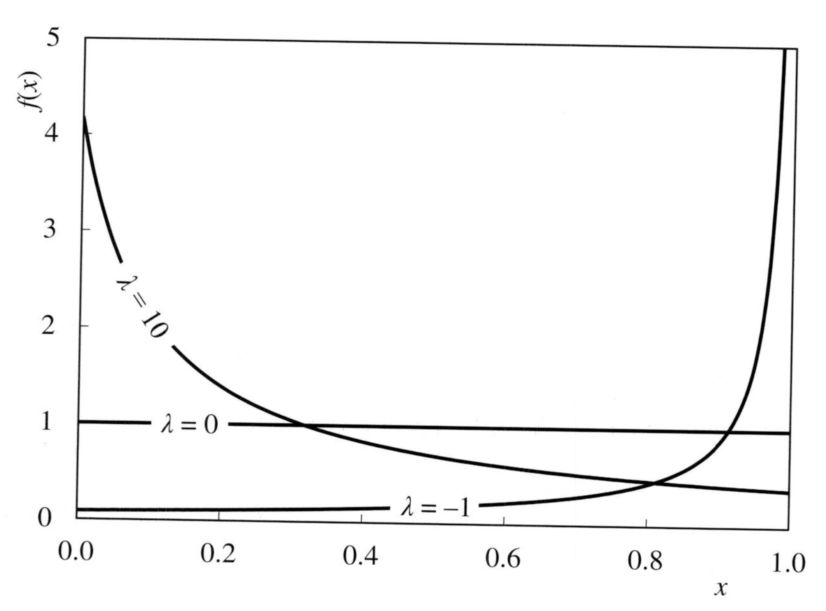

Figure 33.4 Bradford: Effect of λ on shape

$$\sigma^2 = \frac{(x_{max} - x_{min})^2 \{(\lambda+2)\ln(\lambda+1) - 2\lambda\}}{2\lambda \ln^2(\lambda+1)} \tag{33.12}$$

There are formulae published for skewness and kurtosis, but which are far too large to include here.

33.4 Champernowne

The *Champernowne distribution* describes two different distributions. The PDF of the first is

$$f(x) = \frac{\delta_1 \delta_2^{\delta_1} x^{\delta_1 - 1}}{\left(x^{\delta_1} + \delta_2^{\delta_1}\right)^2} \qquad x > 0;\ \delta_1, \delta_2 > 0 \tag{33.13}$$

Figure 33.5 shows the effect of varying δ_1 and δ_2.

The CDF is

$$F(x) = \frac{x^{\delta_1}}{x^{\delta_1} + \delta_2^{\delta_1}} \tag{33.14}$$

Fitting to the C_4 in propane data gives δ_1 as 3.39 and δ_2 as 4.30. *RSS* is poor at 0.0731. There are no formulae published for mean, variance, skewness or kurtosis.

The PDF of the second form of the distribution is

$$f(x) = \frac{\delta_1 \sin(\delta_2)}{\pi(x-\alpha)\left[\left(\frac{x-\alpha}{\beta}\right)^{-\delta_1} + \left(\frac{x-\alpha}{\beta}\right)^{\delta_1} + 2\cos(\delta_2)\right]} \qquad x > 0;\ \beta, \delta_1 > 0;\ 0 < \delta_2 < \pi \tag{33.15}$$

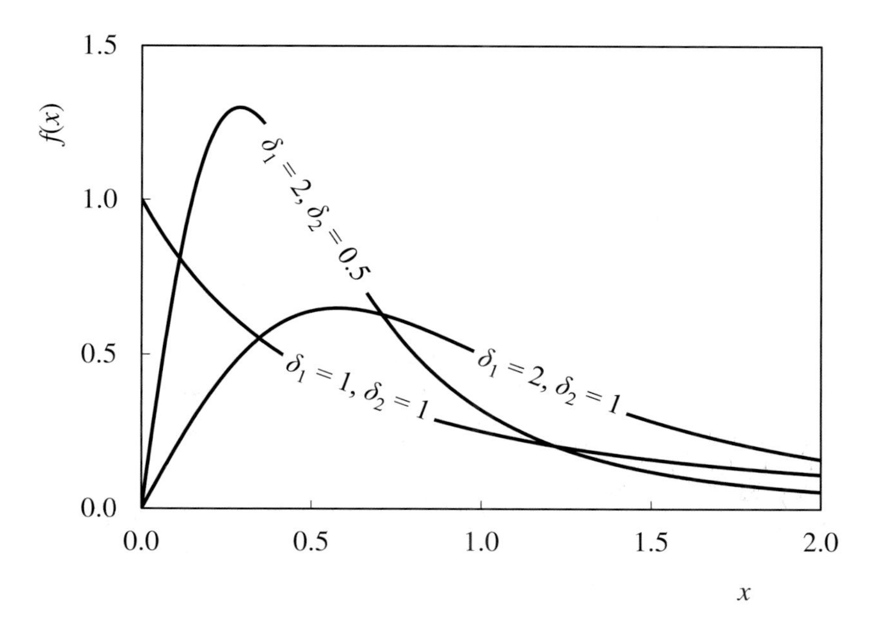

Figure 33.5 Champernowne (1): Effect of δ_1 and δ_2 on shape

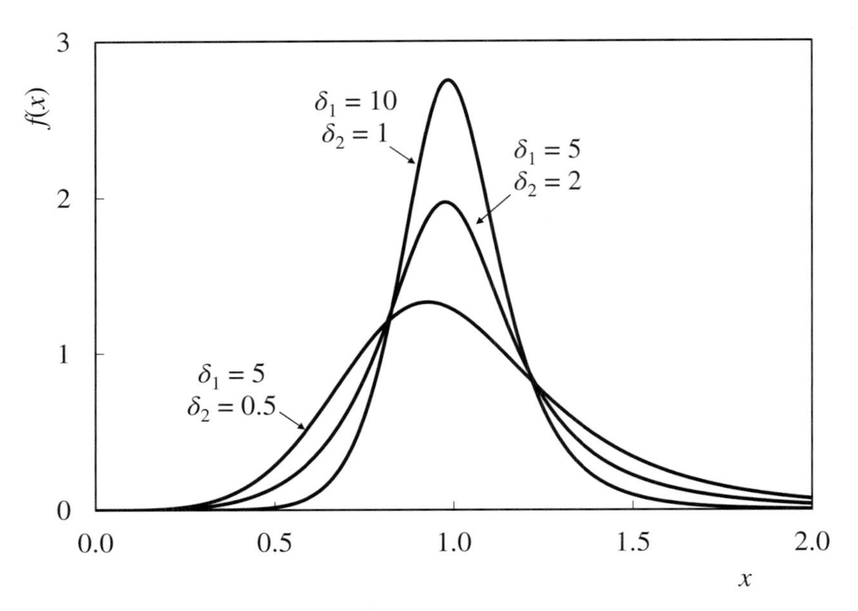

Figure 33.6 Champernowne (2): Effect of δ_1 and δ_2 on shape

Figure 33.6 shows the effect of varying δ_1 and δ_2 with α fixed at 0 and with β at 1.

Fitting to the C_4 in propane data gives α as 1.41, β as 3.00, δ_1 as 2.239, δ_2 as small as possible (set to 0.001) and $F(x_1)$ as 0.0303. *RSS* is mediocre at 0.0441.

Raw moments are given by

$$m_n = \frac{\pi \sin\left(n\frac{\delta_2}{\delta_1}\right)}{\delta_2 \sin\left(n\frac{\pi}{\delta_1}\right)}\beta^n \quad \delta_1 > n \tag{33.16}$$

This gives the mean as 5.69 and the standard deviation as 7.80 – both much larger than the values calculated from the data. Because δ_1 is less than 3 neither skewness nor kurtosis can be calculated.

Figure 33.7 shows the feasible combinations of skewness and kurtosis.

33.5 Davis

The *Davis distribution* is described by the PDF

$$f(x) = \frac{\beta^\delta}{\Gamma(\delta)\zeta(\delta)(x-\alpha)^{\delta+1}}\left[\exp\left(\frac{\beta}{x-\alpha}\right)-1\right]^{-1} \quad \beta,\delta > 0 \tag{33.17}$$

It includes the Riemann zeta function (ζ) as described in Section 11.10. This involves a sum to infinity that only converges if δ is greater than 1. Figure 33.8 shows the effect of varying δ, with α fixed at 0 and β at 1.

Fitting the distribution to the C_4 in propane data gives values of 6.80 for δ, –0.930 for α and 32.9 for β. At 0.0445, *RSS* is bettered by several other, easier to apply, distributions.

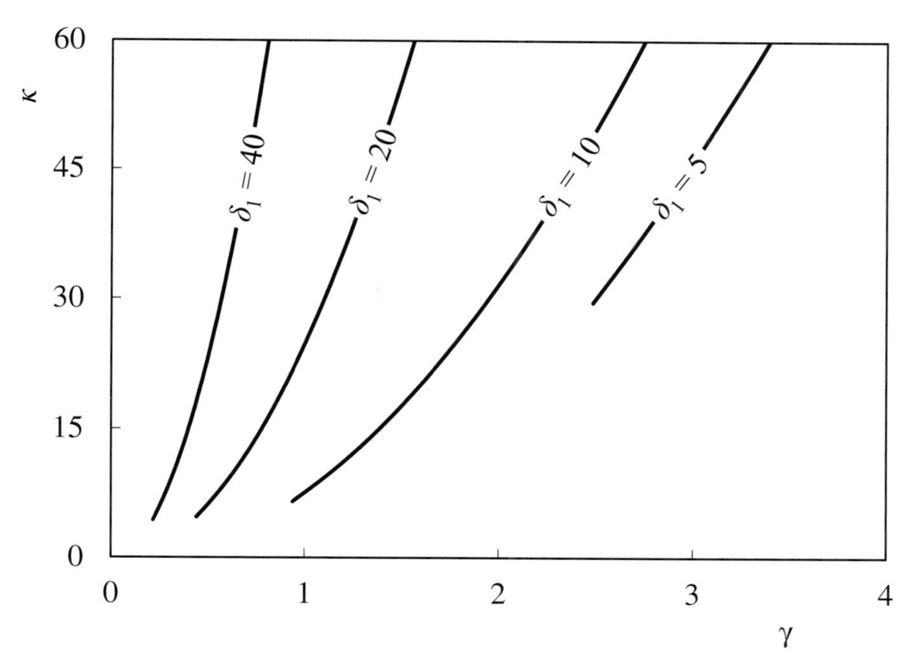

Figure 33.7 Champernowne (2): Feasible combinations of γ and κ

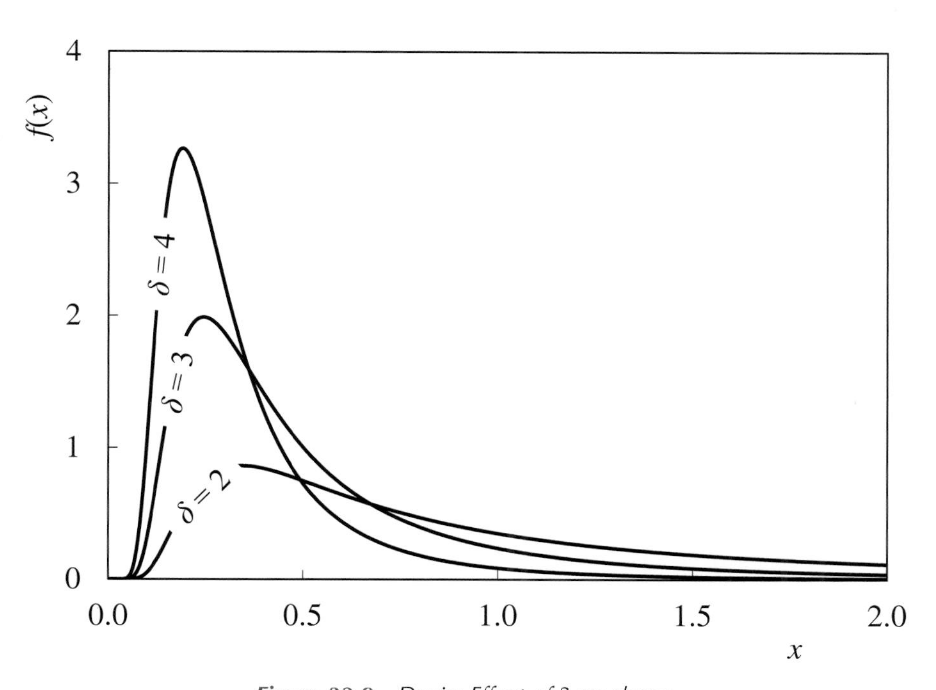

Figure 33.8 Davis: Effect of δ on shape

The mean and variance are

$$\mu = \alpha + \frac{\zeta(\delta-1)}{(\delta-1)\zeta(\delta)}\beta \quad \delta > 2 \tag{33.18}$$

$$\sigma^2 = \frac{(\delta-1)\zeta(\delta-2)\zeta(\delta) - (\delta-2)\zeta^2(\delta-1)}{(\delta-2)(\delta-1)^2\zeta^2(\delta)}\beta^2 \quad \delta > 3 \tag{33.19}$$

These give μ as 4.81 and σ as 2.71.

A special case of the Davis distribution is Planck's Law. This predicts the frequency spectrum of thermal radiation from a body at a given temperature. While expressed in terms of the energy emitted at each frequency, rather than probability, such a spectrum is analogous to a PDF.

33.6 Fréchet

The *Fréchet distribution* is also known as the *inverse Weibull distribution*, the *log-Gompertz distribution* or the *Gumbel-II distribution*. It is described by the PDF

$$f(x) = \frac{\delta}{\beta}\left(\frac{\beta}{x-\alpha}\right)^{\delta+1} \exp\left[-\left(\frac{\beta}{x-\alpha}\right)^{\delta}\right] \quad \beta, \delta > 0 \tag{33.20}$$

Figure 33.9 shows the effect of varying δ, keeping α at 0 and β at 1.

The CDF is

$$F(x) = \exp\left[-\left(\frac{\beta}{x-\alpha}\right)^{\delta}\right] \tag{33.21}$$

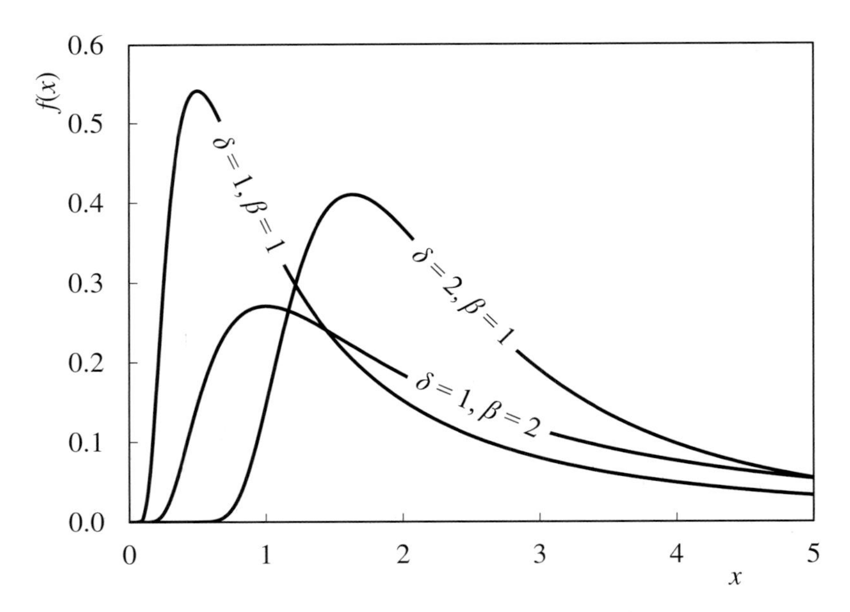

Figure 33.9 *Fréchet: Effect of δ and β on shape*

Fitting the distribution to the C_4 in propane data gives values of 5.09 for δ, −4.87 for α and 8.50 for β. At 0.0391, *RSS* is comparable with many of the well-fitting distributions.

The CDF can be inverted to

$$x(F)=\alpha+\beta[-\ln(F)]^{-1/\delta} \tag{33.22}$$

Raw moments are given by

$$m_n=\Gamma\left(1-\frac{n}{\delta}\right)\beta^n \qquad \delta>n \tag{33.23}$$

And so

$$\mu=\alpha+\beta\Gamma\left(1-\frac{1}{\delta}\right) \qquad \delta>1 \tag{33.24}$$

$$\sigma^2=\beta^2\left[\Gamma\left(1-\frac{2}{\delta}\right)-\Gamma^2\left(1-\frac{1}{\delta}\right)\right] \qquad \delta>2 \tag{33.25}$$

From these equations μ is estimated as 4.99 and σ as 3.03. As Figure 33.10 shows, the fitted skewness and kurtosis are much greater than the calculated values. As Figure 33.11 shows, this arises because the fitted distribution has a longer tail than the data. This would also suggest the fitted standard deviation is an overestimate.

Its long tail is applicable to extreme value analysis that we covered in Chapter 13. It is therefore also known as the *extreme value-II distribution*.

This distribution is not to be confused with the *reverse Weibull distribution*. This is the survival function from Equation (12.72) and is also used in extreme value analysis.

$$S(x)=1-F(x)=\exp\left[-\left(\frac{x-\alpha}{\beta}\right)^{\delta}\right] \tag{33.26}$$

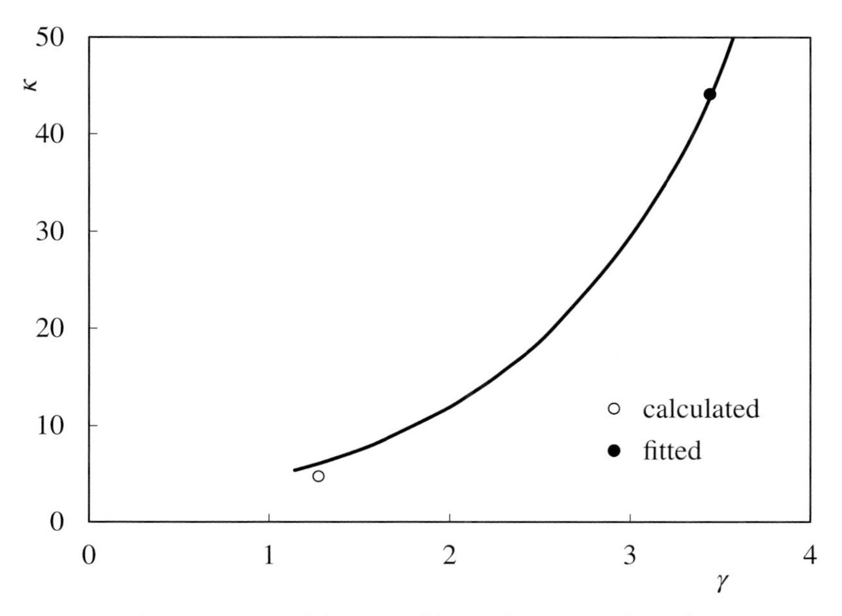

Figure 33.10 *Fréchet: Feasible combinations of γ and κ*

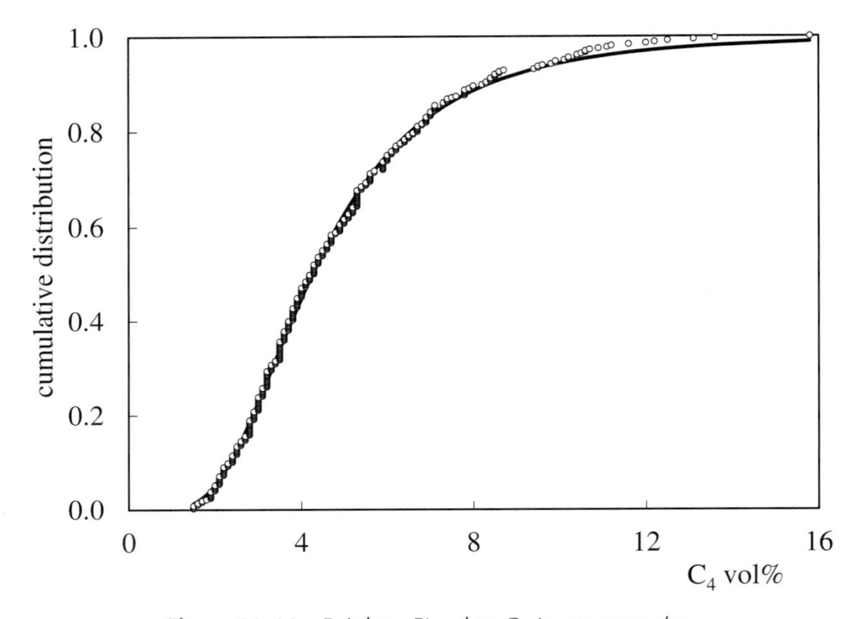

Figure 33.11 Fréchet: Fitted to C_4 in propane data

33.7 Gompertz

The *Gompertz distribution* is described by the PDF

$$f(x) = \alpha \exp(\beta x) \exp\left[-\frac{\alpha}{\beta}[\exp(\beta x) - 1]\right] \quad x \geq 0; \alpha > 0 \tag{33.27}$$

The effect of α and β is shown in Figure 33.12.

Its CDF is

$$F(x) = 1 - \exp\left\{-\frac{\alpha}{\beta}[\exp(\beta x) - 1]\right\} \tag{33.28}$$

Inversion gives the QF

$$x(F) = \frac{1}{\beta} \ln\left[1 - \frac{\beta}{\alpha} \ln(1 - F)\right] \quad 0 \leq F \leq 1 \tag{33.29}$$

Fitting the distribution to the C_4 data gives values for α and β of 0.0482 and 0.436 respectively. With *RSS* at 0.9764, the fit is poor.

Fitting to the absolute values of the NHV disturbances gives α as 32.3 and β as 0.0383, with *RSS* at 0.1074.

The Gompertz distribution is also published in the form

$$f(x) = \alpha \delta^x \exp\left[-\frac{\alpha(\delta^x - 1)}{\ln(\delta)}\right] \quad x \geq 0; \delta > 0 \tag{33.30}$$

$$F(x) = 1 - \exp\left[-\frac{\alpha(\delta^x - 1)}{\ln(\delta)}\right] \tag{33.31}$$

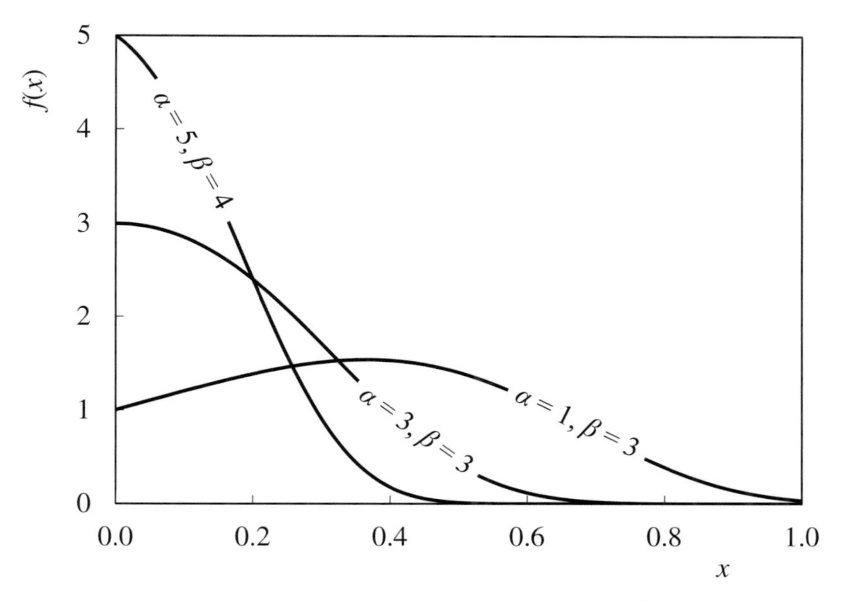

Figure 33.12 Gompertz: Effect of α and β on shape

At first glance this might appear to be a quite different distribution. Understandably the engineer might include them both in the possible prior distributions fitted to the data. Both will give exactly the same *RSS*, giving a very strong indication that they are identical. Indeed, this can often be the easiest way of checking whether distributions sharing the same name are in fact the same. Identical *RSS* would then prompt the engineer to look more closely at the parameters of each CDF to see if those in one can be mapped into those of the other. In this case, β in Equations (33.27) and (33.28) is replaced with $\ln(\delta)$.

$$\ln[\exp(\beta x)] = \beta x = x\ln(\delta) = \ln(\delta^x) \quad \therefore \exp(\beta x) = \delta^x \tag{33.32}$$

The same substitution in Equation (33.29) gives the QF

$$x(F) = \frac{1}{\ln(\delta)}\ln\left[1 - \frac{\ln(\delta)}{\alpha}\ln(1-F)\right] \quad 0 \le F \le 1 \tag{33.33}$$

33.8 Shifted Gompertz

The *shifted Gompertz distribution* is not 'shifted' in the true sense of the word. If it were, then it should be possible, by setting a parameter (such as α), to have it revert to the classic Gompertz distribution.

It is defined by

$$f(x) = \beta\exp[-\beta x - \alpha\exp(-\beta x)]\{1 + \alpha[1 - \exp(-\beta x)]\} \quad x \ge 0;\ \alpha, \beta > 0 \tag{33.34}$$

$$F(x) = \exp[-\alpha\exp(-\beta x)][1 - \exp(-\beta x)] \tag{33.35}$$

Figure 33.13 shows the effect of changing α, with β fixed at 1. Fitting Equation (33.35) to the C_4 in propane data gives α as 6.50 and β as 0.550, with *RSS* at 0.1240.

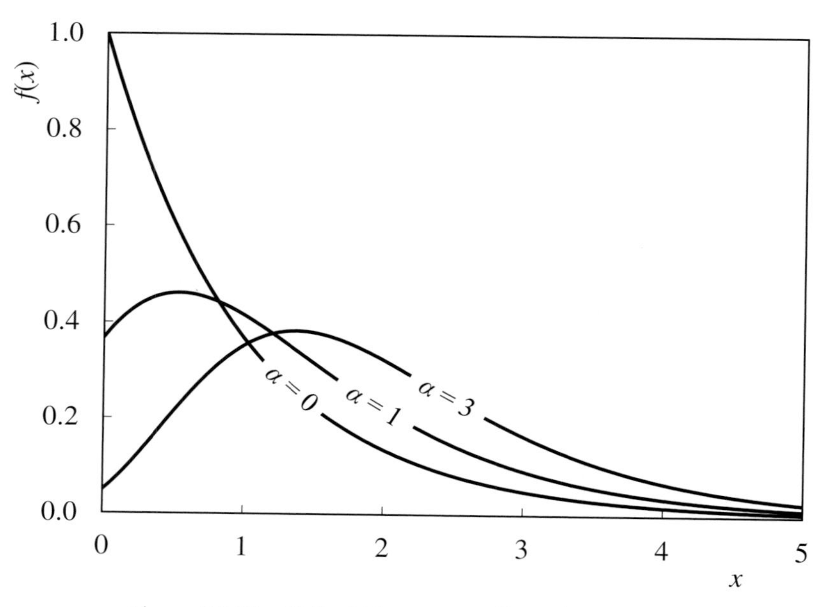

Figure 33.13 Shifted Gompertz: Effect of α on shape

Fitting to the absolute values of the NHV disturbances gives α as −0.862 and β as 0.412, with *RSS* at 0.0718.

33.9 Gompertz–Makeham

The *Gompertz–Makeham distribution*, or sometimes simply the *Makeham distribution*, includes an additional parameter (λ).

$$f(x) = \left(\alpha e^{\beta x} + \lambda\right)\exp\left\{-\frac{\alpha}{\beta}[\exp(\beta x) - 1] - \lambda x\right\} \quad x \geq 0;\ \lambda > 0 \tag{33.36}$$

$$F(x) = 1 - \exp\left\{-\frac{\alpha}{\beta}[\exp(\beta x) - 1] - \lambda x\right\} \tag{33.37}$$

Figure 33.14 illustrates some of what distributions can be represented. Fitting to the C_4 in propane data gives values for α, β and λ of −0.758, −0.227 and 0.643 respectively, with *RSS* at 0.0867.

Fitting to the absolute values of the NHV disturbances gives α as 1.65, β as −0.0798 and λ as −0.677. With *RSS* at 0.0682 this distribution is the best of the Gompertz group.

As with the Gompertz distribution, replacing β with $\ln(\delta)$ gives a different looking version of the same distribution.

$$f(x) = (\alpha\delta^x + \lambda)\exp\left[-\frac{\alpha}{\ln(\delta)}(\delta^x - 1) - \lambda x\right] \tag{33.38}$$

$$F(x) = 1 - \exp\left[-\frac{\alpha}{\ln(\delta)}(\delta^x - 1) - \lambda x\right] \tag{33.39}$$

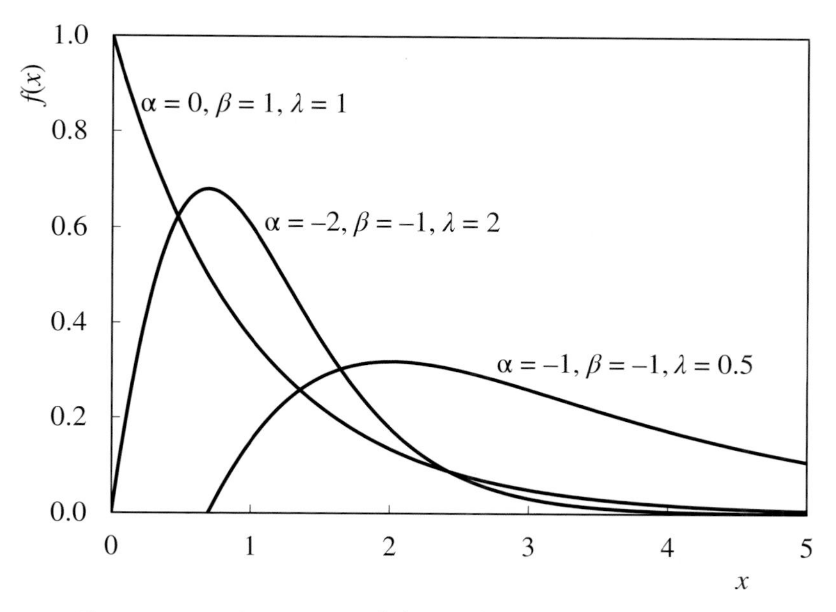

Figure 33.14 *Gompertz–Makeham: Effect of α, β and λ on shape*

33.10 Gamma-Gompertz

The *gamma-Gompertz distribution* is defined by

$$f(x) = \frac{\beta\delta\alpha^{\delta}\exp(\beta x)}{[\alpha - 1 + \exp(\beta x)]^{\delta+1}} \qquad x \geq 0;\ \alpha,\beta,\delta > 0 \tag{33.40}$$

$$F(x) = 1 - \frac{\alpha^{\delta}}{[\alpha - 1 + \exp(\beta x)]^{\delta}} \tag{33.41}$$

Figure 33.15 illustrates some of what distributions can be represented. Of the four versions evaluated for the C_4 data, this distribution fits best. *RSS* is 0.0328 with α set to 229, β to 2.20 and δ to 0.177.

Fitting to the absolute values of the NHV disturbances gives α as 0.228, β as 0.0468 and λ at 4.77, with *RSS* at 0.0732.

For the C_4 in propane data, comparison of the four distributions is shown in Figure 33.16. For all of them, the calculation of mean, variance, skewness and kurtosis involve very complex formulae. Since several, easier to apply, distributions outperform the Gompertz distributions, it is unlikely that their use could be justified in the process industry.

33.11 Hyperbolic

In principle the *hyperbolic distribution* takes the shape of a hyperbola, although this is not obvious from its PDF. This includes the modified Bessel function of the second kind (K), as described in Section 11.8.

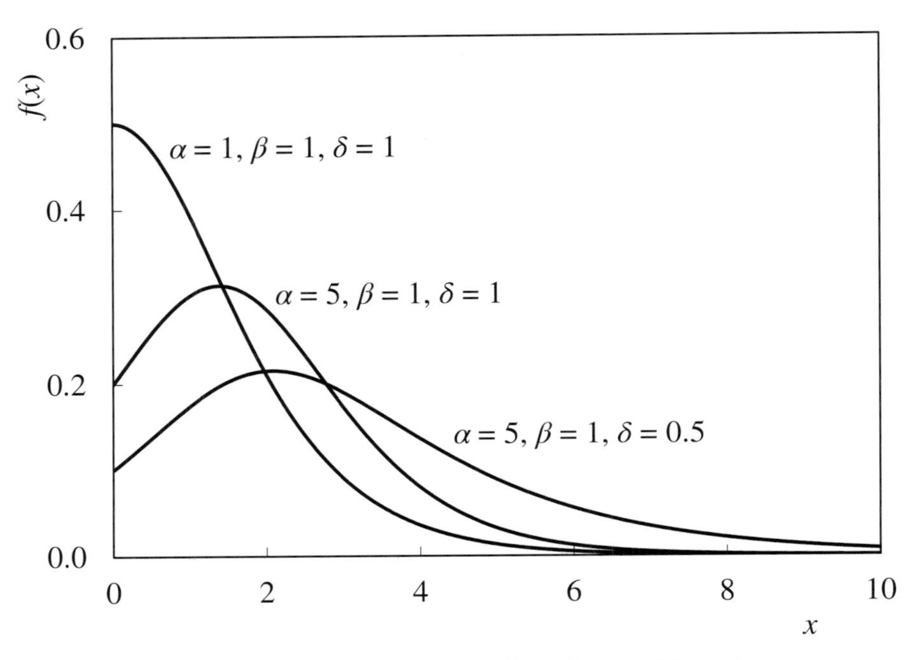

Figure 33.15 Gamma-Gompertz: Effect of α, β and δ on shape

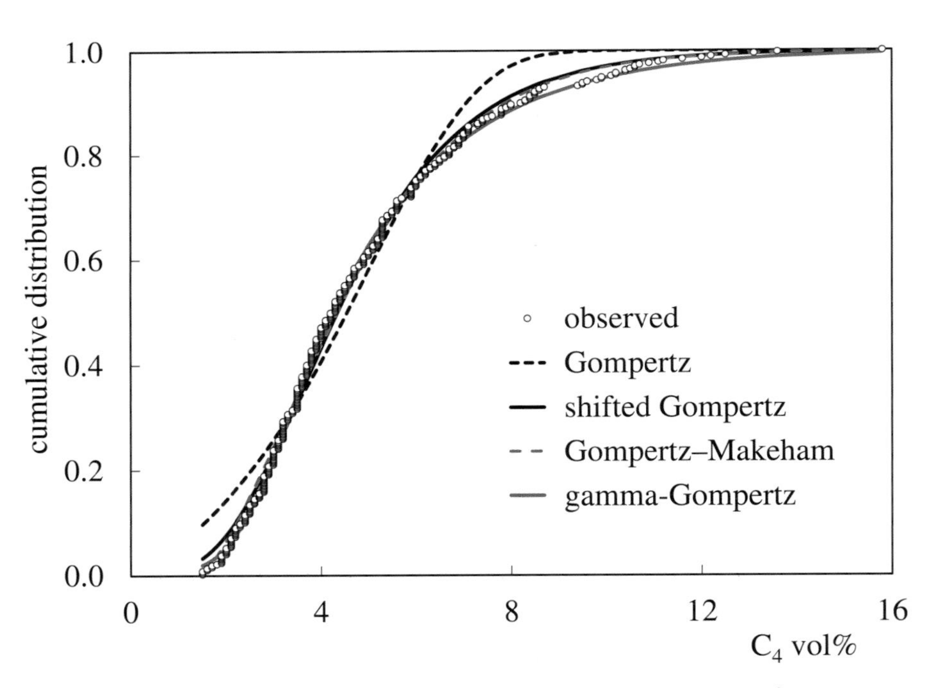

Figure 33.16 Gompertz: Comparison of fits to C_4 in propane data

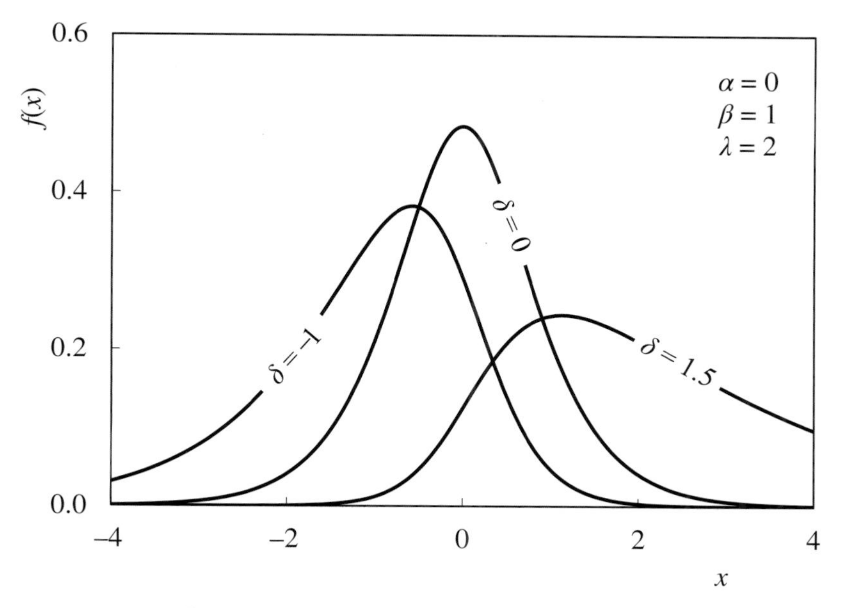

Figure 33.17 *Hyperbolic: Effect of δ on shape*

$$f(x) = \frac{\sqrt{\lambda^2 - \beta^2}}{2\lambda\beta \mathrm{K}_1\left(\beta\sqrt{\lambda^2 - \beta^2}\right)} \exp\left[\delta(x-\alpha) - \lambda\sqrt{\beta^2 + (x-\alpha)^2}\right] \tag{33.42}$$

Figure 33.17 shows how the symmetry parameter (δ) affects skewness. Figure 33.18 shows how λ affects kurtosis.

Fitting to the C_4 in propane data gives α as 2.25, β as 0.744, λ as 1.74, δ as 1.33 and $F(x_1)$ as 0.0182. *RSS* is 0.0279 – the best achieved by any distribution.

The mean and variance are

$$\mu = \alpha + \frac{\beta\delta \mathrm{K}_2\left(\beta\sqrt{\lambda^2 - \delta^2}\right)}{\sqrt{\lambda^2 - \delta^2}\,\mathrm{K}_1\left(\beta\sqrt{\lambda^2 - \delta^2}\right)} \tag{33.43}$$

$$\sigma^2 = \frac{\beta \mathrm{K}_2\left(\beta\sqrt{\lambda^2 - \delta^2}\right)}{\sqrt{\lambda^2 - \delta^2}\,\mathrm{K}_1\left(\beta\sqrt{\lambda^2 - \delta^2}\right)} + \frac{\delta^2\beta^2}{\lambda^2 - \delta^2}\left(\frac{\mathrm{K}_3\left(\beta\sqrt{\lambda^2 - \delta^2}\right)}{\mathrm{K}_1\left(\beta\sqrt{\lambda^2 - \delta^2}\right)} - \frac{\mathrm{K}_2^2\left(\beta\sqrt{\lambda^2 - \delta^2}\right)}{\mathrm{K}_1^2\left(\beta\sqrt{\lambda^2 - \delta^2}\right)}\right) \tag{33.44}$$

These give μ as 4.97 and σ as 2.65 – close to the values determined by calculation from the data.

Fitting to the NHV disturbance data gives α as –0.28, β as 0.0001, λ as 0.970, δ as 0.207 and $F(x_1)$ as 0.0184. Although *RSS* is low at 0.0320, such a low value for β places doubt on the reliability of the fit. Variation in dispersion (quantified by β) can be difficult to distinguish from variation in kurtosis (quantified by λ). The NHV disturbance data is extremely leptokurtic – perhaps aggravating the problem. The estimates of μ as 0.18 and σ as 1.56 are somewhat different from the values calculated from the data – placing further suspicion on the result. This again frustrates the aim of identifying a distribution that works well for all datasets.

The formulae for skewness and kurtosis are too long to be included here but can be derived from the moment generating function.

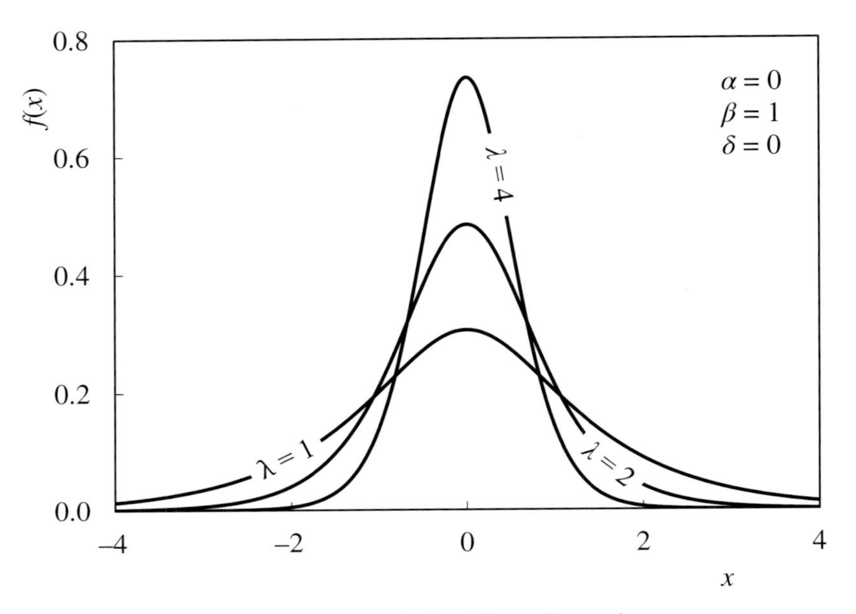

Figure 33.18 Hyperbolic: Effect of λ on shape

$$M(t) = \frac{\sqrt{\lambda^2 - \delta^2}\,\mathrm{K}_1\left(\beta\sqrt{\lambda^2 - (\delta + t)^2}\right)}{\sqrt{\left(\lambda^2 - (\delta + t)^2\mathrm{K}_1\left(\beta\sqrt{\lambda^2 - \delta^2}\right)\exp(\alpha t)\right.}} \tag{33.45}$$

The *generalised hyperbolic distribution* also includes the Bessel function but of an order that need not be an integer. This is not supported by most spreadsheet packages.

33.12 Asymmetric Laplace

The *asymmetric Laplace distribution* is derived from the Laplace distribution described in Section 32.7. It is defined by the PDF

$$f(x) = \frac{\delta}{\beta(\delta^2 + 1)}\exp\left(\frac{x - \alpha}{\beta\delta}\right) \quad x < \alpha;\ \beta, \delta > 0 \tag{33.46}$$

$$f(x) = \frac{\delta}{\beta(\delta^2 + 1)}\exp\left(-\frac{\delta(x - \alpha)}{\beta}\right) \quad x \geq \alpha;\ \beta, \delta > 0 \tag{33.47}$$

Figure 33.19 shows the effect, with α and β constant, of varying δ. Values of δ between 0 and 1 give a positive skewness; values greater than 1 give a negative skewness.

The CDF is

$$F(x) = \frac{\delta^2}{\delta^2 + 1}\exp\left(\frac{x - \alpha}{\beta\delta}\right) \quad x < \alpha \tag{33.48}$$

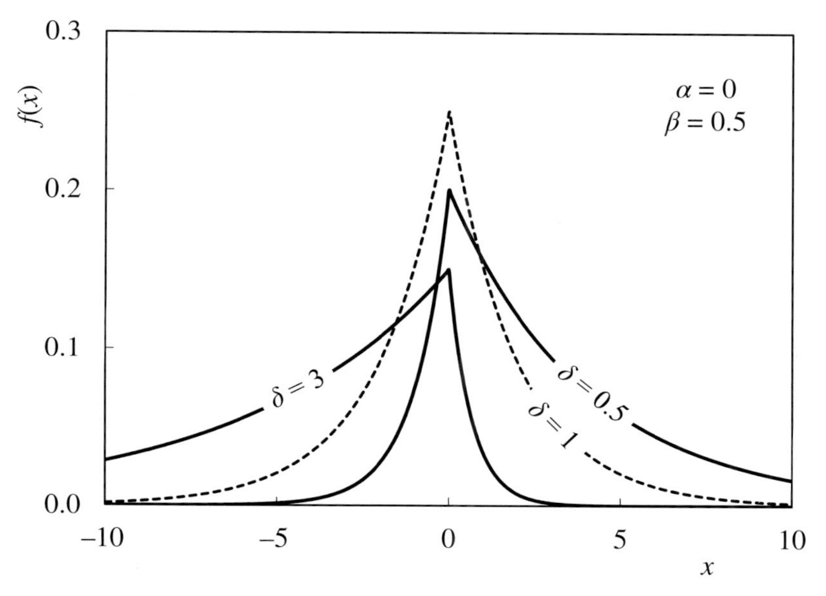

Figure 33.19 *Asymmetric Laplace: Effect of δ on shape*

$$F(x) = 1 - \frac{1}{\delta^2 + 1}\exp\left(-\frac{\delta(x-\alpha)}{\beta}\right) \quad x \ge \alpha \tag{33.49}$$

Fitting to NHV disturbance data gives α as –0.203, β as 1.08 and δ as 0.891. *RSS* is 0.0380, showing that the fit is one of the best.

The mean and variance are

$$\mu = \alpha + \frac{1-\delta^2}{\delta}\beta \tag{33.50}$$

$$\sigma^2 = \frac{\delta^4 + 1}{\delta^2}\beta^2 \tag{33.51}$$

These give values of 0.05 for μ and 1.54 for σ.

Skewness and kurtosis are

$$\gamma = \frac{2\left(1-\delta^6\right)}{\left(\delta^4+1\right)^{3/2}} \tag{33.52}$$

$$\kappa = \frac{3\left(3\delta^8 + 2\delta^4 + 3\right)}{\left(\delta^4+1\right)^2} \tag{33.53}$$

This gives γ as 0.48 and κ as 6.15, both close to the values calculated from the data as 0.62 and 5.87. As Figure 33.20 shows the distribution is restricted to relatively mild skewness and leptokurtosis.

A different presentation of the distribution results from defining β_1 as $\beta\delta$ and β_2 as β/δ. The equations above then become

$$f(x) = \frac{1}{\beta_1 + \beta_2}\exp\left(\frac{x-\alpha}{\beta_1}\right) \quad x < \alpha \tag{33.54}$$

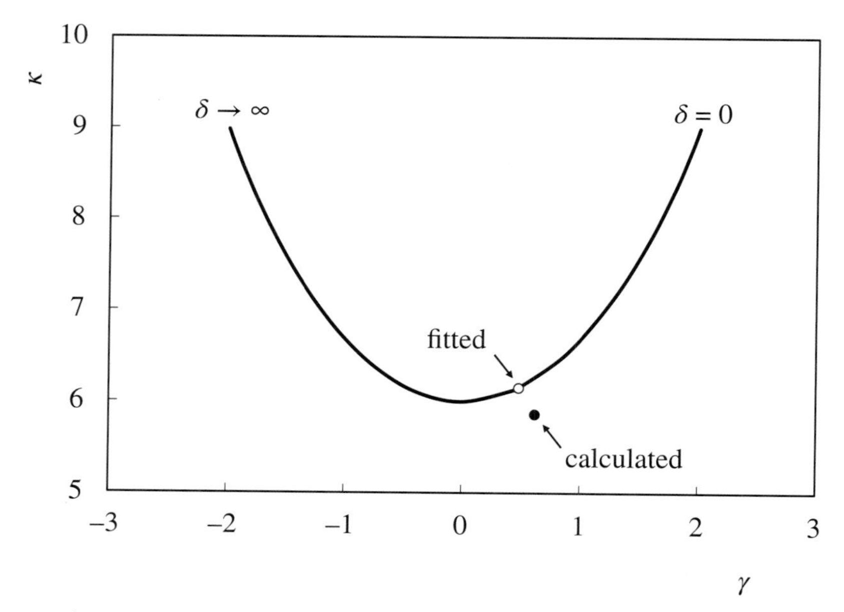

Figure 33.20 Asymmetric Laplace: Feasible combinations of γ and κ

$$f(x)=\frac{1}{\beta_1+\beta_2}\exp\left(-\frac{x-\alpha}{\beta_2}\right) \quad x\geq\alpha \tag{33.55}$$

$$F(x)=\frac{\beta_1}{\beta_1+\beta_2}\exp\left(\frac{x-\alpha}{\beta_1}\right) \quad x<\alpha \tag{33.56}$$

$$F(x)=1-\frac{\beta_2}{\beta_1+\beta_2}\exp\left(\frac{x-\alpha}{\beta_2}\right) \quad x\geq\alpha \tag{33.57}$$

$$\mu=\alpha+\beta_2-\beta_1 \tag{33.58}$$

$$\sigma^2=\beta_1^2+\beta_2^2 \tag{33.59}$$

$$\gamma=\frac{2\left(\beta_1^3-\beta_2^3\right)}{\left(\beta_1^2+\beta_2^2\right)^{3/2}} \tag{33.60}$$

$$\kappa=\frac{3\left(3\beta_1^4+2\beta_1^2\beta_2^2+3\beta_2^4\right)}{\left(\beta_1^2+\beta_2^2\right)^2} \tag{33.61}$$

33.13 Log-Laplace

As with many other distributions, we can replace x with $\ln(x)$ to produce the first of two distributions entitled the *log-Laplace distribution*. While the distribution is symmetrical in $\ln(x)$, it is not so in x, and so might be considered for skewed data. Its PDF is

$$f(x)=\frac{1}{2\beta x}\exp\left(\frac{-|\ln(x)-\alpha|}{\beta}\right) \quad x\geq0;\ \beta>0 \tag{33.62}$$

Its CDF is

$$F(x)=\frac{1}{2}\left[1+\frac{|\ln(x)-\alpha|}{\ln(x)-\alpha}\right]\left[1-\exp\left(\frac{-|\ln(x)-\alpha|}{\beta}\right)\right] \tag{33.63}$$

Figure 33.21 shows the effect of varying α and β. As a lower-bounded distribution it cannot be applied to process disturbances. Because of the discontinuity at $x=e^{\alpha}$, the distribution is unlikely to represent any other set of process data.

Including the same skewness parameter (δ) as used in the asymmetric Laplace distribution gives the *asymmetric log-Laplace distribution*, which is similarly unlikely to be applicable to the process industry.

Another distribution also bears the log-Laplace title and is also known as the *log-double exponential distribution*. Its PDF is

$$f(x)=\frac{\delta}{2\beta}\left(\frac{x-\alpha}{\beta}\right)^{\delta-1} \qquad 0<x<\alpha;\ \beta,\delta>0 \tag{33.64}$$

$$f(x)=\frac{\delta}{2\beta}\left(\frac{x-\alpha}{\beta}\right)^{-\delta-1} \qquad x\geq\alpha;\ \beta,\delta>0 \tag{33.65}$$

If δ is 1, the part of the distribution where x is less than α becomes the uniform distribution. If set to 2 it becomes the triangular distribution. Figure 33.22 shows the effect of increasing δ above this value, with α fixed at 0 and β at 1. The distribution becomes increasingly symmetrical as δ is increased further.

Its CDF is

$$F(x)=\frac{1}{2}\left(\frac{x-\alpha}{\beta}\right)^{\delta} \quad \text{for } 0<x<\alpha \tag{33.66}$$

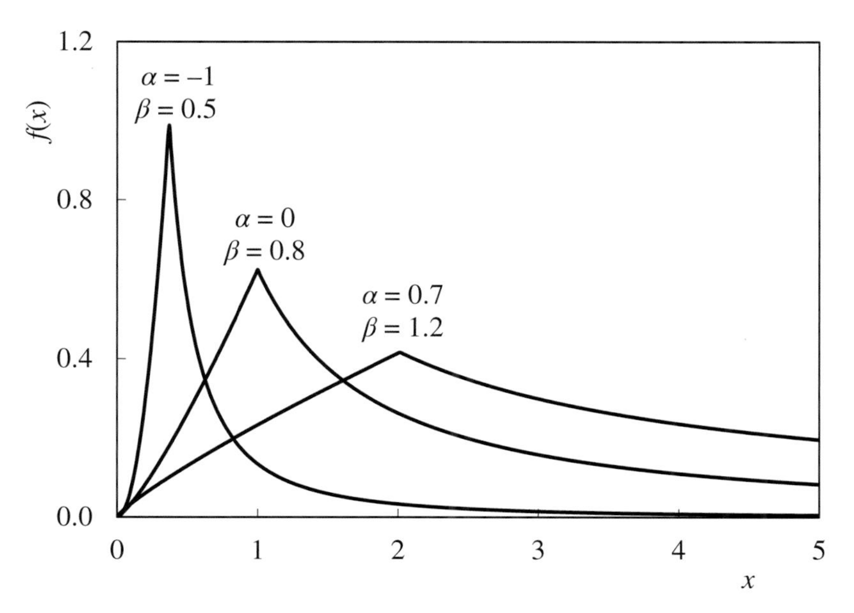

Figure 33.21 *Log-Laplace (1): Effect of α and β on shape*

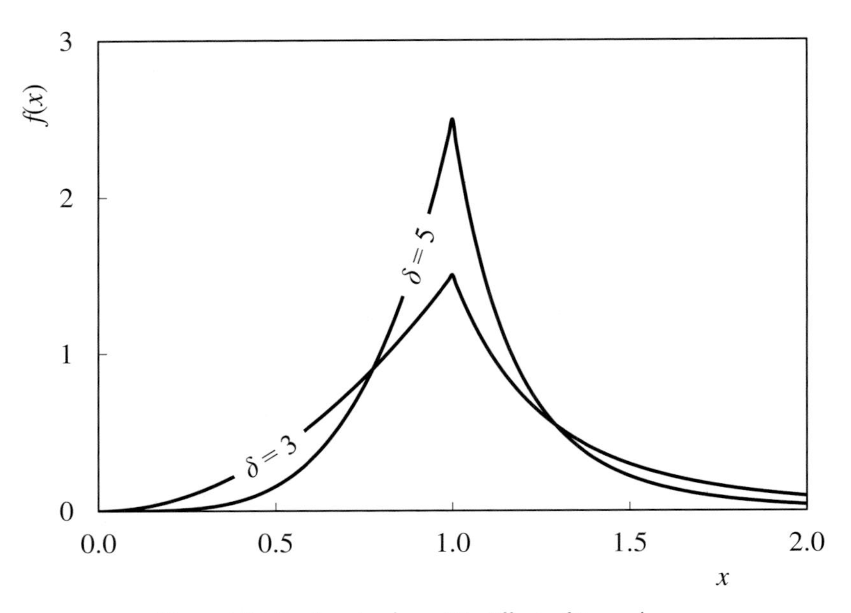

Figure 33.22 Log-Laplace (2): Effect of δ on shape

$$F(x) = 1 - \frac{1}{2}\left(\frac{x-\alpha}{\beta}\right)^{-\delta} \quad \text{for } x \geq \alpha \tag{33.67}$$

This version, too, probably has little application in the process industry.

33.14 Lindley

The *Lindley distribution*[24] has a number of enhanced versions. The basic PDF is

$$f(x) = \frac{\lambda^2}{\lambda+1}(x+1)e^{-\lambda x} \quad x \geq 0, \lambda > 0 \tag{33.68}$$

Figure 33.23 shows the effect of adjusting λ.

Its CDF is

$$F(x) = 1 - \left(\frac{\lambda x}{\lambda+1} + 1\right)e^{-\lambda x} \tag{33.69}$$

Fitting this to the C_4 in propane data gives λ as 0.326 and *RSS* as 3.78. Clearly the fit is very poor – primarily because there is only one parameter to adjust.

Raw moments are given by

$$m_n = \frac{n!(\lambda+n+1)}{\lambda^n(\lambda+1)} \tag{33.70}$$

The mean and variance are therefore

$$\mu = \frac{\lambda+2}{\lambda(\lambda+1)} \tag{33.71}$$

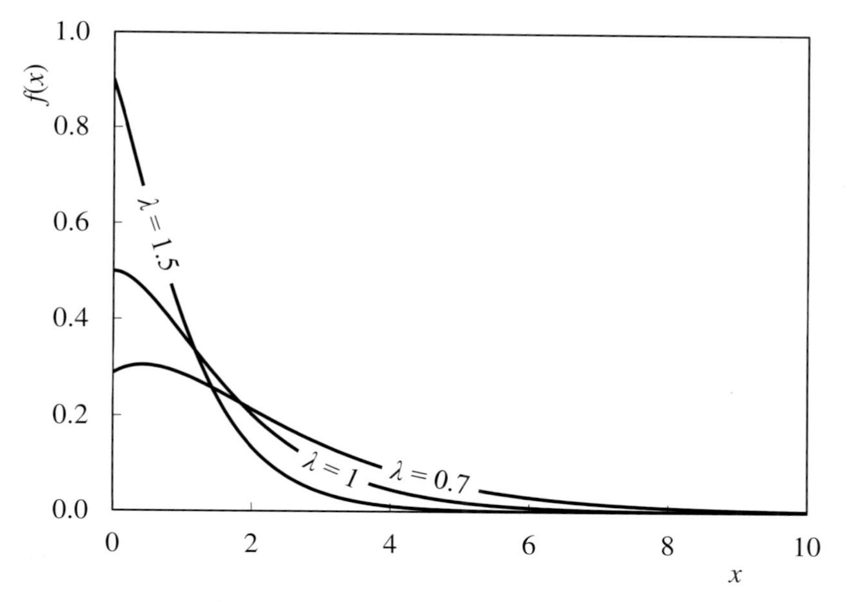

Figure 33.23 indley: Effect of λ on shape

$$\sigma^2 = \frac{\lambda^2 + 6\lambda + 2}{\lambda^2(\lambda+1)^2} \tag{33.72}$$

Figure 33.24 shows the feasible combinations of skewness and kurtosis.

33.15 Lindley-Geometric

One published modification to the Lindley distribution is the addition of another parameter (δ) to give the *Lindley-geometric distribution*.[25] Its PDF is

$$f(x) = \frac{(1-\delta)\dfrac{\lambda^2}{\lambda+1}(x+1)e^{-\lambda x}}{\left[1-\delta\left(\dfrac{\lambda x}{\lambda+1}+1\right)e^{-\lambda x}\right]^2} \qquad x \geq 0;\ \lambda > 0 \tag{33.73}$$

The effect of varying δ, with λ fixed at 1, is shown in Figure 33.25. If δ is set to 0 the PDF reverts to the classic Lindley distribution.

Its CDF is

$$F(x) = \frac{1-\left(\dfrac{\lambda x}{\lambda+1}+1\right)e^{-\lambda x}}{1-\delta\left(\dfrac{\lambda x}{\lambda+1}+1\right)e^{-\lambda x}} \tag{33.74}$$

Fitting to the C_4 in propane data gives 0.861 for λ and −13.1 for δ. The fit, although *RSS* improves to 0.5022, is still poor. There are published formulae for the mean and variance but these are extremely complex.

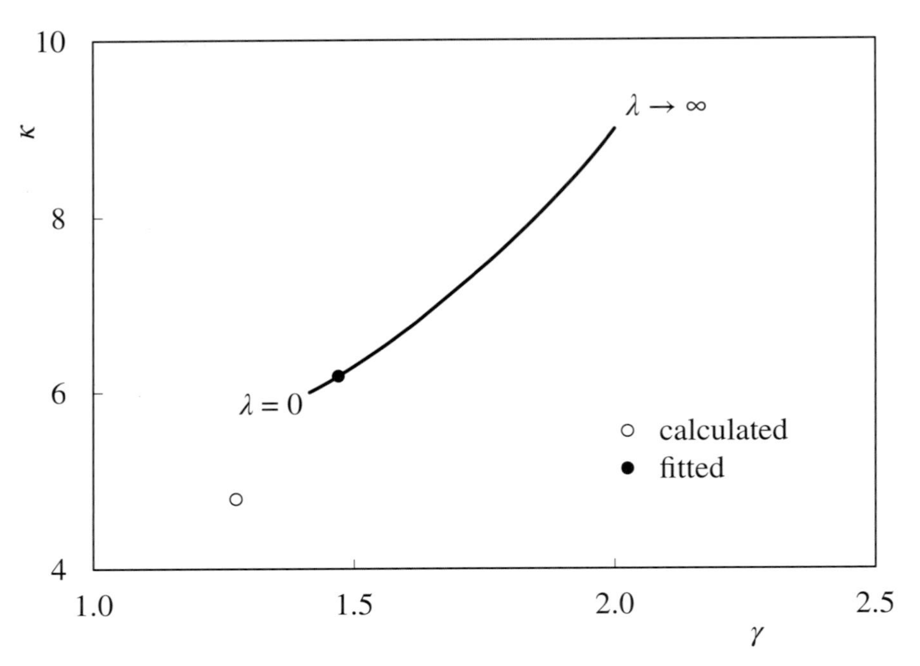

Figure 33.24 Lindley: Feasible combinations of γ and κ

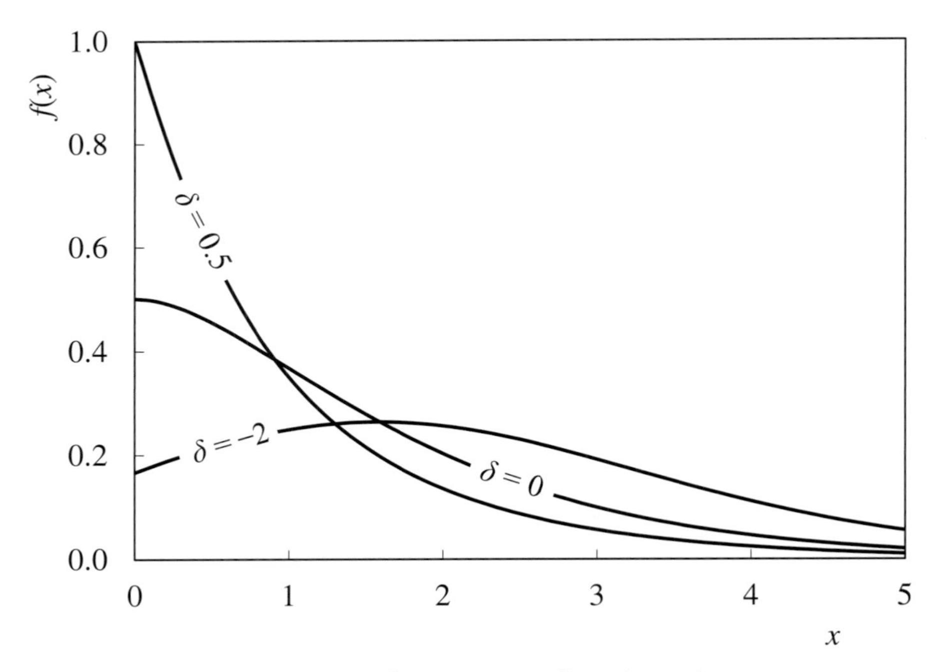

Figure 33.25 Lindley-geometric: Effect of δ on shape

33.16 Generalised Lindley

The PDF of the *generalised Lindley distribution*[26] is

$$f(x) = \frac{\lambda^{\delta_1+1}x^{\delta_1-1}}{\delta_1\Gamma(\delta_1)(\lambda+\delta_2)}(\delta_2 x+\delta_1)e^{-\lambda x} \quad x \geq 0;\ \lambda,\delta_1,\delta_2 > 0 \tag{33.75}$$

This is a compound of the classic Lindley and gamma distributions. Indeed if δ_2 is set to 0 it becomes the classic gamma distribution and if both δ_1 and δ_2 are set to 1 it becomes the classic Lindley distribution. The principle is that, rather than assume λ is a constant, it varies randomly following the gamma distribution.

The effect of varying δ_1 and δ_2, with λ fixed at 1, is shown in Figure 33.26.

There is no published CDF. Fitting the PDF gives values for λ, δ_1 and δ_2 as 0.988, 4.38 and −0.0003. With δ_2 so close to zero, the contribution by the Lindley distribution is negligible. We have effectively selected the gamma distribution. *RSS* is 0.0660, which is a further improvement but still does not approach the accuracy achieved by other distributions. Further there are no published methods of determining the mean, variance, skewness or kurtosis.

Figure 33.27 compares the fits of each of the Lindley distributions.

33.17 Mielke

More fully described as the *Mielke beta-kappa distribution*, its PDF is

$$f(x) = \frac{\frac{\delta_1}{\beta}\left(\frac{x-\alpha}{\beta}\right)^{\delta_1-1}}{\left[1+\left(\frac{x-\alpha}{\beta}\right)^{\delta_2}\right]^{\frac{\delta_1}{\delta_2}+1}} \quad x \geq \alpha;\ \beta,\delta_1,\delta_2 > 0 \tag{33.76}$$

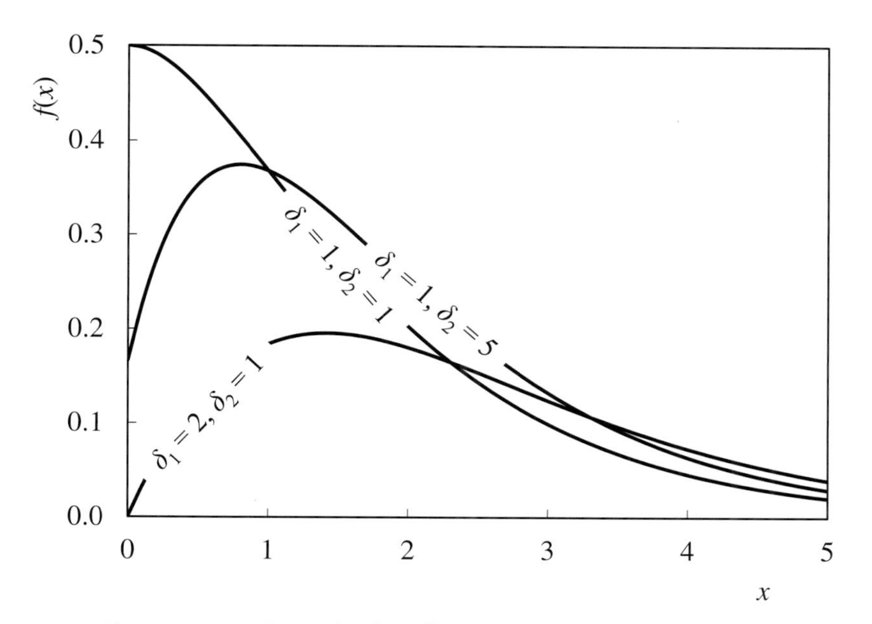

Figure 33.26 Generalised Lindley: Effect of δ_1 and δ_2 on shape

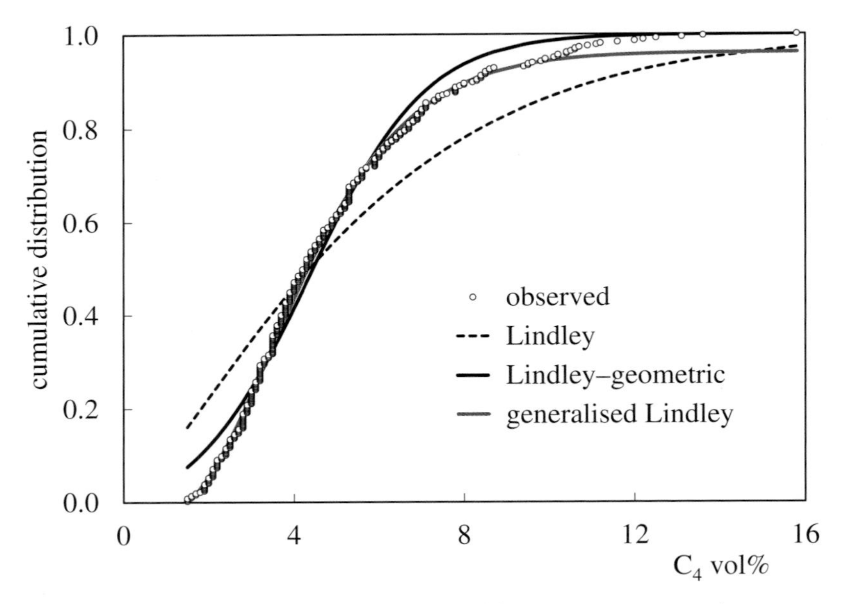

Figure 33.27 *Lindley: Comparison of fits to C_4 in propane data*

Figure 33.28 shows the effect of adjusting δ_1 and δ_2, keeping α at 0 and β at 1.
The CDF is

$$F(x)=\left[\frac{\left(\frac{x-\alpha}{\beta}\right)^{\delta_2}}{1+\left(\frac{x-\alpha}{\beta}\right)^{\delta_2}}\right]^{\frac{\delta_1}{\delta_2}} \tag{33.77}$$

Fitting the distribution to the C_4 in propane data gives α as 1.84, β as 0.0458, δ_2 as 0.999 and δ_1 as 39.2. With *RSS* at 0.4448, this type of distribution would clearly be a poor choice in this case. Further there are no simple formulae to determine mean, variance, skewness or kurtosis.

33.18 Muth

The PDF of the *Muth distribution* is

$$f(x)=\frac{1}{\beta}\left[\exp\left(\frac{\delta(x-\alpha)}{\beta}\right)-\delta\right]\exp\left\{\frac{\delta(x-\alpha)}{\beta}-\frac{1}{\delta}\left[\exp\left(\frac{\delta(x-\alpha)}{\beta}\right)-1\right]\right\} \quad x\geq\alpha;\ \beta,\delta>0 \tag{33.78}$$

Figure 33.29 shows the effect of varying δ with α fixed at 0 and β at 1. The CDF is

$$F(x)=1-\exp\left\{\frac{\delta(x-\alpha)}{\beta}-\frac{1}{\delta}\left[\exp\left(\frac{\delta(x-\alpha)}{\beta}\right)-1\right]\right\} \tag{33.79}$$

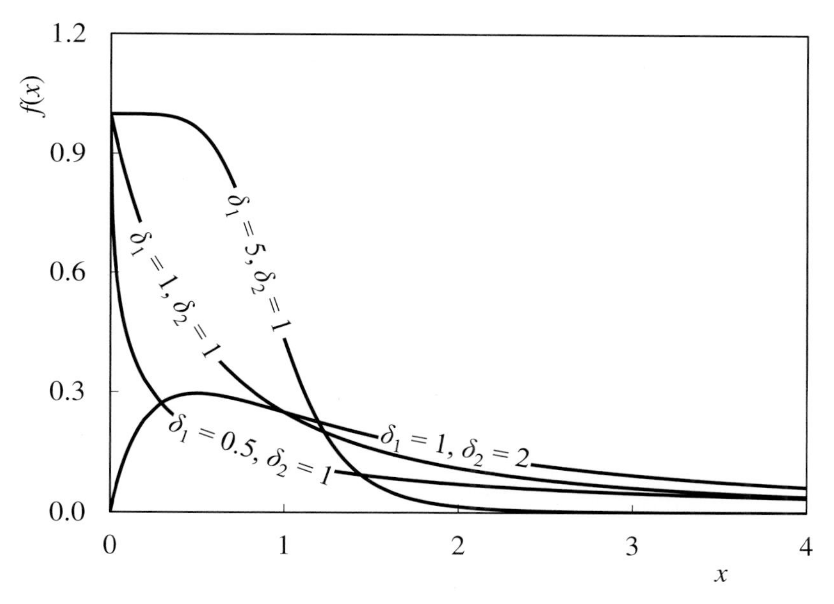

Figure 33.28 *Mielke: Effect of δ_1 and δ_2 on shape*

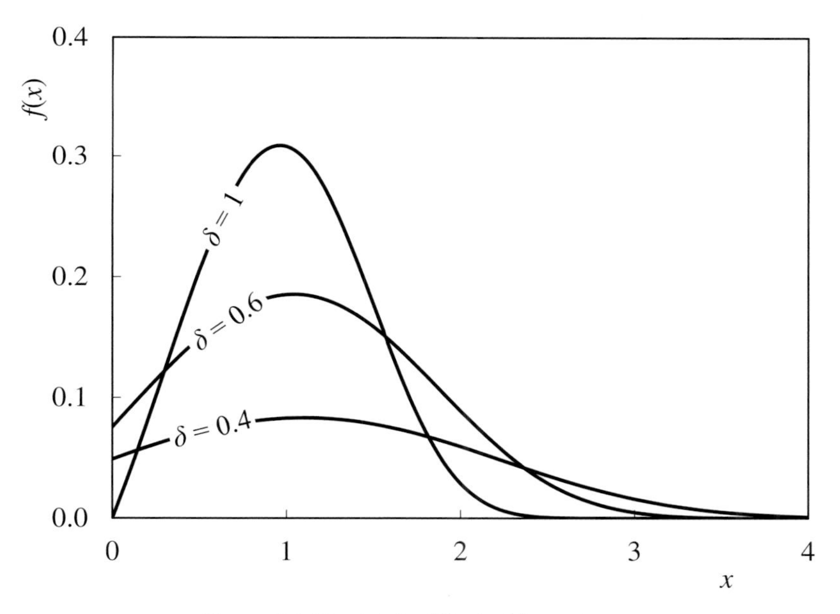

Figure 33.29 *Muth: Effect of δ on shape*

Fitting to the C_4 data gives α as 1.96, β as 2.83 and δ as 0.293. With *RSS* at 0.0850 the fit is considerably better than the normal distribution but many other types of distribution fit the data better. Further there are no published calculations for mean, variance, skewness or kurtosis.

33.19 Nakagami

The PDF of the *Nakagami distribution* is

$$f(x)=\frac{2(x-\alpha)^{\delta-1}}{\beta^{\delta}\Gamma\left(\frac{\delta}{2}\right)}\exp\left[-\left(\frac{x-\alpha}{\beta}\right)^{2}\right] \qquad x\geq\alpha;\ \beta,\delta>0 \tag{33.80}$$

Figure 33.30 shows the effect of changing δ, with α fixed at 0 and β at 1. At higher values, δ has very little effect on shape – mainly affecting location.

Fitting to the C_4 in propane data gives values of 0.720 for δ, 2.35 for α and 5.55 for β. *RSS* is 0.2074 – barely better than the normal distribution.

The equations below then give the mean as 4.83 and the standard deviation as 2.63.

$$\mu=\alpha+\frac{\Gamma\left(\frac{\delta+1}{2}\right)}{\Gamma\left(\frac{\delta}{2}\right)}\beta \tag{33.81}$$

$$\sigma^{2}=\beta\left[\frac{2}{\delta}-\left(\frac{2}{\delta}\right)^{2}\frac{\Gamma^{2}\left(\frac{\delta+1}{2}\right)}{\Gamma^{2}\left(\frac{\delta}{2}\right)}\right] \tag{33.82}$$

As the solid line in Figure 33.31 shows, of concern are the 36 values of x that are less than α and are therefore excluded during fitting. Fixing α at 1.50 (the smallest x) includes all the points and gives values for δ and β of 0.736 and 7.61 respectively. But, as shown by the dashed line, the fit becomes far less accurate with *RSS* at 1.337.

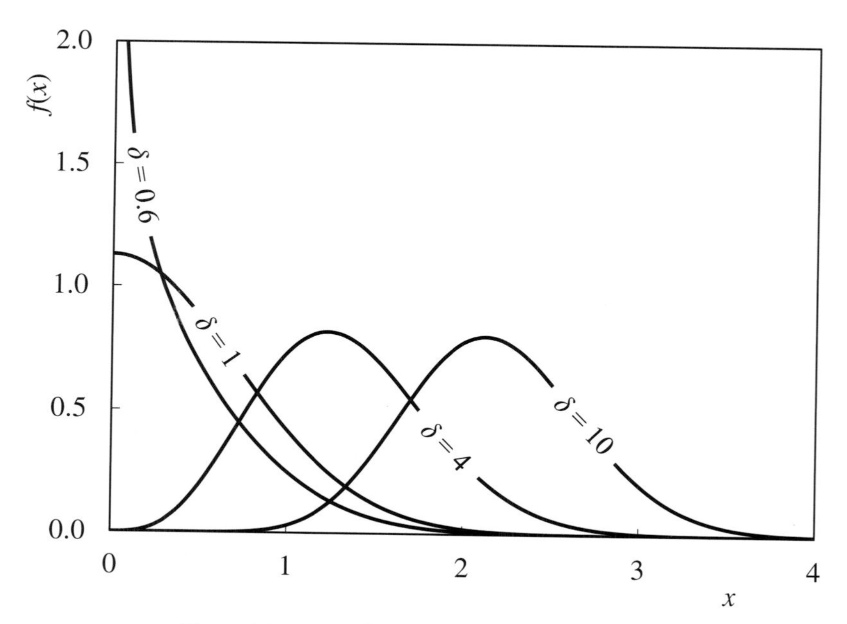

Figure 33.30 Nakagami: Effect of δ on shape

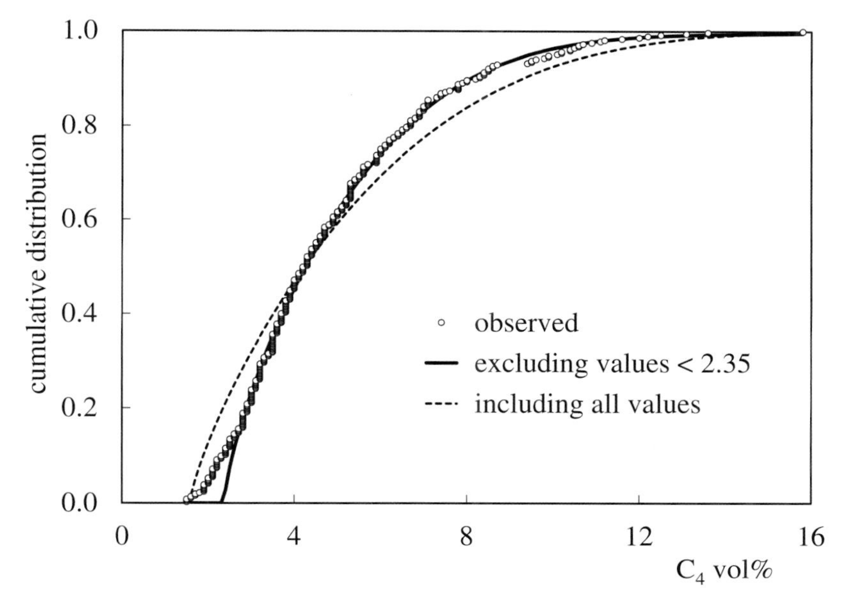

Figure 33.31 Nakagami: Data ignored when fitted to C_4 in propane data

33.20 Power

The *power distribution* is defined by the PDF

$$f(x) = \frac{\delta(x-\alpha)^{\delta-1}}{\beta^{\delta}} \quad \text{for} \quad 0<x<\beta \tag{33.83}$$

Figure 33.32 shows the effect of adjusting δ (with β fixed at 3). Normally β would be set slightly above the highest value in the data. More likely to be applicable to process data are values of δ less than 1. Setting δ to 1 gives the uniform distribution U(0,1/β).

The CDF is

$$F(x) = \left(\frac{x-\alpha}{\beta}\right)^{\delta} \tag{33.84}$$

The CDF can be inverted to give the QF.

$$x(F) = \alpha + \beta F^{1/\delta} \quad 0 \le F \le 1 \tag{33.85}$$

Fitting to the NHV disturbance data gives β as 9.5 and δ as 0.294. With *RSS* at 3.182, the distribution would be a very poor choice.

Fitting to the intervals between the 129 events, of LPG splitter reflux exceeding 70 m^3/hr, gives β as 238 and δ as 0.326. Again the fit is very poor.

The mean and variance are

$$\mu = \alpha + \frac{\beta\delta}{\delta+1} \tag{33.86}$$

$$\sigma^2 = \frac{\delta\beta^2}{(\delta+2)(\delta+1)^2} \tag{33.87}$$

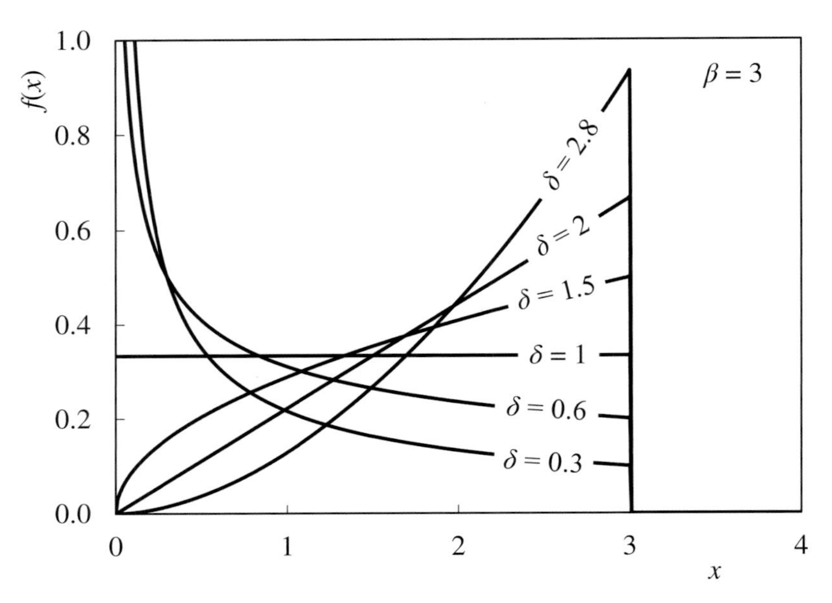

Figure 33.32 *Power: Effect of δ on shape*

Not meriting documenting, the fitted means and standard deviations are well away from those calculated from the data.

Skewness and kurtosis are

$$\gamma = \frac{2(\delta - 1)\sqrt{\delta + 2}}{(\delta + 3)\sqrt{\delta}} \tag{33.88}$$

$$\kappa = \frac{3\left(3\delta^2 - \delta + 2\right)(\delta + 2)}{\delta(\delta + 3)(\delta + 4)} \tag{33.89}$$

Figure 33.33 shows the feasible combinations of skew and kurtosis. The fitted values are well away from all those calculated – showing again the poor choice of prior distribution.

33.21 Two-Sided Power

The *two-sided power (TSP) distribution* is described by the PDF

$$f(x) = \frac{\delta}{x_{max} - x_{min}} \left(\frac{x - x_{min}}{x_{mode} - x_{min}} \right)^{\delta - 1} \quad x_{min} \le x < x_{mode};\ \delta > 0 \tag{33.90}$$

$$f(x) = \frac{\delta}{x_{max} - x_{min}} \left(\frac{x_{max} - x}{x_{max} - x_{mode}} \right)^{\delta - 1} \quad x_{mode} \le x \le x_{max};\ \delta > 0 \tag{33.91}$$

Figure 33.34 shows the effect of varying δ, with x_{min} set to 1, x_{mode} set to 3 and x_{max} set to 8. Setting δ to 1 gives the uniform distribution U(1,8); setting it to 2 gives the triangular distribution.

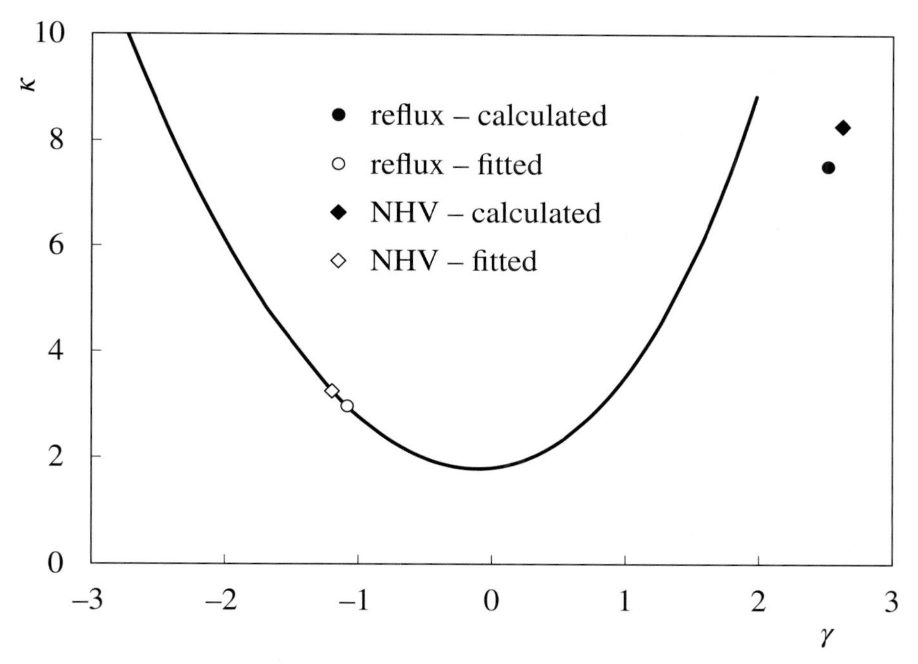

Figure 33.33 *Power: Feasible combinations of γ and κ*

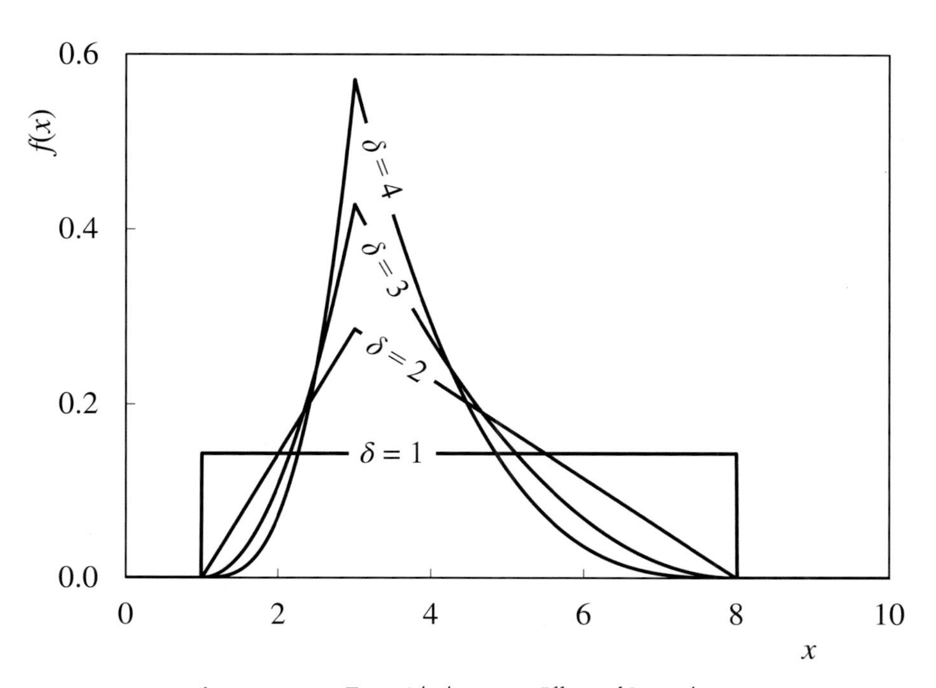

Figure 33.34 *Two-sided power: Effect of δ on shape*

The CDF is

$$F(x)=\frac{x_{\text{mode}}-x_{\min}}{x_{\max}-x_{\min}}\left(\frac{x-x_{\min}}{x_{\text{mode}}-x_{\min}}\right)^{\delta} \qquad x_{\min}\le x<x_{\text{mode}} \tag{33.92}$$

$$F(x)=1-\frac{x_{\max}-x_{\text{mode}}}{x_{\max}-x_{\min}}\left(\frac{x_{\max}-x}{x_{\max}-x_{\text{mode}}}\right)^{\delta} \qquad x_{\text{mode}}\le x\le x_{\max} \tag{33.93}$$

Fitting to the NHV disturbance data, with $x_{\min}$ set to the lowest measurement of −8.42 and $x_{\max}$ set to the highest of 9.51, gives x_{mode} as −0.15 and δ as 7.75. *RSS* is 0.0567.

The mean and variance are

$$\mu=\frac{x_{\min}+(\delta-1)x_{\text{mode}}+x_{\max}}{\delta+1} \tag{33.94}$$

$$\sigma^2=\frac{\delta\left(x_{\min}^2+x_{\max}^2\right)-2x_{\text{mode}}(\delta-1)(x_{\min}+x_{\max}-x_{\text{mode}})-2x_{\min}x_{\max}}{(\delta+1)^2(\delta+2)} \tag{33.95}$$

These give μ as 0.01 and σ as 1.38. The result is reasonably insensitive to the choice of $x_{\min}$ and $x_{\max}$. For example, setting them to −5 and +5 and refitting gives μ as −0.03 and σ as 1.27. Setting them to −15 and +15 gives μ as −0.03 and σ as 1.40. However, the standard deviation is somewhat lower than that calculated from the data. This, and the relatively high *RSS* indicates that the TSP distribution might be a poor choice in this case.

33.22 Exponential Power

The title *exponential power distribution* is given to two quite different distributions. One is less ambiguously known as the *generalised normal distribution*, as covered in Section 20.12. The PDF of the other is

$$f(x)=\frac{\delta\lambda}{\beta}\left(\frac{x-\alpha}{\beta}\right)^{\delta-1}\exp\left[\lambda\left(\frac{x-\alpha}{\beta}\right)^{\delta}\right]\exp\left[1-\lambda\left(\frac{x-\alpha}{\beta}\right)^{\delta}\right] \qquad x\ge\alpha;\ \beta,\delta,\lambda>0 \tag{33.96}$$

Figure 33.35 shows the effect of changing λ and δ, with α fixed at 0 and β at 1.

The CDF is

$$F(x)=1-\exp\left\{1-\exp\left[\lambda\left(\frac{x-\alpha}{\beta}\right)^{\delta}\right]\right\} \tag{33.97}$$

This can be inverted to give the QF.

$$x(F)=\alpha+\beta\left(\frac{\ln(1-\ln(1-F))}{\lambda}\right)^{\frac{1}{\delta}} \qquad 0\le F\le 1 \tag{33.98}$$

Fitting to the C_4 in propane data gives α as 1.97, β as 3.88, λ as 0.826 and δ as 0.905. *RSS* is 0.1102 – a mediocre fit compared to what is achieved by other distributions.

Fitting to the NHV disturbance data gives α as −2.13, β as 3.19, λ as 0.985 and δ as 1.51. *RSS* is 0.2029 – also a mediocre fit.

There are no formulae for mean, variance, skewness or kurtosis.

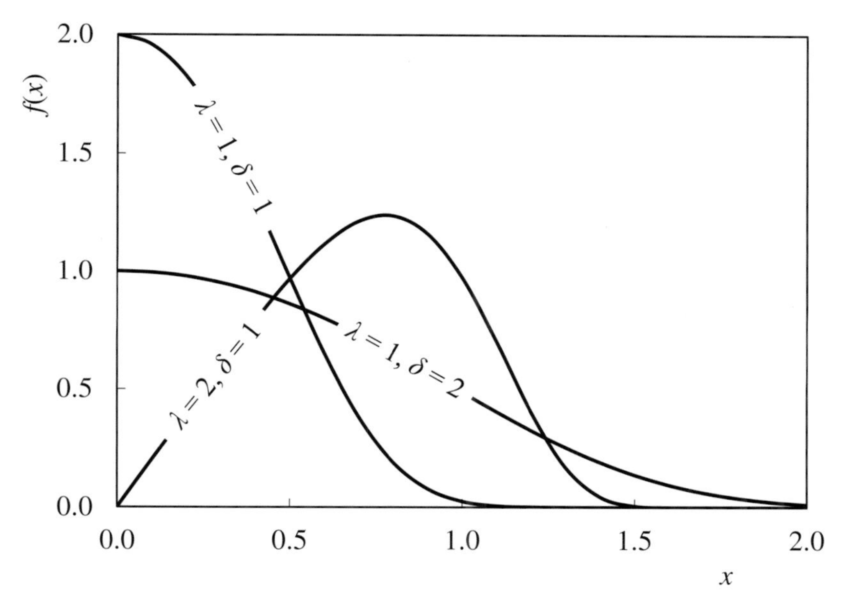

Figure 33.35 Exponential power: Effect of λ and δ on shape

33.23 Rician

The *Rician* (or *Rice*) *distribution* was originally developed to model the fading of a radio signal caused by multi-path interference. It is described by the PDF

$$f(x) = \frac{x}{\beta^2}\exp\left[-\frac{(x^2+\alpha^2)}{2\beta^2}\right]\mathrm{I}_0\left(\frac{\alpha x}{\beta^2}\right) \quad x \geq 0;\ \alpha,\beta > 0 \tag{33.99}$$

I_0 is the zero order modified Bessel function, as described in Section 11.8. Figure 33.36 shows the effect of changing α with β fixed at 1. Values below 2 affect shape; higher values change only location.

The CDF includes the Marcum Q-function (Q_1) described in Section 11.9.

$$F(x) = 1 - \mathrm{Q}_1\left(\frac{\alpha}{\beta}, \frac{x}{\beta}\right) \tag{33.100}$$

For the reasons given, fitting to the C_4 in propane data instead uses the trapezium rule. It gives values of 3.18 for α and 2.28 for β – resulting in 0.1617 for *RSS*. While a better fit than the normal distribution, it is outperformed by many others. The calculations of key parameters such as mean, variance, skewness and kurtosis are complex. In the process industry, the Rician distribution probably has little advantage over other distributions.

33.24 Topp–Leone

The PDF of the *Topp–Leone distribution* is

$$f(x) = \delta(2-2x)\left(2x-x^2\right)^{\delta-1} \quad 0 \leq x \leq 1;\ \delta > 0 \tag{33.101}$$

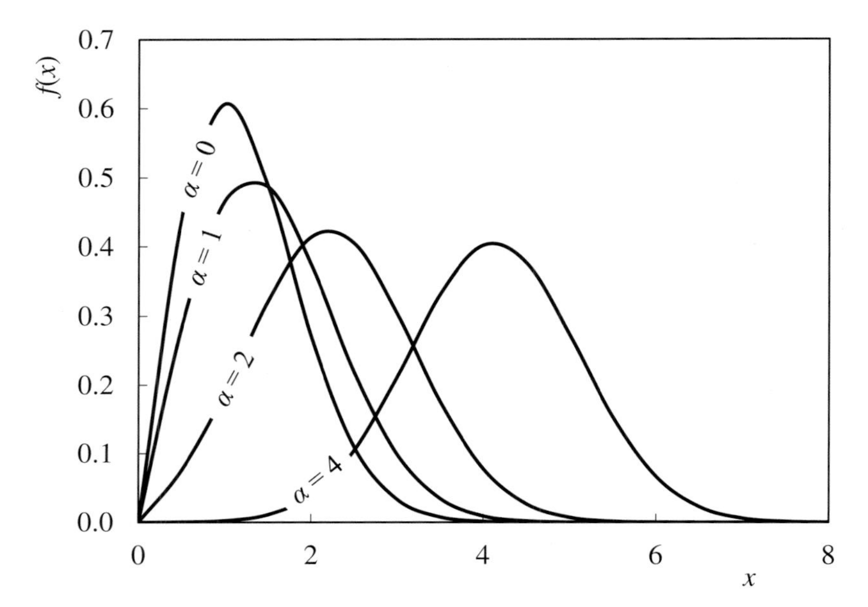

Figure 33.36 *Rician: Effect of α on shape*

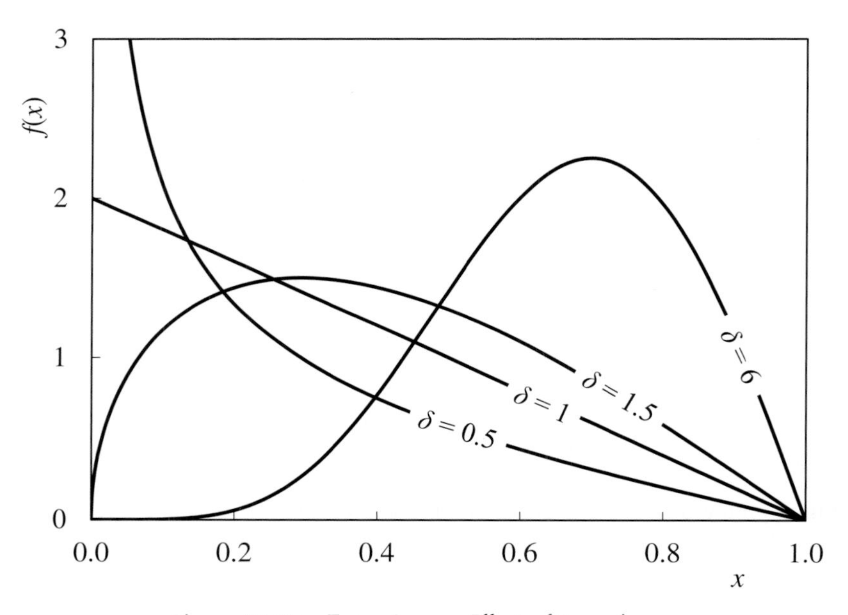

Figure 33.37 *Topp–Leone: Effect of δ on shape*

Figure 33.37 shows the effect of adjusting δ. The CDF is

$$F(x) = \left(2x - x^2\right)^{\delta} \tag{33.102}$$

The CDF can be inverted and solved as a quadratic equation to give the QF.

$$x(F) = 1 - \sqrt{1 - F^{1/\delta}} \quad 0 \leq F \leq 1 \tag{33.103}$$

Including the range of x, so that it can remain in engineering units gives the PDF

$$f(x)=\frac{\delta}{x_{max}-x_{min}}\left[2-2\left(\frac{x-x_{min}}{x_{max}-x_{min}}\right)\right]\left[2\left(\frac{x-x_{min}}{x_{max}-x_{min}}\right)-\left(\frac{x-x_{min}}{x_{max}-x_{min}}\right)^2\right]^{\delta-1} \quad x_{min}\le x\le x_{max};\ \delta>0 \tag{33.104}$$

and the CDF

$$F(x)=\left[2\left(\frac{x-x_{min}}{x_{max}-x_{min}}\right)-\left(\frac{x-x_{min}}{x_{max}-x_{min}}\right)^2\right]^{\delta} \tag{33.105}$$

The QF then becomes

$$x(F)=x_{min}+\left(x_{max}-x_{min}\right)\left(1-\sqrt{1-F^{1/\delta}}\right) \quad 0\le F\le 1 \tag{33.106}$$

Fitting Equation (33.105) to the C_4 in propane data gives values for x_{min}, x_{max} and δ of 1.97, 10.4 and 0.947 respectively. At 0.1520 *RSS* is better than that for the normal distribution, but poor compared to many others. Of greater concern is that range is best fitted by excluding 28 of the more extreme measurements. There are no simple methods for determining mean and variance, let alone skewness and kurtosis.

33.25 Generalised Tukey Lambda

The *generalised Tukey lambda distribution*[27] is described as a QF that cannot be inverted to give a CDF.

$$x(F)=\lambda_1+\frac{1}{\lambda_2}\left[F^{\lambda_3}-(1-F)^{\lambda_4}\right] \quad \lambda_2\neq 0 \tag{33.107}$$

If λ_1 is set to 0 and the other parameters set to λ, this reverts to the symmetrical Tukey distribution we covered in Section 32.11. Examination of Equation (33.107) shows that λ_1 is the location parameter and λ_2 the scale parameter. Shape is determined by λ_3 and λ_4. For example, the distribution can be skewed in either direction depending on which of these is larger. This is illustrated in Figure 33.38.

Fitting Equation (33.107) to the NHV disturbance data gives values for λ_1, λ_2, λ_3 and λ_4 of −0.0604, −0.688, −0.370 and −0.381 respectively. *RSS* is reduced to 8.73 – substantially improving upon the basic Tukey distribution. But, because *RSS* has the units of F not x, we cannot directly compare the value with what is achieved by fitting a CDF. The best of these is the slash distribution but, because it does not have a QF, we cannot show mathematically which gives the better fit. However, Figure 33.39 shows graphically that the generalised Tukey distribution significantly outperforms the slash distribution (shown as the dashed line). In particular it fits the tails very well.

If F is 0.025, from Equation (33.107) we obtain x as −4.29. If F is 0.975 then x is 4.40. The mean is not exactly at 0 but, within the accuracy of the exercise, this shows that 95% of the disturbances are within ±4.3. In other words, there is a probability of 5% that a disturbance will be larger than this. This is detailed in colour on Figure 33.39.

Fitting Equation (33.107) to the C_4 in propane data gives values for λ_1, λ_2, λ_3 and λ_4 of 3.16, −0.00272, −0.00114 and −0.00570 respectively. *RSS* is 15.8. Figure 33.40 plots this

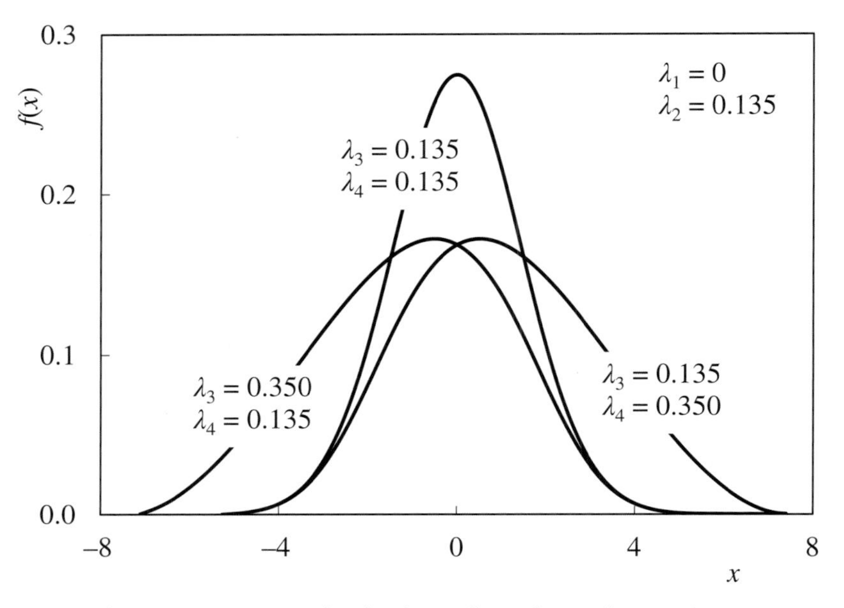

Figure 33.38 *Generalised Tukey: Effect of λ_3 and λ_4 on shape*

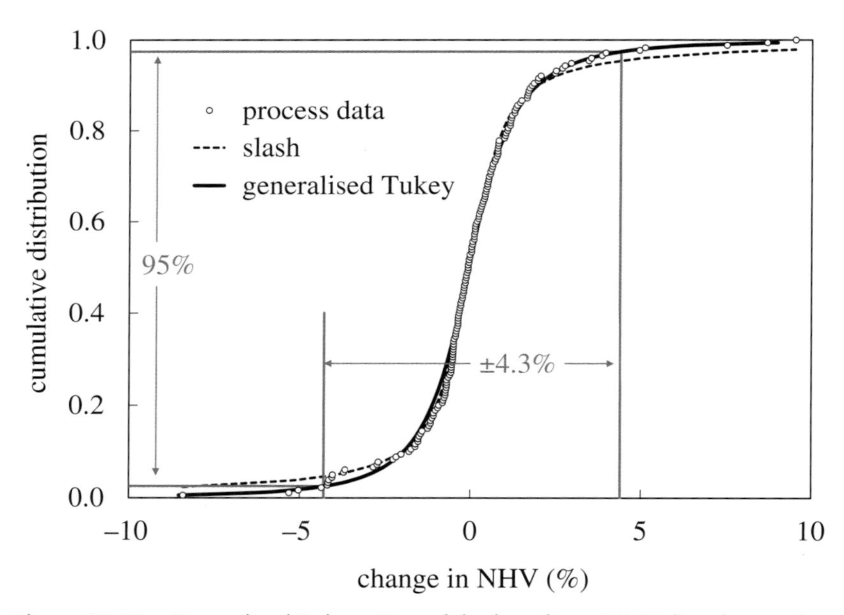

Figure 33.39 *Generalised Tukey: One of the best fits to NHV disturbance data*

distribution against the best-of-the-rest, i.e. the generalised logistic Type IV distribution. The fit is virtually identical. It would appear that we have identified a distribution that might be universally adopted for all datasets. Unfortunately there are no simple formulae for calculating mean, variance, skewness or kurtosis.

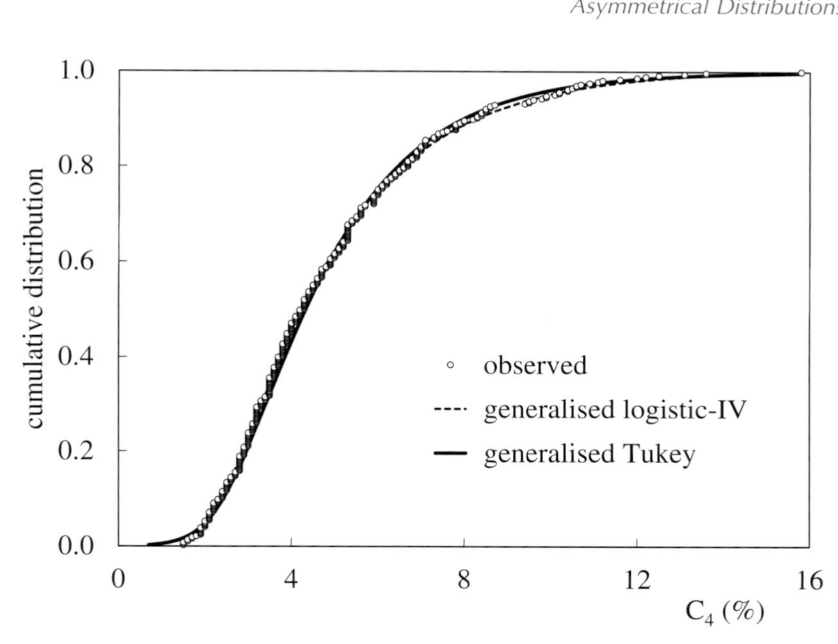

Figure 33.40 Generalised Tukey: One of the best fits to the C_4 in propane data

33.26 Wakeby

The *Wakeby distribution* is another that it is described by a QF that cannot be inverted to a CDF.

$$x(F) = \alpha + \frac{\delta_1}{\delta_2}\left[1-(1-F)^{\delta_2}\right] - \frac{\delta_3}{\delta_4}\left[1-(1-F)^{-\delta_4}\right] \quad 0 \le F \le 1 \tag{33.108}$$

More unusual is its PDF which is written in terms of its CDF (F).

$$f(x) = \frac{(1-F)^{\delta_4+1}}{\alpha(1-F)^{\delta_2+\delta_4} + \delta_3} \tag{33.109}$$

As usual, the coefficient α is the location parameter and, as Figure 33.41 shows, so largely is δ_1. The other shape coefficients can have very similar effects. Figures 33.42, 33.43 and 33.44 show the effect of adjusting δ_2, δ_3 and δ_4. As is common with other distributions that have a large number of coefficients, finding the best combination can be a challenge. Quite different combinations can give almost identically accurate fits.

Fitting to the C_4 in propane data gives values of 2.05 for α, 3.33 for δ_1, 0.188 for δ_2, −0.002 for δ_3 and 0.466 for δ_4. At 7.74, *RSS* is significantly less than that from fitting the QF for the generalised Tukey distribution. The resulting fit is shown as the coloured line in Figure 33.45. Of the distributions that can be expressed as a QF, the Hosking distribution gave one of the best fits. Fitting Equation (12.25) gives α as 2.08, β as 3.44, δ_1 as 0.223 and δ_2 as 0.959. Shown as the black line, only *RSS* (at 9.78) shows it is slightly outperformed by Wakeby distribution. This might be expected from a distribution that has five shape parameters compared to the four of the Hosking distribution.

Also of note are the quite different shape parameters for the Hosking distribution obtained by fitting $x(F)$ compared to fitting $F(x)$. Minimising *RSS* measured in terms of F does not give the same result as minimising it in terms of x. We covered this effect in detail in Chapter 16.

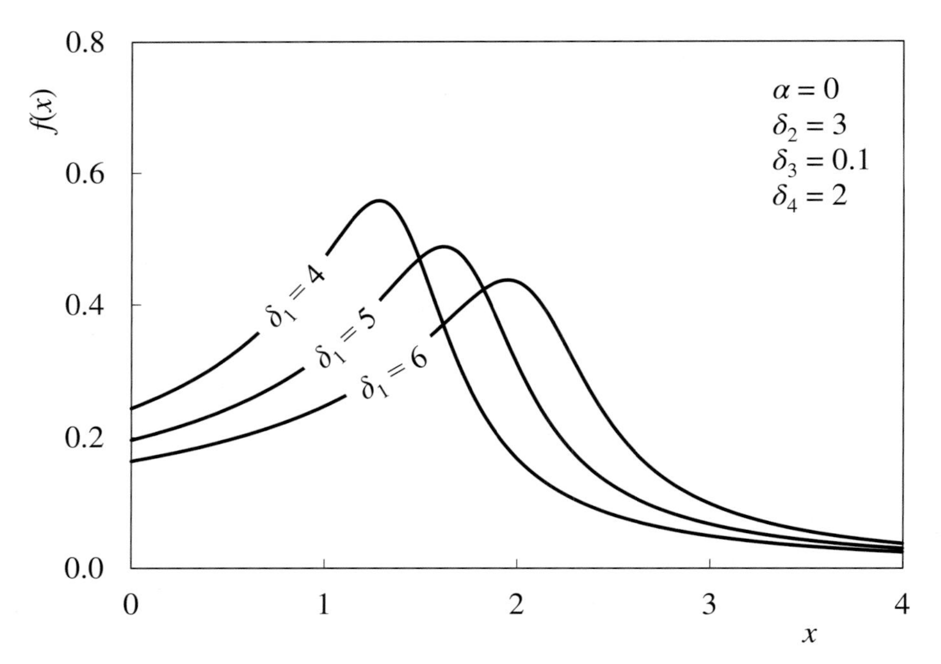

Figure 33.41 *Wakeby: Effect of δ_1 on shape*

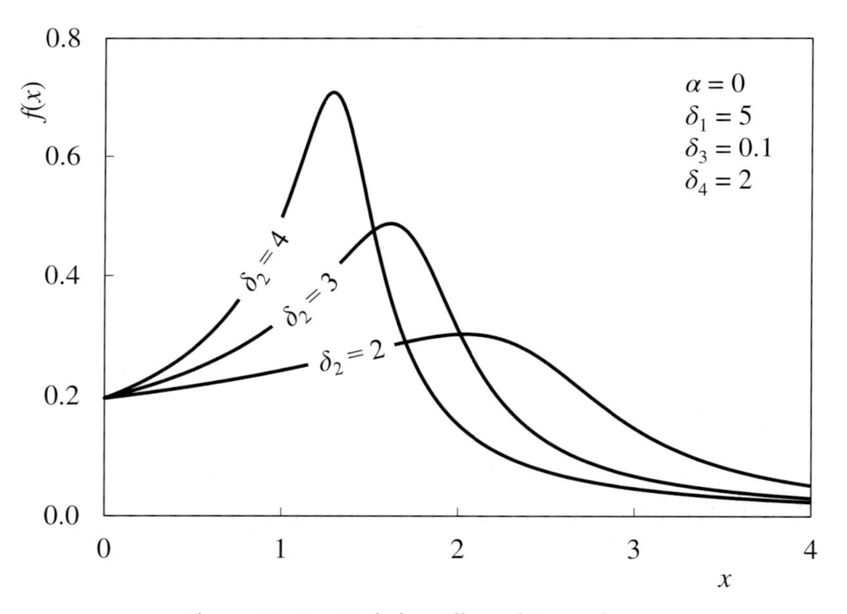

Figure 33.42 *Wakeby: Effect of δ_2 on shape*

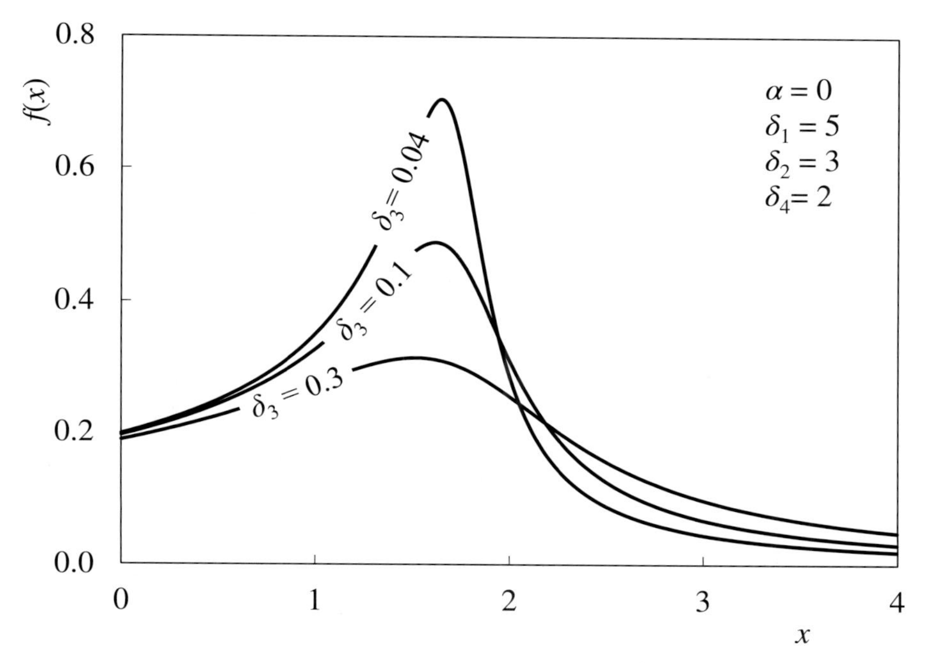

Figure 33.43 *Wakeby: Effect of δ_3 on shape*

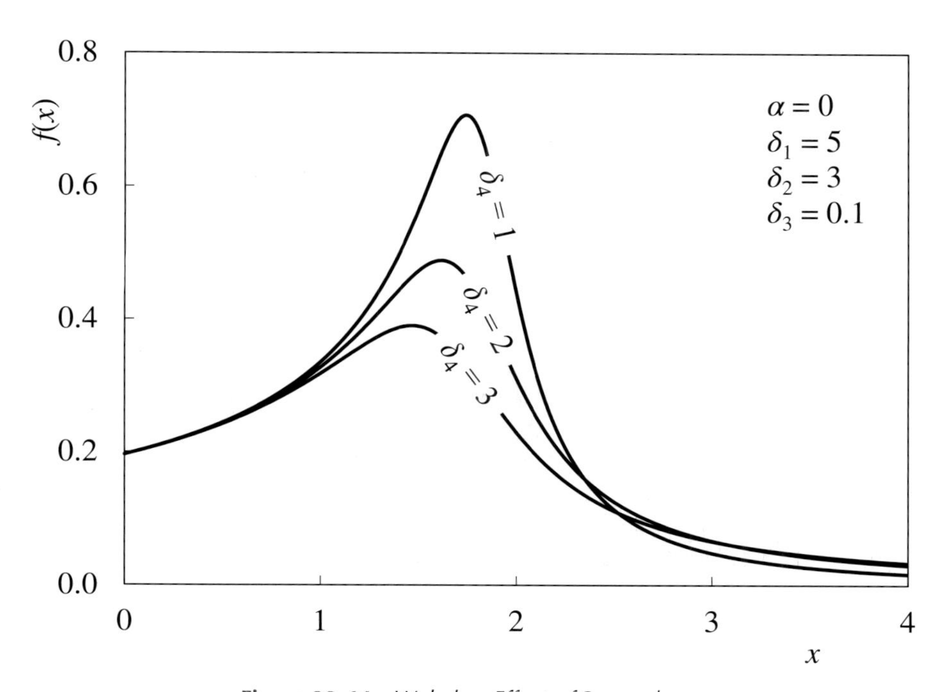

Figure 33.44 *Wakeby: Effect of δ_4 on shape*

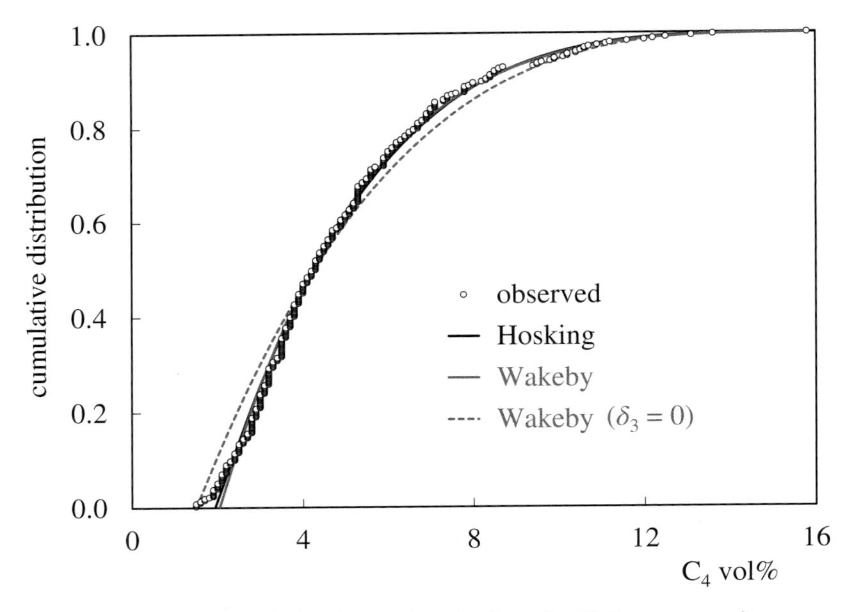

Figure 33.45 Wakeby: Improving the fit to the C_4 in propane data

The mean and variance are

$$\mu = \alpha + \frac{\delta_1}{1+\delta_2} + \frac{\delta_3}{1-\delta_4} \tag{33.110}$$

$$\sigma^2 = \frac{\delta_1^2}{(1+\delta_2)^2(1+2\delta_2)} - \frac{2\delta_1\delta_3}{(1+\delta_2)(1+\delta_2-\delta_4)(\delta_4-1)} - \frac{\delta_3^2}{(\delta_4-1)^2(2\delta_4-1)} \tag{33.111}$$

These give values of 4.85 for the mean and 2.39 for the standard deviation.

Examination of Equation (33.108) shows, if either δ_1 or δ_3 is 0, it can be inverted to a CDF.

$$F(x) = 1 - \left[1 + \frac{\delta_4(x-\alpha)}{\delta_3}\right]^{-\frac{1}{\delta_4}} \qquad \delta_1 = 0 \tag{33.112}$$

$$F(x) = 1 - \left[1 - \frac{\delta_2(x-\alpha)}{\delta_1}\right]^{\frac{1}{\delta_2}} \qquad \delta_3 = 0 \tag{33.113}$$

In our example, δ_3 is very close to 0. Fitting Equation (33.113), to the same C_4 in propane data, gives 1.50 for α, 4.41 for δ_1 and 0.309 for δ_2. With *RSS* at 0.8174, the fit is very poor. The dashed line in Figure 33.45 shows the result, illustrating the benefit of using the full distribution – even if key coefficients are close to 0.

34

Amoroso Distribution

The *Amoroso distribution*[28] is defined by the PDF

$$f(x) = \frac{|\delta_2|}{\beta\Gamma(\delta_1)}\left(\frac{x-\alpha}{\beta}\right)^{\delta_1\delta_2-1}\exp\left[-\left(\frac{x-\alpha}{\beta}\right)^{\delta_2}\right] \quad x \geq \alpha;\ \beta,\delta_1,\delta_2 > 0 \qquad (34.1)$$

Figure 34.1 shows, with α set to 0 and β to 1, the effect of varying δ_1 and δ_2.

As mentioned in Section 31.3, the generalised gamma distribution is a special case of the Amoroso distribution in which α is set to zero. This, however, is just a trivial example of what other distributions can be represented. Table 34.1 lists such distributions that are covered elsewhere in this book, showing how each PDF is derived from Equation (34.1).

In principle, fitting the Amoroso distribution effectively considers every one of these as the prior distribution. However, it is not quite as straightforward. The term δ_2 in Equation (34.1) need not be an integer and so, to avoid a negative number being raised to a non-integer power, x cannot be smaller than α. If the best fit is achieved with a value of α which is larger than some of the lower values of x, for these values $f(x)$ would be set to 0. But, for almost all of the derivable distributions, both δ_2 and the product $\delta_1\delta_2$ are integers and so the restriction on x does not apply. In effect, many of the special cases of the lower-bounded Amoroso distribution are unbounded.

We can illustrate this with an example. Fitting the Amoroso distribution to the C_4 in propane data gives α as 1.22, β as 0.0128, δ_1 as 11.1, δ_2 as 0.43 and $F(x_1)$ as 0.0073. RSS is 0.0302. The Pearson-III distribution can theoretically be fitted by forcing δ_2 to 1. Doing so gives α as 1.60, β as 1.84, δ_1 as 1.78 and $F(x_1)$ as 0.0027. As might be anticipated, with one less shape parameter, RSS increases to 0.0358. This result is very similar to that obtained in Section 27.3, but not identical. Fitting the Pearson-III distribution directly includes the three values of x that are less than a. Fitting the Amoroso equivalent sets $f(x)$ to 0 for these values. As Figure 34.2 shows, on this occasion, the Amoroso distribution is a good fit indistinguishable from that of the Pearson-III. This is because the inclusion, or not, of three very marginal points makes little difference. But, in another example, the effect might be much greater.

Statistics for Process Control Engineers: A Practical Approach, First Edition. Myke King.

Table 34.1 Distributions represented by the Amoroso distribution

distribution	α	β	δ_1	δ_2
shifted exponential	α	β	1	1
standard exponential	0	$\frac{1}{\lambda}$	1	1
Weibull-III	α	β	1	k
chi-squared	0	2	$\frac{f}{2}$	1
gamma	0	β	k	1
Lévy	α	$\frac{\beta}{2}$	$\frac{1}{2}$	−1
inverse gamma	α	β	δ	−1
Pearson Type III	α	β	δ	1
inverse exponential	α	β	1	−1
stretched exponential	0	$\frac{1}{\lambda}$	1	k
chi	α	β	$\frac{f}{2}$	2
inverse Rayleigh	α	β	1	−2
Maxwell	α	$\beta\sqrt{2}$	$\frac{3}{2}$	2
half-normal	α	$\beta\sqrt{2}$	$\frac{1}{2}$	2
inverse chi	α	$\beta\sqrt{2}$	$\frac{f}{2}$	−2
inverse chi-squared	α	$\frac{\beta}{2}$	$\frac{f}{2}$	−1
Rayleigh	α	β	1	2
generalised gamma	0	β	δ_1	δ_2
Fréchet	α	β	1	$-\delta$
Nakagami	α	β	$\frac{\delta}{2}$	2

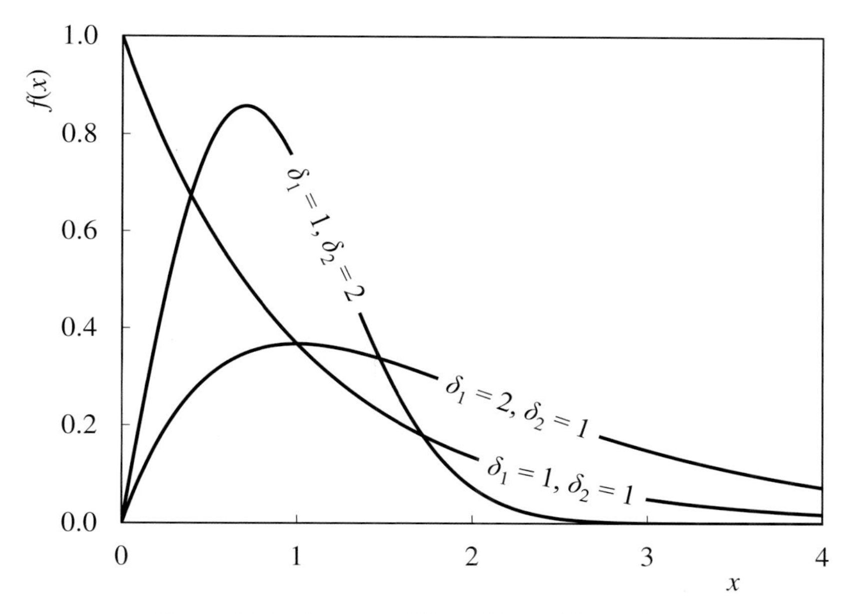

Figure 34.1 Amoroso: Effect of δ_1 and δ_2 on shape

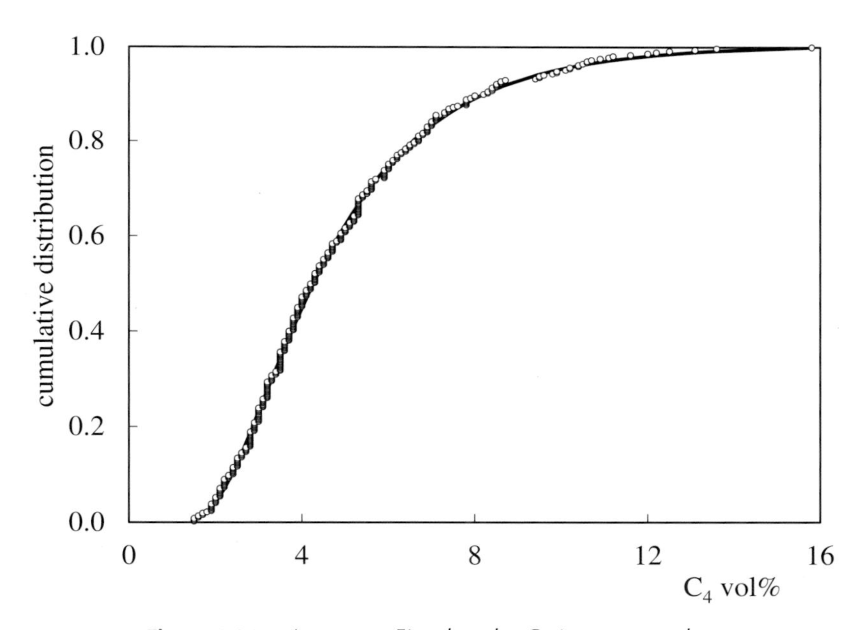

Figure 34.2 Amoroso: Fitted to the C_4 in propane data

The approach should be to compare the shape parameters derived by fitting the Amoroso distribution against those in Table 34.1. If there is a close match to one of the derivable distributions then this should be fitted to the data and the resulting *RSS* compared to that of the Amoroso distribution.

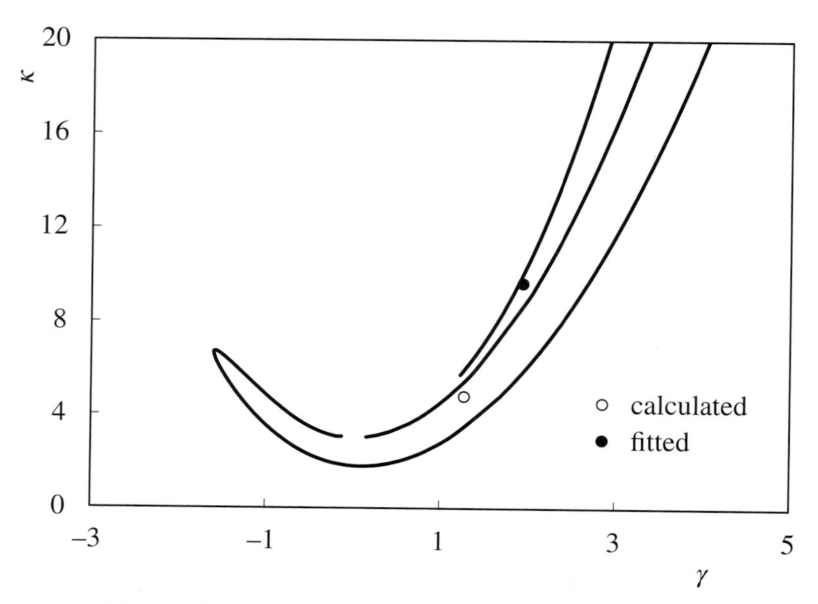

Figure 34.3 Amoroso: Feasible combinations of γ and κ

Moments are given by

$$m_n = \frac{\Gamma\left(\delta_1 + \frac{n}{\delta_2}\right)}{\Gamma(\delta_1)}\beta^n \tag{34.2}$$

From the moments we can derive

$$\mu = \alpha + \frac{\Gamma\left(\delta_1 + \frac{1}{\delta_2}\right)}{\Gamma(\delta_1)}\beta \tag{34.3}$$

$$\sigma^2 = \frac{1}{\Gamma^2(\delta_1)}\left[\Gamma(\delta_1)\Gamma\left(\delta_1 + \frac{2}{\delta_2}\right) - \Gamma^2\left(\delta_1 + \frac{1}{\delta_2}\right)\right]\beta^2 \tag{34.4}$$

These equations give the mean as 5.06 and the standard deviation at 2.77.

Figure 34.3 shows the feasible combinations of skewness and kurtosis covered by the distribution. Platykurtosis and leptokurtosis are covered, as well as skewness in either direction. The fitted values are reasonably close to those calculated from the data.

35

Binomial Distribution

Perhaps the most well known of the discrete distributions, the binomial distribution, we covered in Section 12.11. Equally well known, the Poisson distribution derives from the binomial distribution and was covered in Section 12.12. Here we describe the range of extensions to both.

35.1 Negative-Binomial

The binomial distribution gives the probability of a number of successes; the *negative binomial distribution* gives the probability of a number of failures. Also known as the *Pascal distribution*, it is also a special case of the *Pólya distribution*. It is a compound distribution, starting as the Poisson distribution. But, instead of the expected number of successes (λ) being a constant, it is assumed to follow the gamma distribution.

There are several ways of presenting the PMF. In the first, if p is the probability of success of a single trial, the PMF gives the probability of there being x failures before there are s successes. This means that there must first be $(s-1)$ successes and x failures, followed by a success. Therefore

$$p(x)=\frac{(s+x-1)!}{x!(s-1)!}p^{s-1}(1-p)^{x}\times p=\frac{(s+x-1)!}{x!(s-1)!}p^{s}(1-p)^{x}\quad x\geq 0;\ 0\leq p\leq 1 \tag{35.1}$$

The PMF can also be written in the form that gives the probability of there being x failures in n trials, where n will therefore be the sum of x and s.

$$p(x)=\frac{(n-1)!}{x!(n-x-1)!}p^{n-x}(1-p)^{x}\quad 0\leq x\leq n;\ 0\leq p\leq 1 \tag{35.2}$$

In the batch blending example we require only one success. If the batch is on grade we make no further corrections; so n will be $x+1$. Equation (35.2) therefore reduces to

$$p(x)=p(1-p)^{x}\quad x\geq 0;\ 0\leq p\leq 1 \tag{35.3}$$

Statistics for Process Control Engineers: A Practical Approach, First Edition. Myke King.

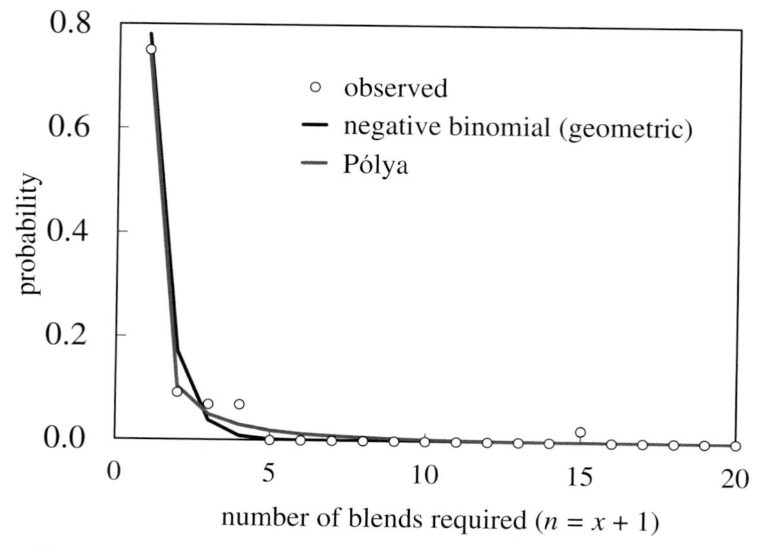

Figure 35.1 *Negative binomial: Fit to batch blending performance*

This is one form of the *geometric distribution* that we will cover later in more detail. In the batch blending example, of 44 completed batches, 33 required no correction. We could therefore assume p is 0.750 and use this value in Equation (35.3) to determine the probability of the need to correct batches in the future.

Alternatively we can fit the distribution to the data. In addition to the 33 batches that needed no correction, 4 required one correction, 3 required two and 3 required three. There is also what appears to be one exceptional case which took 14 corrections. Converting these to observed probabilities gives 0.7500, 0.9909, 0.0682, 0.0682 and 0.0227. Fitting Equation (35.3) to these results gives p as 0.780 – close to that calculated. *RSS* is 0.0124.

Although we might have some confidence in the value of p (because it is based on 44 results), as Figure 35.1 shows, the distribution does not fit the data well. This is because the data includes only 11 cases of batch correction divided between four different values of k. This is far from enough to have confidence that the distribution would predict how many occasions a batch would require two, three, four or more corrections.

The mean can be defined in several different ways. Commonly it is the mean number of trials required to produce s successes, in which case it is

$$\mu = \frac{s}{p} \tag{35.4}$$

It can also be defined as the mean number of successes (μ_S) before there are x failures.

$$\mu_S = \frac{kp}{1-p} \tag{35.5}$$

Or it can be the mean number of failures (μ_K) before there are s successes.

$$\mu_K = \frac{s(1-p)}{p} \tag{35.6}$$

The standard deviation, skewness and kurtosis are the same for all the versions.

$$\sigma^2 = \frac{s(1-p)}{p^2} \tag{35.7}$$

$$\gamma = \frac{2-p}{\sqrt{s(1-p)}} \tag{35.8}$$

$$\kappa = \frac{p^2 + 3(1-p)(s+2)}{s(1-p)} \tag{35.9}$$

35.2 Pólya

The PMF of the Pólya distribution is normally written in the form

$$p(x) = \frac{\Gamma(\alpha+x)\beta^x}{\Gamma(x+1)\Gamma(\alpha)(\beta+1)^{\alpha+x}} \quad x \geq 0;\ \alpha, \beta > 0 \tag{35.10}$$

Fitting this to the batch blending example gives values of 0.169 for α and 4.55 for β. The result is included in Figure 35.1. *RSS* is 0.0032.

The mean is

$$\mu = \alpha\beta \tag{35.11}$$

This gives μ as 0.77. This represents the average number of corrective blends required per batch. In other words, on average, a batch requires 1.77 blends. This result is consistent with a total of 78 blends producing 44 batches. The variance is

$$\sigma^2 = \alpha\beta(\beta+1) \tag{35.12}$$

This gives σ as 2.06, versus the value of 2.21 calculated from the data. As with other distributions, by rearranging Equations (35.11) and (35.12), we can use the calculated values of μ and σ to give initial estimates for α and β.

$$\alpha = \frac{\mu^2}{\sigma^2 - \mu} \tag{35.13}$$

$$\beta = \frac{\sigma^2 - \mu}{\mu} \tag{35.14}$$

Equation (35.10) can be rewritten as

$$p(x) = \frac{(\alpha+x-1)!}{x!(\alpha-1)!}\beta^x(\beta+1)^{-(\alpha+x)} \tag{35.15}$$

This becomes the negative binomial distribution, described by Equation (35.1), if we set

$$\alpha = s \quad \beta = \frac{1-p}{p} \tag{35.16}$$

There is also the *extended negative binomial distribution* but which is too complex to be of practical use in the process industry.

35.3 Geometric

As we saw in Section 35.1, the *geometric distribution* is a special case of the negative-binomial distribution. It is one of the few memoryless discrete distributions. It can be used to estimate the probability that there are x failures before the first success. Its PMF is then

$$p(x) = p(1-p)^x \quad x \geq 0;\ 0 \leq p \leq 1 \tag{35.17}$$

The CDF gives us the probability that there are x or fewer failures.

$$P(x)=1-(1-p)^{x+1} \tag{35.18}$$

This can be inverted to give the QF

$$x(P)=\frac{\ln(1-P)}{\ln(1-p)}-1 \quad 0\leq P\leq 1 \tag{35.19}$$

Remembering that x is an integer, we would have to round the result either up or down. In the example of 'We can be 95% certain that there will be no more than x events', we would round it up.

The mean is

$$\mu=\frac{1-p}{p} \tag{35.20}$$

More commonly we want to estimate the probability that the xth trial is the first success. This version is known as the *shifted geometric distribution* because the minimum value of x is shifted from 0 to 1.

$$p(x)=p(1-p)^{x-1} \tag{35.21}$$

The CDF is

$$P(x)=1-(1-p)^{x} \tag{35.22}$$

The QF is

$$x(P)=\frac{\ln(1-P)}{\ln(1-p)} \quad 0\leq P\leq 1 \tag{35.23}$$

The mean is

$$\mu=\frac{1}{p} \tag{35.24}$$

The variance, skewness and kurtosis are the same for both versions.

$$\sigma^2=\frac{1-p}{p^2} \tag{35.25}$$

$$\gamma=\frac{2-p}{\sqrt{1-p}} \tag{35.26}$$

$$\kappa=\frac{p^2}{1-p}+9 \tag{35.27}$$

The conventional approach to estimating the change achieved by improved control is to apply the Same Percentage Rule as described by Equation (2.1). This assumes that the standard deviation (σ) will be halved and so permit the mean to be moved by half the mean deviation from target. Using the same batch blending data (Table A1.7) we calculate the simple mean as 97.91 and the standard deviation of 9.12. The normal distribution curve is plotted as the 'before' case in Figure 35.2. The 'after' case is therefore based on a mean of 98.96 and a standard deviation of 4.56.

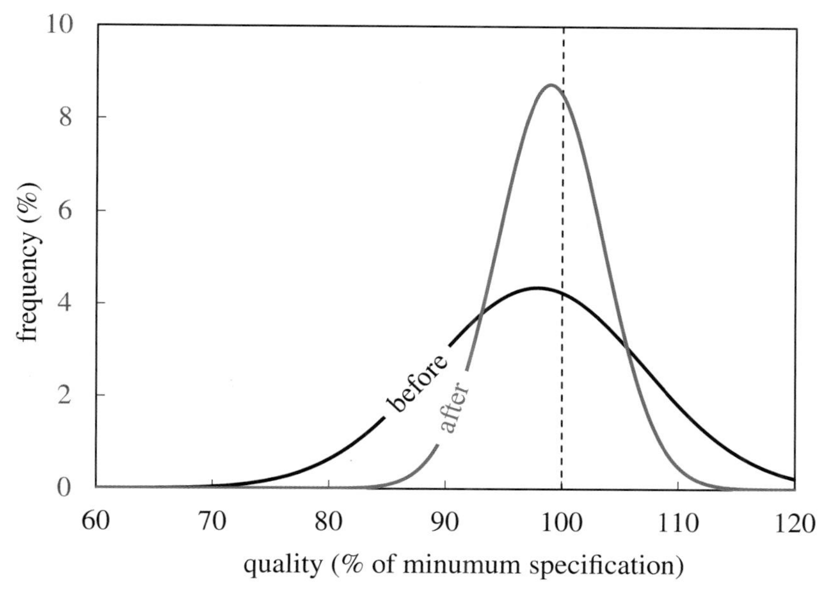

Figure 35.2 Use of same percentage rule to quantify control improvement

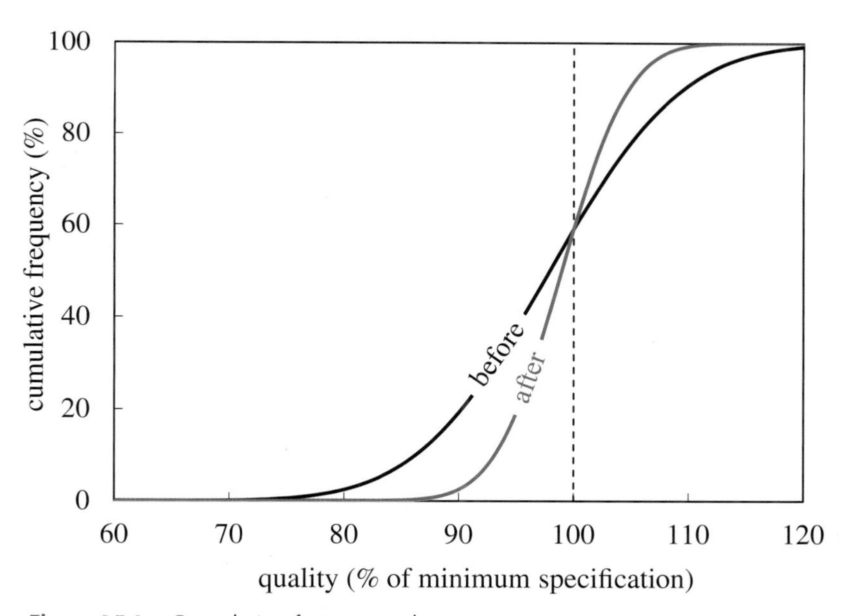

Figure 35.3 Cumulative frequency showing intersection at same percentage

The results are a measure of the quality of 78 blends. Figure 35.3 shows the change in cumulative probability achieved by improved control. There are two problems. Firstly, it would suggest that the probability that a blend is off-grade is 59%, i.e. 46 out of the 78 blends. However, Figure 3.17 shows that, of the 78 results, 34 were off-grade – about 44%. The discrepancy arises because it is not sufficiently accurate to assume that the distribution is normal. Secondly, as its name suggests, applying the Same Percentage Rule results in the same proportion of off-grade results. We need a different approach.

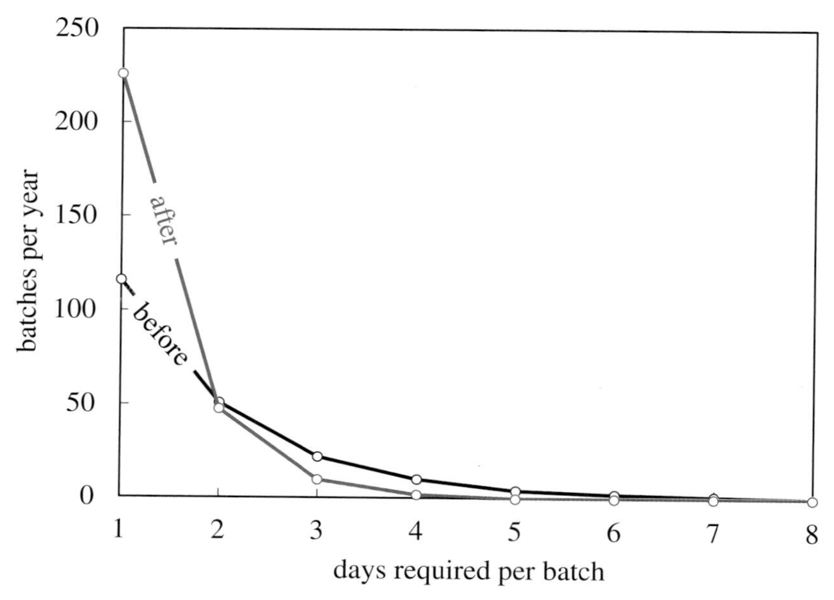

Figure 35.4 *Geometric: Impact of improved control on distribution of days required per batch*

If p is the current probability that a batch of product will be on-grade and p_{new} the probability after implementation of the control improvement then, from Equation (35.25)

$$\frac{\sqrt{1-p_{new}}}{p_{new}} = 0.5\frac{\sqrt{1-p}}{p} \quad \text{or} \quad p_{new} = \frac{2p\left(\sqrt{p^2-p+1}-p\right)}{1-p} \tag{35.28}$$

Of the 78 blends, 44 were on grade. We can therefore assume that the current probability of success (p) is 0.564. From Equation (35.28) we determine that, with improved control, this will increase to 0.788. Out of 78 blends, we would therefore expect the number of off-grade results to fall to 17.

From Equation (35.25), improved control reduces the mean from 1.77 to 1.27 days. In a year, therefore, the number of batches can increase from 206 to 288.

In a year we expect $365p$ finished batches. Equation (35.21) gives the probability that a batch will take x days to complete. Multiplying this by $365p$ gives the expected number of batches per year that take x days. This gives the 'before' frequency distribution shown in Figure 35.4. The 'after' case is plotted by replacing p with p_{new}.

Multiplying each frequency by x and summing the results gives the total number of days per year used in producing the batches. Because each result is rounded off to an integer, the total is 363, not 365. Multiplying each result by 100/363 gives the percentage of the year used. Figure 35.5 plots this as the cumulative allocation of time to batches. This shows that, after the control improvement, we can be 95% certain that the number of blends required for a single batch will not exceed two. Before we were only as certain that it would not exceed five.

While there will be a benefit in reducing the number of trim blends performed, the main benefit would arise if the process is a bottleneck on production. Improved control would then permit a 40% increase in the number of batches produced. A further benefit will be a reduction in quality giveaway. The average quality of the completed batches is 103.8. From Equation (2.1), this will be reduced to 101.9 – so reducing the proportion of the more expensive component used in blending.

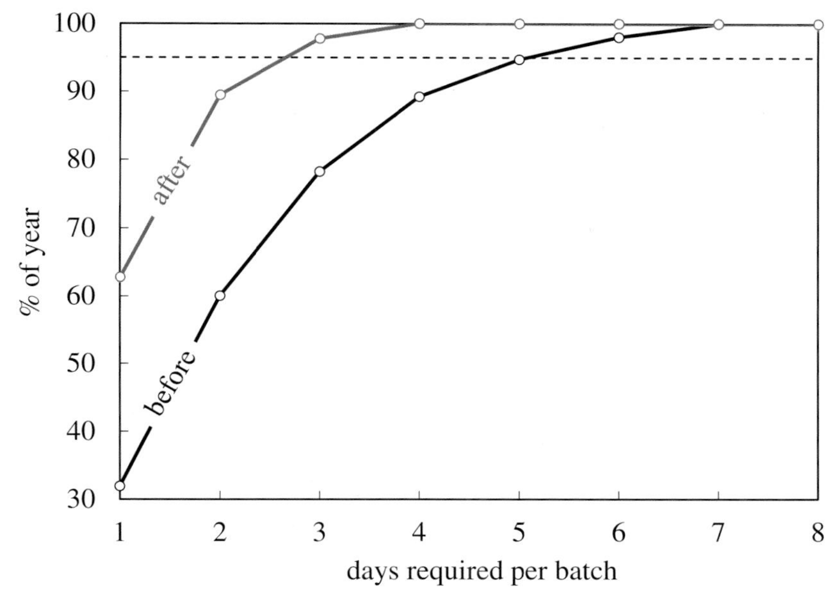

Figure 35.5 *Geometric: Reduction in probability of large number of trim blends*

35.4 Beta-Geometric

The *beta-geometric distribution* is the geometric distribution in which the probability of success, instead of being a constant, varies randomly following the beta distribution. It is another example of a compound distribution. If x is the number of trials to achieve a success, then its PMF is

$$p(x) = \frac{\alpha\Gamma(\alpha+\beta)\Gamma(\beta+x-1)}{\Gamma(\beta)\Gamma(\alpha+\beta+x)} \quad x \geq 1;\ \alpha,\beta > 0 \tag{35.29}$$

Figure 35.6 shows the effect of changing α and β.

Redefining x as the number of failures before a success gives the *shifted beta-geometric distribution*.

$$p(x) = \frac{B(\alpha+1,\beta+x)}{B(\alpha,\beta)} = \frac{\alpha\Gamma(\alpha+\beta)\Gamma(\beta+x)}{\Gamma(\beta)\Gamma(\alpha+\beta+x+1)} \quad x \geq 0;\ \alpha,\beta > 0 \tag{35.30}$$

By then setting β to 1 we obtain the special case of the *Waring–Yule distribution*, sometimes called simply the *Waring distribution*.

$$p(x) = \frac{B(\alpha+1,x+1)}{B(\alpha,1)} = \alpha B(\alpha+1,x+1) = \frac{\alpha\Gamma(\alpha+1)\Gamma(x+1)}{\Gamma(\alpha+x+2)} \quad x \geq 0;\ \alpha > 0 \tag{35.31}$$

The mean and variance are

$$\mu = \frac{\beta}{\alpha-1} \quad \alpha > 1 \tag{35.32}$$

$$\sigma^2 = \frac{\alpha\beta(\alpha+\beta-1)}{(\alpha-1)^2(\alpha-2)} \quad \alpha > 2 \tag{35.33}$$

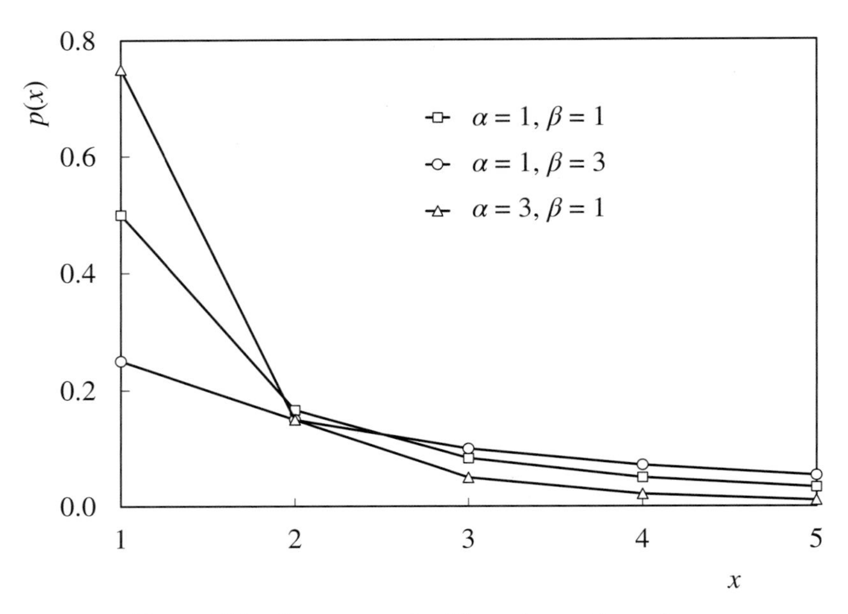

Figure 35.6 Beta-geometric: Effect of α and β on shape

We will use, as an example, the number of days between events – where the event is the LPG splitter reflux exceeding 70 m^3/hr. We can calculate, from the data, the mean (μ) number of days as 1.56. Similarly, from the data, σ is 1.85. Fitting Equation (35.29) to the data gives α as 5.12 and β as 4.75. Equation (35.32) gives the mean as 1.16 and the standard deviation as 2.03 – both reasonably close to the calculated values.

Figure 35.7 confirms that the fit is good. Figure 35.8 plots the same data as a cumulative plot. It shows, for example, that we can be 95% sure that a high reflux event will occur within 5 days of the last event.

35.5 Yule–Simon

By setting β to 1, in Equation (35.29), we obtain a special case of the beta-geometric distribution, known as the *Yule distribution* or *Yule–Simon distribution*. The PMF can be written in several ways.

$$p(x) = \alpha \mathrm{B}(x, \alpha + 1) = \frac{\alpha \Gamma(x) \Gamma(\alpha + 1)}{\Gamma(x + \alpha + 1)} = \frac{\alpha \alpha! (x-1)!}{(x+\alpha)!} \quad x \geq 1;\ \alpha > 0 \tag{35.34}$$

The CDF is

$$P(x) = 1 - x\mathrm{B}(x, \alpha + 1) \tag{35.35}$$

Unlike many other discrete distributions, where x can be 0, in the Yule–Simon distribution it cannot be less than 1. This simply means we have to reformulate the problem. For example, in the batch blending example, we define x as the number of blends required per batch, not the number of trim blends. Fitting it to this example gives α as 3.11. With *RSS* as 0.0066 it is a better fit than the geometric distribution but not as good as the Pólya distribution.

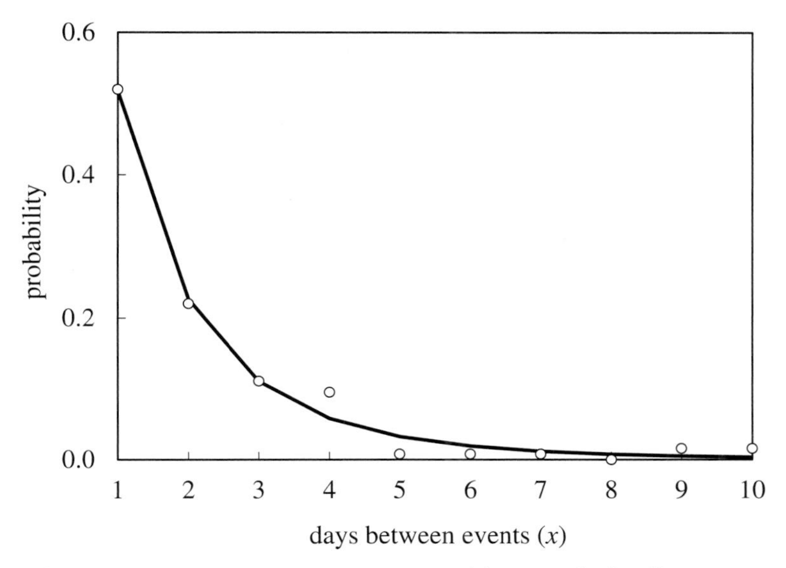

Figure 35.7 Beta-geometric: Fit to interval between high reflux events

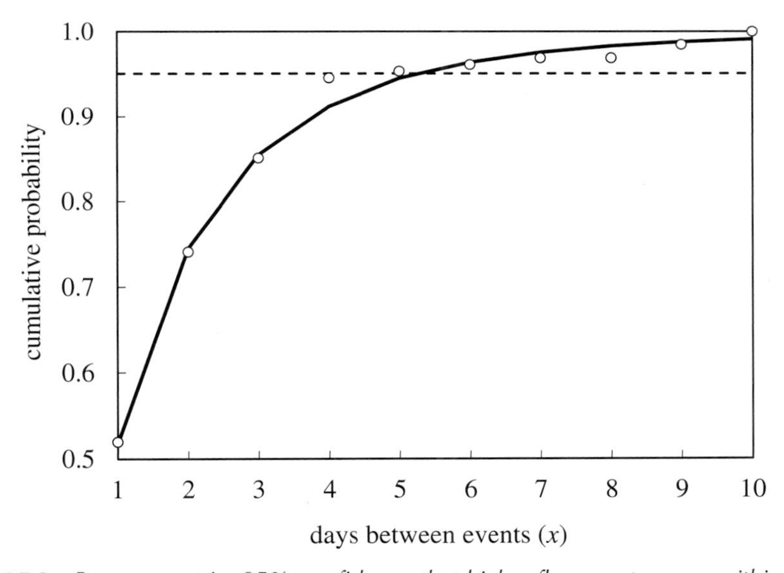

Figure 35.8 Beta-geometric: 95% confidence that high reflux event occurs within 5 days

The mean and variance are

$$\mu = \frac{1}{\alpha - 1} \qquad \alpha > 1 \tag{35.36}$$

$$\sigma^2 = \frac{\alpha^2}{(\alpha-1)^2(\alpha-2)} \qquad \alpha > 2 \tag{35.37}$$

These give μ as 0.47 and σ as 1.40 – both well away from the values calculated from the data and those estimated by the Pólya distribution.

The skewness and kurtosis are

$$\gamma = \frac{(\alpha+1)^2\sqrt{\alpha-2}}{\alpha(\alpha-3)} \qquad \alpha > 3 \tag{35.38}$$

$$\kappa = \frac{\alpha^4 + 10\alpha^3 - 30\alpha^2 + 23\alpha - 22}{\alpha(\alpha-4)(\alpha-3)} \qquad \alpha > 4 \tag{35.39}$$

35.6 Beta-Binomial

The *beta-binomial distribution* is an enhanced form of the binomial distribution. The probability of an event is assumed to be random following the beta distribution. Its PMF is

$$p(x) = \frac{n!}{x!(n-x)!}\frac{\mathrm{B}(x+\alpha, n-x+\beta)}{\mathrm{B}(\alpha,\beta)} \qquad x \geq 0;\ \alpha,\beta > 0 \tag{35.40}$$

Using the same example of the LPG splitter reflux flow, we want to assess the frequency with which the flow exceeds 70 m^3/hr. The 5,000 hourly measurements cover 208 days. Of these there were 117 where the flow never exceeded the limit, 60 where it exceeded it once, 24 where it exceeded it twice and 7 where it exceeded it three times. Converting this values to observed probabilities gives 0.5625, 0.2885, 0.1154 and 0.0337 respectively.

Since we are examining the probability of a number of hourly events occurring in 24 hours, n is 24. Fitting Equation (35.40) to the four data points gives a value of 2.12 for α and 75.8 for β. The result is shown as Figure 35.9, which also shows how the distribution can be adapted by adjusting α and β. Figure 35.10 shows that the fitted distribution closely matches the data – including the zero occasions that there were four or more violations of the limit.

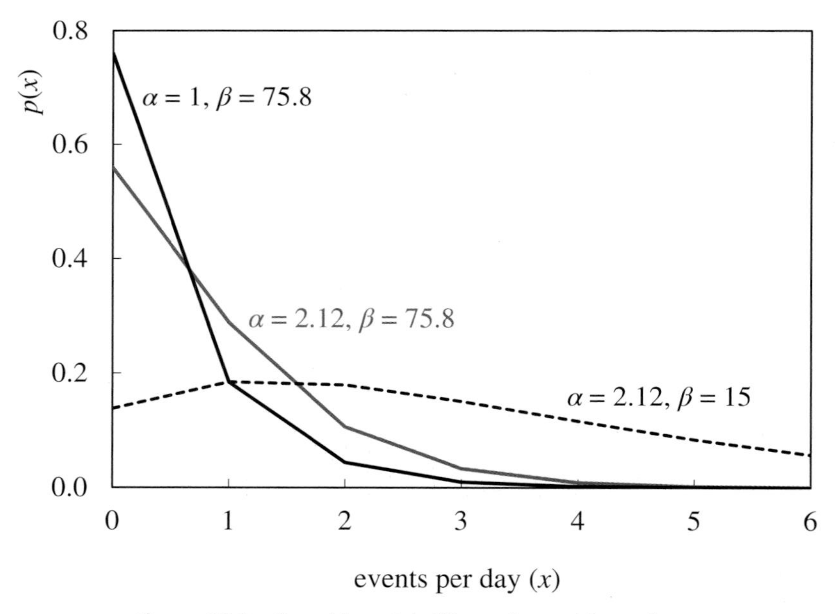

Figure 35.9 *Beta-binomial: Effect of α and β on shape*

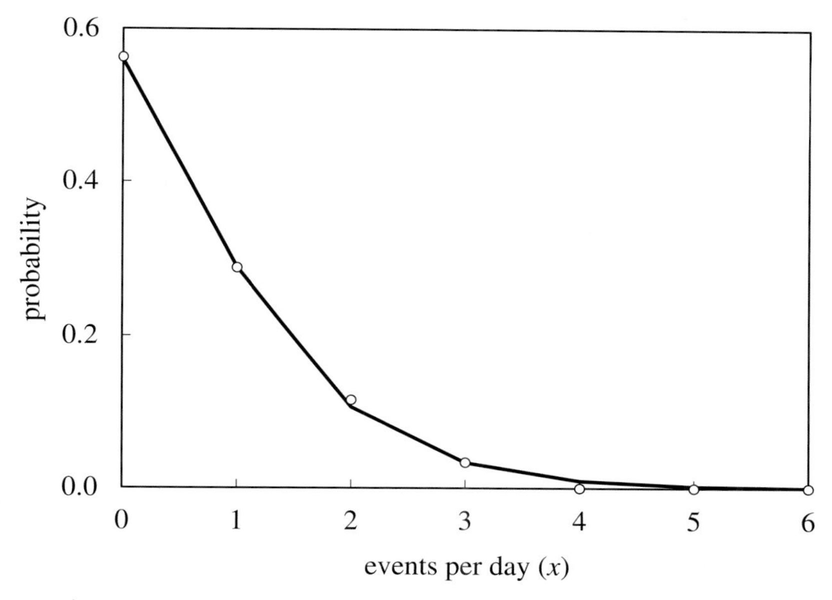

Figure 35.10 *Beta-binomial: Fitted to frequency of high reflux events*

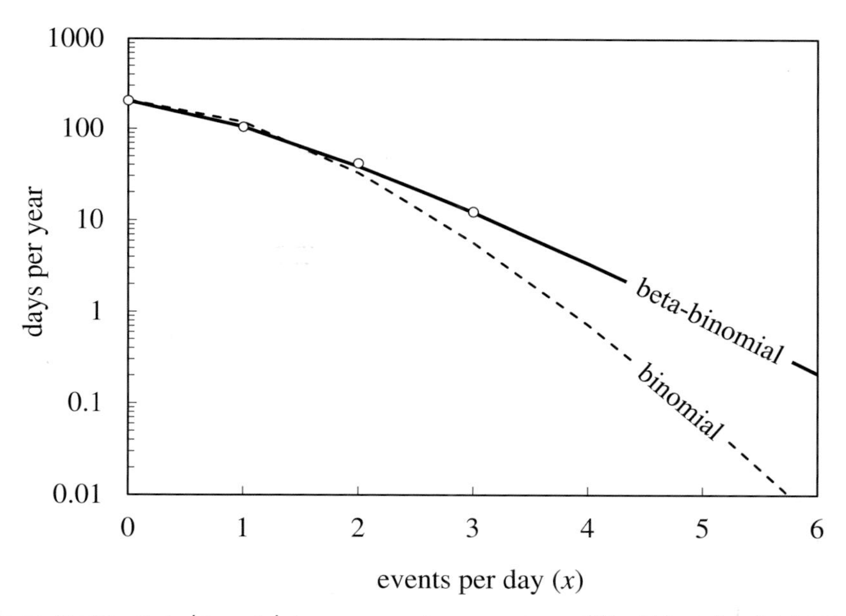

Figure 35.11 *Beta-binomial: Improvement on accuracy of fit of binomial distribution*

Figure 35.11 shows the same data converted to days per year. For example, during 208 days, there were 117 in which there was one event. This is equivalent to 205 events per year. Fitting the distribution to the four data points permits it to be extrapolated to explore the probability of more than three violations of the limit occurring in a day. For example, the probability of four violations in a day can be determined from Equation (35.40) as 0.0093. We would therefore expect this to occur three times per year. Similarly we would expect five violations in day to occur once per year.

From the basic data, we can calculate the mean number of occasions per day that the reflux violated the limit.

$$\mu=\frac{(117\times 0)+(60\times 1)+(24\times 2)+(7\times 3)}{208}=0.62 \tag{35.41}$$

Similarly the standard deviation is

$$\sigma=\sqrt{\frac{117(0-0.62)^2+60(1-0.62)^2+24(2-0.62)^2+7(3-0.62)^2}{208-1}}=0.82 \tag{35.42}$$

These parameters can also be calculated from the fitted distribution.

$$\mu=\frac{n\alpha}{\alpha+\beta} \tag{35.43}$$

$$\sigma^2=\frac{n\alpha\beta(\alpha+\beta+n)}{(\alpha+\beta)^2(\alpha+\beta+1)} \tag{35.44}$$

These equations put μ at 0.65 and σ at 0.91 – reasonably close to the calculated values.

We can compare this distribution against the classic binomial. If the total number of violations of the maximum flow is 129, out of 5,000 measurements this represents an observed probability of 0.0258. Fitting the classic binomial distribution gives a value of p close to this at 0.0234. The dashed line in Figure 35.11 shows how unreliable the extrapolation would be if the probability of the event is assumed to be constant.

Skewness is

$$\gamma=\frac{(\alpha+\beta+2n)(\beta-\alpha)}{\alpha+\beta+2}\sqrt{\frac{\alpha+\beta+1}{n\alpha\beta(\alpha+\beta+n)}} \tag{35.45}$$

There is no formula for kurtosis.

35.7 Beta-Negative Binomial

In the same way that we modify the binomial distribution to form the negative binomial distribution, the *beta-negative binomial distribution* is a modification to the negative-binomial distribution. It is also known as the *generalised Waring distribution* and the *inverse Markov–Pólya distribution*. Like the negative binomial distribution, it gives the probability of a number of failures (x) before there are n successes, but with this assumed to follow the beta distribution. The PMF is

$$p(x)=\frac{(n+x-1)!}{x!(n-1)!}\frac{\mathrm{B}(\alpha+n,\beta+x)}{\mathrm{B}(\alpha,\beta)} \qquad x\geq 0;\ \alpha,\beta>0 \tag{35.46}$$

Figure 35.12 shows the effect of varying α and β with n fixed at 10.

The mean and variance are

$$\mu=\frac{\beta n}{\alpha-1} \qquad \alpha>1 \tag{35.47}$$

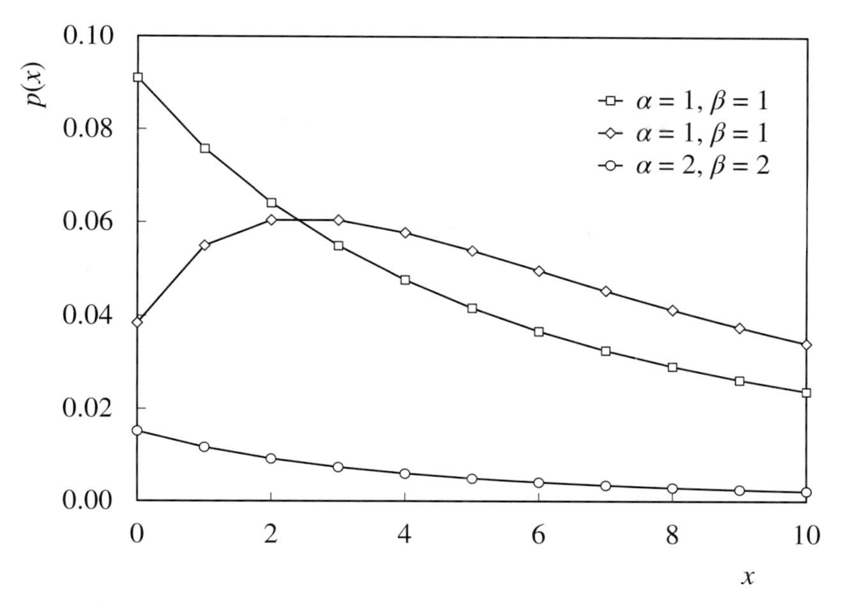

Figure 35.12 Beta-negative binomial: Effect of α and β on shape

$$\sigma^2 = \frac{\beta n(\alpha+\beta-1)(\alpha+n-1)}{(\alpha-2)(\alpha-1)^2} \quad \alpha>2 \tag{35.48}$$

The formulae for skewness and kurtosis are too complex for inclusion here.

The beta-geometric distribution, described in Section 35.4, is a special case of the beta-negative binomial distribution, in which we require only one success after x failures. Therefore n is 1. Making this substitution into Equation (35.46).

$$p(x) = \frac{B(\alpha+1,\beta+x)}{B(\alpha,\beta)} = \frac{\Gamma(\alpha+1)\Gamma(\beta+x)}{\Gamma(\alpha+\beta+x+1)} \frac{\Gamma(\alpha+\beta)}{\Gamma(\alpha)\Gamma(\beta)} = \frac{\alpha\Gamma(\alpha+\beta)\Gamma(\beta+x)}{\Gamma(\beta)\Gamma(\alpha+\beta+x+1)} \tag{35.49}$$

Remembering that x in Equation (35.29) is the number of trials including the success, Equation (35.49) is the same distribution.

The negative binomial is another special case, approached as α and β approach infinity.

35.8 Beta-Pascal

The *beta-Pascal distribution* is a shifted version of the beta-negative binomial distribution. Its PMF is

$$p(x) = \frac{(n-1+x)!}{x!(n-1)!} \frac{B(n+\alpha,\beta+x)}{B(\alpha,\beta)} \quad x \geq 0;\ \alpha,\beta>0 \tag{35.50}$$

Figure 35.13 shows the effect of varying α and β, with n set to 10. The distribution does not fit well the LPG splitter reflux example – requiring impractically large values for α and β.

There are no formulae for mean, variance, skewness or kurtosis.

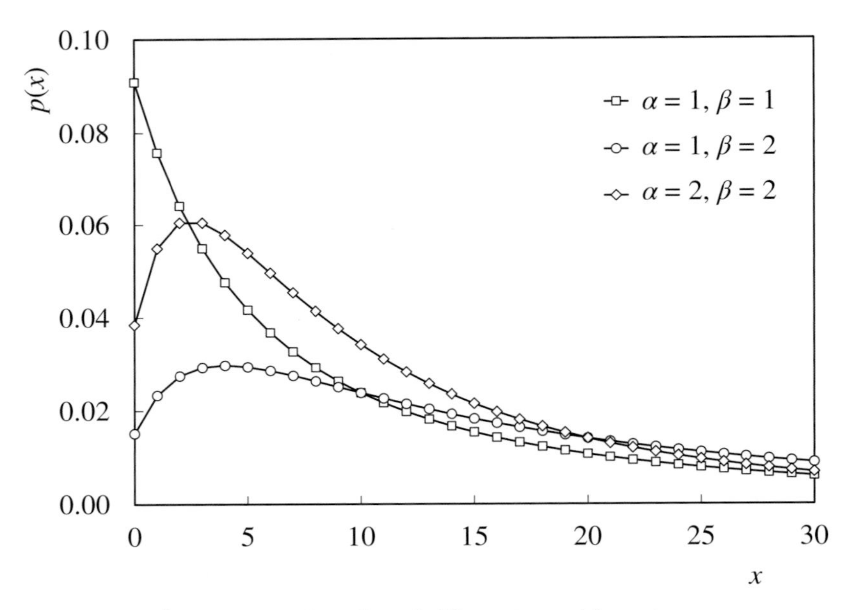

Figure 35.13 *Beta-Pascal: Effect of α and β on shape*

35.9 Gamma-Poisson

Rather than assume the expected number of successes that occur in a chosen time interval (λ) is constant, in the *gamma-Poisson distribution* it is assumed to vary randomly following the gamma distribution. Its PMF is

$$p(x) = \frac{(\beta + x - 1)!}{x!(\beta - 1)!}\left(\frac{1}{\alpha + 1}\right)^{\beta}\left(\frac{\alpha}{\alpha + 1}\right)^{x} \quad x \geq 0;\ \alpha, \beta > 0 \tag{35.51}$$

This becomes the negative-binomial distribution if

$$\alpha = \frac{1 - p}{p} \quad \text{and} \quad \beta = s \tag{35.52}$$

An alternative definition is

$$p(x) = \frac{\Gamma(\beta + x)\alpha^{x}}{\Gamma(\beta)(\alpha + 1)^{\beta + x}x!} \tag{35.53}$$

Figure 35.14 shows the effect of adjusting α and β. The moment generating function is

$$M(t) = (1 - \alpha - \alpha e^{t})^{-\beta} \tag{35.54}$$

leading to

$$\mu = \alpha\beta \tag{35.55}$$

$$\sigma^2 = \alpha\beta + \alpha^2\beta \tag{35.56}$$

$$\gamma^2 = \frac{(1 + 2\alpha)^2}{\alpha\beta(1 + \alpha)} \tag{35.57}$$

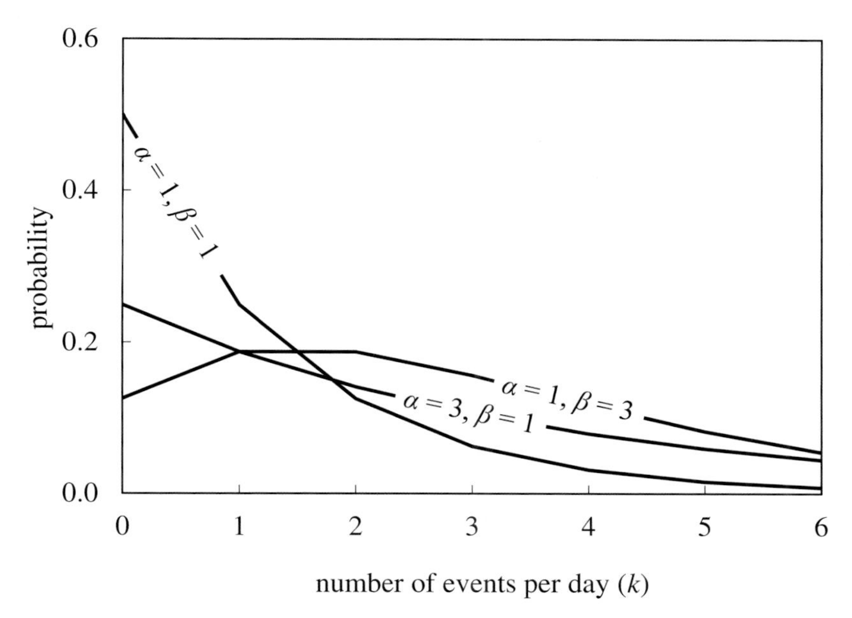

Figure 35.14 Gamma-Poisson: Effect of α and β on shape

$$\kappa = \frac{3\alpha^2\beta + 6\alpha^2 + 3\alpha\beta + 6\alpha + 1}{\alpha\beta(1+\alpha)} \tag{35.58}$$

Fitting Equation (35.53) to the LPG splitter high reflux events gives α as 0.271 and β as 2.42. From Equation (35.55) the best fit value for μ is therefore 0.66 and, from Equation (35.56), for σ it is 0.91.

Comparing again to the classic binomial distribution, we can use the fitted value of 0.0234 for p in Equation (12.95) to plot the dashed line in Figure 35.15. The better fit by the gamma-Poisson confirms that the probability of a high reflux event is not constant. Figure 35.16 shows that the gamma-Poisson distribution accurately models the number of days, per year, that have k events.

35.10 Conway–Maxwell–Poisson

The *Conway–Maxwell–Poisson distribution* is defined by the PMF

$$p(x) = K\frac{\lambda^x}{(x!)^\delta} \qquad x \geq 0;\ \lambda > 0 \tag{35.59}$$

The normalisation constant (K) is determined from

$$K = \frac{1}{\sum_{i=0}^{\infty} \frac{\lambda^i}{(i!)^\delta}} \tag{35.60}$$

While a sum to infinity might seem impractical, it converges very quickly. Even for very small values of δ, fewer than 50 terms are required.

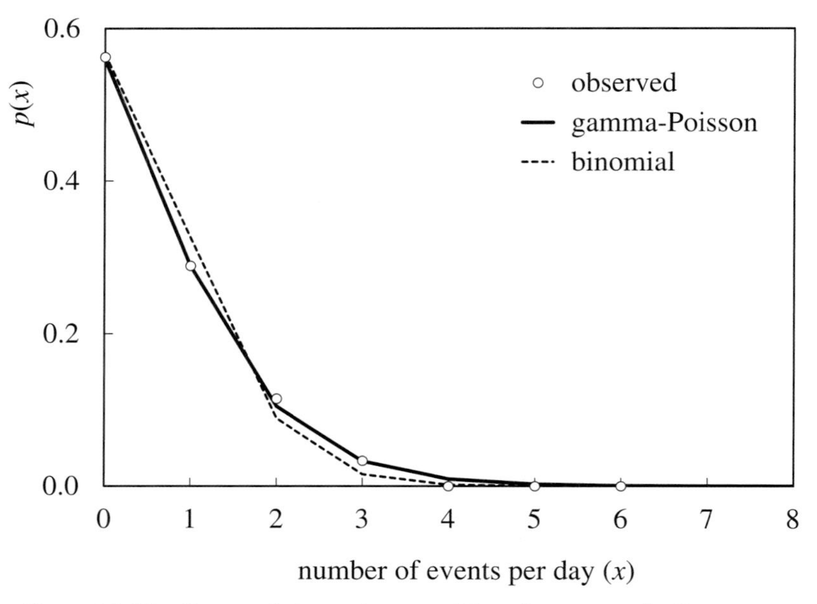

Figure 35.15 Gamma-Poisson: Improved fit to frequency of reflux events

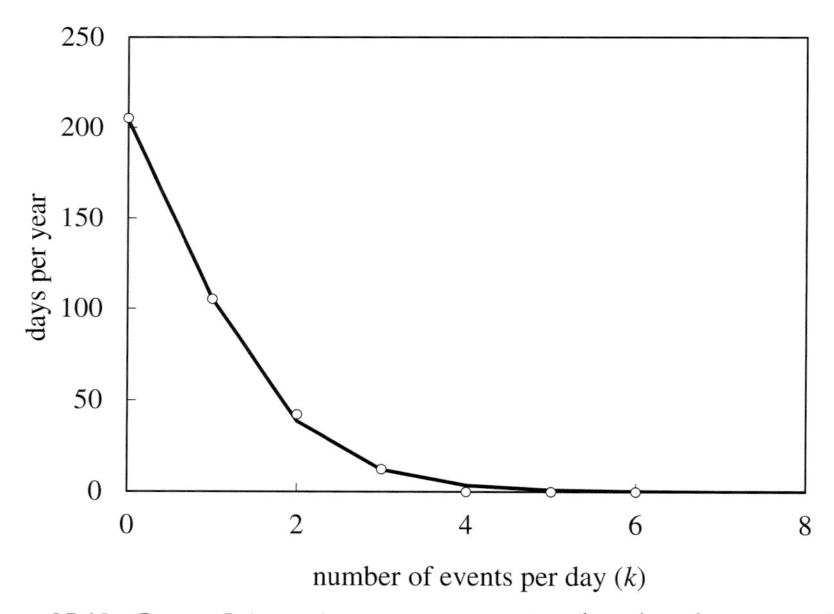

Figure 35.16 Gamma-Poisson: Accurate representation of number of events per day

If δ is set to 0, the distribution becomes the geometric distribution with p equal to $(1 - \lambda)$. If δ is set to 1, it becomes the Poisson distribution. As $\delta \rightarrow \infty$, it becomes the Bernoulli distribution with p equal to $1/(1 + \lambda)$. Figure 35.17 shows the effect of changing δ with λ fixed at 0.8.

Fitting Equation (35.59) to the same reflux example gives λ as 0.516, δ as 0.474 and hence K as 0.560. Compared to the gamma-Poisson distribution RSS is reduced from 0.00021 to 0.00016. Visually the fits would be indistinguishable.

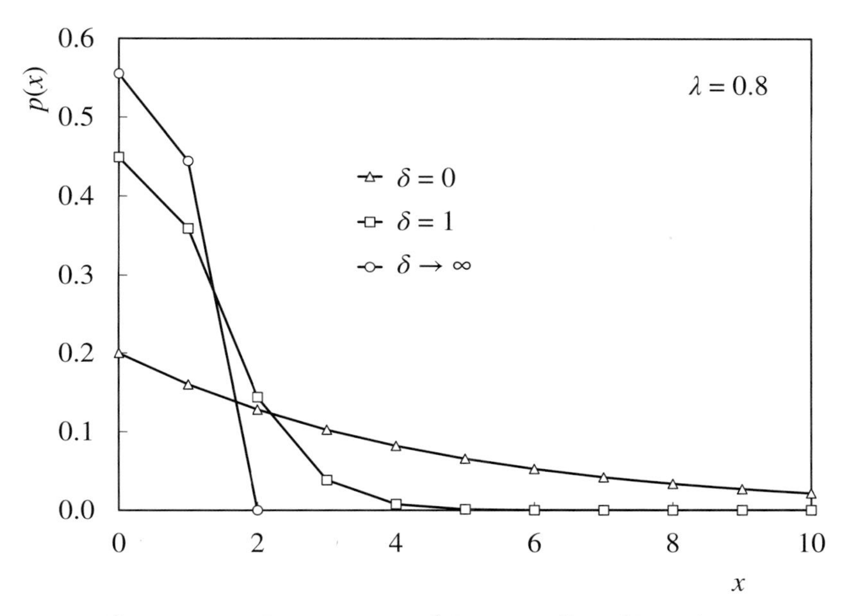

Figure 35.17 Conway–Maxwell–Poisson: Effect of δ on shape

The mean and variance are

$$\mu = K\sum_{i=o}^{\infty}\frac{i\lambda^{i}}{(i!)^{\delta}} \tag{35.61}$$

$$\sigma^{2} = K\sum_{i=o}^{\infty}\frac{i^{2}\lambda^{i}}{(i!)^{\delta}} - \mu^{2} \tag{35.62}$$

These give the mean as 0.65 and the standard deviation as 0.90. These slightly more complex formulae, in this case, put the Conway–Maxwell–Poisson distribution at a slight disadvantage compared to the simpler gamma-Poisson distribution.

The Conway–Maxwell–Poisson does, however, lend itself to a more elegant method of curve fitting. From Equation (35.59)

$$\frac{p(x-1)}{p(x)} = \frac{x^{\delta}}{\lambda} \tag{35.63}$$

$$\therefore \ \log\left(\frac{p(x-1)}{p(x)}\right) = \delta\log(x) - \log(\lambda) \tag{35.64}$$

Figure 35.18 plots Equation (35.64) based on the observed probabilities. The slope of 0.497 is the best estimate for δ. The best estimate of λ is derived from the intercept as $10^{-0.279}$ or 0.526. These values are close to those obtained by conventional least squares curve fitting. However, with only three points, we cannot be certain that the line is approximately straight. With more points the technique can be valuable in assessing whether the distribution is well chosen.

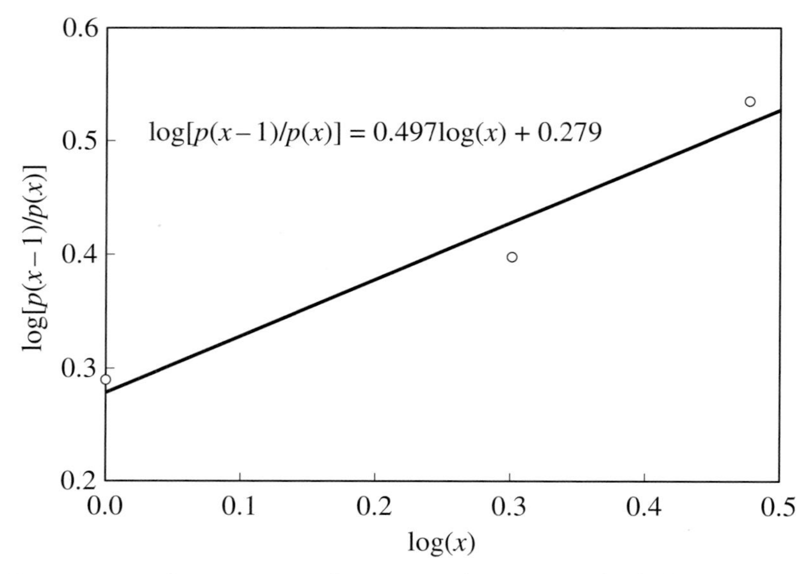

Figure 35.18 *Conway–Maxwell–Poisson: Alternative method of fitting to data*

35.11 Skellam

The *Skellam distribution* gives the probability of the difference between two values selected from two Poisson distributions. To understand this, consider two soccer teams for which we know the average rate of goal scoring per 90 minute game. The Skellam distribution would allow us to determine the probability of a specified difference between their scores when they play each other. Indeed it can be used, albeit naïvely, by gamblers in *spread betting* on sports and stock markets.

The distribution involves the use of the modified Bessel function (I_x), described in Section 11.8. If the expected number of events, in a fixed time, in two Poisson distributions are λ_1 and λ_2 then the PMF of the Skellam distribution is

$$p(x) = \exp(-\lambda_1 - \lambda_2)\left(\frac{\lambda_i}{\lambda_2}\right)^{\frac{x}{2}} I_x\left(2\sqrt{\lambda_1\lambda_2}\right) \quad \lambda_1, \lambda_2 > 0 \tag{35.65}$$

We can again use the stock level example with its five (λ_1) low-stock events per year. We might explore investment in product storage that is anticipated to reduce this to three (λ_2) per year. These values were used to plot the points Figure 35.19. They lie very close to the normal distribution, shown as the coloured line, with

$$\mu = \lambda_1 - \lambda_2 \tag{35.66}$$

$$\sigma^2 = \lambda_1 + \lambda_2 \tag{35.67}$$

Not surprisingly it shows the most likely reduction in the number of events per year is two. However the probability of achieving this is only 0.14. To determine the benefit of the investment we might better examine the probability of any improvement. For this we need the complementary CDF derived from summing the PMF

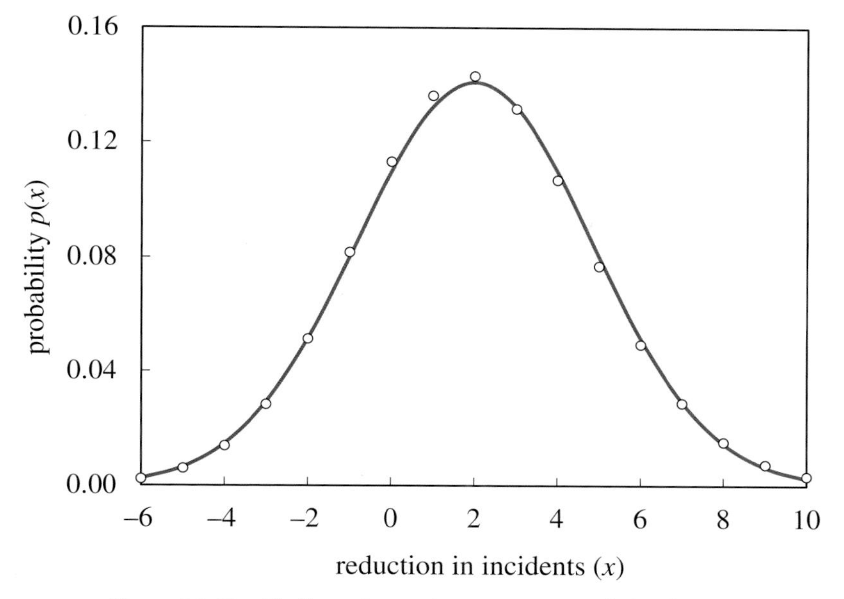

Figure 35.19 *Skellam: Approximation to normal distribution*

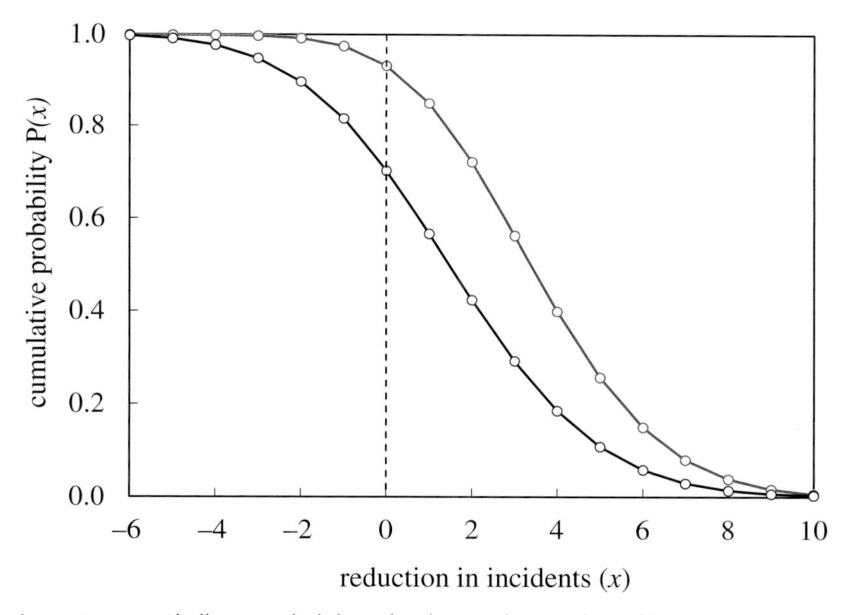

Figure 35.20 *Skellam: Probability of reducing the number of low-stock incidents*

$$P(x) = 1 - \sum_{i=-\infty}^{x} \exp(-\lambda_1 - \lambda_2)\left(\frac{\lambda_i}{\lambda_2}\right)^{\frac{i}{2}} \mathrm{I}_i\left(2\sqrt{\lambda_1 \lambda_2}\right) \tag{35.68}$$

In this example x can be negative; there is a possibility that there could be an increase in the number of events. The CDF has to be derived therefore by integrating the PMF from $-\infty$.

Clearly this is not practical; instead we integrate from a value for which the PMF is close enough to zero. For example, if k is −6, Equation (35.68) puts the probability at 0.002. Excluding lower values of k therefore has very little effect on the CDF.

Figure 35.20 plots (as the black line) the complementary CDF and shows the probability of the reduction in the number of events exceeding zero is 0.70. In other words we can be 70% certain that the investment will be beneficial. If this were considered too low then a more costly solution might reduce the average number of events to one per year. The resulting CDF is plotted as the coloured line, showing that the probability of improvement then increases to 93%.

36

Other Discrete Distributions

There are a number of discrete distributions that have no obvious connection to the binomial distribution or its derivatives. They are included here in alphabetical order. The reader might reasonably question the relevance of some to the process industry. Many have been included because they are well known and some explanation is merited as to why they have little application in the process industry. Others included are here simply because some might consider them interesting and identify a use previously not considered.

36.1 Benford

Benford's Law, also known as the *First Digit Law*, concerns the frequency at which leading digits occur in datasets. For certain datasets, the probability that a member of the set begins with the digits (x) is

$$P_x = \log_{10}\left(1 + \frac{1}{x}\right) \quad x \geq 1 \qquad (36.1)$$

Provided x is an integer the law is applicable to any string of leading digits. More commonly though it is applied to the single leading digit. Figure 36.1 shows the expected *Benford distribution*. It generally requires that the dataset covers several orders of magnitude.

One application is the detection of fraud in company accounts. Data fabricated by fraudsters might be expected to be uniformly distributed. To help understand why the Benford distribution applies to company accounts, consider all the possible additions of three single-digit numbers. There are 729 (9^3) possible results, of which 525 (72%) will have 1 as the leading digit. 120 (16%) will have 2 as the leading digit. We would need to consider a much wider range of possible summations to come close to the Benford distribution, but this example demonstrates the principle. Single entries in company accounts are themselves the results of addition. Incoming and outgoing invoices will each be an addition of several separately priced items plus taxes.

Figure 36.1 also shows the observed distribution of the leading digit in some 2,000 transactions passing through the accounts of the author's consulting company. While generally

Statistics for Process Control Engineers: A Practical Approach, First Edition. Myke King.

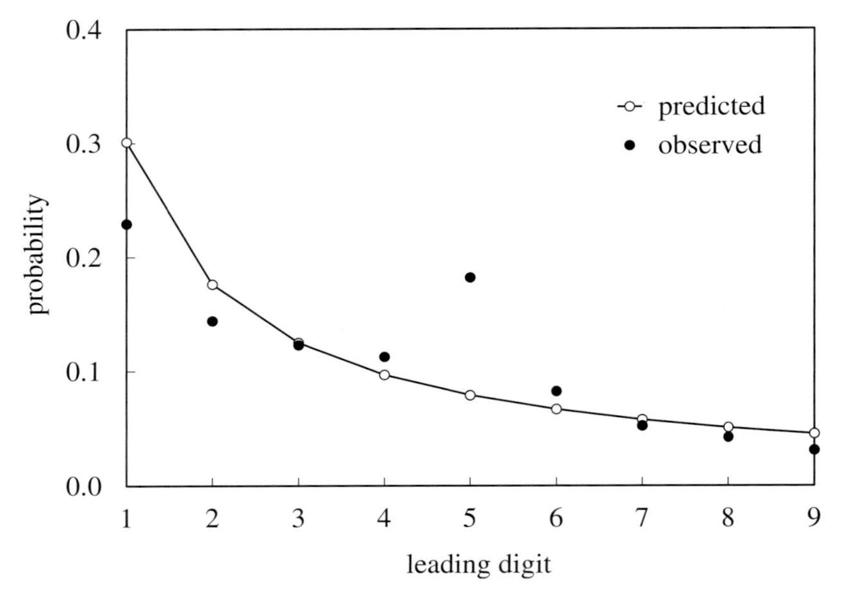

Figure 36.1 *Benford: Expected and actual distribution of leading digit in accounts*

following the expected distribution, most notable is the digit 5. While this could be the work of a very unimaginative fraudster, it is in fact accounted for by a large number of frequent regular transactions of two amounts – both having 5 as their first digit. Figure 36.2 is a P–P plot showing that otherwise the distribution is close to that expected

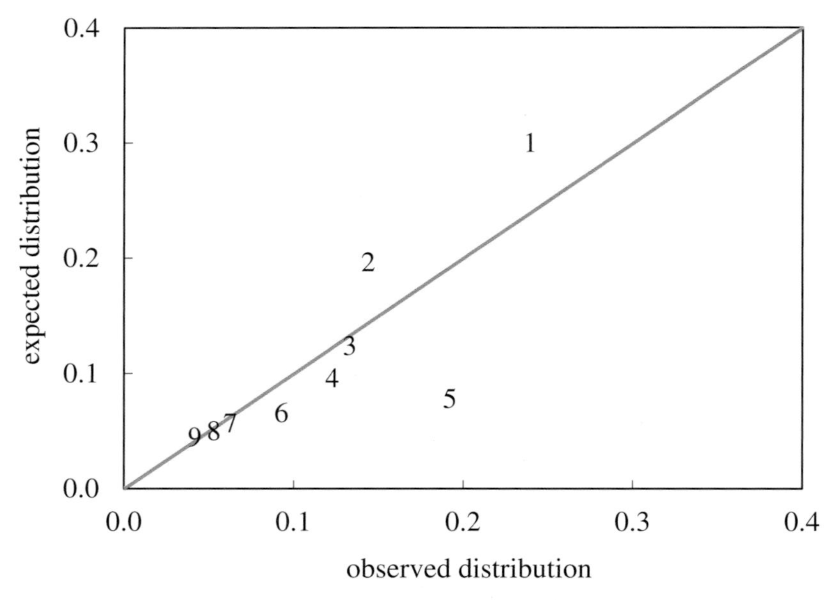

Figure 36.2 *Benford: P–P plot showing exception from expected distribution*

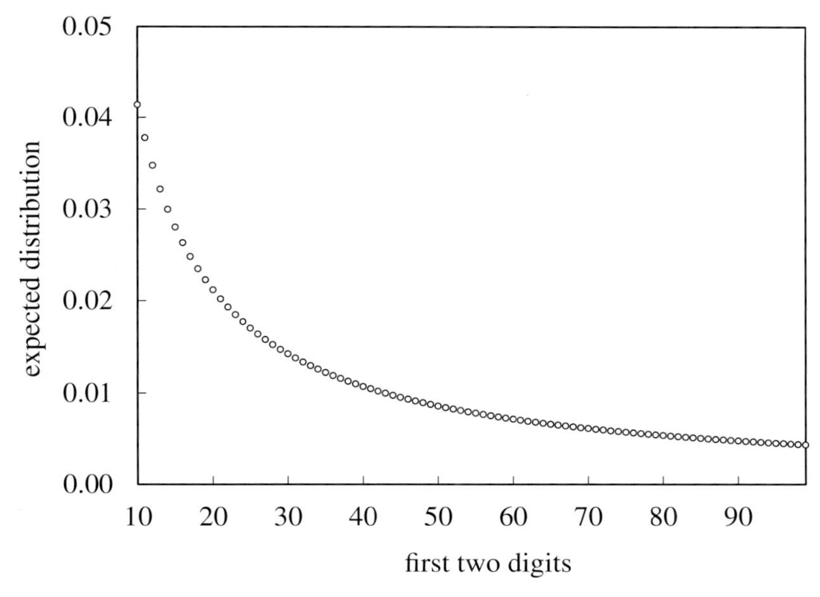

Figure 36.3 Benford: Expected distribution of two leading digits

In practice, such investigations would also explore the distribution of the second and perhaps subsequent digits. For example, Figure 36.3 plots Equation (36.1) for all two-digit values of x. It is also possible to examine the occurrence of digits in other positions. For example, the probability of the second digit being a 5 is given by the sum of the probabilities of the first two digits being 15, 25, … or 95. From Equation (36.1) we can calculate this as

$$P_{x=5} = \log_{10}\left(1 + \frac{1}{15}\right) + \log_{10}\left(1 + \frac{1}{25}\right) \ldots + \log_{10}\left(1 + \frac{1}{95}\right) = 0.0967 \qquad (36.2)$$

In general the probability of the second digit being x is

$$P_x = \sum_{i=1}^{9} \log_{10}\left(1 + \frac{1}{10i + x}\right) \qquad (36.3)$$

Figure 36.4 shows the expected distribution of the second digit.

The Benford distribution might at first appear to be an unlikely choice for process data. However, atomic weights are the result of summing the number of protons and neutrons in the nucleus. Figure 36.5 includes all 108 known elements and shows that they have some tendency to follow the Benford distribution displayed by the solid line. Further, molecular weights are the sum of atomic weights and there are many properties, such as gas density and volumetric NHV, that are directly proportional to molecular weight. In principle these might then follow the distribution although, in practice, it is unlikely that the data include sufficient variation in order of magnitude for it to be observable.

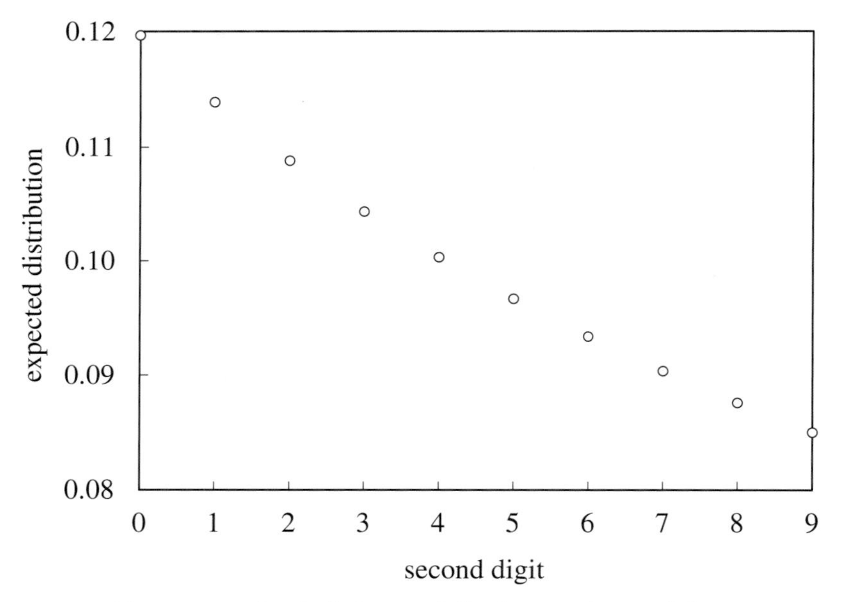

Figure 36.4 Benford: Expected distribution of second digit

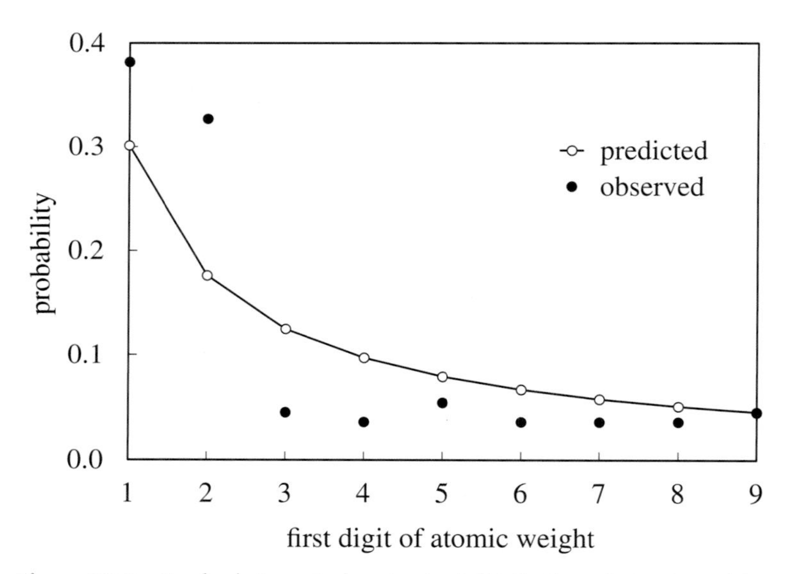

Figure 36.5 Benford: Expected and actual distribution of atomic weights

36.2 Borel–Tanner

The *Borel–Tanner distribution* is defined by the PMF

$$p(x) = \frac{ne^{-\lambda x}(\lambda x)^{x-n}}{x(x-n)!} \qquad x \geq n;\ 0 < \lambda < 1 \tag{36.4}$$

Its main application is to queuing, where n is the number in the queue at the start and k is the number of items dealt with before the queue is empty. The *intensity* (λ) is the average number of items arriving – expressed as a fraction of the number that can be dealt with. Clearly λ must be less than 1 for the queue to be emptied. The distribution assumes that arrivals follow the Poisson distribution and that the time to deal with each item is constant.

Figure 36.6 shows the effect of changing λ with n set to 5. Figure 36.7 shows the effect of changing n with λ set to 0.5.

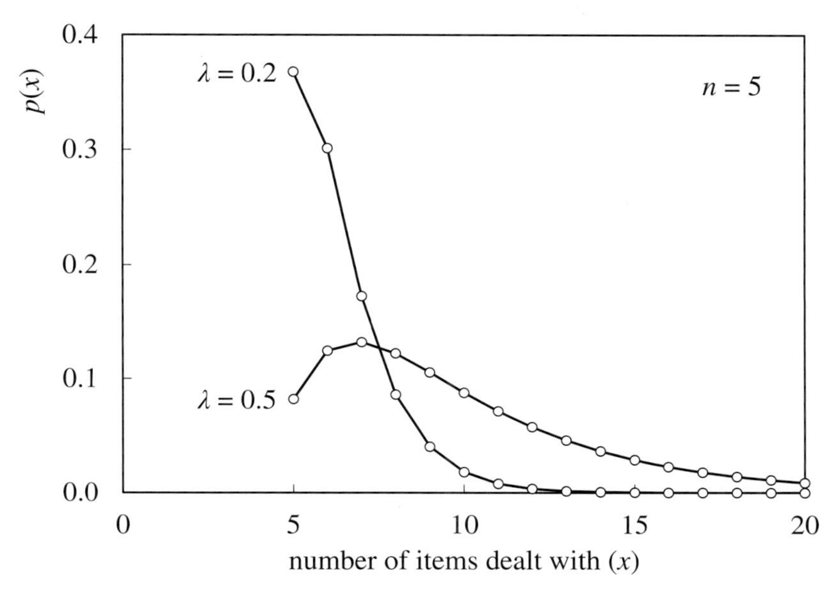

Figure 36.6 Borel–Tanner: Effect of λ on shape

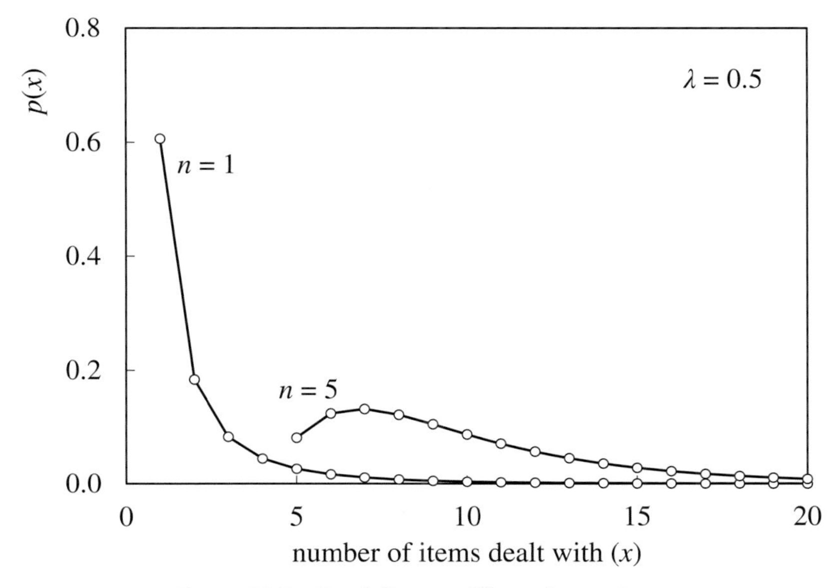

Figure 36.7 Borel–Tanner: Effect of n on shape

As an example, consider a process in which a number of batch reactors are operated simultaneously. On completion, a batch is pumped to its own intermediate storage tank. The material in the tank is later tested and, if on grade, routed to the main shipping tank. Production runs 24 hours a day, every day, and typically produces 20 batches per week. The laboratory staff work 09:00 to 17:00, Monday to Friday. Typically they can test and empty 35 intermediate tanks per week. It is important that, at the end of the laboratory's working day on Friday, there are no outstanding tests. On Monday morning, 64 hours later, the laboratory would expect 8 (n) batches awaiting testing.

The intensity (λ) is 20/35, or 0.571. Figure 36.8 plots the cumulative distribution derived from Equation (36.4). The black curve shows that currently in about 4% of the weeks the number of batches that must be tested exceeds the capacity of 35 and so production has to be reduced.

There is a plan to increase production to 24 batches per week. The intensity will therefore increase to 24/35, or 0.686. There are now likely to be 10 batches waiting for testing on Monday morning. The coloured curve in Figure 36.8 shows that production would have to be reduced in about 31% of the weeks.

To maintain the current availability the capacity of the laboratory must be increased to 42 per week – as shown by the black curve in Figure 36.9. Of course, it is not surprising that a 20% increase in production requires a 20% increase in laboratory capacity. While it is reassuring that the technique confirms this, its value lies in allowing us to explore other options. For example, if the cost of increasing the laboratory capacity beyond 40 was prohibitive then the intensity would reduce to 0.600. The coloured line in Figure 36.9 shows that we would now expect to exceed the capacity in about 7% of the weeks.

The mean and variance are

$$\mu = \frac{n}{1-\lambda} \tag{36.5}$$

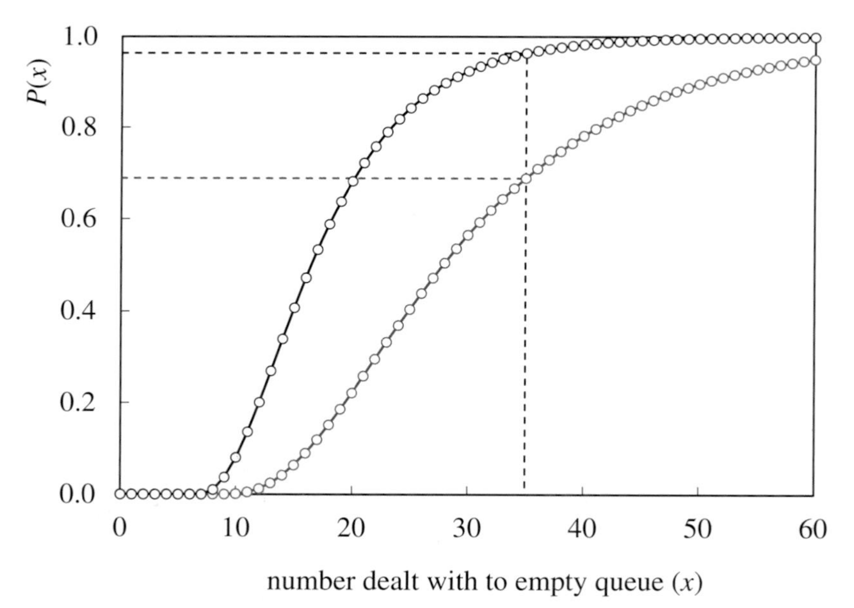

Figure 36.8 *Borel–Tanner: Impact of increased production on likelihood of completion*

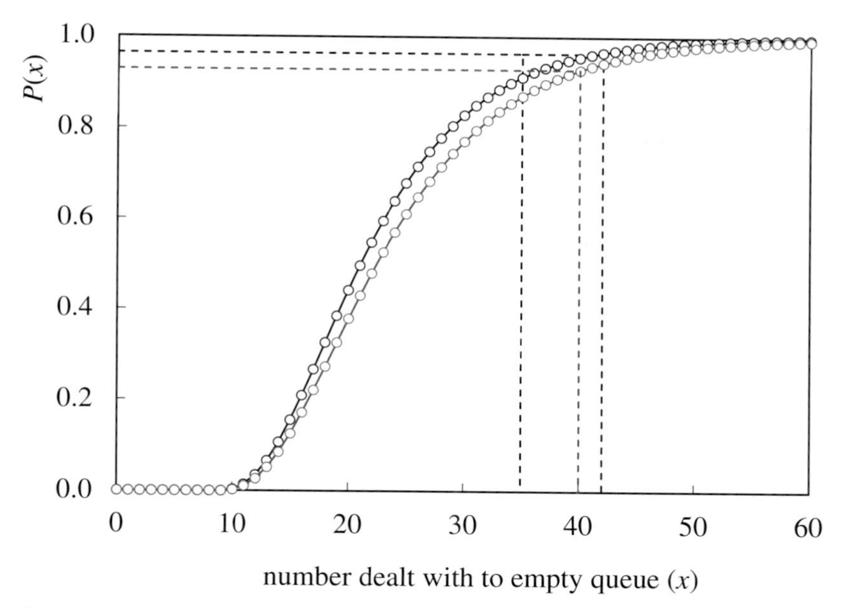

Figure 36.9 Borel–Tanner: Impact of restricted increase in laboratory capacity

$$\sigma^2 = \frac{n\lambda}{(1-\lambda)^3} \tag{36.6}$$

The *Borel distribution* is obtained by setting n to 1.

$$p(x) = \frac{e^{-\lambda x}(\lambda x)^{x-1}}{x!} \tag{36.7}$$

The Borel distribution can be shifted by α, in which case it becomes the *Lagrange–Poisson distribution* or the *Poisson–Consul* distribution.

$$p(x) = \frac{\alpha e^{-(\lambda x+\alpha)}(\lambda x+\alpha)^{x-1}}{x!} \tag{36.8}$$

$$\mu = \frac{\alpha}{1-\lambda} \tag{36.9}$$

$$\sigma^2 = \frac{\alpha}{(1-\lambda)^3} \tag{36.10}$$

36.3 Consul

The *Consul distribution* is described by the PMF

$$p(x) = \frac{(\alpha x)!\beta^{x-1}(1-\beta)^{\alpha x-x+1}}{x!(\alpha x-x+1)!} \qquad x \geq 1;\ 0 \leq \beta < 1;\ 1 \leq \alpha \leq \frac{1}{\beta} \tag{36.11}$$

Figure 36.10 shows the effect of α and β. If α is set to 1, and β replaced with $(1-p)$, the distribution becomes the form of geometric distribution described by Equation (35.21).

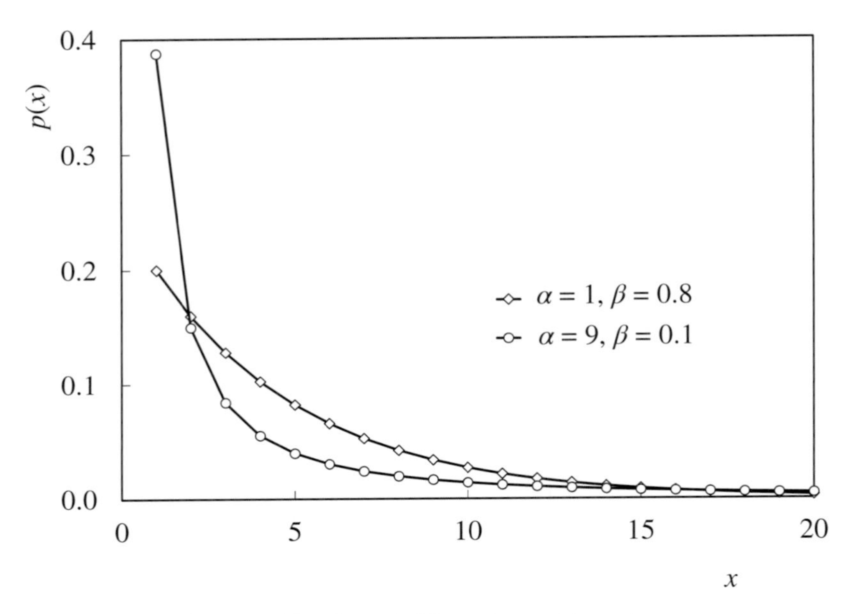

Figure 36.10 Consul: Effect of α and β on shape

The mean and variance are

$$\mu = \frac{1}{1-\alpha\beta} \tag{36.12}$$

$$\sigma^2 = \frac{\alpha\beta(1-\beta)}{(1-\alpha\beta)^2} \tag{36.13}$$

Rearranging Equation (36.12)

$$\beta = \frac{\mu-1}{\alpha\mu} \tag{36.14}$$

Substituting in Equation (36.11) gives an alternative definition of the PMF, based on μ.

$$p(x) = \frac{(\alpha x)!}{x!(\alpha x - x + 1)!}\left(\frac{\mu-1}{\alpha\mu}\right)^{x-1}\left(1-\frac{\mu-1}{\alpha\mu}\right)^{\alpha x - x + 1} \tag{36.15}$$

36.4 Delaporte

The *Delaporte distribution* is somewhat unusual in that the PMF is defined as a summation.

$$p(x) = \sum_{i=0}^{x} \frac{\Gamma(\alpha+i)\beta^i \lambda^{x-i} e^{-\lambda}}{\Gamma(\alpha) i! (\beta+1)^{\alpha+i}(x-i)!} \qquad x \geq 0;\ \lambda \geq 0;\ \alpha,\beta > 0 \tag{36.16}$$

Figure 36.11 shows the effect of varying λ. If λ is zero, the distribution becomes the negative binomial. Figure 36.12 shows the effect of varying α and β. If both are zero, the distribution becomes the Poisson distribution. The fact that the PMF covers other distributions might

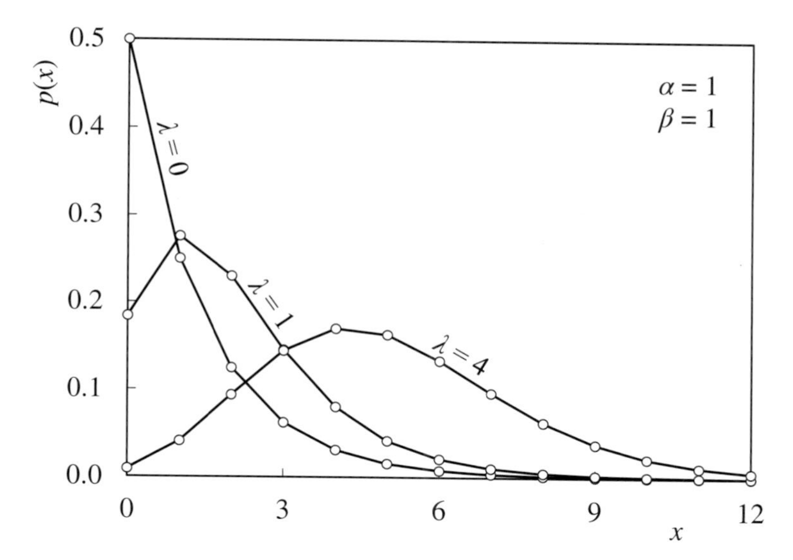

Figure 36.11 Delaporte: Effect of λ on shape

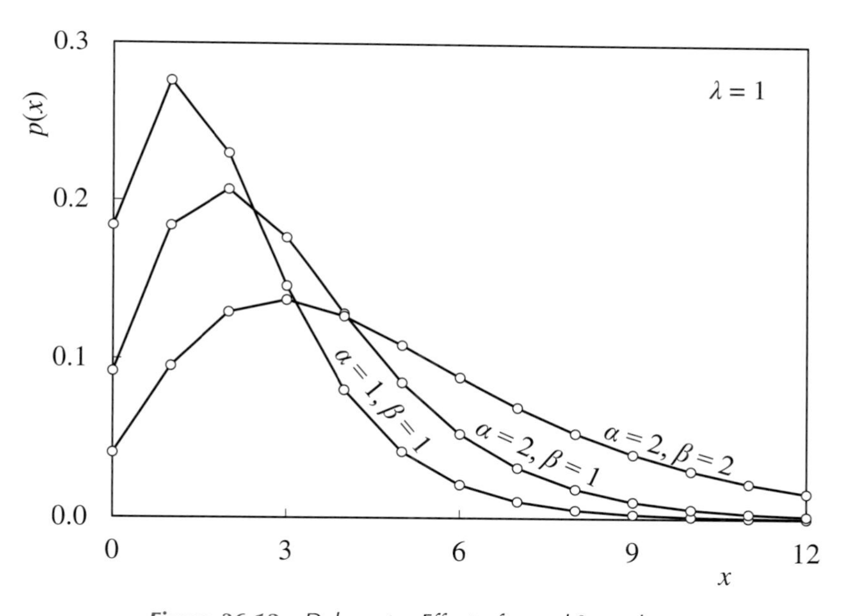

Figure 36.12 Delaporte: Effect of α and β on shape

suggest, for example, that we need not consider the Poisson as the prior distribution. However if α is close to zero, then adjusting β has very little effect and vice versa. So, if the distribution is close to Poisson, there will be a very wide range of values of α and β that give apparently the same fit. Similarly, if both are large, then adjusting λ can have little effect. A search algorithm, if it moves into these regions, can easily become 'lost' and manual intervention will be required to locate the best fit. It is therefore better to try fitting the simpler distributions first and, given the complexity of its PMF, only consider the Delaporte distribution if one of these fails to provide the required fit.

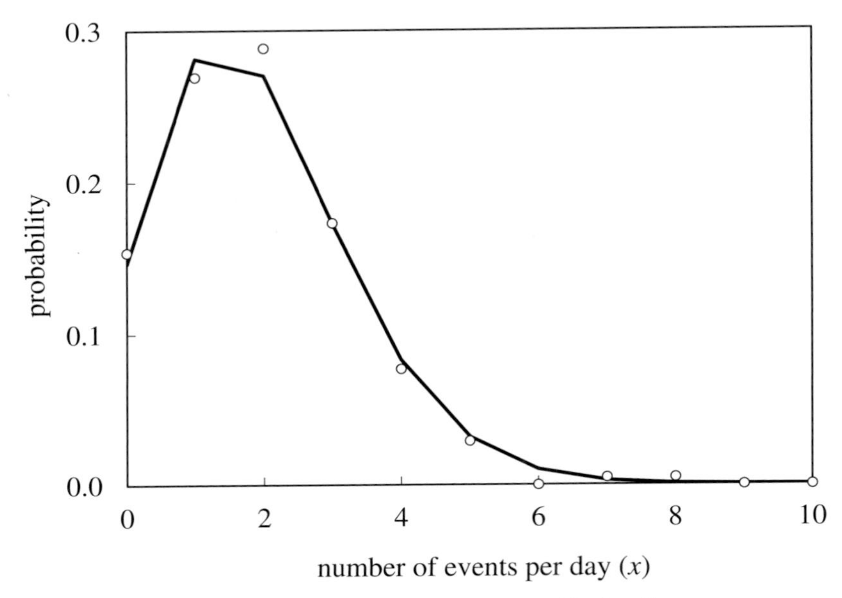

Figure 36.13 Delaporte: Fit to frequency of high reflux events

To illustrate this we can use the LPG reflux flow example but, this time, define an event as the flow exceeding 65 m^3/hr. Figure 36.13 shows the distribution of the number of events that occurred in a 24 hour period, along with the fitted distribution with λ set to 1.92, α and β to 0. The same accuracy of fit can be achieved with a very different values for either α or β.

Fortunately, if the resulting fit does set either α or β to zero, the value of the other parameter has no impact on the mean or variance that are

$$\mu = \lambda + \alpha\beta \tag{36.17}$$

$$\sigma^2 = \lambda + \alpha\beta(1+\beta) \tag{36.18}$$

36.5 Flory–Schulz

The *Flory–Schulz distribution* was originally developed to describe the relative ratios of molecule lengths after a polymerisation reaction. Its PMF is

$$p(x) = p^2 x(1-p)^{x-1} \quad x \geq 1;\ 0 \leq p \leq 1 \tag{36.19}$$

Figure 36.14 shows the effect of varying p. As expected the probability of a molecule having zero length is zero. This makes the distribution applicable to data where we expect P_0 to be zero.

The example of the time between events of the LPG splitter reflux exceeding 65 m^3/hr strictly requires a continuous distribution function. However, because the reflux flow was recorded at hourly intervals, the time between events is an integer number of hours. The data can also therefore be treated as discontinuous. Further there cannot be an interval of zero.

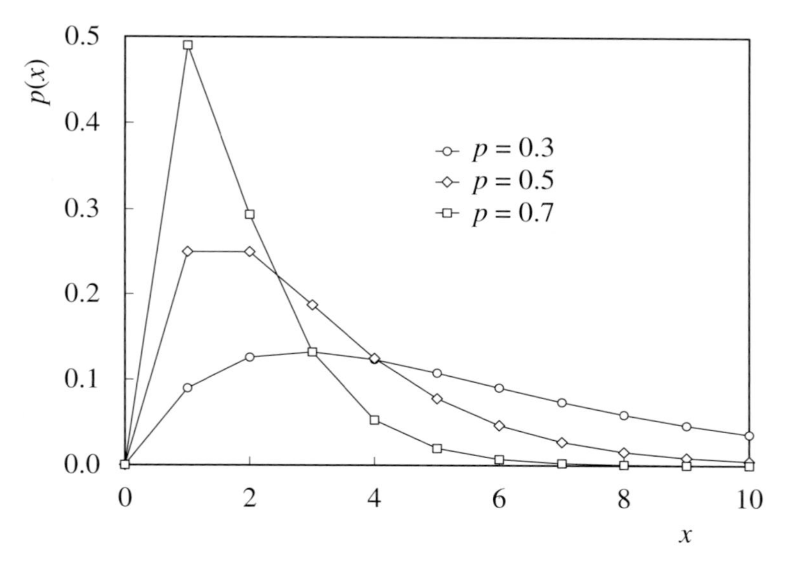

Figure 36.14 *Flory–Schulz: Effect of* p *on shape*

Fitting Equation (36.19) gives p as 0.155. With *RSS* at 0.1458, the fit is quite poor compared to that achieved, for example, by the exponential distribution.

The mean is

$$\mu = \frac{2-p}{p} \tag{36.20}$$

This gives μ as 11.9 hours – close to that estimated using the exponential distribution. The other key parameters are

$$\sigma^2 = \frac{2(1-p)}{p^2} \tag{36.21}$$

$$\gamma = \frac{2-p}{\sqrt{2(1-p)}} \tag{36.22}$$

$$\kappa = \frac{p^2 - 12p + 12}{2(1-p)} \tag{36.23}$$

36.6 Hypergeometric

Included here because it is commonly referred to in text books and spreadsheet packages, the *hypergeometric distribution* has limited application to the process industry. The example normally used is the *urn problem*. This might be a container holding a mixture of two coloured balls, say, black and white. The total number of balls in the container is N which includes K black balls, and hence $(N-K)$ white balls. A number (n) balls are then withdrawn, without returning any to the container. The probability that x black balls, and hence $(n-x)$ white balls, will be withdrawn is

$$p(x)=\frac{\binom{K}{x}\binom{N-K}{n-x}}{\binom{N}{n}} \qquad 0\leq x\leq n;\ 0\leq n\leq N \tag{36.24}$$

With N fixed at 100 and n at 10, Figure 36.15 shows the effect of varying K. The distribution has been applied in situations where a small number (n) of batches of product are randomly selected, for testing, from a much larger number (N). The test would be of the pass/fail type; x would be the number of batches failing the test. Knowing $p(x)$, perhaps from historical data, would allow the total number (K) of off-grade batches to be estimated. If excessive, then this might instigate selection of more batches for testing.

The mean and variance are

$$\mu=\frac{nK}{N} \tag{36.25}$$

$$\sigma^2=\frac{nK(N-K)(N-n)}{N^2(N-1)} \tag{36.26}$$

There are formulae for skewness and kurtosis but which are too complex to merit inclusion here.

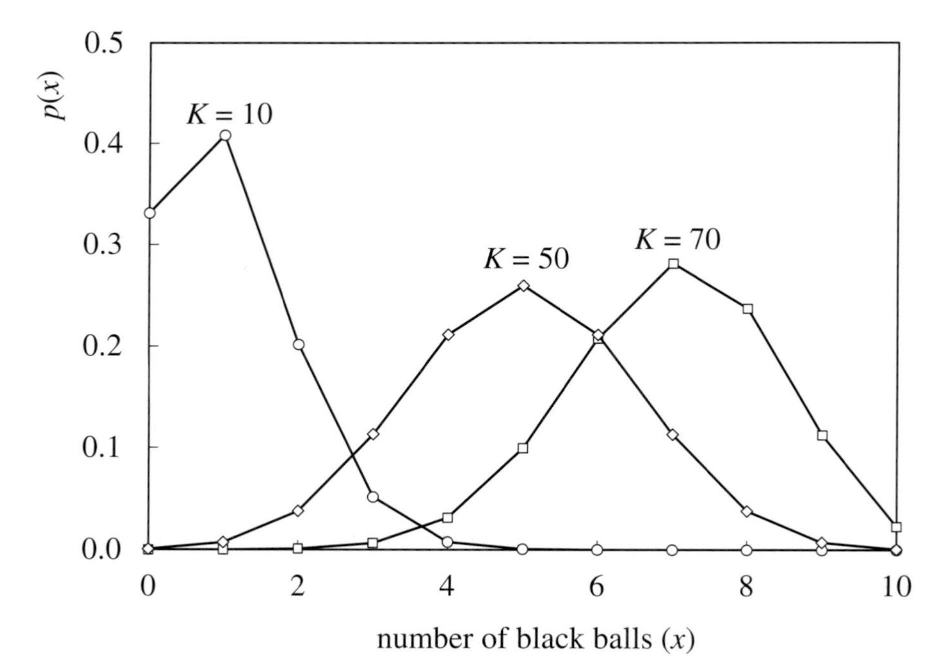

Figure 36.15 Hypergeometric: Effect of K on shape

36.7 Negative Hypergeometric

The PMF of the *negative hypergeometric distribution* is

$$p(x)=\frac{\binom{n_1+x-1}{x}\binom{n_2+n_3-(n_1+x+1)}{n_2-x}}{\binom{n_2+n_3-1}{n_2}} \qquad 0\le x\le n_2;\ n_1,n_2,n_3>0 \tag{36.27}$$

Figure 36.16 shows the effect of varying n_1, with n_2 fixed at 20 and n_3 at 10.

Theoretically the PMF could be fitted to process data although, since n_1, n_2 and n_3 are integers, specialist software would be required.

The mean and variance are

$$\mu=\frac{n_1 n_2}{n_3} \tag{36.28}$$

$$\sigma^2=\frac{n_1 n_2(n_2+n_3)(n_3-n_1)}{n_3^2(n_3+1)} \tag{36.29}$$

36.8 Logarithmic

The title of *logarithmic distribution* is also used for another unrelated distribution and so this one is better described as the *log-series distribution*. It PMF is

$$p(x)=\frac{-p^x}{x\ln(1-p)} \qquad x\ge 1;\ 0\le p\le 1 \tag{36.30}$$

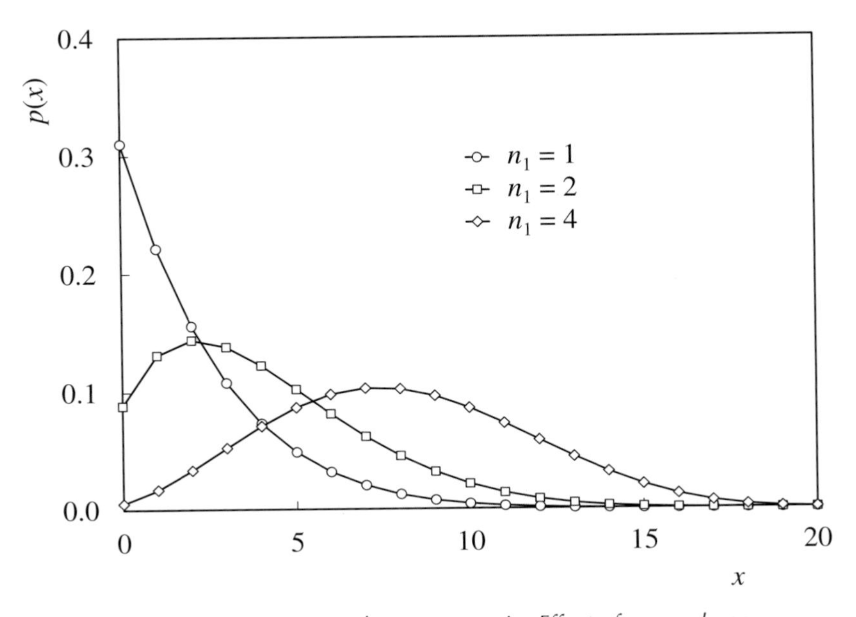

Figure 36.16 Negative hypergeometric: Effect of n_1 on shape

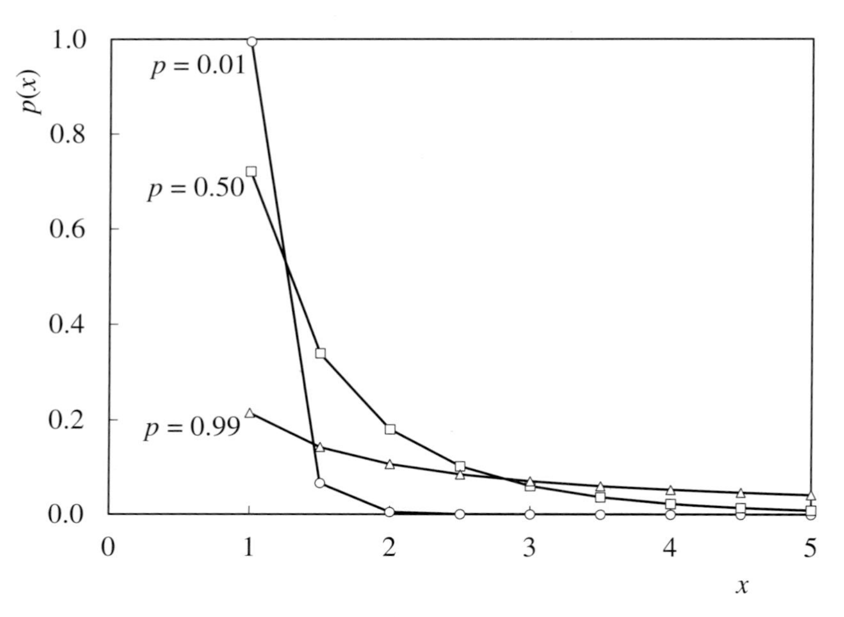

Figure 36.17 *Logarithmic: Effect of* p *on shape*

Figure 36.17 shows the effect of changing p. The mean and variance are

$$\mu = \frac{-p}{(1-p)\ln(1-p)} \tag{36.31}$$

$$\sigma^2 = \frac{-p[p+\ln(1-p)]}{(1-p)^2\ln^2(1-p)} \tag{36.32}$$

Using our batch blending example in Table A1.7, it took 78 blends to make 44 batches; 33 met the specification first time, 4 required two blends, 3 required three and 3 required four. There is also what appears to be one exceptional case that took 15 blends. Converting these to observed probabilities gives 0.7500, 0.9909, 0.0682, 0.0682 and 0.0227. Fitting Equation (36.30) to these results gives p as 0.428. This is the probability that the event will occur, i.e. the probability that a blend will be off-grade. The fitted value is close to that calculated from the data, i.e. 0.436 (1 – 44/78). The fit is shown as Figure 36.18.

The mean number of blends per batch can be calculated as 1.77 (78/44). From Equation (36.31) the estimate is somewhat different at 1.34. Using this result, if a blend takes a day to complete, we would complete 273 batches in 365 days. From Equation (36.32) the standard deviation is 0.74. If improved control halved this variability then iteratively solving Equation (36.32) shows that p would reduce to 0.192. Putting this result into Equation (36.31) shows that the mean reduces to 1.11. This increases, to 328, the number of batches that can be completed in 365 days, i.e. a 20% increase in production.

While this example illustrates the calculation method the volume of data used is too small to have confidence in the result. Apart from the large difference between the calculated and fitted results for the mean, Equation (36.30) would predict a probability of 3.5×10^{-7} that a batch would take 15 blends. In fact the observed probability is 1 in 44 batches. As Figure 36.18 shows there is also a mismatch when x is 2 and 4. Selecting a different distribution is unlikely to adequately improve the fit.

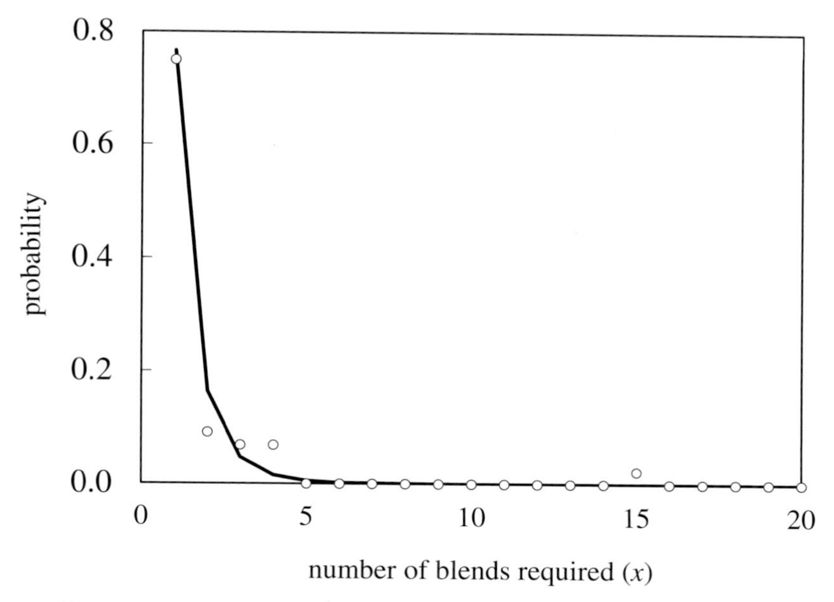

Figure 36.18 Logarithmic: Fit to number of blends require per batch

36.9 Discrete Weibull

The *discrete Weibull distribution* is described by the PMF

$$p(x) = (1-p)^{x^\beta} - (1-p)^{(x+1)^\beta} \quad x \geq 0;\ 0 \leq p \leq 1;\ \beta > 0 \tag{36.33}$$

The definition of p is different from other distributions we have covered; rather it being the average probability of the event occurring, it is the probability that none will occur in the defined period.

Figure 36.19 shows the effect of varying β, with p fixed at 0.2.

There are alternative ways of formulating the PMF. One is to define q as $1-p$.

$$p(x) = q^{x^\beta} - q^{(x+1)^\beta} \tag{36.34}$$

Another is to then replace q.

$$q = \exp\left(-\alpha^{-\beta}\right) \tag{36.35}$$

This results in

$$p(x) = \exp\left[-\left(\frac{x}{\alpha}\right)^\beta\right] - \exp\left[-\left(\frac{x+1}{\alpha}\right)^\beta\right] \tag{36.36}$$

Fitting to the LPG splitter high reflux flow events, gives p as 0.560 and β as 1.20. The accuracy of fit is very similar to that of the gamma-Poisson distribution.

The corresponding CDF are

$$P(x) = 1 - (1-p)^{(x+1)^\beta} \tag{36.37}$$

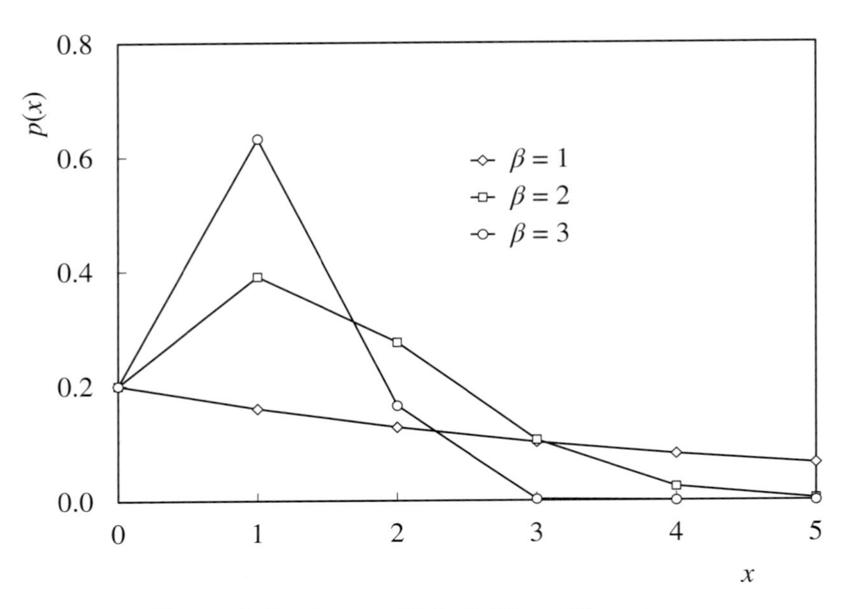

Figure 36.19 Discrete Weibull: Effect of β on shape

$$P(x) = 1 - \exp\left[-\left(\frac{x+1}{\alpha}\right)^{\beta}\right] \tag{36.38}$$

The corresponding QF are

$$x(P) = \left[\frac{\ln(1-P)}{\ln(1-p)}\right]^{\frac{1}{\beta}} - 1 \quad 0 \le P \le 1 \tag{36.39}$$

$$x(P) = \alpha[-\ln(1-P)]^{1/\beta} - 1 \quad 0 \le P \le 1 \tag{36.40}$$

The mean can be determined from

$$\mu = \sum_{x=0}^{\infty}\left[(1-p)^{x^{\beta}} - (1-p)^{(x+1)^{\beta}}\right]x \tag{36.41}$$

While the summation is theoretically to infinity, in practice we need only calculate to the maximum possible value of x. The maximum value in the data is 3. Summing to $x = 20$ gives μ as 0.656 – identical to that derived from fitting the gamma-Poisson distribution.

There are no formulae for variance, skewness or kurtosis.

36.10 Zeta

The *zeta distribution* takes its name from its use of the Riemann zeta function, described in Section 11.10. Its PMF is

$$p(x) = \frac{x^{-\lambda}}{\zeta(\lambda)} \quad x \ge 1;\ \lambda > 0 \tag{36.42}$$

Fitting to batch blending example gives λ as 2.53. *RSS* is 0.0049 and so is bettered by the Pólya distribution. The mean and variance are

$$\mu = \frac{\zeta(\lambda-1)}{\zeta(\lambda)} \quad \text{for } \lambda > 2 \tag{36.43}$$

$$\sigma^2 = \frac{\zeta(\lambda)\zeta(\lambda-2)-\zeta(\lambda-1)^2}{\zeta(\lambda)^2} \quad \text{for } \lambda > 3 \tag{36.44}$$

The mean can be calculated as 1.86, close to that calculated from the data. Because λ is less than 3, the standard deviation cannot be determined.

The skewness and kurtosis can be calculated from the moments.

$$m_n = \frac{\zeta(\lambda-n)}{\zeta(\lambda)} \quad \text{for } \lambda > n+1 \tag{36.45}$$

36.11 Zipf

The *Zipf distribution* was derived from the study of language. This showed that the frequency of any word in a large quantity of text is inversely proportional to its ranking (x) in the frequency table.

$$p(x) \propto \frac{1}{x} \quad x \geq 1 \tag{36.46}$$

Of over 100,000 words in this book, the most frequently used 'the' appears 8.39% of the time. Its observed probability (P_1) is therefore 0.0839. Second most frequent is 'of' ($P_2 = 0.0334$) and third is 'is' ($P_3 = 0.0306$). These points are plotted, along with those for the remaining of the top 30 words, as Figure 36.20.

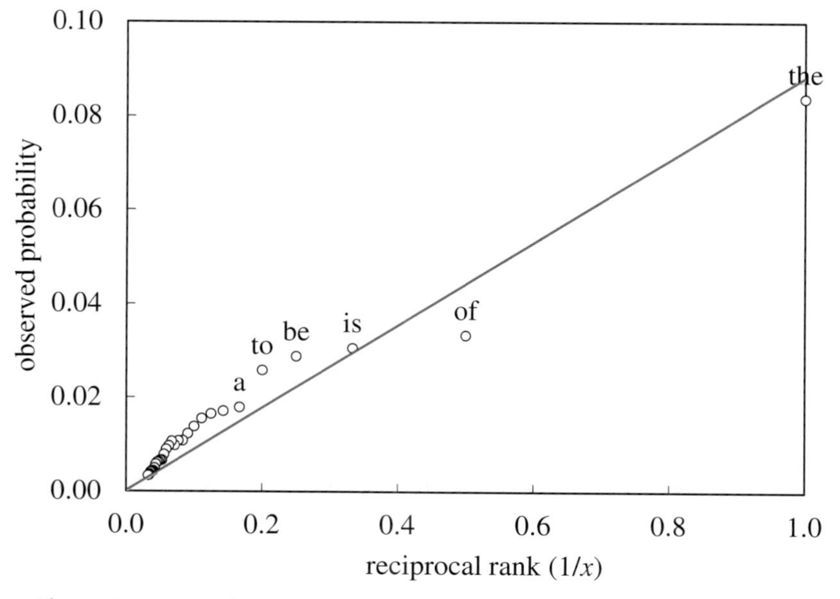

Figure 36.20 *Zipf: Actual distribution against that expected from ranking*

More generally studies show that

$$p(x) = \frac{x^{-\lambda}}{\ln(K.N)} \tag{36.47}$$

The terms λ and K are shape parameters, while N is the total number of words in the text. Fitted to the frequency of the top 100 words in this book, K is 2.05 and λ is 0.865. These parameters can be used to validate the authorship of large texts by comparison with texts whose authorship is certain. As Figure 36.21 shows the author of this book seems to use the word 'of' less than might be expected.

However, if we exclude the words 'the' and 'of' from the fit, K becomes 0.19 and λ, at 0.936, is much closer to the predicted value of 1. The resulting P–P plot is shown as Figure 36.22. As might be expected, the excluded words lie well away from the line. This is an example the *king effect*, in which the one or two elements ranked highest in a dataset do not conform to the distribution followed by all the other elements.

The reason that the Zipf distribution fits to language has never been fully explained. The evidence is largely empirical. It also has been used to describe the distribution of hits over a selection of web sites. So, while it was developed for a very specific application, there is no reason why it should not be considered for any dataset.

The commonly used form of the PMF is similar to that of the zeta distribution.

$$p(x) = \frac{x^{-\lambda}}{H_{N,\lambda}} \tag{36.48}$$

The CDF is

$$P(x) = \frac{H_{x,\lambda}}{H_{N,\lambda}} \tag{36.49}$$

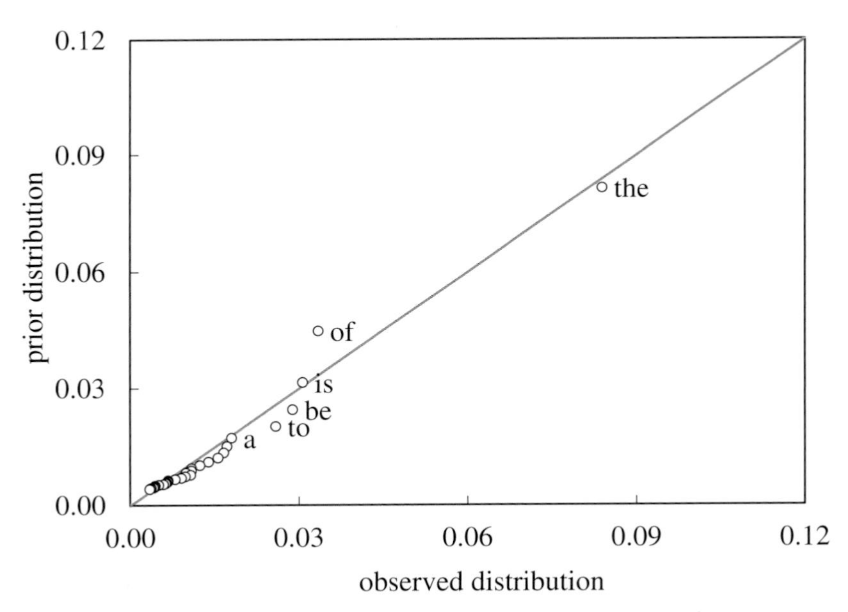

Figure 36.21 *Zipf: P–P plot showing close match for almost all words*

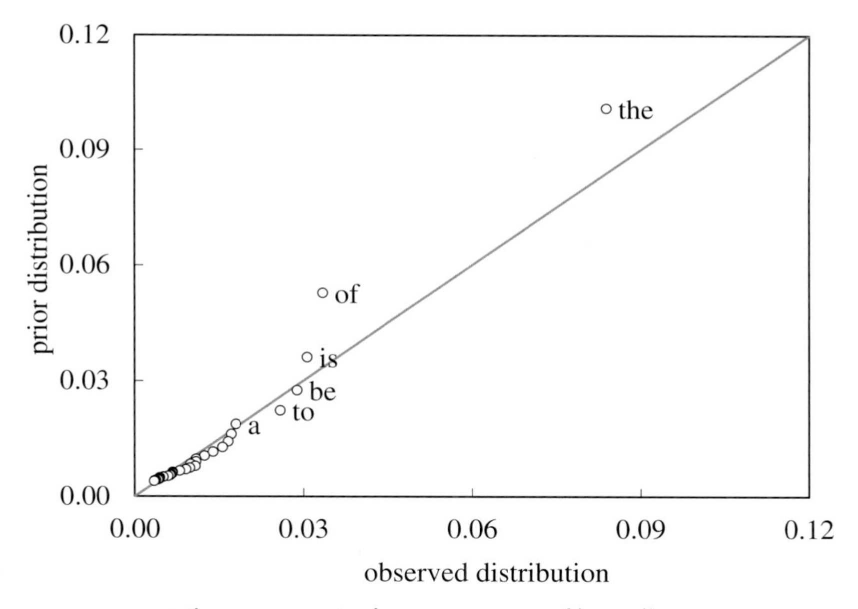

Figure 36.22 *Zipf: Demonstration of king effect*

$H_{N,\lambda}$ is the harmonic number, as described in Section 11.11, and N is the number of elements. Applying it to the batch blending example, we choose N as 44 – the total number of batches. Fitting gives λ as 2.53. This is virtually identical to the zeta distribution because, for large values of N, $H_{N,\lambda}$ approaches $\zeta(\lambda)$. The mean will also be the same as that for the zeta distribution.

$$\mu = \frac{H_{N,\lambda-1}}{H_{N,\lambda}} \tag{36.50}$$

The *Zipf–Mandelbrot distribution* includes an additional parameter (α), where α is an integer. It is defined by

$$p(x) = \frac{(x+\alpha)^{-\lambda}}{H_{N,\alpha,\lambda}} \tag{36.51}$$

$$P(x) = \frac{H_{x,\alpha,\lambda}}{H_{N,\alpha,\lambda}} \tag{36.52}$$

$$\mu = \frac{H_{N,\alpha,\lambda-1}}{H_{N,\alpha,\lambda}} - \alpha \tag{36.53}$$

36.12 Parabolic Fractal

The *parabolic fractal distribution* is the relationship between the logarithm of the probability and a quadratic function of the logarithm of the ranking.

$$\ln[p(x)] = a + b\ln(x) + c\ln^2(x) \qquad x > 0 \tag{36.54}$$

In general, when fitting the PMF to event data, some of the actual frequencies may be zero and so $\ln[p(x)]$ would present a problem. This resolved by rearranging Equation (36.54).

$$p(x) = e^{a} x^{b} \exp\{c \ln^{2}(x)\} \quad x > 0 \tag{36.55}$$

This also allows us to compare the fit with that of other distributions since *RSS* will have the units of $p(x)$ rather than $\ln[p(x)]$.

Fitting this to the frequency of the top 100 words in this book gives a as −2.53, b as −0.777 and c as −0.0278. *RSS* is 0.000286, compared to 0.000304 for the Zipf distribution, and so the fit is slightly less accurate.

An alternative choice is the stretched exponential distribution, described by Equation (28.27). Fitting gives δ as 0.336 and λ as 0.0364. At 0.000272, *RSS* is slightly improved.

Appendix 1

Data Used in Examples

Table A1.1 *Results for vol% C_4 in propane rundown*

Jan	Feb	Mar	Apr	May	Jun	Jul	Aug	Sep	Oct	Nov	Dec
5.3	4.9	4.2	2.4	2.4	2.1	2.8	2.4	4.6	12.5	2.8	5.2
4.3	4.8	2.2	2.6	2.5	2.2	5.0	6.7	4.9	6.4	3.2	10.9
2.5	5.3	2.8	7.0	5.2	9.4	5.3	4.5	5.9	3.8	2.5	5.0
3.9	5.9	4.3	10.2	6.0	3.1	3.1	4.1	11.1	6.2	9.5	3.1
3.6	5.3	10.6	3.5	7.3	3.5	3.8	4.5	1.7	5.3	8.6	2.8
3.8	4.1	7.0	7.4	5.9	3.5	6.5	6.0	4.3	10.1	3.5	3.2
3.8	6.7	2.1	3.1	7.1	3.9	3.0	5.6	2.8	4.2	5.2	4.6
4.9	2.9	5.6	12.2	8.5	6.8	2.4	6.2	4.5	2.0	3.0	3.2
5.3	2.6	1.9	3.5	4.0	5.2	7.0	6.3	5.0	2.4	2.9	5.6
3.2	3.1	6.9	10.7	1.6	6.7	5.6	8.5	8.0	4.3	7.0	8.4
3.6	13.1	9.6	3.6	2.0	6.1	3.6	3.8	5.9	2.7	1.7	2.9
4.0	3.0	3.8	3.7	5.1	4.9	2.8	13.6	3.9	2.0	7.4	2.2
3.9	3.0	10.6	4.3	2.2	5.0	3.2	10.5	3.9	4.7	1.9	5.7
2.8	4.6	6.7	4.7	3.2	3.2	15.8	5.3	6.4	4.1	4.4	5.1
4.1	2.2	6.9	1.9	2.7	4.0	6.6	3.4	3.7	2.2	7.8	5.3
4.0	8.4	3.9	4.7	6.0	3.9	8.7	1.9	3.2	12.0	8.3	1.6
4.4	5.3	3.5	4.4	2.1	3.2	5.1	2.1	4.8	6.1	4.3	2.9
6.9	4.4	8.2	6.9	3.3	1.5	3.7	3.5	9.9	2.5	3.0	3.4
6.2	7.8	2.5	5.9	7.9	4.9	4.7	1.5	3.3	5.1	5.5	2.9
11.6	3.0	6.1	9.8	2.2	3.5	3.7	6.9	4.2	3.5	3.1	7.1
5.7	5.6	3.5	9.5	2.3	3.0	7.3	6.7	7.6	2.9	2.9	7.1
10.4	6.4	6.6	6.3	3.5	6.0	2.0	7.1	2.1	2.1	3.7	7.8
9.9	2.1	2.8	5.4	3.7	2.5	3.8	6.8	3.0	2.0	4.2	3.5
1.5	3.4	5.3	4.0	5.5	4.2	3.6	5.3	4.7	4.5	5.6	3.5
4.4	2.3	3.3	3.2	2.6	5.3	6.2	2.8	2.5	4.3	2.8	4.7
7.1	3.8	2.7	1.8	3.6	5.9	2.8	7.8	3.1	3.5	7.4	6.5
4.4	5.5	2.3	3.8	3.0	8.6	8.4	4.0	4.6	3.0	4.0	5.2
8.0	4.3	4.5	4.0	3.0	5.4	5.9	3.7	11.2	3.6	1.9	6.5
2.4		4.1	5.6	10.4	8.5	2.6	6.0	3.9	8.3	5.3	3.7
2.8		3.3	7.5	5.4	2.7	3.2	4.7	3.2	10.2	4.6	3.5
3.6		7.8		4.9		3.3	3.8		3.2		1.9

Statistics for Process Control Engineers: A Practical Approach, First Edition. Myke King.

Table A1.2 *Results for vol% C_2 in 100 propane cargoes*

3.6	3.7	3.4	4.7	2.5
3.4	2.1	3.8	4.0	4.6
4.8	4.5	2.9	3.0	3.8
2.7	4.1	2.9	3.7	3.5
4.1	4.3	4.0	3.2	3.2
4.8	3.3	3.3	3.5	4.8
4.3	3.1	2.2	3.2	3.2
2.7	2.8	3.6	4.4	5.0
3.2	4.0	3.2	1.9	4.6
3.4	2.1	3.4	4.8	4.7
4.2	3.1	2.9	2.6	2.7
3.4	3.9	3.4	2.1	2.4
2.9	2.8	3.4	3.7	3.8
3.3	4.0	3.2	2.4	3.0
4.3	4.9	4.1	4.1	3.7
3.6	3.1	3.8	2.2	4.4
3.6	4.0	3.7	3.5	3.9
4.2	3.7	3.5	3.0	2.3
3.4	3.1	3.1	4.9	4.8
3.2	3.4	2.6	4.3	4.0

Table A1.3 *Gasoil 95% distillation points (°C)*

Jan	Feb	Mar	Apr
347	338	359	355
364	**377**	360	356
364	366	355	355
362	363	323	352
359	**369**	337	357
366	366	341	355
360	355	353	352
359	341	345	358
358	344	350	350
358	362	355	350
356	353	359	352
354	359	359	363
360	359	366	358
338	338	362	362
347	345	358	363
361	365	361	365
365	353	365	365
362	356	362	363
362	355	364	**367**
366	366	365	365
360	354	363	360
361	351	**368**	
355	355	361	
357	357	355	
355	351	361	
362	353	359	
354	345	357	
365	356	353	
367		354	
336		354	
337		356	

Table A1.4 *Results for fuel gas NHV (MJ/sm^3)*

Jan	Feb	Mar	Apr	May	Jun
37.79	37.77	38.78	38.87	39.87	37.57
37.72	34.59	38.50	38.73	39.93	38.26
37.53	33.15	38.54	39.11	39.74	41.90
37.46	34.84	39.11	38.75	39.60	40.21
37.43	37.45	40.08	37.33	37.50	38.72
37.88	36.85	39.93	35.82	37.79	38.13
38.18	37.28	40.14	36.70	39.09	37.99
38.22	37.02	40.31	38.51	39.06	37.55
38.18	36.86	40.99	38.31	39.37	37.78
38.12	36.97	41.09	38.22	39.46	37.66
38.13	37.23	39.30	38.77	39.44	37.45
37.82	37.40	38.19	38.72	39.23	36.91
38.27	37.52	37.53	38.13	38.93	38.32
38.06	37.31	37.73	38.14	39.13	39.37
38.69	40.55	38.35	38.06	39.38	39.08
38.44	39.98	38.24	38.20	37.40	38.77
38.28	39.97	37.83	37.43	38.87	38.68
38.33	38.30	38.60	37.86	37.26	38.65
38.74	38.24	38.85	38.03	37.56	39.16
38.75	38.22	38.48	37.93	37.63	38.69
38.55	37.83	38.29	38.24	38.14	39.45
38.43	38.84	38.95	38.01	37.73	39.01
38.48	39.49	38.68	38.10	36.88	39.32
38.34	39.53	38.79	38.32	37.96	39.24
38.62	39.52	38.98	38.81	38.35	38.61
38.40	39.03	38.81	39.00	38.51	39.00
38.20	39.53	38.36	40.37	38.38	37.94
38.15	39.85	38.25	40.17	38.40	38.68
38.17		37.97	40.31	38.13	38.41
37.49		37.78	39.71	38.35	38.22
37.34		38.19		37.52	

Table A1.5 *Fuel gas analyses (mol %)*

sample	H_2	CH_4	C_2	C_3	C_4	C_5	C_{6+}	N_2	CO	CO_2	H_2S
1	55.85	15.65	14.02	9.13	3.69	0.54	0.11	0.11	0.65	0.22	0.03
2	54.60	13.06	13.06	9.36	6.89	1.13	0.21	0.87	0.41	0.31	0.11
3	52.54	13.16	12.24	12.44	6.68	1.03	0.21	0.87	0.62	0.21	0.01
4	52.52	13.10	13.00	12.37	5.45	1.36	0.31	1.15	0.52	0.21	0.00
5	52.29	12.48	12.79	10.95	7.16	1.43	0.41	1.76	0.51	0.20	0.01
6	63.19	10.65	10.45	7.33	5.73	0.40	0.40	1.73	0.00	0.10	0.01
7	62.89	10.15	10.85	7.74	5.63	0.40	0.60	1.73	0.00	0.00	0.02
8	64.72	11.56	10.84	6.40	4.75	0.72	0.10	0.38	0.41	0.10	0.01
9	54.25	12.45	13.16	8.91	6.88	1.52	0.51	1.59	0.51	0.20	0.03
10	60.77	12.50	11.78	7.69	5.02	0.82	0.20	0.63	0.31	0.10	0.17
11	65.60	10.88	10.07	6.81	4.68	0.92	0.20	0.40	0.31	0.10	0.03
12	59.71	14.02	12.67	8.00	4.26	0.83	0.10	0.00	0.31	0.10	0.00
13	59.95	12.46	11.75	9.70	4.60	1.12	0.20	0.00	0.20	0.00	0.01
14	60.99	10.42	11.04	8.46	5.99	1.24	0.62	0.89	0.21	0.10	0.04
15	70.49	8.66	7.33	6.01	5.20	0.92	0.20	0.30	0.41	0.31	0.17
16	74.08	8.46	7.52	6.48	2.40	0.31	0.10	0.00	0.52	0.10	0.00
17	57.34	11.63	11.73	9.78	7.82	0.93	0.21	0.15	0.31	0.10	0.01
18	54.28	13.59	13.70	8.38	5.72	1.64	0.51	1.55	0.41	0.20	0.01
19	56.31	14.57	13.43	8.47	4.44	1.03	0.21	0.79	0.41	0.31	0.01
20	52.95	12.76	13.67	9.21	6.68	1.62	0.51	1.89	0.51	0.20	0.01
21	53.05	13.57	13.57	10.02	5.97	1.32	0.61	1.19	0.51	0.20	0.00
22	52.20	14.62	14.00	9.57	6.49	1.24	0.31	0.76	0.51	0.21	0.09
23	52.34	13.66	13.16	9.85	8.34	1.31	0.30	0.34	0.50	0.20	0.00
24	53.93	13.74	12.71	9.53	6.25	1.33	0.41	1.25	0.51	0.21	0.12
25	53.73	14.25	13.64	9.46	5.29	1.32	0.41	1.19	0.51	0.20	0.00
26	53.86	13.90	13.55	9.07	5.17	1.26	0.34	2.00	0.57	0.23	0.03
27	53.35	13.01	13.62	9.76	6.50	1.42	0.41	1.21	0.51	0.20	0.01
28	64.99	9.13	9.24	8.15	5.43	0.98	0.11	1.09	0.54	0.33	0.01
29	48.20	13.62	14.14	10.16	9.22	1.57	0.52	1.72	0.52	0.31	0.01
30	42.37	15.59	13.95	11.08	11.49	1.54	0.31	2.40	0.82	0.41	0.04
31	61.55	12.50	9.91	7.13	6.30	1.03	0.52	0.38	0.41	0.10	0.17
32	58.51	11.96	9.53	10.85	5.88	1.12	0.51	1.09	0.41	0.10	0.05
33	59.37	14.09	10.12	10.23	4.59	0.63	0.10	0.12	0.52	0.21	0.02
34	51.90	14.08	12.33	9.76	7.09	1.44	0.51	1.79	0.72	0.31	0.06
35	56.20	14.62	11.82	11.20	3.94	0.62	0.10	0.61	0.62	0.21	0.06
36	63.42	10.69	9.56	7.30	5.04	1.23	0.62	1.38	0.51	0.21	0.05
37	59.92	12.32	10.56	10.56	4.24	0.72	0.10	0.79	0.21	0.52	0.06
38	68.29	10.98	9.15	5.89	3.76	1.02	0.30	0.00	0.30	0.20	0.10
39	44.61	16.47	16.57	12.38	7.16	1.43	0.31	0.53	0.31	0.20	0.02

Table A1.6 Stock levels

422	1118	620	1250	1485	435	1589	1785	560
520	1380	**260**	1510	1485	625	1470	1920	820
620	1740	420	1510	1485	940	1450	1510	1085
855	2075	620	1395	1485	1000	1680	1155	1365
970	1780	800	1295	1485	1025	2000	1480	1640
1210	1785	1050	1135	1485	960	1705	1780	1900
1475	1940	710	910	1175	1340	1255	2080	1470
1490	1930	1060	850	840	1780	1450	1680	1100
1560	1660	560	845	620	2000	1770	1570	1045
1718	1740	950	740	485	1910	2095	1790	1080
1837	1895	1220	745	445	1575	2090	2025	1410
1820	1785	1105	980	470	1205	1920	1795	1750
1860	1450	1105	1180	560	945	1600	1665	1640
1845	990	750	960	620	885	1630	1540	1320
1560	1260	435	925	680	1180	1950	1100	1365
1465	1490	**75**	960	700	1400	1860	940	1340
1520	1720	360	1060	455	1760	1685	1225	1360
1595	1880	560	1245	780	1770	1690	1560	1420
1500	1600	720	1480	1185	1140	2000	1820	1645
1400	1685	960	1480	1140	470	1485	1995	1850
1450	1495	950	1480	1070	505	1680	1605	
1740	1625	902	1480	700	845	2000	1310	
2040	1450	900	1480	300	1180	1590	960	
1900	1450	970	1480	**280**	1525	1440	690	
1500	1065	1095	1490	340	1865	1470	580	

Table A1.7 *Results for batch blending (including trim blends)*

batch	property
1	107.0
2	104.7
3	100.0
4	102.3
5	111.6
	90.7
	88.4
	90.7
6	104.7
7	107.0
8	102.3
9	100.0
	96.5
	96.5
10	118.6
11	102.3
12	100.0
13	100.0
14	107.0
15	104.7
16	103.5
	88.4
17	102.3
16	102.3
19	104.7
20	109.3
21	107.0
22	102.3
	97.7
23	100.0
24	100.0
25	102.3
	97.7
26	102.3
27	100.0
	70.9
	69.8
28	100.0

batch	property
	76.7
	81.4
	83.7
	83.7
	86.0
	84.9
	77.9
	77.9
	84.9
	91.9
	93.0
	95.3
	96.5
	97.7
29	110.5
30	107.0
31	102.3
	97.7
32	103.5
33	102.3
	97.7
	96.5
34	104.7
	97.7
	97.7
	95.3
35	104.7
36	104.7
37	101.2
38	102.3
39	107.0
40	102.3
	95.3
	97.7
	98.8
41	104.7
42	100.0
43	100.0
44	100.0
	98.8

Table A1.8 Two-tailed confidence intervals

n	confidence (%)
0.0	0.0
0.5	38.3
1.0	68.3
1.5	86.6
2.0	95.4
2.5	98.76
3.0	99.73
3.5	99.953
4.0	99.9937
4.5	99.99932
5.0	99.999943
5.5	99.9999962
6.0	99.99999980
6.5	99.999999992
7.0	99.9999999997

confidence (%)	n
50	0.6745
60	0.8416
70	1.0364
80	1.2816
85	1.4395
90	1.6449
95	1.9600
98	2.3263
99	2.5758
99.5	2.8070
99.9	3.2905
99.99	3.8906
99.999	4.4172
99.9999	4.8916
99.99999	5.3267

Table A1.9 One-tailed confidence intervals

n	confidence (%)
0.0	0.0
0.5	69.1
1.0	84.1
1.5	93.3
2.0	97.7
2.5	99.38
3.0	99.87
3.5	99.977
4.0	99.9968
4.5	99.99966
5.0	99.999971
5.5	99.9999981
6.0	99.99999990
6.5	99.999999996
7.0	99.9999999999

confidence (%)	n
50	0.0000
60	0.2533
70	0.5244
80	0.8416
85	1.0364
90	1.2816
95	1.6449
98	2.0538
99	2.3263
99.5	2.5758
99.9	3.0893
99.99	3.7005
99.999	4.1860
99.9999	4.7534
99.99999	5.1993

Appendix 2
Summary of Distributions

Included here are two tables, Table A2.1 for continuous distributions and Table A2.2 for discrete distributions. They list, in alphabetical order, all the distributions described in this book. Their purpose is to assist the reader in shortlisting the distributions that might be considered for a given dataset.

Where distributions share the same name, different versions are numbered as $^{(1)}$, $^{(2)}$, etc. This should not be confused with distributions that are numbered as part of their title, e.g. Pearson-I, Pearson-II, etc.

A blank row in the table indicates that the details will be found under the distribution's alternative name.

A ✓ for the lower or upper bound means that it exists only under certain circumstances or is determined from a formula too complex to show in the table. Full details will be found in the detailed description of the distribution.

For the PDF (or PMF), CDF and QF a ✓ shows that the function is included in the book. A blank generally implies that it is mathematically too complex to be included. A CDF, if not presented as a function, can always be developed by applying the trapezium rule to the PDF (or PMF). And the lack of a QF is usually only a minor inconvenience that, if needed, can be overcome by iterative or graphical methods. However, if required for Monte Carlo simulation, its existence becomes more important. A few distributions have only a QF. This requires a slightly different method of fitting and should not, in itself, be a reason for rejecting a distribution.

Distributions with no shape parameters are highly unlikely to be of practical use. Similarly distributions with a large number of parameters may provide a better fit but are less likely to be robust. They are also unlikely to have simple formulae for mean, variance, etc.

For mean (μ), standard deviation (σ), skewness (γ) and kurtosis (κ) a ✓ shows that a formula is included in this book. In some cases it may be presented as a general formula for the raw moments. A ✗ means that the parameter cannot be defined. A blank means either that it is too complex for inclusion or has not been published. Most distributions include shape parameters that give location and dispersion. Depending on the application, these might be used instead of mean and standard deviation. Further, moments can usually be derived using the trapezium rule.

Statistics for Process Control Engineers: A Practical Approach, First Edition. Myke King.

Excel supports many of the commonly used distributions. Where relevant, they are shown in the tables. In Excel 2013 their names take the form '.DIST'. This provides both the PDF and CDF. For each there is '.INV' which is the QF. For the majority of the distributions, which are not specifically included, the PDF, CDF and QF can be defined as a calculation. Many distributions use the ERF function. Used by several distributions, all four kinds of Bessel function (I, J, K and Y) are supported, although only for integer orders. Not shown in the table, Excel also supports the GAMMA, $\Gamma(x)$, and FACT, $x!$, functions. Also not shown are the common mathematical functions such as EXP, e^x; LN, $\ln(x)$; etc. There is also, of course, the full range of trigonometric functions, such as SIN, $\sin(x)$; ASIN, $\sin^{-1}(x)$; SINH, $\sinh(x)$; etc. Common statistical functions such as AVERAGE, STDEV, SKEW, VAR, COVAR.S and PEARSON are based on the formulae given in this book. However, note that KURT gives excess kurtosis and so must have 3 added to give kurtosis.

Table A2.1 Summary of continuous distributions

distribution	lower bound	upper bound	PDF	CDF	QF	shape parameters	μ	σ	γ	κ	alternative names	Excel function
Amoroso	α	∞	✓			$\alpha, \beta, \delta_1, \delta_2$	✓	✓	✓	✓		
anglit	$\mu - \pi\beta/4$	$\mu + \pi\beta/4$	✓	✓	✓	μ, β	μ	✓	0	✓		
arcsine (first kind)	0	1	✓	✓	✓		0.5	0.125	0	−1.5		
arcsine (second kind)	−1	1	✓						0			
arctangent	α	∞	✓	✓	✓	α, β, δ			0			
asymmetric Laplace	$-\infty$	∞	✓	✓		α, β, δ	✓	✓	✓	✓		
Balding–Nichols	0	1	✓			μ, p	μ	✓	✓	✓		
Bates	0	1	✓			n	✓	✓	0	✓		
Benini	x_{min}	∞	✓	✓		x_{min}, α, β						
Benktander-I	0	∞	✓	✓		λ, δ	✓					
Benktander-II	0	∞	✓	✓		λ, δ	✓					
beta rectangular	$-\infty$	∞	✓			$\delta_1, \delta_2, \delta_3$						
beta-I	0	1	✓			δ_1, δ_2	✓	✓	✓	✓	Feller–Pareto	BETA.DIST
beta-II	0	β	✓			$\beta, \delta_1, \delta_2, \delta_3$	✓	✓	✓	✓	beta prime inverted beta	
beta-IV	x_{min}	x_{max}	✓			$x_{min}, x_{max}, \delta_1, \delta_2$	✓	✓	✓	✓	beta subjective PERT	BETA.DIST
Birnbaum–Saunders	α		✓	✓		α, β, δ	✓	✓	✓	✓	fatigue life	
bounded Pareto-I	x_{min}	x_{max}	✓	✓	✓	x_{min}, x_{max}, δ	✓	✓				
Bradford	x_{min}	x_{max}	✓	✓		$x_{min}, x_{max}, \lambda$	✓	✓				
Burr-I	$-\infty$	∞									uniform	
Burr-II											generalised logistic	
Burr-III	α	∞	✓	✓	✓	$\alpha, \beta, \delta_1, \delta_2$						

(continued overleaf)

Table A2.1 (*continued*)

distribution	lower bound	upper bound	PDF	CDF	QF	shape parameters	μ	σ	γ	κ	alternative names	Excel function
Burr-IV	$-\alpha$	$\alpha+\beta\delta_1$	✓	✓	✓	$\alpha, \beta, \delta_1, \delta_2$						
Burr-V	$\alpha-\pi/2$	$\alpha+\pi/2$	✓	✓	✓	$\alpha, \beta, \delta_1, \delta_2$						
Burr-VI	$-\infty$	∞	✓	✓	✓	$\alpha, \beta, \delta_1, \delta_2$						
Burr-VII	$-\infty$	∞	✓	✓	✓	α, β, δ						
Burr-VIII	$-\infty$	∞	✓	✓	✓	α, β, δ						
Burr-IX	$-\infty$	∞	✓	✓	✓	$\alpha, \beta, \delta_1, \delta_2$						
Burr-X	$-\infty$	∞	✓	✓	✓	α, β, δ						
Burr-XI	$-\infty$	∞	✓	✓	✓	α, β, δ						
Burr-XII	α	∞	✓	✓	✓	$\alpha, \beta, \delta_1, \delta_2$	✓	✓	✓	✓	Pareto-IV Singh–Maddala	
cardioid	0	2π	✓	✓		μ, β	μ					
Cauchy	$-\infty$	∞	✓	✓	✓	α, β	✗	✗	✗	✗	McCullagh Breit–Wigner Lorenz Cauchy–Lorenz	
Champernowne[1]	0	∞	✓	✓		δ_1, δ_2						
Champernowne[2]	0	∞	✓			$\alpha, \beta, \delta_1, \delta_2$	✓	✓	✓	✓		
chi	$-\infty$	∞	✓			α, β, f	✓	✓	✓	✓		
chi-squared (standard)	$-\infty$	∞	✓			f	✓	✓	✓	✓		CHISQ.DIST
chi-squared (shifted and scaled)	$-\infty$	∞	✓			α, β, f	✓	✓	✓	✓		
cosine	$\mu-\pi\beta/2$	$\mu+\pi\beta/2$	✓	✓	✓	μ, β	μ	✓	0			
Dagum-II	α	∞	✓	✓	✓	$\alpha, \beta, \delta_1, \delta_2, \delta_3$						

Dagum-III	✓	∞	✓	✓	✓	$\alpha, \beta, \delta_1, \delta_2, \delta_3$						
Davis	$-\infty$	∞	✓			α, β, δ	✓	✓				
double exponential	$-\infty$	∞	✓			α, β	✓	✓			Laplace	
erf	$-\infty$	∞	✓	✓		α	0	✓	0	3		
exponential (standard)	0	∞	✓	✓	✓	λ	✓	✓	2	9		EXPON. DIST
exponential (shifted)	α	∞	✓	✓	✓	α, β	✓	✓	2	9	Pearson-X	
exponentiated Kumaraswamy–Dagum	0	∞	✓	✓	✓	$\lambda, \delta_1, \delta_2, \delta_3, \delta_4$					EKD	
exponential logarithmic	0	∞	✓	✓	✓	λ, δ						
exponential power	α	∞	✓	✓	✓	α, β, δ						
exponentially modified Gaussian	$-\infty$	∞	✓	✓		α, β, λ	✓	✓	✓	✓	EMG	ERF
extreme value-I	$-\infty$	∞	✓	✓	✓	α, β, δ						
Fisher	$-\infty$	∞	✓			f_1, f_2	✓	✓	✓	✓	Fisher–Snedecor Snedecor-F	F.DIST
folded normal	$-\infty$	∞	✓	✓		α, β						ERF
Fréchet	$-\infty$	∞	✓	✓	✓	α, β, δ	✓	✓	✓	✓	Gumbel-II log-Gompertz inverse Weibull	
gamma	0	∞	✓			β or λ, k	✓	✓	✓	✓		GAMMA. DIST
gamma-Gompertz	0	∞	✓	✓		α, β, δ						
generalised beta	0	β	✓			$\beta, \delta_1, \delta_2, \delta_3$	✓	✓	✓	✓		
generalised beta prime[1]	α	∞	✓			$\alpha, \beta, \delta_1, \delta_2, \delta_3$	✓	✓	✓	✓	transformed beta	

(continued overleaf)

Table A2.1 (*continued*)

distribution	lower bound	upper bound	PDF	CDF	QF	shape parameters	μ	σ	γ	κ	alternative names	Excel function
generalised beta prime[(2)]	0	1	✓			$\beta, \delta_1, \delta_2, \delta_3, \delta_4$						
generalised exponential	α	∞	✓	✓	✓	$\alpha, \beta, \lambda, \delta$						
generalised extreme value	α	∞	✓			α, β, δ					GEV	
generalised gamma	$-\infty$	∞	✓			$\beta, \delta_1, \delta_2$	✓	✓	✓	✓	transformed gamma Stacy–Mihram	
generalised inverse Gaussian	0	∞	✓			α, β, δ	✓	✓	✓	✓	GIG Sichel	BESSELK
generalised Lindley	0	∞	✓			$\lambda, \delta_1, \delta_2$						
generalised logistic	$\alpha - \beta/\delta$ ($\delta < 0$)	$\alpha - \beta/\delta$ ($\delta > 0$)	✓	✓	✓	α, β, δ	✓	✓			Burr-II	
generalised logistic-I	α	∞	✓	✓	✓	α, β, δ						
generalised logistic-II	$-\infty$	∞	✓	✓	✓	α, β, δ						
generalised logistic-III	α	∞	✓			α, β, δ						
generalised logistic-IV	α	∞	✓			$\alpha, \beta, \delta_1, \delta_2$					exponential general beta-II	
generalised log-logistic	α	∞	✓			$\alpha, \beta, \delta_1, \delta_2$						
generalised normal[(1)]	$-\infty$	∞	✓			μ, β, λ	μ	✓	0	✓	generalised Gaussian exponential power generalised error	
generalised normal[(2)]	$-\infty$	$m + \beta/\delta$	✓	✓		m, β, δ	✓	✓	✓	✓	generalised Gaussian	
generalised normal[(3)]	α	∞	✓			$\alpha, \beta, \delta_1, \delta_2$	✓	✓	✓	✓	generalised Gaussian	
generalised Pareto[(1)]											Stoppa	
generalised Pareto[(2)]	α	∞	✓	✓	✓	α, β, δ	✓	✓	✓	✓		

generalised Pareto[3]	α	∞	✓	✓		α, β, δ						
generalised Tukey lambda	$-\infty$	∞			✓	$\lambda_1, \lambda_2, \lambda_3, \lambda_4$						
Gibrat	$-\infty$	∞	✓				~1.65	~4.67	~6.19	~114		
Gompertz	0	∞	✓	✓	✓	α, β						
Gompertz–Makeham	0	∞	✓	✓		α, β						
Gompertz–Verhulst	✓	∞	✓	✓	✓	$\alpha, \beta, \delta_1, \delta_2$						
Gumbel-I	$-\infty$	∞	✓	✓							extreme value max extreme value min	
Gumbel-II											Fréchet	
half-logistic	α	∞	✓	✓		α, λ	✓	✓				
half-normal	$-\infty$	∞	✓			α, β	✓	✓	~1.00	~0.87		
Hjorth	0	∞	✓	✓		$\beta, \delta_1, \delta_2$					IDB	
Hosking	$-\infty$	∞	✓	✓	✓	$\alpha, \beta, \delta_1, \delta_2$						
hyperbolic	$-\infty$	∞	✓			α, β, λ	✓	✓	✓	✓		BESSELK
hyperbolic secant	$-\infty$	∞	✓	✓	✓	μ, σ	μ	σ	0	5	sech	
hyperexponential	0	∞	✓	✓		$p_1, p_2 \ldots, \lambda_1, \lambda_2 \ldots$	✓	✓	✓	✓		
hypoexponential	0	∞	✓			λ_1, λ_2 ..	✓	✓	✓		generalised Erlang	
inverse Burr	α	∞	✓	✓	✓	$\alpha, \beta, \delta_1, \delta_2$	✓	✓	✓	✓	Dagum-I	
inverse chi	$-\infty$	∞	✓			α, β, f	✓	✓	✓	✓		
inverse chi-squared	$-\infty$	∞	✓			α, β, f	✓	✓	✓	✓		
inverse exponential	α	∞	✓	✓	✓	α, β	✗	✗	✗	✗		
inverse gamma											Pearson-V	
inverse Gaussian (two parameter)	0	∞	✓	✓		μ, β	μ	✓	✓	✓		
inverse Gaussian (three parameter)	α	∞	✓			α, μ, σ	μ	σ				

(continued overleaf)

Table A2.1 (*continued*)

distribution	lower bound	upper bound	PDF	CDF	QF	shape parameters	μ	σ	γ	κ	alternative names	Excel function
inverse paralogistic	α	∞	✓			α, β, δ						
inverse Pareto	α	∞	✓	✓		α, β, δ	✓	✓	✓	✓		
inverse Rayleigh	$-\infty$	∞	✓			α, β	✓	✗	✗	✗		
Irwin–Hall	0	∞	✓			n	✓	✓	1	✓		
Johnson S_B	$-\infty$	∞	✓	✓		$\alpha, \beta, \mu_J, \sigma_J$						ERF
Johnson S_L	μ_J	∞	✓	✓		$\alpha, \beta, \mu_J, \sigma_J$	✓	✓				
Johnson S_N	$-\infty$	∞	✓	✓		$\alpha, \beta, \mu_J, \sigma_J$	✓	✓	0	3		
Johnson S_U	$-\infty$	∞	✓	✓		$\alpha, \beta, \mu_J, \sigma_J$	✓	✓				
kappa	$-\infty$	∞	✓			K, α, β, δ			0			
Kumaraswamy	0	1	✓	✓	✓	δ_1, δ_2	✓	✓	✓	✓	minimax	
Laplace	$-\infty$	∞	✓	✓	✓	μ, σ	μ	σ	0			
Lévy	α	∞	✓	✓		α, β	✓				reciprocal Gaussian	ERF
Lindley	0	∞	✓	✓		λ	✓	✓	✓	✓		
Lindley-geometric	0	∞	✓	✓		λ, δ						
log-Cauchy	0	∞	✓	✓	✓	α, β						
log F	$-\infty$	∞	✓			α, β, f_1, f_2	✓	✓			Fisher z	
log-gamma[(1)]	$\alpha-1$	∞	✓			$\alpha, \delta_1, \delta_2$	✓	✓	✓	✓		
log-gamma[(2)]	0	1	✓			δ_1, δ_2						
log-gamma[(3)]	$\alpha-1$	∞	✓			$\alpha, \delta_1, \delta_2$						
log-gamma[(4)]	$-\infty$	∞	✓			$\alpha, \delta_1, \delta_2$						
log-gamma[(5)]	$-\infty$	∞	✓			α, β, δ						
logistic	$-\infty$	∞	✓	✓	✓	μ, β	μ	✓	0	4.2		

logistic exponential	0	∞	✓	✓	✓	λ, δ						
logit-normal	0	1	✓	✓		α, β						ERF
log-Laplace[(1)]	0	∞	✓	✓		α, β						
log-Laplace[(2)]	0	α	✓	✓		α, β					log-double exponential	
log-logistic	α	∞	✓	✓	✓	α, β, δ	✓	✓	✓	✓	Fisk	
lognormal	0	∞	✓	✓		α, β	✓	✓	✓	✓		LOGNORM.DIST ERF
log-Student *t*	0	∞	✓			α, β, f						
Lomax	0	∞	✓	✓	✓	β, δ	✓	✓	✓	✓		
Maxwell	−∞	∞	✓	✓		α, β	✓	✓	~0.49	~0.11	Maxwell-Boltzman	
Maxwell–Jüttner	0	∞	✓	✓		λ, α						
Mielke	α	∞	✓	✓		$\alpha, \beta, \delta_1, \delta_2$						
minimax odds	0	∞	✓	✓	✓	δ_1, δ_2						
Moyal	−∞	∞	✓	✓		α, β	✓	✓			Landau (approx.)	ERF
Muth	α	∞	✓	✓		α, β, δ						
Nakagami	α	∞	✓			α, β, δ						
noncentral beta	0	1	✓	✓		α, β, δ						BESSELI
noncentral chi-squared	−∞	∞	✓	✓		α, β, δ	✓	✓	✓	✓		BESSELI
noncentral F	−∞	∞	✓			δ, f_1, f_2	✓	✓				
noncentral Student *t*	−∞	∞	✓			δ, f	✓	✓				
normal	−∞	∞	✓	✓		μ, σ	μ	σ	0	3	Gaussian Pearson-0	NORM.DIST ERF
normal inverse Gaussian	−∞	∞	✓			$\alpha, \beta, \delta, \lambda$	✓	✓	✓	✓	NIG	BESSELK

(*continued overleaf*)

Table A2.1 (continued)

distribution	lower bound	upper bound	PDF	CDF	QF	shape parameters	μ	σ	γ	κ	alternative names	Excel function
Nukiyama–Tanasawa	0	∞	✓			$\beta, \delta_1, \delta_2$						
paralogistic	α	∞	✓	✓	✓	α, β, δ	✓	✓	✓	✓		
Pareto-I	β	∞	✓	✓	✓	β, δ	✓	✓	✓	✓	Pearson-XI	
Pareto-II	α	∞	✓	✓	✓	α, β, δ	✓	✓	✓	✓		
Pareto-III	α	∞	✓	✓	✓	α, β, δ	✓	✓	✓	✓		
Pearson-I	α	$\alpha+\beta$	✓			$\alpha, \beta, \delta_1, \delta_2$	✓	✓	✓	✓	beta-I	BETA.DIST
Pearson-II	α	$\alpha+\beta$	✓			α, β, δ	0.5	✓	0	✓	symmetric-beta	BETA.DIST
Pearson-III	$-\infty$	∞	✓			α, β, δ	✓	✓	✓	✓		
Pearson-IV	$-\infty$	∞	✓			$\alpha, \beta, \delta_1, \delta_2$	✓	✓	✓	✓		
Pearson-V	$-\infty$	∞	✓			α, β, δ	✓	✓	✓	✓		
Pearson-VI	$-\infty$	∞	✓			$\alpha, \beta, \delta_1, \delta_2$	✓	✓	✓	✓		
Pearson-VII	$-\infty$	∞	✓			α, β, δ	✓	✓	0	✓		
Pearson-VIII	$-\infty$	∞	✓			K, α, β						
Pearson-IX	$-\infty$	∞	✓			K, α, β						
Pearson-X											exponential	
Pearson-XI											Pareto-I	
Pearson-XII	$-\infty$	∞	✓			$K, \alpha, \beta_1, \beta_1$						
PERT	x_{min}	x_{max}	✓			$x_{min}, x_{max}, x_{mode}, \lambda$	✓	✓	✓	✓	beta-IV	
power	0	β	✓	✓	✓	α, β, δ	✓	✓	✓	✓		
power lognormal	0	∞	✓	✓		α, β, p					Marshall-Olkin	
q-exponential	0	∞	✓	✓	✓	λ, q	✓	✓	✓	✓		
q-gamma	0	∞	✓			λ, δ, q	✓	✓	✓	✓		
q-Gaussian	✓	✓	✓			μ, β, q	✓	✓	0	✓		
q-Weibull	0	∞	✓	✓	✓	α, β, δ, q						

raised cosine	$\mu-\beta$	$\mu+\beta$	✓	✓		μ, β	μ	✓	0	~2.41		
Rayleigh	$-\infty$	∞	✓	✓	✓	α, β	✓	✓	~0.63	~0.25		
reciprocal inverse Gaussian	0	∞	✓			α, λ	✓	✓	✓	✓		
reverse exponential	$-\infty$	∞										
reverse Weibull	$-\infty$	∞	✓	✓	✓	α, β, δ						
Rician	0	∞	✓	✓		α, β					Rice	BESSELI
shifted Gompertz	0	∞	✓	✓		α, β						
skew-logistic(1)	$-\infty$	∞	✓	✓		α, β, δ						
skew-logistic(2)	$-\infty$	∞	✓	✓		α, β, δ						
skew-normal	$-\infty$	∞	✓			α, β, δ	✓	✓	✓	✓		
slash	$-\infty$	∞	✓	✓		μ, σ	μ	σ	0	✗		ERF
Stoppa-I											power	
Stoppa-II	β	∞	✓	✓	✓	$\beta, \delta_1, \delta_2$	✓	✓	✓	✓	generalised Pareto(1)	
Stoppa-III											generalised exponential	
Stoppa-IV	$\alpha+\beta/2$	$\alpha+\beta$	✓	✓	✓	α, β, δ						
Stoppa-V											Burr-V	
stretched exponential	0	∞	✓	✓	✓	λ, δ	✓	✓	✓	✓		
Student *t*	$-\infty$	∞	✓			f	0	✓	0	✓		T.DIST
Topp–Leone	0	1	✓	✓	✓							
triangular	x_{min}	x_{max}	✓	✓		$x_{min}, x_{max}, x_{mode}$	✓	✓		2.4	lack of knowledge	
Tukey lambda	$-\infty$	∞			✓	λ	0	✓	0			
two-sided power	x_{min}	x_{max}	✓	✓		x_{min}, x_{max}, δ	✓	✓			TSP	

(*continued overleaf*)

Table A2.1 (continued)

distribution	lower bound	upper bound	PDF	CDF	QF	shape parameters	μ	σ	γ	κ	alternative names	Excel function
uniform	$x_{\min}$	$x_{\max}$	✓	✓	✓	$x_{\min}, x_{\max}$	✓	✓	0	1.8	rectangular Burr-I	RAND
von Mises	$\mu-\pi$	$\mu+\pi$	✓			μ, β	μ				Tikhonov	BESSELI
Wakeby	$-\infty$	∞			✓	$\delta_1, \delta_2, \delta_3, \delta_4$	✓	✓				
Wald	0	∞	✓	✓		β	1	✓	✓	✓		ERF
Weibull-I	0	∞	✓	✓	✓	δ	✓	✓	✓	✓		
Weibull-II	0	∞	✓	✓	✓	β, δ	✓	✓	✓	✓	Rosin-Rammler	WEIBULL. DIST
Weibull-III	α	∞	✓	✓	✓	α, β, δ	✓	✓	✓	✓		
Wigner semicircle	$-r$	r	✓	✓		r	0	✓	0	2		

Table A2.2 Summary of discrete distributions

distribution	lower bound	upper bound	PMF	CDF	QF	shape parameters	μ	σ	γ	κ	alternative names	Excel function
Benford	1	∞	✓								first digit	
Bernoulli	0	1	✓			p	✓	✓	✓	✓		
beta-binomial	0	∞	✓			α, β	✓	✓	✓	✗		
beta-geometric	1	∞	✓			α, β	✓	✓				
beta-negative binomial	0	∞	✓			α, β, n	✓	✓			generalised Waring inverse Markov-Pólya	
beta-Pascal	0	∞	✓			α, β, n						
binomial	0	n	✓			n, p	✓	✓	✓	✓		BINOM.DIST
Borel	1	∞	✓			λ	✓	✓				
Borel–Tanner	n	∞	✓			n, λ	✓	✓				
Consul	1	∞	✓			α, β	✓	✓				
Conway–Maxwell–Poisson	0	∞	✓			λ	✓	✓				
Delaporte	0	∞	✓			α, β, λ	✓	✓				
Erlang	0	∞	✓			β or λ, k	✓	✓	✓	✓		
Flory–Schulz	1	∞	✓			p	✓	✓	✓	✓		
gamma-Poisson	0	∞	✓			α, β	✓	✓	✓	✓		
geometric	0	∞	✓	✓	✓	p	✓	✓	✓	✓		NEGBINOM.DIST
hypergeometric	0	n	✓			N, K, n	✓					HYPGEOM.DIST
Lagrange–Poisson	n	∞	✓			n, α, λ	✓	✓			Poisson–Consul	
logarithmic	1	∞	✓			p	✓	✓			log-series	
multinomial	0	n	✓			$n, p_1, p_2 \ldots$						
negative binomial	0	∞	✓			n, p	✓	✓	✓	✓	Pascal	NEGBINOM.DIST
negative hypergeometric	0	n_2	✓			n_1, n_2, n_3	✓	✓				

(*continued overleaf*)

Table A2.2 (*continued*)

distribution	lower bound	upper bound	PMF	CDF	QF	shape parameters	μ	σ	γ	κ	alternative names	Excel function
parabolic fractal	0	∞	✓			a, b, c						
Poisson	0	∞	✓			λ	✓	✓				POISSON.DIST
Poisson binomial	0	n	✓			n, p						
Pólya	0	∞	✓			α, β	✓	✓				
Skellam	$-\infty$	∞	✓			λ_1, λ_2	✓	✓				BESSELI
uniform	x_{min}	x_{max}	✓	✓		x_{min}, x_{max}	✓	✓	0	✓	rectangular	
Weibull (discrete)	0	∞	✓	✓	✓	p, β	✓					
Yule–Simon	1	∞	✓	✓		α	✓	✓	✓	✓	Yule	
zeta	1	∞	✓			λ	✓	✓	✓	✓		
Zipf	1	∞	✓	✓		λ, K	✓					
Zipf–Mandelbrot	1	∞	✓	✓		λ, K, α	✓					

References

1. Martin G.D., Turpin L.E., Cline R.P. (1991) Estimating control function benefits. *Hydrocarbon Processing*, **70**(6), 68–73.
2. Martin G.D. (2004) Understand control benefits estimates. *Hydrocarbon Processing*, **83**(10), 43–46.
3. EEMUA (2013) Alarm systems, a guide to design, management and procurement. Publication No. 191, third edition.
4. Horwitz W. (2003) *Journal of the Association of Official Analytical Chemists*, **86**, 109.
5. Boyer K.W., Horwitz W., Albert R. (1985) *Analytical Chemistry*, **57**, 454–459.
6. Hosking J.R.M. (1994) The four-parameter kappa distribution. *IBM Journal or Research and Development*, **38**(3), 251–258.
7. Box G.E.P., Hunter J.S., Hunter W.G. (2005) Statistics for Experimenters. *John Wiley & Sons, Inc.*, **53**.
8. Anderson T.W. (1994) *The Statistical Analysis of Time Series*. John Wiley & Sons, Inc., 186–187.
9. Good I.J. (1953) The population frequencies of species and the estimation of population parameters. *Biometrika*, **40**, 237–260.
10. Tsallis C. (1988) *Journal of Statistical Physics*, **52**, 479.
11. Burr I.W. (1942) Cumulative frequency functions. *Annals of Mathematical Statistics*, **13**, 215–232.
12. Dagum C. (2006) Wealth distribution models: analysis and applications. *Statistica*, **LXVI**(3), 235–268.
13. Sastry D.V.S., Bhati D. (2016) A new skew logistic distribution: properties and applications. *Brazilian Journal of Probability and Statistics* **30**(2), 248–271.
14. Huang S., Oluyede B.O. (2014) Exponentiated Kumaraswamy-Dagum distribution with applications to income and lifetime data. *Journal of Statistical Distributions and Applications*, **1**, 8.
15. Stoppa G. (1990) Proprietà campionarie di un nuovo modello Pareto generalizzato. *XXXV Riunione Scientifica della Società Italiana di Statistica*, Padova, 137–144.
16. McDonald J.B. (1984) Some generalized functions for the size distribution of income. *Econometrica*, **52**, 647–664.
17. Johnson N.L. (1949) Systems of frequency curves generated by methods of translation. *Biometrika*, **36**, 149–176.
18. Pearson K. (1893) Contributions to the mathematical theory of evolution. *Proceedings of the Royal Society*, **54**, 329–333.
19. Pearson K. (1895) Contributions to the mathematical theory of evolution, II: Skew variation in homogeneous material. *Philosophical Transactions of the Royal Society. Series A*, **186**, 343–414.
20. Pearson K. (1901) Mathematical contributions to the theory of evolution, X: Supplement to a memoir on skew variation. *Philosophical Transactions of the Royal Society. Series A*, **197**, 443–459.
21. Pearson K. (1916) Mathematical contributions to the theory of evolution, XIX: second supplement to a memoir on skew variation. *Philosophical Transactions of the Royal Society of London, Series A*, **216**, 429–457.
22. Gupta R.D., Kundu D. (2007) Generalized exponential distribution: existing results and some recent developments. *Journal of Statistical Planning and Inference*, **137**(11), 3537–3547.

Statistics for Process Control Engineers: A Practical Approach, First Edition. Myke King.

23. Venter G.G. (1983) Transformed beta and gamma distributions and aggregate losses. *Proceedings of the Casualty Actuarial Society*, **LXX**, 62.
24. Lindley D.V. (1958) Fiducial distributions and Bayes' theorem. *Journal of the Royal Statistical Society: Series B: Methodological*, **20**(1), 102–107.
25. Zakerzadeh H., Mahmoudi E. (2012) A new two parameter lifetime distribution: model and properties. arXiv:1204.4248.
26. Zakerzadeh H., Dolati A. (2009) Generalized Lindley distribution. *Journal of Mathematical Extension*, **3**(2), 13–25.
27. Chalabi Y., Scott D.J., Würtz D. (2012). Flexible distribution modeling with the generalized lambda distribution. Part of PhD thesis, Swiss Federal Institute of Technology, Zurich.
28. Amoroso L. (1925) Ricerche intorno alla curva dei redditi. *Annali di Mathematica*, **IV**(2), 123–159.

Index

Statistics for Process Control Engineers: A Practical Approach, First Edition. Myke King.